Global Blue Economy

The blue economy is an economic arena that depends on the benefits and values realized from the coastal and marine environments. This book explains the 'sustainable blue economy' as a marine-based economy that provides social and economic benefits for current and future generations. It restores, protects, and maintains the diversity, productivity, and resilience of marine ecosystems, and is based on clean technologies, renewable energy, and circular material flows.

Features

- Illustrates the fundamental concepts, tools, techniques, and details of a global blue economy
- Describes the scale and scope of the global blue economy and the role that observations, measurements, and forecasts play in supporting the safe and effective use of the ocean and its resources
- Includes many case studies from different countries and explores energy demands with emphasis on offshore oil and gas exploration methods and techniques
- Stimulates the political will and actions of governments and other partners for activities that effectively shape the framework of blue economy developments in many countries
- Clarifies the links among blue economy, sustainable development, and economic growth, and recognizes the importance of sustainable development goals for enhancing the economic benefits from the sustainable uses of marine resources
- Investigates the problems that threaten marine ecosystems and presents a set of management toolboxes and models for solving the issues of the blue economy in selected countries

This book provides a survey of the current state of understanding, activities, and policies related to the blue economy as it is being pursued in different industries and countries. A comprehensive resource for anyone interested.

Applied Ecology and Environmental Management – A Series

Series Editor: **Steven M. Bartell**, *Oak Ridge Associated Universities;*
Sven E. Jorgensen, *Copenhagen University, Denmark*

Global Blue Economy: Analysis, Developments, and Challenges
Md. Nazrul Islam and Steven M. Bartell

Managing Environmental Data: Principles, Techniques, and Best Practices
Gerald A. Burnette

Environmental Management Handbook, Second Edition – Six Volume Set, Second Edition
Edited by Brian D. Fath, Sven Erik Jorgensen

Sustainable Development Indicators: An Exergy-Based Approach
Søren Nors Nielsen

Environmental Management of Marine Ecosystems
Edited by Md. Nazrul Islam, Sven Erik Jorgensen

Ecotoxicology and Chemistry Applications in Environmental Management
Sven Erik Jorgensen

Ecological Forest Management Handbook
Edited by Guy R. Larocque

Integrated Environmental Management: A Transdisciplinary Approach
Sven Erik Jørgensen, Joao Carlos Marques, Søren Nors Nielsen

Ecological Processes Handbook
Luca Palmeri, Alberto Barausse, Sven Erik Jorgensen

Handbook of Inland Aquatic Ecosystem Management
Sven Erik Jorgensen, Jose Galizia Tundisi, Takako Matsumura Tundisi

Eco Cities: A Planning Guide
Edited by Zhifeng Yang

For more information on this series, please visit: www.routledge.com/Applied-Ecology-and-Environmental-Management/book-series/CRCAPPECOENV?pd=published,forthcoming&pg=2&pp=12&so=pub&view=list

Global Blue Economy
Analysis, Developments, and Challenges

Edited by
Md. Nazrul Islam and Steven M. Bartell

CRC Press
Taylor & Francis Group
Boca Raton London New York

CRC Press is an imprint of the
Taylor & Francis Group, an **informa** business

First edition published 2023
by CRC Press
6000 Broken Sound Parkway NW, Suite 300, Boca Raton, FL 33487-2742

and by CRC Press
4 Park Square, Milton Park, Abingdon, Oxon, OX14 4RN

CRC Press is an imprint of Taylor & Francis Group, LLC

© 2023 Taylor & Francis Group, LLC

ISBN: 9781032012728 (hbk)
ISBN: 9781032026251 (pbk)
ISBN: 9781003184287 (ebk)

DOI: 10.1201/9781003184287

Typeset in Times
by Newgen Publishing UK

Dedication

To
SAHANAJ TAMANNA
(Wife of Prof. Dr. Md. Nazrul Islam)
&
SABABA MOBASHIRA ISLAM
(Daughter of Prof. Dr. Md. Nazrul Islam)

Contents

Preface

Key components of international cooperation for the blue economy approach is innovative research and application. A science technology policy and nexus-based approach is essential for the development of the global blue economy, commencing with an initial assessment, and critically, the evaluation of the blue capital at our disposal. This book provides a basis for informed decision-making and adaptive management. This major undertaking must be addressed and continually refined and upgraded in line with changing circumstances, evolving technologies, and our increasing understanding, otherwise the blue economy approach will flounder. This underlines the importance of technical assistance, technology transfer, and capacity building in the pursuit of sustainable development of a blue economy.

This book has elaborated on the existing economic importance of maritime economic activities that establish the scope, aims, technology, and policy integration for a future blue economy. It also looks at the current state of research and technology development in these economic activities based on an extensive patent and publication analysis on global blue economy sectors and their future directions. This book will provide a suite of opportunities for sustainable and equitable blue growth in both traditional and emerging sectors including shipping and port facilities, fisheries, aquaculture, tourism, energy, and biotechnology. Oceans are critical to satisfying Earth's life support systems and the billions of people who are reliant on oceans for livelihoods, food security, and economic development. Some of the chapters of this book will discuss the marine-based economic development that leads to enhanced human well-being and social equity, while explicitly reducing environmental risks and ecological insufficiencies. The concept of a blue economy is a topic open to several interpretations based on the geographical locations and sectors discussed in the available literature. Representative sectors and the activities are described and discussed in *Global Blue Economy: Analysis, Developments, and Challenges*.

The book focuses on fundamental concepts, tools, and pillars of the blue economy, the generation of new resources to supplement a blue economy, and offers several case studies that investigate selected global scenarios and the technology-policy nexus of a global blue economy. The book gives an indication of how to manage complex and dynamic blue economy sectors in an ecological and sustainable manner within the domain of international maritime negotiations. It also investigates the problems that threaten marine ecosystems and presents a set of management toolboxes and models for solving the issues that challenge a sustainable blue economy. *Global Blue Economy: Analysis, Developments, and Challenges* consists of 17 chapters emphasizing the fundamental concepts, tools, and pillars of the blue economy, including marine ecosystem services, the prevailing blue economy and SDG linkages, sustainable fisheries management, aquaculture engineering, effective and modern seafood cultivation techniques to increase global food security, and exploration technologies for deep sea mining and drilling. The book will explore the energy demands, off-shore oil and gas exploration methods and techniques, identify the opportunities of off-shore wind power, wave, and tide energy, ship building, submarine, and underwater robotics vehicles, and examine the productivity of hydrocarbons of blue-green algae, seaweed, and biofuels, and address ocean health and pollution.

The blue economy emphasizes conservation and sustainable management, which are fundamental to managing the ocean in a sustainable manner. The concept seeks to promote economic growth, social inclusion, and preservation or improvement of livelihoods, while ensuring environmental sustainability. Realizing the importance of the blue economy globally, Chapter 2 aims to analyze the potential for a blue economy in Malaysia in various sectors related to the ocean, namely, ports and shipping, offshore oil and gas, fisheries, coastal tourism, and maritime transport. The chapter also identifies the opportunities and challenges drawn from the blue economy concept, in general. Finally, the chapter concludes by highlighting future plans in Malaysia for adapting the blue economy concept in the maritime sector, which will contribute to the sustainable social and economic growth of the country.

An interesting case study on salmon farm operators seeks to meet this need either by the establishment of new locations, or by the expansion of existing sites. Finding space for new sites can be challenging, due to interactions between them and the surrounding environment. As a result, operators are looking toward more physically exposed areas in which to place sites. Parasitic sea lice pose a particular environmental challenge and public concern. The enhanced availability of host fish in farms can allow lice to reach much greater numbers than they would in the absence of farms. Operationally, threshold abundances are used to define when management action should be taken, but there is no consensus approach to determining whether a particular proposed development will present an unacceptable risk in terms of heightened sea lice infestation levels and consequent risk to wild fish. We demonstrate the use of biophysical models in assessing physical suitability and parasite risk for two hypothetical farm sites. In a broader context, we demonstrate how infestation pressure is likely to decrease with farm isolation, and to a lesser extent with wave exposure.

Fish farming operators are seeking suitable offshore sites as an inevitable choice for sustainable and high-quality fish production. However, offshore fish farming has its challenges due to a relatively high energy environment with poor accessibility in the more remote sites. Nevertheless, combinations of fish farming with other marine activities are desirable from an economic viewpoint. The overall infrastructure and operational procedure will no doubt be more complex, and the increased functionalities will bring more risks and require more rigorous assessments for warrants and insurance coverage than solely fish farming activity. More research and developments are needed in this area.

One of our goals with this book was for it to have broad application to fundamentals concepts, pillars, and approaches of the blue economy. Sustainable blue economy approaches will change current methodologies used for resource assessment and future regulation of marine resources. Conventional fishery economic theory focuses on single species, hence neglecting externalities from the fisheries on marine ecosystems, for example, the food web and bottom structures. It is expected to realize environmental risk finance for raising this huge cost. We report a practical theory that utilizes statistical, ecological, and financial models to allow projection onto challenges to water environment preservation under anthropogenic pressure. The key component of our theory is the physical-biogeochemistry model that elaborates the mechanism of water environment adaptation, no matter whether it be estuarine, inland sea, or lacustrine environment, to aggressive biotic and abiotic stresses in a deterministic fashion. Numerical simulations, which are based on future scenarios of meteorological variables generated by the stochastic time series model, allow us to calculate the likelihood that environmental risks materialize. Risk quantification enables environmental risks to be tradable in the financial markets; thus, we can cover any costs associated with them through a risk finance framework. Our approach is to propose ecosystem-based adaptations to such potential disasters within the scope of risk finance. A case study of Biwa Lake shows how such a cutting-edge theory has advanced the understanding of resilience to water environment challenges and contributed to more informed decision-making about the blue economy.

The Central Indian Ocean Basin (CIOB) is one of the largest and richest polymetallic nodule-bearing areas in the world's oceans. The basin has varied morpho-structures, sediments, and rocks and other materials that are 'seeds' for growth of nodules through hydrogenetic and diagenetic processes. The abundance of $5\,kg/m^2$ of nodules on the seafloor, metal grade of 2% (copper, nickel, and cobalt), and areas of low slope angles collectively make the CIOB a potential target for nodule mining. For more than four decades, India has been carrying out a nodule programme that largely encompassed exploration and collection of baseline data for environmental impact assessment studies. As a contractor with the International Seabed Authority, India has certain obligations to fulfil prior to the mining of the nodules. In Chapter 7, we provide a gist of the different investigations conducted by India in the CIOB and this is followed by processes developed for metal beneficiation from the nodules, technological progress that is underway to mine the nodules, and the economic viability of mining the nodules. There exist a number of reports that pertain to the geological and geophysical characteristics of the CIOB, but the implementation of a blue economy has never been discussed with all the seriousness that it deserves. Hence, we address this important concern and the ways in which India could move the different sectors of the blue economy toward a profitable and sustainable mining of the nodules.

The major seaweed culturing methods that are followed by farmers in the coastal regions of Tamil Nadu and other parts of India are (i) the single rope floating raft method, (ii) the fixed bottom long line/monoline method, (iii) the bamboo raft method, and (iv) the net bag method. In India, there is a strong need to expand coastal seaweed cultivation to offshore large-scale farming. Currently, the major problems associated with the seaweed industry are overexploitation of raw materials, low quality of stocks, and lack of labor. Besides, most seaweeds are prone to epiphytism, and they are colonized by epibionts such as bacteria, protists, algae, and invertebrates. High-resolution hydro-dynamic modelling should be carried out before and after constructing farming structures for higher productivity. Furthermore, surveys need to be conducted to identify the seaweed cultivation zones to promote livelihood activities in coastal areas, with advanced seaweed mechanized vehicles within large spatial scales to identify proper localities for large-scale seaweed culture.

Take, for example, our ability to assess the fact that the increasing world population cannot be supported without new infrastructures based on the utilization of the oceans, which store almost all water, carbon, and energy resources on Earth. It is important, however, to assess the associated technologies and systems from the viewpoint of sustainability since the goal is to develop a sustainable blue economy. It is a way of deriving the economic growth of a nation through its contribution from the ocean and coastal-based activities while assuring environmental sustainability and livelihood development. Sri Lanka is an island nation of the Indian Ocean with an EEZ of approximately eight times its terrestrial extent together with a continuous coastline, where the concept of the blue economy is vital for the development of the country. Therefore, Chapter 8 discusses the development and challenges of the Indian Ocean Blue Economy and opportunities for Sri Lanka and its relationship with the other nations of the region. In relation to its geographical significance through the connectivity of the island to the east-west maritime route, where half of the world's trade is taking place through this region, the country has had a history in international maritime trading since ancient times. As reflected by the Linear Shipping Index, Sri Lanka shows its importance by being ranked fifth among the other Indian Ocean Rim Countries. Coastal waters of the island are rich in marine biological resources, indicated by more than six hundred species, while this diversity, together with natural beaches and tropical climates are of service to the blue economy through tourism. The coastal, offshore, and deep-sea fisheries of the country play a significant role in economic terms via value addition to national income, employment, and foreign exchange through exports. The world's capture fisheries and aquacul-ture production have reached its peak in Sri Lanka though the policymakers doesn't create too much space for enhancing blue economy. But Sri Lanka has still great potential to expand its aquaculture and fisheries.

The Indian Ocean with its four major ridge systems collectively forms the Indian Ocean Ridge System (IORS). The hydrothermal fields along these ridges, namely the Carlsberg Ridge (CR), the Central Indian Ridge (CIR), the South West Indian Ridge (SWIR), and the South East Indian Ridge (SEIR) are slow to ultra-slow spreading ridges. The IORS are geologically, tectonically, and petro-logically different among themselves and from the global mid-ocean ridges. Unlike the other slow spreading and well-studied Mid-Atlantic Ridge, the number of hydrothermal vent sites that host seafloor massive sulfide (SMS) deposits is few and far between along the IORS. Yet, in the future it may be viable to mine the SMS deposits because of their economic potential and the need for them in hi-tech industries. We present an overview of the hydrothermal sites and associated SMS that occur along the IORS as reported by several researchers. The inception of and investigations of the

IORS by India are detailed. This is followed by the importance of applying the paradigm of the blue economy and its various facets by India in the exploration, exploitation, and allied activities for the SMS resources. In the long run it is perhaps feasible to profitably recover the SMS with minimal harm to the environment by having various mitigation measures in place.

The deep-sea mining of mineral resources is moving toward reality. This blue economy sector has attracted attention because of the discovery of deposits presenting significant amounts of metals of economic interest, such as nickel, copper-cobalt, and rare earth elements. The global transition to clean energy with a low-carbon economy increased the demand for such metals, which serve as raw materials for renewable energy infrastructure and novel technology. However, knowledge gaps about life in the deep-sea, uncertainty, and doubt about the methods and techniques of deep-sea mining and their potential risks and impacts on marine ecosystems are the main challenges facing commercial activities of these minerals. Chapter 12 analyses the main opportunities and challenges facing deep-sea mining. For this reason, in this chapter, we used a political, economic, social, technological, legal, and environmental (PESTLE) analysis tool combined with a SWOT analysis (the Strengths, Weaknesses, Opportunities, and Threats) powered by a literature review of available data and information related to deep-sea mining.

Within the last few decades, the industry of seafood production/harvesting has seen a massive uplift. Farm-based seafood production is increasing day by day to meet the high consumption numbers as it is becoming popular among consumers. Numerous aquaculture methods are available according to the condition of the harvesting area and harvested species. These methods have their own pros and cons. In enhancing the blue economy these methods play a vital role. Seafood imports from developing countries are also increasing. Countries like Bangladesh are playing vital roles in exporting seafood to the global market and making sure that the demand is fulfilled. After the maritime boundary dispute settlement on the Bay of Bengal, more doors of opportunity in this sector have opened. However, this is still lacking in sustainable management and the production of seafood. Seafood for sustainable development will help Bangladesh to achieve a sustainable production and import hub for the whole world.

In recent years, the ocean plastic waste problem has become a worldwide concern, especially after many saw pictures of straws stabbed in sea turtles' noses or dead whales with massive amounts of plastic packaging in their stomachs. The US and Japan consume massive amounts of disposable plastic products every year, and most of the waste plastic products are exported to China instead of being recycled domestically. From January 2018, affected by China's ban on waste plastic importation, the US and Japan had to learn how to deal with waste plastic products domestically. During the G7 summit meeting held in Canada in 2018, participants adopted the 'Blueprint for Healthy Oceans, Seas and Resilient Communities' that outlines commitments related to resilient coasts and coastal communities to solve ocean plastic waste problems. Moreover, England, France, Germany, Italy, Canada, and the EU also signed the 'Ocean Plastic Charter' to further strengthen plastic management. Meanwhile, it is noticeable that neither the US nor Japan signed the charter.

Scientific research upholds the notion that the world is rapidly recognizing the value of the blue economy. Concerns about the ocean and the coastal regions have encouraged policymakers and academic institutions throughout the world to seek more and better understanding of the blue economy and ocean-based activity. Every nation makes significant decisions in terms of management operations, data mining, data analysis, monitoring, and product creation. In one chapter, these judgments are taken and carried out based on the current situation and future goals. Because there are some inconsistencies in perception, additional dialogue and discourse, including this case study of the ocean-based economy, is required. Another chapter also covers global marine pollution scenarios, blue economy and ocean health perceptions, pollution and their impact on marine ecosystems and blue economies, and marine pollution management approaches for guiding nations in their future planning, leading the world to a sustainable future and strengthening the global blue economy. There are several management ideas and alternative management options for the growth of the blue economy which include identifying and stopping pollution, enhancing global interaction, evaluating and promoting developmental accomplishments, and promoting ocean health development and the blue economy. Furthermore, in terms of the scope and richness of research and assessment, there remains a long way to go. The chapters of this book have been compiled and analyzed and contain contrasting scenarios in the Indian Ocean region including India, Malaysia, Japan, Sri Lanka, and Bangladesh. This book explores case study research methods and models to help readers categorize the problem, identifies global marine regime creation challenges for a blue economy, and presents both conceptual and simulation models to predict the future technology-policy nexus of the global blue economy.

Each chapter of this book brings fresh ideas to this new, emerging scientific frontier of *Global Blue Economy: Analysis, Developments, and Challenges.* The book presents viewpoints of the authors on the challenges involved in the design and implementation of sound environmental management of marine ecosystems and is valuable to both academics and practitioners wishing to deepen their knowledge in the field of marine ecosystems and approaches to their management. We offer a formal and heartfelt thank you to all the authors for providing their collaborative insights and for putting up with us during the editorial phase of producing this book. We greatly appreciate the superb editorial work and patience of Irma Britton and others.

This book would not have been written without decades of collegial interactions and community engagement with our peers, students, and mentors and our forward-thinking community, stakeholders and researchers who have advanced the concepts of environmental management of marine ecosystems for marine science, the environment, and society.

Md. Nazrul Islam
Steven M. Bartell

Acknowledgments

The editors would like to acknowledge the help of all the people involved in this book project and, more specifically, to the authors and reviewers that took part in the review process. Without their support, this book would not have become a reality. The editors are grateful to express their gratitude to the many people who provided support, offered comments, allowed them to quote their remarks, data, and information, and assisted in the editing, proofreading, and design.

First, the editors would like to thank each one of the authors for their contributions. Our sincere gratitude goes to all chapter authors who contributed their time and expertise to this book. Second, the editors wish to acknowledge the valuable contributions of the reviewers regarding the improvement of quality, coherence, and content presentation of chapters. Most of the authors also served as referees and we highly appreciate their twofold task. The late Prof. Sven Eric Jorgensen was the higher-ranking mentor of the first editor of this book and deserves much more credit than he generally receives. His creative and supportive influence was felt strongly and was hugely useful to encourage us to write this book series.

I would like to thank Irma Britton, Senior Editor, Environmental and Engineering, CRC Press/Taylor & Francis Group for enabling me to publish this book. I would like to thank Michele Dimont, Project Editor, CRC Press/Taylor & Francis Group for helping me in the process of selection, editing, and production.

My father had a dream that I could cross the border of Bangladesh and contribute to the world through my research and scientific writing. Today my father would have been very happy to see the publication of this excellent book. Unfortunately, I lost him a few years ago. In this beautiful moment today, I pray to the Almighty Allah (SW) that He might provide my father a heavenly place. My mother is very happy to know about the publication of this book. She has prayed for my good health so that in the future I can do many more significant research projects and write more high-quality academic books for the welfare of future generations.

Above all, I would like to thank my beloved wife, Sahanaj Tamman, my loving daughter, Sababa Mobashira Islam, and the rest of my family, who supported and encouraged me despite the time it took me away from them. Especially my beloved wife and adorable daughter, their genuine support and love helped me to work so hard. I couldn't give them the precious time and love they deserve in many cases because I was busy writing this book. I remain eternally grateful to them as it was a long and difficult journey for them.

Last and not least, I beg forgiveness to all those who have been with me over the course of the years and whose names I have failed to mention.

I hope you like what we have done.

Md. Nazrul Islam
Dhaka, Bangladesh

About the Editors

Md. Nazrul Islam is a permanent professor in the Department of Geography and Environment at Jahangirnagar University, Savar, Dhaka, Bangladesh. Prof. Islam earned his PhD from the University of Tokyo, Japan. In addition, he has completed a two-year standard JSPS postdoctoral research fellowship from the University of Tokyo, Japan. Prof. Islam's fields of interest are: environmental systems modeling, climate change and risk modeling, modeling of phytoplankton transition, harmful algae, and marine ecosystems with regard to dealing with hydrodynamic ecosystems coupled models on coastal seas, bays and estuaries, application of computer-based programming for numerical simulation modeling, and more. Prof. Islam is an expert on scientific research techniques and methods to develop the models for environmental systems analysis research. Prof. Islam has also visited as an invited speaker in several foreign universities in Japan, US, Australia, UK, Canada, China, South Korea, Germany, France, the Netherlands, Taiwan, and Vietnam. Prof. Islam has been awarded the Best Young Researcher Award by the International Society of Ecological Modeling (ISEM) for his outstanding contribution to the ecological modeling fields, 2013, Toulouse, France. Prof. Islam has made more than 40 scholarly presentations in more than 20 countries around the world, authored more than 150 peer-reviewed articles, and authored 15 books and research volumes. Currently, Prof. Islam has jointly published an excellent textbook entitled, *Environmental Management of Marine Ecosystems* with the late Prof. Sven Erik Jorgensen by CRC Press/Taylor & Francis. He has also currently published some excellent series of books entitled, *Climate Change Impacts, Mitigation and Adaptation in Developing Countries* (case studies on Bangladesh I, Bangladesh II, India I, India II and India III, Springer Publication, the Netherlands, Germany, and USA). Prof. Islam is currently serving as Executive Editor-in-Chief of the journal, *Modeling Earth Systems and Environment*, Springer International Publications.

Steven M. Bartell has extensive experience and technical skills in quantitative ecosystem analysis, ecological modeling, and ecological risk assessment. He contributed extensively to the development of the *USEPA Framework and Guidelines for Ecological Risk Assessment*. Dr. Bartell has applied his modeling skills in assessing ecological risks posed by eutrophication, ionizing radiation, chemical contaminants, invasive species, habitat degradation, and altered hydrology. Dr. Bartell has professional working knowledge on the use of toxicity data for assessing human and ecological risks. He has developed quantitative methods for extrapolating toxicity benchmarks for use in forecasting ecological risks to populations, communities, and ecosystems. Dr. Bartell has also developed complex aquatic ecosystem models in support of coastal marine ecosystem management and restoration.

Contributors

Nagi Abdussamie
University of Tasmania
Australia

Thomas Adams
Scottish Sea Farms Limited, South Shian
Argyll, UK

Md. Wahidul Alam
University of Chittagong, Chittagong,
Bangladesh

Dmitry Aleynik
Scottish Marine Institute,
Oban, UK

Mir Mohammad Ali
Sher-e-Bangla Agricultural University,
Dhaka, Bangladesh

Ankeeta A. Amonkar
Dnyanprassarak Mandal's College and
    Research Centre,
Mapusa, India

Keerthi Sri Senarathna Atapaththu
University of Ruhuna,
Matara, Sri Lanka

Joerg Baumeister
Griffith University
Australia

Md. Simul Bhuyan
Bangladesh Oceanographic Research Institute,
Cox's Bazar, Bangladesh

Yunil Chu
University of Queensland St Lucia,
Queensland, Australia

Nawalage S. Cooray
International University of Japan,
Japan

Mohan Kumar Das
National Oceanographic And Maritime
    Institute, Dhaka, Bangladesh

Monika Das
Matshya Bhaban, Dhaka,
Bangladesh

Keith Davidson
Scottish Marine Institute,
Oban, UK

Ramadoss Dineshram
CSIR - National Institute of Oceanography
Goa, India

Hanizah Idris
University of Malaya
Malaysia

Ryo Ikeda
Tohoku University
Japan

Temjensangba Imchen
CSIR - National Institute of
    Oceanography
Goa, India

Md. Nazrul Islam
Jahangirnagar University, Savar, Dhaka,
Bangladesh

Md. Shahriar Islam
Jahangirnagar University, Savar, Dhaka,
Bangladesh

S. M. Rashedul Islam
Jahangirnagar University, Savar, Dhaka,
Bangladesh

Sridhar D. Iyer
CSIR-National Institute of Oceanography,
Dona Paula, Goa, India

Muthuswamy Jaikumar
Gujarat Institute of Desert Ecology
Gujarat, India

Dong-Sheng Jeng
Griffith University
Australia

Niyati Gopinath Kalangutkar
Goa University,
Taleigao Plateau, India

Khaled Mahamud Khan
Jahangirnagar University
Dhaka, Bangladesh

Kentaro Kikuchi
Faculty of Economics, Shiga University
Japan

Daisuke Kitazawa
Institute of Industrial Science (IIS)
The University of Tokyo, Japan

Hideya Kubo
The Organising Committee of the World
Masters Games Kansai,
Japan

Xiaoyue Liu
Tohoku University
Japan

Gaku Manago
Tohoku University
Japan

Sourav Mandal
CSIR - National Institute of Oceanography
Goa, India

Nezha Mejjad
Faculty of Sciences, Ben M'sik, University
   Hassan II
Casablanca, Morocco

Istiak Ahamed Mojumder
University of Chittagong, Chittagong,
Bangladesh

Sobnom Mustary
Birkbeck, University of London, UK

Md. Noman
Jahangirnagar University, Savar, Dhaka,
Bangladesh

Kazuaki Okubo
Tohoku University
Japan

Shiori Osanai
Tohoku University
Japan

Upul Premarathna
University of Ruhuna, Wellamadama,
Matara, Sri Lanka

Tilak Priyadarshana
University of Ruhuna
Matara, Sri Lanka

Kannan Rangesh
Madurai Kamaraj University,
Tamil Nadu, India

Md. Rashed-Un-Nabi
University of Chittagong, Chittagong,
Bangladesh

Marzia Rovere
National Research Council,
Bologna, Italy

Ranjan Roy
Sher-e-Bangla Agricultural University,
Dhaka, Bangladesh

Kevin Roy B. Serrona
Prince George's County Government
USA

Al Rabby Siemens
Jahangirnagar University
Dhaka, Bangladesh

Sahanaj Tamanna
Bangladesh Environmental Modeling Alliance
   (BEMA), Mirpur, Dhaka, Bangladesh

Tadao Tanabe
Shibaura Institute of Technology
Japan

Kosuke Toshiki
University of Miyazaki
Japan

Chien Ming Wang
University of Queensland St Lucia,
Queensland, Australia

Shuoyao Wang
Shanghai SUS Environment Co. Ltd,
China

Takero Yoshida
Tokyo University of Marine Science and
    Technology
Japan

Jeongsoo Yu
Tohoku University
Japan

Hong Zhang
Griffith University
Australia

Jinxin Zhou
The University of Tokyo
Japan

1 Concepts, Tools, and Pillars of the Blue Economy

A Synthesis and Critical Review

Md. Nazrul Islam
Department of Geography and Environment, Jahangirnagar University,
Savar, Dhaka, Bangladesh
E-mail: nazrul_geo@juniv.edu

CONTENTS

1.1 CONCEPTUAL PARADIGM OF THE BLUE ECONOMY

The term 'blue economy' denotes a new idea about the better management of marine ecosystem services, also called 'blue' resources (Kathijotes, 2013; Islam et al. 2018; Keen et al. 2018; Bir et al. 2020; Martínez-Vázquez et al. 2021). The International Maritime Organization (IMO) said that it was essential to properly utilize water resources, specifically marine water, as a sustainable resource for future generations (Bigg et al. 2003; Karani and Failler, 2020; Kabil et al. 2021; Kabil et al. 2021). The blue economy aspires to promote better well-being and radical egalitarianism while simultaneously lowering environmental risks and environmental inadequacies in ocean biodiversity. As with the 'green economy', the blue economy aspires to consider human and social development in the context of the environmental and ecological degradation of the ocean (Behnam, 2012; IFFO, 2013; Ebarvia, 2016; Voyer, 2018; Garland et al. 2019; Lee et al. 2020). The economic activities of coastal dwellers, such as fishing, shipbuilding, maritime transportation, coastal tourism, and so on, are directly and indirectly included in the blue economy. The term 'blue economy' is a relatively more modern concept than the idea of water resources and water awareness (Raakjaer et al. 2014; Hussain et al. 2017; Sarker et al. 2018; Lee et al. 2020).

The blue economy sector is currently an emerging initiative led by Small Island Developing States (SIDS), but it applies to all coastline states or countries with an interest in waters beyond their borders (Pinto et al. 2015; Hadjimichael, 2018; Bennett et al. 2019; Lu et al. 2019). The SIDS are a distinct group of 38 UN Member States and 20 Non-UN Members/Associate Members of United Nations regional commissions that face unique social, economic, and environmental vulnerabilities. However, the SIDS have traditionally relied on maritime resources for growth. The blue economy, although including the notion of ocean-based economies, encompasses much more (World Bank, 2017; Kabil et al. 2021). Oceans are treated as 'Dynamic Spaces' in the blue economy, with spatial planning incorporating resource usage, conservation, and long-term use, as well as oil and mineral extraction, nanoparticle production, renewable energy generation, and maritime transit (IGBP, 2013; Bennett et al. 2019; Nagy and Nene, 2021). In economic modeling and decision-making processes, the blue economy currently covers accountancy, appraisals, and solutions for coastal resources (Lu et al. 2019; Kabil et al. 2021). The blue economy is a model for poor nations to achieve long-term progress by addressing fairness in entrance to, expansion of, and sharing of wealth from ocean biodiversity (Behnam, 2012; Ebarvia, 2016; Martnez-Vázquez et al. 2021).

The blue economy is gaining popularity in many nations throughout the world, with some even putting it on their national agendas to strengthen their marine policy (Behnam, 2012; Kathijotes, 2013; Silver et al. 2015; Keen et al. 2018). Many nations have taken steps to strengthen their action plans by focusing on the sustainable exploitation of aquaculture resources, global warming, and protection of the environment. In addition, national policies have been updated to increase awareness of, and to take some steps towards reducing poverty among the people by creating alternative jobs that employ marine habitats (Ebarvia, 2016; Bennett et al. 2019; Nagy and Nene, 2021). To support the blue economy, several governments have produced national policies and action plans. Many poor nations, on the other hand, are still disregarding the blue economy sectors in order to benefit from these new marine resources (Islam et al. 2018; Nagy and Nene, 2021). Several nations have developed national strategies and initiatives to promote the blue economy. However, many developing nations are still ignoring the blue economy industries in order to make use of these new marine riches (Islam et al. 2018; Nagy and Nene, 2021).

By connecting vendors and buyers, seas are becoming increasingly crucial in promoting global trade (Brodie et al. 2020) (Figure 1.1). As the significance and relevance of the relationship between land and water grows, the patterns of behavior of such trade organizations on the waters are gaining greater regulatory and commercial attention (Bigg et al. 2003; Karani and Failler, 2020; Kabil et al. 2021). The notion of the blue economy was formed against this backdrop. The term 'ocean economy' or 'blue economy (BE)' is indeed a novel idea born out of the 2012 United Nations Conference on Sustainable Development in Rio de Janeiro (Martinez-Vázquez et al. 2021). The blue economy

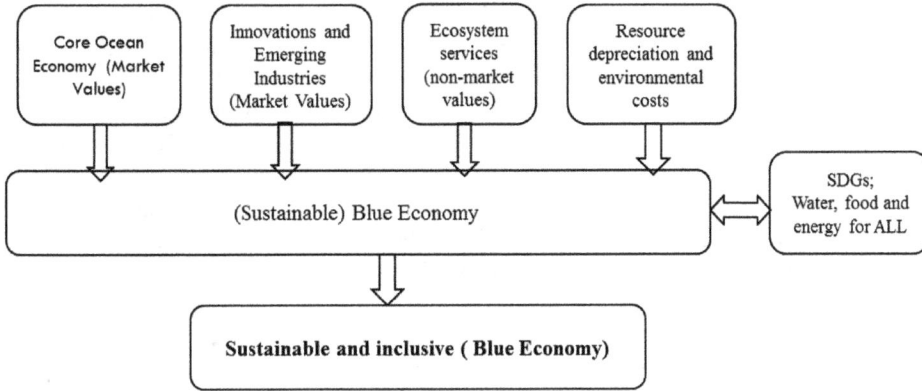

FIGURE 1.1 The blue economy is promoted with the goal of improving human well-being and social equality to achieve long-term prosperity.

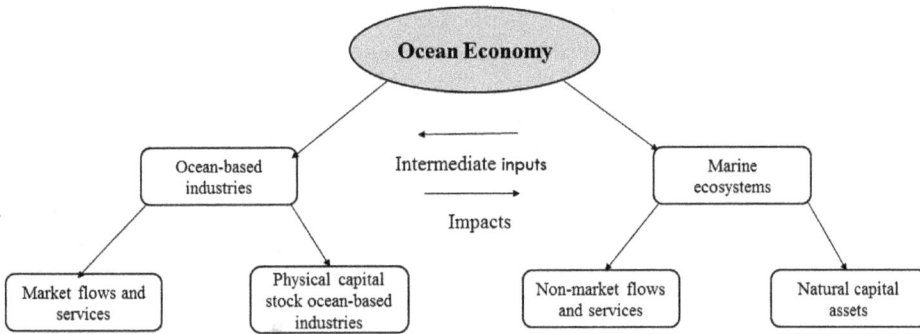

FIGURE 1.2 Fundamentals of blue economy (OECD, 2016; the Ocean Economy in 2030).

is gaining popularity as a strategy for safeguarding the world's oceans and water resources (Bigg et al. 2003; Karani and Failler, 2020; Kabil et al. 2021). Whenever economic growth is aligned with the long-term ability of marine ecosystems to maintain the activity, blue economy considerations may well emerge (Smith-Godfrey, 2016). More particularly, the notion of blue economy suggests inherent contradictions between two narratives: the development, growth and preservation of ocean resources (Sarker et al. 2018), and the protection of ocean resources (Sarker et al. 2018).

The notion of the blue economy is currently associated with business and economic actions, and it stems from the necessity to incorporate restoration and endurance into the calculations in maritime administrator's accounts (Ebarvia, 2016; Voyer and Leeuwen, 2019). The ecosystem or biodiversity of the water could also be expanded (Kathijotes, 2013; Keen et al. 2018; Brodie et al. 2020).

The environmental sustainability plugin enables the incorporation of consumption and refilling, whilst providing low or no greenhouse gas (GHG) outcomes, which is critical to the process of engaging in various activities (Smith-Godfrey, 2016; Kathijotes and Sekhniashvili, 2017; Sarker et al. 2018). Another component of durability also refers to the long-term viability of the sea as a food source for both humans and animals (Zhang et al. 2004; Lu et al. 2019).

As per the World Bank's Refinement in Island Emerging States concept document (Patil, 2016), 'Blue Economy is an undersea economic boom that results in improved people happiness and social fairness while minimizing overall implications and ecological resource scarcity' (Kathijotes, 2013; Keen et al. 2018) (see Figure 1.2).

In 2009, Maria Cantwell, United States Senator of Washington State, pointed out in the opening statement of the hearing on "The Blue Economy: The Role of the Oceans in our Nation's Economic Future" that "The "Blue Economy" – the jobs and economic opportunities that emerge from our oceans, Great Lakes, and coastal resources – is one of the main tools to rebuilding the United States economy.": The Participation of the Ocean waters in our Nation's Financial Future' (World Bank, 2017; Wenhai et al. 2019; Martnez-Vázquez et al. 2021). Similar international organizations, such as the United Nations Environment Program (UNEP), distinguish between blue-green economy (Johnson et al. 2018). They encourage reduced-carbon, asset transportation, fisheries, marine tourism, and marine energy firms as a method for addressing climate change (UNEP et al. 2012; Ebarvia, 2016; Johnson et al. 2018).

Whenever reference is made to the ocean economy, it is commonly believed that it is a fluid concept that is implemented in various settings and by various players. Silver et al. (2015) looked at why the word was utilized during the Rio+20 World Conference, noting how well the 'financial sector' was a concept employed by several parties throughout the negotiations to promote certain concepts and initiatives (Voyer et al. 2018).

There were four dominating discourses identified:

- Ocean as Environmental Equity: a concept used by quasi ecological groups to recommend that the essential services supplied by aquatic habitats be acknowledged and receive greater appreciation.
- Oceans as Good Business: marine industries such as fishing and transportation, as well as development banks, lobbied for widespread awareness of ocean-based enterprises and their global benefits (Martinez-Vázquez et al. 2021).
- Oceans as a Critical Part of Western Growth: Pacific SIDS have been deeply involved in developing the economic base to meet their particular lives and quality objectives.
- Seas as Comparatively Tiny Fisheries' Future Prosperity: this topic focused on poverty reduction and the importance of small scale fisheries (SSFs) in providing protein and economic chances for the world's poor. Its promotion was greatly supported by SSF groups and supporters, including institutions worldwide.

The blue economy, in a nutshell, is the concept of 'achieving maximum natural oceanic resources and increase within ecological restrictions' and 'bifurcation of monetary success from environmental contamination' (Wenhai et al. 2019; Martnez-Vázquez et al. 2021). The blue economy, as shown in Table 1.1, is made up of a variety of interconnected industries that use the richness of the waters to stimulate economic growth via environmental sustainability.

1.2 UNITED NATIONS (UN) CONCEPT OF THE BLUE ECONOMY

The blue economy, as per a UN spokesperson, ' "A blue economy is a long-term strategy aimed at supporting sustainable economic growth through oceans-related sectors and activities, while improving human well-being and social equity and preserving the environment," (World Bank, 2017; Heidkamp et al. 2021). There are a number of obstacles and challenges that must be resolved in order to strengthen the global blue economy. Defining and effectively managing the multiple components of maritime sustainability, from commercial fishing to organism integrity and pollution avoidance, is a big challenge for the blue economy (World Bank, 2017; Lee et al. 2020). Furthermore, we must understand that long-term ocean capacity planning will be implemented jointly with local ecosystem based management approach (Sharafuddin and Madhavan, 2020). This is a massive challenge, especially given the limited resources of SIDS and least developed countries (LDCs).' According to the UN, the blue economy will help the UN's future developments, one of which is 'Life Below Water' (Kathijotes, 2013; Keen et al. 2018; Bennett et al. 2019). The United Nations

TABLE 1.1
The Blue Economy Is Made Up of Interrelated Industries That Use the Resources of the Waters to Drive Economic Progress via Sustainability

Activity Type	Service to the Sea	Factory	Growth Drivers	References
Living resource harvesting	Seafood	Fisheries	Food Security (demand for food, nutrition and protein)	Sarker et al. (2018)
		Trade of seafood products	Food demand, nutrition and protein	Wessells and Wallström (2019)
		Trade of non-edible seafood products	Cosmetics, pet food, and pharmaceuticals products are in high demand.	Ferdouse et al. (2018)
		Aquaculture	Demand for food, nutrition and protein	Frankic and Hershner (2003)
	Ocean biotechnology	Pharmaceuticals and chemicals	Necessary resources for healthcare and industry	Rasmussen and Morrissey (2007)
Extraction of nonliving resources and creation of fresh resources are two different things.	Minerals	Mining on the seafloor	Mineral demand	Mitra et al. (2021)
	Energy	Gas and oil Renewables	Alternative energy sources are in high demand.	Galván et al. (2016) Mitra et al. (2021)
	Pure water	Desalination	Fresh water demand	Holland et al. (2015)
Recurrent and quasi natural energies	Renewable energy generation (off-shore)	Renewables	Demand for alternative energy sources	Chu et al. (2015)
Economic activities and trade in and around the oceans	Transport and trade	Shipping Port infrastructure and services	Growth in seaborne trade; International regulations	Ebarvia (2016) Ebarvia (2016)
	Tourism and recreation	Tourism Coastal Development	Growth of global tourism Coastal urbanization Domestic regulations	Lenzen et al. (2018) Martins et al. (2012) Ebarvia (2016)
Indirect contribution to economic activities and environments	Ocean governance and protocol	Technology, Resources and Development	Resources and development in ocean technologies	Voyer and Leeuwen, (2019)
	Carbon control	Blue Carbon	Increased coastal marine preservation and conservation efforts	Baral and Guha (2004)
	Coastal Protection	Habitat protection and restoration	Conserve and prevent the marine species extinction, fragmentation, or reduction	Blumm (2017)
	Waste arrangement	Assimilation of nutrients and wastes	The process of nutrients being absorbed by each cell of the body in the form of energy	Gichana et al. (2018)

Source: Modified and adopted from Keen et al. 2018; Islam et al. 2018; Lee et al. 2020.

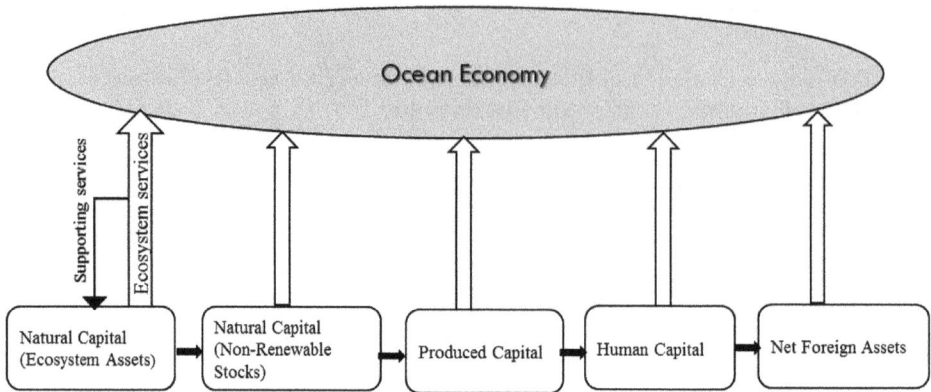

FIGURE 1.3 Ecosystems services and assets of blue economy visions (modified from the concept and definitions of Lange et al. (2018) and Brown et al. (2016)).

hosted its first conference on environmental stewardship in Stockholm in 1972, with the purpose of strengthening city planners (Lu et al. 2019; Lee at al. 2020). The next UN Convention, held in Rio in 1992 and focusing on the competitive advantage for the resilient company, was followed in 2002 by the Johannesburg symposium, which concentrated on the social aspect of sustainable development (Lee at al. 2020).

In the last decade, a range of stakeholders have widely pushed the blue economy or 'ocean/marine economics' as a notion or approach for safeguarding the nearby seas and coastal organisms (Pauli, 2010; Lee at al. 2020). The concept of the blue economy was born in Rio de Janeiro in 2012 as part of the concept of environmental sustainability (UNCTAD, 2014; Voyer et al. 2018). In the absence of precise definitions, words such as 'ocean economy' or 'marine economy' are being used (Johnson et al. 2018) (see Figure 1.3). The 'blue economy,' as defined by the United Nations, is an ocean economic model that aims to 'improve human well-being and radical equality while drastically decreasing environmental risks and natural resource shortages' (UN Report 2014, p. 2; Silver et al. 2015; Voyer et al. 2018).

The World Bank describes the blue economy as 'the responsible use of ocean diversity for economic progress, improved livelihoods, and creating jobs whilst maintaining the health of the oceanic ecosphere' (World Bank, 2017, p. 6; Lee at al. 2020). The definition provided by the global bank encompasses a wide variety of characteristics of marine resilience, from fishery resources to environmental quality and pollution avoidance (World Bank, 2017; Johnson et al. 2018; Lu et al. 2019). Importantly, the notion itself necessitates cross-sector and inter-sector collaboration across a wide range of stakeholders and associations. Therefore, tailored to their needs, different stakeholders will support different goals or concepts. It indicates that certain potential conflicts or challenges may arise from diverse participants' decisions or desires.

1.3 HOW CAN BUILDING A BLUE ECONOMY HELP US ACHIEVE THE SUSTAINABLE DEVELOPMENT GOALS (SDGs)?

The blue economy is a melting pot of development opportunities, as well as sensitive and fragile ecosystems that require conservation (Kathijotes, 2013; Keen et al. 2018; Lee at al. 2020). Because of the inherent inconsistencies between these two ideas, solutions must embrace the benefits of the ocean economy while simultaneously recognizing and reducing its drawbacks (Jones et al. 2020). According to the UN Development Strategy, income growth in the blue economy is both egalitarian and environmentally responsible, and it emphasizes the significance of balancing the commercial, social, and physical components of sustainable development in connection to the ocean (Griggs et al. 2013; Bennett et al. 2019). The UN has designated the years 2021 to 2030 as the 'Decade of Ocean Studies for Sustainability Goals,' for the purpose of encouraging measures to reverse the deterioration of aquatic biota and gathering ocean specialists from around the world under one roof (World Bank, 2017; Potgieter, 2018). This framework is designed to guarantee that marine science can adequately help developing nations in their long-term efforts to expand the ocean. In respect to oceans, the World Bank highlights 'managing the triple bottom lines of environmental protection' as a critical component of the BE (World Bank, 2017, p.4). In reality, however, achieving a balance is challenging due to the fact that ocean conditions have worsened significantly with the addition of industrial and human activities, with frequently conflicting aims such as contamination, unsuitable fishing, and biological degradation (United Nations, 2016).

The UN defines the blue economy as an ocean enterprise that aspires to 'increase human well-being and social fairness while considerably minimizing environmental hazards and ecological scarcity,' as articulated by the UN in 2014. UNCTAD (2014), p. 3. As per 'Resources Are taken for Nature' (2015), the blue industry is a saltwater economy that:

- Addresses poverty, livelihood possibilities, revenue, job prospects, health, safety, equality, and good governance for present and future generations through promoting food security, social equality, welfare, revenue, employment creation, wellness, protection, fairness, and democratic reform (Potgieter, 2018).
- Aims to ensure that marine ecosystems, which are the economy's natural capital, are repaired, protected, and appreciated for their variety, productivity, resilience, crucial services, and intrinsic value.
- Is based on green technology, sustainable sources, and cyclical material flows to assure long-term macroeconomic stability while staying within the natural limits of One Earth (Baltic Sea Action Plan (2013) Baltic Eco-region Programme, 2015, p. 1).

It is observed that development concepts span ecological and planetary borders, with turning points generating a fresh concern for re-evaluating the economy and revaluing the maritime economic link globally (Voyer and Leeuwen, 2019; Lee at al. 2020). Attempting to relate the blue economy to the UN's development interests is very problematic, especially when household or manufacturing goals such as lowering fossil-fuel-based carbon emissions, or supplying power, begin to compete or fight (World Bank, 2017; Lee et al. 2020; Lewis et al. 2021). The SDGs, together with 17 goals, 169 objectives, and 232 targets, are the product of a multi-stakeholder agreement among countries to minimize unsustainability and foster sustained growth (Sarker et al. 2018; Bebbington, 2018; Alexander and Delabre, 2019). However, establishing the size and scope of the blue economy in conformity with the UN's sustainable development goals is difficult, if not impossible. Furthermore, the relevant players in the ocean economy, as well as their respective interests and functions, are unclear (World Bank, 2017; Potgieter, 2018).

1.4 SCOPE AND ECONOMIC OPPORTUNITIES OF THE BLUE ECONOMY

Governments all over the world have been very much engaged in efforts to achieve self-determination over marine domains since the signing of the UN International Maritime Law Treaty in 1982 (Kildow and McIlgorm, 2010; Brodie Rudolph et al. 2020). Those maritime domains are sometimes vast, perhaps larger than the total area of a country, and they include a varied spectrum of biological and non-living elements (Nagy and Nene, 2021). The stagnation of traditional dry land industries, along with the depletion of earth resources, has sparked an interest in the economic prospects that exist beneath the ocean (OECD, 2016; Voyer et al. 2018). While maritime commerce and business is not new, recent developments point to a trend towards a more organize marine sector, one that combines conflicting uses, allocates 'property,' and provides procedures and management systems to maintain national assets under state control (Winder and Le Heron, 2017). In places outside national authority, including the open seas, UN-led negotiations are taking place to establish how deep marine resources should be used and controlled to safeguard species and to provide new prospects (Warner, 2009). As a result, the oceans have grown into development zones, creating new opportunities for coastal people and governments with maritime holdings to build and enhance their businesses (United Nations, 2014).

The blue economy is increasing in popularity as a new governance tool for addressing optimal ocean usage at the global, provincial, and national levels. Apart from the lack of a universally agreed definition, there is much uncertainty over the scope of governance of a blue economy (Behnam, 2012; Voyer et al. 2018). Based on the industries analyzed, the magnitude of the blue economy/blue development varies. Industries such as fisheries, aquaculture, ecotourism, transport, bioengineering, maritime security, quarrying, oil and gas, and sustainable sources use the seas and inland waters (Schutter and Hicks, 2019). These numerous sectors, and the flora and fauna that they support, have a direct influence on the marine environment and the flora and fauna that it supports. The purpose of an overall blue economy plan is to assess solutions for reducing the combined impact of diverse economic sectors on live water resources, biodiversity, and natural ecosystems (Cervigni and Scandizzo, 2017). Blue economy strategies have elevated food security, aquaculture, eco-system assistance, marine and coastal tourism, and respectable livelihood opportunities in a number of coastal developing countries, together with the SIDS, with the goal of progressively incorporating other key sectors based on their circumstances (Voyer and Leeuwen, 2019).

The blue economy idea is built on the separation of socioeconomic success and environmental degradation (Behnam, 2012; Sarker et al. 2018). The blue economy method is based on evaluating and integrating the true quality of natural (blue) equity into all areas of economic action (worldview, making plans, infrastructural facilities, trade, travel, renewable energy resource exploitation, and power supply) to achieve this (Table 1.2). While environmental and ecological standards must be

---

TABLE 1.2

Scope and Economic Opportunities of Marine Resources which Contribute to the Ocean Economy

Non-living Resource Extraction or Resource Creation	Exploitation of Live Resources	Trade and Commerce on and Near the Seas	Conservation and Management of Ecosystems
Mining on the seafloor/deep seafloor	Fish stocks	Shipping (marine transportation)	Blue carbon
Petroleum and natural gas	Aquaculture	Construction and maintenance of ships	Surveillance and marine security are two crucial features of marine safety
Water (desalinization)	Marine biotechnology	designing a jetty	Habitat protection/restoration
Dredging	Recreational fishing and boating	Port infrastructure and services	Hazard protection
Tidal/wave energy; coastal/ offshore wind) are examples of sustainable energy sources.	Seafood processing	Ocean-related services and on seas, research and development, and teaching coastline development Oceans and terrestrial tourism are being defended.	Ecological/ecosystem research Waste treatment and disposal

Source: After modified from the Economist, 2015; Voyer et al. 2018; Voyer and van Leeuwen, 2019.

respected, performance and resource optimization are critical (Voyer and Leeuwen, 2019). This means using local materials wherever feasible, as well as blue, low-energy solutions to provide utilization and advantages rather than the brown, high fuel, underemployment, and industrialized development models (Schutter and Hicks, 2019). Equity mainstreaming at the regional and international level allows developing nations to generate more money from their assets, enabling them to support their people, improve the environment, decrease budget deficit, and assist with the abolition of hunger and poverty.

1.4.1 Fundamentals of Blue Economy and So-called Lenses

In 2012, during in the UN Convention on Sustainability, often known as Rio+20, the phrase 'blue economy' was coined. The goal is to build on territories' green energy principles by stressing the development that is possible with appropriate ocean biodiversity management (Kathijotes, 2013; Ebarvia, 2016; Keen et al. 2018). Whereas the blue economy is mostly based on ecological sustainability principles, there is no commonly agreed definition of it yet. In practice, a wide range of players have utilized the term for a wide range of objectives. Voyer et al. (2018) confirmed four common perceptions of the maritime sector in current conversations, drawing on previous research by Silver et al. (2015). These four 'lenses' (Table 1.3) are as follows: (Voyer and van Leeuwen, 2019).

TABLE 1.3
Fundamentals of Blue Economy and So-Called Lenses

Fundamentals of Blue Economy Lenses	
Lens-01: Ocean as Natural Capital	**Lens-02: Ocean as Drivers of Innovation**
1) Attention on marine ecosystem services	1) Focus on technological and technical fixes
2) Environmental NGOs and marine economy	2) Underwater vehicles and surveying
3) Eco-tourism, shipping and MPAs	3) Including innovation hubs and others
4) Marine services valuations and utilization	4) Companies, ministries, and favored by several research organizations
5) Industries that use a lot of carbon (like oil and gas)	5) New sectors, such as renewable energy
6) Shoreline and coastal resources	6) Bioscience and deep ocean mining
7) Significant source of animal protein	7) Systematic investigation of marine resources
Lens-03: Ocean as Livelihoods	**Lens-04: Ocean as Trade and Commerce**
1) Alternative food security and poverty alleviation	1) Focus on economic growth and environment
2) Poor and underdeveloped states, proponents for small-scale fishing, and favored by development organizations	2) Favored by industries and large global economies (EU, OECD etc.)
3) Oceanography, aquaculture, and environmentalist on a small scale	3) Foreign businesses, shipping, and petroleum and gas
4) From the production of seafood through freight transportation	4) Sustainable energy and deep sea miners
5) Precautionary approach to deep sea mining	5) Helping to achieve sustainable development
6) Influences our weather and climate	6) Focus on petroleum dominates commerce
7) Biochemical and medicinal plants and animals	7) Offshore mining and power generation

Source: (After modification from Voyer et al. 2018; Voyer and van Leeuwen, 2019).

1.4.2 Differentiate between Green and Blue Economy

The UN Assemblies on Environmental Sustainability 'Rio +20,' which took place in Rio de Janeiro from June 20 to 22, 2012, focused on two primary themes: expanding the concept of green economy and building and upgrading the organizational system for nature conservation (Pretorius and Henwood, 2019; Nagy and Nene, 2021). The incorporation of the idea of maritime sector, which is gleaned from green economy, provided the impetus for 'sustainable'. Sustainable is taken in the context of the stability of both behavior (institutional, trade, legislative, and compliance), the revitalized dedication (ecosystem surroundings), and the cultures that rely on it (livelihood opportunities and food) (Danovaro et al. 2017). For selecting the optimal balance, this equilibrium may be analyzed and translated into an efficacy metric.

Throughout the Rio+20 planning process, several coastal nations questioned the green economy's emphasis and applicability to them. Strong arguments were made for a more prominent blue economy plan to be addressed during the Rio +20 preparation phase (Kathijotes, 2013; Keen et al. 2018; Sharafuddin and Madhavan, 2020). This approach is crucial since the oceans, particularly the shared heritage of the high seas, represent in many ways the last frontier for humanity's ambition for long-term progress. Efforts were undertaken at the institutional level to increase the blue component of the sustainable future, as mentioned in the paper 'Sustainable Future in a Blue World,' but global momentum has accelerated much faster (Garlock et al. 2020). Throughout the Rio +20 process, it became clear that the world's major oceans and seas demand more in-depth attention and concerted effort (Colgan, 2018). The Associations' expert panel symposium on oceans, seas, and sustainability, the Global Water Council's efforts, the Worldwide Coalition for Ocean waters, and the UN's five Strategic Agendas 2012–2016, which set a major priority on seas and oceans, all highlight this.

In 2012 and 2018, the European Union announced its 'Blue Growth Project' for the gradual development of the marine and maritime sectors, with the objective of advancing the Europe 2020 vision of smart, sustainable, and equitable growth (Guerreiro, 2021). The blue economy has been illustrated by APEC, the East Asia Summit (EAS), the South Asian Association for Regional Cooperation (SAARC), and the Indian Ocean Edge Association (IORA), all of which have inspired intergovernmental approaches, developed cooperative strategies and action plans, and progressed toward fish farming exploitation of resources (UNEP, 2013; Lu et al. 2019).

1.4.3 EU's Blue Growth Strategy and Blue Economy Innovation Plan

The European Commission will unveil its new plan for a sustainable blue economy in May 2021. Because both the European Green Deal and the SDG agenda demand the transformation of the EU's economy, the SBE Strategy takes a comprehensive approach to the EU's blue economy (Gureva, 2018). In 2012, the European Union unveiled the 'Blue Growth Plan', stating that blue growth will be at the heart of all maritime policy and pinpointing important development areas and particular initiatives for the future (Pretorius and Henwood, 2019; Voyer et al. 2018; Garlock et al. 2020). The Blue Business Strategy has launched initiatives in a wide range of policy areas relating to Europe's oceans, seas, and coasts, enabling cross-border and cross-sector cooperation among seafaring businesses, government bodies, and stakeholders to ensure the marine environment's long-term viability. The Blue Economy Development Plan was released in 2014, with the plan stating that it will be executed in three ways. As an example:

- Develop sectors with a high potential for long-term job creation and growth
- Essential components will give knowledge, legal clarity, and security in the blue economy
- Sea basin plans will ensure tailor-made measures and encourage international collaboration.

In 2017, the European Union issued their study on the Blue Growth Agenda for More Responsible Growth and Jobs inside the Blue Economy. This study examines what has been learned and achieved since 2012, as well as what is taking place currently, but also what is missing (Mulazzani and Malorgio, 2017; Potgieter, 2018). The study covers five topics: (i) promoting growth in five focus areas, including blue energy, aquaculture, coastal areas and seafaring tourism, blue biotech, and seafloor mineral resources; (ii) the advantages of ocean data, road networks, and seagoing surveillance to facilitate blue growth of the economy; (iii) trying to promote a partnership approach; (iv) growing incentive to invest; and (v) having to adapt blue business strategy to new challenges.

As per the Food and Agriculture Organisation (FAO), the commercial fishing sectors employ almost 60 million individuals worldwide, the highest numbers of workers are in Asia (85 percent), followed by Africa (9 percent), the Americas (4 percent), and Europe and Oceania (1 percent each) (FAO, 2020). Over 350 million people are employed in fishing, aquaculture, seashore and marine tourism (Chan et al. 2021). There were also 5 million people working in the blue economy sector in 2018, representing a significant increase of 11.6% compared to the year before (Dinati et al. 2021). Although sectors such as coastal and marine tourism, as well as fisheries and aquaculture are severely affected by the coronavirus pandemic, the blue economy as a whole presents a huge potential in terms of its contribution to a green recovery. (The Blue Economy Report, 2020; FAO (2020), Donati (2021)).

1.5 MAJOR PILLARS OF THE GLOBAL BLUE ECONOMY

The blue economy encompasses not just economic viability, but also the conservation and augmentation of intangible blue resources such as local traditions, carbon capture, and coastal resiliency to help vulnerable governments mitigate the impacts of global warming (Attri, 2016; Wenhai et al. 2019). The blue economy concept highlights the significance of norms in reducing poverty, guaranteeing food security and nutrition, reducing the problem of global warming, and building equitable and resilient employment (Voyer and Leeuwen, 2019).

As a consequence, the blue economy mixes economic expansion and long-term preservation. As a result, the blue economy aims to inspire economic development, connectedness, and the retention or development of livelihood strategies while also preserving the sustainability goals of the seas and coastlines (World Bank Group, 2017; Schutter and Hicks, 2019), a concept known as 'sustainable usage.' Many forward-thinking firms, industry associations, scientists, government officials, and marine activists are looking for ways to make the oceans economy a reality. As depicted in Figure 1.4, the fundamental pillars of the blue economy are: a) government accountability, b) vision

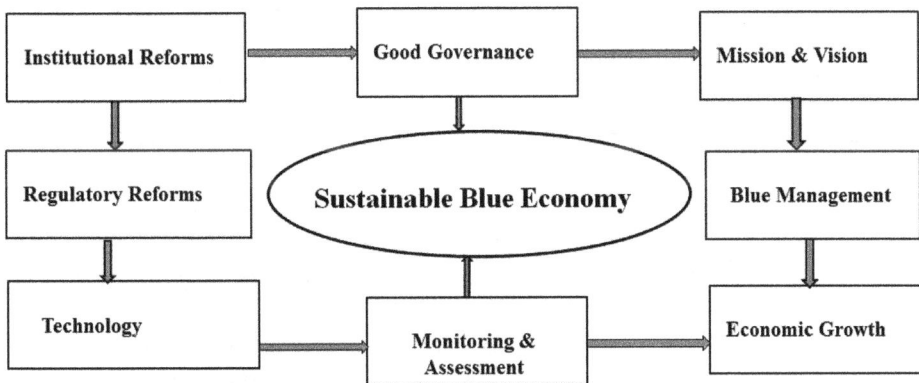

FIGURE 1.4 Major pillars of a sustainably blue economy and economic growth possibilities to foster economic growth, social inclusion, and the conservation or enhancement of livelihoods (Adapted from Attri, 2016).

and mission, c) technologies, d) blue administration, e) inspection, and f) legal and governance reforms. (Attri, 2016)

The sea is a rich form of renewable energy and non-renewable environmental assets that have provided different economic, societal, and cultural benefits throughout history and which have huge future potential (Voyer and Leeuwen, 2019). These benefits are typically obtained through sectors of the economy that have increased in number and activity during the previous 50 years (Attri, 2016; Nagy and Nene, 2021). Improved access to, usage of, and efficiency from marine environmental assets has resulted in growth in the sea's economic sectors (Klinger et al. 2018). Simultaneously, disagreements over marine resources have emerged across sectors (for example, tourism vs. off-shore hydrocarbon production), at a corporate level (for example, people, groups, and governments), over geographical areas (for example, local rivers, regional seas, and the global oceans), and time periods (for example, between current and future uses). Continued economic expansion in the use of the seas is projected to create cross-sector conflicts, as well as the possibility of environmental degradation, wasteful natural resource usage, and other morally reprehensible outcomes (Klinger et al. 2018). The ocean's dynamic character, which varies over a wide range of geographical and temporal dimensions, complicates cross-sector administration (Klinger et al. 2018). Due to global warming and natural variability, the ocean is experiencing rising heat transfer, increased acidity, and changes in other physiochemical oceanographic features (Harley et al. 2006). Melting sea ice, increasing sea levels, and changing organisms (for example, changes in abundance, diversity, and variety) are all consequences of these changes, all of which have various implications for the ability of mankind to benefit from the ocean. Table 1.4 depicts the connections between significant sectors,

TABLE 1.4
Linkages Between Major Sectors Their Characteristics and Management Frameworks to Respond to Developing a Global Blue Economy

S. L.	Sectors	Characteristics	References
1	Maritime Tourism	Nautical tourism includes sea-based activities such as rowing, windsurfing, cruises, and marine sports, as well as the land-based social infrastructure that support them.	Ecorys (2013)
2	Beach Tourism	Beach tourism utilizes the aesthetic and environmental values of the beach.	Matthews et al. (2021)
3	Cruise Ships	The cruise business is one of the fastest-growing segments of a tourist industry that is already rising at a rapid pace.	Wood (2007)
4	Yachting	Yachting is the use of sailing boats for sporting purposes; yacht races.	Philpott (2005)
5	Aquaculture and Fisheries	Marine aquaculture and fisheries present an opportunity for increasing seafood production	Gentry (2017)
6	Recreational / subsistence	Marine recreational and semi-subsistence activities have a high economic value	Gomez et al. (2021)
7	Marine Sport fishing	Sport or offshore fishing, as well as large game fishing, are terms used to describe this type of fishing.	Cox et al. (2002)
8	Trade and Commercial	Maritime shipping is the fundamental part of world trade.	Ojala and Tenold (2017)
9	Mariculture	Seaweeds, mussels, oysters, shrimp, prawns, salmon, and other fish species are all part of the mariculture industry. Mariculture provides opportunities for long-term protein-rich food production as well as community economic growth.	Neori et al. (2004)

TABLE 1.4 (Continued)
Linkages Between Major Sectors Their Characteristics and Management Frameworks to Respond to Developing a Global Blue Economy

S. L.	Sectors	Characteristics	References
10	Marine Biodiversity	As a result, marine biodiversity refers to the diversity and quantity of organisms found in the world's oceans and seas.	Cheung et al. (2009)
11	Marine protected areas	A marine protected area (MPA) is a portion of the ocean where human activity is restricted by the government.	De Santo (2013)
12	Goods and services	Ecological goods and services are the economic benefits (goods and services) arising from the ecological functions of marine ecosystems.	Dias (2011)
13	Amenity values	Marine users' perceptions of a location's aspects that provide a good, delightful benefit are referred to as amenity.	Frampton (2010)
14	Maritime transport and ports	Maritime transport is a mode of transportation in which products (or persons) are carried by sea.	Maiola et al. (2009)
15	Maritime industry	Since the dawn of recorded history, the marine industry has been the focal point for global goods transportation.	Czachorowski et al. (2019)
16	International marine shipping	International maritime shipping is a well-known forwarder that specializes in roll-on/roll-off goods.	Andrews (2015)
17	Port and Harbour facilities	A port is a seaside business district with infrastructure including cranes, warehouses, and docks that assist trade and transit. A port is a spot on the coast where watercraft can be stored or parked.	Santos et al. (2019)
18	Offshore Petroleum	Offshore oil and gas refer to operations in the oil and gas business that take place along a coastline.	Nguyen et al. (2014)
19	Marine Crude oil	Microbes that utilise crude oil's rich source of energy and carbon can be found in saltwater, sediments, and shorelines all over the world, from the tropics to the poles.	McGenity et al. (2012)
20	Natural Gas	The remains of sea algae and land plants have generated natural gas and mineral oil, with substantial amounts collecting in specific rock strata.	Staplin (1969)
21	Dredging and aggregate extraction	Dredging in the sea has various advantages for transportation, building, and other enterprises. Marine aggregates are sands and gravels that occur naturally here on inner continental shelves.	Velegrakis et al. (2010)
22	Sand Mining	Sand mining is thus a lucrative business and fuels illegal extraction.	Sonak et al. (2006)
23	Sea Port dredging	In the maritime sector, port dredging is a critical activity. It has the following functions: It contributes to safer trips by reducing underwater traffic and ensuring adequate bottom clearance.	Grech et al. (2013)
24	Marne Environmental protection	The practice of individuals, organizations, and governments protecting the natural environment is known as marine environmental protection.	Christie et al. (2017)

their characteristics, and frameworks for adapting to the developing global maritime economy. Leadership must be flexible and adaptive, enabling single or multi-management frameworks to adjust to changing socioeconomic and social conditions in real time.

1.6   CHALLENGES OF SUSTAINABLE OR BLUE ECONOMY GROWTH

A healthy blue economy aims to boost economic growth, protect and improve livelihoods across a variety of industries while also ensuring the long-term usage of ocean biodiversity (Alhaddi, 2014). Cooperation, adaptability, possibility, and interconnection are all hallmarks of the circular economy (Alhaddi, 2014; Garlock et al. 2020). Blue growth has a lot of promise in terms of ensuring long-term ocean usage, but it also has several drawbacks. The lack of a single, agreed-upon blue growth goal is one of the most evident impediments (Pretorius and Henwood, 2019). Some people define blue as getting shared prosperity from aquatic and marine assets while preventing destruction of blue natural capital, while others define it as obtaining maximum economic growth from aquatic and marine resources whilst avoiding deterioration of blue capital assets (Colgan, 2018). This lack of shared understanding might explain the lack of comprehensive blue growth plans and of more precise and equitable goals and objectives that cut across industries.

Another issue is transdisciplinary learning and training to 'speak the same language' (Eikeset et al. 2018; Eikeset et al. 2018). Not only do scientists need to interact across professions, but they also need to engage with policy experts and other stakeholders who might have even more diverse perceptions of blue development and other essential terms (Eikeset et al. 2018). To guarantee that study educates and promotes viable, integrated, and complete solutions and their implementation, close engagement with stakeholders is required. This appears to be achievable in theory, assuming that the data are being collected and that the results are unambiguous and conveyed to policymakers and regulators.

To address stakeholder conflicts, Figure 1.5: a theoretical model for establishing a sustainable blue economy, is complex and requires a holistic approach to governance. Another challenge lies in addressing potential conflicts, which mainly arise from tradeoffs between different uses of ocean territory but also commonly include who decides what should be open to public debate (Alhaddi, 2014; Pretorius and Henwood, 2019). Salmon farming, for example, has become an important part of Norway's economy, and the industry has set the standard for feeding practices, economization,

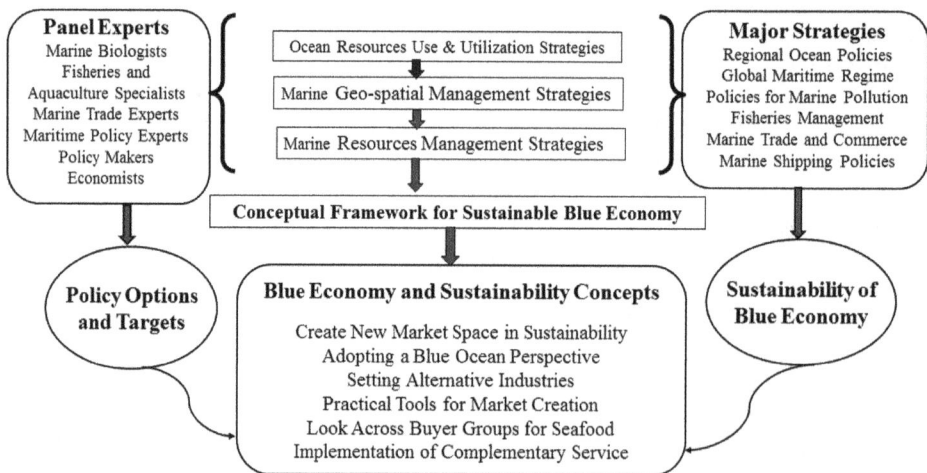

FIGURE 1.5   Conceptual framework for developing sustainable blue economy to resolve the conflicts between stakeholders is difficult and requires holistic approach to governance.

and sustainability practices per unit of output. Alternatively, fish farming can have significant environmental and biological effects in the ocean, influencing other users of ocean space. Comprehensive examination of tradeoffs between various ocean uses necessitates collaboration and coordination across a wide range of scientific disciplines and stakeholders. Stakeholder conflict resolution is difficult and demands a holistic approach to governance.

1.6.1 Challenges and Opportunities of Global Blue Economy

Despite several problems and opportunities, the global blue economy sectors have the potential to enhance collaboration and communication among scientists, industry, and politics, resulting in a coordinated effort to battle climate change (Exner-Pirot, 2012). To overcome these concerns, more research is needed, and collaboration with stakeholders would be advantageous. Before tackling these challenges, it's necessary to grasp the major strengths, limitations, opportunities, and threats facing the current global blue economy scenarios (Beveridge et al. 2020). It is critical to develop policies at the global, regional, and local levels as soon as possible to address the difficulties.

1.6.2 Sustainable Blue Economy Finance Principles

Every marine business investment is based on the Principles of Responsible Blue Economy Finance. These were introduced in 2018 as the world's largest first global guiding framework for bankers, insurers, and financiers to establish a sustainable blue economy. They advocate the attainment of SDG 14 (Life Below Waters) and the adoption of sea regulations, which will allow the financial world to incorporate sustainability into the ocean-based economy (Table 1.5) (Sumaila et al. 2021).

1.6.3 Economic Sustainability of the Blue Economy

Many countries employ the economy as a policy instrument or method to boost economic growth and employment creation (Wenhai et al. 2019). The blue economy must adhere to SDG 14, which stresses the responsible protection and use of oceans, seas, and coastal habitats (Spalding, 2016). The objective is to accomplish social and economic growth while also maintaining a dynamic resource and environmental balance. In their second planning section the UN Commission on Sustainable Development's Preliminary Committee underlined methods to embrace the blue economy, believing it to be congruent with the Rio+20 Summit's basic elements (Ioc-Unesco, 2011. As outlined in the Rio+20 debates, the green economy represents a paradigm change in economic development. In the development and administration of ocean and coastal zones, international society frequently refers to the blue economy as the green economy or the green development model (Rio+20 Pacific Preparatory Meeting, 2011).

We should maintain a healthy coastal and agricultural ecosystem based on an assessment of maritime industrialization and the integrity of the aquatic organisms (Spalding, 2016; Hasan et al. 2018). The combating of pollution problems, such as marine transportation wastes, plastic litter, and microplastics, reducing the effects of global warming, and developing a blue economy based on a sustainable model centered on supporting a healthy environment, is needed (Thushari and Senevirathna, 2020). Fisheries, aquaculture, tourism, shipping, seafloor mining, oil and natural gas, renewables, and transport are all profitable frontiers for so many governments and businesses aspects (Hassan et al. 2014). Several policies have been established by the Environmental Protection Agency (EPA) to assist in the protection of beaches, the limitation of pollution from ships, the reduction of marine debris, and the prohibition of ocean dumping. Off-shore drilling can be reduced by depending on renewable energy sources such as wind and solar power (Madara and Perera, 2020).

TABLE 1.5

The Sustainable Blue Economy Finance Principles

S. L.	Principles	Major Initiatives and Ideas	References
1	Protective	Marine ecosystem variety, productivity, resilience, value, and general health, as well as the livelihoods and communities that rely on them, must be protected or maintained.	Binet et al. (2015) EIB (2018)
2	Compliant	To comply to applicable legal and regulatory frameworks at the international, regional, national, and other levels that support development and ocean health.	Bennett et al. (2019) EIB (2018)
3	Risk-aware	To assess the risks and systemic implications so that decision-making processes and actions can be adjusted.	UNDP (2012)
4	Systemic	To determine the overall and cumulative effects of our investments, operations, and initiatives throughout value chains.	Shiiba et al. (2021) EIB (2018)
5	Inclusive	To improve local lives and effectively interact with key stakeholders.	Uy and Tapnio (2021)
6	Cooperative	To spread ocean knowledge, best practices for a sustainable blue economy, lessons gained, viewpoints, and ideas in order to promote and apply these principles.	Bennett (2020) EIB (2018) Uy and Tapnio (2021)
7	Transparent	With due regard for confidentiality, assure the social, environmental, and economic repercussions (both positive and negative).	Bovino and Niesten (2021); EIB (2018)
8	Purposeful	Making a direct role in the achievement of Sustainable Development Goals Goal 14: Preserve and use oceans, oceans, and coastal habitats in a sustainable manner for slow gestation.	Sumaila et al.(2021) EIB (2018)
9	Impactful	To provide funds and help so that current and future generations can benefit from our ocean's social, environmental, and economic advantages.	Shiiba et al. (2021) EIB (2018)
10	Precautionary	The principle will take precedence when scientific data are unavailable.	Child and Hicks (2019)
11	Diversified	To reach a larger range of sustainable infrastructure projects, including those in conventional and non-traditional marine industries, as well as smaller and big efforts.	Huwyler et al. (2014) Uy and Tapnio (2021)
12	Solution-driven	Identifying and promoting the economic case for such programs, as well as encouraging the dissemination of best practices produced in this manner.	Ram and Kaidou-Jeffrey (2020)
13	Partnering	To accelerate the change for a better blue economy, particularly via the establishment and operation of coastal and marine spatial planning ideas.	EIB (2018) Uy and Tapnio (2021)
14	Science-led	To promote the Blue Economy's long-term financial prospects and to disseminate scientific data and information about the marine environment.	Fenichel et al. (2020) EIB (2018)
15	Reporting	Publicly report (annually) on how the institution is working to implement the Principles	EIB (2018)

Source: Founders of the Sustainable Blue Economy Finance Principles. The European Commission (EC) and the European Investment Bank (EIB) as well as by the Swedish Government via the UN Environment Programme Sustainable Blue Economy Initiatives on oceans and seas, 2018.

The effective utilization of marine resources in order to promote long-term economic growth has received a lot of attention in recent years all around the world (Voyer et al. 2018). The blue economy, according to the World Wildlife Fund (2015), is a marine-based economy that:

- Contributes significantly to food security, poverty eradication, income, jobs, medicine, safety, fairness, and stability for the present generation, as well as providing social and economic advantages.

- Ensures that the diversity, productivity, resilience, essential functions, and the inherent worth of marine ecosystems, the ecological integrity on which the world's economy is founded, will be restored, conserved, and sustained.
- Is based on clean technology, renewable energy, and cyclical material flows to assure long-term financial stability while staying constrained by the restrictions of one planet (Voyer et al. 2018).

Finally, the financial sector as an economic term embraces all sectors of the global economy. Further, as a macroeconomic concept, the blue economy encompasses all aspects of national and international governance, economic growth, environment protection and durability, and intercultural understanding (Wenhai et al. 2019). The financial sector combines green growth and sustainable development. It emphasizes the need for long-term management and cooperation between the economic systems of the ocean and coastal zones, as well as the marine ecology (Okafor-Yarwood et al. 2020). Taking account of the above features, we describe the blue economy as sustainable services, and all other related activities that utilize and conserve coastal and marine resources (Kadagi et al. 2020). Several obstacles must be addressed, ranging from private industry to research and innovation, to non-governmental organizations, and to governmental policies.

1.6.4 Environmental Sustainability of the Blue Economy

As per the World Bank, the blue economy is the 'responsible use of marine biodiversity for income progress, better lifestyles, and jobs while increasing the safety of the ocean ecosystem' (Abhinav et al. 2020). From the European Commission, 'any commercial activity related to ports, oceans, and coasts.' The world's seas and beaches, together known as the Blue World, are a veritable harvest for humanity (Fang et al. 2020). Food, oxygen, and a means of subsistence are all provided by them. The Earth is a blue planet, with oceans spanning 71% of its 510 million square kilometers surface area and an average depth of four times that of land (Bennet et al. 2019). The marine biosphere is thus the world's largest ecosystem, despite the fact that it has a great diversity of species, ranging from almost barren to fertile hotspots (Gamage, 2016; Blumm, 2017; Hossain, 2020). 'Academics, philosophers, corporate leaders, and governments alike are increasingly concerned about environmental deterioration and the imbalance between man and nature' (Zhongming et al. 2019). Many people are concerned about the growing gap between affluent and poor people, as well as the ongoing inability to provide the basic requirements of all (not just humans) (Gamage, 2016). The mass extinction and our incapacity to overcome poverty appears to be the sole long-term trends of our time (Kathijotes, 2013). Although we're really looking reality in the face, it looks like we lack that perspective and talent needed to make a contribution and steer our extravagant consumption culture and aggressive business world towards sustainably.

The blue economy aims to restore, retain, and maintain the diversity, productivity, resilience, vital functions, and inherent worth of marine ecosystems (Brears, 2021). Economic activity should take place in the ocean, and ocean ecosystems should be able to support it in the long run while staying robust and healthy (Kathijotes and Sekhniashvili, 2017). It is widely seen as a long-term strategy aiming at promoting long-term economic growth through ocean-related sectors and activities while also protecting the environment (UNCTAD, 2014). In this context, ecosystem-based ocean management has been shown to maintain ocean productivity through time while also ensuring long-term economic growth. The blue economy attempts to restore, preserve, and protect marine ecosystems' diversity, productivity, resilience, important functions, and inherent value (Voyer, et al. 2018). It is usually regarded as a long-term strategy aimed at encouraging long-term economic growth through ocean-related industries and activities while simultaneously safeguarding the environment (Winther, et al. 2020). In this context, it has been demonstrated that ecosystem-based ocean management may preserve ocean productivity throughout time while simultaneously ensuring long-term economic growth.

Ecosystem-based management outlines the decision-making process as well as the goals that must be met, together with precise rules for economic and environmental preservation (O'Hagan, 2020). It can assist in the coordination of numerous rules affecting the coastal zone and marine enterprises, ranging from traditional ocean industries to emerging ocean health firms (Gamage, 2016; Blumm, 2017; Hossain, 2020). The goal of ecosystem-based management is to promote 'ocean resource protection and sustainable usage' (Wenhai, et al. 2019). Ecosystem-based management protects not just the ocean's environmental features, but also the blue economy concept's approach, goals, and implementation.

1.6.5 Social Sustainability of the Blue Economy

Identifying and regulating the positive and negative impacts of companies on people is critical to social sustainability. It is critical that a company's stakeholder relationships and engagement are of high quality (Cisneros-Montemayor et al. 2021). Employees, value chain workers, customers, and local communities all have a direct or indirect impact on businesses, and it is vital to handle these implications ahead of time. Social sustainability has gained significantly less attention in public debate than economic and environmental sustainability (Voyer and Leeuwen, 2019).

Environmental justice, human health, resource security, and education are only a few of the important social factors that influence social sustainability (Manikarachchim, 2014). Efforts to promote social sustainability should also aim to develop economic and environmental benefits, according to the three pillars idea (Purvis et al. 2019). In the blue economy, measures to promote social sustainability could include focusing corporate efforts on employee retention rather than economic priorities (Wenhai et al. 2019). Employee happiness, for example, is likely to benefit the company financially by increasing employee motivation (Jamshidi and Jafari, 2021). Efforts to improve social sustainability may also have a positive impact on the environment. People's nutrition choices, for example, can have a significant impact on both human and environmental health, therefore, advocating healthy eating can also benefit the environment.

1.6.6 Linking SDGs and Policies to the Global Blue Economy

Sensible policy creation, funding, and execution at the municipal, regional, and international levels are critical to society's capacity to achieve a certain goal (SDGs) (Scharlemann et al. 2020). As a result, policymakers must now direct activities to advance the SDGs while minimizing negative repercussions (Novaglio et al. 2021). In order to attain the intended outcomes, decision-makers must comprehend the feedbacks and interrelations between stakeholders in the blue economy, the national economy, and environmental sustainability, prompting reviews to better comprehend how blue economy sectors influence our capabilities to attain the 2030 agenda (McKinley et al. 2019; Nash et al. 2020). Although the increase in goal interaction research is a great step forward, trade-offs and synergy are not a fundamental part of the SDG review process. Good leadership is necessary for long-term stewardship of aquatic ecological sustainability, as well as ensuring biodiversity and ecological resilience, all of which contribute to human resilience in the face of a range of problems, including changing climate (Folke, 2016). By minimizing risks and providing incentives for innovation, effective governance will also assist in the establishment of an enabling climate for responsible private sector investments throughout the value chain (Gamage, 2016; Blumm, 2017; Hossain, 2020). Finally, improved governance will increase the macroeconomic contribution of fisheries, aquaculture, and mariculture, raising the sector's visibility and, as a result, resource allocation.

1.6.7 BLUE ECONOMY AND MARINE POLLUTION ISSUES

The chemical and biological factors that make up marine pollution include plastic waste, petroleum-based pollutants, dangerous metals, synthetic compounds, pharmaceuticals, pesticides, and a foul stew of nitrogen, phosphorus, fertilizer, and sewage (Landrigan et al. 2020). Marine pollution is becoming more of a problem in today's world. Increased amounts of chemicals in coastal water, such as nitrogen and phosphorus, enhance the growth of algal blooms, which can be harmful to people and dangerous to other species (Wurtsbaugh et al. 2019). Algal blooms have major health and environmental effects, as well as a negative impact on the fishing and tourism industries in the area (Grattan et al. 2016). Marine waste includes a variety of plastic goods such as shopping bags and beverage bottles, as well as cigarette butts, bottle caps, food wrappers, and fishing gear (Pasternak et al. 2017). Plastic waste is a particularly harmful contaminant due to its long lifespan. Plastic items could take hundreds of years to decompose.

Around 80% of ocean pollution originates on land, with the remaining 20% coming from discharges from maritime ships, offshore industrial operations, and at-sea rubbish disposal (Daoji and Daler, 2004). In-shore contamination is caused by wastewater discharges, industrial releases, agricultural runoff, and riverine pollution along coastlines and in bays, ports, and estuaries (Landrigan et al. 2020). Along the coasts of quickly developing countries, some of the world's worst ocean pollution can be found (Nellemann, et al, 2008). For this reason, cultivating aquaculture and enhancing blue economy currently face difficulties.

Nonetheless, many nations are acting, however preventing marine pollution alone will not be enough to boost developing countries' blue economy sectors. More than sixty countries have implemented legislation banning or forbidding the use of throwaway plastic goods, according to United Nations research from 2018 (Schnurr, et al. 2018). Some key ways that governments can combat marine pollution and promote a healthy, productive, and resilient sea are as follows:

- Reduce plastic use and implement litter control policy
- Reduce or recycle plastic
- Diminish discharge of untreated sewage
- Control chemical and industrial pollution
- Increase money for the prevention and treatment of water debris.
- Strengthen laws on marine litter
- Incorporate preventative and control programs into national policy
- Develop local competence and technological abilities
- Raise public awareness
- Create alliances to combat marine pollution
- Evaluate pollution problems in a systematic manner
- Assess the economic impacts

1.6.8 TOOLS AND OPPORTUNITIES FOR A BLUE ECONOMY

Diet and health security from the fisheries sector, social and economic development from the fisheries sector, aquatic and marine tourist industry, shipping, mining, and energy, as well as natural ecosystems like sequestering carbon, water filtration, atmospheric and heat regulation, erosion safeguards, and severe weather event protection are all provided by ocean and inland waters (Johnson et al. 2018). Issues and difficulties bring both challenges and opportunities, and the blue economy offers a wide range of solutions for sustainable, clean, and fair blue growth in both established and emerging industries (Gamage, 2016; Blumm, 2017; Hossain, 2020). From bio-prospecting to seabed mineral mining, technological advancements are paving the way for new horizons in marine resource development (Behnam, 2012; Hossain et al. 2014; Chowdhury et al. 2015). Wind, wave, and tide, as well as thermal and biomass sources, provide a lot of promise for renewable blue energy generation at sea (Hoegh-Guldberg et al. 2015).

1.6.8.1   Fisheries, Aquaculture and Seafood Production

Around the world, capture fisheries continue to be in great demand. The importance of fisheries and aquaculture in supplying food, nutrition, and jobs is demonstrated in this paper (Tacon et al. 2009; Belton and Thilsted, 2014). The fisheries and aquaculture sectors have grown quickly in recent decades, with worldwide output, trade, and consumption reaching new highs in 2018 (Shamsuzzaman, et al. 2020; FAO, 2020). However, since the early 1990s, aquaculture has accounted for the largest share of growth in the industry as a whole, while capture fisheries production has remained largely stable, with some growth primarily relating to inland capture. Sustainable fishing may be an essential aspect of a successful blue economy, with marine fisheries contributing more than US$270 billion to world GDP each year (World Bank, 2017b. Marine fisheries provide a large source of animal protein, essential minerals, and omega-3 fatty acids to the 300 million inhabitants who work in the industry. They also help to address the nutritional demands of the three billion people who rely on fish for protein, trace nutrients, and omega-3 fatty acids (FAO, 2016). Fishing is important in many of the world's poorest locations, where fish is a critical source of nutrition and the industry functions as a social security net (Dulvy et al. 2011). The removal of fishery subsidies that cause overexploitation, as well as the deployment of integrated, ecosystem-based techniques based on the best available knowledge in a precautionary environment, offer the potential of restoring critical stocks and increasing catches (Bunnefeld et al. 2011; Cohen et al. 2019). Implementing solid management practices promises improved sustainable catches, reduced energy use, and lower prices, securing livelihoods and improving food security.

Aquaculture, like catch fisheries, has the potential to expand in importance for the blue economy. While small-scale aquaculture has a significant impact on food security and employment in underdeveloped nations, high operating expenses and economies of scale necessitate a culture of high-value commodities oriented towards local tourist and export markets (Chuenpagdee et al. 2008; Farmery et al. 2021). Given the SIDS' reputation as tourist destinations, it is vital to decrease the environmental effect of aquaculture (Chuenpagdee et al. 2005). As a result, the sector plan will detail the technologies and innovations necessary to sustain ocean health and guarantee that aquaculture products, namely, fish raised in a clean, well-managed environment, compete on a global scale with the tourist business.

1.6.8.2   Oil and Gas, Deep-sea Mining and the Blue Economy

Mineral reserves on and under the seabed are being studied and exploited all over the world. The technique of obtaining minerals from the seabed is known as seabed mining (Hunter and Taylor, 2014; Cuyvers et al. 2018). Deep-sea mining has become more economically viable due to growing availability of metals and relatively unusual commodities, as well as technical advancements (Biswas et al. 2015). Digging has been proposed on the aquila plains, at depths, and near hydrothermal systems. The bulk of seabed mining is now restricted to shallow water zones, but technology is improving, and experimental deeper bottom dredging may be possible in the future (Frankic and Hershner, 2003). Extrinsic advantages to the price of mined commodities are expected to occur largely on the seafloor and at drill sites as a consequence of substrate surface water drainage produced by both drilling and returning seawater (Montserrat et al. 2019). Environment impacts can arise at any stage of the manufacturing process as a result of unforeseen events or natural disasters, with possible ramifications including changes in kinesiology, biological modifications, and perhaps exacerbating the transaction's effects on natural environments (Klinger et al. 2018).

1.6.8.3   Offshore Wind, Wave and Tide Energy Production

Incoming solar heat sustains the bulk of renewable ocean energy forms, putting them indirectly into the category of solar energy (Häyhä and Franzese, 2014; Bennett et al. 2019). The fluctuating gravitational pulls of the moon and sun on the planet and its oceans produce tidal force (Butikov, 2002). This is sustainable due to its ability to fulfill the world's energy needs while lowering carbon emissions in the long run. While some renewable marine energy projects are still in the planning phase, others have been in operation for a while with variable degrees of both technical and economic success.

1.6.8.4 Offshore Wind Energy

Offshore wind power is the installation of wind turbines in large bodies of water (Byrne and Houlsby, 2003). Winds are blowing more quickly and more uniformly at sea than on land, implying less wear on engine components and more power generated by every rotor (Musial et al. 2006). Wind potential energy is typically proportional to the cube of wind speed. As a result, even a little increase in flow speed causes a huge increase in energy production (Pishgar-Komleh et al. 2015). A turbine at a location with a wind speed of 25 km/h, for instance, could produce around 50% more electricity than a turbine in a location with a mean wind speed of 23 km/h (Hoover et al. 2005).

Offshore wind is also the most established form of maritime renewable energy in terms of technological progress, regulations, and generating capacity (Appiott et al. 2004). Research and expertise with both territory power generation and offshore oil and gas production have greatly aided offshore wind power designs and other project features (Afewerki, 2022). This is currently a viable sustainable material in many areas, and it is gaining global attention because of its enormous resources, and is typically located near large electricity concentrations in coastal towns (Creutzig et al. 2014). Based on these features, offshore wind energy appears to have the most immediate potential for energy generation, grid integration, and combating climate change (Stephens et al. 2009).

1.6.8.5 Offshore Tidal Energy and the Blue Economy

Wind and solar energy are less predictable than tides. Tide mills were originally used to crush grain, but today they're utilized to generate energy at tide power plants (Neill et al. 2018). Traditional tidal power has had high prices and a limited number of locations with acceptable tidal ranges or flowrate, limiting its total availability among alternative energy sources (Rashid and Barua, 2016). New technological improvements, on the other hand, show that marine power's overall energy production might be higher than previously estimated, cutting costs (Roberts et al. 2016). Only a few tidal power plants are still operational today, among of which is the La Rance tidal power station in France, which has been in operation since 1966 (Hammons, 1993). To collect energy from tidal flows, the prospective energy generated when two bodies of water isolated by a dam or inundation have different altitudes is used (Rourke et al. 2010; Toupin, 2016). Immersed in water tidal turbines rely exclusively on the kinetics of free water movement. Tidal turbines must be significantly more durable than wind turbines because water is 800 percent denser and more destructive than air (Tong, 2019; Behera, 2022).

1.6.8.6 Shipping, Port and Maritime Logistics with the Blue Economy

Maritime shipping accounts for more than 80% of all global commodity trade in 2015, and this ratio is substantially higher in most impoverished nations (Warren, 2007). In terms of value, some researchers, such as Lloyd's List Intelligence, claim that marine seaborne trade accounted for 55% of all international trade in 2013, while others say it was closer to 70% (UNCTAD 2016). The effects of climate change (such as rising sea levels, increased temperatures, and more frequent and/or intense storms) pose significant threats to crucial transportation systems, services, and processes, particularly in SIDS and coastal LDCs, requiring an insight into the underlying threats and hazards, as well as the application of improved adaptation strategies (Smith-Godfrey, 2016; Kathijotes and Sekhniashvili, 2017; Sarker et al. 2018).

Given the critical role of ports in the worldwide economic system, environmental and resilience solutions for ports are an urgent need. The primary environmental implications of maritime traffic include marine and air pollution, marine debris, underwater noise, and the introduction and spread of exotic species (Gamage, 2016; Blumm, 2017; Hossain, 2020). New international rules compel the shipping sector to spend heavily on environmental technology such as emissions control, waste disposal, and bilge water purification. Some of the expenditure is not only beneficial to the environment, but could also save money in the long term, for example, by boosting fuel economy.

1.6.8.7   Marine Manufacturing and Ship-building

Marine shipping is an important part of global trade and commerce at present. It deals with nearly 90% of all goods transported in one form or another through shipping. The value of the worldwide maritime product transportation business is in the trillions of dollars, and it is steadily expanding (Pretorius and Henwood, 2019). Approximately 76% of all trade involves some type of maritime transportation (Notteboom, 2004). Despite the fact that the shipping business has seen its fair share of ups and downs due to economic crises and the industry's financially fragile character, worldwide demand appears to be growing at a steady pace (Sanusi, 2011). The ability of shipping to convey goods and resources from their point of origin to their final destination is essential to modern life (Lipton et al. 1990). For an economic region such as the European Union, shipping accounts for 80% of all exports and imports by volume and 50% by value Hoekman and Djankov, 1997).

Shipping, on the other hand, is reliant on the availability of ships and boats capable of delivering a wide range of commodities across vast distances (Moutoukias, 1988; Jacks and Pendakur, 2010). Shipbuilding is a lengthy process that might take up to 1.5 years for medium to large vessels. As a result, regardless of how strong the maritime transport business is, the shipbuilding industry controls and restricts commerce, based on the number of vessels built. The construction of cruise liners and pleasure vessels is another prominent area where shipbuilding plays a significant role (Johnson et al. 2018). Several countries throughout the world offer such facilities that are accessible to both residents and visitors. It benefits the tourism and coastal sectors in various countries. For example, to stimulate the economy, India, a country with no prior cruise or pleasure vessel history, has lately established cruises along its coast. As a result, we can see that passenger shipbuilding is just as vital as commercial cargo shipbuilding.

1.6.8.8   Enhancing Marine Commerce, Tourism and Leisure

Maritime tourism is just one of several types of tourism that help countries along the coast (Moreno and Amelung, 2009). Tourists and visitors participate in active and passive leisure and vacation pursuits or excursions on (or in) coastal seas, shorelines, and their immediate environs in Maritime tourism (Higham, 2017). Marine leisure refers to a wide range of activities or interests that locals, tourists, and day visitors engage in when visiting various marine-related destinations (Smith-Godfrey, 2016; Kathijotes and Sekhniashvili, 2017; Sarker et al. 2018). Because of its diversity, tourism is one of the industries that might gain the most from the blue economy. The ocean environment has always been one of the most appealing tourist destinations (Mertha et al. 2017; Wiarti et al. 2017; Lapa et al. 2021). All beach activities, sea kayaking, excursions to fishing communities and lighthouses, maritime museums, sailing and motor yachting, maritime festivals, Arctic and Antarctic tourism, and many more activities are all covered (Sankrusme, 2017).

1.6.9   Implementation of Science-Policy Nexus for Developing Blue Economy

Healthy marine ecosystems offer a home for a wide variety of marine species while also supplying essential human necessities such as food, medicine, and alternative energies, as well as climate regulation, ecotourism, entertainment, coastal protection, and job generation (Potts et al. 2014; Smith-Godfrey, 2016; Kathijotes and Sekhniashvili, 2017; Sarker et al. 2018). Despite the fact that oceans encompass over 70% of our world and provide such vital services and goods, there is still a lot we don't know about them (Laffoley et al. 2019). Kenya hosted and supported the inaugural Worldwide Construction Blue Economy Conference in November 2018, which again was founded by Canada and Japan, in recognition of science's critical role in leveraging the Economic Base and the reality that humankind's existence is strongly reliant on a functioning ocean (Farmery et al. 2021).

Many barriers to the sustainable economic perspective exist in developing countries, including waterlogging, marine pollution, including acidification of the oceans and blue carbon, a shortage of trained professionals, synchronizing sectoral initiatives, plans, and laws, poor seafloor stewardship, and popular backing, to name a few (Kalam et al. 2018). The exploitation of maritime resources whilst avoiding pollution, overuse, and mismanagement, is needed. For the above contexts the following would be recommended for developing a sustainable blue economy for both developed and developing nations:

- **Ensuring education and awareness for all:** all people in developing countries should get proper education and awareness to enhance the blue economy sectors globally.
- **Exchanging technological knowledge and research:** the governments of both developed and developing countries should invest more in new better technologies to achieve the sustainable blue economy that will help future generations.
- **Strong environmental law and maritime policies:** the policymakers and the stakeholders both should implement more effective marine environmental law and ensure better law enforcement. That will ensure less environmental damage to marine resources.
- **Important to focus on blue economy:** every developed country is focusing more on the blue economy. This is the unlimited source of wealth which will help to improve people's standard of living. Similarly, it is important for the governments of developed countries to play a proper role in supporting the blue economy sectors in developing countries.
- **Stable political environment:** a stable political situation allows a country to develop the blue economy sectors more quickly. People will invest more for their benefit in this sector.
- **Mostly in Energy sector:** Most of the developed countries energy sectors are well organized. They have many alternative energy production plants and systems. Electricity is the main indicator for developing a country economy. No country will develop without energy sector improvement. So, the ocean would be the best source of energy to resolve the future energy crisis in developing nations.
- **Better policy for the blue economy:** policy makers should create effective policies to develop the blue economy, including South Asian and African countries.
- **Awareness in people:** countries should inform their citizens about damage to the marine environment and resources. They should use the natural resources properly and not cause any damage to environment that cannot be reversed. They should also aim to use renewable resources, which is environment friendly.

1.7   CONCLUSIONS

The oceans are a significant carbon sink, a critical home for millions of species, and an important component in environmental and human health. The blue economy is a concept that highlights how the seas serve as a crucial food supply and a worldwide commercial facilitator. The blue economy refers to the long-term use of marine habitats for economic expansion, increased revenue, and employment generation while conserving the sustainability of ocean ecosystems. However, regular floods, maritime pollution such as ocean algal blooms and salinity, and blue carbon, as well as a shortage of qualified personnel, interacting policy positions, plans, and regulations, poor ocean leadership, and public influence, to name a few challenges, all present difficulties to the blue economy standpoint. Furthermore, our human footprint is harming the health of the seas because of continuous man-made stresses. To have clean seas, recover marine life, and create the long-term conditions for resilient and functional oceans, we must drastically reduce these stresses. To do so, we must make significant adjustments in policies, institutions, and practices that are not currently in place to improve the global economy's sustainability.

ACKNOWLEDGMENTS

I would like to express my appreciation to all of the writers and contributors whose work I reviewed in order to compile the current chapter in this book. In order to examine the information relating to this chapter, I also used several websites, free domains, blogs, and other sources. I would like to express my respect and appreciation for the unnamed authors. In addition, I would like to express my thanks to the SUMITOMO Foundation, Tokyo, Japan for providing financing assistance for this study on the global blue economy and seafood production practices observed in Japanese tradition.

REFERENCES

Abhinav, K. A. Collu, M. Benjamins, S. Cai, H. Hughes, A. Jiang, B. ... and Zhou, B. Z. (2020). Offshore multi-purpose platforms for a Blue Growth: A technological, environmental and socio-economic review. *Science of the Total Environment, 734*, 138256.

Alexander, A. and Delabre, I. (2019). Linking sustainable supply chain management with the sustainable development goals: Indicators, scales and substantive impacts. In *Sustainable Development Goals and Sustainable Supply Chains in the Post-global Economy* edited by Natalia Yakovleva; Regina Frei; Sudhir Rama Murthy (pp. 95–111). Springer, Cham.

Alhaddi, H. (2014). Blue ocean strategy and sustainability for strategic management. *International Proceedings of Economics Development and Research, 82*, 125.

Andrews, G. (2015). *Performance and Prospects of Roll-On/Roll-Off Service: A Study with Special Reference to Cochin Port Trust (Doctoral Dissertation)*. Indian Maritime University, India.

Appiott, J. Dhanju, A. and Cicin-Sain, B. (2014). Encouraging renewable energy in the offshore environment. *Ocean and Coastal Management, 90*, 58–64.

Atkins, D. E. (2003). *Revolutionizing Science and Engineering Through Cyberinfrastructure: Report of the National Science Foundation Blue-Ribbon Advisory Panel on Cyberinfrastructure*. National Science Foundation.

Attri, V. N. (2016). *An Emerging New Development Paradigm of the Blue Economy in IORA; A Policy Framework for the Future*. Chair Indian Ocean Studies, Indian Ocean Rim Association (IORA), University of Mauritius.

Baltic Sea Action Plan (2013) Baltic Sea Action Plan is it on track? WWF Baltic Ecoregion Programme, page 1-2; https://wwf.fi/app/uploads/3/g/u/mrfzp55gwhbujpoptha5z5a/wwf_balticseaactionplan_2013-final.pdf

Baral, A. and Guha, G. S. (2004). Trees for carbon sequestration or fossil fuel substitution: the issue of cost vs. carbon benefit. *Biomass and Bioenergy, 27*(1), 41–55.

Bebbington, J. and Unerman, J. (2018). Achieving the United Nations Sustainable Development Goals: an enabling role for accounting research. Accounting, Auditing and Accountability Journal.

Behera, A. (2022). Energy Harvesting and Storing Materials. In *Advanced Materials*: An Introduction to Modern Materials Science, by Ajit Behera (pp. 507–555). Springer, Cham.

Behnam, A. (2012). Building a blue economy: strategy, opportunities and partnerships in the Seas of East Asia. In *The East Asian Seas Congress 2012*, Changwon.

Belton, B. and Thilsted, S. H. (2014). Fisheries in transition: food and nutrition security implications for the global South. *Global Food Security, 3*(1), 59–66.

Bennett, N. J. Cisneros-Montemayor, A. M. Blythe, J. Silver, J. J. Singh, G. Andrews, N. Calò, A. Christie, P. Di Franco, A. Sumaila, U.R. et al. (2019). Towards a sustainable and equitable blue economy. *Nat. Sustain.* 2, 991–993.

Bennett, V. (2020). EBRD signs up to Sustainable Blue Economy Finance Principles. *The European Bank for Reconstruction and Development*. https://ebrd.com/news/2020/ebrd-signs-up-to-sustainable-blue-economy-finance-principles.html

Beveridge, C. Hossain, F. Biswas, R. K. Haque, A. A. Ahmad, S. K. Biswas, N. K. ... and Bhuyan, M. A. (2020). Stakeholder-driven development of a cloud-based, satellite remote sensing tool to monitor suspended sediment concentrations in major Bangladesh Rivers. *Environmental Modelling and Software*, 133, 104843.

Bigg, G. Jickells, T. Liss, P. and Osborn, T. (2003). The role of the oceans in climate. *International Journal of Climatology*, 23, 1127–1159. doi: 10.1002/joc.926

Binet, T. Diazabakana, A. and Hernandez, S. (2015). *Sustainable Financing of Marine Protected Areas in the Mediterranean: A Financial Analysis* (Vertigo Lab, MedPAN, RAC/SPA, WWF Mediterranean).

Bir, J. Golder, M. R. Zobayer, M. F. A. Das, K. K. Chowdhury, S. Z. Das, L. M. and Paul, P. C. (2020). A review on blue economy in Bangladesh: prospects and challenges. *International Journal of Natural and Social Sciences*, 7(4): 21–29. doi: 10.5281/zenodo.4270719

Biswas, A. K. M. A. A. Sattar, M. A. Hossain, M. A. Faisal, M. and Islam, M. R. (2015). An internal environmental displacement and livelihood security in Uttar Bedkashi Union of Bangladesh. *Science and Education*, 3(6), 163–175.

Blue Economy Report (2020) Blue-Cloud has received funding from the European Union's Horizon programme call BG-07-2019-2020, topic: [A] 2019 - Blue Cloud services, Grant Agreement No.862409. https://blue-cloud.org/news/eu-blue-economy-report-2020

Blumm, M. C. (2017). Indian treaty fishing rights and the environment: Affirming the right to habitat protection and restoration. *Wash. L. Rev.* 92, 1.

Bovino, R. and Niesten, E. (2021). *Production of a Baseline Inventory of Existing and Potential Sustainable Blue Finance Investors to Support the CLME+ Vision*. EcoAdvisors, Inc. February 2021. Mr. Robbie Bovino, robbie@ecoadvisors.org.; Dr. Eduard Niesten, eddy@ecoadvisors.org. All rights reserved. Cover photo by Rj lerich, (1).

Brears, R. C. (2021). *Developing the Blue Economy*. Palgrave Macmillan.

Brodie Rudolph, T. et al. (2020). A transition to sustainable ocean governance. *Nat Commun*, 11, 3600 (2020). https://doi.org/10.1038/s41467-020-17410-2.

Browman, H. I. Stergiou, K. I. Cury, P. M. Hilborn, R. Jennings, S. Lotze, H. K. and Mace, P. M. (2004). Perspectives on ecosystem-based approaches to the management of marine resources. *Marine Ecology—Progress Series*, 274, 269–303.

Bunnefeld, N. Hoshino, E. and Milner-Gulland, E. J. (2011). Management strategy evaluation: a powerful tool for conservation? *Trends in Ecology and Evolution*, 26(9), 441–447.

Butikov, E. I. (2002). A dynamical picture of the oceanic tides. *American Journal of Physics*, 70(10), 1001–1011.

Byrne, B. W. and Houlsby, G. T. (2003). Foundations for offshore wind turbines. *Philosophical Transactions of the Royal Society of London. Series A: Mathematical, Physical and Engineering Sciences*, 361(1813), 2909–2930.

Cervigni, R. and Scandizzo, P. L. (2017). *The Ocean Economy in Mauritius: Making It Happen, Making It Last*. Washington, DC: The World Bank.

Cheung, W. W. Lam, V. W. Sarmiento, J. L. Kearney, K. Watson, R. and Pauly, D. (2009). Projecting global marine biodiversity impacts under climate change scenarios. *Fish and Fisheries*, 10(3), 235–251.

Childs, J. R. and Hicks, C. C. (2019). Securing the blue: political ecologies of the blue economy in Africa. *Journal of Political Ecology*, 26(1), 323–340.

Chowdhury, S. R. Hossain, M. S. Sharifuzzaman, S. M. and Sarker, S. (2015). *Blue Carbon in the Coastal Ecosystems of Bangladesh. Project Document, Support to Bangladesh on Climate Change Negotiation and Knowledge Management on Various Streams of UNFCCC Process Project, Funded by DFID and Danida*, implemented by IUCN Bangladesh Country Office.

Christie, P. Bennett, N. J. Gray, N. J. Wilhelm, T. A. Lewis, N. A. Parks, J. ... and Friedlander, A. M. (2017). Why people matter in ocean governance: incorporating human dimensions into large-scale marine protected areas. *Marine Policy*, 84, 273–284.

Chu, K. C. Kaifuku, K. and Saitou, K. (2015). Optimal integration of alternative energy sources in production systems with customer demand forecast. *IEEE Transactions on Automation Science and Engineering*, 13(1), 206–214.

Chuenpagdee, R. Degnbol, P. Bavinck, M. Jentoft, S. Johnson, D. Pullin, R. and Williams, S. (2005). *Challenges and concerns in capture fisheries and aquaculture. In Fish for Life: Interactive Governance for Fisheries (pp. 25–40)*. Amsterdam University Press, Amsterdam, The Netherlands.

Chuenpagdee, R. Kooiman, J. and Pullin, R. (2008). Assessing governability in capture fisheries, aquaculture and coastal zones. *Journal of Transdisciplinary Environmental Studies*, 7(1), 1–20.

Cisneros-Montemayor, A. M. Moreno-Báez, M. Reygondeau, G. Cheung, W. W. Crosman, K. M. González-Espinosa, P. C. ... and Ota, Y. (2021). Enabling conditions for an equitable and sustainable blue economy. *Nature*, 591(7850), 396–401.

Cohen, P. J. Allison, E. H. andrew, N. L. Cinner, J. Evans, L. S. Fabinyi, M. ... and Ratner, B. D. (2019). Securing a just space for small-scale fisheries in the blue economy. *Frontiers in Marine Science*, 6, 171.

Colgan, C. S. (2018). *The Blue Economy Handbook of the Indian Ocean Region*, edited by by V.N. Attri (Editor), Narnia Bohler-Mulleris (Editor), 38.

Cox, S. P. Beard, T. D. and Walters, C. (2002). Harvest control in open-access sport fisheries: hot rod or asleep at the reel? *Bulletin of Marine Science*, 70(2), 749–761.

Creutzig, F. Goldschmidt, J. C. Lehmann, P. Schmid, E. von Blücher, F. Breyer, C. ... and Wiegandt, K. (2014). Catching two European birds with one renewable stone: mitigating climate change and Eurozone crisis by an energy transition. *Renewable and Sustainable Energy Reviews*, 38, 1015–1028.

Cuyvers, L. Berry, W. Gjerde, K. Thiele, T. and Wilhem, C. (2018). *Deep Seabed Mining, a Rising Environmental Challenge*. IUCN, Gland, Switzerland.

Czachorowski, K. Solesvik, M. and Kondratenko, Y. (2019). The application of blockchain technology in the maritime industry. In *Green IT Engineering: Social, Business and Industrial Applications*, written by Karen V. Czachorowski; Solesvik Marina; Yuriy Kondratenko (pp. 561–577). Springer, Cham.

Danovaro, R. Aguzzi, J. Fanelli, E. Billett, D. Gjerde, K. Jamieson, A. Van Dover, C. L. (2017). An ecosystem-based deep-ocean strategy. *Science*, 355: 452–454.

Daoji, L. and Daler, D. (2004). *Ocean Pollution from Land-Based Sources: East China Sea*. Ambio, China. 107–113.

De Santo, and E. M. (2013). Missing marine protected area (MPA) targets: how the push for quantity over quality undermines sustainability and social justice. *Journal of Environmental Management*, 124, 137–146.

Dias, V. (2011). *Values of Ecological Goods and Services Provided by Wetland for Policy Development in Saskatchewan* (Doctoral dissertation, University of Saskatchewan).

Donati, A. Hallet, L. L. and Tortora, J. J. (2021). Space and economic development on Earth: the case of blue economy. In *A Research Agenda for Space Policy*. Edward Elgar Publishing.

Dulvy, N. K. Reynolds, J. D. Pilling, G. M. Pinnegar, J. K. Phillips, J. S. Allison, E. H. and Badjeck, M. C. (2011). *Fisheries Management and Governance Challenges in a Climate Change*.

Ebarvia, M. C. M. (2016). Economic assessment of oceans for sustainable blue economy development. *Journal of Ocean and Coastal Economics*, 2(2), 7.

Ecorys, I. E. S. (2013). IRS (2013). *Apprenticeship and Traineeship Schemes in EU27: Key Success Factors*, 6.

Eikeset, A. M. Mazzarella, A. B. Davíðsdóttir, B. Klinger, D. H. Levin, S. A. Rovenskaya, E. and Stenseth, N. C. (2018). What is blue growth? The semantics of 'Sustainable Development' of marine environments. *Marine Policy*, 87, 177–179.

European Investment Bank (EIB) (2018) European Investment Bank statistical report 2018, European Investment Bank, 2019, https://data.europa.eu/doi/10.2867/187290

Exner-Pirot, H. and Gulledge, J. (2012). Climate change and international security: The Arctic as a bellwether. *Center for Climate and energy solutions*.

Fang, X. Zou, J. Wu, Y. Zhang, Y. Zhao, Y. and Zhang, H. (2021). Evaluation of the sustainable development of an island 'Blue Economy': a case study of Hainan. *China. Sustainable Cities and Society*, 66, 102662.

FAO (2020). *The State of World Fisheries and Aquaculture 2020. Sustainability in action*. Rome.

Farmery, A. K. Allison, E. H. Andrew, N. L. Troell, M . Voyer, M. Campbell, B. ... and Steenbergen, D. (2021). Blind spots in visions of a "blue economy" could undermine the ocean's contribution to eliminating hunger and malnutrition. *One Earth*, 4(1), 28-38.

Fenichel, E. P. Addicott, E. T. Grimsrud, K. M. Lange, G. M. Porras, I. and Milligan, B. (2020). Modifying national accounts for sustainable ocean development. *Nature Sustainability*, 3(11), 889–895.

Ferdouse, F. Holdt, S. L. Smith, R. Murúa, P. and Yang, Z. (2018). The global status of seaweed production, trade and utilization. *Globefish Research Programme*, 124, I.

Folke, C. Biggs, R. Norström, A. V. Reyers, B. and Rockström, J. (2016). Social-ecological resilience and biosphere-based sustainability science. *Ecology and Society, 21*(3).

Frampton, A. P. (2010). A review of amenity beach management. *Journal of Coastal Research*, 26(6), 1112–1122.

Frankic, A. and Hershner, C. (2003). Sustainable aquaculture: developing the promise of aquaculture. *Aquaculture International*, 11(6), 517–530.

Galván, J. Cantillo, V. and Arellana, J. (2016). Factors influencing demand for buses powered by alternative energy sources. *Journal of Public Transportation*, 19(2), 2.

Garland, M. Axon, S. Graziano, M. Morrissey, J. and Heidkamp, C.P. (2019). The blue economy: identifying geographic concepts and sensitivities. *Geogr. Compass.13*(7), e12445.

Garlock, T. Asche, F. and erson, J. Bjørndal, T. Kumar, G. Lorenzen, K. .and Tveterås, R. (2020). A global blue revolution: aquaculture growth across regions, species, and countries. *Reviews in Fisheries Science and Aquaculture*, 28(1), 107–116.

Gamage, R. N. (2016). Blue economy in Southeast Asia: oceans as the new frontier of economic development. *Maritime Affairs: Journal of the National Maritime Foundation of India*, 12(2), 1–15.

Gentry, R. R. Froehlich, H. E. Grimm, D. Kareiva, P. Parke, M. Rust, M. ... and Halpern, B. S. (2017). Mapping the global potential for marine aquaculture. *Nature Ecology and Evolution*, 1(9), 1317–1324.

Gichana, Z. M. Liti, D. Waidbacher, H. Zollitsch, W. Drexler, S. and Waikibia, J. (2018). Waste management in recirculating aquaculture system through bacteria dissimilation and plant assimilation. *Aquaculture International*, 26(6), 1541–1572.

Gómez, S. Carreño, A. and Lloret, J. (2021). Cultural heritage and environmental ethical values in governance models: conflicts between recreational fisheries and other maritime activities in Mediterranean marine protected areas. *Marine Policy,* 129, 104529.

Grech, A. Bos, M. Brodie, J. Coles, R. Dale, A. Gilbert, R. ... and Smith, A. (2013). Guiding principles for the improved governance of port and shipping impacts in the Great Barrier Reef. *Marine Pollution Bulletin*, 75(1–2), 8–20.

Grattan, L. M. Holobaugh, S. and Morris Jr, J. G. (2016). Harmful algal blooms and public health. *Harmful Algae,* 57, 2–8.

Griggs, D. et al. (2013). Sustainable development goals for people and planet. *Nature*, *495*(7441), 305-307.

Guerreiro José (2021) The Blue Growth Challenge to Maritime Governance, Frontiers in Marine Science, 8, 10.3389/fmars.2021.681546, https://www.frontiersin.org/article/10.3389/fmars.2021.681546

Gureva, Maria. (2018). Sustainable leadership and the development of lgreenr economy in the European Union. 10.2991/icseal-18.2018.40.

Habib, K. A. and Islam, M. J. (2020). An updated checklist of marine fisheries in Bangladesh. *Bangladesh Journal of Fisheries*, 32, 357–367.

Hadjimichael, M. A. (2018). Call for a blue degrowth: Unravelling the European Union's fisheries and maritime policies. *Mar. Policy,* 94, 158–164

Hammons, T. J. (1993). Tidal power. *Proceedings of the IEEE*, 81(3), 419–433.

Harley, C. D. Randall Hughes, A. Hultgren, K. M. Miner, B. G. Sorte, C. J. and Thornber, C. S. (2006). The impacts of climate change in coastal marine systems. *Ecology Letters,* 9, 228–241.

Hasan, M. M. Hossain, B. M. S. Alam, M. J. Chowdhury, K. M. A. Al Karim, A. and Chowdhury, N. M. K. (2018). The prospects of blue economy to promote bangladesh into a middle-income country. *Open Journal of Marine Science*, 8, 355–369.

Hassan, S. R. Hassan, M. K. and Islam, M. S. (2014). Tourist-group consideration in tourism carrying capacity assessment: A new approach for the Saint Martin's island, Bangladesh. *Journal of Economics and Sustainable Development*, 5, 150–158.

Häyhä, T. and Franzese, P.P. (2014). Ecosystem services assessment: a review under an ecological-economic and systems perspective. *Ecological Modelling,* 289, 124–132.

Heidkamp, C. P. Garland, M. and Krak, L. (2021). Enacting a just and sustainable blue economy through transdisciplinary action research. *The Geographical Journal.* 10.1111/geoj.12410.

Hoegh-Guldberg, O. et al. (2015). *Reviving the Ocean Economy: Action Agenda for 2015*. WWF International, Geneva. https://wwf.de/fileadmin/fm-wwf/PublikationenPDF/WWF-Report-Reviving-the-Ocean-Economy-Summary.pdf

Hoekman, B. and Djankov, S. (1997). Determinants of the export structure of countries in Central and Eastern Europe. *The World Bank Economic Review*, 11(3), 471–487.

Holland, R. A. Scott, K. A. Flörke, M. Brown, G. Ewers, R. M. Farmer, E. ... and Eigenbrod, F. (2015). Global impacts of energy demand on the freshwater resources of nations. *Proceedings of the National Academy of Sciences,* 112(48), E6707–E6716.

Hoover, S. L. and Morrison, M. L. (2005). Behavior of red-tailed hawks in a wind turbine development. *The Journal of Wildlife Management,* 69(1), 150–159.

Hossain, F. (2020.) Adaptation measures (AMs) and mitigation policies (MPs) to climate change and sustainable blue economy: a global perspective. *Journal of Water and Climate Change.* 12(5), 1344-1369.

Hossain, M. S. Chowdhury, S. R. Navera, U. K. Hossain, M. A. R. Imam, B. and Sharifuzzaman, S. M. (2014). Opportunities and strategies for ocean and river resources management. *Background Paper for Preparation of the 7th Five Year Plan* (p. 61) **edited by Maud Bennett.** FAO, Dhaka, Bangladesh.

Hunter, T. and Taylor, M. (2014). *Deep Sea Bed Mining in the South Pacific: A Background Paper.* Centre for International Minerals and Energy Law.

Hussain, G. M. Failler, P. Karim, A. A. and Alam, M. K. (2017). Review on opportunities, constraints and challenges of blue economy development in Bangladesh. *Journal of Fisheries and Life Sciences,* 2(1), 45–57.

Huwyler, F. Käppeli, J. Serafimova, K. Swanson, E. and Tobin, J. (2014). *Conservation Finance: Moving Beyond Donor Funding Toward an Investor-Driven Approach.* Credit Suisse, WWF, McKinsey and Company. *Zurich, Switzerland.*

IFFO (2013). Linkage between farmed fish and wild fish—fishmeal and fish oil as feed ingredients in the context of sustainable aquaculture. *Jonathan Shepherd, Director General, International Fishmeal and Fish Oil Organisation, OECD Conference Paris 15th–16th April 2010.*

IGBP (2013). *Ocean Acidification Summary for Policymakers—Third Symposium on the Ocean in a High-CO$_2$ World.* International Geosphere-Biosphere Programme, Stockholm, Sweden.

Islam, M. K. Rahaman, M. and Ahmed, Z. (2018). Blue Economy of Bangladesh: Opportunities and Challenges for Sustainable Development. *Advances in Social Sciences Research Journal,* 5(8): 168-178. https://doi.org/10.14738/assrj.58.4937

Jacks, D. S. and Pendakur, K. (2010). Global trade and the maritime transport revolution. *The Review of Economics and Statistics,* 92(4), 745–755.

Jadot, C. (2020). How can Small Islands harness the Blue Economy to build Climate Resilience and Protect Biodiversity?¿ Cómo pueden las Islas Pequeñas aprovechar la Economía Azul para generar Resiliencia Climática y Proteger la Biodiversidad? Comment les Petites Îles peuvent-elles exploiter l'Économie Bleue pour Renforcer la Résilience. *Proceedings of the 73rd Gulf and Caribbean Fisheries Institutes November* 2, 6.

Jafrin, N. Saif, A. N. M. and Hossain, M. I. (2016). Blue economy in Bangladesh: proposed model and policy recommendations. *Journal of Economics and Sustainable Development,* 7(21): 131-134. www.iiste.org ISSN 2222–1700 (Paper) ISSN 2222–2855 (Online).

Jamshidi, S. and Jafari, M. (2021). Environmental impacts of physical and dynamical characteristics of the southern coastal waters of the Caspian Sea. *Earth and Environmental Science Transactions of the Royal Society of Edinburgh,* 112(2), 111–124.

Johnson, K. Dalton, G. and Masters, I. (2018). *Building Industries at Sea: 'Blue Growth' and the New Maritime Economy.* River Publishers **River Publishers,** Alsbjergvej 10, 9260 Gistrup, Denmark.

Jones, K. M. Briney, K. A. Goben, A. Salo, D. Asher, A. and Perry, M. R. (2020). A comprehensive primer to library learning analytics practices, initiatives, and privacy issues. *College and Research Libraries,* 81(3), 570–591.

Kabil, M. Priatmoko, S. Magda, R. and Dávid, L. D. (2021). Blue economy and coastal tourism: a comprehensive visualization bibliometric analysis. *Sustainability,* 13, 3650. https://doi.org/10.3390/su13073650

Kadagi, N. Okafor-Yarwood, I. Uku, J. Adewumi, I. Miranda, N. and Elegbede, I. (2020). *The Blue Economy— Cultural Livelihood—Ecosystem Conservation Triangle: The African Experience. Frontiers in Marine Science,* p.586.

Kalam, A. (2018). Ocean Economy Planning: Adaptability of the 'Blue Economy' model in Bangladesh Maritime Context. *Institute of Regional Studies, Islamabad,* 36(1), 3–45.

Karani, P. and Failler, P. (2020). Comparative coastal and marine tourism, climate change, and the blue economy in African large marine ecosystems. *Environ. Dev.* 36, 100572.

Kathijotes, N. (2013). Keynote: blue economy—environmental and behavioural aspects towards sustainable coastal development. *Procedia Soc. Behav. Sci.* 101, 7–13. doi: 10.1016/j.sbspro.2013.07.173

Kathijotes, N. and Sekhniashvili, D. (2017). Blue economy: technologies for sustainable development. *Zbornik Radova Građevinskog Fakulteta*, 33(30), 621–624.

Keen, M. R. Schwarz, A.-M. and Wini-Simeon, L. (2018). Towards defining the Blue Economy: practical lessons from Pacific Ocean governance. *Mar. Policy,* 88, 333–341. doi: 10.1016/j.marpol.2017.03.002

Kildow, J. T. and McIlgorm, A. (2010). The importance of estimating the contribution of the oceans to national economies. *Marine Policy*, 34(3), 367-374.

Klinger, D. H. Eikeset, A. M. Davíðsdóttir, B. Winter, A. M. and Watson, J. R. (2018). The mechanics of blue growth: management of oceanic natural resource use with multiple, interacting sectors. *Marine Policy,* 87, 356–362.

Konar, M. and Ding, H. (2020). A sustainable ocean economy for 2050: approximating its benefits and costs. World Resources Institute, Washington, D.C. United States.

Laffoley, D. Baxter, J. M. Day, J. C. Wenzel, L. Bueno, P. and Zischka, K. (2019). Marine protected areas. In *World Seas: An Environmental Evaluation* (pp. 549–569). Academic Press. Cambridge, MA, USA

Landrigan, P. J. Stegeman, J. J. Fleming, L. E. Allemand, D. anderson, D. M. Backer, L. C. ... and Rampal, P. (2020). Human health and ocean pollution. *Annals of Global Health*, 86(1): 151

Lapa, K. Gorica, K. and Proko, E. (2021). Development of new tourism models for blue economy using internet of things (IOT)-challenges for vlora region. *SMATECH 2021 Proceedings 21–22 October 2021*, 148.

Lee, K.-H. Noh, J. and Khim, J. S. (2020). The blue economy and the united nations' sustainable development goals: challenges and opportunities. *Environ. Int.* 137, 105528.

Lenzen, M. Sun, Y. Y. Faturay, F. Ting, Y. P. Geschke, A. and Malik, A. (2018). The carbon footprint of global tourism. *Nature Climate Change*, 8(6), 522–528.

Lewis, D. J. Yang, X. Moise, D. and Roddy, S. J. (2021). Dynamic synergies between China's belt and road initiative and the UN's sustainable development goals. *Journal of International Business Policy*, 4(1), 58–79.

Lipton, D. Sachs, J. Fischer, S. and Kornai, J. (1990). Creating a market economy in Eastern Europe: The case of Poland. *Brookings Papers on Economic Activity*, 1990(1), 75–147.

Lu, W. Cusack, Caroline, Baker, Maria, Tao, Wang, Mingbao, Chen, Paige, Kelli, Xiaofan, Zhang, Levin, Lisa, Escobar, Elva, Amon, Diva, Yue, Yin, Reitz, Anja, Neves, Antonio Augusto Sepp, O'Rourke, Eleanor, Mannarini, Gianandrea, Pearlman, Jay, Tinker, Jonathan, Horsburgh, Kevin J. Lehodey, Patrick, Pouliquen, Sylvie, Dale, Trine, Peng, Zhao, Yufeng, Yang (2019). Successful blue economy examples with an emphasis on international perspectives. *Frontiers in Marine Science,* 6, 261, doi: 10.3389/fmars.2019.00261

Madara, K. H. and Perera, L. A. S. (2020). *A Study on Blue–Green Economy: Evidence from Sri Lanka.* Available at SSRN 3844251.

Mahmood, S. and Rashid, H. (2018). *The Conception, Planning and Implementation of Integrated Coastal and Ocean Management for Sustainable Blue Economy in Bangladesh.* World Maritime University Dissertations.

Manikarachchim, I. (2014). *Stepping Up from Green Revolution to Blue Economy: A New Paradigm for Poverty Eradication and Sustainable Development in South Asia.* World Maritime University Dissertations

Martínez-Vázquez, R. M. Milán-García, J. and de Pablo Valenciano, J. (2021). Challenges of the blue economy: evidence and research trends. *Environ Sci Eur,* 33, 61. https://doi.org/10.1186/s12302-021-00502-1

Martins, C. D. Arantes, N. Faveri, C. Batista, M. B. Oliveira, E. C. Pagliosa, P. R. ... and Horta, P. A. (2012). The impact of coastal urbanization on the structure of phytobenthic communities in southern Brazil. *Marine Pollution Bulletin*, 64(4), 772–778.

Matthews, L. Scott, D. and Andrey, J. (2021). Development of a data-driven weather index for beach parks tourism. *International Journal of Biometeorology*, 65(5), 749–762.

McGenity, T. J. Folwell, B. D. McKew, B. A. and Sanni, G. O. (2012). Marine crude-oil biodegradation: a central role for interspecies interactions. *Aquatic Biosystems*, 8(1), 1–19.

McKinley, E. Aller-Rojas, O. Hattam, C. Germond-Duret, C. San Martín, I. V. Hopkins, C. R. ... and Potts, T. (2019). Charting the course for a blue economy in Peru: a research agenda. *Environment, Development and Sustainability*, 21(5), 2253–2275.

Mertha, I. W. Wiarti, L. Y. and Tirtawati, N. M. (2017). Sustainable marine tourism: its existence and rolefor local communities in Bali. *WCBM,* 387. Miola, A. Paccagnan, V. Mannino, I. Massarutto, A. Perujo, A. and Turvani, M. (2009). *External Costs of Transportation Case Study: Maritime Transport.* JRC, Ispra.

Mitra, A. Gobato, R. Saha, A. Pramanick, P. Dhar, I. and Zaman, S. (2021). Non-living resources of the sea as the pillars of blue economy. Parana Journal of Science and Education, Vol. 7, No. 3, 2021, pp. 28-34. http://tiny.cc/PJSE24476153v7i3p028-034

Montserrat, F. Guilhon, M. Corrêa, P. V. F. Bergo, N. M. Signori, C. N. Tura, P. M. ... and Turra, A. (2019). Deep-sea mining on the Rio Grande Rise (Southwestern Atlantic): A review on environmental baseline, ecosystem services and potential impacts. *Deep Sea Research Part I: Oceanographic Research Papers,* 145, 31–58.

Moutoukias, Z. (1988). Power, corruption, and commerce: the making of the local Administrative structure in seventeenth-century Buenos Aires. *Hispanic American Historical Review*, 68(4), 771–801.

Mulazzani, L. and Malorgio, G. (2017). Blue growth and ecosystem services. *Marine Policy*, 85, 17-24.

Musial, W. Butterfield, S. and Ram, B. (2006, May). Energy from offshore wind. In *Offshore technology conference*. OnePetro.

Nagy, H. and Nene, S. (2021). Blue gold: advancing blue economy governance in Africa. *Sustainability,* 13, 7153. https://doi.org/10.3390/su13137153

Nash, K. L. Blythe, J. L. Cvitanovic, C. Fulton, E. A. Halpern, B. S. Milner-Gulland, E. J. ... and Blanchard, J. L. (2020). To achieve a sustainable blue future, progress assessments must include interdependencies between the sustainable development goals. *One Earth*, 2(2), 161–173.

Neill, S. P. Angeloudis, A. Robins, P. E. Walkington, I. Ward, S. L. Masters, I. ... and Falconer, R. (2018). Tidal range energy resource and optimization—Past perspectives and future challenges. *Renewable Energy,* 127, 763–778.

Nellemann, C. Hain, S. and Alder, J. (2008). *In dead water: merging of climate change with pollution, overharvest, and infestations in the world's fishing grounds*. UNEP/Earthprint.

Neori, A. Chopin, T. Troell, M. Buschmann, A. H. Kraemer, G. P. Halling, C. ... and Yarish, C. (2004). Integrated aquaculture: rationale, evolution and state of the art emphasizing seaweed biofiltration in modern mariculture. *Aquaculture*, 231(1–4), 361–391.

Nguyen, T. V. Voldsund, M. Elmegaard, B. Ertesvåg, I. S. and Kjelstrup, S. (2014). On the definition of exergy efficiencies for petroleum systems: Application to offshore oil and gas processing. *Energy,* 73, 264–281.

Notteboom, T. E. (2004). Container shipping and ports: an overview. *Review of Network Economics*, 3(2): 86-106

Novaglio, C. Bax, N. Boschetti, F. Emad, G. R. Frusher, S. Fullbrook, L. ... and Fulton, E. A. (2021). Deep aspirations: towards a sustainable offshore blue economy. *Reviews in Fish Biology and Fisheries*, 32, 1–22.

O'Hagan, A. M. (2020). Ecosystem-based management (EBM) and ecosystem services in EU law, policy and governance. In *Ecosystem-based management, ecosystem services and aquatic biodiversity* (pp. 353-372). Springer, Cham.

OECD, I. (2016). Energy and air pollution: world energy outlook special report 2016.

Ojala, J. and Tenold, S. (2017). Maritime trade and merchant shipping: the shipping/trade ratio since the 1870s. *International Journal of Maritime History*, 29(4), 838–854.

Okafor-Yarwood, I. Kadagi, N. I. Miranda, N. A. Uku, J. Elegbede, I. O. and Adewumi, I. J. (2020). The blue economy—cultural livelihood—ecosystem conservation triangle: the African experience. *Frontiers in Marine Science*, 7, 586.

Patil, P. G. Virdin, J, Diez, S. M. Roberts, J. Singh, A. (2016). Toward A Blue Economy: A Promise for Sustainable Growth in the Caribbean; An Overview. The World Bank, Washington D.C.

Pasternak, G. Zviely, D. Ribic, C. A. Ariel, A. and Spanier, E. (2017). Sources, composition and spatial distribution of marine debris along the Mediterranean coast of Israel. *Marine Pollution Bulletin, 114*(2), 1036-1045.

Pauli, G. A. (2010). *The Blue Economy: 10 Years, 100 Innovations, 100 Million Jobs*. Paradigm Publications. aradigm Publications, 202 Bendix Drive Taos NM 87571

Philpott, A. B. (2005). Stochastic optimization and yacht racing. In *Applications of Stochastic Programming* (pp. 315–336).

Pinto, H. Cruz, A. R. and Combe, C. (2015). Cooperation and the emergence of maritime clusters in the atlantic: analysis and implications of innovation and human capital for blue growth. *Mar. Policy*, 57, 167–177.

Pishgar-Komleh, S. H. Keyhani, A. and Sefeedpari, P. (2015). Wind speed and power density analysis based on Weibull and Rayleigh distributions (a case study: Firouzkooh county of Iran). *Renewable and Sustainable Energy Reviews*, 42, 313–322.

Potgieter, T. (2018). Oceans economy, blue economy, and security: notes on the South African potential and developments. *Journal of the Indian Ocean Region*, 14(1), 49–70.

Potts, T. Burdon, D. Jackson, E. Atkins, J. Saunders, J. Hastings, E. and Langmead, O. (2014). Do marine protected areas deliver flows of ecosystem services to support human welfare? *Marine Policy, 44,* 139–148.

Pretorius, R. and Henwood, R. (2019). Governing Africa's blue economy: the protection and utilisation of the continent's blue spaces. *Studia Univ. Babes-Bolyai Studia Eur.* 64, 119–148.

Purvis, B. Mao, Y. and Robinson, D. (2019). Three pillars of sustainability: in search of conceptual origins. *Sustainability science, 14*(3), 681-695.

Raakjaer, J. van Leeuwen, J. van Tatenhove, J. and Hadjimichael, M. (2014). Ecosystem-based marine management in european regional seas calls for nested governance structures and coordination: a policy brief. *Mar. Policy, 50,* 373–381.

Ram, J. and Kaidou-Jeffrey, D. (2020). Financing the blue economy in the Wider Caribbean. In *The Caribbean Blue Economy By Justin Ram, Donna Kaidou-Jeffrey* (pp. 210–225). Publication, Taylor and Francis, 270 Madison Ave, New York, NY, United States.

Rashid, M. M. and Barua, B. (2016). Tidal energy and its prospects in Bangladesh. In *Proceedings of the 2016 International Conference on Industrial Engineering and Operations Management Kuala Lumpur,* Malaysia.

Rasmussen, R. S. and Morrissey, M. T. (2007). Marine biotechnology for production of food ingredients. *Advances in Food and Nutrition Research, 52,* 237–292.

Rio+20 Pacific Preparatory Meeting (2011). *The 'Blue Economy': A Pacific Small Island Developing States Perspective Apia, Samoa.* Apia: Rio+20 Pacific Preparatory Meeting.

Roberts, A. Thomas, B. Sewell, P. Khan, Z. Balmain, S. and Gillman, J. (2016). Current tidal power technologies and their suitability for applications in coastal and marine areas. *Journal of Ocean Engineering and Marine Energy, 2*(2), 227–245.

Rourke, F. O. Boyle, F. and Reynolds, A. (2010). Tidal energy update 2009. *Applied Energy,* 87(2), 398–409.

Sankrusme, S. (2017). *Potential Development Strategies on Marine and Beach Tourism.* Anchor Academic Publishing. Imprint of Bedey and Thoms Media GmbH, Hermannstal 119 k, 22119 Hamburg, Germany

Santos, M. Radicchi, E. and Zagnoli, P. (2019). Port's role as a determinant of cruise destination socio-economic sustainability. *Sustainability*, 11(17), 4542.

Sanusi, S. L. (2011). Global financial meltdown and the reforms in the Nigerian banking sector. *CBN Journal of Applied Statistics (JAS)*, 2(1), 7.

Sarker, S. Bhuyan, M. A. H. and Rahman, M. M. (2018). From science to action: exploring the potentials of Blue Economy for enhancing economic sustainability in Bangladesh. *Ocean Coast. Manag.* 157, 180–192.

Scharlemann, J. P. et al. (2020). Towards understanding interactions between Sustainable Development Goals: The role of environment—human linkages. *Sustainability Science*, 15(6), 1573–1584.

Schnurr, R. E. Alboiu, V. Chaudhary, M. Corbett, R. A. Quanz, M. E. Sankar, K. ... and Walker, T. R. (2018). Reducing marine pollution from single-use plastics (SUPs): A review. *Marine pollution bulletin, 137,* 157-171.

Schutter, M. S. and Hicks, C. C. (2019). Networking the Blue Economy in Seychelles: pioneers, resistance, and the power of influence. *Journal of Political Ecology*, 26(1), 425–447.

Sharafuddin, M. A. and Madhavan, M. (2020). Thematic evolution of blue tourism: a scientometric analysis and systematic review. *Global Business Review*, 0972150920966885.

Sharafuddin, M. A. and Madhavan, M. (2020) 'Thematic Evolution of Blue Tourism: A Scientometric Analysis and Systematic Review', Global Business Review. doi: 10.1177/0972150920966885.

Silver, J. J. Gray, N. J. Campbell, L. M. Fairbanks, L. W. and Gruby, R. L. (2015). Blue economy and competing discourses in international oceans governance. *J. Environ. Dev.* 24, 135–160.

Smith-Godfrey, S. (2016). Defining the blue economy. *Maritime Affairs: Journal of the National Maritime Foundation of India,* 12, 1–7. doi: 10.1080/09733159.2016.1175131.

Sonak, S. Pangam, P. Sonak, M. and Mayekar, D. (2006). Impact of sand mining on local ecology. In *Multiple Dimensions of Global Environmental Change by* Sangeeta Sonak (pp. 101–121). Teri Press, New Delhi.

Spalding, Mark J. (2016). The new blue economy: the future of sustainability. *Journal of Ocean and Coastal Economics,* 2(2), 8.

Staplin, F. L. (1969). Sedimentary organic matter, organic metamorphism, and oil and gas occurrence. *Bulletin of Canadian Petroleum Geology*, 17(1), 47–66.

Stephens, J. C. Rand, G. M. and Melnick, L. L. (2009). Wind energy in US media: a comparative state-level analysis of a critical climate change mitigation technology. *Environmental Communication*, 3(2), 168–190.

Sumaila, U. R. et al. (2021). Financing a sustainable ocean economy. *Nat Commun,* 12, 3259. https://doi.org/10.1038/s41467-021-23168-y

Tacon, A. G. Metian, M. Turchini, G. M. and De Silva, S. S. (2009). Responsible aquaculture and trophic level implications to global fish supply. *Reviews in Fisheries Science,* 18(1), 94–105.

Thushari, G. G. N. and Senevirathna, J. D. M. (2020). Plastic pollution in the marine environment. *Heliyon*, 6(8), e04709.

Tong, C. (2019). Advanced materials and devices for hydropower and ocean energy. Edited by Colin Tong, In *Introduction to Materials for Advanced Energy Systems* (pp. 445–501). Springer, Cham.

Toupin, M. (2016). *Scientific Validation of Standards for Tidal Current Energy Resource Assessment* (Doctoral dissertation, Université d'Ottawa/University of Ottawa).

UNCTAD W. (2014). United nations conference on trade and development. *Review of Maritime Transport.*

UNDP (2012). *Catalysing Ocean Finance*. Volume 1: Transforming Markets to Restore and Protect the Global Ocean (United Nations Development Programme).

UNEP, F. (2012). Principles for sustainable insurance.

UNEP (2013): United Nations Environment Programme, www.un.org/youthenvoy/2013/08/unep-united-nations-environment-programme/

United Nations Report (2016) The Potential of the Blue Economy: Increasing Long-term Benefits of the Sustainable Use of Marine Resources for Small Island Developing States and Coastal Least Developed Countries, World Bank Group, United Nations.

UN Report (2014) The Potential of the Blue Economy: Increasing Long-term Benefits of the Sustainable Use of Marine Resources for Small Island Developing States and Coastal Least Developed Countries, World Bank Group, United Nations.

Uy, N. and Tapnio, C. (2021). Turning blue, green and gray: opportunities for blue-green infrastructure in the Philippines. In *Ecosystem-Based Disaster and Climate Resilience* edited by • Mahua Mukherjee, Rajib Shaw (pp. 161–184). Springer, Singapore.

Velegrakis, A. F. Ballay, A. Poulos, S. E. Radzevičius, R. Bellec, V. K. and Manso, F. (2010). European marine aggregates resources: origins, usage, prospecting and dredging techniques. *Journal of Coastal Research,* Special Issue No. 51: 1–14.

Voyer, M. (2018). Shades of blue: what do competing interpretations of the blue economy mean for oceans governance? *J. Environ. Policy Plan.* 2018, 23, 595–616.

Voyer, M. and van Leeuwen, J. (2019). 'Social license to operate' in the blue economy. *Resour. Policy,* 62, 102–113. doi: 10.1016/j.resourpol.2019.02.020

Voyer, M. Quirk, G. McIlgorm, A. and Azmi, K. (2018). Shades of blue: what do competing interpretations of the blue economy mean for oceans governance? *Environ. Policy Plan,* 20, 595–616.

Wallace, S. W. and Ziemba, W. T. (Eds.). (2005). *Applications of stochastic programming.* Society for Industrial and Applied Mathematics. 3600 Market Street, 6th Floor Philadelphia, PA 19104 USA

Warner, M. (2009). *The letters of the republic: Publication and the public sphere in eighteenth-century America.* Harvard University Press.

Warren, J. F. (2007). *The Sulu Zone, 1768–1898: The Dynamics of External Trade, Slavery, and Ethnicity in the Transformation of a Southeast Asian Maritime State.* Nus Press: NUS Press Pte Ltd; AS3 #01-02; 3 Arts Link; National University of Singapore Singapore; 117569

Wenhai, L. Cusack, C. Baker, M. Tao, W. Mingbao, C. Paige, K. and Yufeng, Y. (2019). Successful blue economy examples with an emphasis on international perspectives. *Frontiers in Marine Science,* 6, 261.

Wessells, C. R. and Wallström, P. (2019). New dimensions in world fisheries: implications for US and EC trade in seafood. In *Agricultural Trade Conflicts and GATT,* Edited ByGiovanni Anania, Colin A. Carter, Alex F. McCalla (pp. 515–535). Routledge. USA, New York.

Wiarti, L. (2017). The impact of marine tourism in Lovina, Bali: the perspective of local community. *Asean Journal on Hospitality and Tourism,* 15, 139–150.

Winder, G. M. and Le Heron, R. (2017). Assembling a Blue Economy moment? Geographic engagement with globalizing biological-economic relations in multi-use marine environments. *Dialogues in Human Geography,* 7(1), 3-26.

Winther, J. G. Dai, M. Douvere, F. Fernandes, L. Halpin, P. Hoel, A. H. ... and Whitehouse, S. (2020). Integrated ocean management. *World Resources Institute*, 2020-09.

Wood, R. E. (2007). Cruise ships: Deterritorialized destinations. In *Tourism and Transport* (pp. 148–161). Routledge, USA, New York

World Bank (2017). United Nations Department of Economic and Social Affairs. 2017. *The Potential of the Blue Economy: Increasing Long-term Benefits of the Sustainable Use of Marine Resources for Small Island Developing States and Coastal Least Developed Countries.* World Bank, Washington, DC. https://openknowledge.worldbank.org/handle/10986/26843. License: CC BY 3.0 IGO.

Wurtsbaugh, W. A. Paerl, H. W. and Dodds, W. K. (2019). Nutrients, eutrophication and harmful algal blooms along the freshwater to marine continuum. *Wiley Interdisciplinary Reviews: Water, 6*(5), e1373.

Zhang, Y.-g. Dong, L.-j. Yang, J. Wang, S.-y. and Song, X.-r. (2004). Sustainable development of marine economy in China. *Chinese Geographical Science*, 14(4), 308–313.

Zhongming, Z. Linong, L. Xiaona, Y. Wangqiang, Z. and Wei, L. (2019). Maritime Cooperation in SASEC. South Asia Subregional Economic Cooperation.

2 Realizing Blue Economy Potential in Malaysia, Opportunities and Challenges

Hanizah Idris
Department of Southeast Asian Studies, Faculty of Arts and
Social Sciences, University of Malaya, Kuala Lumpur, Malaysia
E-mail: wafa@um.edu.my

CONTENTS

2.1 INTRODUCTION

The fact that oceans and seas matter for sustainable development is undeniable. Oceans and seas cover over two-thirds of the Earth's surface and contribute to livelihoods, provide food and minerals, decent work, generate oxygen, determine weather patterns and temperature and serve as highways for seaborne international trade. According to the Economist Intelligence Unit (EIU), the ocean may also be a new economic frontier, driven by a growing population in search of new sources of growth, and rapid technological advances, making new resources accessible (Economist Intelligence Unit 2015).

As the ocean is inherently fluid in nature, the compartmentalization of ocean, coastal and marine industries from its operating environment of watersheds and ecosystems, to the harmonization of traditional economic activities with sustainable economic values, becomes a challenging activity (Smith-Godfrey 2016). Measuring the ocean economy gives a country a first order of understanding of the economic importance of the sea (Economist Intelligence Unit 2015). Undoubtedly, ocean resources generate numerous benefits to the world economy and offer essential opportunities for transportation, food production, energy, mineral extraction, biotechnology, human settlement in coastal areas, tourism and recreation, and scientific research (Kaczynski 2011, pp. 21–32).

A blue economy is a long-term strategy aimed at supporting sustainable economic growth through ocean related sectors and activities, while improving human well-being and social equity and pre-serving the environment (World Bank and UN DESA 2017). The blue economy is a relatively a new term that has been used in various world studies to refer to a comprehensive set of economic activ-ities concerning the seas and promoting the context of the sustainable development of a country or a region (Nikcevic and Skuric 2021, p. 2). The established blue economy sector contributed 1.3 per-cent to the European Union (EU) economy and 1.8 percent to EU employment in 2017 (European Commission 2019). At the Mediterranean level 'Blue Economy' constitutes an advantage for the development of the region. Amongst the uses with the highest importance for the Mediterranean blue economy is maritime transport. The sector is an essential element of the economy and for job creation across the Mediterranean, but it is also exposed to market fluctuation and international crises. This makes it a relatively volatile source of growth and jobs in a world increasingly exposed to shocks (Union for the Mediterranean 2021).

With increasing recognition of the importance of the world's oceans and coasts, and realizing that natural resources are declining, many countries, including Malaysia, are focusing on reassessing the value of oceans and coasts and actively establishing strategies to develop a sustainable blue economy. The need to balance the economic, social, and environmental dimensions of sustainable development in relation to the oceans is a key component of the blue economy. Realizing the full potential of the blue economy also requires the effective inclusion and active participation of all societal groups. In this context, traditional knowledge and practices can also provide culturally appropriate approaches for supporting improved governance (World Bank and UNDESA 2017).

For a country like Malaysia, the seas surrounding the country contain productive and diverse habitats with the major ecosystems being mangroves, coral reefs, and seagrasses, amongst others. These are productive natural ecosystems that contribute significantly to human, food, economic and environment security. Malaysia is located in the Indo-Pacific region where its coastlines border the Andaman Sea, the Straits of Malacca and Singapore, the Gulf of Thailand, and the South China Sea, to name but a few. As seen in Figure 2.1 below, Malaysia's territory consists of Peninsular Malaysia which is part of the mainland of Southeast Asia and Sabah and Sarawak (East Malaysia), separated by the South China Sea. The coast and seas are always part of nation's social, economic, security and culture and natural parameters, which are interlinked and influenced by internal as well as external factors. These sectors are dynamic and continuously changing, providing goods and ser-vices and in turn being affected by its utility (Kaur 2018).

2.2   DEFINING THE BLUE ECONOMY

How to stimulate economic growth in the ocean areas is widely understood, but it is not clear what the sustainable ocean economy or blue economy should look like, and under what policies, conditions and pathways is it most likely to develop (Patil et al 2016). The term 'Blue Economy' has been used in different ways and similar terms such as 'ocean economy' or 'marine economy' are used without clear definitions (Lee, Noh, and Khim 2020). The urgency of the ocean health challenge is becoming more prominent in the global policy discourse. The importance of oceans for sustainable development is widely recognized by the international community and was embodied in Chapter 7 of the Agenda 21, the Johannesburg Plan of Implementation, the Rio+20 outcome document *The Future We Want*, and the 2030 Agenda for Sustainable Development, where various actions taken by global institutions on sustainable development are identified (Sustainable Development Goals Report 2017).

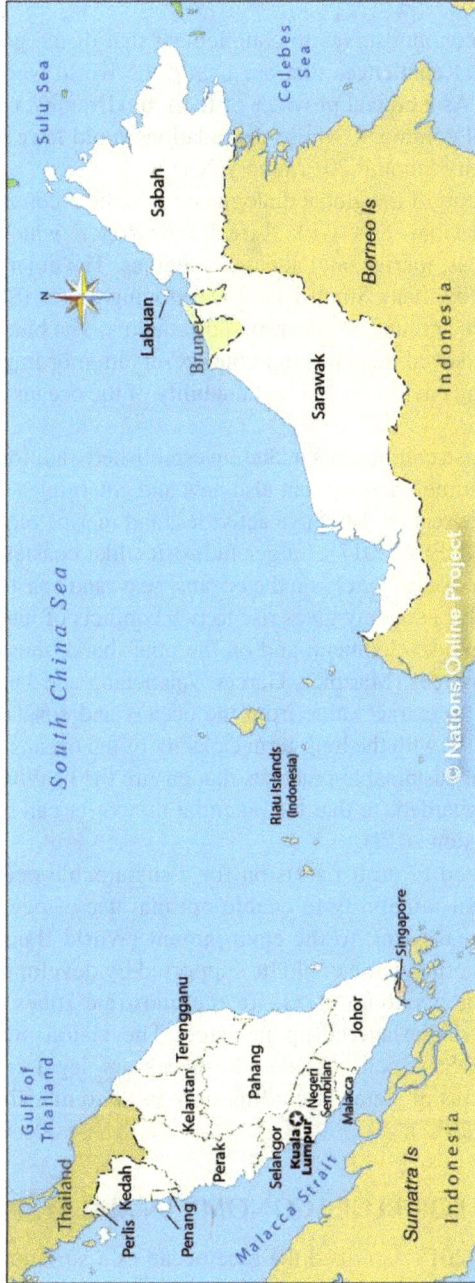

FIGURE 2.1 Map of Malaysia (nationonline.org).

The 'Rio+20' United Nations Conference on Sustainable Development (UNCSD), held in Rio De Janeiro 2012, focused on two key themes: the further development and the refinement of the Institutional Framework for Sustainable Development and the advancement of the 'Green Economy' (UNEP 2012). The 2012 Rio+20 first raised the notion of a blue economy and the need to stimulate 'blue growth', particularly for island nations and developing countries with significant coastlines and/or maritime areas.

The concept of the blue economy serves to complement that of the green economy by rightly highlighting the environmental challenges that are facing the world's oceans which account for two thirds of our blue planet. As a crucial provider of food, biodiversity, carbon sequestration, and energy, oceans are true global commons, whose degradation would have disastrous consequences for all mankind. (Blue Economy Summit 20th January 2014)

The blue economy is thus part of the global dialogue on specific recommendations for achieving the Sustainable Development Goal (SDG) 14: 'Life Below Water' which seeks to conserve and sustainably use the oceans, seas, marine, and aquatic resources. The emerging concept of the blue economy has been embraced by many Small Island Developing States (SIDS) as a mechanism to realize sustainable growth based around an ocean-based economy. The blue economy concepts seek to improve economic growth, social inclusion, and the preservation or improvement of livelihoods while at the same time ensuring environmental sustainability of the oceans and coastal areas (World Bank and UNDESA 2017).

The blue economy has diverse components including established traditional ocean industries such as fisheries, tourism, and maritime transport, but also new and emerging activities, such as offshore renewable energy, aquaculture, seabed extractive activities, and marine biotechnology and bioprospecting (World Bank and UNDESA 2017). Larger industries like coastal development, shipping, and port infrastructure and services also rely on the oceans, seas, and coasts. It is important to highlight that the concept of the blue economy gives rise to two conflicts of interest: on one hand, those linked to economic growth and development, and on the other hand, those linked to safeguarding and protecting the oceans resources (Martinez, Garcia, Valenciano and Jamie 2021). A sustainable blue economy allows society to extract value from the oceans and coastal regions. However, this extraction needs to be in balance with the long-term capacity of the oceans to sustain such activities through the implementation of sustainable practices that ensure the health of the oceans and where productivity economy is safeguarded, so that the potential they offer can be realized and sustained over time (European Commission 2019).

Each country will thus need to draft its vision for a sustainable ocean economy, including how to balance growth and sustainability to enable optimal use of ocean resources with maximum benefit (or least minimal harm) to the environment (World Bank and UNDESA 2017). According to the World Bank, the vision could be supported by development plans and policies, sometimes referred to as blue economy plans, for the maritime zones of each country, which will support the attainment of the agreed-upon vision. The vision must further be anchored in the provisions of UNCLOS, which provides the necessary legal certainty with respect to maritime rights and obligations of states, including in regards to maritime space and resources (World Bank and UNDESA 2017).

2.3   PRIORITIES AREAS FOR BLUE ECONOMY IN MALAYSIA

The World Ocean Summit in 2015 described the blue ocean as a sustainable ocean economy that emerges when economic activity is in balance with ocean ecosystems' support to remain resilient and healthy. For Malaysia the concept of the blue economy is relatively new although the country has been engaging with sustainable development for more than a decade. Malaysia is among the countries with the richest marine biodiversity in the world, and the marine resources provide essential contributions to the livelihood and sustenance of the people.

Various studies have been carried out on the concept of the blue economy and its importance to Malaysia's long term economic growth as well as ocean sustainability. For example, in cooperation with the Partnerships in Environmental Management for the Seas of East Asia (PEMSEA), an assessment of the state of the country's oceans and coasts was carried out by the Maritime Institute of Malaysia (MIMA), an agency under the Ministry of Transport, Malaysia, between 2015 and 2017 to provide:

a) Better understanding of Malaysia's coasts and seas, as well as the threats from human activities, and climate change, and the challenges of managing these areas.
b) Aid policymaking, planning and management of the coastal and marine areas of the country, including the resources, environment, economic activities, and investment, and foster inter-agency collaboration.
c) Contribute to the blue economy assessment, and to monitoring the implementation of the Sustainable Development Goals (SDGs), Sustainable Development Strategy for the Seas of East Asia (SDS-EA), other international agreements, as well as national laws and policies (Kaur 2020)

At the regional level, PEMSEA in its 2018 report identified the following eight key industries of the blue economy:

- Fisheries and aquaculture
- Ports, shipping and marine transport
- Tourism, resorts and coastal development
- Oil and gas
- Coastal manufacturing
- Seabed mining
- Renewable energy
- Marine biotechnology, marine technology and environmental services.

The Regional Network of Local Governments Implementing Integrated Coastal Management (RNLG) was officially launched by the GEF/UNDP/IMO PEMSEA in March 2001 as a forum for exchanging information and experiences in integrated coastal management (ICM) practices among local governments in the region (PEMSEA Charter 2005). Following the Fourth Ministerial Forum on the Sustainable Development Strategy for the Seas of East Asia in Changwon City, ROC Korea in 2012, the committee understood the blue economy to be the practical ocean-based economic model using green infrastructure and technologies, innovative financing mechanisms and proactive institutional arrangements for meeting the twin goals of protecting our oceans and coasts and enhancing their potential contribution to sustainable development, including human well-being, and reducing environmental risks and ecological scarcities (Ebarvia 2016). The declaration enables the development of an ocean-based blue economy in the region through strengthened support for SDC-SEA implementation and other relevant initiatives.

According to PEMSEA, the Malaysian ocean economy contributes 4 percent in terms of employment and is the second largest after Indonesia. In terms of value, the size of the ocean economy gross value added for Malaysia was US$63 billion and US$17.7 billion in terms of the value of ecosystems service in 2015. Table 2.1 shows the size of the ocean economy in East Asia and Southeast Asia in 2015. For Malaysia, the discourse on SDG14, which has been guided by conservation objectives and large economic interests is shifting and greater emphasis has now been placed on the blue economy, generating employment, eradicating poverty and hunger, encouraging entrepreneurship, and creating opportunities for men, women, youth and indigenous people.

TABLE 2.1
Size of Ocean Economy (Gross Value in US$ Billion in 2015)

Ocean Economy	Gross Value (US$ billion)
Cambodia	2.39
China	1,041.92
Indonesia	182.54
Malaysia	63.00
Philippines	11.81
Korea (ROK)	43.53*
Singapore	20.78
Thailand	118.19
Timor-Leste	1.97
Vietnam	38.53
Total	1,512.85

* 2013 latest data available.

Source: Maria Corazon Ebarvia, Blue economy: initiative in the East Asian Seas. *PEMSEA* (2018) https://unescap.org/sites/default/files/02_04_G_Blue_economy_PEMSEA_1-3Aug2018.pdf.

In the Tenth Malaysia Plan (2010), the government recognized the importance of sustainability as part of a comprehensive socio-economic development plan (Tenth Malaysia Plan 2010). In 2009, Malaysia set a voluntary target for reducing greenhouse gas (GHG) emission resulting from its GDP by up to 40 percent, compared to 2005 levels, by 2020. In 2013, Malaysia had already achieved a 33 percent reduction. The country continues developing green growth in the 11th Malaysia Plan which will ensure that socio economic development is pursued more sustainably, beginning at the planning stage, and continuing throughout the implementation and evaluation stages (Eleventh Malaysia Plan 2016). The energy sector, which is a major contributor to national GHG emissions, has undertaken steps to increase the use of clean and environmentally friendly sources.

Malaysia is well-endowed with natural and marine resources. In 2020, petroleum products and liquefied natural gas (LNG) contributed 6.3 percent or RM61.9 billion and 2.0 percent or RM28.8 billion respectively to Malaysia's total exports of RM981 billion (MIDA 2020). The maritime trade and shipping industry have been the backbone of the country, and more than 90 percent of goods are exported by sea. As the coronavirus (COVID-19) pandemic has resulted in massive economic consequences, it is perhaps timely for Malaysia to focus on its ports and shipping sector which form its maritime industry. With the country's growing population, demand for seafood will increase, shipping traffic and tourism will continue to grow, and new ocean industries will expand. The maritime industry contributes about 40 percent of the country's gross economic product. Over the last ten years, before the pandemic, Malaysian ports have recorded an average trade growth of 3 percent in compound cargo throughput. Table 2.2 shows the five potential maritime sectors and their subsectors for the implementation of the blue economy in Malaysia.

2.3.1 PORT ACTIVITIES

2.3.1.1 Development

The rapid growth and development of the Malaysian economy over the past decades cannot be viewed apart from the country's location alongside the world's most important trade routes, namely the Straits of Malacca (Jeevan 2015, Idris and Ramli 2018). The Straits of Malacca has been a strategic waterway for centuries and still retains its importance today. The strategic location and good

TABLE 2.2
The Five Potential Maritime Sectors and Their Sub-sectors for the Implementation of the Blue Economy in Malaysia

Sector	Sub-sector
1. Port activities	Warehousing and storage
	Cargo handling Construction of projects
	Service activities incidental to water transportation
2. Marine Non-living resources oil and gas	Extraction of Crude petroleum
	Extraction of natural gas
	Extraction of marine aggregates
	Supporting activities for petroleum and natural gas
	Supporting activities for other mining
3. Marine Living resources	Capture fishing
	Aquaculture sector
	Processing and distribution
4. Maritime Tourism Coastal and island Tourism Cruise tourism	Accommodation
	Transport
	Other expenditure
5. Maritime transport shipbuilding and repairing	Inland passenger water transport
	Inland freight water transport
	Marine machinery
	Marine equipment
	Repair and maintenance of ships and boats
	Sea and coastal passenger water transport
	Sea and coastal freight water transport

Source: Author's observation.

connectivity make Malaysia one of the preferred countries to enter the Southeast Asian market. Currently, Malaysia has a total of seven main federal ports, namely Port Klang, Penang Port, Johor Port, Kemaman Port, Kuantan Port, Port of Tanjung Pelepas (PTP) and Bintulu Port. Figure 2.2 shows seaports in Malaysia including Kuching port in Sarawak, Labuan port and Kudat, Sandakan, Lahad Datu, Kunak and Tawau ports in Sabah.

Meanwhile, all ports in Sabah and Sarawak (except Bintulu Port and Labuan Port) are under the jurisdiction of the State Governments of Sabah and Sarawak respectively (Ministry of Transport 2021). The administration of ports was legislated under the respective port act. Port authorities were established to govern the federal ports including Port Klang, Johor Port, Penang Port, Kuantan Port and Bintulu Port (Jeevan 2015). The major ports in Malaysia had experienced a privatization process since 1986 (Idris 2000). Following the Port Privatization Act in 1990, the port authorities have been transferring the operation activities to private parties, whereby establishing their role as facilitators, regulators and owners of their designated port area (Egide van der Heide 2020). The privatization Master Plan (PMP) was introduced in 1991 to guide the implementation of the privatization pro-gramme. The PMP contained, among others, a broad policy framework for privatization, procedures for implementation, and assignment of priorities between projects to be prioritized (Mohd Rahim 2017). The privatization programme had five main objectives namely:

  i. To facilitate economic growth
 ii. Relieve the financial and administrative burden of the Government
iii. Improve efficiency and productivity

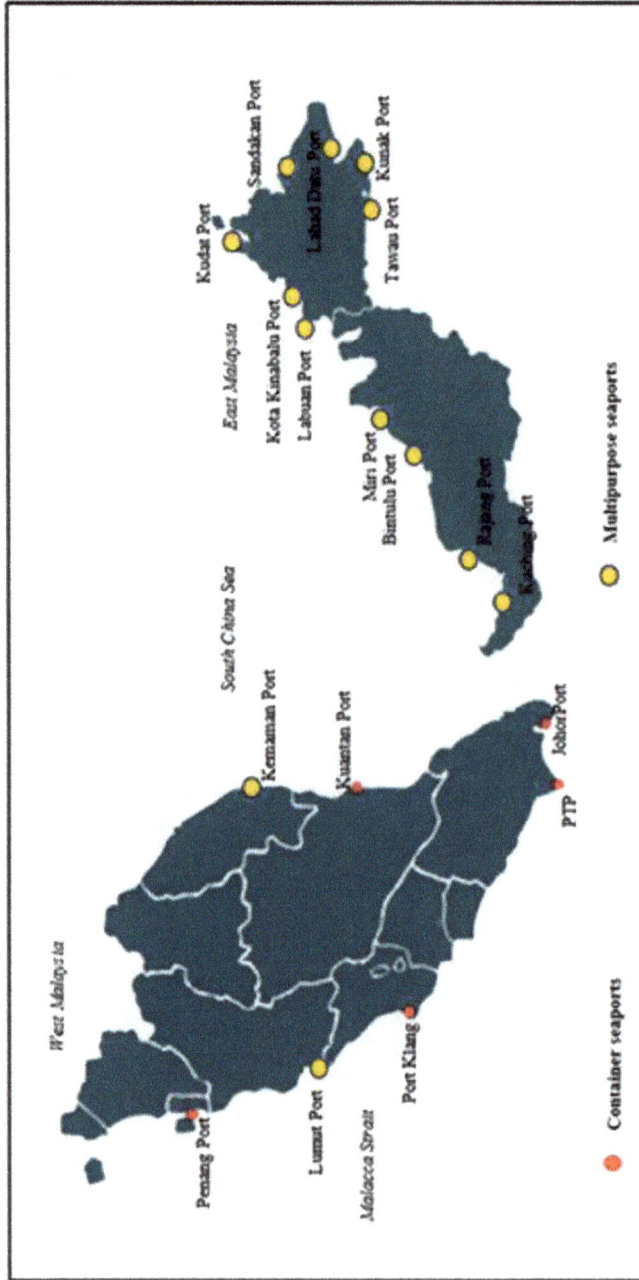

FIGURE 2.2 Malaysian seaports (Jagaan Jeevan, Yeng, C. K. and Othman, M. R. (2021) Extension of the seaport life cycle (SLC) by utilizing existing land capacity for current and future trade preparation, *The Asian Journal and Shipping Logistics*, 37: 45–60).

iv.   Reduce the size and presence of the public sector in the economy
v.   Help to meet the restrictive objectives of the National Development Policy.

Malaysia's premier maritime hub, Port Klang (Wesports and Northport) became the first port to be privatized, followed by other major ports in the country including Penang Port, Johor Port, PTP, Kuantan Port, and Kemaman Port. In terms of types of cargo, about 70 percent of the cargo is containerized. Supporting the growth are Malaysia's premier Port Klang, the country's national load center and PTP, the second largest port in the country. Both Port Klang and PTP handled 64 percent of total cargo throughput by Malaysian ports in 2018. PTP began its operation in 1999 and it is one of the fastest growing container ports in Southeast Asia. PTP, a joint venture between the Malaysia based MMC Group and the Netherlands based APM Terminals, registered a strong growth after accomplishing a record-breaking 9.8 million TEUS total throughput in 2020 despite global economic uncertainties and the health pandemic (PTP 2021). MMC Corporation Berhad handles port operations in Malaysia namely PTP, Johor Port Berhad, Northport (Malaysia) Berhad, Penang Port Sdn. Berhad, and Tanjung Bruas Port Sdn. Berhad. MMC Corporation Berhad is one of the 10th largest port operators in the world with a total container handling capacity across all ports in Malaysia of 21.5 million TEUs (MMC Annual Report 2020).

Container trade development in Asia is faster in comparison to other countries in the world (Jeevan et al 2019). In 2019, Asia contributed almost 65 percent of total world container traffic (UNCTAD 2020). Since 2000, Malaysia has recorded a 400 percent growth in container throughput, now taking almost a quarter of all containers handled in the region (Egide van der Heide 2020). In 2016 Port Klang managed to overtake Rotterdam as the eleventh leading port worldwide (UNCTAD 2017). With a total throughput of 24.9 million in 2018, Malaysian ports handled as many containers as the ports of Rotterdam and Antwerp combined. According to the United Nations Conference on Trade and Development (UNCTAD), Malaysia is the world's fifth best connected country in terms of shipping and line connectivity as shown in Table 2.3.

Malaysian ports have invested heavily in the port infrastructure and port capacity expansion projects (Jeevan 2015). In future planning, Port Klang intends to increase its capacity by 50 percent to 30 million twenty-equivalent units (TEUs) per annum by 2040. Similarly, PTP is developing a new berth that will add 3.5 million TEUs to its current capacity by 2025 (Idris and Ramli 2018).

TABLE 2.3
UNCTAD Maritime Connectivity Index

Best Connected Countries and/or Territories

 1. China
 2. Singapore
 3. Korea (Rep)
 4. Hong Kong (China)
 5. Malaysia
 6. Netherlands
 7. Germany
 8. United States
 9. United Kingdom
10. Belgium

Source: Egide van der Heide (2020). *Port Development in Malaysia*, Report by the Embassy of the Kingdom of Netherland in Malaysia.

Apart from trade and shipping, the ports are also important for cruise tourism. The cruise terminals are located in Penang, Pulau Indah in Selangor, Kota Kinabalu in Sabah and Langkawi in Kedah.

2.3.1.2   Potential

2.3.1.2.1   Sustainable Maritime Ports and Shipping

According to analysis from the Malaysian Investment Development Authority (MIDA 2021) of all the key industries listed, a promising prospect for Malaysia to focus on is 'ports, shipping and marine transport' which form a maritime industry. The maritime industry needs to be more technologically driven for better stewardship of the ocean, or 'blue' resources. The industry is continuously under pressure to meet commercial marketplace needs and carbon emissions regulations set by the International Maritime Organization (IMO).

Although global ocean-related risks such as illegal, unreported, unregulated (IUU) fishing, and marine plastic pollution are difficult to control, opportunities in an area such as ocean renewable energy are rapidly emerging (MIDA 2021). With Malaysia's dependence on international trade, staying aligned to global trends is important, with the shift to sustainability being one of the key trends of the past decade. Malaysian exporters have taken steps to adopt the United Nations-supported SDG through implementing environmental, social, and governance (ESG) standards that meet the requirements for ethical impact and sustainability (MIDA 2020).

2.3.1.2.2   Green Port Policy and Environmental Initiatives

Malaysian ports have taken the initial steps towards the development of a Green Port Policy, intended to guide port operators into becoming both environmentally-friendly and commercially viable. The policy includes the study of the fuel quality of ships in ports, and environmental as well as energy, electricity and fuel saving initiatives. Malaysia is committed to reducing its greenhouse gas emission by up to 45 percent in terms of emissions resulting from its GDP, by 2030. Under the National Automotive Policy 2014, Malaysia had determined to reduce energy consumption and emissions of harmful gases by encouraging the use of energy-efficient vehicles (EEV) (National Automotive Policy 2014). As the nation's economy and trade grow, ports are required to minimize their emissions and pollutions. Port Klang, the premier port in Malaysia aims to be Malaysia's premier maritime hub in terms of sustainable port development in line with the National Transport Policy (NPC). In 2021, Port Klang received international recognition from the Asia Pacific Economic Cooperation (APEC) Port Service Network (APSN) under the Green Port Award System (GPAS) (Westports 2021).

The achievement was made possible through the efforts of the Port Klang Authority and both its terminal operators Northport (Malaysia) and Westports Malaysia Berhad whereby various green initiative programs were initiated and implemented related to alternative energy via solar LED lighting, investing in more energy efficient environmentally friendly cargo handling fleets, monitoring carbon emissions from port equipment and vehicles, marine water quality monitoring and treatment before being released to port waters, paperless transactions, establishing waste management standard operating procedures and other ongoing green initiatives (Westports 2021). PTP, the second major port in Malaysia has stepped up measures to become a green port by offering an advanced vessel traffic monitoring process by introducing the Marine Resource Management System (MRMS) and Vessel Traffic Monitoring and Information System (VTMIS).

PTP, the country's largest container terminal is the first port in Malaysia to attain the VTIMS in 2016 (PTP 2016). VTMIS is a service that primarily provides improvement in terms of the efficiency of vessel traffic movement and navigation within port approaches or hazardous areas. Both the latest systems have the ability to reduce the company's carbon footprint, specifically paper usage. It also further strengthened the safety and security of vessel navigation within the PTP terminal (PTP 2016). Other environmental initiatives taken by the port community included adopting a marine sanctuary area, collaborating with the Malaysian Nature Society, beach cleaning activities, mangrove planting and environmental monitoring and waste management. Malaysia and Singapore were also involved in joint cooperation in tackling chemical oil spills: The Emergency Response Plan.

In recent years, environmental sustainability has become a major policy concern in global maritime transport (UNCTAD 2019). Ports are increasingly expected to meet other performance criteria by ensuring highest service reliability and standards relating to quality, security, safety, financial sustainability, resource conservation, environmental protection, and social inclusion, many of which are linked to key SDGs (UNCTAD 2017). Achieving environmental sustainability, including in maritime transport, is an imperative of the 2030 Agenda for Sustainable Development. The new IMO 2020 regulation, bringing the sulfur cap in fuel oil ships down from 3.50 percent to 0.50 percent, is expected to bring significant benefits for human health and the environment (UNCTAD 2019).

More stringent environmental requirements continue to shape the maritime sector (UNCTAD 2020). Carriers need to maintain services levels, reduced costs, and at the same time ensure sustainability in operations. Furthermore, by promoting liquid natural gas-powered ships, the industry can reduce costs and use a cleaner source of energy, in line with energy and climate related targets under SDG 7 (on energy) and 13 (on climate change) (UNCTAD 2017).

2.3.1.2.3   Port Interaction with Hinterland

Malaysian ports need to innovate and modernize their operating facilities through digitalization and automation. As a large transshipment hub, Malaysia should benefit from technological innovation at the port in terms of cargo handling and terminal design intended to increase efficiency and capacity in a sustainable way. The future port is a smart port, fully automated with a single integrated port community system but there is a concern about the impact on the jobs of port workers if full scale automation were to be implemented. Digitalization and automation are transforming the shipping sector and requiring new skills. The latest technologies provide new opportunities to achieve greater sustainability in shipping and ports, as well as enhanced performance and efficiency (UNCTAD 2019).

Ports activities also provide basic infrastructure for many other sectors including fishing, transport, marine extraction of minerals, oil and gas, maritime tourism and marine renewable energy. In this case, ports act as facilitators of economic and trade development for their hinterland. On the other hand, there will be a competition in terms of land use with respect to other developments such as coastal tourism and aquaculture. As an attraction for coastal and marine tourism, cruise tourism has become one of the sectors identified in the national key economic areas. The Cruise and Integrated Seaport Infrastructure Blueprint for Malaysia has been prepared as an outline for the cruise industry in Malaysia to achieve international standards.

2.3.2   MARINE NON-LIVING RESOURCES: THE OIL AND GAS INDUSTRY

2.3.2.1   Development

Malaysia is the second largest oil exporter and natural gas producer in Southeast Asia and is the fifth largest exporter of liquified natural gas (LNG) in the world in 2019. Malaysia's energy industry is an important sector of growth of the economy (U.S. Energy Information Administration 2021). Malaysia's national oil and natural gas company Petroliam Nasional Berhad (PETRONAS) holds exclusive ownership rights to all oil and natural gas exploration and production projects in Malaysia. PETRONAS has an upstream presence that extends across 20 countries globally, with 245 producing fields, 429 offshore platforms and 30 floating facilities (PETRONAS Annual Report 2019). Malaysia exported 276,000 b/d of crude oil in 2019 and shipped almost all its crude oil exports to Asia Pacific countries including Australia, India, Thailand and Singapore. PETRONAS is also a dominant player in the natural gas sector.

Malaysia's marketed dry natural gas production has risen during the past decade reaching 2.5 trillion cubic feet (Tcf) in 2018. The increase since 2016 has been a result of projects that have come online in the past few years. Besides natural gas, Malaysia has shipped about 1.2 Tcf of LNG, and this accounted for 7 percent of LNG exports worldwide in 2019. Major importers of Malaysia's LNG in 2019 were all in the Asia Pacific region, namely Japan, China, and South Korea. The world's first floating LNG facility, the PFNG SATU successful loaded 10 LNG cargoes in 2019, after its relocation from the Kanowit gas field in Sarawak to the Kebabangan gas field in Sabah. These facilities which have the capacity to produce 1.2 million tonnes of LNG per year, are capable of extracting natural gas from gas fields in water depths up to 200 meters via a flexible subsea pipeline for the liquefaction, production, storage and offloading processes of LNG at the offshore gas field (PETRONAS Annual Report 2019).

The development of Pengerang Integrated Petroleum Complex (PIPC) with oil storage terminal in Johor has completed phase 2 of its construction and this has increased its crude oil storage capacity to 20.8 million barrels for crude oil and petroleum product storage. Phase 3 construction is expected to complete by 2021 and to add about 2.7 million barrels of storage of clean petroleum product. It will become the largest commercial oil storage facilities in Southeast Asia, a joint venture consisting of Vopak, Dialog Group and the state government of Johor.

2.3.2.2 Potential

Petroleum and other liquids and natural gas are the primary energy sources consumed in Malaysia, with estimated shares of 37 percent and 36 percent respectively in 2019. Coal provides 21 percent of the country's energy consumption. Renewable energy accounts for 6 percent of total consumption. The high dependence on petroleum products will be reduced by promoting the use of alternative fuels. The use of biofuel will be promoted while research and development efforts (RandD) into the production of biodiesel will be given full support (Ninth Malaysia Plan 2006). Gas will continue to play a crucial role in powering Malaysia as it is the cleanest burning fossil fuel and reliable source of energy, making it a complementary partner to renewable energy, moving towards low carbon renewable energy in the future.

Other renewable energy sources such as solar will continue to be developed. Efforts have been made to promote the development of biofuel using palm oil as a renewable source of energy during the Ninth Malaysia Plan, (Ninth Malaysia Plan 2006). Studies will be conducted to explore new renewable energy sources such as wind, geothermal and ocean energy. The government will provide 1,740 personnel to the Sustainable Energy Development Authority (SEDA), creating experts in the field of biomass, mini hydro and solar PV (Eleventh Malaysia Plan 2016). Therefore, increased effort is needed in a collaboration with the government and with agencies to shape the long-term energy policy as part of the national development agenda.

2.3.3 Maritime Tourism

2.3.3.1 Development

Tourism the single largest industry in the world of today. Tourism is an agenda issue in the social, environmental and economic levels of many governments (Bhuiyan, Siwar and Ismail 2013). Maritime or ocean-related tourism, as well as coastal tourism, are vital sectors of the economy in many countries. Globally, coastal tourism is the largest market segment and is growing rapidly, becoming less sustainable (Kathijotes 2013). Coastal and maritime tourism has been identified as one of the sectors with high potential for sustainable jobs and growth in the Blue Growth Strategy in the EU (European Commission 2019).

Coastal and ocean-related tourism comes in many forms and includes dive tourism, maritime archaeology, surfing, cruises, ecotourism, and recreational fishing operations (World Bank and

Blue Economy Potential in Malaysia

TABLE 2.4
The Total Number of Tourist Arrival in Malaysia and Total Tourist Receipts, 2015–2019 (Billion RM)

Year	Tourist Arrival (million)	Tourist Receipts (billion RM)
2015	25.72	69.1
2016	26.76	82.1
2017	25.95	82.1
2018	25.83	84.1
2019	26.10	86.1

Source: Ministry of Tourism, Arts and Culture, Malaysia , 2020.

UNDESA 2017). With 48,000 km of coastline, Malaysia boasts some of the most beautiful islands and beaches in Asia with the Straits of Malacca in the west, the South China Sea in the east and the Andaman Sea to the Northwest. The Malaysian government has given emphasis to tourism since the middle of the 1980s. Today, tourism is the third biggest contributor to Malaysia's GDP after manufacturing and commodities. It contributes RM86.14 billion to Malaysia's economy with 26.1 million tourists in 2019 as shown in Table 2.4, ranked number two in the region and ranked 77 in the world when considering tourist numbers in relation to the population of Malaysia. The top ten international tourist arrivals are from Singapore, Indonesia, China, Thailand, Brunei, India, South Korea, Japan, The Philippines, and Vietnam.

In the Ninth Malaysia Plan, greater emphasis has been given to ecotourism and sustainable development through the protection of natural resources such as recreational forests, parks, beaches, islands, and lakes (Ninth Malaysia Plan 2006). The nine key initiatives for tourism development in this period, amongst others, are to attract private sector investment for the growth of the industry, to develop the quality of tourism services including improving the direction and coordination of tourism activities, to improve the diversification of tourism products and activities, to give emphasis to ecotourism, agro-tourism, culture and heritage tourism, the homestay program and thematic events, to cooperate with the private sector for the marketing and promotion of tourism, an integrated approach has been taken for ensuring sustainable development in tourism industry in planning and implementation.

2.3.3.2 Potential

2.3.3.2.1 Ecotourism and Marine Parks

Sustainable development can be part of the blue economy, promote conservation and sustainable use of marine environments and species, and generate income for local communities (Kathijotes 2013, World Bank and UNDESA 2017). Ecotourism is one of Malaysia's biggest tourist attractions. Malaysia is a hotspot for tourists seeking to experience flora and fauna, ancient rainforests, beautiful beaches and reefs, spectacular natural formations, and unparalleled biodiversity (Idris 2021). Marine parks (other protected areas) have been established for the protection, conservation, and management of the marine environment. Currently there are 42 islands gazetted as marine parks in Malaysia managed by the Department for Marine Parks. The principal goals are to protect, conserve and manage in perpetuity representative ecosystems, particularly coral reefs and their respective flora and fauna, so that they remain undamaged for future generations (Department of Marine Park 2021). Other goals are to promote scientific research and to inculcate public understanding, appreciation and enjoyment of Malaysia's marine heritage.

The National Tourism Plan (NTP) was formulated in 1992 by the Ministry of Tourism, Arts and Culture, to develop the tourism industry. The policy incorporates necessary guidelines and management practices for tourism destination development. The policy is based on community-based tourism, cooperation and coordination in tourism development, identified potential tourism assets, and diversification of new products. Ecotourism has been identified as one of the sustainable forms in this plan. NTP has emphasized sustainable ecotourism development in natural islands such as highlands, coastal areas, marine parks, islands, national and state parks, geological sites and, wetland and RAMSAR sites, turtle landing sites and fire-fly sites.

Figure 2.3 shows the Tun Mustapha Park (TMP), the largest marine conservation area in Malaysia, covering almost 1 million hectares with a stunning biodiversity of marine life (WWF-Malaysia 2017). It also incorporates more than 50 islands and islets off the northern coast of Sabah, which is home to the districts of Kudat, Kota Marudu and Pitas. A hugely important area for conservation, it is one of the conservation areas of the Sulu Sulawesi Marine Ecoregion (SSME) and a priority seascape within the Coral Triangle, which is itself acknowledged as the centre of the world's marine biodiversity (WWF-Malaysia 2017). TMP boasts more than 250 species of hard corals, around 360 species of fish, endangered green turtles and dugongs as well as significant primary rainforest, seagrass beds supporting food security and the livelihoods thousands of people.

TMP gazettement comes after more than 13 years of preparatory work led by Sabah parks with government agencies, local communities, international partners, and from support from nongovernmental organizations (NGOs), including the WWF. Fishing has been banned in some areas of the park to better protect the endangered species, including sharks, but it is still allowed with limitations in certain designated zones. It will be an International Union for Conservation of Nature (IUCN), Category VI Park where sustainable uses are still allowed and local communities living within it are able to continue their activities within designated zones. According to WWF-Malaysia, the park is an important part of Sabah's fisheries industry, which produces about 100 tonnes of fish (worth some of RM700,000) each day. Some 80,000 people make their living from the sea along the coast, and many are fishermen. Plans are now underway to increase ecotourism in the area, which is expected to generate income of RM343.4 million over the next two decades for the park.

FIGURE 2.3 The Tun Mustapha Park (WWF Malaysia 2016).

The Malaysian government has continued to promote sustainable tourism development in the National Physical Plan (NPP)-2. The plan prioritizes several sustainable tourism approaches including ecotourism, culture tourism, tangible and intangible zones and assets. Malaysia has also introduced new programs such as Eco-Host Malaysia to enhance public awareness of the need to preserve and conserve the natural environment. It also aims to educate the public on the current environmental situation and how human activities effect ecosystems, including the local community (Ministry of Tourism, Arts and Culture, Malaysia 2021). Other programs such as Ecoteer, a volunteer-based ecotourism company, allows travelers to experience and protect nature simultaneously through restorative ecotourism in Malaysia.

To harness the competitiveness, Malaysia's tourism industry, the Ministry of Tourism, Arts and Culture has established a National Tourism Policy, 2020–2030. The aim is to transform the tourism industry by harnessing public-private sector partnerships and embracing digitalization to drive innovation and competitiveness towards sustainable and inclusive development in line with the United Nations Sustainable Development Goals (National Tourism Policy 2020–2030). The National Tourism Policy has outlined six transformational strategies, first to strengthen governance capacity, second to create special tourism investment zones, third to embrace smart tourism, fourth to enhance demand sophistication, fifth to practise sustainable and responsible tourism and finally, sixth to upskill human capital.

2.3.3.2.2 Cruise Tourism

Malaysia is one of the region's most attractive cruise destinations. Malaysian ports are fully equipped with state-of-the art facilities, offering convenient berthing spots for cruise ships from all over the world (Ministry of Tourism, Arts and Culture, Malaysia 2021). The Cruise and Ferry Integrated Seaport Infrastructure Blueprint for Malaysia was commissioned by the Economic Planning Unit in 2011 in collaboration with the Ministry of Transport and the Ministry of Tourism, Arts and Culture (MOTAC), detailing the visions and policy for cruise industry development in Malaysia until 2020. The blueprint identified six ports as having potential to contribute significantly to the Malaysian cruise industry. The six ports were Penang, Port Klang, Kota Kinabalu, Langkawi, Melaka and Kuching in Sarawak (PEMANDU Associates 2021). In 2013, Kuantan Port joined the line-up of a primary ports due to its growing strategic importance for international cruise lines developing their East Asia sectors (PEMANDU Associates 2021). Kuantan emergence as a core port supports projected traffic increase of cruises to countries along the eastern coast of Southeast Asia, Hong Kong and Taiwan.

Over the last few years, Malaysia has seen an encouraging number of cruise ships calling at ports in various countries port, particularly Port Klang, Penang Port, Melaka and Langkawi. From January to May 2017, a total of 253 international cruise ships called at the ports in the 11 countries, thus indicating an increase of 9.48 percent compared to the same period in 2016. During the same period, the country received 405,554 cruise passengers compared to a previous 330,473 passengers, an increase of 18.51 percent. Malaysian cruise terminals such as Langkawi, Penang Port, Melaka and Port Klang are located close to local attractions, offering cruise passengers an experience of a big-city atmosphere and easy access to ecotourism attractions, beaches, authentic culture, and exotic cuisine. Meanwhile, Kota Kinabalu which is the capital city of Sabah is a gateway to the natural beauties of the rainforest and orangutan watching tends to be popular among many segments of the cruise market.

2.3.4 FISHERIES SECTOR

2.3.4.1 Development

Global production of fish and seafood has quadrupled over the past 50 years. According to the Fisheries Development Authority of Malaysia (FDAM), the international seafood trade ensures food security and contributes to Malaysia's economic growth. As for 2018, the amount of seafood imported and exported to/from Malaysia was 1,249,539 t (1.4 billion USD) and 232,156 t (632.08 billion USD), with a net import quantity of 1,017,023 t. As COVID-19 impacted the global demand for seafood products also plunged. Malaysia covers a total area of 329,847 sq km with a population of 32.06 million in 2020 (DOSM 2020). The countries share maritime borders with the Philippines, Vietnam, and Singapore in the south. The fishery sector has for decades been playing an important role as a major supplier for animal protein to the Malaysian population.

In 2015, the fisheries sector has provided employment to 175,980 people and its contribution to the GDP was at 1.1 percent. The Ministry of Agriculture and Agro-Based Industry, Malaysia is responsible for setting up management policies relating to agriculture and fisheries. The Department of Fisheries, Sabah (DOFS) is responsible for fisheries matters in Sabah. Two other important agencies that deal with fisheries matters in Malaysia with exception of Sabah are the Department of Fisheries, Malaysia (DOF) and FDAM. DOF and DOFS assess and periodically update data on fishing capacity and establish management measures to ensure that Malaysian capture fisheries are functioning in a sustainable manner (National Plan of Action for the Management of the Fishing Capacity in Malaysia, PLAN 2 2015).

Various other legislation and regulations have been put in place to establish controls for related dimensions of fisheries management, import and export, health and safety and these complement the fisheries Act (USAID 2018). DOF released a National Plan of Action (NAP) for the Management of Fishing Capacity in Malaysia for the period 2007–2010 known as PLAN 1 to fulfill a commitment undertaken by the country as set forth in the 1999 FAO International Plan of Action for the Management of Fishing Capacity (IPOA-Capacity). The original PLAN 1 focused on strategies related to the effective management of fishing capacity for the sustainable exploitation of fishery resources in Malaysia. The revised plan, known as PLAN 2, reviewed the achievements of PLAN 1 and to meet the long-term objectives of NPOA which is to achieve an efficient, equitable and transparent management of fishing capacity (National Plan of Action for the Management of the Fishing Capacity in Malaysia, PLAN 2 2015).

The development of the fishing industry in Malaysia followed the guidelines of National Agriculture Policies, with the latest being the National Agro-Food Policy 2011–2020 (NAP). Amongst others, the policies are intended to sustainably modernize and transform the fisheries industry in Malaysia. The fisheries policy currently adopted is targeted towards the exploration and exploitation of resources in new areas of the offshore waters on a large-scale commercial basis. At the same time, Malaysia will continue to give great importance to the sustainable management of coastal fisheries. Important challenges in the fisheries industry in Malaysia are listed below:

- Resources being overfished
- Overcapacity
- Inadequate updated data on fisheries resources
- Incomplete gear specification documentation
- Inadequate capacity and capability of monitoring and surveillance
- Insufficient public awareness and participation
- Decisions inconsistent with current policies
- Foreign fishermen working on board local fishing vessels
- Incentives
- Lack of political will for and awareness of conservation and management
- Encroachment of local fishing vessels into prohibited zones
- Encroachment of foreign fishing vessels.

Marine Protected Areas (MPAs) have been suggested as useful tools to conserve marine habitats and biodiversity, particularly the fisheries and coral reefs (Gazi Md Nurul Islam et al 2017). MPA was first initiated by DOF in the 1980s. Subsequently, most coral reef islands were gazetted as Marine Parks by 1994 under the Fisheries Act 1985 which was amended to the Marine Parks and Marine Reserves Order 1994. The management of MPAs under the DOF was less effective due to encroachment by trawler nets. In dealing with the problems of encroachment and overexploitation of fisheries, the government created the Marine Park Department Malaysia in 2004 under the Ministry of Natural Resources and Environment (Gazi Md Nuzul Islam et al 2017). In peninsular Malaysia, the main regulatory method implemented for managing the fisheries resources is limited entry licensing through zoning. The DOF initiated fisheries conservation and introduced the Fisheries Prohibited Area Regulations in 1983 under the Fisheries Act 1985. In Malaysia, the regulations that govern the Marine Parks are still provided by the Fisheries Act 1985. The MPAs have been declared as protected areas of sea where a variety of uses are permitted such as snorkeling, diving, boating, and beach use but where fishing is prohibited within the MPA.

2.3.4.2 Potential

2.3.4.2.1 Aquaculture

In 2004, agriculture including aquaculture development, has been given top priority by the government to ensure food security and to reduce the food import bill. The aquaculture in Malaysia began way back in 1920 with the extensive poly-culture of Chinese-carp in pools created in water-filled mines and has continued to develop from 1930 until the present day. Given the long coastline of about 4,780 kilometers, brackish water aquaculture dominates the agriculture industry in Malaysia. Brackish water aquaculture is categorized by the extensive culture of the bivalve mollusk, mostly in the western coastal water where there is an abundance of mud flats suitable for the culture of blood cockles. Land-based earthen ponds have spread throughout the country, with the biggest area in the state of Sabah. The culture of marine fish in floating net-cages in lagoons and sheltered coastal waters is concentrated mainly in peninsular Malaysia.

Aquaculture is becoming economically more important as well as increasing the variety of local fish products for food security. Although aquaculture production is still small compared to fish catches, it has been identified as one of the critical activities to ensure food security since the Seventh Malaysia Plan (1996). Further impetus has been given to enhancing agriculture development as a third engine of growth in the Eighth Malaysia Plan (2001–2005). To chart the development of the Malaysian agriculture sector, the government formulated the first National Agricultural Policy (NAP) in the early 1980's. In the third NAP (1998–2010), the promotion of sustainable aquaculture development was one of the priorities, where the aims are to increase aquaculture production.

Sustainable fisheries development has become increasingly important as it is also an important generator of economic and social progress for the rural poor. Various management strategies have been formulated and implemented to control fishing activity and to promote sustainability and conservation of marine resources and ecosystems. The government introduced the National Agrofood Policy (NAP) in 2010 as a replacement for the third NAP. The blueprint sets a direction to transform the agricultural sector to become more dynamic, progressive and sustainable. The NAP covers a period between 2011 to 2020.

2.3.4.2.2 Marine Protected Areas (MPAs)

To have full integration of aquatic resource management in Malaysia, the management of marine parks which is one of MPAs has been placed under the DOF. To date there are several types of MPAs such as Marine Parks, Refugia and Conservation Zones (The New Straits Times, September 9, 2019). MPAs in Malaysia consist of four different types of protecting areas namely marine parks, fishing prohibited areas, wild life reserves and turtle sanctuaries (The International Coral Reef Initiative 2020). The different types of marine managed areas are due to the different objectives, namely biodiversity which include the management of fisheries, turtles and habitats. There are five management authorities that managed the marine areas in Malaysia, the Department of Fisheries under the

Ministry of Agriculture, and Department of Marine Parks, Malaysia under the Ministry of Natural Resources and Environment are responsible for the management of marine areas in peninsular Malaysia. The other managing agencies are Sarawak Forestry Department, Sabah Parks and Sabah Wildlife Department. They are responsible for the management of marine protected areas in the state of Sarawak and in the state of Sabah respectively (The International Coral Reef Initiative 2020).

Malaysia is supposed to set aside at least 10 percent of the coastal and marine environments as MPAs by the end of 2020 to meet one of the Aichi's Biodiversity Targets (WWF 2017). However, with 25,357.9 sq km currently gazetted for MPAs, the country is only at 5.3 percent. According to the WWF, the world had lost 27 percent of its coral reefs and if the present rate continued, 60 percent will die over the next 30 years. According to WWF-Malaysia, the reefs in the Straits of Malacca have an economic value of RM2.3 billion. Malaysia's marine parks managed to attract an average of 650,000 visitors annually, showing the great potential of the marine ecotourism sector in Malaysia. For example, Pulau Tioman marine park rated by the International Coral Reef Triangle Initiative (ICTI) as a Category Four Flagship Site, is identified as a 'large, effectively managed site with regional ecological, governance and socioeconomic importance' and which meets the highest level criteria for management effectiveness (The Star 21 Jan 2020). Pulau Tioman is a prime example of working with the local communities, with an estimated 3,500 residents having undergone major socioeconomic changes with the gazetting of the island and its waters as marine parks. Therefore, under the National Policy for Biological Diversity 2016–2025, the Fisheries Department is obliged to protect, and where necessary restore, ecosystems and habitats such as mangroves, seagrasses, limestone hills, wetlands and coral reefs.

Aside from MPAs, other methodologies adopted for conservation are known as refugia. The refugia in Malaysia are based on four main components (Mohd. Ghazali Manap 2018):

a) Identification and management of fisheries and critical habitat linkages at two priority fisheries refugia
b) Improvement of the management of critical habitats of fish stocks of trans-boundary significance via national actions to strengthen the enabling environment and knowledge-base for refugia management
c) Information management and dissemination of the fisheries refugia concept
d) National coordination for integrated fish stocks and critical habitat management.

In 2000 the DOFS introduced one concept that had been rejuvenated from traditional practices, called the *Tagal* System. The aim of the *Tagal* System is to strengthen the smart partnership between government agencies and local communities in protecting and reviving depleted river fish populations and harvesting such resources in a sustainable manner for the benefit of the local communities. The system which started with 30 areas in 2001 had expanded to 487 areas by 2014 and benefited 170,000 local communities in various economic activities. Seventeen districts in the West of Sabah comprising of 192 rivers have already been involved in *Tagal* Programs (Mohd. Ghazali Manap 2018).

Finally, the Ministry of Agriculture and Agro based Industry realised a new direction for the Ministry that covers the period 2019–2020. The document highlights the five-point food security plan, 18 strategies and 51 initiatives that aim to strengthen the implementation of the NAP-2011–2020. This action plan is geared towards ensuring national food security and rural economic development, as well as spurring domestic investment and international trade with the intention of lifting farmers, breeders, and fishermen out of poverty. Challenges to the sustainable management of fisheries included attention to trawling in the vulnerable reef areas, overfishing, and the use of destructive and illegal fishing methods. In addition, encroachment by commercial fishing vessels into traditional fishing areas and the use of excessive lights on the purse seiners (which attracted excess and often juvenile fish), also lead to overfishing (Wild Life Fund Malaysia 2017). The result was declining fish stocks and habitat degradation which threatened the very survival of fisheries, including high value fish species (Wild Life Fund Malaysia 2017).

2.3.5 MARITIME TRANSPORT—SHIP BUILDING AND REPAIR

2.3.5.1 Development and Potential

Ship building and ship repairing (SBSR) is a strategic industry that contributes to Malaysia's economic growth. It facilitates shipping which carries an estimated 95 percent of national trade, and the offshore oil and gas (OandG) which remain the backbone of the Malaysian economy. There are about 100 shipyards in Malaysia which build and undertake maintenance, repair and overhaul vessels of various types and sizes. Most of the shipyards are concentrated in the cities of Sibu and Miri in the state of Sarawak and specialize in building and repairing small to medium sized vessels such as tugboats, offshore support vessels (OSVs), barges, anchor handlers, and passenger boats.

Shipyards in the west coast of peninsular Malaysia focus on building and repairing naval patrol vessels, and fabricating offshore structures, while most in the east coast concentrate on fishing vessels. The SBSR industry contributes significantly to national economy and trade. In 2015 the industry generated an estimated RM6.4 billion worth of revenue and provided employment to 35,000 people (not including sub-contractors at shipyards) (Khalid 2021).

2.4 OPPORTUNITIES AND CHALLENGES

An important challenge of the blue economy is thus to understand and better manage the many aspects of oceanic sustainability, ranging from sustainable fisheries to ecosystem health to pollution. Bannet et al (2019) suggested that the global rush to develop the blue economy risks harming both the marine environment and human wellbeing with bold policies and actions being urgently needed as ocean development proliferates within Economic Exclusive Zones (EEZ) and in areas beyond national jurisdiction. As Voyer et al (2018) pointed out, the blue economy operates in two competing camps: a) opportunities for growth and development and; b) threatened and vulnerable spaces in need of protection.

At the same time, the ocean economy or blue economy can contribute to addressing some of the concerns associated with economic and environment vulnerability, including those associated with remoteness, by fostering international and regional cooperation under an 'open space approach', which is expressed in the literature as marine spatial planning (UNCTAD 2014, Smith-Godfrey 2016). The concept of ocean economy also embodies economy and trade activities that integrate conservation, sustainable use, and management biodiversity, including maritime ecosystems and genetic resources (UNCTAD 2014). Thus, the ocean economy offers significant development opportunities but raises challenges, not only to SIDS and LDCs but to all countries.

Despite the wide range of economic activity dependent on and shaped by the ocean as a discrete and unique segment of the global economy and its potential for growth, measures for 'ocean economy' have historically not been available (Patil et al 2016). The contribution of the ocean economy to global GDP has rarely been measured although a number of recent estimates have been made, hampered by a lack of comprehensive data made available for analysis. So far, the most extensive effort is the development of an Ocean Economy Database by the OECD, suggesting a contribution in 2010 on the order of US$1.5 trillion in added value (OECD 2016). This value is growing rapidly, and prior to the COVID-19 pandemic, it was projected to increase to USD 3.0 trillion in 2030 (Sumaila et al 2021). Note that this is likely to be an underestimate since many evaluations do not include benefits that lack a market value, such as cultural, social, and ceremonial values (Sumaila et al 2021).

'Measurement is difficult, not least because the lines between coastal and ocean economies are often blurred. And comparisons between countries are complicated by differences in measurement systems, income, and geography. It is also likely that the economic contribution of the ocean is undervalued in many countries. National accounting systems often treat large sectors such as oil and gas and coastal real assets separately. Meanwhile, few estimates give any sense of the value of non-market goods and services such as carbon sequestration, to the ocean economy'. (Economist Intelligence Unit 2015).

According to Smith-Godfrey (2016), the definition of blue economy could be: 'Blue Economy is the sustainable industrialization of the oceans to the benefit of all'. The motivation for including 'benefit' is to balance improvements in equity and wellbeing of both humankind and the environment with reduction in ecological scarcities, bringing the elements of resource efficiency and a low-carbon footprint. Benefit, as defined in this context, allows for the measurement of all the included elements, which may be interpreted as a measurement of effectiveness (Smith-Godfrey 2016). By adopting the definition of the blue economy as 'the sustainable industrialization of the oceans to the benefit of all', together with the criteria linked to it and the balance it seeks to strike between activity and value, several aspects need to be analyzed.

The ocean activities include harvesting of living resources, extraction of non-living resources, generation of new resources, and trade of resources and resource health, which were established by applying a value chain method on the oceans (Smith-Godfrey 2016, p. 4). To discuss the opportunities and challenges posted by the blue economy, it is essential to identify ocean activity and its value. For example, different activities will have different values that contribute to the economic growth of a country, including the establishment of new emerging industries, as shown in Table 2.5. This method was applied on various types of activity using a life-cycle analysis and a cradle-to-grave approach (Smith-Godfrey 2016).

Here the drivers of change vary according to the different activities, for example, for ports and shipping activities the drivers include consumer demands, international regulations for trade and transport, and globalization. For tourism and recreation activities, the drivers are growth in demand for tourism and urbanization, and increases in mobility, accessibility and the demand for

TABLE 2.5
Ocean's Value Chain

Type of Activity	Related Industries	Providing	Emerging Industries
1. Harvesting of living resources	Fisheries	Food security for humans and animals	Aquaculture and mari-culture Pharmaceutical and chemical industries
2. Extraction of non-living resources	Mining oil and gas exploration Alluvial mining	Energy	Seabed mining
3. Generation of new resources	Oil and gas supply Water desalination	Energy and water	Water desalination Renewable energy such as wave energy, ocean wind and solar farming
4. Trade in resources such as transportation of resources Trade tourism and recreation (services)	Shipping, ports and infrastructure supporting the services and coastal development for tourism		Shipping and ports Cabotage Eco-marine tourism Marine real estate development Maritime and marine culture and heritage
5. Resource health	Ocean surveillance and monitoring coastal governance and ocean management		Ocean technologies blue carbon habitat protection rehabilitation and restoration pollution and waste technologies

Source: Adapted from Simone Smith-Godfrey (April 2016). Defining the blue economy. *Maritime Affairs: Journal of the National Maritime Foundation of India*, 1: 58–64.

conservation activities (Smith-Godfrey 2016). For resource health activities, the drivers of change include research and development in technologies, developments in carbon regulations, political stability from a security perspective, urbanization, preservation, and conservation demand.

However, many aspects of the current ocean resource use pattern make it unsustainable and therefore difficult to measure. For example, human transformation of marine ecosystems has resulted in widespread biodiversity lost and habitat damage. This is not new phenomenon since intense exploitation of ocean resources, overfishing, and pollution are major anthropogenic threats to the future sustainability of the oceans and their resources. Unsustainable use of oceans and its resources has reduced the ecosystems' resilience and increased mankind's vulnerability to future global change, thereby incurring a huge economic cost to society. The cost of inaction in the conservation and sustainable use of the ocean are high (Sumaila et al 2021).

Clearly, there is need to change existing practices to ensure that they are compatible with that ocean economy in a way that is sustainable. There is a need to overcome current economic trends that are rapidly degrading ocean resources through unsustainable extraction of marine resources, physical alterations and the destruction of marine and coastal habitats and landscapes, climate change, and marine pollution (World Bank and UNDESA 2017). Sustainable used of marine resources is a complex phenomenon that requires an interdisciplinary approach. In order to overcome the gap between economic prosperity and the need for the sustainability of marine resources, the ecological dimension is significant because it emphasizes the establishment of efficient sea resource management that includes a number of activities. In the first instance there is a need for the prevention of marine pollution, protection of the marine and coastal ecosystems, encouragement of regulated fishing, conservation of coastal and marine areas, the employment of the new marine technologies and the acknowledgment of the scientific background (Nikcevic and Skuric 2021).

Addressing sustainability and equity demands attention to governance. However, ocean governance is highly complex and often lacks coherence and coordination. Bennet et al (2019) pointed out that blue economy governance focuses on how the ocean will be developed and by whom, how and to whom benefits will be distributed, how damage will be minimized, and who will bear responsibility for environmental and social outcomes. Inclusive governance requires that decision making structures and processes are representative of diverse factors from civil society, the private sectors and the governments as shown in Figure 2.4. Ocean spaces and resources are often shared and

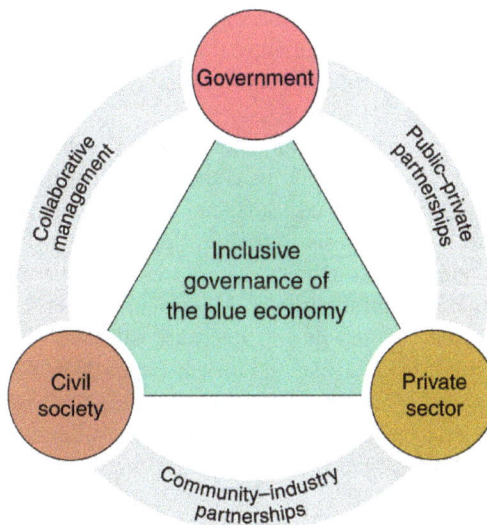

FIGURE 2.4 Inclusive governance of the blue economy (Nathan, J. B. et al. (October 2019). Towards a sustainable and equitable blue economy. *Nature Sustainability*, 2(11): 1–3).

accessed by numerous users including coastal communities, small-scale fisheries and indigenous people who should have a right to participate in the decision regarding allocation of property rights, resources and benefits from, and management of the blue economy (Cohen et al 2019).

One of the challenges to the process of implementation of the blue economy is financial in nature. Financial plans must be put in place to guarantee long term operation. Multisector stakeholder agreements between government and its public utilities, private organizations, and environmental protection groups can be drafted to shoulder long-term expenses with the commitment to share data and equipment utilization (Bethel, Buravleva and Tang 2021). For ocean economy to be sustainable, ocean finance must be adequate and directed to sustainable use and governance of the ocean and its resources (Sumaila et al 2021). This encompasses local, national, and international level financial instruments that are provided by individuals, public and private companies, governments and other non-governmental/inter-governmental institutions.

2.5   CONCLUSION

It is clear that from the above discussion, Malaysia is still in the middle of implementing sustainable development and at the same time promoting green growth in several sectors including the maritime port and shipping sector, fisheries, maritime tourism and the oil and gas industry. While the 'blue economy' concept is very new and not widely used yet in Malaysia, the term has gained enormous momentum at the regional and international levels. Realizing the potential of the blue economy and its benefits in Malaysia, the ports and shipping sector and the tourism sector could be the central sectors of the blue economy concept in the country. Currently, ports are facing increased competitive pressure. Optimization of operations, cost reduction and trade promotion could help improve productivity. Recently, many stakeholders have been calling for an integrated approach to sustainable development; a 'blue economy' that builds an understanding of the world's ocean that not only accounts for more than 70 percent of the planet's surface area but also forms a foundation for global economic sustainability.

Apart from the ports and shipping sector, the maritime tourism sector will be a booming sector of the blue economy. There are approximately 3.2 million jobs in the maritime tourism sector in Europe alone. For a country like Malaysia, tourism, including maritime tourism, is a generator for economic growth and further priority will be given by the government to develop ecotourism that will reduce the negative impact on the environment. Coastal and maritime tourism depends greatly on good environmental conditions and a good water quality in particular. Cruising can also be considered as part of coastal tourism and maritime tourism and the industry grows extremely fast.

The fishing industry has not changed a lot in recent years. The importance of fishing for feeding the world's population has undoubtedly remained the same. Unfortunately, the negative effects of the fishing industry have hardly changed. Sustainable fishing has huge potential in Malaysia. Malaysia should continue to give emphasis to the development of MPAs. Malaysia as one of the twelve leading biodiverse countries in the world and places a lot of emphasis on conservation and sustainable utilization of its rich natural heritage. Malaysia achieves success in managing marine parks and has been recognized for this by the International Coral Triangle Initiative (ICTI), a multilateral partnership involving Malaysia, Indonesia, Papua New Guinea, the Philippines, the Solomon Islands and Timor Leste. However, the rise of population and rapid development of the MPA areas has made biodiversity conservation increasingly challenging. Malaysia's biodiversity riches must be managed and conserved properly so that the benefits can be passed to the next generations.

Finally, the greatest challenges in implementing the blue economy concept in most countries is that the ocean is still not a priority and the lack of a national ocean policy and institutional arrangements. There is an urgent need to have a common understanding of the blue economy and link it to the

SDGs and other regional and international agreements. Yet, neither an obvious coordinating body nor a comprehensive set of blue economy guidelines currently exist (Bennet et al 2019). Thus, Bennet et al (2019) recommended that the UN establish or designate a commission or agency within the Economic and Social Council system to be responsible for developing best practices and establish international guidelines for the implementation, monitoring and management of blue economy activities. Guidelines would provide a foundation for international deliberation and multilateral discussion, as well as guidance on national policies and corporate activities (Bennet et al 2019).

Moving forward, the ocean sector which is crucial for Malaysia's economy through its resources and ecosystems services that support trade and industries, requires proper management and conservation strategies to achieve the best economic, environment and social outcomes. There should be changes in the role of governance and the encouragement of more private sector participation to ensure sustainable and inclusive well-being and to promote a circular economy and climate resilience. Some of the initiatives that could be undertaken include developing a blue economy profile and conducting pilot studies of the ocean to help define and refine Malaysia's concept of a blue economy and promoting the use of ocean economy data in marine planning and at the regional level to facilitate further engagement of Malaysia with other countries in the region on related matters (The Borneo Post 2018). It is envisaged that the blue economy initiative would further drive sustainable development at the national level (The Borneo Post 2018).

REFERENCES

Audrey Dermawan. "Fishermen urged to venture into aquaculture," *The New Straits Times*, September 9, 2019, https://nst.com.my/news/nation/2019/09/519960/fishermen_urged_venture_aquaculture

Bannet, N. J. et al. October 2019. Towards a sustainable and equitable blue economy. *Nature Sustainability* 2(11): 1–3. https://researchgate.net/publication/336532839

Bethel, B. J. B. Buravleva, Y. and Tang, D. May 2021. Blue economy and blue activities: opportunities, challenges and recommendations for the Bahamas. *Water* 13: 1399. https://mdpi.com/journal/water

Bhuiyan, M. A. H. Siwar, C. and Ismail, S. M. June 2013. Tourism development in Malaysia from the perspective of development plans. *Asian Social Science* 9:11–18. https://researchgate-net/publication/286432146_Tourism_Development_in_Malaysia_from_the_Perspective_Development_Plans

Cohen, P. J. et al. April 2019. Securing a just space for small-scale fisheries in the blue economy. *Front. Mar. Sci* 6: 1–8.

Department of Marine Park, 2021. Malaysia. http://marinepark.dof.gov.my

Department of Statistics Malaysia (DOSM), 2020. . https://dosm.gov.my

Ebarvia, Maria Corazon, M. February 2016. Economic assessment of oceans for sustainable blue economy development. *Journal of Ocean and Coastal Economics* 2(2): 1–29. https://cbe.miis.edu/joce/vol2/iss2/7

Economist Intelligence Unit. 2015. The blue economy: growth, opportunity and challenges. https://eiuperspectives.economist.com/sustainability/blue-economy

Egide van der Heide. 2020. *Port development in Malaysia*. Report by the Embassy of the Kingdom of Netherland in Malaysia.

Eleventh Malaysia Plan. 2016. Economic Planning Unit (EPU). https://epu.gov.my/sites/default/files/2021-05/Chapter%201.pdf

European Commission. 2019. The EU blue economy report 2019. Publication office of the European Union, Luxembourg. https://prod5.assets-cdn.io/event/3769/assets/8442090163-fc038d4d6f.pdf

Gazi, M. N. I. et al. 2017. Community perspective of governance for effective management of marine protected areas in Malaysia. *Ocean and Coastal Management* 135: 34–42.

Idris, H. December 2000. The expansion of the Southeast Asian ports: prospects and challenges. *Jati-Journal of Southeast Asian Studies* 5: 69–78.

Idris, H. 2021. The social-cultural aspects of maritime Malaysia. In *Malaysia as a Maritime Nation,* ed. Harun, R. and Ja'afar, S. 247-278. Kuala Lumpur: Maritime Institute of Malaysia (MIMA).

Idris, H. and Ramli, M. F. December 2018. Southeast Asian region maritime connectivity and the potential development of the Northern Sea route for commercial shipping. *Jati-Journal of Southeast Asian Studies* 23(2): 25–46.

Jeevan, J. et al. December 2015. The implication of the growth of port throughput on the port capacity: the case of Malaysian major container seaports. *International Navigational of E-navigation and Maritime Economy* 3: 84–98. https://sciencedirect.com

Jeevan, J. Yeng, C. K. and Othman, M. R. 2021. Extension of the seaport life cycle (SLC) 2015 by utilizing existing land capacity for current and future trade preparation. *The Asia Journal and Shipping Logistics* 37: 45–60.

Jeevan, J. et al. 2019. The impact of dry port operations on container seaports competitiveness. *Maritime Policy and Management* 46:4-23

K-H, Lee, Junsung, N. and Jong, S. K. 2020. The blue economy and the United Nations' sustainable development goals: challenges and opportunities. *Environmental International* 137: 105528. 1–6.

Kaczynski, W. 2011. The future of the blue economy: lessons for the European Union. *Found Manag* 3(1): 21–32.

Kathijotes, N. 2013. Keynote: Blue economy-environmental and behaviorial aspects towards sustainable coastal development. *Procedia-Social and Behaviorial Sciences* 101: 7–13.

Kaur, C. R.. "Time to focus on blue economy," *The Star*. 2018, May 7, https://thestar.com.my/opinion/letters/2018/05/07

Kaur, C. R. . Towards a blue economy initiative:assessment and way forward for Malaysia on the environment and resources management aspects. No.9/2016, 17 June 2016. SEA Views MIMA'S online commentary on maritime issues, 1-4. https://researchgate.net/publication/31192481_Towards_a_blue_economy_initiative_Assessment_and_way_forward_for_Malaysia_on_the_environment_and_resources_management_aspects

Khalid. N. June 29, 2021. Upskilling the shipbuilding industry. *The Malaysian Reserve*. https://themalaysianreserve.com/2017/04/03/upskilling-the-shipbuiding

Malaysian Investment Development Authority (MIDA). 2020. https://mida.gov.my

Martínez-Vázquez, Rosa María, Milan-Garcia, Juan, and de Pablo Valenciano, Jamie. 2021. Challenges of the blue economy: evidence and research trends. *Environmental Science Europe* 33(61): 1–7. http://researchgate.net/publication/349294317_Challenges_of_the_Blue_Economy_Evidence_and_Research_Trends

Ministry of Transport, Malaysia. 2020. https://mot.gov.my/en

Ministry of Tourism, Arts and Culture, Malaysia. 2020. https://motac.gov.my

Ministry of Tourism, Arts and Culture, Malaysia. 2021. https://motac.gov.my

MMC Corporation Annual Report. 2020. https://mmc.com.my

Mohd. Ghazali Bin Manap. 2018. Establishment of marine *refugia* in Malaysia: conservation and wild panaeid shrimp stock in Baram, Sarawak and wild lobster population in Tanjung Leman, Johor. Department of Fisheries, Malaysia (Policy paper).

Mohd Rahim, F.A. et al. 2017. *Public-Private Partnership and Private Finance for Infrastructure Projects*. Kuala Lumpur: University of Malaya Press.

National Automotive Policy. 2014. https://maa.org.my/pdf/NAP_2014_policy.pdf

National plan of action for the management of fishing capacity in Malaysia (PLAN 2). 2015. Department of Fisheries, Malaysia, Putrajaya.

National Tourism Policy-Summary 2020–2030. 2020. Ministry of Tourism, Culture and Arts, Malaysia.

Nikcevic, J. and Skuric, M. 2021. A contribution to the sustainable development of mtritime transport in the context of the blue economy: the case of the Montenegro. *Sustainability* 13: 3079. https://mdpi.com/journal/sustainability

Ninth Malaysian Plan, 2006–2010. 2006. Economic Planning Unit (EPU). https://epu.gov.my

Ocean Economic Co-operation and Development (OECD) Report 2016. *Development Co-operation Report 2016: The Sustainable Development Goals as Business Opportunities*. Paris: OECD Publishing. http://dx.doi.org/10.1787/dcr-2016-en

Patil, P. G. Virdin, J. Diez, S. M. Roberts, and J. Singh, A. 2016. *Toward a Blue Economy: A Promise for Sustainable Growth in the Caribbean*. Washington DC: World Bank, http://documents.worldbank.org/curated/en/965464147344986 1013/main-report

PEMANDU Associates. 2021. https://pemandu.org

PETRONAS Annual Report. 2019. http://PETRONAS-Annual Report-2019-v2.pdf

PEMSEA Charter 2005. Agreements and Declarations. https://pemsea.org/publications/agreements-and-declarations/charter-pemsea-network-local-governments-sustainable

Port of Tanjung Pelepas. 2021. Port of Tanjung Pelepas remains resilient amidst global economic uncertainties. www.ptp.com.my/media-hub/news/maritime-fairtrade

Port of Tanjung Pelepas. 2016. PTP embarks on mrm and vtmis systems. https://ptp.com.my/media-hub/news/ptp-embarks-on-mrms-vtmis-systems

Smith-Godfrey, S. April 2016. Defining the blue economy. *Maritime Affairs*: *Journal of the National Maritime Foundation of India* 1: 58–64. http://tandfonline.com/loi/rnmf20

Seventh Malaysian Plan. 1996. https://epu.gov.my/en/economic-developments/development-plans/rmk/seventh-malaysia-plan-1996-2000

Sumaila, U. R. et al. 2021. Financing a sustainable ocean economy. *Nature Communications* 12(3259), 1–11. https://doi.org/10.1038/s41467–02–23168-y; www.nature.com/naturecommunications

Tenth Malaysia Plan. 2010. Economic Planning Unit (EPU). https://epu.gov.my

The Borneo Post. 2018, December 9. Understanding the blue economy and its growing importance. *The Borneo Post*. https://pressreader.com/malaysia/the-borneo-post-sabah/20181209/281728385599443 https://eiuperspectives.economist.com/sustainability/blue-economy

The International coral reef initiative. 2020. http://icriforum.org/about

The sustainable development goals report. 2017. United Nations. Sustainable Development Goals Report 2017 | United Nations

UNCTAD. 2014. *The Oceans Economy: Opportunities and Challenges for Small Islands Developing States*. United Nations: Geneva. https://unctad.org/system/files/official-document/ditcted2014d5_en.pdf

UNCTAD. 2017. *Review of Maritime Transport*. https://unctad.org/system/files/official-document/rmt2017_en.pdf

UNCTAD. 2019. *Review of Maritime Transport*. https://unctad.org/system/files/official-document/rmt2019_en.pdf

UNCTAD.2020. Review of Maritime Transport. http://unctad.org/system/files/official-document/rmt2020_en.pdf

UNEP. 2012. Blue economy concept paper:Rio+20 United Nations Conference Sustainable Development. https://unep.org/resources/report/rio20-outcome-document-future-we-want

Union for the Mediterranean (UfM). 2021. *Towards a Sustainable Blue Economy in the Mediterranean Region*. The UfM Secretariat. Towards a Sustainable Blue Economy in the Mediterranean region – 2021 Edition | EU Neighbours

USAID. 2018. *Malaysia CDT Gap Analysis and Partnership Appraisal*. The Oceans and Fisheries Partnership (USAID Oceans). USA. Malaysia CDT Gap Analysis and Partnership Appraisal – The Oceans and Fisheries Partnership (seafdec-oceanspartnership.org)

U.S Energy information administration. 2021. https://oilandgasmalaysia.pdf

Voyer, M. Quirk, G. McIlgorm, A. Azmi, K. 2018. Shades of blue: What to competing interpretations of the blue economy means for ocean governance? *J. Environment Pol. Plan* 20(5): 565–616. www.tandfonline.com/loi/cjoe20

Westports Holdings Sdn Bhd. 2021. https://westportsholdings.com

Wild Wild Life Fund-Malaysia. 2017. The Tun Mustapa Park case study. https://wwfint.awsassets.panda.org/downloads/tun_mustapha_park_case_study.pdf

World Bank and United Nation Department of Economic and Social Affairs. 2017. *The Potential of the Blue Economy: Increasing Long-Term Benefits of the Sustainable Use of Marine Resources for Small Island Developing States and Coastal Least Developing Countries*. Washington DC: World Bank. http://hdl.handle.net/10986/26843

3 Optimizing the Connectivity of Salmon Farms

Role of Exposure to Wind, Tides, and Isolation

Dmitry Aleynik,[1]* Thomas Adams,[1,2] and Keith Davidson[1]
[1] Scottish Association for Marine Science, Oban, UK
[2] Scottish Sea Farms Limited, South Shian, Connel, Argyll, UK
*Corresponding author: Dmitry Aleynik
E-mail: Dmitry.Aleynik@sams.ac.uk; Tom.Adams@scottishseafarms.com;
Keith.Davidson@sams.ac.uk

CONTENTS

DOI: 10.1201/9781003184287-3

3.1 INTRODUCTION

Growth in consumer demand for fish and other marine food products requires a corresponding increase in production over the coming decades. This is particularly true in high-latitude countries where salmon farming is practised. In Scotland, for example, government is targeting a sustained increase in production to 2030 (Scotland Food and Drink 2017). To meet this demand, farm operators have two main options: the establishment of new sites, or the expansion of the existing ones. Preferred locations for farms are generally found in sheltered fjordic areas. However, as the industry is now well established, such areas are already considered to have reached their capacity in terms of numbers of sites. Furthermore, sites cannot be increased in size without regulatory established limits. A maximum permitted biomass is applied to each site when its licence is granted (based upon the predicted level of released depositional matter such as feed and faeces and its impact on the benthic environment; (Scottish Environmental Protection Agency 2019b), and operators seek to work as closely as possible to this level to maximise profits. To overcome such spatial and size based restrictions on capacity, operators are therefore looking for new (often more exposed) locations in which to site farms.

One of the principal environmental impacts of salmon farming relates to sea lice *Lepeophteirus salmonis* (Kroyer, 1837) (Bron et al. 1993). These are parasitic copepods that feed on the soft tissues of salmonid fish, causing a range of issues including lesions, loss of appetite, osmoregulatory imbalance and increased susceptibility to other challenges such as Amoebic Gill Disease, harmful algal blooms, and jellyfish stings. These serve to reduce fish health and welfare and can, in severe cases, lead to mortality of infested fish. Sea lice were present prior to the establishment of salmon aquaculture, but the enhanced (and year-round) availability of host fish on farms allows lice to reach much greater numbers than they would naturally (Costello 2006; Heuch et al. 2005). Even historically, when production levels were much lower, it was estimated that farmed fish accounted for 78-97% of the larval sea lice in Scottish waters (Butler 2002).

Sea lice infestations have generally been treated through the application of chemicals, both in fish feed and in topical bath treatments. Due to their impacts on the environment (in particular benthic organisms (Veldhoen et al. 2012), the chemicals used to treat and remove sea lice are strictly regulated (Scottish Environmental Protection Agency 2019b), with limits placed on the frequency, number and intensity of treatments. Other approaches to dealing with sea lice include fresh and warm water baths (Grøntvedt et al. 2015), physical barriers (Oppedal et al. 2017), cleaner fish such as wrasse (Leclercq, Davie, and Migaud 2014; Skiftesvik et al. 2013) and coordinated local management strategies (T. P. Adams, Aleynik, and Black 2016; Murray and Gubbins 2016).

Farmers wishing to extend existing sites or establish new ones must demonstrate a commitment to limiting impacts on wild fish. In Scotland, such a planning process is administered at a local council level. Council planners are more experienced at dealing with applications where complaints are well defined and objective (oppositions to building developments, for example). In the case of assessing the environmental impacts of proposed aquaculture developments, they are often faced with the challenge of vocal objectors (the concerns of whom may be justified or unjustified), and a lack of informative and impartial information upon which to gauge the likely impacts of a site development. These might be impacts in absolute, relative, or cumulative terms.

Despite limited physical space in fjordic coastal waters, sustainable development of the industry should be possible. It will however involve consideration of locations that, by virtue of their level of physical exposure, might have been previously considered somewhat less suitable for farming operations. Investigating such options requires coordination between stakeholders at a regional level, appropriate regulation and standardised methods for the appraisal of parasite pressure and wild fish impacts which take into account the interconnected nature of the environment in which farms are located.

Computational biophysical models are ideal for such an application. Over the last decade, methodology for the modelling of sea lice dispersal in the coastal marine environment has reached a consensus in terms of parameterization and constituent processes (Johnsen et al. 2016; Salama and Murray 2013). Advances in computing facilities and hydrodynamic modelling software now allow the simulation of a large spatial domain, incorporating fine resolution in key areas with acceptable levels of computational overhead (Adams et al. 2016; Aleynik et al. 2016; 2018; Wolf et al. 2016). Given appropriate environmental forcing and using existing industry data sources, models now allow the derivation of clearly defined and comprehensible quantitative results. This enables appraisal of individual sites in the context of existing local developments and making scientifically robust comparisons between different options.

To illustrate this approach, we focus on a comparative case study of two hypothetical new salmon aquaculture sites, both of which are more exposed than most existing sites. We use biophysical model simulations to investigate and describe how these sites affect and are affected by their surroundings, in terms of the impact of sea lice populations. We discuss model parameterization, simulation timeframes, variation in environmental conditions, and options for the integration of available data sources relating to production statistics and local sea lice counts. Key outputs relating to the spread of lice from sites are described, including larval density, prevalence and connectivity. We consider how our new sites fit within the region, assessing connectivity across the full network of sites, in the context of wave exposure and physical isolation from established sites.

3.2 METHODS

3.2.1 Summary

A combined hydrodynamic and biological model was used to simulate the spread of 'sea lice' larval particles from existing salmon aquaculture sites, and from two locations where there are currently no fish farms. Simulations covered a continuous period of twelve complete months, incorporating a representative range of varying tides and weather conditions, which together dominate the flow pattern in this region (Edwards 2016). Particle release rates were linked to site biomass and lice treatment threshold levels. A range of metrics including mapped dispersal patterns, dispersal kernels and between-site connectivity were considered to characterize the new sites in the context of their surroundings.

The variability in connectivity of all existing sites was also assessed, in addition to the infection pressure at locations spanning the full coastline. This was related to a wave exposure index and the sum of inverse distances to existing sites.

3.2.2 Scoping Sites

The characteristics of two sites in the Scottish west coast region were investigated. These sites were chosen such that they were relatively nearby to one another, allowing their comparison and a study of interconnectivity between them, and in locations which could potentially be used for future aquaculture development (though no development or plan for development presently exists). The model sites were located at Loch Buie on the south coast of Mull ('Site 1': lon/lat -5.9102°, 56.3339°), and Minard Point on the coast of the mainland, south of Oban ('Site 2': lon/lat -5.5459°, 56.3503°) (stars on Figures 3.1, 3.3, 3.5, 3.6).

Both sites are relatively exposed in comparison with most farms in the region, but they are somewhat sheltered in comparison to completely open water and hence characteristic of the water bodies that aquaculture operators now wish to exploit. Site 1 is in a relatively open loch (fjord) which faces SW, and it is exposed to prevailing winds but sheltered from most other directions. Located by an island, it is expected to have relatively low influence of freshwater. It is relatively isolated from other fish farm sites.

Site 2 is located near a headland at the mouth of a loch and is more sheltered from the prevailing wind direction. Tidal flows in the area around this site are strong and fairly complex, due to the array of islands and narrow straits in the vicinity. The site is likely to be subject to pulses of freshwater emanating from the neighboring loch. Site 2 is less isolated from other existing farms, with these being located both to the north and south of the site.

3.2.3 Physical Models

The underlying models for this study are based on the coupled operational ocean-atmosphere WeStCOMS (West Scotland Coastal Ocean Modelling System), comprised of the WRF meteorological model (Skamarock et al. 2008) and the unstructured FVCOM ocean circulation model (Chen et al. 2013). This utilizes a variable resolution triangular prismatic mesh to simulate water currents, temperature and salinity (among other quantities), representing detailed coastal and bathymetric features where required, whilst retaining computational efficiency. Horizontal resolution ranges between 130 m to 4.5 km at some open boundary locations. Vertically, the model resolves 10 terrain following sigma-layers, with higher density of layers near the surface and seabed with most locations close to the coastlines being shallower than 100 m. WeStCOMS-FVCOM is nested (one-way) into the regional operational North-East-Atlantic ROMS model (Dabrowski et al. 2014), which provides ocean temperature and salinity fluctuations at the open lateral boundary. A range of external forcing processes include tidal excursions driven at the model boundaries with 11 major tidal constituencies computed with Oregon State University OSU inverse tidal solution (Egbert et al. 2010), and the Multi-Satellite Sea-Surface Temperature MUR-SST product available from JPL-NASA (Armstrong et al. 2012). Meteorological forcing data, derived from the high-resolution (2 km) WeStCOMS-WRF model domain, includes cross-sea-surface shortwave and net radiation fluxes, winds and fresh water supply via discharges from major rivers (estimated as the accumulated WRF rainfall over their individual catchment areas).

Specific implementations of these models were developed for the Scottish west coast region, details of which have been described previously (Adams et al. 2016; Aleynik et al. 2016; 2018). The model domain is shown in Figure 3.1 and covers most of the west coast of Scotland, an area that contains a large proportion of the salmon aquaculture sites in the UK (156 sites). The WeStCOMS model produces hindcast and five-day forecasts on a weekly basis over the full period from July 2013 to the present date.

The meteorological model was compared to data from weather stations within the domain, including Dunstaffnage, Tiree and Machrihanish (UK Met Office 2019), with a particular focus on accuracy of wind speed and direction, which have a large impact on the surface water layers in which sea lice generally reside (Amundrud and Murray 2009). The hydrodynamic model was validated against a range of data sources, including tidal pressure gauges, ocean flow parameters obtained using in-situ Acoustic Doppler Current Profilers deployed at moorings, and floating drifter tracks (Aleynik et al. 2016).

The predicted characteristics related to each site included the residual currents, de-tided by removing the synthetic time-series built with principal tidal components using T_TIDE toolbox (Pawlowicz et al. 2002), temperature, salinity and wind components at 10 m height, using an analysis of the model outputs from June 2013 - June 2018. In the planning process for an actual site, a current meter deployment was used to provide a basis for the depositional modelling required to meet regulatory criteria. Instrumental records of the time-series of currents were also used for hydrodynamic model validation at the site. We compared predicted current speed and direction against equivalent ADCP (600 kHz Teledyne RDI Sentinel WH) records collected for submission to SEPA for a proposed site 'West Jura', to the south of two specific sites considered (Location: -5.903137°, 56.064578° marked with a square on Figure 3.1a). The measurements represent currents at the sea surface, at middle depth 10.6 m, and near the seabed 34.2 m over the period between 28th July and 15th August 2016, courtesy of (Kames Fish Farming Ltd 2016).

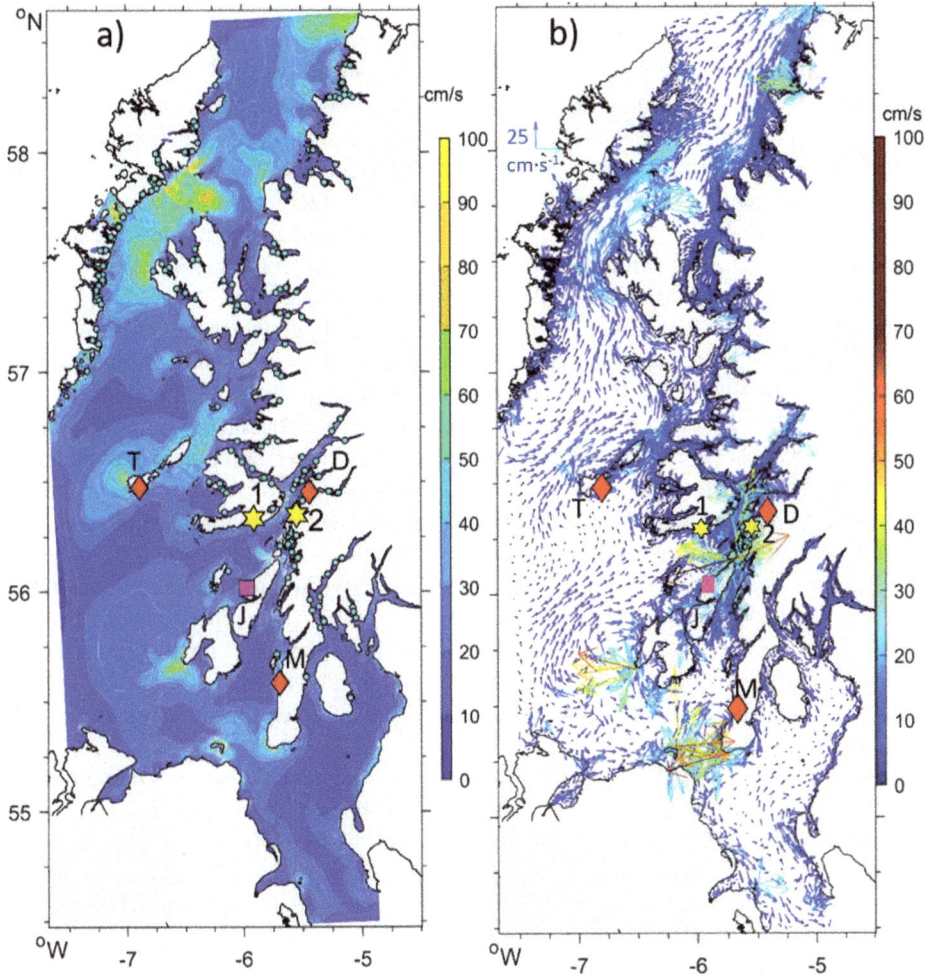

FIGURE 3.1 *Hydrodynamic model domain.* (a) The long-term (2013–2018) maximum current speed in cm·s^{-1}, computed with the WeStCOMS model over the west Scotland region. Yellow stars indicate new sites 1 and 2, and cyan discs indicate all other licensed salmon aquaculture sites. Red diamonds **T, D** and **M** refer to Tiree, Dunstaffnage and Machrihanish weather stations respectively, and the magenta square indicates the Isle of Jura ADCP current meter location. (b) The same long-term mean residual de-tided sea-surface velocity vectors. https://doi.org/10.6084/m9.figshare.15029112.v1

3.2.4 BIOLOGICAL MODEL

Output from the hydrodynamic model was used to drive a particle tracking model, which was parameterized to represent the characteristics of sea lice larvae.

The particle tracking model has been described previously (Adams et al. 2016; Adams et al. 2012). Particles moved horizontally subject to the water currents predicted by the hydrodynamic model, in addition to random turbulent diffusion. Larval particles inhabited the upper layer of the water column, and were not allowed to move vertically between layers (Cantrell et al. 2020), although response to salinity gradient is included by some authors (Johnsen et al. 2014). Particles experience a constant rate of mortality of 0.01 h^{-1}, and are removed from the simulation after 21 days (parameters after (Amundrud and Murray 2009; Salama et al. 2013; Stien et al. 2005). Stage durations are dependent on water temperature (which typically varies between 8 – 14°C in this locality), with particles moving from the non-infective nauplii stage to the infective copepodid stage after accumulating 40 degree-days (1 day at 10 degrees C equates to 10 degree-days). Particles are removed from the simulation after 150 degree days (Johnsen et al. 2014; Samsing et al. 2019). Particles are viewed as 'super-particles'. This means that, in contrast to our previous work, but in line with other studies (Johnsen et al. 2016; Salama et al. 2013), lice particles are able to infect multiple sites, in other words, they do not end their movement when an infection/arrival event occurs. Particles also have a density (reduced over time by mortality) which governs the predicted overall spatial density and connection probabilities. This gives two options for computing connections between sites: records of particle arrivals, and computation of particle density over an appropriate space-time window. The particle tracking code is available in an online repository (Adams 2019).

The model was run for a continuous 380 day period to simulate the spread of sea lice from the 196 existing sites within the model domain, and from the new sites. Presented simulations covered the period 01/01/2016 – 31/12/2016, incorporating a representative balance of periods where winds were very light (May, June), northerly (April), south-westerly (March, July, August), and southerly (September, October). This allows assessment of the likely variation in spread patterns from the sites. Every hour of the model run, we released five particles from each fish farm site. For the estimates of larval density, we applied such scaling that it provided a realistic estimate of the larval release from each site, as described in Equation 1.

The estimated number of nauplii released from a site within a given hour t was $(B \times F \times L \times f)/(24 \times 5)$, where B is the biomass, and F is the number of fish per tonne of farmed biomass. L is the number of lice per fish and f is the estimated number of larvae released per adult female louse per day. To obtain an hourly rate this value is divided by 24 and by the number of particles released per hour (5) to rescale model particle counts to an estimated number of nauplii released.

In order to provide a conservative 'worst case scenario', each operating site's permitted biomass was used, and fixed at the same level throughout the simulations with the range $B_i = 0 - 2649$ tonnes (Scottish Government 2019); table 3.3: 'Scottish west coast salmon farm locations' (accessible at https://doi.org/10.6084/m9.figshare.14997870.v3). For the new sites, the biomass was assumed to be equal to the maximum presently permissible (2500 tonnes). F_i was fixed at 240 fish per tonne, approximating the midpoint of a production cycle (though this is a substantial underestimate at early points in the production cycle). We assumed that the number of lice per fish at all sites was on the threshold of treatment of 0.5 adult female lice pre fish. The estimated number of larvae released per louse per day was $f = 28.2$ (Heuch et al. 2000), and assumed to be the same for all sites, at all times. With the intention to provide a baseline worst case scenario, we ignore any variability in lice counts, stocking patterns, or environmental conditions, which can affect larval release rates over space and time (Stien et al. 2005).

Models we run in a sequential off-line hindcast fashion, with the hydrodynamic model driven by the output from the meteorological model, and the output from the hydrodynamic model used to drive the biological model.

3.2.5   Outputs and Analysis

To assess the influence of the new sites on sea lice spread patterns within their respective surrounding areas, we considered a range of metrics. We mapped weekly and overall average density of larval lice (non-infective juveniles and infective copepodids), separating the contribution of existing and new sites. We calculated the lice density (in number of lice per m^3) by dividing the estimated number of larvae in each model element by its horizontal area (in m^2) on the assumption that the vertical portion of the water column (where lice reside) did not exceed 5 m. Lice are most often found close to the surface (Heuch et al. 2005; Penston et al. 2008). Simulated vertical movements of sea lice usually result in the majority being found in the upper few metres (Johnsen et al. 2016).

The prevalence of lice particles over time (the proportion of weeks in which they were present) and the number of distinct source sites, was calculated throughout the spatial domain. These metrics assess the level of influence existing and new sites have on particular areas, and the potential difference made by the new sites.

Pairwise connectivity (Adams et al. 2012) between existing sites and the new sites (both towards them, and from them) was calculated, in addition to connectivity between existing sites (identified by their allocated reference ID: e.g. 'FFMC53', and so forth). The connection probability was calculated by summing particle density values or particles moving within a radius of 500 m of a site location, and dividing by the number of particles released from each site over a time window of one week (Adams et al. 2012). This allows appraisal of how the networks, created by between-site dispersal of larval sea lice, may be altered by the addition of new sites, and allows for the location of the new site to be reviewed for potential impacts with respect to existing Farm Management Areas (FMAs), within which spatially coordinated management for sea lice is carried out (Adams 2019; Code of Good Practice Management Group 2011). Connections for each site were classified into incoming (sum = 'influx') and outgoing (sum = 'outflux') connections and plotted on maps and as a matrix.

As an alternative measure of connections to sites, infection pressure was calculated by summing element densities, multiplied by inverse distance to the element centroid from the focal location, within a certain radius. That is, $IP_i = \sum_{j, d_{ij} < r} \rho_j \big/ d_{ij}$, where i is the focal location index, j is the hydrodynamic model element index, ρ_j is the particle density in element j, d_{ij} is the distance between the focal location i and the element j, and r is the maximum radius used to select elements for the summation. Value ρ_j is a sum of particle counts and densities over a time window of one week, and tested values of r covered the range 500-5000 m.

Wave exposure index ('wave fetch') was derived from (Burrows et al. 2008), which summed the linear distance to coastline in 16 equally spaced directions spanning the compass (22.5° intervals), as measured using digital mapping. For the wave fetch calculation, we used a 200 m grid, covering all coastal locations within the hydrodynamic model domain. To calculate wave exposure for fish farm sites we identified the nearest grid cell to the recorded farm centre point. An index of fish farm isolation was calculated by summing inverse distances to all other farms in the model domain ($Site\,Density_i = \sum_j 1 \big/ d_{ij}^2$) giving a higher value for locations with more nearby farms (Figure 3.2). We also calculated the isolation/site density index for all coastal points in the fetch calculation grid.

Influx and outflux for all sites within the model domain were plotted against site wave fetch and site isolation metrics, and Generalised Additive Models (GAMs) were fitted to site mean (over time) values (Poisson family, multiplying response values by 10 and rounding to the nearest integer, log link function) in order to describe the relationships more formally. This regression technique allows fitting of non-parametric smoother curves between predictor and response variables and is particularly useful where non-linear responses are expected. Complete details of the method are given in Pedersen et al. (2018).

Where required, geographical coordinate transformations were computed using the SEPA OS to WGS84 Matlab toolbox (Berkeley and Schlicke 2018).

3.3 RESULTS

3.3.1 PHYSICAL MODELS–VALIDATION

Nearest to the sites, the current meter observations cover a two week period between 28th July and 15th August 2016. Predicted winds over that time (Figure 3.3) indicate Pearson correlation coefficients of 0.72 with the time series at the relatively sheltered Dunstaffnage weather station, and 0.88 at more exposed Tiree airport MetOffice weather station (Table 3.1). The WRF model replicates all

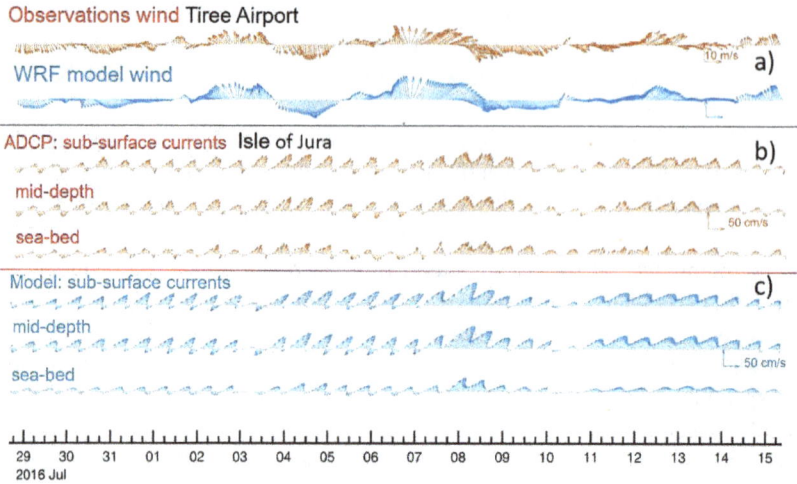

FIGURE 3.2 Observed and modelled wind and currents vectors. Fine-scale, short duration physical model validation over the period between 28th July and 15th August 2016. (a) Hourly wind vectors: observed (top, red) and modelled (below, blue) at Tiree airport weather station. Hourly current vectors at three layers (top, mid-depth, seabed) (b) observed (red) and (c) predicted (blue) at the west Jura survey site. https://doi.org/10.6084/m9.figshare.15029181.v1

key temporal patterns in observed wind speeds over the period of interest. The best fit between short term observed and modelled wind speed at 10 m height (and direction) we found at Tiree weather station with the Willmott Index of Agreement values 0.91 (0.93) respectively (Table 3.1).

The hydrodynamic model clearly indicates a residual poleward averaged flow between mainland Scotland and the Outer Hebrides archipelago (Figure 3.1b) and predominantly outward residual surface currents inside the narrow firths and sea-lochs, as described in previous local empirical studies (Inall et al. 2009). In terms of current speeds, the Pearson correlation coefficient between hydrodynamic model and observed records at the west Jura location (Figure 3.3) varies between 0.82 (sea surface) and 0.75 (seabed) over the 14 day record period (Table 3.1). The highest Index of Agreement (Willmott 1982) between measurements and model was detected for sea-surface and mid-depth layers at hourly and at daily averaged time-series ($I_A = 0.90$, same location). The averaged difference between measured and model surface current speeds does not exceed $2\,cm\,s^{-1}$, which is five times better than the Scottish regulatory requirements of $10\,cm\,s^{-1}$ (Scottish Environmental Protection Agency 2019a). Daily averaged surface current direction was captured even more reliably, with the values of the integrated performance indices Taylor Score, (Taylor 2001) $S_T = 0.93$ and Willmott Index of Agreement $I_A = 0.97$.

3.3.2 PHYSICAL MODELS—SITE CHARACTERISTICS

A statistical analysis of long-term hindcast model runs (June 2013 – April 2019), demonstrates that the two new sites differ in their flow characteristics, but have very similar temperature and salinity properties. Site 2 displayed a faster averaged (residual) surface current speed ($4.15\,cm\,s^{-1}$) than Site 1 ($2.85\,cm\,s^{-1}$). During the winter storms, peak flow exceeded $70\,cm\,s^{-1}$ at Site 2, but only reached $47\,cm\,s^{-1}$ at Site 1, which is slightly more sheltered by the surrounding terrain as shown in Figures 3.4c, d. The averaged wind speed was close to $3\,m\,s^{-1}$ at both sites, and the peak wind speed was $24\,m\,s^{-1}$. The prevailing wind direction at both sites was SW (210°), with stronger winds during winter storm days arriving from WSW (241°) (Table 3.2). The storminess index (days per month with winds exceeding 10 or $15\,m\,s^{-1}$) was on average 1.5–2 days higher at Site 1 than at Site 2. Current and wind speeds in the area around the sites is shown in Figure 3.4.

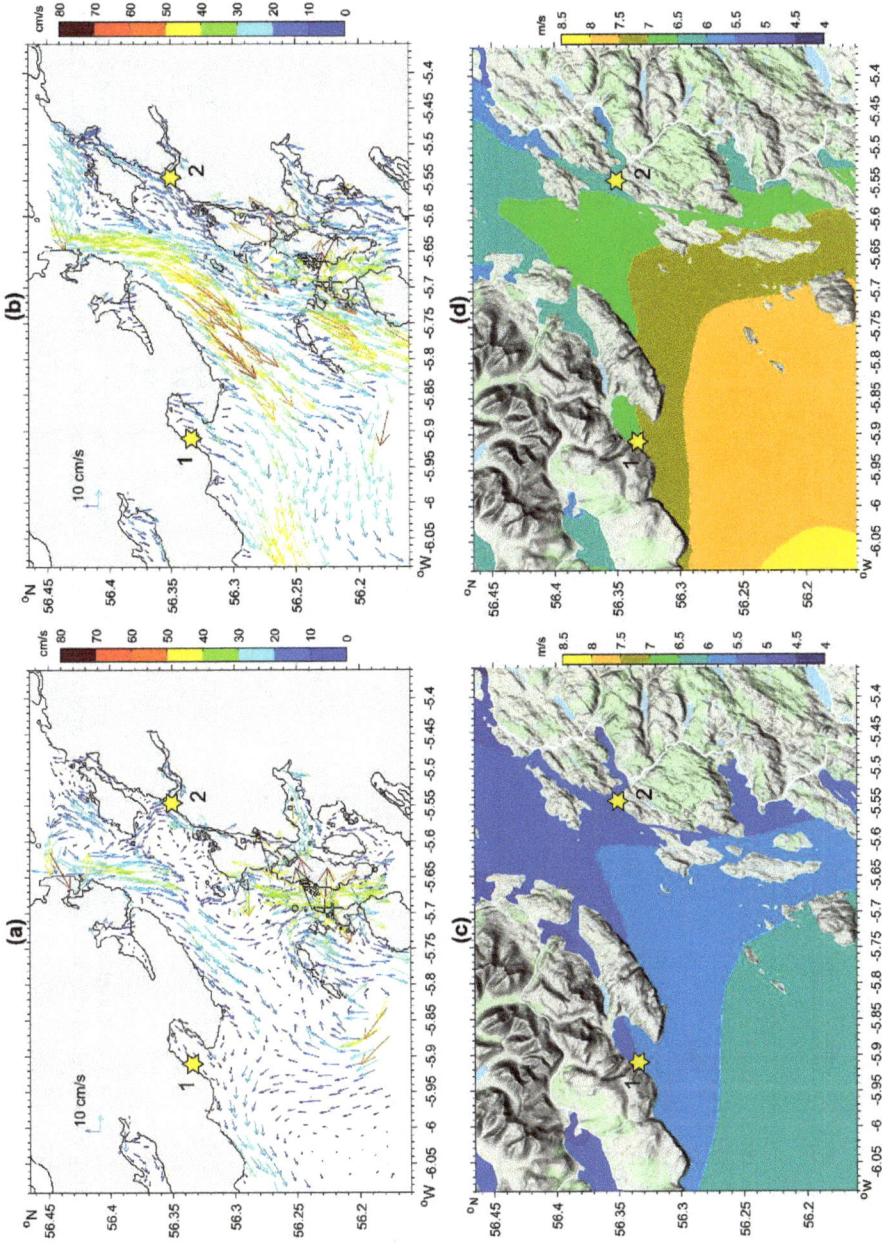

FIGURE 3.3 Model sea-surface currents and wind speeds. (a) Residual (de-tided) currents averaged over 5 years between June 2013 and June 2018 with arrows (spatially sub-sampled for clarity) showing flow direction. (b) Peak sea-surface residual currents: current direction shown with arrows, colours indicate the magnitude in cm·s-1. Peak current may occur at different times at different locations. (c) Mean wind speed (ms-1) in June-August over 5 years. (d) Mean wind speed (ms-1) in all months over 5 years. New sites shown by yellow stars. Mountainous terrain topography included to indicate wind sheltering effects. https://doi.org/10.6084/m9.figshare.15029181.v1

TABLE 3.1

Short Term Statistics (<mean>, STD σ_O and σ_M, Centred RMSD Difference and RMSE Error, Pearson Correlation Coefficients R, Taylor Score S_T (Taylor 2001), Willmott Index of Agreement I_A (Willmott 1982), Peak Values, Kurtosis K_O, K_M and Skewness S_O, S_M) of the WeStCOMS-FVCOM Model (m) Performance against Observations (o)

Currents speed	Depth m	<o>	<m>	σ_O cm s⁻¹	σ_M cm s⁻¹	RMSD	RMSE	Corr. Coef.	Taylor score	Willmott Index	Peak$_O$ cm s⁻¹	Peak$_M$ cm s⁻¹	K_O	K_M	S_O	S_M
aver-aged	0	18.72	17.33	12.92	12.35	7.63	7.75	0.82	0.68	0.90	56.73	63.65	2.84	3.30	0.81	0.83
aged	10.6	17.95	16.45	12.35	11.48	7.49	7.64	0.80	0.66	0.89	55.93	62.31	3.22	3.84	0.92	0.87
<1h>	34.2	14.17	7.60	9.08	4.76	6.34	9.13	0.75	0.40	0.67	47.00	27.97	4.20	5.23	1.11	1.18
aver-aged	0	18.86	17.18	6.63	6.77	3.98	4.33	0.82	0.69	0.89	33.94	28.61	2.65	1.89	0.71	0.39
aged	10.6	17.94	16.37	5.91	6.40	4.08	4.37	0.78	0.63	0.87	34.35	27.42	4.68	2.08	1.14	0.39
<24h>	34.2	14.16	7.57	4.72	2.82	3.22	7.33	0.75	0.45	0.54	28.19	14.70	5.89	3.92	1.43	0.78
Wind speed	Duns	402	565	238	311	216	270	0.72	0.51	0.77	1337	1429	3.38	3.20	0.80	0.71
at MO sta	Tiree	645	752	310	321	157	188	0.88	0.78	0.91	1595	1552	2.85	2.94	0.59	0.38
	Mach	582	720	266	296	166	215	0.83	0.69	0.85	1492	1771	3.00	3.22	0.34	0.37

Current direction	m	<o>	<m>	σ_O cm s⁻¹	σ_M cm s⁻¹	RMSD	RMSE	R	S_T	I_A	Peak$_O$ °	Peak$_M$ °	K_O	K_M	S_O	S_M
aver-aged	0	33.1	46.7	95.9	100.1	63.9	65.4	0.79	0.64	0.89	305	349	2.28	2.82	1.00	1.17
aged	10.6	36.3	45.2	91.6	88.1	53.2	53.5	0.83	0.69	0.91	337	243	2.24	2.25	0.96	1.04
<1h>	34.2	29.8	51.7	107.2	91.2	71.2	70.5	0.75	0.58	0.88	356	328	2.39	2.14	0.86	0.87
aver-aged	0	33.1	46.5	66.1	66.6	17.9	23.0	0.96	0.93	0.97	167	187	2.62	3.17	0.00	0.21
aged	10.6	36.4	45.4	61.4	61.1	17.1	18.4	0.96	0.92	0.98	163	161	2.81	2.11	0.31	-0.20
<24h>	34.2	30.4	51.9	71.6	64.3	33.2	32.2	0.89	0.78	0.95	148	170	2.20	1.97	-0.19	-0.24
Wind direction	St. ID															
	918	247	246	71.7	68.0	53.7	53.4	0.71	0.53	0.84			4.00	3.14	-1.24	-0.64
at MO sta	18974	274	262	79.7	69.1	40.3	39.2	0.86	0.74	0.93			3.53	3.48	-0.97	-0.89
	908	265	267	74.1	66.9	50.7	50.9	0.75	0.58	0.86			3.04	4.01	-1.03	-1.21

Note: The Current Speed (|v|, cm·s⁻¹), Their Directions (°, degrees) at Three Layers (top middle, and near sea-bedseabed) at the 'West Jura' site, Measured with ADCP between 28th July and 15th August 2016. Wind Speed and Direction at 10m Height at Three Nearby Met Office (MO) Weather Stations (Dunstaffnage, Tiree and Machrihanish, not shown in figures) are also Included.

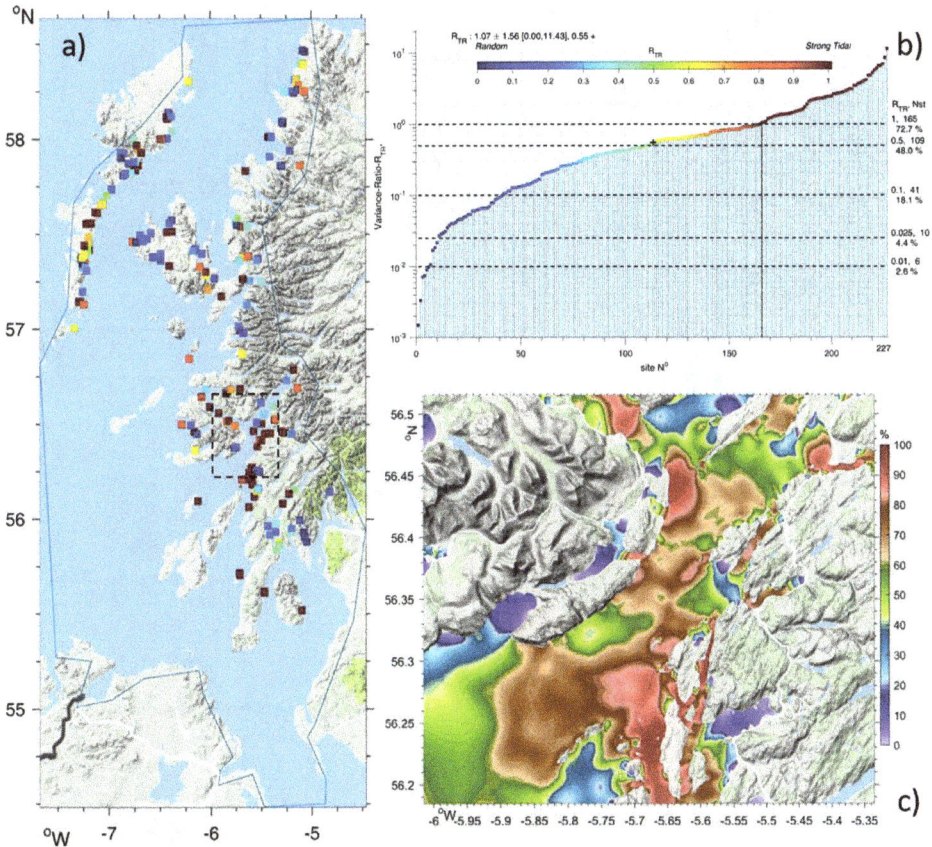

FIGURE 3.4 (a) Spatial distribution of tidal-random ratio RTR between the variances of the synthetic tidal currents time series (computed with 8 major harmonics) and the random (not-tidal) variance of the model currents speed at sea surface for 227 (active and closed) SEPA fin-fish farm sites. (b) The ratio RTR as a sorted distribution, both are shown with same coloured scale. (c) Spatial distribution of the ratio between the variances of the synthetic currents time series, computed with 8 major tidal harmonics, and the total variance of the model currents speed at the sea surface in the area of interest in 2016.
https://doi.org/10.6084/m9.figshare.15029289.v1

The annual average temperature was close to 11°C at both locations, with peak values slightly below 16°C during late summer 2013, and minimum surface values of 5.83°C detected on 19th March 2018. Average (and peak) salinity at the sea surface were 32.9 (34.3) PSU at Site 1, and only 0.1 units lower at Site 2, with a similar small difference between the sites at mid-depth and near the sea-bed (Table 3.2). The seasonal cycle in sea-surface salinity was less pronounced than for temperature, which dominates the evolution of vertical density stratification at both sites during the warmer season.

Coastal ocean flows sustained a balance between tidal and a wide range of non-tidal currents. That include internal wave motion generated by tidal interaction with the topography, various coastal trapped waves, turbulence, wind-driven and inertial oscillations. Baroclinic circulation evolves in response to uneven sea-water density stratification, associated with heat and mass exchanges via the interface between ocean and atmosphere and the freshwater discharge. To enable a scientifically informed decision-making process in practical applications we estimated the relative contributions of two major drivers (tidal and non-tidal) in regional coastal current dynamics at the highest spatial

TABLE 3.2

Long-term WeStCOMS–FVCOM Model Statistics (<mean>, std σ_X, Peak Values and Dates) of a Several Physical Environmental Parameters Such as Temperature (T), Salinity (S), Potential Density (ρ), Eastern and Northern Currents Components (u, v), Velocity Magnitude $|v|$ and Direction (Dir) at Three Layers (surface, mid-depth, sea-bed) at Three Sites (Site 1, Site 2 and West coast of Jura)

Site	Depth	<t>	<s>	<|r|>|r|>	VarT	T	S	ρ	Peak T	Peak S	Peak ρ
Units		C	psu	kg m⁻³	C	C	psu	kg m⁻³	C	psu	kg m⁻³
Site 1	2	11.04	32.90	25.10	0.80	2.69	0.94	0.89	15.82	34.31	26.65
	15	11.02	33.34	25.45	0.52	2.66	0.70	0.72	15.75	34.36	26.70
	32	10.98	33.66	25.70	0.43	2.60	0.60	0.66	15.49	34.52	26.79
Site 2	2	11.00	32.82	25.04	2.02	2.77	1.71	1.42	15.72	34.20	26.65
	20	11.04	33.18	25.31	0.78	2.75	0.93	0.88	15.71	34.20	26.65
	41	11.03	33.30	25.41	0.65	2.73	0.81	0.80	15.71	34.24	26.67
West Jura	2	11.42	34.04	25.91	0.40	2.81	0.46	0.64	16.61	34.79	27.01
	18	11.28	34.12	26.00	0.35	2.69	0.42	0.59	15.99	34.79	27.03
	37	11.13	34.16	26.06	0.32	2.60	0.40	0.57	15.94	34.79	27.04

Site	depth	<u>	<v>	<|v|>	<Dir>	σ_u	σ_v	$\sigma_{\|v\|}$	σ_{Dir}	Peak |V|	Peak |V|
Units	m	cm s⁻¹	cm s⁻¹	cm s⁻¹	°	cm s⁻¹	cm s⁻¹	cm s⁻¹	°	cm s⁻¹	Dir °
Site 1	2	-2.47	-1.43	2.85	239.98	5.37	4.75	3.91	89.12	47.06	43.44
	15	-0.36	-0.28	0.46	232.83	3.46	3.59	2.86	97.20	46.64	43.54
	32	0.03	-0.14	0.14	165.85	1.29	1.67	1.11	108.11	16.57	232.28
Site 2	2	-2.82	3.05	4.15	317.20	9.91	4.27	6.23	117.98	70.92	278.71
	20	0.25	1.55	1.57	9.12	7.58	3.60	4.82	108.68	65.61	61.61
	41	-0.28	0.00	0.28	270.99	3.27	1.49	1.64	94.14	22.18	251.99

W. Jura										
2	9.57	10.01	13.85	43.72	15.98	12.80	14.61	101.50	137.41	50.04
18	8.53	8.66	12.16	44.56	14.08	13.28	13.85	89.80	135.92	50.04
37	4.63	3.96	6.10	49.45	5.40	7.34	5.91	71.69	55.09	31.51

Met Office station	Wind # at 10 m Height	<uw> ms^{-1}	<vw> ms^{-1}	<|vw|> ms^{-1}	<Dir> °	σ_{uw} ms^{-1}	σ_{vw} ms^{-1}	$\sigma_{\|vw\|}$ ms^{-1}	σ_{Dirw} °	Peak wind ms^{-1}	Peak wind Dir °
Dunstaffnage	918	1.07	2.25	2.49	205.52	4.50	4.22	3.33	81.77	23.78	202.51
Tiree	18974	2.09	2.83	3.51	216.42	6.05	6.14	4.14	84.53	28.36	203.54
Machrihanish	908	1.54	1.80	2.37	220.59	5.68	4.57	3.56	81.89	24.57	213.87
Site 1	at	1.51	2.62	3.03	209.9	5.58	4.98	3.58	78.35	24.06	241.63
Site 2	10 m	1.54	2.63	3.05	210.4	5.60	4.99	3.60	78.32	24.16	241.60
West Jura	Height	1.56	2.65	3.07	210.5	5.62	4.99	3.60	78.04	24.19	241.77

Note: The Estimates Obtained for Model Hindcast Runs over the 5 Years 10 Months Period Between June 2013 and April 2019. WeStCOMS WRF Model Winds Statistics over Same Sites and at Three Nearby the Met Office Weather Stations Are Also Included.

and temporal scale available within the WeStCOMS modelling framework. Providing an accurate initial state of the coastal sea and up-to-date boundary conditions for solving equations of motion in the hydrodynamic models is the crucial element to reduce uncertainty in such highly chaotic and stochastic systems as air flows (Lorenz 1969) and coastal currents.

An extensive set of current records collected at aquaculture sites over several decades indicate that *non-tidal* variance could exceed tidal at 75% of sites (Edwards 2015; 2016). We performed similar discrete frequency tidal analyses of the WeStCOMS model surface velocities time series over several years (2013–2018). For the time series $A_{j=o,m,T}$ respectively $\underline{o}$bserved, $\underline{m}$odel and $\underline{T}$idal currents, the variance Var_j over their mean $\bar{m}$ and the Tidal to Random ratio R_{TR} were defined as:

$$Var_j = \frac{1}{N-1}\sum_{i=1}^{N}[(A_i - \bar{m})]^2 ; R_{TR} = \frac{\left(Var_{U_T} + Var_{V_T}\right)}{\left(Var_{U_R} + Var_{V_R}\right)}$$

The results indicate that in the open sea areas outside the narrow channels (such as the Firth of Lorn), modelled tidal forcing accounts for 60 - 70% of the overall sea surface current variance (Figure 3.5c), while near the seabed the tidal contribution in the flow variance increased to over 90%. The non-harmonic contributions are therefore substantial, supplying up to 30 - 40% of the kinetic energy at the sea surface, but less than 10-15% near the seabed. In many side sea-lochs non-tidal impact could be amplified, becoming a dominant factor (80-90% of variance) over the whole water column.

The analysis of the spatial distribution of the ratio between tidal and non-tidal variance allow advance assessment and reduced uncertainty levels in the expected average state, extremes and potentially severe current predictions of the flow patterns for already existing fin-fish and shell-fish aqua-farm sites (Figure 3.5a, b). Enhanced non-tidal variance values near the perspective site locations for developing aqua-farms could indicate their exposure to aperiodic and less predictable high flow speed events.

3.3.3 SEA LOUSE DISPERSAL

In Figures 3.6 and 3.7, nine sites with non-zero biomass during the study period are marked by small dots. New sites are indicated by stars. Lice values shown in the plots in this section are mean values that take the dispersal patterns predicted over the full simulated time-period, into account. The density and distribution of lice dispersing from existing sites varies on a weekly basis as shown using the animations at https://doi.org/10.6084/m9.figshare.14975010 (Adams and Aleynik 2021).

3.3.4 DISPERSAL OF LICE FROM EXISTING SITES

Lice from existing sites were predicted to be present at most locations within the focal domain, throughout much of the year (Figure 3.6a, b). The highest modelled density and modelled prevalence of sea lice was generally observed closer to shore, with declining density and prevalence at locations away from the coastline. However, it is also worth noting a lower predicted density and prevalence of lice in many of the enclosed sea lochs, due to the dominant outward residual (de-tided) flow in the surface waters of these environments (Figure 3.4a, b). The density and prevalence of lice dispersing from existing sites was predicted to be highest in the locality of New Site 2.

3.3.5 DISPERSAL OF LICE FROM NEW SITES

Particles released from both sites spread over fairly large distances to cover much of the focal area (Figure 3.6c–f). Against a backdrop of lice from existing sites, lice from the new sites increase the

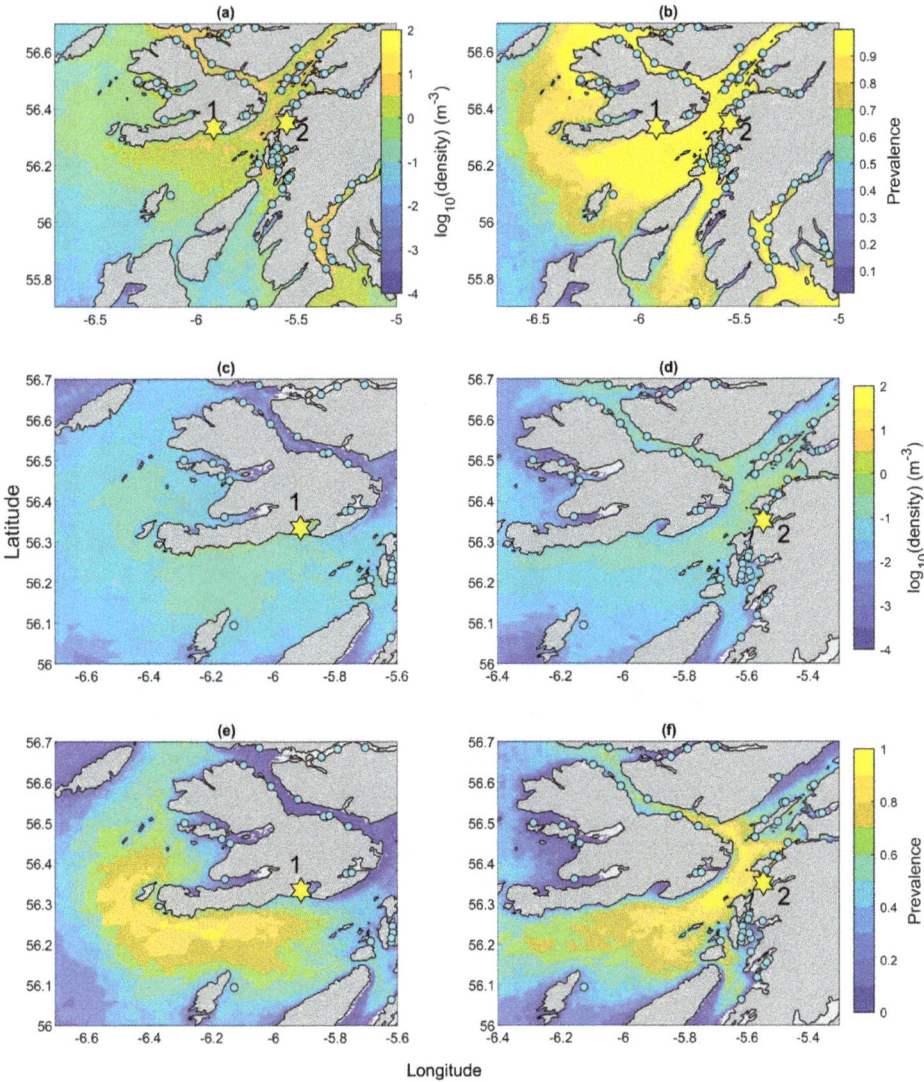

FIGURE 3.5 Predicted sealice dispersal density and prevalence from existing sites. (a) Mean density (over 1 year) of copepodid (infective) lice spread from existing sites, based on particle densities weighted by consented site biomass. (b) Prevalence of lice released from existing sites (proportion of weeks when lice were present). New sites are marked by yellow stars, and existing sites by blue dots. (c) Predicted mean density of lice dispersing from new site 1. (d) Predicted mean density of lice dispersing from new site 2. (e) Prevalence of infective copepodid lice larvae over the simulation period, from new site 1, and (f) new site 2. https://doi.org/10.6084/m9.figshare.15029313.v1

density but do not significantly alter overall spread. The plume of lice particles dispersing from Site 2 was at its highest density between the mainland and the southern entrance to Sound of Mull, covering a number of other fish farm sites and blocking two channels which might act as important corridors for wild fish inhabiting the many rivers and streams feeding this body of water. The plume from Site 1 on the southern coast of the Island of Mull exists in a more open environment, and areas of high density were found near to the south-west coast of Mull, away from existing sites and major rivers.

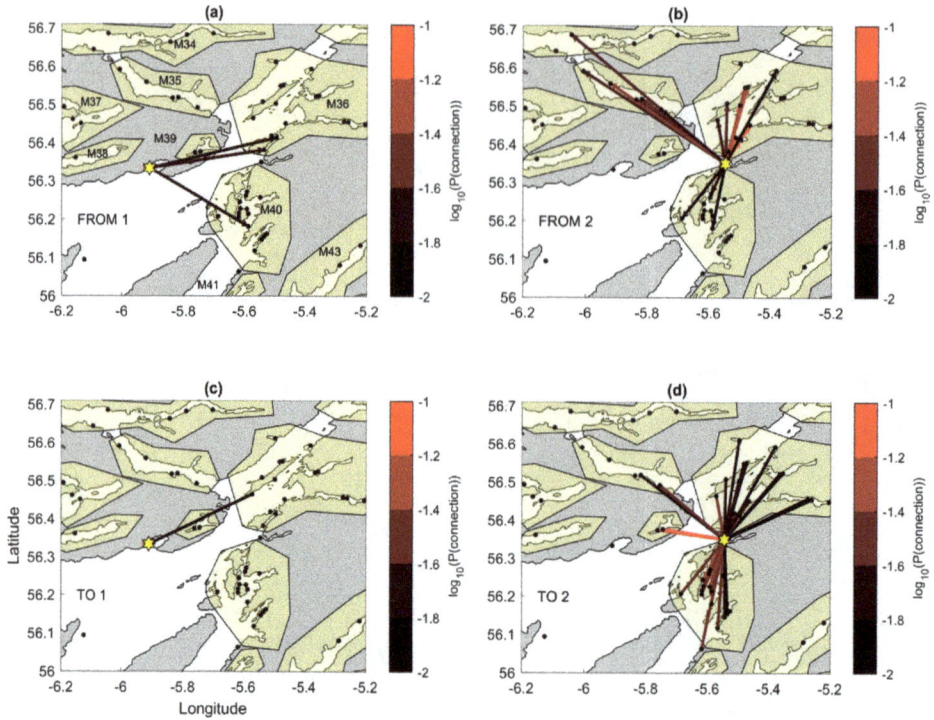

FIGURE 3.6 Site connectivity between existing and new sites. Connections are indicated by arrows, with colour relating to connection strength (on a logarithmic scale) from sites 1 and 2 (a, b) and from existing farms toward sites 1 and 2 (c, d). All sites exhibited 'self-connections' (the potential for larval particles to re-infect the same site), but these are omitted from this diagram for clarity. Yellow shaded regions indicate approximate Farm Management Area boundaries (labelled M in panel (a)). https://doi.org/10.6084/m9.figshare.15029322

3.3.6 CONNECTIVITY

Figure 3.7 maps the connections made by dispersing model sea lice particles either departing from or arriving at the two new sites. Owing to the strong currents in the area and variable conditions due to meteorology, the area containing the sites is generally well connected (Figures 3.7 and 3.8), with many sites connected sufficiently to impact population dynamics. Previous work indicated a threshold total incoming connection probability of 0.01 as being sufficient for population expansion (Toorians and Adams 2020).

Both sites were predicted to self-infect at a rate exceeding the threshold stated above. Site 1 was predicted to supply particles to three other sites (two to the north and one to the south), and to be supplied with particles by one other site, at rates higher than the threshold (Figure 3.6c, e). In contrast, Site 2 (which sits closer to a region of complex tidal currents) was predicted to have stronger connections with many sites to both the south and the north (Figure 3.6d, f).

The connections at Site 1 could potentially introduce a link between two management areas (FMAs M36 and M40). However, the links created between the same two FMAs (and additionally M34 and M35) by Site 2 were much stronger and more temporally persistent. Moreover, stronger links already exist between these two FMAs, even excluding the connections to and from the two new sites.

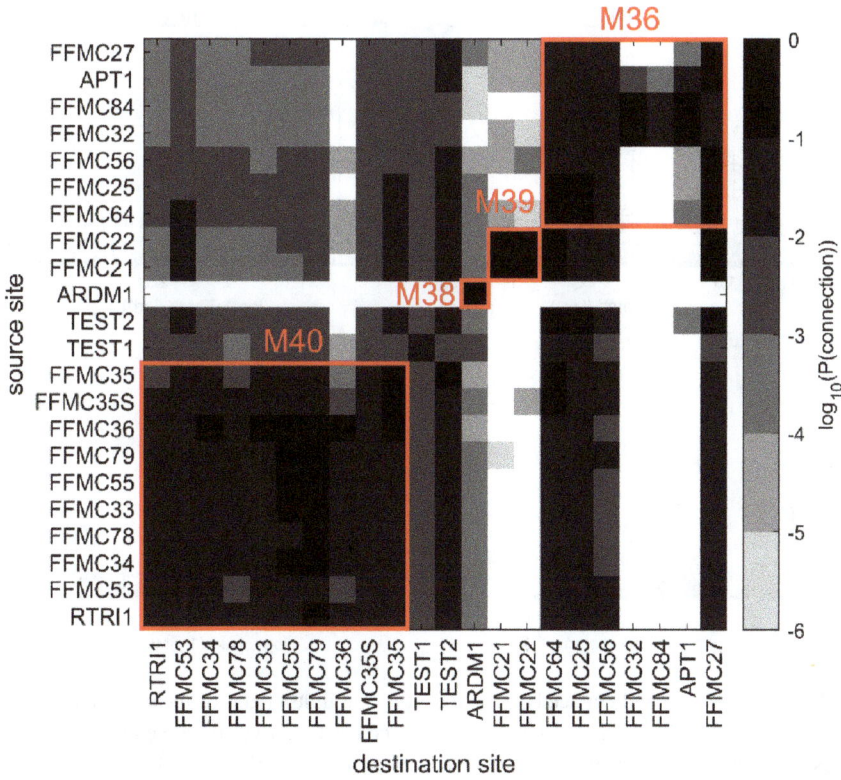

FIGURE 3.7 Connectivity matrix. A representation of the strength of connections between sites in the locality of the new sites, plotted as log-transformed pairwise values. The new sites are identified as 'TEST1' and 'TEST2'. Farm Management Area (FMA) groupings (matching Figure 3.6) are identified by red boxes and names. https://doi.org/10.6084/m9.figshare.15029364

3.3.7 EXPOSURE AND ISOLATION

The influx and outflux values for the sites, derived from the biophysical model, were fairly strongly positively correlated (Spearman's rank correlation = 0.60). However, the physical explanatory variables, fetch and isolation were weakly negatively correlated (Spearman's rank correlation = -0.30) (Figure 3.9). Influx and outflux for all sites are plotted against site wave fetch and isolation indices in Figure 3.10, highlighting the two new sites. Much variability between sites was apparent, but more wave exposed sites tend to have lower levels of influx and outflux (or more consistently low levels over time). However, it remains possible for very sheltered sites to have low influx and outflux, as indicated by the leftmost point in Figure 3.10a, b. More isolated sites had lower influx and outflux values. Visually, the most connected sites in the model domain appear to be those with intermediate levels of isolation. Fitting GAMs confirmed these relationships, with strongly significant smoothers describing a decline in median influx with fetch, a humped relationship with isolation (lowest influx at highest isolation) (Figure 3.11a, b; overall deviance explained 26.5%). Similar strongly significant smoothers were fitted for median outflux (Figure 3.11c, d) and overall deviance explained 25.8%.

Connectivity influx and infection pressure (when each are plotted on a logarithmic scale) follow a positive and approximately linear relationship, indicating a power-law relationship (not shown).

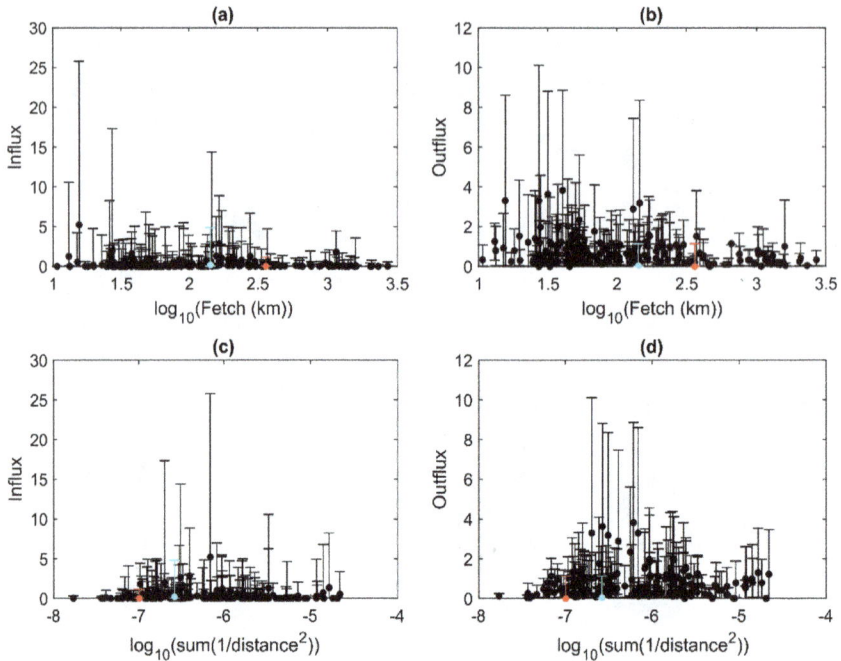

FIGURE 3.8 Site exposure and connectivity metrics. (a) Site influx (sum of incoming connections) versus wave exposure index. (b) Site outflux (sum of incoming connections) versus wave exposure index. (c) Site influx versus isolation (sum of inverse squared distances to fish farm sites; small value on x-axis indicates more isolated). (d) Site outflux versus isolation. Points show median weekly value for a site over the simulated period, and error bars indicate 10 and 90 percentile values. Values for Site 1 (red) and Site 2 (cyan) are identified in each plot. https://doi.org/10.6084/m9.figshare.15029370

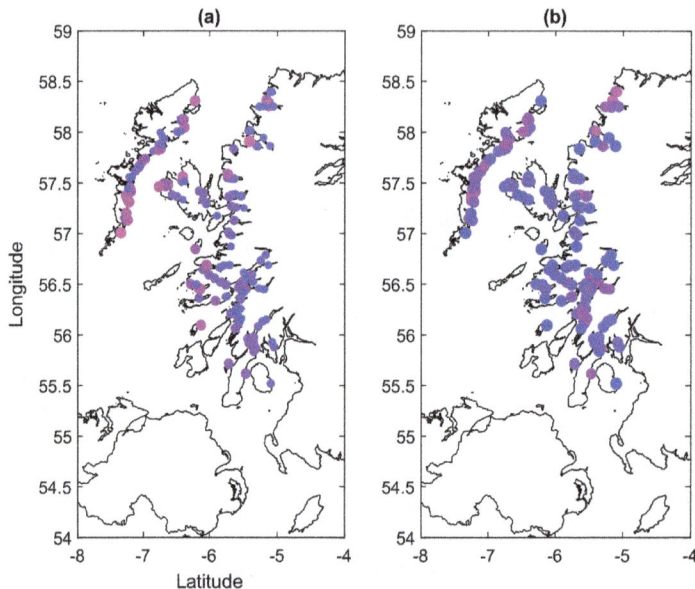

FIGURE 3.9 (a) Wave fetch (log-transformed) for salmon aquaculture sites on the west coast of Scotland (pink = higher wave fetch). (b) Local site density for salmon aquaculture sites (pink = higher local density of sites). https://doi.org/10.6084/m9.figshare.15029385

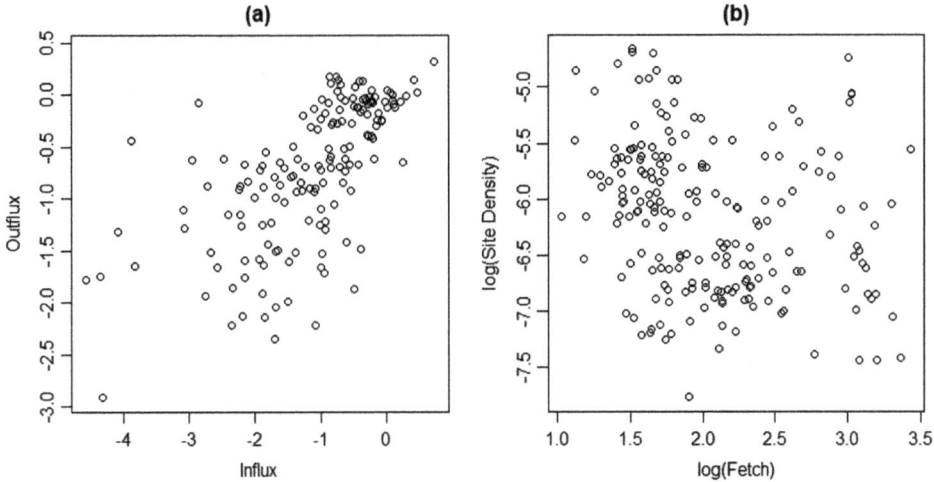

FIGURE 3.10 Site influx and outflux characteristics for all sites within the model domain. (a) Scatterplot of mean influx and outflux. (b) Scatterplot of site fetch and isolation. https://doi.org/10.6084/m9.figshare.15029388

For the non-site coastal locations given in the fetch grid, it is therefore possible to derive similar relationships between infection pressure and wave fetch, or isolation from existing fish farm sites. There is a limited tendency towards lower infection pressure at more wave exposed sites, while very isolated locations far from fish farms display the lowest infection pressure values (Figure 3.12a, b).

3.4 DISCUSSION

In this chapter, a model-based comparison of the sea-lice dispersal characteristics of two hypothetical new salmon aquaculture sites has been discussed. These are sites which possess certain characteristics that require development if the industry is to expand in Scotland. We sought to understand what impact such sites would have, and how the network of existing sites would affect them. By taking a broader look at these new and existing sites together, it was possible to investigate the general characteristics of the dispersal and connectivity pattern, which allowed us to provide insights into the potential benefits offered by moving operations to more exposed and / or isolated locations. Results such as these are important to inform the decision making process of site selection. However, such comparative information is currently often lacking in the site planning process, which may force local planners to make judgements based on subjective or generic statements about potential site impacts. The analysis determines the preferred site from the available options based on a number of quantitative criteria.

Both sites (and particularly Site 1) are relatively more exposed to external forcing impact than the majority of existing sites, which tend to be in more sheltered fjords and sea lochs. Hence, although the physical characteristics of Site 1 were less energetic than Site 2 in terms of current speed, the (modelled) currents at both sites are within a range suitable for adult salmon (Johansson et al. 2014), and therefore, both sites would be suitable for offshore expansion under this criterion. However, the mean wind speed is lower at Site 2, and it is less exposed to storms than Site 1; and hence a trade-off in wind would be required if an operator were to decide which of the sites to develop.

The strength of connectivity with other sites in the network will determine the number of lice infesting each site, and the extent to which outbreaks of lice spread between sites. Threshold values of between-site connectivity may play a role in the occurrence of outbreaks (Toorians and Adams 2020) and higher connectivity implies a greater risk of spatially distributed lice outbreaks. In addition to the magnitude of connectivity, directionality of connection is also important. Sites may be a source of lice for other sites, a destination, a self-infector, or some combination of the three. Both

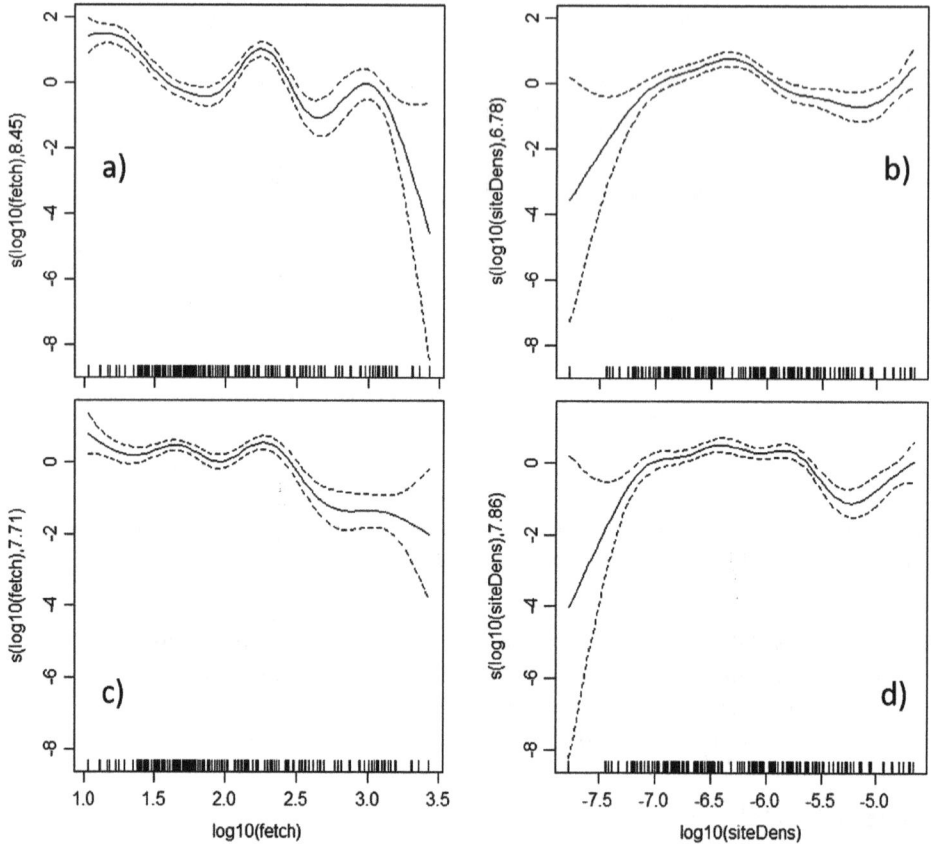

FIGURE 3.11 *Generalised Additive Model* (GAM) for site influx (a, b) and outflux (c, d). Fitted smoother response to wave fetch (a, c). Fitted smoother response to isolation (sum of inverse squared distances to fish farm sites (b, d). https://doi.org/10.6084/m9.figshare.15029400

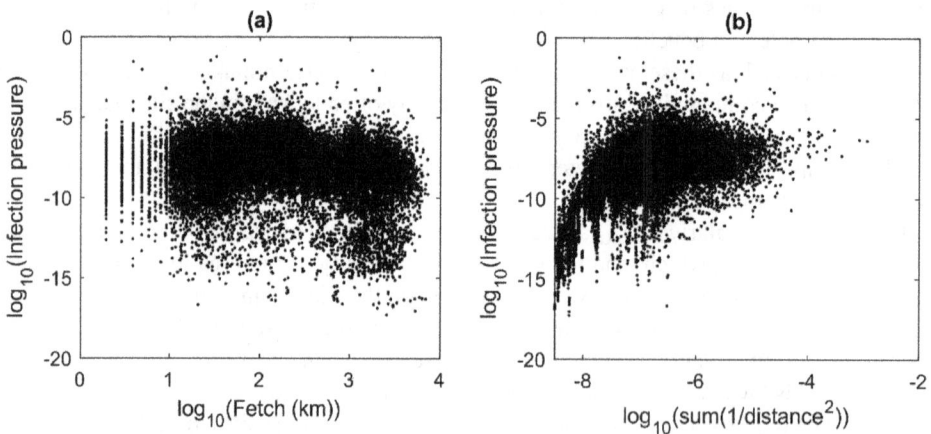

FIGURE 3.12 *Mean infection pressure* (inverse distance-squared weighted particle density within 2000 m; mean over time) for coastal points within the model domain, on a 200 m grid: (a) Infection pressure versus wave fetch (log-log), (b) infection pressure versus isolation (sum of inverse squared distances to fish farm sites). https://doi.org/10.6084/m9.figshare.15029415.v1

the considered sites created new connections between management areas, one (Site 2) much more strongly than the other (Site 1). Connections already exist between sites in these two areas, but it would be important to consider the approach taken to management of the new Site 2, which would ideally match that of the connected areas (FMAs M36 and M40).

The spatial extent and distribution of dispersal from sites is also of key interest, in particular of infective copepodids. In addition to dispersing to other sites, lice can infect wild fish moving freely in the water column. In terms of isolating a site from its neighboring environment, a dispersal kernel focused on the site itself is ideal. The predicted focal area of infective stage lice produced by a site should be as far as possible from embayment or other habitat known to be used by sea trout, which have been shown to experience mortality rates elevated by around 50% at locations up to 15 km from farm sites (Shephard et al. 2016). Ideally, it should not present a complete barrier to likely migratory routes of salmon (covering shore to shore in a sea loch, for example). Long-range, low-density dispersal may be tolerated, although the nature of the impact on wild fish depends on their characteristics and behaviour (Moriarty and Murray, unpublished). In this context, Site 1 has dispersal focused in open areas of water, away from existing sites and river mouths. On the other hand, the plume of Site 2 covers the entrance to Loch Linnhe (a likely passage for migrating salmon) at a high density. Based on this, and the connectivity computation, Site 1 appears preferable in this sense.

Considering existing and new sites together, our study suggests that more exposed sites are likely to have slightly lower connectivity with existing sites, as are sites that are physically isolated by distance from others. However, there is a high level of variability between sites, and the statistical models partially explain this variability. It is also worth bearing in mind that exposure and isolation do not necessarily go hand in hand, and therefore it is likely to be useful to investigate sites which are both exposed and isolated. These potential benefits of considering more exposed sites may be affected by other factors, such as currents flow regime, temperature or salinity profiles, which may favour or limit lice settlement (Bricknell et al. 2006), survival (Groner et al. 2016) or development (Stien et al. 2005; Samsing et al. 2016) in different localities and at different times of year.

Over the period presented here, lice counts were only available in an aggregated form, grouped within Fish Health Management Reporting Areas. From 2018, counts have become available for each individual farm on a monthly basis, with a three-month delay (Scottish Salmon Producers Organisation, 2018). This makes it possible to look back at the patterns of infestation that have occurred and assess the risks that would be posed to particular sites more precisely than was possible here. For more immediate operational purposes, Norway operates a system that provides up to date spatial information on a weekly basis (BarentsWatch 2019), which has allowed the testing of a louse forecasting system (Sandvik et al. 2016). A final factor, limiting assessment of lice challenge on a regional scale (between operators), is the need to translate between lice numbers (which are reported on a 'per fish' basis) and sites' current stocking level (which is reported in kilograms). The relationship between stocking and biomass clearly varies over time, depending on fish size, and at an early stage in the production cycle the number of fish per tonne of biomass will be many times the value used here.

3.5 CONCLUSIONS

Finding space for aquaculture in complex and often crowded coastal environments is challenging, requiring balanced analysis of competing concerns (Leith et al. 2014). The move to more exposed locations involves a range of considerations, incorporating potential operational challenges (stress on infrastructure, difficulties in access, risks to fish health, slowing feed-conversion and fish growth rates under increased flow dynamics or higher wave fetch exposure environments). The potential wide-range of benefits include the reduced intensity of benthic impacts (Scottish Environmental Protection Agency 2017) or risk of harmful algal blooms (Smayda 2006; Davidson et al. 2016; 2021), and a reduced impact on local communities due to their isolation. This work demonstrates how biophysical models can support the decision making process for salmon farm development, particularly into more exposed locations, and how they enable evaluation and quantification of potential benefits in terms of decreased parasite connectivity. Sites established in more exposed or physically isolated locations could offer benefits in terms of parasite population connectivity, which would have consequent benefits in controlling the broader meta-population.

ACKNOWLEDGEMENTS

We thank Kames Fish Farming limited for permission to use the current meter record collected during the process of planning for a potential new site. Development of the models described here were funded by UK Research and Innovation (BBSRC and NERC) under the project 'Evaluating the Environmental Conditions Required for the Development of Offshore Aquaculture' (grant number BB/S00419X/1), the European Union Interreg VA projects 'Collaborative Oceanography and Monitoring for Protected Areas and Species' (COMPASS) and 'Combining Autonomous observations and Models for Predicting and Understanding Shelf seas' (CAMPUS – grant number NE/R00675X/1). The UKRI ARCHER computing facilities (https://www.archer2.ac.uk) and NERC–funded HPC cluster Samhanach at SAMS helped to advance and to run WeStCOMS physical model.

SUPPLEMENTARY MATERIALS

The following supplementary materials are available online to complement the main chapter:

Appendix S3.1. Scottish west coast salmon farm locations (S3.1_WestCoastFarmLocations.csv, https://doi.org/10.6084/m9.figshare.14997870.v3)

CSV file containing details of the 196 existing salmon aquaculture sites used in this study, including the SEPA reference identifier, Eastings and Northings in metres on the Ordnance Survey UK National Grid, the maximum biomass in tonnes over the study period, and the Farm Management Area (FMA), Disease Management Area (DMA) and Fish Health Management Reporting Area (FHMRA) to which each site is allocated (areas described in more detail by Adams et al. 2016).

Appendix S3.2: Animation of weekly lice densities (S2_weeklyLiceDensity.mp4 https://doi.org/10.6084/m9.figshare.14975010.v2, T. Adams and Aleynik 2021). An animation showing spatiotemporal variation in model predicted lice densities over the study period, based on releases from existing site locations.

REFERENCES

Adams, T. P. 2019. *BioTracker—Biological Particle Tracking in Unstructured and Structured Hydrodynamic Grids* (version 1). Java. Oban, UK: Scottish Association for Marine Science. https://bitbucket.org/tomadams1982/biotracker/.

Adams, T. and D. Aleynik. July 12. 2021. 'S2_weeklyLiceDensity. Media Animation.' *Figshare*. https://doi.org/10.6084/M9.Figshare.14975010.

Adams, T. Black, K. MacIntyre, C. MacIntyre, I. and Dean, R. 2012. 'Connectivity Modelling and Network Analysis of Sea Lice Infection in Loch Fyne, West Coast of Scotland.' *Aquaculture Environment Interactions* 3 (1): 51–63. https://doi.org/10.3354/aei00052.

Adams, T. P. Aleynik, D. and Black, K. D. 2016. 'Temporal Variability in Sea Lice Population Connectivity and Implications for Regional Management Protocols.' *Aquaculture Environment Interactions* 8: 585–96. https://doi.org/10.3354/aei00203.

Aleynik, D. Dale, A. C. Porter, M. and Davidson, K. 2016. 'A High Resolution Hydrodynamic Model System Suitable for Novel Harmful Algal Bloom Modelling in Areas of Complex Coastline and Topography.' *Harmful Algae* 53: 102–17. https://doi.org/10.1016/j.hal.2015.11.012.

Aleynik, D. Adams, T. Davidson, K. Dale, A. Porter, M. Black, K. and Burrows, M. 2018. 'Biophysical Modelling of Marine Organisms: Fundamentals and Applications to Management of Coastal Waters.' *Environmental Management of Marine Ecosystems*. Edited by Md. Nazrul Islam Sven Erik Jørgensen, CRC Press. https://lccn.loc.gov/2017020721 (ISBN 9781498767729)

Amundrud, T. L. and Murray, A. G. 2009. 'Modelling Sea Lice Dispersion under Varying Environmental Forcing in a Scottish Sea Loch.' *Journal of Fish Diseases* 32 (1): 27–44. https://doi.org/10.1111/j.1365–2761.2008.00980.x.

Armstrong, E. M. Wagner, G. Vazquez-Cuervo, J. and Chin, T. M. 2012. 'Comparisons of Regional Satellite Sea Surface Temperature Gradients Derived from MODIS and AVHRR Sensors.' *International Journal of Remote Sensing* 33 (21): 6639–51. https://doi.org/10.1080/01431161.2012.692832.

BarentsWatch. 2019. 'Norwegian Fish Health. BarentsWatch Web Portal.' *Environmental Modelling Data Portal.* Accessed 2022/07/18, https://barentswatch.no/fiskehelse/settings.

Berkeley, A. and Schlicke, T. 2018. *A MATLAB Toolbox for Reading, Processing and Manipulating the Particle Tracking Models AutoDepomod and NewDepomod* (version 1). Matlab. UK: Scottish Environmental Protection Agency. https://github.com/OceanMetSEPA/depomod_toolbox.

Bricknell, I. R. Dalesman, S. J. O'Shea, S. Pert, C. C. and Mordue Luntz, A. J. 2006. 'Effect of Environmental Salinity on Sea Lice Lepeophtheirus Salmonis Settlement Success.' *Diseases of Aquatic Organisms* 71: 201–12. https://doi.org/10.3354/dao071201.

Bron, J. E. Sommerwille, C. Wootten, R. and Rae, G. H. 1993. 'Fallowing of Marine Atlantic Salmon, Salmo Salar L. Farms as a Method for the Control of Sea Lice, Lepeophtheirus Salmonis (Kroyer, 1837).' *Journal of Fish Diseases* 16 (5): 487–93. https://doi.org/10.1111/j.1365-2761.1993.tb00882.x .

Burrows, M. T. Harvey, R. and Robb, L. 2008. 'Wave Exposure Indices from Digital Coastlines and the Prediction of Rocky Shore Community Structure.' *Marine Ecology Progress Series* 353: 1–12. https://doi.org/10.3354/meps07284.

Butler, J. R. A. 2002. 'Wild Salmonids and Sea Louse Infestations on the West Coast of Scotland: Sources of Infection and Implications for the Management of Marine Salmon Farms.' *Pest Management Science* 58 (6): 595–608. https://doi.org/10.1002/ps.490.

Cantrell, D. Filgueira, R. Revie, C. W. Rees, E. E. Vanderstichel, R. Guo, M. Foreman, M. G. G. Wan, D. and Grant, J. 2020. 'The Relevance of Larval Biology on Spatiotemporal Patterns of Pathogen Connectivity among Open-Marine Salmon Farms.' *Canadian Journal of Fisheries and Aquatic Sciences* 77 (3): 505–19. https://doi.org/10.1139/cjfas-2019-0040.

Chen, C. Beardsley, R. and Cowles, G. 2013. *An Unstructured Grid, Finite-Volume Coastal Ocean Model: FVCOM User Manual, 4th Edition. SMAST/UMASSD-13-0701.* MEDM, University of Massachusetts-Dartmouth. http://fvcom.smast.umassd.edu/fvcom/.

Code of Good Practice Management Group. 2011. *A Code of Good Practice for Scottish Finfish Aquaculture.* Scottish Salmon Producers' Organisation. https://scottishsalmon.co.uk/code-of-good-practice.

Costello, M. J. 2006. 'Ecology of Sea Lice Parasitic on Farmed and Wild Fish.' *Trends in Parasitology* 22 (10): 475–83. https://doi.org/10.1016/j.pt.2006.08.006.

Dabrowski, T. Lyons, K. Berry, A. Cusack, C. and Nolan, G. D. 2014. 'An Operational Biogeochemical Model of the North-East Atlantic: Model Description and Skill Assessment.' *Journal of Marine Systems* 129: 350–67. https://doi.org/10.1016/j.jmarsys.2013.08.001.

Davidson, K. Whyte, C. Aleynik, D. Dale, A. Gontarek, S. Kurekin, A. A. McNeill, S. et al. 2021. 'HABreports: Online Early Warning of Harmful Algal and Biotoxin Risk for the Scottish Shellfish and Finfish Aquaculture Industries.' *Frontiers in Marine Science* 8: 631732. https://doi.org/10.3389/fmars.2021.631732.

Davidson, K. Anderson, D. M. Marcos, M. Reguera, B. Silke, J. Sourisseau, M. and Maguire, J. 2016. 'Forecasting the Risk of Harmful Algal Blooms.' *Harmful Algae* 53: 1–7. https://doi.org/10.1016/j.hal.2015.11.005.

Edwards, A. 2015. *A Technical Standard for Scottish Finfish Aquaculture, Appendix A6.3 A Note on the Estimation of Extreme Currents in West Scottish Coastal Waters.* 00479005. The Scottish Government, Edinburgh, UK: Marine Scotland Science. https://gov.scot/publications/technical-standard-scottish-finfish-aquaculture/.

Edwards, A. 2016. 'The Balance of Tidal and Wind-Driven Currents in West Scottish Coastal Waters.' In: *MASTS 6th Annual Science Meeting.* Glasgow, UK: MASTS, Ed. by Dr Bee Berx. P1. http://masts.ac.uk/media/36074/abstracts-for-phys-ocean-session-web.pdf.

Egbert, G. D. Erofeeva, S. Y. and Ray, R. D. 2010. 'Assimilation of Altimetry Data for Nonlinear Shallow-Water Tides: Quarter-Diurnal Tides of the Northwest European Shelf.' *Continental Shelf Research* 30 (6): 668–79. https://doi.org/10.1016/j.csr.2009.10.011.

Groner, M. L. McEwan, G. F. Rees, E. E. Gettinby, G. and Revie, C. W. 2016. 'Quantifying the Influence of Salinity and Temperature on the Population Dynamics of a Marine Ectoparasite.' *Canadian Journal of Fisheries and Aquatic Sciences* 73 (8): 1281–91. https://doi.org/10.1139/cjfas-2015-0444

Grøntvedt, R. N. Nerbøvik, I. K. G. Viljugrein, H. Lillehaug, A. Nilsen, H. and Gjevre, A. G. 2015. 'Thermal De-Licing of Salmonid Fish—Documentation of Fish Welfare and Effect.' 13. Norwegian Veterinary Institute's Report Series. Norwegian Veterinary Institute. https://vetinst.no/rapporter-og-publikasjoner/ rapporter/2015/thermal-de-licing-of-salmonid-fish-documentation-of-fish-welfare-and-effect/_/attachm ent/download/3130da85-cba1-4773-8e2e-8dc9d89fe2d1:51e2b38493e6fde1e5340541b9053b00d6e90 9dd/Report%2013_15%20-%20Thermal%20de-licing%20of%20salmonid%20fish%20-%20docume ntation%20of%20fish%20welfare%20and%20effect%20-%20english.pdf

Heuch, P. A. Nordhagen, J. R. and Schram, T. A. 2000. 'Egg Production in the Salmon Louse [Lepeophtheirus Salmonis (Krøyer)] in Relation to Origin and Water Temperature.' *Aquaculture Research* 31 (11): 805–14. https://doi.org/10.1046/j.1365-2109.2000.00512.x.

Heuch, P. A. Bjørn, P. A. Finstad, B. Holst, J. C. Asplin, L. and Nilsen, F. 2005. 'Corrigendum to 'A Review of the Norwegian 'National Action Plan Against Salmon Lice on Salmonids': The Effect on Wild Salmonids' [Aquaculture 246 (2005) 79–92].' *Aquaculture* 250 (1–2): 535. https://doi.org/10.1016/j.aquaculture.2005.10.003.

Inall, M. Gillibrand, P. Griffiths, C. MacDougal, N. and Blackwell, K. 2009. 'On the Oceanographic Variability of the North-West European Shelf to the West of Scotland.' *Journal of Marine Systems* 77 (3): 210–26. https://doi.org/10.1016/j.jmarsys.2007.12.012.

Johnsen, I. A. Asplin, L. C. Sandvik, A. D. and Serra-Llinares, R. M. 2016. 'Salmon Lice Dispersion in a Northern Norwegian Fjord System and the Impact of Vertical Movements.' *Aquaculture Environment Interactions* 8: 99–116. https://doi.org/10.3354/aei00162.

Johnsen, I. Fiksen, A, Ø Sandvik, A. D. and Asplin, L. 2014. 'Vertical Salmon Lice Behaviour as a Response to Environmental Conditions and Its Influence on Regional Dispersion in a Fjord System.' *Aquaculture Environment Interactions* 5 (2): 127–41. https://doi.org/10.3354/aei00098.

Kames Fish Farming Ltd. 2016. *Jura West Hydrographic and Site Survey Report.* 1. Connel, Argyll. UK: TransTech Limited.

Leclercq, E. Davie, A. and Migaud, H. 2014. 'Delousing Efficiency of Farmed Ballan Wrasse (Labrus Bergylta) against Lepeophtheirus Salmonis Infecting Atlantic Salmon (Salmo Salar) Post-Smolts.' *Pest Management Science* 70 (8): 1274–82. https://doi.org/10.1002/ps.3692.

Leith, P. Ogier, E. and Haward, M. 2014. 'Science and Social License: Defining Environmental Sustainability of Atlantic Salmon Aquaculture in South-Eastern Tasmania, Australia.' *Social Epistemology* 28 (3–4): 277–96. https://doi.org/10.1080/02691728.2014.922641.

Lorenz, E. N. 1969. 'Atmospheric Predictability as Revealed by Naturally Occurring Analogues'. *J. of Atmos. Science,* 26 (4): 636–46. https://doi.org/10.1175/1520-0469(1969)26%3C636:APA RBN%3E2.0.CO;2.

Murray, A. G. and Gubbins, M. 2016. 'Spatial Management Measures for Disease Mitigation as Practised in Scottish Aquaculture.' *Marine Policy* 70: 93–100. https://doi.org/10.1016/j.marpol.2016.04.052.

Oppedal, F. Samsing, F. Dempster, T. Wright, D. W. Bui, S. and Stien, L. H. 2017. 'Sea Lice Infestation Levels Decrease with Deeper 'snorkel' Barriers in Atlantic Salmon Sea-Cages.' *Pest Management Science* 73 (9): 1935–43. https://doi.org/10.1002/ps.4560.

Pawlowicz, R. Beardsley, R. and Lentz, S. 2002. 'Classical Tidal Harmonic Analysis Including Error Estimates in MATLAB Using T_TIDE.' *Computers and Geosciences* 28 (8): 929–37. https://doi.org/10.1016/S0098-3004(02)00013-4.

Pedersen, E. J. Miller, D. L. Simpson, G. L. and Ross, N. 2018. *Hierarchical Generalized Additive Models: An Introduction with Mgcv.* PeerJ. https://doi.org/10.7287/peerj.preprints.27320.

Penston, M. J. Millar, C. P. Zuur, A. and Davies, I. M. 2008. 'Spatial and Temporal Distribution of Lepeophtheirus Salmonis (Krøyer) Larvae in a Sea Loch Containing Atlantic Salmon, Salmo Salar L. Farms on the North-West Coast of Scotland.' *Journal of Fish Diseases* 31 (5): 361–71. https://doi.org/10.1111/j.1365-2761.2008.00915.x.

Salama, N. K. G. and Murray, A. G. 2013. 'A Comparison of Modelling Approaches to Assess the Transmission of Pathogens between Scottish Fish Farms: The Role of Hydrodynamics and Site Biomass.' *Preventive Veterinary Medicine* 108 (4): 285–93. https://doi.org/10.1016/j.prevetmed.2012.11.005.

Salama, N. K. G. Collins, C. M. Fraser, J. G. Dunn, J. Pert, C. C. Murray, A. G. and Rabe, B. 2013. 'Development and Assessment of a Biophysical Dispersal Model for Sea Lice.' *Journal of Fish Diseases* 36 (3): 323–37. https://doi.org/10.1111/jfd.12065.

Samsing, F. Oppedal, F. Dalvin, S. Johnsen, I. Vågseth, T. and Dempster, T. 2016. 'Salmon Lice (*Lepeophtheirus Salmonis*) Development Times, Body Size, and Reproductive Outputs Follow Universal Models of Temperature Dependence.' *Canadian Journal of Fisheries and Aquatic Sciences* 73 (12): 1841–51. https://doi.org/10.1139/cjfas-2016–0050.

Samsing, F. Johnsen, I. Treml, E. A. and Dempster, T. 2019. 'Identifying 'firebreaks' to Fragment Dispersal Networks of a Marine Parasite.' *International Journal for Parasitology* 49 (3–4): 277–86. https://doi.org/10.1016/j.ijpara.2018.11.005.

Sandvik, A. D. Bjørn, P. A. AAdlandsvik, B. Asplin, L. Skarðhamar, J. Johnsen, I. A. Myksvoll, M. and Skogen, M. D. 2016. 'Toward a Model-Based Prediction System for Salmon Lice Infestation Pressure.' *Aquaculture Environment Interactions* 8: 527–42. https://doi.org/10.3354/aei00193.

Scotland Food and Drink. 2017. 'Aquaculture Growth to 2030: A Strategic Plan for Farming Scotland's Seas.' *Scotland Food and Drink*, Edinburgh, UK. https://foodanddrink.scot/resources/sector-strategies/aqua culture-growth-to-2030-a-strategic-plan-for-farming-scotland-s-seas/.

Scottish Environmental Protection Agency. 2017. 'Fish Farm Manual.' *SEPA*. https://sepa.org.uk/regulations/ water/aquaculture/fish-farm-manual/.

———. 2019a. 'Aquaculture Modelling: Regulatory Modelling Guidance for the Aquaculture Sector.' *SEPA*. https://sepa.org.uk/media/433436/regulatory-modelling-process-and-reporting-guidance-for-the-aqua culture-sector_v10_june2019.pdf. (Accessed 2020/08/17)

———. 2019b. 'Protection of the Marine Environment: Discharges from Marine Pen Fish Farms, a Strengthened Regulatory Framework.' *SEPA*. https://sepa.org.uk/media/433439/finfish-aquaculture-annex-2019_3 1052019.pdf.

Scottish Government. 2019. *Scotland's Aquaculture*. Scottish Government. aquaculture.scotland.gov.uk.

Scottish Salmon Producers Organisation. 2018. 'Scottish Salmon Farm Sea Lice Reporting.' http://scottishsal mon.co.uk/monthly-sea-lice-reports/.

Shephard, S. MacIntyre, C. and Gargan, P. 2016. 'Aquaculture and Environmental Drivers of Salmon Lice Infestation and Body Condition in Sea Trout.' *Aquaculture Environment Interactions* 8: 597–610. https://doi.org/10.3354/aei00201.

Skamarock, W. Klemp, J. Dudhia, J. Gill, D. Barker, D. Wang, W. Huang, X-Y. and Duda, M. 2008. 'A Description of the Advanced Research WRF Version 3.' Application/pdf. UCAR/NCAR. https://doi.org/10.5065/D68S4MVH.

Skiftesvik, A. B. Serra-Llinares, R. M. Durif, C. M. F. Johansen, I. S. and Browman, H. I. 2013. 'Delousing of Atlantic Salmon (Salmo Salar) by Cultured vs. Wild Ballan Wrasse (Labrus Bergylta).' *Aquaculture* 402–403: 113–18. https://doi.org/10.1016/j.aquaculture.2013.03.032.

Smayda, T. J. 2006. 'Harmful Algal Bloom Communities in Scottish Coastal Waters: Relationship to Fish Farming and Regional Comparisons—A Review.'Natural Scotland, Scottish Executive. Paper 2006/3. https://www.webarchive.org.uk/wayback/archive/20160124215502mp_/http://www.gov.scot/Resource/ Doc/92174/0022031.pdf

Stien, A. Bjørn, P. A. Heuch, P. A. and Elston, D. A. 2005. 'Population Dynamics of Salmon Lice Lepeophtheirus Salmonis on Atlantic Salmon and Sea Trout.' *Marine Ecology Progress Series* 290: 263–75. https://doi.org/10.3354/meps290263.

Taylor, K. E. 2001. 'Summarizing Multiple Aspects of Model Performance in a Single Diagram.' *Journal of Geophysical Research: Atmospheres* 106 (D7): 7183–92. https://doi.org/10.1029/2000JD900719.

Toorians, M. E. M. and Adams, T. P. 2020. 'Critical Connectivity Thresholds and the Role of Temperature in Parasite Metapopulations.' *Ecological Modelling* 435: 109258. https://doi.org/10.1016/j.ecolmodel.2020.109258.

UK Met Office. 2019. 'Met Office Integrated Data Archive System (MIDAS) Land and Marine Surface Stations Data (1853-Current).' On-line catalogue. https://catalogue.ceda.ac.uk/uuid/220a65615218d5c9cc9e4785a3234bd0.

Veldhoen, N. Ikonomou, M. G. Buday, C. Jordan, J. Rehaume, V. Cabecinha, M. Dubetz, C. et al. 2012. 'Biological Effects of the Anti-Parasitic Chemotherapeutant Emamectin Benzoate on a Non-Target Crustacean, the Spot Prawn (*Pandalus Platyceros* Brandt, 1851) under Laboratory Conditions.' *Aquatic Toxicology* 108: 94–105. https://doi.org/10.1016/j.aquatox.2011.10.015.

Willmott, C. J. 1982. 'Some Comments on the Evaluation of Model Performance.' *Bulletin of the American Meteorological Society* 63 (11): 1309–13.
https://doi.org/10.1175/1520-0477(1982)063<1309:SCOTEO>2.0.CO;2

Wolf, J. Yates, N. Brereton, A. Buckland, H. de Dominicis, M. Gallego, A. and O'Hara Murray, R. 2016. 'The Scottish Shelf Model. Part 1: Shelf-Wide Domain.' *Scottish Marine and Freshwater Science* 7 (3): 1. Marine Scotland Science. ISSN: 2043–7722. https://data.marine.gov.scot/dataset/scottish-shelf-model-part-1-shelf-wide-domain/resource/559a1e03–8fc4–4b9a-bb03-e363517d7f8d.

4 Offshore Fish Farming
Challenges and Developments in Fish Pen Designs

Chien Ming Wang,[1]* Yunil Chu,[2] Joerg Baumeister,[2]
Hong Zhang,[2] Dong-Sheng Jeng,[2] and Nagi Abdussamie[3]
[1] The University of Queensland, Australia
[2] Griffith University, Australia
[3] University of Tasmania, Australia
* Corresponding author: C.M. Wang
E-mail: cm.wang@uq.edu.au; y.chu@griffith.edu.au; j.baumeister@griffith.
edu.au; hong.zhang@griffith.edu.au; d.jeng@Griffith.edu.au; nagi.
abdussamie@utas.edu.au

CONTENTS

DOI: 10.1201/9781003184287-4

4.1 INTRODUCTION

4.1.1 Background

Captured fisheries have become unsustainable because most of the wild captured species have been overfished or fully fished with no potential for increase in production. On the other hand, in recent decades, farmed aquaculture has taken an increasing role in filling the gap between seafood supply and rising demand as shown in Figure 4.1 (FAO Report, 2020). A recent report published by DNV (DNV, 2021) gave an estimate of marine aquaculture production by 2050. According to the report, among aquatic animals, farmed finfish will dominate marine aquaculture production (more than 50%) when considering edible weight versus total live weight. The edible weight for finfish is expected at 14 million tonnes, with crustaceans taking up 7 million tonnes and molluscs at 6 million tonnes by 2050. However, farmed fish production has been slowing down due to less nearshore sheltered sea space being licensed for fish farming, plus the emergence of the sea lice problem (especially in Norwegian, Scottish and Chilean farm sites) plus public and environmental opposition to the expansion of nearshore fish farms. If this trend is not reversed, farmed fish will not be able to meet global seafood demand.

To date, almost all marine water fish farms are located at nearshore sites where they are sheltered (in bays, coves and fjords), have a shallow water depth and are hugging the shorelines, mainly for safe operation and easy access to service facilities such as power supply, feed, hatchery, storage, maintenance, and fish processing. With an increasing demand for a higher production target and cost-effective operation for fish production, many suitable nearshore sites have already been fully exploited and most farming pens have reached their allowable fish stock density (Huguenin, 1997; Stickney, 2002). The current nearshore fish farming practice has led to conflicts with local communities, conservation and environmental groups. The criticisms voiced against nearshore fish farming are environmental degradation due to water pollution, noise pollution and their unsightly appearance (Colbourne, 2005; Noroi et al. 2011; Shainee et al. 2013; Tidwell, 2012). The competition for common sea space in coastal areas has intensified, not only amongst fish farmers but also with other marine sectors such as shipping, tourism, conservation and recreation. Moreover, incidents of farmed fish escaping and the spread of diseases have seriously threatened the native sea life population (Beveridge, 2008; Huguenin, 1997; Taranger et al. 2015; Tidwell, 2012; Verhoeven et al. 2018).

In response to environmental concerns and pressures from regulatory authorities, fish farming companies have started exploring offshore sites in their quest to expand fish production in a more

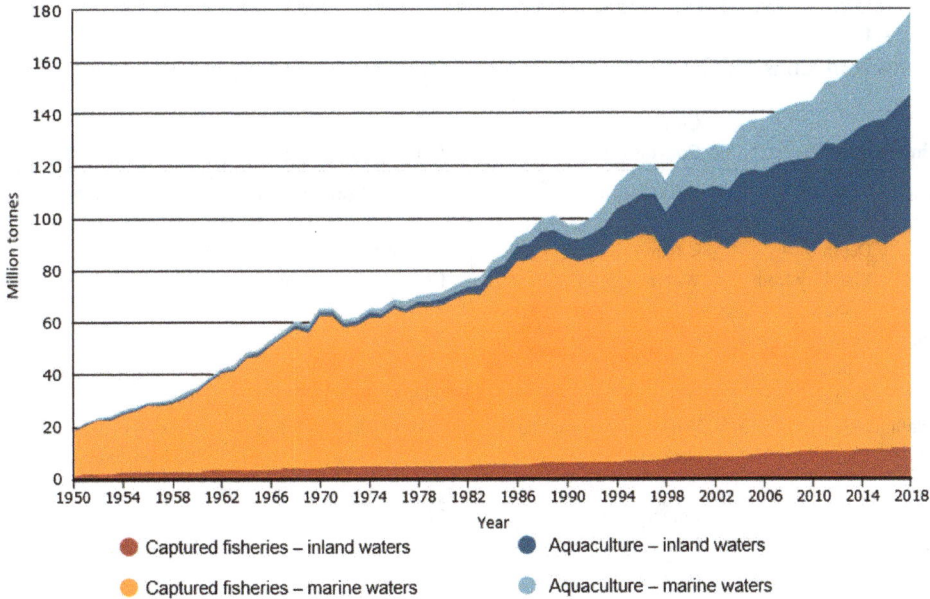

FIGURE 4.1 World captured fisheries and aquaculture production (FAO, 2020).

sustainable and environmental friendly way (Bjelland et al. 2016; Buck, 2007; Holm et al. 2017; Kankainen and Mikalsen, 2014; Kapetsky et al. 2013). Offshore sites offer more space and pristine sea water with less contest with other sea space users (Holm et al. 2017; Huguenin, 1997; Kankainen and Mikalsen, 2014; Tidwell, 2012). The offshore environment with stronger waves, currents and deeper waters helps with waste dispersal and preventing the accumulation of fish wastes (such as, uneaten feed or faeces) under fish pens as well as containing less parasites and diseases. Consequently, fish farming operators are seeking suitable offshore sites as a location of choice for sustainable and high-quality fish production. However, offshore fish farming has its challenges due to a relatively high energy environment with poor accessibility in the remote sites. These challenges will be discussed further in Section 4.2.

4.1.2 DEFINITION OF OFFSHORE FOR FISH FARMING

The definition of '*offshore*' for fish farming takes on different forms according to different stakeholders who include fish farmers, legislators, government agencies, researchers, technology providers and classification societies. The terms of reference for the definition of '*offshore*' for fish farming vary and include various parameters such as met-ocean conditions, bathymetry, geographical distances, technologies used and any combinations of these. In addition to the word '*offshore*', such words as '*exposed*', '*high energy*' and '*remote*' are also used as synonyms (Morro et al. 2021).

According to Drumm (2010), offshore aquaculture is defined as taking place in the open sea with significant exposure to wind and wave action, and requires equipment and servicing vessels to survive and operate in severe sea conditions. Spanish law defines offshore as the sea area outside the straight line joining two major capes or promontories. The sea space within these capes is correspondingly defined as inshore waters (Cabello, 2000). Holmer (2009) defined three classes for fish farming sites: Class 1 - coastal farming; Class 2 off-coast farming; and Class 3 - offshore farming, based on physical and hydrodynamic settings as shown in Table 4.1. It can be seen that distance from shoreline, water depth, and significant wave height are the key-parameters for defining offshore fish

TABLE 4.1

Definitions of Coastal, Off-Coast and Offshore Farming According to Holmer (2009)

Defining Setting		Class 1 Coastal Farming	Class 2 Off-coast Farming	Class 3 Offshore Farming
Physical setting	Distance	< 500 m from shore	500 m to 3 km from shore	> 3 km from shore
	Depth	< 10 m	10 m to 50 m	> 50 m
	Visibility from shore	Within sight of shore users	Usually within sight	Not visible from shore
Exposure	Significant wave height	< 1 m	3 m to 4 m	Up to 5 m
	Accessibility	100%	90%	80%
Legal definitions		Within costal baseline National waters	Within coastal baseline National waters	Outside coastal baseline National/international waters
Major countries with fish farming under the various classes		China, Chile, Norway	Chile, Norway, Mediterranean	USA (Hawaii), Spain (Canaries)

Note: Accessibility <100% refers to limitations in access to the farm due to weather conditions.

TABLE 4.2

Classification of Wave by Wave Height According to NS 9415

Wave Classes	Significant Wave Height H_s (m)	Peak Wave Period T_p (s)	Current Classes	Current Velocity (m/s)	Degree of Exposure
A	0.0-0.5	0.0–2.0	a	0.0–0.3	Low
B	0.5–1.0	1.6–3.2	b	0.3–0.5	Moderate
C	1.0–2.0	2.5–5.1	c	0.5–1.0	Large
D	2.0–3.0	4.0–6.7	d	1.0–1.5	High
E	> 3.0	5.3–18.0	e	>1.5	Severe

Source: NS 9415, 2009.

farming. Also shown in Table 4.2 as referred to by the Norway Standard of Marine fish farms (that is, NS 9415 (2009)), Holmer's Class 1 sites can be regarded as moderate exposure sites with respect to the significant wave height, whilst Classes 2 and 3 can be regarded as severe exposure sites. Figure 4.2 presents the definition of inshore and offshore waters based on the Spanish law together with Holmer's three classes for fish farming.

There are also some restrictions related to the selection of offshore sites for fish farming. For example, Cardia and Lovatelli (2016) recommended that the water depth should be at least three times deeper than the open net pen depth and no less than 15 m between the pen bottom and the seabed. In addition, Kapetsky et al. (2013) pointed out realistic restrictions imposed on conditions for offshore fish farming such as:

- Offshore fish farming should take place within Exclusive Economic Zones (EEZ) (that is, up to 200 nautical miles or 370.4 km from the low water mark) in order to ensure national

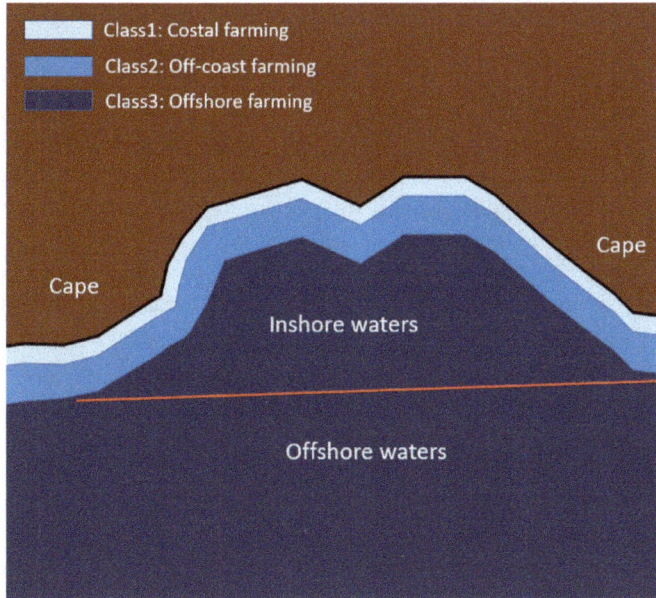

FIGURE 4.2 Definition of inshore and offshore waters with Holmer's classes for fish farming sites.

governance and to provide for the legal protection of investors. The maximum distance from the coastline to an offshore site is recommended to be 25 nautical miles (46.3 km) for economic feasibility, taking into account installation and operation as reported by Jin (2008).

- The depth threshold for conventional sea pens is about 25 m to 100 m based on actual practice and feasible mooring methods and costs.
- Current velocity is within 0.1 m/s to 1 m/s for cultured fish in the confined open net pen.
- The operation will be dependent on onshore facilities to support offshore grow-out installation (for example, feed, holding seed, storage, maintenance, set-up for processing and transporting harvested fish).

After studying the various definitions, parameters and viewpoints on offshore fish farming, it can be seen that offshore fish farming might be categorized by:

(i) an unsheltered site which is seaward of a straight line joining the closest two major capes or promontories, and at least about 3 km seaward of the shoreline but within the EEZ,
(ii) a water depth greater than 50 m,
(ii) current velocity within 0.1 m/s to 1 m/s, and
(iv) significant wave heights exceeding 3 m.

Note that the above definition of offshore fish farming is meant for a preliminary engineering design of the offshore fish pen. There are other factors such as environmental, ecological, other regulatory issues and fish health that have to be considered in the final offshore fish pen design.

4.2 CHALLENGES FACED BY OFFSHORE FISH FARMING

There are many challenges associated with moving fish farms to offshore sites which have not been clearly identified and rigorously studied. As a result, fish farmers are not fully confident about moving to offshore fish farming. It is crucial to identify the challenges in offshore fish farming as they affect running costs, productivity, fish mortality, and HSE (health, safety and environment) for workers.

An appropriate offshore fish pen design should consider environmental risks associated with the selected farming site such as waves, currents and wind. In addition, the offshore fish farming system should provide not only sufficient farming space but also allow stable positioning for easy operation and maintenance as well as a guarantee of the fish's well-being for optimum growth (Shainee et al. 2013). Occurrences of structural damage, sinking of pens and failures of the mooring system can bring massive fish escapes that threaten biosecurity and the profitability of the farming business. Moreover, a relatively poor accessibility in the remote sites puts farming operators in a difficult situation maintaining facilities, monitoring fish behaviour, and carrying out planned feeding and anti-parasite/disease treatments (Morro et al. 2021; Taranger et al. 2015).

After a thorough literature review, the following environmental and operational challenges together with design challenges have been identified:

(1) Environmental challenges:
   • water depth
   • current velocity
   • wave action
   • seabed condition
   • adverse weather and storms
(2) Operational challenges:
   • conducive environment for fish welfare
   • vessel collision with fish pens
   • marine animal invasion
   • infrastructure for offshore fish farming (for example, utility vessels, power supply)
   • economic sustainability of operations (including material selection)
(3) Design challenges:
   • lack of experience in designing mega/submerged offshore fish pens
   • lack of standardized and comprehensive design guidelines/codes

4.2.1 Environmental Challenges

4.2.1.1 Water Depth

Water depth directly affects installation and maintenance costs for the anchoring and mooring system. The length of the mooring lines is usually three to five times the water depth. Therefore, deeper water means more costs for anchoring and mooring systems (Cardia and Lovatelli, 2016; Forster, 2013). The cost for surveying the seabed by using autonomous or remote vehicles at such large water depths will also be high. In addition, a greater water depth may influence the fish growth rate due to a deficiency of illumination and oxygen saturation, or a wide variation of water temperature.

On the flip side, a large water depth can lessen the concentration of waste sediment in the area around fish pens (see Figure 4.3). The deeper water allows for greater dilution potential for dissolved waste and time for detrital consumers to act on solid waste. Since water gets into the pen not only through the sides but also through the bottom, keeping the pen bottom clear is essential to ensure pristine water for the fish (Chacon-Torres et al. 1988). Moreover, deeper water allows a much taller fish pen that gives more space for fish movement and lessens the probability of fish disease. It also allows fish to swim to deeper and calmer water zones during a storm.

4.2.1.2 Current Velocity

Although an adequate current flow is essential for farming fish in pens for the replenishment of oxygen and removal of organic waste, high flow rates may have a detrimental impact on both the

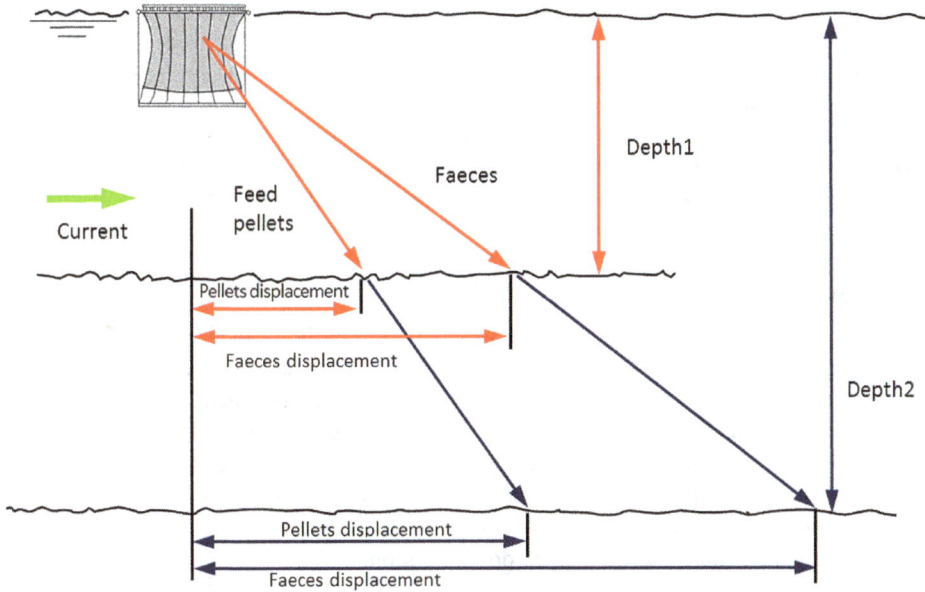

FIGURE 4.3 Influence of depth in solid waste displacement on seabed below pens (Cardia and Lovatelli, 2016).

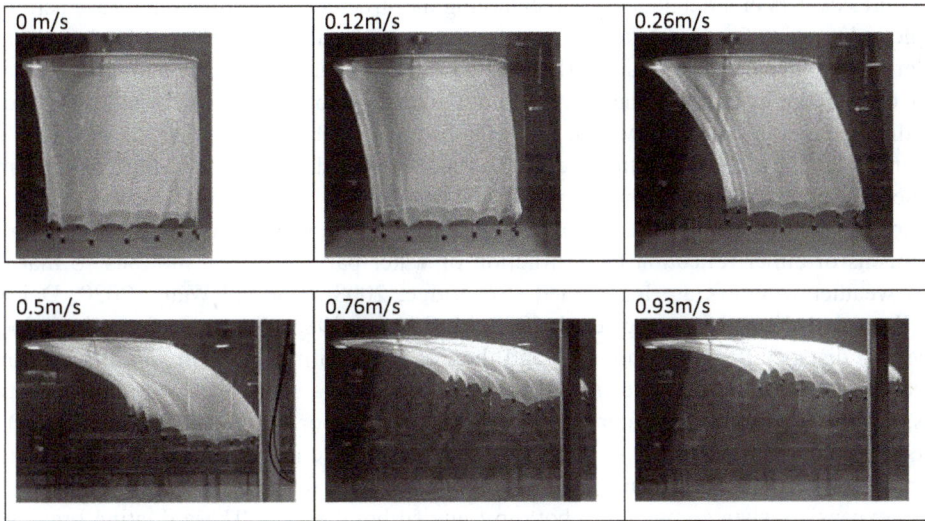

FIGURE 4.4 Net pen models subjected to increasing flow velocity (Moe-Føre et al. 2016).

pen system and on the fish. The maximum drag loads on nets are in most cases caused by current and not by waves. Especially for a flexible open net pen system, horizontal drag forces exerted by current on the pen can reduce the internal volume of the pen. These cause excessive strain on the pen collar and increases tension on mooring lines. Moe-Føre et al. (2016) conducted experimental tests and observed significant volume reduction in fish pens with increasing current velocity as shown in Figure 4.4. Klebert et al. (2015) performed a full-scale field measurement for the current flow field with multiple fish pens. The test results show that the maximum reduction in volume of the fish pen is up to 30% when the current velocity exceeds 0.6 m/s.

Moreover, under an excessive current flow, fish may spend too much energy on swimming, as well as suffering from unacceptable losses of feed (Beveridge, 2008). Consequently, fish growth is curbed and the risk of mortality increases. For example, Solstorm et al. (2015) tested post-smolts of Atlantic salmon (98.6 g, 22.3 cm) in water velocities ranging from slow (0.04 m/s) to fast (0.33 m/s) over six weeks. They found that fish subjected to fast water velocity showed 5% lower weight gain when compared to fish subjected to moderate and slow velocities. In practice, current velocities in the range of 0.1 m/s to 0.6 m/s have been found to be satisfactory for salmon fish farming (Beveridge, 2008; Faltinsen, 2015; Gowen and Edwards, 1990; Hvas et al. 2017; Hvas and Oppedal, 2019; Kapetsky et al. 2013; Oldham et al. 2019; Remen et al. 2016; Yuen et al. 2019).

4.2.1.3 Wave Action

Waves play a significant role in determining whether fish can be farmed in offshore sites. Some farmed fish species, whose habitat is in sheltered sites in nature, are not well adapted to living in high energy wave conditions as they prefer a calm and peaceful environment. Although fish can dive in deeper water where it is relatively calm, they still prefer to be near to the surface of the water for sunlight, oxygen saturation, lower static sea pressure, nutrients/plankton and surface air that is necessary for fish with swim bladders. Moreover, excessive wave action in offshore sites not only harms fish well-being, but it can damage pen structures and moorings, interrupt a worker's routine operation or even place the worker in a hazardous situation.

Recently, several approaches for fish farming have been proposed to control the risk associated with wave action. These approaches include having a flexible structure that moves with the waves, submerging parts or the whole structure, altering the environmental condition by using floating barriers/breakwaters or strengthening and enlarging the structure to withstand the wave action. For example, submerging fish pens take an evasive action to reduce the effect of wave load on the structure during bad weather. Liu et al. (2019) carried out physical experiments and showed that the tension of the mooring rope and the movement of the floating collar were also significantly reduced as the diving depth increases. However, when the fish pen reached a certain depth, the attenuation trend tends to stabilize. Based on these results, it was established that about one third of the depth of the water is the optimal submergence depth for the fish pen.

A floating breakwater with a sufficient draught can attenuate wave transmission through mechanisms of either reflection or destruction of water particle orbital motions so that operational weather windows are lengthened (Beveridge, 2008; Chu and Wang, 2020; Dai et al. 2018; Kato et al. 1979; Matsunaga et al. 2002; McCartney, 1985). When the farm site is located at a certain distance from the shoreline and there is a prevailing wave direction, the farm can be placed in the breakwater's lee side. If the fish farm is in a sea space with multi-directional waves, a floating closed breakwater (in other words, circular or octagonal shaped barriers) may be used to attenuate wave forces in order to create a calm internal water basin for fish farming. Floating breakwaters are excellent for sheltering offshore fish farms from waves as they are relatively inexpensive when compared to bottom founded breakwaters. These floating breakwaters may be moored by using catenary chains in relatively deep water and such breakwaters do not interfere with currents. They can be readily reconfigured as farms expand or pens are removed (Kato et al. 1979).

Figure 4.5 shows the use of the Bridgestone floating breakwater system to protect fish farms in Japan. Figure 4.6 shows a porous collar barrier for COSPAR design which was introduced in a numerical study done by Chu and Wang (2020). According to the study, the porous collar barrier can be a part of semi-submersible floating fish pens so that it can reduce wave transmission by at least 60% inside the fish pen.

4.2.1.4 Seabed Condition

The seabed condition in particular affects the mooring system and selection of the anchorage method. A good anchor provides reliable holding power. Therefore, it is important to know the

FIGURE 4.5   Bridgestone breakwater systems, Japan (Kato et al. 1979).

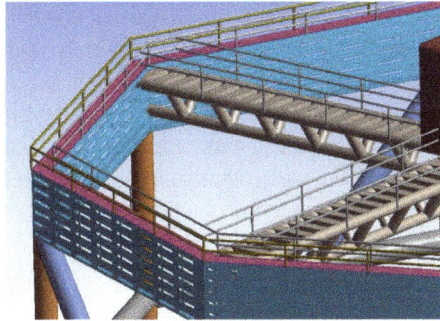

FIGURE 4.6   Porous collar barrier (Chu and Wang, 2020).

sea bottom conditions in order to select the correct type of anchoring system. The continuous action of a seawater dynamic load on the seabed causes the accumulation of pore pressure in the submarine soil layer. When the accumulated pore pressure exceeds the initial stress, the seabed will be liquefied and eventually instability and destruction of the soil layer will occur (Jeng, 2018; Wang et al. 2014). Not only might shells and seagrass on upper layers prevent an anchor from taking hold, but bottom layers with sand, mud, peat or clay require different anchoring mechanisms (Cardia and Lovatelli, 2016). Apart from these, the seabed might include submarine fiber optic cables, telephone lines or pipelines, explosive areas, or historical shipwreck sites (Cardia and Lovatelli, 2016). These limitations should be indicated and considered for pen designs and mooring systems.

Detailed seabed analysis is needed for determining what kind of anchorage method would be suitable for the site. For example, traditional anchorage (gravity method, see Figure 4.7) is suitable at sites where there is adequate deep sediment layer (Kankainen and Mikalsen, 2014). If the sea bottom is rocky, drilling would be a better method to keep the mooring system at the site. Echo sounding and sea bottom samples are methods used to evaluate anchorage (Kankainen and Mikalsen, 2014). Seismic survey will also be required for seabed soil investigation.

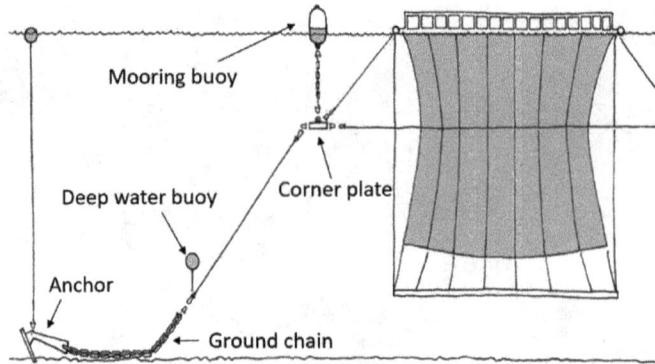

FIGURE 4.7 Components of anchoring and mooring system (Cardia and Lovatelli, 2016).

4.2.1.5 Adverse Weather and Storms

Infrastructure and equipment failures caused by extreme environmental conditions not only bring huge losses to investors, but they also create a high-risk working environment for workers (Jensen et al. 2010; Mapes, 2017). Storms (hurricanes, or cyclones or typhoons) are meteorological phenomena that pose a risk to offshore fish farms due to associated strong winds, resultant waves and currents generated (Beveridge, 2008; Kankainen and Mikalsen, 2014; Kapetsky et al. 2013; Tidwell, 2012). They mostly occur in the tropical-equatorial zones, namely in the region between the Tropic of Cancer and the Tropic of Capricorn, but their incidence can extend to the North Atlantic and North Pacific (Cardia and Lovatelli, 2016).

A storm surge is a long-term 'wave' that can maintain a water level above normal levels for hours or even days. A storm surge can combine with the astronomical tide to create a storm tide. The magnitude of the surge is affected by several factors, such as storm intensity, magnitude, wind speed, approach to the coast and coastal bathymetry (Wamsley et al. 2010). Although there are few studies on fish pens under extreme sea conditions, we can still draw some information about storms from other offshore structures.

The occurrence of storms should be rigorously analysed so as to detect an appropriate offshore fish farming site and to predict the environmental forces for pen designs (Cardia and Lovatelli, 2016; Huguenin, 1997). Submerged pens are more suitable for areas where there is a high incidence of storms and extreme weather conditions since wave forces and associating pitching and surging motion during storms reduce significantly with increasing water depth. Therefore, the submergibility of offshore pens provides an excellent protection for pens and fish against destructive storms.

4.2.2 Operational Challenges

4.2.2.1 Conducive Environment for Fish Welfare

Fish demand the best environmental conditions for growth. Shainee et al. (2013) listed the key parameters for the survival needs of fish (see Table 4.3) under five factors (water quality, stocking density, feed conversion, less motion and smaller net deflection). These factors and parameters are very important for the design of fish pens.

The best quality of water for fish farming is species-dependent, since each type of fish thrives in a particular water temperature, salinity, dissolved oxygen, pH and turbidity (Pillay, 2004; Stien et al. 2013). However, the optimum design parameters for fish pen designs are unknown for many fish species (Shainee et al. 2013).

The next important parameter is the stocking density which is dependent on the net water volume. Sufficient living space is essential for the life of fish. A good pen design should provide sufficient net

TABLE 4.3
Design Parameters for Fish Living

	Factors	Parameters
1	Water quality	Dissolved oxygen, salinity, temperature, pH, turbidity, pollution, infestation, biofouling
2	Stocking density	Net volume, dissolved oxygen
3	Feed conversion	Motion, stocking density, feeding frequency, feed type
4	Less motion	Waves, current, wind, pen design
5	Smaller net deflection	Waves, current, pen design

Source: Shainee et al. 2013.

volume under a strong current. Therefore, the minimum pen net deformation due to current loads should be considered in the design to provide sufficient fish pen volume (Faltinsen, 2015). Huang et al. (2008) found that the current-induced effects on the net-pen system were more important than those due to waves only. So, they concluded that farming sites should not be situated in areas where the current velocity exceeds 1 m/s, unless engineering solutions are found to overcome serious net-pen volume deformation.

Feeding of fish is done daily from dawn to dusk by using a feed distributor that is remotely controlled. Fish pens must be designed to securely hold the feed distributors so that they do not get dislodged in the event of storms. In salmon farms, fish gather at the water surface to consume dry pellets quickly which would otherwise become moist and sink fast and out of reach from the fish. Note that dry pellets normally contain a high level of fish meal with enriched nutrients and must be kept at less than 10% moisture level and supplied at the water surface (Lovell, 1989; Pandey, 2018). Therefore, it is important for a fish pen to have calm surface water in order to reduce feed wastes and keep fish growth at an acceptable level. A new fish pen design for deployment in energetic offshore sites may thus require an engineering solution to reduce wave transmission inside the pen.

The environmental forces of the selected site can influence the fish welfare and the integrity of the pen system. Hence, pen designs must not only be robust enough to survive the strong environmental forces, but they should also have the means to avoid or dissipate the excess energy in order to provide a stable and relatively quiet environment for fish to grow well. Therefore, the challenge is to design a system that copes best with the environmental forces by means of advanced technology and economically affordable methods (Shainee et al. 2013).

Table 4.4 presents the optimal growing conditions for various commonly farmed marine fish species, from the species factsheets given in the Appendix of Le François et al. (2010).

4.2.2.2   Vessel Collision with Fish Pens

Large fish farming facilities in offshore sites are exposed to collision with non-aquaculture vessels, or aquaculture support vessels such as well-boats and feed-barges (see Figure 4.8). The aquaculture support vessels for offshore sites will be much larger than the nearshore fish farms for their required fish capacity. Therefore, collision accidents in the offshore sites can be more catastrophic than those that happen in the nearshore sites. The main consequence of the collision is not only failure of the aquaculture system, but massive fish escapes that can threaten biosecurity and the profitability of the farming business. A strengthening method such as berthing reinforcement for ships and/or a back-up system are required to protect the fish farming facilities from such vessel collision accidents. Alternatively, a dynamic positioning system may be applied in order to prevent accidents associated with interactive motions between vessels and pens during harvesting or bathing operations.

TABLE 4.4
Optimal Farming Condition for Farming Marine Fish Species

Farming Marine Fish Species	Rainbow Trout	European Whitefish	Atlantic cod	Barramundi	Atlantic Salmon	Atlantic, Southern, Pacific Bluefin tuna
Commercial size	0.6 to 0.9 kg	0.8 to 1 kg	3 to 5 kg	0.4 to 3 kg	3 to 7 kg	80 kg
Years to reach commercial size	10 to 13 months	2 to 3 years from hatching	2 to 3 years from hatching	1.5 to 2 years from hatching	2 to 3 years from 50 to 100 g smolt	1.5–2.5 years from 15 kg fish
Open pen culture	Yes (Seawater pen)	Yes	Yes	Yes ($2 \times 2 \times 2$ m to $16 \times 16 \times 8$ m size of pen)	Yes (5 to 20 m deep sea water)	Yes (40 m diameter floating ring)
Close containment culture	Yes (Indoor/ outdoor tanks)	Yes (Pond, plastic tanks)	Yes (land-based tanks)	Yes (Pond <2 m deep)	N.A	N.A
Rearing density	20 to 40 kg/m³	20 to 30 kg/m³	20 to 25 kg/m³ (open pen) 40 to 50 kg/m³ (closed containment)	15 to 25 kg/m³	<25 kg/m³ Norway: 5 to 15 kg/m³, Australia and Chile: 8 to 10 kg/m³	2 kg/m³ to 5 kg/m³
Optimal temperature	15°C	17 to 19°C	<15°C	27 to 36°C	15°C	N/A

Source: Le François et al. 2010.

FIGURE 4.8 Steel pen damaged due to ship collision (Photo from: https://salmonexpert.cl/article/sernape
sca-y-transmarko-activan-protocolos-de-accin-por-accidente-en-centro/).

FIGURE 4.9 Whales enter the net pen (https://fiskeridir.no/Akvakultur/Erfaringsbase/Knoelhval-i-merd).

4.2.2.3 Marine Animal Invasion

Containing a large number of fish in offshore sites, net pens can attract wild predators such as
whales and sharks which normally do not venture into nearshore sites. To gain access, such massive
predators can damage parts or entire fish pens, thereby causing fish escapees. Figure 4.9 shows an
invasion of a whale (9 m in length) in a Norwegian fish farm. At about 2 m below the water surface,
a hole in the net was observed that was damaged by the whale. The hole was covered immediately
to prevent salmon from escaping and the whale was towed out of the pen by lowering part of the
pen wall.

Offshore fish pens must therefore be designed to keep out predators by introducing methods such
as acoustic deterrent systems (Croix, 2008) and employing more durable nets (such as the EcoNet)
or double net pen (for example, Huon's Fortress Pen). For example, EcoNet (see Figure 4.10(a)) was
developed by the AKVA Group who used it for Ocean Farm 1 project for prevention of net wear and
tear. The EcoNet is made from very strong but lightweight PET (Polyethylene Terephthalate) and it
has been certified under the Norwegian fish farming standard, NS 9415 (2009), to have a lifetime in
the water for up to 14 years. Figure 4.10(b) shows Huon Aquaculture's patented Fortress Pen. This
fish pen was developed in response to a need to keep out predators (such as seals) by employing a
durable double net system. The net material is made from ultra-high-molecular-weight polyethylene

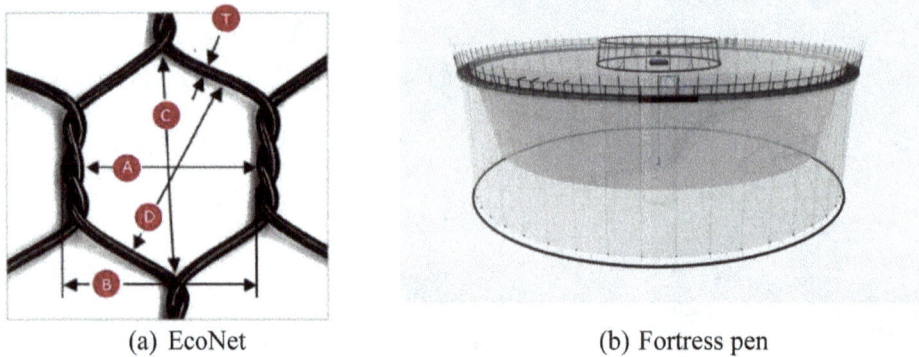

| (a) EcoNet | (b) Fortress pen |

FIGURE 4.10 (a) AKVA Group's EcoNet, (b) Huon Aquaculture's Fortress Pen (photo from: (a) https:// akvagroup.com/pen-based-aquaculture/pens-nets/nets-/econet, (b) https://huonaqua.com.au/wp-content/uplo ads/2017/08/Huon-Fortress-Brochure.pdf).

(UHMWPE), the same material that is also used in bullet-proof vests, that can withstand extremely high current flows.

4.2.2.4 Infrastructure for Offshore Fish Farming

This infrastructure includes utility vessels and the power supply. Fish farming has to cater for all stages of production from spawning, rearing fries and fingerlings, producing mature fish, harvesting and packing. The distance between the farm site and necessary dry land support facilities directly affects running costs (Cardia and Lovatelli, 2016). Therefore, Kapetsky et al. (2013) considered 25 nautical miles (46.3 km) as the limit for economical offshore site development. Aquaculture Forum Bremerhaven reported the urgent need to plan for a more comprehensive development of water-based infrastructure for offshore fish farming (Kapetsky et al. 2013; Rosenthal et al. 2012).

Although there are many novel fish pen designs that aim to operate in offshore sites, the global vessel fleet for offshore fish farming is not yet sufficient to be fully viable for the offshore fish farming operation so that there is a delay in industry maturity. Moreover, there is no international code of practice for aquaculture vessels operating in offshore sites.

Power supply is needed for electrical devices and equipment for monitoring and for automated processes that have become essential for offshore fish farming. However, the power supply is not cheap and easily accessible at offshore sites. The fish farming industry currently relies heavily on diesel to power operations such as ventilation, feeding, lighting, net cleaning, fish bathing and harvesting. Moreover, some submerged fish pen designs require a substantial ballast mass to keep the fish pens in deep water. It will consume a lot of power to fill and empty the ballast water for draught control or water exchange.

4.2.2.5 Economic Sustainability for Operation (Including Material Selection)

California Environmental Associates presented their global review of offshore fish aquaculture in 2018 (CEA, 2018). In the report, it is highlighted that small-scale offshore farming projects will face a challenge in order to become a profitable operation. As it is, these small-scale offshore projects have high capital costs, have to contend with intense oceanographic conditions, and have an unclear path to economies of scale. Although massive industrialization and automation could provide a more profitable business model, the current offshore farming projects have yet to prove their economic sustainability (CEA, 2018).

Ellingsen and Aanondsen (2006) emphasize that it is not only the quality of fish that is important to consumers, but also the environmental impact of farming, processing and transportation are becoming important issues. The performance of fish pen products is an important part of the design for environmental sustainability. The reliability of products is directly related to safety of life and property. Poor-quality materials and facilities can cause potential safety hazards in equipment, resulting in hazards such as broken fish nets. On the other hand, construction cost is also a consideration for engineering design. From the perspective of the structure of circular full-floating HDPE (High-density polyethylene) pen facilities, the main factors affecting the price of pen facilities are the materials and prices of pen frames and nets, as well as the mooring systems. Balancing safety and engineering cost are a key part of sustainable development.

4.2.3   Design Challenges

4.2.3.1   Lack of Experience in Designing Mega/Submerged Offshore Fish Pens

Unlike nearshore fish pen designs which have matured over decades, there are very few offshore fish pen designs for reference. Most offshore fish pen designs are recently developed and remain in a conceptual design stage. Only a few of them (such as Ocean Farm 1 and Havfarm 1) are in the trialing phase. There is little data on their long-term survivability, durability, maintenance, and repair methods, and guarantees for fish well-being. Therefore, there are still many unknowns associated with the new offshore fish pen designs in particular with mega/submerged designs. As the development and operation of these offshore fish pen designs are still in their infancy, offshore and marine engineers face unknown challenges which can only be revealed in a long-term operation of these fish pens.

4.2.3.2   Lack of Standardized and Comprehensive Design Guidelines/Codes

There is lack of design guidelines/codes which could help to create a path forward for building offshore fish farms that are storm proof, ensure fish well-being and growth, and which are safe for workers performing the farming activities.

Within the aquaculture industry, approval of a new fish pen design is necessary before clients have the confidence to invest in building the fish pens and also for commercialization. The design approval may be given by maritime classification societies in accordance with their certifying rules for offshore fish farming. In recent years, some classification societies such as ABS and DNV officially published certifying rules for offshore fish farming installations in the following documents:

- ABS, 2018, Guide for building and classing: Offshore fish farming installations
- DNV, 2017, Rules for classification: Offshore fish farming units and installations (DNVGL-RU-OI-0503)

These certifying rules, however, only cover hull structures, onboard machinery, and equipment that is not part of the aquaculture systems as is common practice in certifying offshore oil and gas units. The guidelines do not cover primary aquaculture elements such as floating collars and net pens (made of polymers, concrete or equivalent) and their associated equipment, feeding and production facilities, feedstock facilities and fish escape prevention devices. They indicate that aquaculture systems will be assessed under the jurisdiction of local authorities which may not exist in some countries.

Owing to a lack of standardized and comprehensive guidance for the offshore fish farming industry, a wide range of fish pen designs has appeared; thereby making it difficult to establish a single strategy for achieving a more cost-effective business model. A design guidance/code for offshore fish pen designs will include many considerations such as maintenance methods, manned or unmanned facilities, risks associated with fish farm sites, marine warranty, special requirements from operators and legislative issues for the utilization of common ocean space.

TABLE 4.5
Flexible Collar Pens—Pros and Cons and Suitability for Application in Offshore Sites

Pros	Cons	Application for Offshore Sites
• high resilience to wave forces with a long service life (>10 years), • high resistance to rotting, weathering and biofouling, • easily formed into various configurations and relatively cheap when ordered in large volumes, • easily constructed in-land and towed by boats to install.	• problems with deformation of the net due to strong waves and currents, • twisting and turning problems of stanchions, • limited walkway access putting workers in danger during bad weather, • difficulty in placing feed systems due to space constraint, • needs large service vessels.	• some have shown to survive storms with significant wave height (H_s) of 10 m, • Possible alterations for offshore use by featuring submergibility, • little empirical or theoretical evidence to offer complete confirmation of the extreme sea state and survivability on a long-term basis.

FIGURE 4.11   Tubenet pen (Photo from: https://fishfarmingexpert.com/article/mowi-goes-deep-to-beat-lice-problem/).

4.3   RECENT DEVELOPMENTS IN OFFSHORE FISH PEN DESIGNS

4.3.1   Modified Flexible Collar Pens

Flexible collar pens have been widely used for fish farming in Japan, Western Europe, North America, South America, New Zealand, and Australia. High-density polyethylene (HDPE) is the most commonly used material in modern industrial fish farming. The main structural elements of these pens are floatable pipes, which can be assembled in various ways to produce the floating collar. The pipes are held together by a series of brackets with stanchions and distributed throughout the entire boundaries to suspend the fish net (Cardia and Lovatelli, 2016). Table 4.5 summarizes the flexible fish pens in terms of pros, cons and their suitability for application in offshore sites.

4.3.1.1   Tubenet

The most recent development of the flexible collar pen by the AKVA Group ASA is the Tubenet as shown in Figure 4.11. The Tubenet system uses a net to keep salmon below the sea lice layer (top 10 m

TABLE 4.6
Submerged Pens—Pros and Cons and Suitability for Application in Offshore Sites

Advantages	Disadvantages	Application for Offshore Sites
• either be unattended by surface units, accessed only when needed, or remotely controlled, • best features to avoid surface debris and effects of storms, • structural strength does not need to be as great as surface structures.	• a lack of visibility in normal operation, • relatively complex to operate due to its submerged mode and maintenance and operating services are difficult, • operating costs may be relatively higher than surface mode structures.	• unknown submerged pens being deployed in offshore sites, • submerged operation is in question to compete surface operation.

of water layer) and protect salmon from strong waves. A large cylindrical and tarpaulin-walled passageway, called the 'snorkel', in the centre of the pen protects salmon from sea lice when they swim to the surface to fill their swim bladders. Feed is delivered by way of subsurface feeding tubes. Only the centre section, where the salmon surface to refill their swim bladders, requires bird netting. In the outer ring the salmon are kept 14 m below the surface so that they can be protected from strong waves. The tarpaulin 'tube' extends to a depth of 14 m and the feeders are placed at 13 m. The inner cylinder is 60 m in circumference. Mowi ASA (Norway) and AKVA have successfully trialled the Tubenet system, and it is expected to be commercially adopted by fish farmers in Scotland and Norway.

4.3.2   SUBMERGED FISH PENS

In order to avoid strong surface waves, submerged fish pens are proposed. The pens are submerged to a suitable water depth below the hazardous upper water column. A hypothesis of the submerged fish pens is that fish welfare and production efficiency will be as good or even better in the submerged position as in the surface position where high energetic waves are present. These submerged pens may be raised temporarily to the surface for necessary maintenance requirements and for fish harvesting. Table 4.6 summarizes the pros and cons of submerged pens as well as their suitability for deployment in offshore sites.

Below, is some discussion of various submerged fish pen designs that have been proposed or built.

4.3.2.1   Atlantis

Atlantis Subsea Farming (see Figure 4.12), the cooperative AKVA group, has developed a submerged offshore fish pen with flexible collars and a compressed air chamber for large-scale salmon production. This submerged pen has a circumference of up to 160 m. Air and fish feed can be added through hoses from a supply vessel.

4.3.2.2   Giant Offshore

Giant Offshore pens are designed for exposed localities with the aim of minimizing the risk of escape, better protection against salmon lice and reducing point discharges of nutrients. Operations and monitoring are performed in full, from integrated base vessels. The pen is designed from a flexible material, with the strength to withstand being more exposed than today's conventional nearshore fish farms.

Giant Offshore is a 500 m long cylindrical construction with pointed ends, where the middle section is 300 m long and 40 m in diameter. The middle section has five large net bags which together make up the farming volume of 290,000 m³, where it is possible to produce 2.2 million salmon. The

FIGURE 4.12 Atlantis subsea farming (Photo from: https://atlantisfarming.no/nyheter/atlantis-subsea-farming-a-farming-concept-designed-for-the-future).

FIGURE 4.13 Giant offshore (Photo from: http://giganteoffshore.no/).

construction is designed to keep a large distance between the net bags and mooring bodies, with the intention of reducing the risk of damage to the net (Figure 4.13).

4.3.2.3 AquaPod

AquaPod (see Figure 4.14) was developed by Ocean Farm Technologies in the United States. It has a two-point anchor for mooring and some operational advances such as net cleaning and removal of dead fish. It is located 8 miles offshore in a water depth of 45 m, with strong currents and waves. It has already been proven to withstand waves of up to 10 m high. The structure is made from recycled polyethylene plastic with fiberglass reinforcement. The reason for its geodesic shape is that this has the least surface area possible compared to its volume, and this helps to make the pen predator proof. Future pods are equipped with a propeller mechanism and GPS so that they can be used as transporting vessels that carry juvenile fish and arrive at the desired location with fish that are ready to harvest.

4.3.2.4 NSENGI Fish Pen

NSENGI (Nippon Steel and Sumikin Engineering Co. Ltd) has investigated the co-location of an existing offshore platform and a fish farm. It has carried out offshore verification testing of large-scale seabed

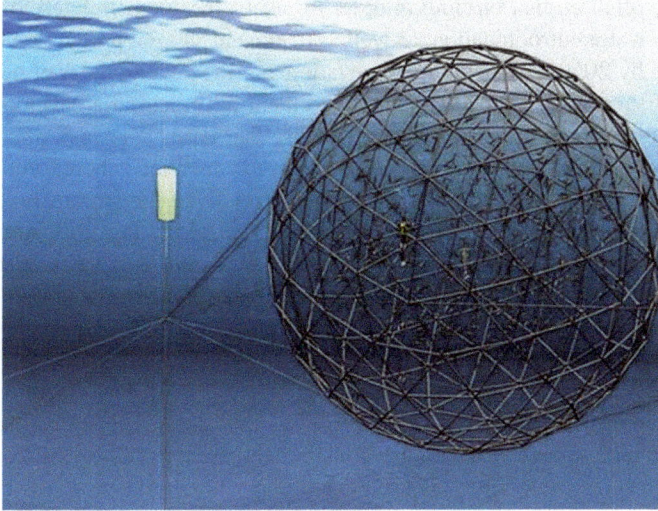

FIGURE 4.14 Submerged aquapod pen from ocean farm technologies (Photo from: https://wired.co.uk/article/aquapod-sustainable-fish-farm).

FIGURE 4.15 Sinking fish pens (Photo from: https://eng.nipponsteel.com/english/news/2016/20161003.html).

resting pens at a salmon farm which is 3 km from the shoreline of Sakaiminato, Tottori Prefecture, Japan. Each pen has a volume of 50,000 m³. The fish pen is designed for the following environmental conditions: significant wave height of 7 - 9 m corresponding to a wave period of 10 to 16 s, a current velocity of 2 knots and a water depth of 60 m. The pens are suitable for farming fish species such as Seriola (Japanese amberjack) and Coho salmon. The pens are serviced by a jacket platform that houses the equipment and feedstock storage facility for automated feeding of the fish (Figure 4.15).

4.3.3 NOVEL OFFSHORE FISH PEN DESIGNS

In recent years, a few global fish farming companies have developed novel designs for offshore fish pens and built them for trial tests at selected offshore sites. Norwegian fish farm operators are leading the way to offshore fish farming in order to resolve problems related to parasites (such as sea lice) caused by water eutrophication, as well as to expand salmon production that is expected to be five times more by 2050. To encourage innovation and the development of next generation fish pen designs for offshore use, the Norwegian government has released a development scheme that is financially lucrative for fish farm operators. Chinese fish farms also seek offshore sites because of the need to expand seafood production to meet their exponentially increasing domestic seafood consumption. China has the advantage of possessing a relatively good construction infrastructure that allows it to fabricate mega-size fish pen projects at a relatively low cost. According to China's Ministry of Agriculture 'National Marine Ranch Demonstration Zone Construction Planning (2017–2025)', China plans to develop 2,500 km² of national fish farming waters by 2025. Several hundreds of offshore fish farming facilities are expected to be deployed in the Chinese offshore sites. Construction of these offshore facilities will reduce the scale of commercial fishing and restore ocean resources.

Offshore fish pen designs may be divided into two main streams; open net pen system and the closed containment tank system based on fish containment methods (Chu et al. 2020). Both systems will be discussed in detail in the following sections.

4.3.3.1   Open Net Pen System

The open net pen system is the most widely used for marine fish farming. These net-based pens are generally slender structures, with a low mass in comparison to the size of the structure. They have a large damping-to-mass ratio, and this can effectively eliminate resonance problems. Depending on net holding methods, open net pen systems can be either deformable (flexible type) or robust (rigid type). However recent offshore fish pen designs mostly adopted the rigid type to withstand strong waves and currents whilst providing a sufficient farming volume. Table 4.7 presents features of the rigid open net pen system in terms of its pros and cons, and its application for offshore sites.

Some examples of the open net pen system for offshore use are described and shown in Figures 4.16–4.24.

TABLE 4.7
Rigid Open Net System—Pros and Cons and Suitability for Application in Offshore Sites

Pros	Cons	Application for Offshore Sites
• stable working platform for all husbandry and management operations, • potential for integral feeding and harvesting systems, • construction and repair facilities may be done in conventional shipyards, • if rigid frames are used, they maintain farming volumes and keep fish in place, • relatively small vertical motion due to large mass, • low natural frequency avoiding wave resonance.	• need for large and heavy structures, • requires good port facilities and/or expensive towing to install, • susceptibility to structural failure in extreme conditions, • large masses require heavier mooring systems, • involves relatively high capital costs, • rigorous engineering analyses, design and high-quality control in construction is essential to ensure safety in offshore operation.	• deployed at some exposed offshore sites where an occurrence of extreme storms is rare, • proto-type testing performed in deployed cage of Ocean Farm1 and Shenlan1, • despite not enough operating records, it is the most progressed type for farming fish in offshore sites, • reported some failures related to human error and adverse weather.

FIGURE 4.16 Ocean farm 1 (Photo courtesy of Charles Lim).

FIGURE 4.17 Smart fish farm (Photo from: https://salmonbusiness.com/gustav-witzoe-our-biggest-challenger-is-land-based/).

4.3.3.1.1 Ocean Farm 1

SalMar Group, a Norwegian fish farm operator, developed Ocean Farm 1 (see Figure 4.16). Ocean Farm 1 is a result of robust technology and principles used in submersible offshore units. It is a full-scale pilot offshore fish pen deployed about 5 km off the coast of central Norway. With a diameter of 110 m and volume of 250,000 m^3, the pen is able to accommodate 1.5 million salmon (Zhao et al. 2019). It is intended for offshore installation in water 100 m – 300 m in depth with a 25-year life-span. It has more than 20,000 sensors and over 100 monitors and control units. It can be immersed in deep water by filling the ballast tanks and is moored by eight lines tied at fairleads placed on the lower parts of the side vertical columns. It uses Eco-net, developed by AKVA Group by using PET (polyethylene terephthalate), which has a hard surface that resists marine fouling and makes it easy to clean in the water, as well as improving durability and preventing fish escape.

FIGURE 4.18   Shenlan 1 (Photo from: https://swissre.com/reinsurance/property-and-casualty/reinsurance/marine/offshore-fish-farming-facilities-challenges-marine-insurers.html).

FIGURE 4.19   Havfarm1(Photofrom:https://fishfarmingexpert.com/article/in-the-shadow-of-the-giant-watch-a-video-of-the-havfarm/).

FIGURE 4.20   Zhenyu 1 aquaculture platform (Photo from: http://cn.zpmc.com/news/cont.aspx?id=1296).

FIGURE 4.21 Viewpoint seafarm (Photo from: https://viewpointaqua.no/seafarm/).

FIGURE 4.22 Spider cage (Photo from: https://viewpointaqua.no/spidercage/).

FIGURE 4.23 De Maas SSFF150 pen (Photo from: http://demaas-smc.com/).

4.3.3.1.2   Smart Fish Farm

SalMar group is planning to build an upgrade version of Ocean Farm 1 which is an even larger fish pen with a diameter of 160 m and can accommodate three million salmon and produce 23,000 tonnes round weight. It has been named, Smart Fish Farm (see Figure 4.17) and will have five to ten units in the first phase. The first unit is expected to have an investment level of EUR 225 million, falling to EUR 147-200 million for additional units. The product cost is expected at EUR 3.6/kilo of salmon for the first unit, falling to EUR 3.3 when several units are in operation. SalMar expects production start-up of Smart Fish Farm in the second quarter of 2024, based on approval of sites and volumes by summer 2021 and expects to enter into a construction contract during the fourth quarter of 2021.

4.3.3.1.3   Shenlan

Shenlan 1 and Shenlan 2 were developed for salmon and trout farming about 130 nautical miles off the shore of Rizhao in east China's Shandong province. Shenlan 1 has already been deployed at the site and it has a diameter of 60 m and is 35 m in height; it is able to culture 300,000 salmon. A noticeable feature, that differentiates it from Ocean Farm 1, is the presence of a centre oscillating buoy to generate power for fish farm operations (Figure 4.18). Shenlan 2 is in the planning stage. It will have a 60 m diameter and a height of 80 m, and it will be able to accommodate about 1 million salmon.

4.3.3.1.4   Havfarm 1

Havfarm 1 is the world's longest fish farm, 385 m long and 59.5 m wide and has a capacity for 10,000 tonnes of salmon (over 2 million fish). Constructed in the Yantai Shipyard, China for Norwegian farmer Nordlaks, Havfarm 1 comprises a steel frame for six 47m x 47m pens made of HDPE mixed with copper (from Garware, India), with open nets at 60 m depth (Figure 4.19). The facilities are designed to withstand 10m significant wave height. Havfarm 1 is moored with 11 anchors, each weighing 22 tonnes and each anchor has a holding power between 300 and 450 tonnes. It is sited 5 km south-west of Hadseloya in Hadsel municipality in Vesteralen, Norway. A 7 km long subsea power cable supplies power to the Havfarm 1 from the shore.

Havefarm 1 is a vessel-shaped fish farm concept which intends to minimize the wave loads coming from the bow in open sea sites. A single-point mooring system is employed for the vessel-shaped fish farm concept to allow the entire fish farm to rotate freely in order to reduce the environmental loads on the structure and improve fish wastes dispersion. In Li et al. (2017) study, the hydrodynamic properties of a basic geometry of the vessel hull were obtained from the frequency-domain analysis. Time-domain simulations were performed by coupling the hull with the mooring system. Mobron et al. (2020) developed a method to estimate the fatigue utilization by using Dynamic Amplification Factor for a fatigue assessment of Havfarm 1 design.

4.3.3.1.5   Zhenyu 1

Zhenyu 1 aquaculture platform (Figure 4.20) was constructed in Lianjiang County, Fuzhou City, Fujian Province and launched in 2019. It is located in the sea area of Dinghai Bay, Fujian Province. The platform is olive-shaped, with a total length of 60m and a width of 30m. The aquaculture water body reaches 130 million cubic meters. It is expected to produce 120 tonnes of high-quality commercial marine fish annually. It is mainly composed of a floating body structure, an aquaculture frame, and a rotating mechanism. Zhengyu 1 was designed based on a concept for preventing biofouling. The pen can be rotated about the horizontal axis so that the underwater bio-fouled part of the pen can be brought above the water surface for cleaning. There is a wind turbine to supply power.

4.3.3.1.6   Viewpoint Seafarm and Spider Cage

Nova Sea AS, a Norwegian company, designed two innovative concepts: the Viewpoint Seafarm (see Figure 4.21) and the Spider cage (see Figure 4.22) for offshore fish farming based on semi-submersible technology. Viewpoint Seafarm comprises a hub on a semi-submersible platform and four floating net pens connected to the semi-submersible by a hinge system. Each floater has a projected area of 50 m × 35 m. Scale testing has been done with 11 m significant wave height and the system showed a stable motion response (Lindeboom, 2018).

The Spider Pen has a dedicated 100 m diameter circular barrier, with an outer steel ring and another ring inside for heave compensation. It is designed to shield the actual fish pen from heavy sea conditions and sea lice. The design has been tested up to a significant wave height of 11 m with and without current, where general motion, acceleration, load and sloshing have been accessed (Lindeboom, 2018).

4.3.3.1.7 De Mass SSFF150 Pen

De Maas SMC, a firm operating in the offshore oil and gas services industry, is partnering the Chinese government to build a $151 million deep-water aquaculture farm off the coast of China. De Maas will design and build five SSFF150 (Semi-submersible Spar Fish Farm) pens. Each is 139 m in diameter and 12 m high (see Figure 4.23). The central tower will house machinery spaces, feed storage and provides accommodation for operators. By submerging underwater, the pen can be protected from storms.

4.3.3.1.8 Impact-9 Submersible Salmon Pen

The Irish company Impact-9 invented a submersible fish pen design which can survive storms based on an innovative flexible design that allows salmon production at offshore sites (see Figure 4.24(a)). The fish pen, 90 m in diameter and 125,000 m^3 in volume, can produce 3,000 tonnes of salmon in every 12 month grow-out cycle. The design considers the environmental conditions of Scottish marine sites that are over 60 m deep. Impact-9's salmon pen design is based around a central, strong backbone structure, which is centrally moored with a single-point mooring. It will provide a safe working platform and a home for feed and farming equipment. Each net is suspended from the backbone and includes an inflatable collar near the base. The pen would alleviate sea lice and algal bloom problems by keeping salmon 6 m below the surface or more if required. The adoption of inflatable beam collars (made of HDPE) into the structure allows operational innovations and flexibility (see Figure 4.24(b)) so that the design will achieve inexpensive access to offshore zones, a lot less operational cost on divers, less dependence on well boats, and resistance to seal ice and algal blooms.

4.3.3.2 Closed Containment Tank System

In order to control the water quality and the production process, closed containment fish tanks were introduced in the 1990s (Beveridge, 2008). These closed containment fish tanks, also known as recirculating aquaculture systems (RAS), are generally sited on land and involve a recirculation system for the water where internal water is constantly filtered (Tidwell, 2012). Such a system is efficient with freshwater, but when dealing with saltwater, there is a production of sulfates at the tank bottom that kill fish if the tank is not properly cleaned.

Floating closed containment tanks for offshore fish farming are recent developments prompted by the need to protect the internal environment from negative external factors. These tanks contain water that is constantly refreshed using a flow-through system. The water that is pumped into the containment is controlled up to the desired requirements of the internal environment. This may include, extracting water from greater depth, filtering, oxygenation or other water treatments. Water that flows out of the system may also be required to be treated before discharging. So far, floating closed containment tanks for fish farming have been deployed in benign waters. For use in exposed sites where there are stronger waves and currents, the closed containment tanks have to be designed to mitigate sloshing of the water in the tank to ensure the well-being of the fish. Current developments are presented below regardless whether they are used for nearshore or offshore fish farming.

(a)

(b)

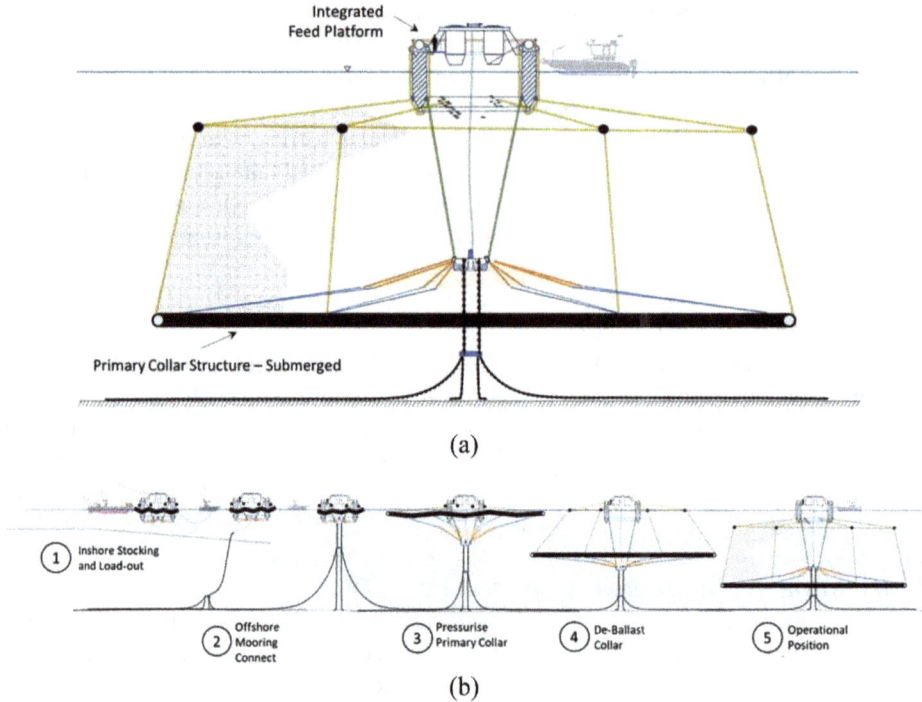

FIGURE 4.24 (a) Impact-9 submersible salmon pen, (b) diagram showing deployment of Impact-9 pen system (photo from: https://fishfarmingexpert.com/article/taking-the-plunge-a-submersible-cage-for-scotland/, https://thefishsite.com/articles/novel-offshore-fish-farm-edges-closer-to-commercial-reality).

Table 4.8 summarizes the floating closed containment tank systems in terms of pros and cons and their applications in offshore sites.

Figures 4.25 to 4.34 show some recent developments of the floating closed containment tanks for fish farming.

4.3.3.2.1 Fish Farm Egg

The fish farm egg concept (see Figure 4.25), developed by 'Hauge Aqua' uses a fully enclosed egg-shaped structure. The water flow enables the system to draw inlet water segregated from where outlet water is released. Water enters by the use of two main pumps that suck water from 20 m below the bottom of the structure. The water quality and volume can be controlled, ensuring steady oxygen levels. It is estimated to cost about NOK 600 million (about USD 60 million).

4.3.3.2.2 Neptun

The Neptun was developed by Aquafarm Equipment (Figure 4.26(a)). The tank has an internal diameter of 40 m, a circumference of 126 m, a depth of 22 m and a gross volume of 21,000 m³. The tank is designed against a wind speed of 30 m/s, and a current velocity of 1.0 m/s and its design life is 25 years. Figure 4.26(b) shows an internal view of the tank with inlet and outlet holes for water circulation. The tank is made from Glass Fiber Reinforced Polymer (GFRP) elements and reinforced with steel in areas that bear the most stress. The design also includes a pump system to extract large volumes of water from a depth of 25 m or more. As the concept of the containment tank is to collect dead fish, fish waste and uneaten fish feed from the sloped bottom, there is a flexible pipeline that connects the lowest point to the waste separator.

TABLE 4.8
Floating Closed Containment Tanks—Pros and Cons and Suitability for Application in Offshore Sites

Advantages	Disadvantages	Application for Offshore Sites
• has control over water replacement so that water can be constantly disinfected to remove pathogenic organisms, • external environmental events like algal blooms is no longer a problem, • organic waste can be removed by biofiltration system before discharging the water back to the sea, • threat of predators (such as sharks and seals) is completely eliminated, • achieve a higher production rate when compared to the open pen system.	• requires a power supply system to deploy in offshore sites, • may be too expensive to bring power from land for offshore fish farming, • requires significant construction and equipment costs, more management demands for monitoring and intervention, • detrimental sloshing effect to both structure and fish by the contained water.	• it is still unknown as to whether a closed containment tank can be deployed in offshore sites, • several challenging issues are raised such as sloshing, swirling and power resources.

FIGURE 4.25 Closed fish farm concept 'fish farm egg' (Photo from: http://sysla.no/fisk/skal-bruke-600-mill-pa-lukkedeoppdrettsegg/).

(a)

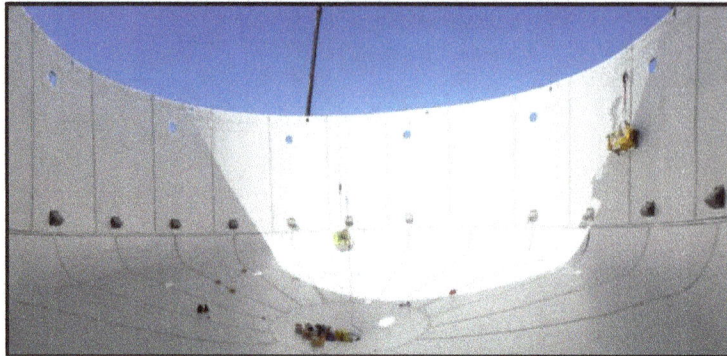

(b)

FIGURE 4.26 (a) Neptun closed containment fish tank, (b) internal view of Neptun semi-closed containment fish tank (photo from: http://aquafarm.no/closed-pen/).

4.3.3.2.3 Salmon Home 1

Dr. Techn. Olav Olsen, a Norway based marine technology consulting company, proposed a closed containment concrete tank, named Salmon Home 1, for offshore farming (see Figure 4.27). The cylindrical concrete tank has a 14.8 m inner diameter, 16.5 m outer diameter and 6 m height providing 1,000 m³ for salmon farming. It uses an existing mooring system such as a system of 2 × 4 or 2 × 8 nets in a rectangular array. It has a sloping bottom for easy collection of organic waste (Olsen, 2020). For a larger tank, of about 16,000 m³, in a water depth of 80 m and significant wave height of 2.5 m, it is proposed that the mooring system comprises of eight 48 mm chains.

4.3.3.2.4 Eco-Ark

AME2 Pte Ltd, a Singapore based company, has developed a closed-containment flow-through floating fish farm called the Eco-Ark as shown in Figure 4.28(a). It has four large tanks of 500 m³ and is about the size of an Olympic swimming pool. It can produce 166 tonnes of fish annually and enables sustainable farming out at sea in volumes that are 20 times more than average minimum production levels at traditional coastal farms. It has a flow-through water supply system. It has a roof equipped with solar panels to supply electricity for the fish farm (Leow and Tan, 2019).

The following environmental conditions were used for designing the Eco-Ark: wind speed 15 m/s, significant wave height 0.5 m, current velocity 1.2 m/s, and water depth 10 m. The mooring system

(a)

(b)

FIGURE 4.27   (a) Salmon Home 1, (b) Salmon Home 1 on site (Photo courtesy of Tor Ole Olsen).

adopted is a spud housing on the port and starboard of the hull. The spud housing allows for self-installation and removal for quick mobilization and demobilisation when required. A total of six spuds with a diameter of 762 mm and length of 25 m are installed at the site in order to keep the Eco-Ark in position. The total weight of Eco-Ark is 5300 tonnes.

The Eco-Ark was constructed in Batam Island, Indonesia and was deployed in the northern coast of Singapore near Pulau Ubin in September 2019. Eco-Ark was designed, constructed and surveyed according to the rules of the classification society, Bureau Veritas as a special service floating fish farm and is fully insured for Hull and Machinery (H&M) and also has third party liability insurance for up to $500 million.

The Eco-Ark allows augmentation and integration by forming a fleet connected to a lift-dock facility that enables the cultivation and processing of fish on site (see Figure 4.28 (b)).

4.3.3.2.5   Marine Donut

The Norwegian salmon farmer, Marine Harvest ASA now known as Mowi ASA, developed a closed containment design, named the Marine Donut. The concept is owned by ODP Donut Solutions (see Figure 4.29). The Marine Donut is a tube made out of HDPE with a volume of 20,000 m^3 and has a capacity of approximately 1000 tonnes of biomass. The design should be able to withstand

(a)

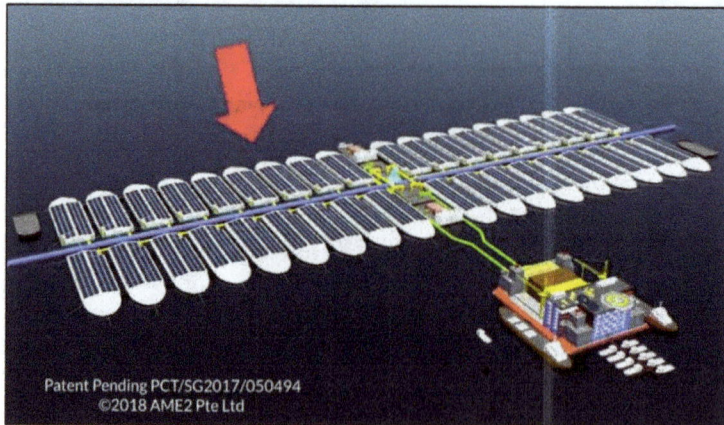

Patent Pending PCT/SG2017/050494
©2018 AME2 Pte Ltd

(b)

FIGURE 4.28 (a) Eco-Ark closed containment system, (b) Eco-Ark fleet connected to lift-dock (photo courtesy of Mr. Ban Tat Leow).

a significant wave height of 3 m. Water is taken in via six inlet pipes that reach below the sea lice barrier and water is continuously circulated inside the donut. In 2019, Norway's Directorate of Fisheries granted permission for 1,100 tonnes of biomass to be used to test the design.

4.3.3.2.6 Stadion Laks

Stadion Laks, is a bathtub-shaped floating aquaculture system made out of reinforced concrete that holds 34,000 m³ of water. It is designed for a stock density of 50–75 kg/m³ and should be able to accommodate smolts and post-smolts up to harvest sized 4–6 kg fish. Water is pumped from below the sea lice barrier, from at least 20 m depth, and is circulated. Various installations and management systems are in place to secure water quality to achieve the best health for the fish (see Figure 4.30). It is looking forward to commencing the construction in 2022 and its operation in 2023.

4.3.3.2.7 Preline

Preline Fish farming Systems AS has developed a concept that is closely related to a raceway system as described by Tidwell (2012). Water is not circulated but runs though the containment. It is a 40 m

FIGURE 4.29 Marine donut—Close containment concept design of Marine harvest (Photo from: http:// marin.bergen-chamber.no/en/teknologi/Growth-through-innovation/).

FIGURE 4.30 Stadion Laks concept (Photo from: https://stadionlaks.no/en/home/).

long oval HDPE tube with a volume of 2000 m³ that is suspended at the water surface. It is designed to hold 100,000–200,000 salmon of up to 1 kg in weight. Slots in the tube that extend above the free surface allow the fish to reach for air. At the bottom, waste collection traps are installed. On the two far ends, it holds two tubes that reach down below the sea lice barrier. Figure 4.31 shows the artist impression of the preline. The right-hand tube is the inlet and the other the outlet. Pumps ensure a water velocity of roughly 0.15 m/s through the system. Fish are kept in the horizontal part of the system and internal net boundaries are put in place to ensure fish do not escape via the inlet or outlet.

4.3.3.2.8 FishGLOBE

The FishGLOBE fish pen has been developed to reduce the required sea space, minimize production costs and have minimal negative impact on the environment. When viewed from the water surface, the FishGLOBE appears as a floating iceberg (see Figure 4.32). The latest model is the FiskGLOBE K10 with a total volume of 30,000 m³ and a biomass capacity of 2000 tonnes. The structure has a total height of 35 m with a cylindrical shape. The inlet and outlet pipes together with the central pipe are a part of the load bearing structure by increasing its stiffness. The FishGLOBE can withstand waves with a significant wave height of up to 2.5 m and currents of up to 1.0 m/s. Oxygen and large quantities of water are supplied by pumping from below the structure where it is free from sea lice.

FIGURE 4.31　Artist impression of the Preline (Photo from: http://preline.no/).

FIGURE 4.32　Fish globe deployed in Norway (Photo from: https://fishglobe.no/).

The interior can be airtight. By increasing internal air pressure, water can be pressed out to gently transport the fish out of the tank. An internal current can be introduced to converge the waste as shown in a study by Gorle et al. (2018) through CFD simulations.

4.3.3.2.9　Eco Cage

Serge Ferrari, a France based company, has developed a composite fabric that is suitable for flexible bag-type closed containment aquaculture systems and is called Biobrane Aqua. They have a range available to suit different purposes and environmental loads. The Biobrane Aqua 2050 is applied in the Eco Cage, which is produced by EcoMerden AS in Norway (see Figure 4.33). The construction consists of three main parts: (1) a steel circular shaped collar ring which holds the heavy-duty flexible wall; (2) the flexible wall creating an enclosed internal environment; and (3) an internal net inside the fabric where the fish reside.

4.3.3.2.10　FiiZK

Another flexible closed-containment aquaculture system that uses the Biobrane Aqua 2050 is a concept developed by the Norwegian-based company FiiZK AS. They deliver a flexible closed-containment solution for the aqua culture industry. Designs in different sizes are possible (see Figure 4.34). Water is drawn from greater depths by extended pipes reaching below the sea lice barrier. Before the water is pumped into the containment area it is oxygenated. Waste is collected at the bottom of the pen, from where it is removed and processed externally.

FIGURE 4.33 Eco Cage – a flexible type closed containment aqua culture systems (Photo from: https://ecomerden.com/).

FIGURE 4.34 Flexible closed containment aqua culture systems designed by FiiZK (Photo: https://fiizk.com/en/product/closed-cage/).

4.3.4 KEY OBSERVATIONS WITH REGARD TO RECENT OFFSHORE FISH PEN DEVELOPMENTS

- Most offshore fish pen designs are at the conceptual stage. There are, however, a few real scale proto-type designs which have been built (examples are, Ocean Farm 1, Shenlan 1, and Havfarm 1).
- To date, the operation of offshore fish farms is still in its infancy. For example, Ocean Farm 1 and Shenlan 1 started operation only a few years ago!
- Instead of an array of small pens in nearshore fish farms, offshore fish farms tend to have a single 'mega' fish pen that can accommodate more than a million fish.
- Instead of flexible pens and fabric nets, offshore fish pens involve the use of rigid frames and stiffer nets (such as PET or metal) for robustness.

- A submerged pen is a good solution for offshore fish farms as it avoids strong surface waves, and its structural strength does not need to be as great as surface structures.
- Offshore fish pens are commonly equipped with remote and autonomous devices for operation, maintenance, monitoring and surveillance (feeding, ventilation, lighting, cleaning, and the removal of waste).
- Offshore fish pens that are sited at a considerable distance from the shoreline will have to tap on wind, wave, and solar energy for their power supplies.
- The existing submerged pens and closed containment tank designs are small in size.
- Closed containment tank designs have only been deployed in nearshore sites.

4.4   INTEGRATION OR CO-LOCATION OF OFFSHORE FISH FARM WITH OTHER MARINE SECTORS

Offshore fish farming would have to generate high economic value in a short period for a quick return of investment. Therefore, additional functionalities of the fish pens are necessary. These include integrating or co-locating the offshore fish pens with renewable energy production, or involving cruise tourism, maritime transport and leisure and entertainment activities that can bring added income streams.

The renewable energy sector is actively seeking offshore sites for capturing strong and sustainable natural energy resources and expecting less societal impact from inconspicuous offshore locations than from nearshore locations. In general, offshore renewable energy refers to the generation of electricity from ocean-based resources including winds, waves, tides, and salinity and thermal properties as well as the conversion of generated electricity to hydrogen (Wiersma and Devine-Wright, 2014). In order to succeed in the offshore renewable energy business, it is necessary to suggest a stable but cost-effective floating substructure and mooring system that should be robust enough to withstand the highly energetic environments and save capital and operating costs including the cost of power delivery from the long-distant sites to the end users.

There is a great potential for collaboration between the aquaculture and renewable energy sectors as they share a common challenge to explore more energetic and exposed localities to sustain production growth (Weiss et al. 2018). By having an integrated and/or co-located solution for offshore fish farming and renewable energy harvest, not only can the fish farms be remotely and autonomously operated via clean energy, but the renewable energy partners can also save capital costs by sharing the substructure and mooring system and reduce power delivery cost through direct power supply to the end users (namely, fish farms). Both industry partners can leverage their profit significantly and synergetic benefits are expected to attract other marine sectors by suggesting better utilization of sea space, and reduction of service and maintenance costs by sharing labour, transportation, monitoring and operational control (Sakellariadou and Kostopoulou, 2015).

CAPEX (Capital Expenditures) and OPEX (Operational Expenditure) for offshore fish farms are certainly much higher than for nearshore fish farms. Co-location or integration of offshore fish farm with other marine sectors will allow the sharing of maintenance and support ships (for example, utility vessels, work boats, general cargo ships and barges) that are required to operate multiple activities (Holm et al. 2017; Kaiser et al. 2011). Moreover, offshore renewable energy facilities can provide freshwater via a desalination process, and hydrogen power and oxygen for fish pens via water splitting (Papandroulakis et al. 2017; Wang et al. 2019). The floating platforms will be able to accommodate fish feed silos and equipment so that transportation costs and the numbers of service vessels for fish feed delivery can be reduced.

Below are some examples of an integrated offshore fish farm and renewable energy production facility.

4.4.1   PLOCAN

The Oceanic Platform of the Canary Islands (PLOCAN) as shown in Figure 4.35 is sited in water depths ranging from 40 m to 200 m and 1.5 km from the coast of the island of Gran Canaria, Spain.

FIGURE 4.35 Oceanic platform of the Canary Islands (PLOCAN) (photo from: https://steemit.com/steems tem/@geronimo14/plocan-a-boost-to-the-blue-economy).

FIGURE 4.36 Blue Growth Farm project's Multipurpose floating platform (photo courtesy from Prof Maurizio Collu of Strathclude University).

The PLOCAN comprises multidisciplinary laboratories for analyzing bio-geo-chemical variables in the water column. It has a main deck area with an office building and a helideck. It carries equipment for loading and unloading the material for experimental testing, and other general-purpose facilities such as workshops, cranes and other equipment providing basic operational support. This facility also aims to encourage development of offshore aquaculture, offshore wind turbines and other marine structures.

4.4.2 BLUE GROWTH FARM

The Blue Growth Farm project is EU's ambitious project to produce advanced industrial knowledge in a fully integrated and efficient offshore multipurpose floating platform. This platform provides a central protected pool to farm fish, as well as large storage and deck areas to host a commercial 10 MW wind turbine and a number of wave energy converters (WEC). Figure 4.36 shows a 1/15 scale of the Blue Growth Farm deployed at a site near the port of Reggio Calabris in Italy.

4.4.3 GIEC'S SEMI-SUBMERSIBLE WAVE POWERED AQUACULTURE PEN

An integrated offshore renewable energy facility, water desalination plant and fish farms has been implemented by the Guangzhou Institute of Energy Conversion (GIEC). Figure 4.37 shows GIEC's semi-submersible wave powered aquaculture pen with seawater desalination plant on board and solar panel roof. This multipurpose open sea aquaculture platform is named, the Penghu platform.

4.4.4 HEX BOX

Ocean Aquafarms developed a new concept of offshore salmon farm named the Hex Box as shown in Figure 4.38. The concept can operate at sites with a significant wave height of up to 10 m, a water depth greater than 100 m and a wind speed of up to 100 knots, thereby allowing salmon farming in areas that are inaccessible with today's technology. The Hex Box uses a ballasting system to be able to change the draft from 4 m to 30 m for inspection and replacement. It has a 275 m circumference providing a net submerged volume of 430,000 m³. The net bag can be suspended in winches a further 20 m below the frame structure. For Australia and New Zealand, the pen can be armed with a full

FIGURE 4.37   Penghu open sea aquaculture platform (photo courtesy of Mr. Ban Tat Leow).

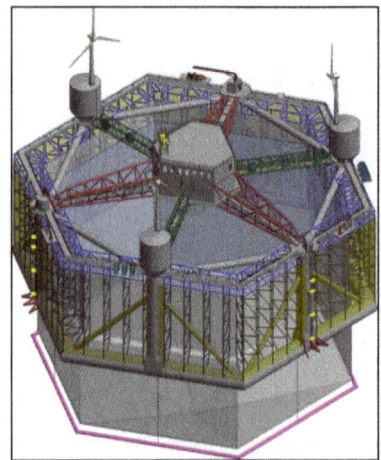

FIGURE 4.38   Hex box offshore salmon farm by ocean aquafarms (photo from: http://oceanaquafarms.com/product/hex-box-norway-2/).

FIGURE 4.39   Illustration of concept of FOWT-SFFC (Zheng and Lei, 2018).

double net against predators. It is equipped with two or three deck cranes for operation. In addition to two diesel generators for power supply, the Hex Box carries three wind turbines (3x100 kW) with batteries to reduce usage of hydrocarbon fuel. The mooring system comprises of 6 to 9 mooring ropes with fixed anchor points in the seabed. The overall mooring system, including ropes, has been demonstrated to meet Norwegian requirements by over three times. A scaled model has been tested and showed promising results (SalmonBusiness, 2020). The construction cost of the Hex Box is approximately USD70 to 90 million.

4.4.5   FOWT-SFFC

Zheng and Lei (2018) presented an integrated design of a Floating Offshore Wind Turbine and a conical Steel Fish-Farming Pen (FOWT-SFFC) as shown in Figure 4.39. The integrated pen design can generate multi-megawatt power and encloses a 200,000 m³ volume of seawater for farming fish. The inner space of the pen can be subdivided into eight sectors to raise a variety of fish. To balance the gravity with buoyancy, high density concrete is placed in the radial and ring pontoons for ballast. The bottom net is attached to lifting devices inside the pen so that can be moved vertically from the bottom to the surface of the water, for harvesting. Nets are made of copper alloy to resist seawater corrosion and biofouling.

4.4.6   COSPAR

Chu and Wang (2019) proposed a combined design of a spar platform and a fish pen with a partially porous collar barrier, named, the COSPAR fish pen (see Figure 4.40). The pen design features an octagonal shape with a partially porous collar barrier to attenuate wave energy for a calmer water environment inside the pen. The pen has a diameter of 80 m, a height of 39 m and encloses a water volume of about 180,000 m³. The deep draught spar is 82 m in height and is made from concrete for its bottom half and from steel for its top half. The spar carries a wind turbine and a control unit. The pen is connected by four truss girders (above water) and 16 girders at the base of the spar so that both pen and spar work as a monolithic rigid body. The four top girders form walkways to access the control unit and the wind turbine. For mooring, four catenary chains are attached to the spar 38 m under the water surface (outside the fish pen) so as to mitigate the tension force in the mooring lines and to reduce the benthic footprint.

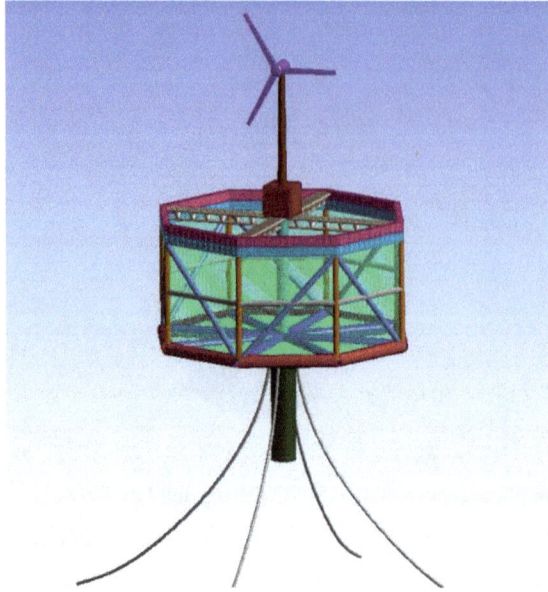

FIGURE 4.40 COSPAR fish pen design (Chu and Wang, 2020).

FIGURE 4.41 Genghai No.1 (Photo from: https://swissre.com/reinsurance/property-and-casualty/reinsura nce/marine/offshore-fish-farming-facilities-challenges-marine-insurers.html).

4.4.7 Genghai No.1

Another interesting design combines aquaculture production with leisure/entertainment activities. Swimming pools, scuba diving facilities and hotels can be placed onto offshore fish farming platforms. Figure 4.41 shows Genghai No.1, which is an aquaculture farm, ocean monitoring centre, as well as a leisure centre. It has an aquaculture volume of $27,000\,m^3$ equivalent to 14 standard swimming pools, and it can accommodate 300 visitors at any given time.

4.5 CONCLUSION

In summary, there is an increasing interest in the fish farming industry to move offshore for sustainable farming, larger sea space and higher fish production. However, offshore operation generally requires higher capital and production costs (Jansen et al. 2016) and therefore rigorous research and development must be carried out urgently to seek cost effective solutions. Also, fish pen designs should consider the health of the fish, fish diseases, exposure to toxicity, fish growth, harvesting of fish, transportation to the market, and environmental issues. Feasibility in offshore fish farming may be achieved through the adoption of new developments of multi-functional, modularity for ease of scaling the farm sizes and an autonomous infrastructure that has been validated by the oil and offshore industry (Dalton et al. 2019; Grinham et al. 2020). By co-locating offshore renewable energy systems (wind turbines, wave energy converters) and floating platforms (that can accommodate fish feed silos, feeding equipment, harvesting cranes and nets, fish processing and packaging plants, waste treatment plants, desalination plants) with offshore fish farms, it is possible to leverage the benefits of collocation, vertical integration and shared services and to reduce operating times and costs (Chen et al. 2020). Also, the use of offshore renewable energy helps to decarbonize the fish farming industry.

Nevertheless, a combination of fish farming with other marine activities is desirable from an economic point of view. The overall infrastructure and operational procedure will no doubt be more complex, and the increased functionalities will bring more risk and require more rigorous assessments for warrants and insurance coverage than would be the case with solely fish farming activities. More research and development are needed in this area.

ACKNOWLEDGEMENT

The authors acknowledge the financial support of the Blue Economy Cooperative Research Centre, established and supported under the Australian Government's Cooperative Research Centers Programme.

REFERENCES

Beveridge, M. C. M. 2008. *Cage Aquaculture, Aquaculture*. John Wiley and Sons, Hoboken, New Jersey, US.

Bjelland, H. Fore, V. Lader, M. Kristiansen, P. Holmen, D. Fredheim, I. M. Grotli, A. Fathi, E. I. Oppedal, V. F. Utne, I. B. and Schjolberg, I. 2016. *Exposed Aquaculture in Norway*. Ocean 2015—MTS/IEEE Washington. 1–10.

Buck, B. H. 2007. Farming in a high energy environment: potentials and constraints of sustainable offshore aquaculture in the German Bight (North Sea). *ber. Polarforsch. Meeresforsch* 543, 238.

Cabello, L. 2000. Production methods for offshore fish management. *FAO Tech. Rep.* 191–202.

Cardia, F. and Lovatelli, A. 2016. Aquaculture operations in floating HDPE cages. *Fish. Aquac.* 593, 176.

CEA, 2018. Offshore Finfish Aquaculture [WWW Document]. https:// eaconsulting.com/wp.../CEA-Offshore-Aquaculture-Report-2018.pdf (accessed 7.26.2019).

Chacon-Torres, A. Ross, L. G. and Beveridge, M. C. M. 1988. The effects of fish behaviour on dye dispersion and water exchange in small net cages. *Aquaculture* 73, 283–293.

Chen, P. Cahoon, S. Bhaskar, P. Abdussamie, N. Adams, L. Fernado, I. Lee, K. H. Wu, Y. Gunarathne, N. Balk, D. Shepherd, T. Dutton, I. Thornton, S. Corden-McKinley, B. and McGookin, B. 2020. *Logistics Challenges to Offshore/High Energy Co-location of Aquaculture and Energy Industries, P.5.20.003.* Launceston, Tasmania, Australia.

Chu, Y. I. and Wang, C. M. 2019. Combined spar and partially porous wall fish cage for offshore Site. *16th East Asian-Pacific Conf. Struct. Eng. Constr.* 569–581.

Chu, Y.I. and Wang, C. M. 2020. Design development of porous collar barrier for offshore floating fish cage against wave action, debris and predators. *Aquac. Eng.* 92, 102137.

Chu, Y. I. Wang, C. M. Park, J. C. and Lader, P. F. 2020. Review of cage and containment tank designs for offshore fish farming. *Aquaculture* 519, 734928.

Colbourne, D. B. 2005. Another perspective on challenges in open ocean aquaculture development. *IEEE J. Ocean. Eng.* 30, 4–11.

Croix, R.D. La, 2008. Preventing damage to fish by seals and sea lions. *Hydroacoustics Inc*, 23.

Dai, J. Wang, C. M. Utsunomiya, T. and Duan, W. 2018. Review of recent research and developments on floating breakwaters. *Ocean Eng.* 158, 132–151.

Dalton, G. Bardócz, T. Blanch, M. Campbell, D. Johnson, K. Lawrence, G. Lilas, T. Friis-Madsen, E. Neumann, F. Nikitas, N. Ortega, S.T. Pletsas, D. Simal, P.D. Sørensen, H.C. Stefanakou, A. and Masters, I. 2019. Feasibility of investment in Blue Growth multiple-use of space and multi-use platform projects; results of a novel assessment approach and case studies. *Renew. Sustain. Energy Rev.* 107, 338–359.

DNV, 2021. *Oceans' Future to 2050, Marine Aquaculutre Forecast*. Hovik, Norway.

Drumm, A. 2010. *Evaluation of the Promotion of Offshore Aquaculture Through a Technology Platform (OATP)*, 46.

Ellingsen, H. and Aanondsen, S .A. 2006. Environmental impacts of wild caught cod and farmed salmon—A comparison with chicken. *Int. J. Life Cycle Assess.* 11, 60–65.

Faltinsen, O. M. 2015. Hydrodynamics of marine and offshore structures. *J. Hydrodyn.* 26, 835–847.

FAO, 2020. *The State of World Fisheries and Aquaculture 2020, Food and Agriculture Org. of the United Nations*. Rome.

Forster, J. 2013. A review of opportunities, technical constraints, and future needs of offshore mariculture— Temperate waters, in: *Technical Workshop Proceedings: Expanding Mariculture Farther Offshore: Technical, Environmental*, Spatial, and Governance Challenges, 77–100.

Gorle, J. M. R. Terjesen, B. F. Holan, A. B. Berge, A. and Summerfelt, S. T. 2018. Qualifying the design of a floating closed-containment fish farm using computational fluid dynamics. *Biosyst. Eng.* 175, 63–81.

Gowen, R. J. and Edwards, A. 1990. The interaction between physical and biological processes in coastal and offshore fish farming: an overview. *Eng. Offshore Fish Farming*, 39–47.

Grinham, A. Marouchos, A. Martini, A. Fisher, A. Seet, B. C. Guihen, D. Williams, G. Symonds, J. Ross, J. Soutar, J. Heasman, K. Huang, L. Lea, M. A. Leary, M. Sikka, P. King, P. Cossu, R. Adams, S. Arachchillage, S. J. Albert, S. Cahoon, S. Bird, S. Connolly, R. Edwards, S. and Bannister, R. 2020. *Autonomous Marine Systems at Offshore Aquaculture and Energy Sites, P.1.20.002—Final Project Report*. Launceston, Tasmania, Australia.

Holm, P. Buck, B. H. and Langan, R. 2017. Introduction: New approaches to sustainable offshore food production and the development of offshore platforms. *Aquaculture Perspective of Multi-Use Sites in the Open Ocean Untapped Potential for Marine Resources in the Anthropocene.* 1–20.

Holmer, M. 2009. Environmental issues of fish farming in offshore waters: perspectives, concerns and research needs. *Aquac. Environ. Interact.* 1, 57–70.

Huang, C. C. Tang, H. J. and Liu, J. Y. 2008. Effects of waves and currents on gravity-type cages in the open sea. *Aquac. Eng.* 38, 105–116.

Huguenin, J. E. 1997. The design, operations and economics of cage culture systems. *Aquac. Eng.* 16, 167–203.

Hvas, M. and Oppedal, F. 2019. Influence of experimental set-up and methodology for measurements of metabolic rates and critical swimming speed in Atlantic salmon Salmo salar. *J. Fish Biol.* 95, 893–902.

Hvas, M. Folkedal, O. Solstorm, D. Vågseth, T. Fosse, J. O. Gansel, L. C. and Oppedal, F. 2017. Assessing swimming capacity and schooling behaviour in farmed Atlantic salmon Salmo salar with experimental push-cages. *Aquaculture* 473, 423–429. https://doi.org/10.1016/j.aquaculture.2017.03.013

Jansen, H. M. Van Den Burg, S. Bolman, B. Jak, R. G. Kamermans, P. Poelman, M. and Stuiver, M. 2016. The feasibility of offshore aquaculture and its potential for multi-use in the North Sea. *Aquac. Int.* 24, 735–756.

Jeng, D. S. 2018. *Mechanics of Wave-Seabed-Structure Interactions*. Cambridge University Press, Cambridge, UK.

Jensen, Ø. Dempster, T. Eb, T. Uglem, I. and Fredheim, A. 2010. Escapes of fish from Norwegian sea-cage aquaculture: causes, consequences and methods to prevent escape. *Aquac. Environ. Interact.* 12, 71–83.

Jin, D. 2008. Economic models of potential U.S. offshore aquaculture operations, in: *Offshore Aquaculture United States: Economic Considerations, Implications and Opportunities*, 117–140.

Kaiser, M. J. Snyder, B. and Yu, Y. 2011. A review of the feasibility, costs, and benefits of platform-based open ocean aquaculture in the Gulf of Mexico. *Ocean Coast. Manag.* 54, 721–730.

Kankainen, M. and Mikalsen, R. 2014. Offshore fish farm investment and competitiveness in the Baltic sea. *Reports of Aquabest Project*, 2.

Kapetsky, J. M. Aguilar-Manjarrez, J. and Jenness, J. 2013. A global assessment of offshore mariculture potential from a spatial perspective. *FAO Fisheries and Aquaculture Techical Paper,* 181.

Kato, J. Noma, T. and Uekita, Y. 1979. Design of floating breakwaters. *Advanced in Aquaculture.* Fishing News Books, Oxford.

Klebert, P. Patursson, Ø. Endresen, P. C. Rundtop, P. Birkevold, J. and Rasmussen, H. W. 2015. Three-dimensional deformation of a large circular flexible sea cage in high currents: field experiment and modeling. *Ocean Eng.* 104, 511–520.

Le François, N. R. Jobling, M. Carter, C. Blier, P. U. and Savoie, A. 2010. Finfish aquaculture diversification. *Finfish Aquaculture Diversification* 1–681.

Leow, B. T. and Tan, H. K. 2019. Technology-driven sustainable aquaculture for eco-tourism, in: *WCFS2019,* 209–218. Springer, Singapore. https://doi.org/10.1007/978-981-13-8743-2_11

Li, L. Jiang, Z. and Ong, M. C. 2017. A preliminary study of a vessel-shaped offshore fish farm concept. *Int. Conf. Offshore Mech. Arct. Eng.* 57724, V006T05A006.

Lindeboom, R. 2018. Future proofing fish farming. MARIN's news magazine for the maritime industry [WWW Document]. https://content.yudu.com/web/1r3p1/0A3a046/MRN125/html/index.html?origin=reader (accessed 1.7.2019).

Liu, S. Bi, C. Yang, H. Huang, L. Liang, Z. and Zhao, Y. 2019. Experimental study on the hydrodynamic characteristics of a submersible fish cage at various depths in waves. *J. Ocean Univ. China* 18, 701–709.

Lovell, T. 1989. *Nutrition and Feeding of Fish, Aquaculture.* Van Nostrand Reinhold, New York.

Mapes, L. 2017. State approves 1 million more farmed fish for Puget Sound, despite escape [WWW Document]. https://seattletimes.com/seattle-news/environment/state-approves-1-million-more-farmed-fish-for-puget-sound-despite-escape/ (accessed 2.1.2020).

Matsunaga, N. Hashida, M. Uzaki, K. I. Kanzaki, T. and Uragami, Y. 2002. Performance of Wave Absorption by a Steel Floating Breakwater with Truss Structure. *Proc. Twelfth Int. Offshore Polar Eng. Conf.* 12, 768–772.

McCartney, B. L. 1985. Floating breakwater design. *J. Wterway, Port, Costal, Ocean Enginneering* 111, 304–318.

Mobron, E. Torgersen, T. Zhu, S. Riis, J. and Bye, M. 2020. Design of Havfarm 1. *Proc. Pave Waves* 26, 503–516.

Moe-Føre, H. Lader, P. F. Lien, E. and Hopperstad, O. S. 2016. Structural response of high solidity net cage models in uniform flow. *J. Fluids Struct.* 65, 180–195.

Morro, B. Planellas, S. R. Davidson, K. Adams, T. P. Falconer, L. Holloway, M. Dale, A. Aleynik, D. Thies, P. R. Khalid, F. Hardwick, J. Smith, H. Gillibrand, P. A. and Rey-Planellas, S. 2021. Offshore aquaculture of finfish: big expectations at sea. *Rev. Aquac.* 14, 791-815.

Noroi, G. Á. Glud, R. N. Gaard, E. and Simonsen, K. 2011. Environmental impacts of coastal fish farming: Carbon and nitrogen budgets for trout farming in kaldbaksfjørour (Faroe Islands). *Mar. Ecol. Prog. Ser.* 431, 223–241.

NS 9415, 2009. *Marine Fish Farms Requirements for Design, Dimensioning, Production, Installation and Operation.* Norwegian Standard.

Oldham, T. Nowak, B. Hvas, M. and Oppedal, F. 2019. Metabolic and functional impacts of hypoxia vary with size in Atlantic salmon. *Comp. Biochem. Physiol.—Part A Mol. Integr. Physiol.* 231, 30–38.

Olsen, T. O. 2020. Fish Farming in Floating Structures, in: *Proceedings of the World Conference on Floating Solutions,* Lecture Notes in Civil Engineering, 191–208.

Pandey, B. 2018. *Pellet Technical Quality of Feeds for Atlantic Salmon.* Master's thesis, Norwegian University of Life Sciences.

Papandroulakis, N. Thomsen, C. Mintenbeck, K. Mayorga, P. and Hernandez-Brito, J. J. 2017. The EU-Project 'TROPOS,' *Aquaculture Perspective of Multi-Use Sites in the Open Ocean.*

Pillay, T. V. R. 2004. *Aquaculture and the Environment, 2nd ed, Freshwater Fisheries Ecology.* Blackwell Publishing Ltd. Oxford.

Remen, M. Solstorm, F. Bui, S. Klebert, P. Vågseth, T. Solstorm, D. Hvas, M. and Oppedal, F. 2016. Critical swimming speed in groups of Atlantic salmon Salmo salar. *Aquac. Environ. Interact.* 8, 659–664.

Rosenthal, H. Costa-Pierce, B. Krause, G. and Buck, B. H. 2012. *Bremerhaven Declaration on the Future of Global Open Ocean Aquaculture-Part II: Recommendations on Subject Areas and Justifications.*

Sakellariadou, F. and Kostopoulou, E. 2015. Marine Ecotourism from the Perspective of Blue Growth. *7th iConEc Conf. Compet. Stab. Knowledge-based Econ.* 1–27.

Salmon Business, 2020. 'Hex Box' Offshore salmon farm: Today, we are in dialogue with several parties who are interested in the concept [WWW Document]. https://salmonbusiness.com/hex-box-offshore-salmon-farm-today-we-are-in-dialogue-with-several-parties-who-are-interested-in-the-concept/ (accessed 4.24.2020).

Shainee, M. Ellingsen, H. Leira, B. J. and Fredheim, A. 2013. Design theory in offshore fish cage designing. *Aquaculture 392–395*, 134–141.

Solstorm, F. Solstorm, D. Oppedal, F. Fernö, A. Fraser, T. W. K. and Olsen, R. E. 2015. Fast water currents reduce production performance of post-smolt Atlantic salmon Salmo salar. *Aquac. Environ. Interact.* 7, 125–134.

Stickney, R. R. 2002. Impacts of cage and net-pen culture on water quality and benthic communities. *Aquac. Environ.* United States 105–118.

Stien, L. H. Bracke, M. B. M. Folkedal, O. Nilsson, J. Oppedal, F. Torgersen, T. Kittilsen, S. Midtlyng, P. J. Vindas, M. A. Øverli, Ø. and Kristiansen, T. S. 2013. Salmon Welfare Index Model (SWIM 1.0): a semantic model for overall welfare assessment of caged Atlantic salmon: review of the selected welfare indicators and model presentation. *Rev. Aquac.* 5, 33–57.

Taranger, G. L. Karlsen, Ø. Bannister, R. J. Glover, K. A. Husa, V. Karlsbakk, E. Kvamme, B. O. Boxaspen, K. K. Bjørn, P. A. Finstad, B. Madhun, A. S. Morton, H. C. and Svåsand, T. 2015. Risk assessment of the environmental impact of Norwegian Atlantic salmon farming. *ICES J. Mar. Sci.* 72, 997–1021.

Tidwell, J. H. 2012. *Aquaculture Production Systems, Wiley-Blackwell.* Wiley-Blackwell, Oxford, UK.

Verhoeven, J. T. P. Salvo, F. Knight, R. Hamoutene, D. and Dufour, S. C. 2018. Temporal bacterial surveillance of salmon aquaculture sites indicates a long lasting benthic impact with minimal recovery. *Front. Microbiol.* 9, 3054.

Wamsley, T. V. Cialone, M. A. Smith, J. M. Atkinson, J. H. and Rosati, J. D. 2010. The potential of wetlands in reducing storm surge. *Ocean Eng.* 37, 59–68. https://doi.org/10.1016/j.oceaneng.2009.07.018

Wang, C. M. Chu, Y. I. and Park, J. C. 2019. Moving offshore for fish farming. *J. Aquac. Mar. Biol.* 8, 38–39.

Wang, L. Z. Shen, K. M. Li, L. L. and Guo, Z. 2014. Integrated analysis of drag embedment anchor installation. *Ocean Eng.* 88, 149–163.

Weiss, C. V. C. Ondiviela, B. Guinda, X. del Jesus, F. González, J. Guanche, R. and Juanes, J. A. 2018. Co-location opportunities for renewable energies and aquaculture facilities in the Canary Archipelago. *Ocean Coast. Manag.* 166, 62–71.

Wiersma, B. and Devine-Wright, P. 2014. Public engagement with offshore renewable energy: a critical review. *Wiley Interdiscip. Rev. Clim. Chang.* 5, 493–507.

Yuen, J. W. Dempster, T. Oppedal, F. and Hvas, M. 2019. Physiological performance of ballan wrasse (Labrus bergylta) at different temperatures and its implication for cleaner fish usage in salmon aquaculture. *Biol. Control* 135, 117–123.

Zhao, Y. Guan, C. Bi, C. Liu, H. and Cui, Y. 2019. Experimental investigations on hydrodynamic responses of a semi-submersible offshore fish farm in waves. *J. Mar. Sci. Eng.* 7, 238.

Zheng, X. Y. and Lei, Y. 2018. Stochastic response analysis for a floating offshore wind turbine integrated with a steel fish farming cage. *Appl. Sci.* 8, 1229.

5 Risk Finance for Natural Disaster in Lakes and Coastal Seas Using Modeling Techniques

Jinxin Zhou,[1] Kentaro Kikuchi,[2] Hideya Kubo,[3] Takero Yoshida,[4] Md. Nazrul Islam,[5] and Daisuke Kitazawa[1]*
[1] Institute of Industrial Science, The University of Tokyo, Japan
[2] Faculty of Economics, Shiga University, Japan
[3] The Organising Committee of the World Masters Games Kansai, Japan
[4] Department of Ocean Sciences, Tokyo University of Marine Science and Technology, Japan
[5] Department of Geography and Environment, Jahangirnagar University, Savar, Dhaka, Bangladesh
* Corresponding author: Daisuke Kitazawa
Email: jxzhou@iis.u-tokyo.ac.jp; kentaro-kikuchi@biwako.shiga-u.ac.jp; hkubo2019@rikkyo.ac.jp; tyoshi3@kaiyodai.ac.jp; nazrul_geo@juniv.edu.bd; dkita@iis.u-tokyo.ac.jp

CONTENTS

DOI: 10.1201/9781003184287-5

5.1 INTRODUCTION

Humans, from a global perspective, first survived on freshwater resources, for example, lakes (Bouch and Jones 1961), and the growth in population and technology has driven the exploitation of marine resources (Erlandson 2008). Nowadays, both lakes and seas are subject to equally high expectations. Reynaud and Lanzanova (2017) initially conducted a meta-analysis with a worldwide dataset of 699 records from 133 studies and estimated the average economic value provided by lakes to be over US$100 per respondent per year and a maximum of US$403 per property per year for hedonic price studies. On the other hand, reports from the Organisation for Economic Cooperation and Development (OECD), an organization at the heart of international cooperation, estimated the total value brought by ocean economic activities to be worth US$1.5 trillion per annum and claimed that the annual blue growth would outpace the growth in terrestrial activities for the next few decades (OECD 2016; OECD 2019). However, those expectations are challenged by continuously emerging uncertainties, such as eutrophication and global warming (lakes: Kumagai et al. 2003; coastal seas: Rabalais et al. 2009). Taking global warming as an example, according to the report by the Intergovernmental Panel on Climate Change (Pörtner et al. 2019), it has been estimated that climate-induced ocean degradation will cost the global economy US$428 billion per annum by 2050 and US$1.98 trillion per annum by 2100.

Concerns about the undesirable effects on aquatic environment are growing around the world and consequently, countermeasures have been taken. The foremost thing is to monitor and assess water quality or, indirectly, meteorological data in terms of climate change. For example, the Global Environment Monitoring System (GEMS), initiated in 1972 after the United Nations Stockholm Conference on the Environment, aims to conduct climate-related monitoring and ocean monitoring throughout the world (Gwynne 1982). Five years later in 1977, the GEMS/Water Programme was inaugurated, especially aiming at freshwater monitoring and already covering 103 countries in 2001 (Robarts et al. 2003). The second approach is to restore the affected aquatic environment. Taking the global restoration of eutrophic lakes as an example, Jeppesen et al. (2003) summarized the efforts worldwide to combat eutrophication by 1) reducing the phosphorus input, 2) removing planktivorous and benthivorous fish but stocking piscivorous fish, 3) protecting or planting submerged macrophytes, and 4) introducing artificial structures. However, both monitoring and restoring programs are capital-intensive and need bilateral and multilateral cooperation. For example, the above-mentioned GEMS/Water Programme was led by the United Nations Environment Programme and the World Health Organization, and assisted by the United Nations Educational, Scientific and Cultural Organization and the World Meteorological Organization (Robarts et al. 2003).

Risk finance is an effective way to quickly secure the large amounts of funds needed when risks materialize. Risk finance is a risk transfer mechanism in which a third party compensates for damages on behalf of the entity holding the risk if the risk materializes. In addition to conventional insurance, securitization and financial derivatives may be used as risk transfer instruments. Risk finance has been used as a means of risk transfer for entities that are exposed to natural disasters such as typhoons, hurricanes, earthquakes, and droughts. For example, in 2010, Sompo Japan developed a product, called 'weather index insurance', for rice farmers in northeastern Thailand who are at risk of reduced yields due to drought. Under this product, Sompo Japan's local subsidiary will pay an insurance claim to the Bank for Agricultural and Agricultural Cooperatives (BAAC) in the event of a drought event in which the cumulative precipitation (weather index) falls below a predetermined threshold during a certain period (such as, July to September), and BAAC will pay the equivalent amount to the farmer. Sompo Japan is also developing weather index products for rice and sesame farmers in Myanmar and longan (a type of fruit) farmers in Thailand.

The risks posed to countries and regions by natural disasters under climate change have become an international concern and consequently, international conferences, such as workshops held by the United Nations Framework Convention on Climate Change and the Rio+20 Conference in 2012, have discussed the development of systems to reduce disaster risks. Through these international discussions, it has been widely recognized that risk finance is an effective means of securing funds quickly before and after disasters. There is also a growing trend in efforts toward internationally coordinated risk finance to address the impacts of climate change risks. An example of a risk finance initiative involving governments and international organizations is the Caribbean Catastrophe Risk Insurance Facility (CCRIF), established in 2007 through a partnership between 18 Caribbean countries and the World Bank. The Caribbean is a region that is prone to hurricanes and earthquakes, and in the event of these natural disasters, CCRIF provides compensation according to the amount of damage. CCRIF Segregated Portfolio Company, which was restructured from CCRIF in 2014, has received financial support from the Government of Japan, the U.S. State Department, and a multi-donor trust fund by the Government of Canada, the European Union, the World Bank, and so forth.

As described above, risk finance has contributed to the risk transfer of natural disasters associated with climate change. However, to the best of the authors' knowledge, there are still no risk finance initiatives targeting the risk of degradation of marine and lake environments. In this chapter, we introduce two models as tools for modeling the risk of environmental degradation: a statistical model and a physical-biogeochemical model. In addition, taking the environmental degradation risk of Lake Biwa in Japan as an example, we explain how to calculate the probability of environmental degradation using these methods in implementing risk financing for marine and lake environmental risks.

5.2 STATISTICAL MODEL

Towards realizing risk finance for environmental risks in lakes and seas, it is necessary to 1) identify the target risk, 2) model the stochastic fluctuations of the indicators (risk indices) that represent the status of the identified risks, and 3) calculate the probability of occurrence of the risks based on the model.

Variables that affect environmental risks in lakes and seas, such as water temperature and wind speed, are often seasonal in nature. Unseasonal terms obtained by applying seasonal adjustment to these variables depend on past levels or are determined stochastically in some cases. Statistical approaches such as time series models are effective in capturing the patterns of these fluctuations. The autoregressive model is comparatively simple and is popular among time series models. Hence, we present an overview of an autoregressive model in this section. Furthermore, we explain how to estimate model parameters and calculate the probability of risk occurrence based on time series models using the Monte Carlo method.

5.2.1 Model Structure and Parameter Estimation

5.2.1.1 Autoregressive Model

The autoregressive (AR) model is commonly adopted to extract the fluctuation characteristics of time-varying data, especially in the fields of statistics, economics, and signal processing. The AR model, as its name explains, is based on a regressive analysis and, for example, the present value of water temperature is supposed to be dependent on its historical values. The AR model specifies a linear regression with a stochastic term. Moreover, the AR model is capable of flexibly adjusting the time range by defining the order p, because the time of auto-dependence is different among indices in nature, with the notation AR(p) indicating an AR model of order p, as defined in Eq. 1.

$$X_t = c + \sum_{i=1}^{p} \rho_i X_{t-i} + \varepsilon_t \tag{1}$$

where X_t is the seasonally adjusted risk index at the time t, ρ_i is the ith parameter of the model with i ranging from 1 to p, $\varepsilon_t \sim i.i.d.N(0,1)$ are the independent and identically standard normal distributed random variables, and c is a constant.

5.2.1.2 Parameter Estimation

In estimating the model parameters of statistical models, maximum likelihood estimation (MLE) is widely applied (Rossi 2018). MLE selects a parameter set when the likelihood $L(\theta \mid X)$, which represents the probability that data set X occurs given the parameter set θ, is maximized. The form of the likelihood $L(\theta \mid X)$ is determined dependent on the density function of stochastic terms ($\varepsilon_t \sim i.i.d.N(0,1)$ in Eq.1) of a statistical model. Once we have the likelihood $L(\theta \mid X)$, we search for a set of parameters that maximizes the likelihood (Eq. 2) through analytical or numerical methods:

$$L(\hat{\theta}) = L(\hat{\theta} \mid X) = maxL(\theta \mid X), \theta \in \Theta \tag{2}$$

where $\hat{\theta}$ is the best parameter obtained by MLE and Θ is the parameter space that contains all possible parameter values.

5.2.1.3 Model Selection

In general, increasing the degree p of the AR(p) model increases the goodness of fit to the data. However, as the degree of the selected AR model increases, the model becomes more complex and is prone to overfitting. Information criteria such as the Akaike Information Criteria (AIC, Akaike 1973) and the Schwarz Bayesian Information Criteria (SBIC or BIC (Schwarz 1978) are used to select a model that can predict unknown situations without making the model too complex. In section 5.4.3, we adopt BIC as statistical model selection criteria; thus, we here present the formulation of BIC (Eq. 3).

$$BIC = -2\ln\left(L(\hat{\theta})\right) + k\ln(n) \tag{3}$$

where k is the number of model parameters, and n is the number of data points or the sample size. The model with the lowest BIC value is selected.

5.2.2 Monte Carlo Simulation

After selecting the best AR(p) model based on information criteria, we perform the Monte Carlo simulation of risk variables using the selected AR(p) to calculate the probability of risk occurrence. Although there is no consensus on the definition of Monte Carlo simulation (Sawilowsky 2003; Ripley 2009), the algorithm of Monte Carlo simulation generally aims to solve problems through random sampling from a probability distribution (Kroese et al. 2014). We show below a method of conducting the Monte Carlo simulation by the repeated sampling of the random variables (ε_t) in the AR model of Eq.1 to calculate the probability of risk occurrence.

Let us take an example of a risk finance product that aims to hedge the risk of facing less rainfall during the rainy season in one city, to better understand the risk calculation process by the Monte Carlo simulation. The observed precipitation Y_t at time t is decomposed into the seasonal part S_t and the unseasonal part X_t, namely, $Y_t = S_t + X_t$. Suppose that the insurance company must pay for the drought crisis when the average precipitation during $t + (T- l)$ and $t + T$ falls below a predetermined value, for example, a. To calculate the premiums for this risk finance product, we must estimate the drought crisis probability in advance. In the following, we explain the method of calculating drought probability based on a Monte Carlo simulation.

First, we suppose that parameters in the unseasonal component (X_t) have been estimated by the AR(1) model from its historical data. Then, the Monte Carlo simulation must cover the whole insurance span and therefore is conducted from $t + 1$ to $t + T$ with t being the present time. Finally, this simulation is repeated N times. The random effects for the kth $(1 \le k \le N)$ repetition can be labeled accordingly, for example, ε_{t+1}^k at time $t + 1$, and when their probability distribution follows the standard normal distribution, the precipitation in the future, for example, Y_{t+1}^k, can be calculated accordingly (Eq. 4) using Eq. 1.

$$Y_{t+1}^k = S_{t+1} + X_{t+1}^k = S_{t+1} + c + \rho_1 X_t^k + \varepsilon_{t+1}^k \tag{4}$$

Back to the definition in the risk finance product, the drought crisis depends on the average precipitation during $t + (T - l)$ and $t + T$, which can be calculated by Eq. 5 for the kth repetition. By comparing the value of this calculated mean value $(\bar{Y})$ and the predetermined value (a), the drought crisis is quantified and judged.

$$\bar{Y} = \frac{1}{l+1} \sum_{i=0}^{l} Y_{t+(T-i)}^k \tag{5}$$

In this way, after completing the whole N repetitions, the occurrence of this drought crisis is counted as M, and therefore the risk of drought crisis under the scope of this risk finance product is M/N.

5.3 PHYSICAL-BIOGEOCHEMICAL MODEL

Another modeling technique is to apply process-based theoretical models to investigate the fluctuation of the aquatic environment under uncertainties. The aquatic environment can be roughly divided into two components: water hydrodynamics like flow velocity and water density, and water quality composed of the abiotic environment as in the concentration of nutrients, and the biotic environment as in the concentration of phytoplankton. Accordingly, most physical-biogeochemical models consist of hydrodynamic and ecosystem submodels. Among many models that have been developed worldwide, the Marine Environmental Committee (MEC) ocean model (MEC 2000) is chosen in this chapter. Because the MEC ocean model has been practised and validated for both lakes (for example, Kitazawa et al. 2018) and coastal seas (for example, Zhou et al. 2021), this model is readily prepared to serve current risk finance for natural disasters in lakes and coastal seas.

5.3.1 HYDRODYNAMIC SUBMODEL

In this submodel, the water movement driven by either wind or water density, together with water temperature and salinity that determine the water density, is described by the governing equations (Eqs. 6–12). However, especially for lakes and coastal seas where the effects of natural disasters are enormous and concerning, two assumptions are adopted for practical simplicity: a hydrostatic approximation where the weight of the water identically balances the pressure, and a Boussinesq approximation where variations in density have no effects on the flow field, except that they give rise to buoyancy forces. The equations are described in the Cartesian coordinate system, where the x, y, and z axes point eastward, northward, and vertically upwards, respectively.

(The motion equations)

$$\frac{\partial u}{\partial t}+\frac{\partial(uu)}{\partial x}+\frac{\partial(vu)}{\partial y}+\frac{\partial(wu)}{\partial z}=-\frac{1}{\rho_0}\cdot\frac{\partial p}{\partial x}+fv+\frac{\partial}{\partial x}\left(A_M\frac{\partial u}{\partial x}\right)+\frac{\partial}{\partial y}\left(A_M\frac{\partial u}{\partial y}\right)+\frac{\partial}{\partial z}\left(K_M\frac{\partial u}{\partial z}\right) \quad (6)$$

$$\frac{\partial v}{\partial t}+\frac{\partial(uv)}{\partial x}+\frac{\partial(vv)}{\partial y}+\frac{\partial(wv)}{\partial z}=-\frac{1}{\rho_0}\cdot\frac{\partial p}{\partial y}-fu+\frac{\partial}{\partial x}\left(A_M\frac{\partial v}{\partial x}\right)+\frac{\partial}{\partial y}\left(A_M\frac{\partial v}{\partial y}\right)+\frac{\partial}{\partial z}\left(K_M\frac{\partial v}{\partial z}\right) \quad (7)$$

$$0=-\frac{1}{\rho}\cdot\frac{\partial P}{\partial z}-g \quad (8)$$

(The continuity equation)

$$\frac{\partial u}{\partial x}+\frac{\partial v}{\partial y}+\frac{\partial w}{\partial z}=0 \quad (9)$$

(The advection-diffusion equation of water temperature and salinity)

$$\frac{\partial T}{\partial t}+\frac{\partial(uT)}{\partial x}+\frac{\partial(vT)}{\partial y}+\frac{\partial(wT)}{\partial z}=\frac{\partial}{\partial x}\left(A_H\frac{\partial T}{\partial x}\right)+\frac{\partial}{\partial y}\left(A_H\frac{\partial T}{\partial y}\right)+\frac{\partial}{\partial z}\left(K_H\frac{\partial T}{\partial z}\right)+R_{TMP} \quad (10)$$

$$\frac{\partial S}{\partial t}+\frac{\partial(uS)}{\partial x}+\frac{\partial(vS)}{\partial y}+\frac{\partial(wS)}{\partial z}=\frac{\partial}{\partial x}\left(A_H\frac{\partial S}{\partial x}\right)+\frac{\partial}{\partial y}\left(A_H\frac{\partial S}{\partial y}\right)+\frac{\partial}{\partial z}\left(K_H\frac{\partial S}{\partial z}\right)+R_{SAL} \quad (11)$$

$$\rho=1028.14-0.0735T-0.00469T^2+(S-35.0)\cdot(0.802-0.002T) \quad (12)$$

where t (s) is time, u, v, and w (m s^{-1}) the x, y, and z components of flow velocity, ρ_0 (kg m^{-3}) the reference density of water, ρ (kg m^{-3}) the density of water, p (N m^{-2}) the pressure, g (m s^{-2}) the acceleration due to gravity, f (s^{-1}) the Coriolis parameter, A_M (m^{2}s^{-1}) the horizontal eddy viscosity coefficient, K_M (m^{2}s^{-1}) the vertical eddy viscosity coefficient, T (°C) the water temperature, S (psu) the water salinity, A_H (m^{2}s^{-1}) the horizontal eddy diffusivity coefficient, K_H (m^{2}s^{-1}) the vertical eddy diffusivity coefficient, R_{TMP} (°C s^{-1}) the heat flux through rivers, and R_{SAL} (psu s^{-1}) the salt flux through rivers. Note that water salinity in lakes is low and its advection-diffusion equation is generally unnecessary.

5.3.2 Ecosystem Submodel

The ecosystem submodel is based on the observed food web in the aquatic environment at the eutrophic level (Figure 5.1). The number of state variables changes due to the range of complexity of the ecosystem submodel from a NPZ (Nutrient-Phytoplankton-Zooplankton) model to a pelagic, benthic, and higher-trophic coupled model. A relatively simplified model includes seven state variables—phytoplankton (*PHY*), zooplankton (*ZOO*), particulate and dissolved organic carbons (*POC* and *DOC*), dissolved inorganic phosphorus and nitrogen (*DIP* and *DIN*), and dissolved oxygen (*DO*) (Figure 5.1). Phosphorus and nitrogen are two limiting nutrients that control primary

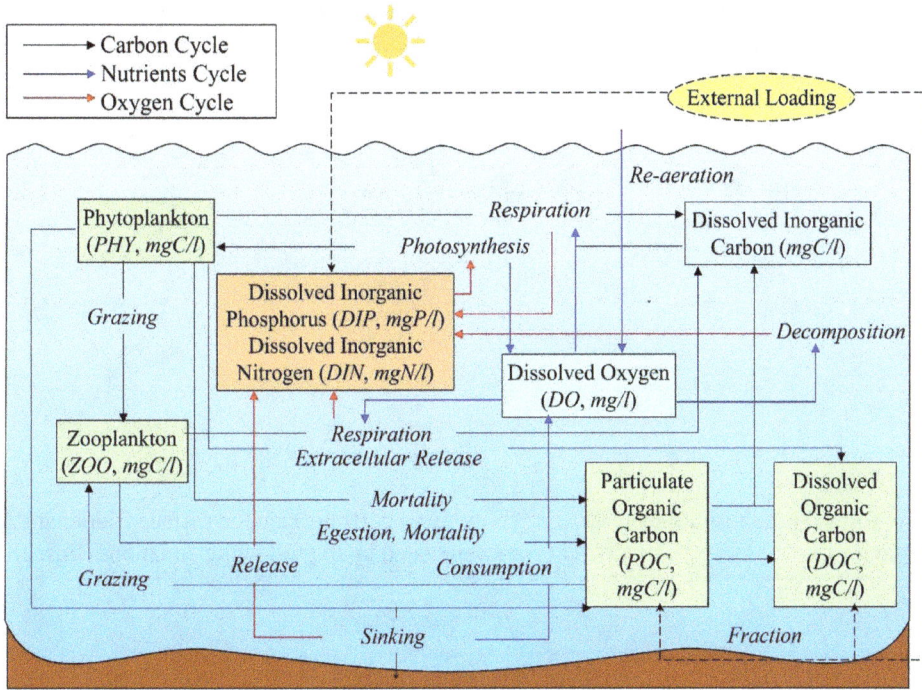

FIGURE 5.1 Material cycle in ecosystem submodel. The variables enclosed by boxes indicate state variables in the ecosystem submodel (not including inorganic carbon), and descriptions by arrows indicate chemical-biological processes.

production. Silicate is sometimes ignored in the model because it is either in relatively sufficient supply, such as in lakes or in a completely negligible concentration such as in the seas.

With an assumption that the transportation of plankton and nutrients is based on the surrounding current field, the governing equation for each state variable can be described by an advection-diffusion equation, similarly to that practised by water temperature and salinity.

$$\frac{\partial B}{\partial t} + \frac{\partial (uB)}{\partial x} + \frac{\partial (vB)}{\partial y} + \frac{\partial (wB)}{\partial z} = \frac{\partial}{\partial x}\left(A_H \frac{\partial B}{\partial x}\right) + \frac{\partial}{\partial y}\left(A_H \frac{\partial B}{\partial y}\right) + \frac{\partial}{\partial z}\left(K_H \frac{\partial B}{\partial z}\right) + \left(\frac{\partial B}{\partial t}\right)^* + R_B \quad (13)$$

where B denotes each state variable and R_B represents the effects of the river on the corresponding state variables. The item $\left(\frac{\partial B}{\partial t}\right)^*$ summarises the effects on corresponding state variables due to the chemical and biological interactions, with more details given in Zhou (2020).

5.3.3 BOUNDARY CONDITIONS

Boundary conditions refer to the theoretical incorporation of those boundaries that surround the calculation domain. Typically, they are conditions of the land, river, water bottom, water surface, and open boundary for seas. At the land and water bottom boundary, flow velocity is usually affected by boundary friction. The river and open boundary generate the water exchange and its accompanying heat, salt, and nutrient flux. However, at the water surface, when considering the effects of meteorological conditions, the interaction is much more complicated.

First, consider the momentum condition at the water surface. Usually, the water surface is regarded as the free surface with frictional stresses by wind (Eqs. 14–17).

$$p = p_a,$$ (14)

$$-\frac{\partial \zeta}{\partial t} - u\frac{\partial \zeta}{\partial x} - v\frac{\partial \zeta}{\partial y} + w = 0,$$ (15)

$$K_M \frac{\partial u}{\partial z} = \frac{\tau_x}{\rho_0},$$ (16)

$$K_M \frac{\partial v}{\partial z} = \frac{\tau_y}{\rho_0}$$ (17)

where p_a ($N\,m^{-2}$) is atmospheric pressure, τ_x ($N\,m^{-2}$) and τ_y ($N\,m^{-2}$) are the surface frictional stresses in the x and y directions, respectively. They are calculated by the following equations (Eqs. 18–22).

$$\tau_x = C_d \rho_a W_x \sqrt{W_x^2 + W_y^2},$$ (18)

$$\tau_y = C_d \rho_a W_y \sqrt{W_x^2 + W_y^2},$$ (19)

$$\rho_a = 1.293 \frac{273.15}{273.15 + T_a} \frac{P_a}{1013.25} \left(1 - 0.378\frac{E_a}{P_a}\right),$$ (20)

$$E_a = E_s \cdot humd,$$ (21)

$$E_s = 6.1078 \times 10^{7.5T_a/(237.3+T_a)}$$ (22)

where C_d is the coefficient of wind friction, W_x ($N\,m^{-2}$) and W_y ($N\,m^{-2}$) are wind velocities in the x and y directions, respectively. E_a (hPa) is the atmospheric vapor pressure and E_s (hPa) is the saturated vapor pressure. T_a (°C) is the atmospheric temperature and $humd$ is the relative humidity ranging from 0 to 1.

Second, consider the heat and salt fluxes through the water surface. Affected by the interactions with the atmosphere, water temperature and salinity have a fluctuation that can't be ignored (Eqs. 23–24).

$$-K_H \frac{\partial T}{\partial z} = \frac{Q_T}{\rho_0 C_p},$$ (23)

$$-K_H \frac{\partial S}{\partial z} = \frac{Q_S}{\rho_0}$$ (24)

where C_p ($J\,kg^{-1}\,K^{-1}$) is the specific heat at constant pressure, Q_T ($J\,m^{-2}\,s^{-1}$) and Q_S ($kg\,m^{-2}\,s^{-1}$) are the heat and salt fluxes through the water surface, respectively.

The heat transfer is due mainly to the solar radiation from air and heat transport from water (Eq. 25), and we adopt the bulk equation (Eq. 26) for practical convenience.

$$Q_T = Q_r - Q_l - Q_h - Q_e \tag{25}$$

$$Q_f = \alpha_f \rho_a \left(T - T_a \right) \sqrt{W_x^2 + W_y^2} \tag{26}$$

where Q_r (J m^{-2}s^{-1}) is the short-wave solar radiation, Q_l (J m^{-2}s^{-1}) is the net long-wave radiation from water, Q_h (J m^{-2}s^{-1}) is the sensible heat transport due to convection, and Q_e (J m^{-2}s^{-1}) is the latent heat transport due to evaporation. Q_f (J m^{-2}s^{-1}) represents the transferred heat, and α_f is the parameter representing the characteristics of the fluid and requires manual specification. For more details on applying the bulk equation to heat flux calculation, refer to Zhou (2020).

On the other hand, the salt flux is caused by precipitation and water evaporation. Here, depending on the wind speed, we adopt different equations for the salt flux calculation.

(wind speed slower than 1.0 m s^{-1})

$$Q_S = S \left[\rho_a C_E' \left\{ \left(T - T_a \right) + 0.61 \left(T_a + 273.15 \right) \left(q_s - q_a \right) \right\}^{1/3} - P_r \right] \tag{27}$$

(otherwise)

$$Q_S = S \left\{ \rho_a C_E \left(q_s - q_a \right) \sqrt{W_x^2 + W_y^2} - P_r \right\} \tag{28}$$

where q_s is the saturated specific humidity, q_a is the specific humidity, and P_r (kg m^{-2}s^{-1}) is the atmospheric precipitation. C_E and C_E' are bulk coefficients for latent heat transfer.

5.4 CASE STUDY IN LAKE BIWA, JAPAN

Because impacts of global warming on aquatic environment share a common mechanism (Yanik and Aslan 2018), we take Lake Biwa, Japan – one of the world's ancient lakes in the archaeological and cultural sense (Kawanabe 2019) – for an example to elucidate how the above mentioned two modeling techniques are applied to risk finance for natural disasters in lakes and seas.

5.4.1 LAKE BIWA AND ENVIRONMENTAL PROBLEMS

Lake Biwa, the largest holomictic lake in Japan, is in Shiga Prefecture and is the source of drinking water for about 11 million people living in Osaka, the largest city in western Japan and its suburbs. Lake Biwa consists of two basins: Northern Basin and Southern Basin (Figure 5.2). The Northern Basin is relatively large and deep, with an area of 612 km² and a maximum depth of 104 m. The Southern Basin is small and shallow with an area of 58 km² and a maximum depth of 8 m (Kumagai et al. 2003). The total lake volume is 27.5 billion m³, where the Northern Basin has a volume of 27.3 billion m³ and water exchange there takes 5.5 years. Water exchange in the other basin, which has a volume of 0.2 billion m³, takes 15 days. One of the factors affecting the water exchange is stratification, typically during the spring and autumn seasons. The water temperature begins to increase from 8°C in late March and reaches about 28°C at the water surface in August. However, the water temperature keeps about 8°C at the water bottom throughout the year, resulting in the temperature difference and consequent stratification. As the surface water cools down during autumn and winter, the stratification gradually weakens and the water is more readily mixed up; typically,

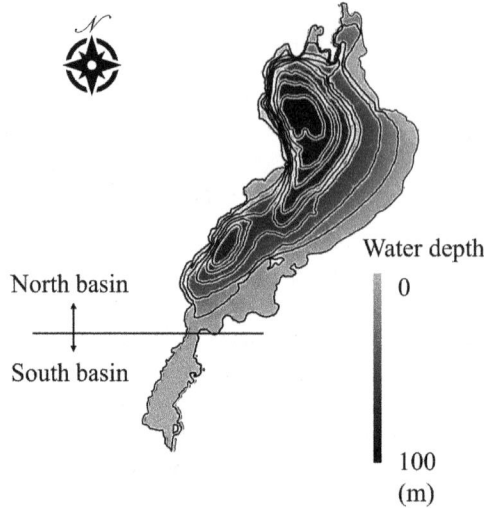

FIGURE 5.2   Bathymetry of Lake Biwa and discrimination between Northern and Southern Basins.

the stratification completely disappears around February. Accompanied by the disappeared stratification and water mixture, also known as overturn, saturated dissolved oxygen above the thermocline is transported to the deep lake and supports the activities of aquatic organisms below the thermocline, for example, the decomposition of organic matter by bacteria. The dissolved oxygen tension marks the minimum value (about $4\,mg\,L^{-1}$) around December at the deepest point of the Northern Basin. Then the concentration of dissolved oxygen is recovered to the saturated level at $11\,mg\,L^{-1}$ when the vertical mixing of waters occurs around February.

However, since the 1980s the atmospheric temperature around Lake Biwa rose by 1°C and precipitation decreased by 100 mm, according to the meteorological observations (www.data.jma. go.jp/multi/index.html?lang=en; Accessed on June 8, 2021). With the warming weather, Woolway and Merchant (2019) indicated a less frequent mixture in lake waters by summarizing the climate change impacts on mixing regimes in 635 lakes around the world, and Woolway et al. (2021) further pointed out the effects of this prolonged stratification on lake deoxygenation. Lake Biwa, like any other of Earth's largest lakes, for example, Lake Michigan (Anderson et al. 2021), is suffering from global warming (Kumagai et al. 2003). The minimum concentration of dissolved oxygen at the deepest point of the Northern Basin kept decreasing and was sometimes less than $2\,mg\,L^{-1}$ (Figure 5.3), which is a general criterion of hypoxic water. Hypoxia has implications on the Lake Biwa environment, especially habitat loss and the mass death of aquatic species, and the lifestyle of the local population (Kawanabe 2019). As a result, the occurrence of hypoxic water in Lake Biwa has attracted public attention, especially in local government.

5.4.2   RISK FINANCE PRODUCT FOR THE LOSS OF OVERTURN IN LAKE BIWA

If we face the loss of overturn in Lake Biwa in winter, the local governments concerned will have to take immediate measures, such as purchasing oxygen injectors for the bottom layer of Lake Biwa. Risk finance is an effective way to secure funds for these countermeasures as soon as possible. Kubo et al. (2021) proposed a risk finance product for the risk of loss of the overturn in Lake Biwa using options, a type of financial derivative. Compared to conventional insurance, risk finance with options can be used to raise funds quickly when risks materialize. An option is a contingent claim that gives its buyer the right to receive a certain amount of money from its seller if a predetermined condition is met on the maturity date. On the contract date, the buyer pays the seller an option

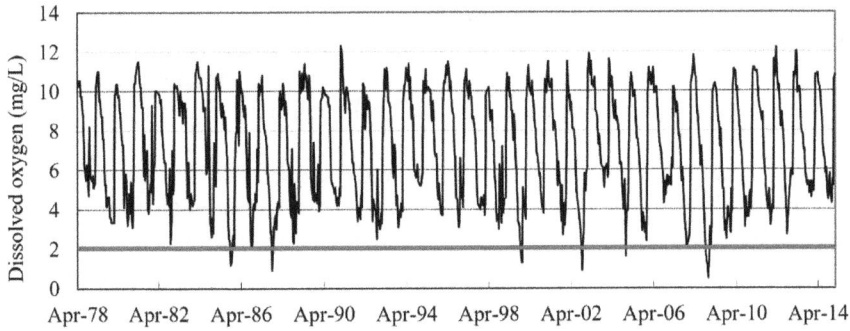

FIGURE 5.3 Annual change in the minimum concentration of dissolved oxygen at the deepest point of Lake Biwa, with the red line representing the general criterion of hypoxic water. Data are provided by Lake Biwa Environmental Research Institute.

premium. Let us take as an example the option against the loss of Lake Biwa overturn proposed in Kubo et al. (2021) as follows.

- Buyer: a local government, Seller: X company
- Date of the contract: October 1, 2014
- Duration of the contract: 196 days
- Date of maturity: April 15, 2015
- Amount of premium to be paid on the contract date: A Japanese Yen (Yen)
- Terms and conditions: If the amount of dissolved oxygen at the bottom layer of Lake Biwa does not reach $8.0 \, \mathrm{mg \, L^{-1}}$ on the maturity date, the seller of this option is obligated to pay C Yen to the buyer on the day following the maturity date.

The target of the risk faced by local governments is the loss of the overturn in Lake Biwa. The loss of the overturn in Lake Biwa means that dissolved oxygen at the bottom layer of Lake Biwa remains at a low level even at the beginning of spring. For this reason, in the above option, we see this risk as the amount of dissolved oxygen at the bottom layer of Lake Biwa falling below a certain threshold ($8.0 \, \mathrm{mg \, L^{-1}}$ in the above example) even after the winter is over. The reason why we here set a threshold at $8.0 \, \mathrm{mg \, L^{-1}}$ is due to the historical fact that DO levels after the occurrence of overturn are always over $8.0 \, \mathrm{mg \, L^{-1}}$. With this option, when the loss of overturn occurs, the local government, as the buyer, can quickly receive C Yen from the seller and take actions to prevent further aggravation of the hypoxia situation in Lake Biwa. When an overturn occurs, the seller makes a profit of A Yen from this option contract. Here, the problem is how to set the level A Yen of the option premium. According to a financial theory, the option premium A depends not only on C but also on the probability of loss of the overturn in Lake Biwa. To estimate the probability of the loss of overturn, we need some quantitative models. Kubo et al. (2021) present the two models mentioned above, statistical and physical-biogeochemical approaches, to estimate the probability.

5.4.3 Application of Statistical Model

In the previous section, we exemplified an option against the loss of Lake Biwa's overturn. Calculating the probability that the amount of dissolved oxygen at the bottom of Lake Biwa falls below a certain level in early spring allows us to determine the level of the option's premium. Thus, in this section, we will explain how to calculate the probability by constructing a time series model of dissolved oxygen fluctuations with observation data since 1979 in Lake Biwa.

In the current analysis, we use data on dissolved oxygen at the bottom of Lake Biwa from April 1978 to March 2015. Because dissolved oxygen at the bottom of Lake Biwa has been measured twice a month, the time unit (t) is taken for 15 days, and every month is approximated as 30 days for simplicity. Following this data standardization, each observation record is utilized at an even time interval.

The change of dissolved oxygen, similar to the water temperature, has a seasonal pattern, and therefore its fluctuation is divided into two components—seasonal variation and irregular variation. The irregular DO variation (U_t) can be calculated with the following Eq. 29.

$$U_t = \ln DO_t - \overline{\ln DO_t} \tag{29}$$

where DO_t ($mg\,L^{-1}$) is the concentration of bottom dissolved oxygen, $\ln x$ is the natural logarithm of x, and $\overline{\ln DO_t}$ represents the mean value of each natural logarithm of observed DO values for the corresponding month, the seasonal level of the natural logarithm of DO concentration.

The time series fluctuation of irregular DO variation is modeled by the following extended version of AR(k) we explained in section 5.2.1.1:

$$U_t = \sum_{i=1}^{k} \rho_i U_{t-i} + \sigma_t \varepsilon_t, \varepsilon_t \sim i.i.d.N(0,1) \tag{30}$$

where σ_t denotes the amplitude of the random effects and it has the seasonal variation as modeled by a sine function in Eq. 31.

$$\sigma_t = \sigma_0 - \sigma_1 |\sin(\pi t + \alpha)| \tag{31}$$

where σ_0 is the constant, σ_1 defines the amplitude of variation by a sine function, and α represents the phase difference. Hereafter, we also denote this extended version of AR(k) as AR(k).

We estimated the model parameters based on the maximum likelihood method using the DO data from 1979 to 2012. Table 5.1 presents the parameter estimation results and the values of BIC for AR(1), AR(2), and AR(3). AR(2) has the lowest BIC value. Figure 5.4 compares the observed and estimated values based on AR(2) of DO and shows a good fit between the model estimates and the observed data.

TABLE 5.1
Results of Parameter Estimation and BIC Values for Three AR Models

Items	AR(1)	AR(2)	AR(3)
σ_0	0.4680 (0.0026)	0.4647 (0.0028)	0.4638 (0.0029)
σ_1	0.4096 (0.0036)	0.4063 (0.0039)	0.4058 (0.0040)
α	0.3756π (0.0123π)	0.4151π (0.0103π)	0.4566π (0.0094π)
ρ_1	0.5225 (0.0141)	0.4751 (0.0170)	0.4697 (0.0166)
ρ_2		0.09571 (0.0171)	0.06811 (0.0167)
ρ_3			0.05636 (0.0132)
$\ln L(\hat{\theta})$	319.58	324.44	326.70
BIC	−612.01	−614.95	−612.69

Note: The numbers in parentheses show the standard error of each parameter.

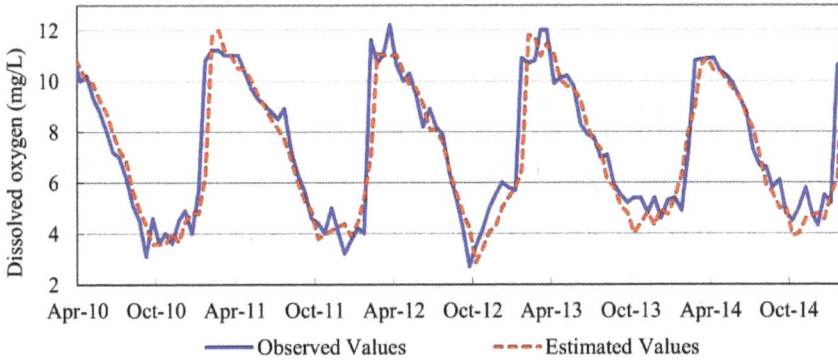

FIGURE 5.4 Comparison of observed (solid line) and estimated (dashed line) DO concentration. These estimated values are based on the AR(2) model with parameters shown in Table 5.1.

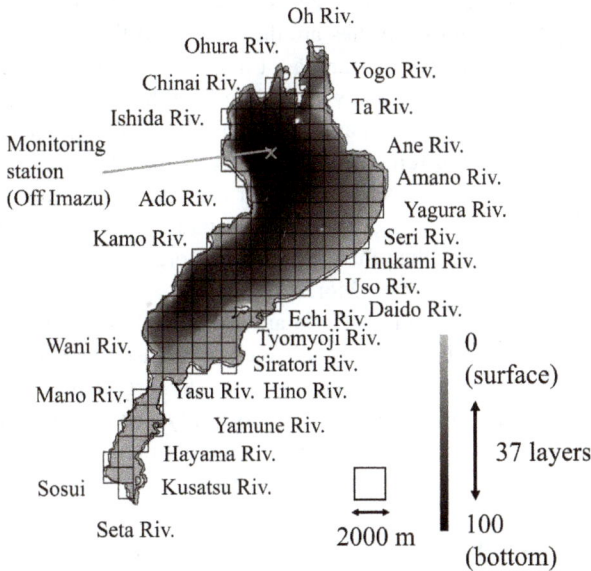

FIGURE 5.5 Grid system in Lake Biwa, and the monitoring station off Imazu, and the locations of rivers.

To calculate the probability of the loss of the overturn in Lake Biwa towards setting the premium of the option presented in section 5.4.2, we perform the Monte Carlo simulation of DO based on AR(2). Specifically, we run the Monte Carlo simulation of DO to obtain 65,000 sample paths from October 1, 2014, to April 15, 2015, and judge whether the DO value of each simulated path on April 15, 2015, is below $8\,mg\,L^{-1}$ or not. According to our calculations, the DO level on April 15, 2015, was below $8\,mg\,L^{-1}$ in 169 sample paths. Hence, we can estimate the probability of the loss of the overturn in Lake Biwa as 169/65000 = 0.26%.

5.4.4 APPLICATION OF PHYSICAL-BIOGEOCHEMICAL MODEL

5.4.4.1 Computational Conditions

The computational domain and grid system of Lake Biwa for this analysis are shown in Figure 5.5. The area is latticed by the structured grids, and its horizontal size is 1,000 m for model validation, and 2,000 m for long-term simulation, with a size of 2.5 m in the vertical direction. For numerical practicality, a staggered grid system is adopted to arrange the evaluation points of flow velocity and pressure, water temperature, density, and state variables.

With the given input of computational conditions, water surface elevation is first calculated, followed by horizontal and vertical flow velocities, water temperature, and each state variable. As for the finite difference scheme, an explicit time integration method of Euler, Quadratic Upstream Interpolation for Convective Kinematics, and a second-order central difference method were adopted for the time derivative term, the advection term, the eddy viscosity, and the diffusivity terms, respectively. Then, water density is calculated by the state equation as a function of only water temperature. The time step is set to be 20 seconds for 1,000 m mesh, and 40 seconds for 2,000 m mesh.

5.4.4.2 State Variables

In Lake Biwa, phytoplankton blooms start from the rapid growth of diatoms such as *Melosira* in early spring. Then it is followed by the dinoflagellate *Uroglena americana* in late spring, and in sequence, green algae and blue-green algae such as *Microcystis* and *Anabena* during summer and autumn. The dominant phytoplankton species are due largely to the seasonal variation in nutrient levels, and the life cycles of these species share a lot in common. As a result, it is robust to model these species with one state variable for analysis. Similarly, although zooplankton is dominated by many species depending on feeding habitat and food abundance, it is also summarized by one state variable. Particulate organic carbon refers to fecal pellets, dead phytoplankton, zooplankton, and bacteria. Dissolved organic carbon is composed of an extracellular release of phytoplankton and fraction production with the bacterial decomposition of particulate organic carbon. Therefore, its decomposition rate is much smaller than that of particulate organic carbon, but their accumulation readily raises the oxygen depletion level. Moreover, special attention should be paid to the fact that microbial loop is not considered and bacteria are not treated as a state variable, although their decomposition of organic carbon is implicitly modeled to release nutrients and consume oxygen. Finally, the effects of upper trophic level predators, such as large fish, birds, and human fishing activities, are neglected on the assumption that they have only a secondary contribution to the food web system in Lake Biwa, and therefore they exert a limited influence on oxygen depletion as well as the loss of overturn.

5.4.4.3 Boundary Conditions

On the sidewalls, the velocity component normal to the walls is zero, and a free-slip condition is applied to the velocity component parallel to the walls. The normal gradients of water temperature and of each state variable are also zero so that there is no advective or diffusive heat, chemical considerations, or plankton through the walls. The effects of fluxes of heat and chemical matter through rivers are considered in this case, with the layout of the river (Figure 5.5).

The boundary conditions of kinematics and bottom stress are assumed at the bottom surface, without consideration of the heat flux through it. The sinking organic matter is assumed to be decomposed immediately with oxygen consumption and nutrient release, while some nutrients are fossilized in the sediment. Finally, at the water surface, the dynamics, kinematics, wind stress, and heat flux are all modeled. The wind stress and heat flux through the water surface are estimated by the bulk formula.

5.4.4.4 Parameter Estimation and Initial Values

All parameters are determined in accordance with the physical and biogeochemical conditions in Lake Biwa, and here, thinking about parameters in physical conditions as an example, the horizontal eddy viscosity and diffusivity are determined by the grid size following Richardson's 4/3 power law. The vertical components, when considering the dominant stratification, are determined by the

TABLE 5.2
Parameter Values in the Physical Submodel

Symbol	Value	Unit	Reference
ρ_0	1000	$\mathrm{kg\,m^{-3}}$	
G	9.8	$\mathrm{m\,s^{-2}}$	
Cd	0.0015	–	Kondo (1975)
Cp	3.93×10^3	$\mathrm{J\,kg^{-1}\,K^{-1}}$	
C_E	1.1×10^{-3}	–	
C'_E	1.1×10^{-3}	–	

stratification function with additional assumptions, such as an instantaneous mixture of water. In addition, the value of the Coriolis parameter depends on the latitude of Lake Biwa. Values for basic or above-mentioned parameters are listed in Table 5.2. For more details about the parameter estimation, refer to Zhou (2020).

The calculation of the water current and the water level starts from the value of zero for the model stabilization and spin-up. The initial values of water temperature, salinity, and each pelagic state variable are consistent with the empirical values for all grids.

5.4.4.5 Scenarios and Threshold for Overturn

In this section, in accordance with Kubo et al. (2021), we present a methodology for determining the probability of the loss of the overturn in Lake Biwa by combining a physical-biogeochemical model with a statistical model for several meteorological variables.

Before creating scenarios about climate change, the risk factors dominating the loss of overturn should be identified from the perspective of numerical efficiency. There are mainly two approaches that lead to the same conclusion: statistical (for example, Kubo et al. 2021) or mechanistic (Yoshida et al. 2018). Atmospheric temperature, solar radiation, and wind speed are the top three risk factors.

First, we model the time series of these risk factors by the multivariate vector regressive model, the extension of the AR model, and estimate model parameters using their historical data. Next, we generate tens of thousands of scenarios for a group of atmospheric temperature, solar radiation, and wind speed using a Monte Carlo simulation based on the estimated model. Theoretically speaking, feeding all the generated scenarios into the physical-biogeochemical model and calculating the ratio of the number of scenarios, in which the level of dissolved oxygen output in early spring remains low to the total number of simulations (namely, the loss of overturn), allows us to calculate the probability of the loss of overturn. In practice, however, the computational load of the physical-biogeochemical model is so high that it is not feasible to input all tens of thousands of meteorological scenarios and perform numerical simulations of the model.

To solve this problem, Kubo et al. (2021) proposed an alternative methodology to calculate the probability of the loss of the overturn in Lake Biwa. The idea was to search for a 'threshold zone' of meteorological conditions where the overturn may or may not occur. If it were possible to identify such a threshold zone, it would be possible to judge whether the overturn in Lake Biwa occurs by giving one meteorological condition: temperature, wind speed, and total solar radiation. Kubo et al. (2021) select dozens of scenarios, including those in which the loss of overturn is expected, from tens of thousands of scenarios for a group of the atmospheric temperature, solar radiation, and wind speed generated by the Monte Carlo simulation based on the estimated model. For each of these selected scenarios, they perform numerical calculations of the physical-biogeochemical model and check the amount of dissolved oxygen at the bottom layer in early spring to determine the occurrence or loss of the overturn in Lake Biwa for each scenario. The results obtained are summarized in

TABLE 5.3
The Threshold for the Loss of the Overturn in Lake Biwa

		Wind speed in January (ms⁻¹)					
		1.75	1.8	1.85	1.9	1.95	2.0
Atmospheric Temperature in December (°C)	10						
	9						
	8						
	7						
	6						
	5						

Note: Colour in grey shows the loss of overturn and the bold line shows the threshold value.

a matrix with the average atmospheric temperature in December and the average wind speed in January as axes (Table 5.3). If the cells in Table 5.3 are colored gray, it means the loss of overturn in the numerical simulation; otherwise, it means that the overturn occurs. From Table 5.3, it can be seen that the weaker the wind speed and the higher the temperature in winter, the less likely it is that the overturn in Lake Biwa will occur. The bold line indicated in Table 5.3 seems to show a threshold zone that separates whether the overturn occurs. Once we obtain a threshold zone, we can judge the occurrence or loss of overturn for a scenario without performing numerical calculations of the physical-biogeochemical model by checking whether the scenario is in the gray zone in Table 5.3. This leads to the calculation of the probability of the loss of the overturn in Lake Biwa.

As a matter of fact, according to Kubo et al. (2021), when the physical-biogeochemical model is calculated in some scenarios with an average wind speed of $1.7\,\mathrm{m\,s^{-1}}$ in January, all results show the occurrence of overturn. For this reason, it should be noted that additional detailed analyses for the threshold zones obtained in Table 5.3 are needed.

As discussed above, we believe that a methodology that combines a statistical model with natural science models that are deterministic models based on the principles of scientific phenomena, such as the physical-biogeochemical model, is effective in quantifying environmental risks.

5.5   FUTURE APPLICATION

The number of natural catastrophes and the resultant loss are continuously rising on a global scale (Alexander 2018). Besides the example of Lake Biwa in this chapter, similar threats have been discovered by a meta-analysis of lakes in Europe (Blenckner et al. 2007). Therefore, risk finance has a high potential to mitigate the deterioration risk of the marine and lake environment. The environment is becoming more conducive to implementing risk finance for marine and lake environmental risks. First, environment-related issues are under the spotlight nowadays, especially with the adoption of Sustainable Development Goals at the UN General Assembly in September 2015, and interest in investing in environment-related issues has simultaneously grown. Second, from an asset management perspective, risk finance products targeting environmental risks have a low correlation with financial instruments such as stocks and bonds and contribute to investors' portfolio diversification.

To calculate the cost of risk transfer in risk finance, we need to quantitatively model the targeted risk variable's fluctuation using a statistical model or a natural science model proposed in this chapter. These two techniques have their advantages and disadvantages. The statistical model is characteristically able to handle the risk variable directly, for example, the hypoxia situation. Modeling the risk variable based on a statistical model, such as the AR model, has the following three strengths: 1) ease of the data collection process of the risk variables; 2) user-friendly model operation without special background knowledge; and 3) compact output without data post-processing. Although the statistical model has these advantages, it should be noted that the model is estimated only from the past patterns of changes in the risk variables. Since the statistical model does not consider mechanisms of natural scientific phenomena, it cannot capture rare events that deviate from past fluctuation patterns, which leads to misinterpretation of the probability of risk occurrence. Hence, we need to consider this when determining the cost level of risk transfer in risk finance.

On the other hand, the process-based physical-biogeochemical models are deterministic; however, they take into account the mechanisms of natural scientific phenomena. They can obtain more robust results than the statistical models in environmental risk analysis, but they have disadvantages. First, because they consider physical, chemical, and biological processes, a wide range of knowledge is necessary to understand and debug models. Second, they require data with high quantity and quality. The data needs to support the parameter estimation, initial condition, boundary condition, and model validation. Third, the whole analysis takes a long time and requires more computer performance than statistical models.

Since the physical-biogeochemical model is deterministic, we cannot use it to calculate the probability of risk occurrence in implementing environmental risk finance. Thus, in section 5.4.4.5, taking the environmental risk of Lake Biwa as an example, we have presented a methodology for calculating an environmental risk probability by combining a physical-biogeochemical model with a statistical model of risk variables. This methodology needs further verification, but once established, it should be a powerful tool for calculating environmental risks accurately.

To increase the use of environmental risk finance and to reduce environmental risks in various parts of the world, it goes without saying that it is essential to calculate the probability of risk occurrence with high accuracy. We believe that the modeling techniques presented in this chapter help us to calculate the risk probability accurately and implement environmental risk finance.

REFERENCES

Akaike, H. 1973. Information Theory and an Extension of the Maximum Likelihood Principle. *Proceedings of the 2nd International Symposium on Information*, ed. B.N. Petrov, and F. Csaki, 267–281. Akademiai Kiado, Budapest.

Alexander, D. 2018. *Natural Disasters*. Routledge, London, UK.

Anderson, E. J. Stow, C. A. Gronewold, A. D. Mason, L. A. McCormick, M. J. Qian, S. S. Ruberg, S. A. Beadle, K. Constant, S. A. and Hawley. N. 2021. Seasonal Overturn and Stratification Changes Drive Deep-Water Warming in One of Earth's Largest Lakes. *Nature Communications* 12: 1–9.

Blenckner, T. Adrian, R. Livingstone, D. M. Jennings, E. Weyhenmeyer, G. A. George, D. G. Jankowski, T. Järvinen, M. Aonghusa, C. N. and Noges, T. 2007. Large-scale Climatic Signatures in Lakes across Europe: A Metaanalysis. *Global Change Biology* 13: 1314–1326.

Bouch, C. M. L. and Jones, G.P. 1961. *A Short Economic and Social History of the Lake Counties*, 1500–1830. Manchester University Press, Manchester, UK.

Gwynne, M. D. 1982. The Global Environment Monitoring System (GEMS) of UNEP. *Environmental Conservation* 9: 35–41.

Jeppesen, E. Søndergaard, M. Jensen, J. and Lauridsen, T. 2003. Restoration of Eutrophic Lakes: A Global Perspective. In *Freshwater Management: Global versus Local Perspectives*, ed. M. Kumagai, and W. F. Vincent, 135–151. Springer, Tokyo, Japan.

Kawanabe, H. 2019. *Lake Biwa: Interactions between Nature and People*. Springer Nature.

Kitazawa, D. Yoshida, T. Zhou, J. and Park, S. 2018. Comparative Study on Vertical Circulation in Deep Lakes: Lake Biwa and Lake Ikeda. In *2018 OCEANS—MTS/IEEE Kobe Techno-Oceans (OTO)* 1–4, IEEE.

Kondo, J. 1975. Air-Sea Bulk Transfer Coefficients in Diabatic Conditions. *Boundary-Layer Meteorology* 9: 91–112.

Kroese, D. P. Brereton, T. Taimre, T. and Botev, Z. I. 2014. Why the Monte Carlo Method is so Important Today. *Wiley Interdisciplinary Reviews: Computational Statistics* 6: 386–392.

Kubo, H. Kikuchi, K. Kitazawa, D. and Yoshida, T. 2021. Proposal of Environmental Risk Finance for the Risk of Overturn Failure in Lake Biwa (in Japanese). *Journal of Insurance Science* 653: 1–30.

Kumagai, M. Vincent, W. F. Ishikawa, K. and Aota, Y. 2003. Lessons from Lake Biwa and Other Asian Lakes: Global and Local Perspectives. In *Freshwater Management: Global versus Local Perspectives*, ed. M. Kumagai, and W. F. Vincent, 1–22. Springer, Tokyo, Japan.

MEC. November 2000. *The Society of Naval Architects of Japan: MEC Model Workshop*. Vol. 1.

OECD. 2016. The Ocean Economy in 2030. https://www.oecd-ilibrary.org/content/publication/9789264251 724-en.

———. 2019. Rethinking Innovation for a Sustainable Ocean Economy. https://www.oecd-ilibrary.org/cont ent/publication/9789264311053-en.

Pörtner, H. O. Roberts, D. C. Masson-Delmotte, V. Zhai, P. Tignor, M. Poloczanska, E. Mintenbeck, K. Nicolai, M. Okem, A. and Petzold, J. 2019. IPCC Special Report on the Ocean and Cryosphere in a Changing Climate. *IPCC Intergovernmental Panel on Climate Change: Geneva, Switzerland*.

Rabalais, N. N. Turner, R. E. Díaz, R. J. and Justić, D. 2009. Global Change and Eutrophication of Coastal Waters. *ICES Journal of Marine Science* 66: 1528–1537.

Reynaud, A. and Lanzanova, D. 2017. A Global Meta-Analysis of the Value of Ecosystem Services Provided by Lakes. *Ecological Economics* 137: 184–194.

Rick, T. C. and Erlandson, J. M. 2008. *Human Impacts on Ancient Marine Ecosystems: A Global Perspective*. Univ of California Press.

Ripley, B. D. 2009. *Stochastic Simulation*. John Wiley and Sons, Hoboken, USA.

Robarts, R. D. Fraser, A. S. Hodgson, K. M. and Paquette, G. M. 2003. Monitoring and Assessing Global Water Quality-the GEMS-Water Experience. In *Freshwater Management: Global versus Local Perspectives*, ed. M. Kumagai, and W. F. Vincent, 23–40. Springer, Tokyo, Japan.

Rossi, R. J. 2018. *Mathematical Statistics: An Introduction to Likelihood Based Inference*. John Wiley and Sons, Hoboken, USA.

Sawilowsky, S. S. 2003. You Think You've Got Trivials? *Journal of Modern Applied Statistical Methods* 2: 21.

Schwarz, G. 1978. Estimating the Dimension of a Model. *Annals of Statistics* 6: 461–464.

Woolway, R. I. and Merchant, C. J. 2019. Worldwide Alteration of Lake Mixing Regimes in Response to Climate Change. *Nature Geoscience* 12: 271–276.

Woolway, R. I. Sharma, S. Weyhenmeyer, G. A. Debolskiy, A. Golub, M. Mercado-Bettín, D. Perroud, M. et al. 2021. Phenological Shifts in Lake Stratification under Climate Change. *Nature Communications* 12: 1–11.

Yanik, T. and Aslan, I. 2018. Impact of Global Warming on Aquatic Animals. *Pakistan Journal of Zoology* 50: 353–363.

Yoshida, T. Kitazawa, D. Zhou, J. Park, S. Kubo, H. Kikuchi, K. and Yoshiyama, K 2018. Numerical Simulation of Overturn in Lake Biwa and Its Relation to Climate Change (in Japanese). *SEISAN KENKYU* 70: 25–28.

Zhou, J. 2020. Numerical Analysis of the Variation in Ecosystem with the Recovery of Aquaculture after Great Earthquake Disaster in Coastal Seas. Ph.D. Thesis.

Zhou, J. Kitazawa, D. Yoshida, T. Fujii, T. Zhang, J. Dong, S. and Li, Q. 2021. Numerical Simulation of Dissolved Aquaculture Waste Transport Based on Water Circulation around Shellfish and Salmon Farm Sites in Onagawa Bay, Northeast Japan. *Journal of Marine Science and Technology* 26: 812–827.

6 Blue Economy Prospects, Opportunities, Challenges, Risks, and Sustainable Development Pathways in Bangladesh

Md. Simul Bhuyan,*[1] Md. Nazrul Islam,**[2] Mir Mohammad Ali,[3] Md. Rashed-Un-Nabi,[4] Md. Wahidul Alam,[5] Monika Das,[6] Ranjan Roy,[7] Mohan Kumar Das,[8] Istiak Ahamed Mojumder,[9] and Sobnom Mustary[10]

[1] Bangladesh Oceanographic Research Institute (BORI), Cox's Bazar, Bangladesh

[2] Department of Geography and Environment, Jahangirnagar University, Savar, Dhaka, Bangladesh

[3] Department of Aquaculture, Sher-e-Bangla Agricultural University, Dhaka, Bangladesh

[4] Department of Fisheries, Faculty of Marine Sciences and Fisheries, University of Chittagong, Chittagong, Bangladesh

[5] Department of Oceanography, Faculty of Marine Sciences and Fisheries, University of Chittagong, Chittagong, Bangladesh

[6] Department of Fisheries, Matshya Bhaban, Dhaka, Bangladesh

[7] Department of Agriculture Extension and Information System, Sher-e-Bangla Agricultural University, Dhaka, Bangladesh

[8] National Oceanographic and Maritime Institute, Dhaka, Bangladesh

[9] Department of Zoology, University of Chittagong, Chittagong, Bangladesh

[10] Department of Biological Sciences, Birkbeck, University of London, UK

* Corresponding author: Md. Simul Bhuyan. Email: simulbhuyan@gmail.com

**Corresponding author: Md. Nazrul Islam. Email: nazrul_geo@juniv.edu

CONTENTS

DOI: 10.1201/9781003184287-6

6.1 INTRODUCTION

Sustainable Development Goal 14 emphasizes the life below water. SDG 14 (Life Below Water) has strong synergies with SDG 6 (Clean Water and Sanitation) and SDG 15 (Life On Land) and a potential trade-off with SDG 12 (Responsible Consumption and Production) (Lee et al. 2020). The blue economy is the use of ocean resources in a sustainable way. The European Union (2018) defined the blue economy as the 'all-economic activities related to oceans, seas, and coasts that cover a wide range of interlinked established and emerging sectors' (European Union, 2018) (Figure 6.1).

Roughly 40% of the world's population lives within 100 km of the coast (UN, 2017) and three-quarters of the world's large cities are located on the coast (Rayner et al. 2019). Marine ecosystems along the coast are valuable ecological and socio-economic sectors that deliver products and services (Seitz et al. 2013). Three billion people rely on oceans for their livelihood. 350 million employees are interconnected with the ocean and developing countries account for 97% of all fishermen.

The blue economy (BE) is a blessing/crucial factor for ocean-based countries like Bangladesh, for them to become developed countries. Gunter Pauli (a Belgian economist) reported that the ocean will provide 10 million jobs through 100 innovations that will take 10 years. The world economy is 88 trillion USD, of which 24 trillion USD solely comes from the sea. The world population is increasing and it will be 900 crores by 2050 (Bhuyan et al. 2020). This huge population will feed largely from the sea. Oceans provide 15% of protein for humans worldwide. 32% of CO_2 is absorbed by the ocean which has a great role in climate change mitigation (Parletta, 2019). 30% of oil and gas, 50% of magnesium, and many life-saving medicines are extracted from the sea (Bhuyan et al. 2020). By 2030, the global added value of marine equipment is projected to contribute US$ 300 billion to the global economy (BALance Technology Consulting, 2014). The contribution of oceans and their coasts in global GDP is depicted in Figure 6.2.

The blue economy is currently worth US$ 1.5 trillion and provides employment for 31 million people around the world (OECD, 2016). This amount will be US$ 3 trillion by 2030 which will

Established Industries	Emerging Industries
Capture fisheries and aquaculture	Open water aquaculture
Seafood processing	Deep- and ultra-deep-water oil and gas
Shipping	Offshore wind energy
Ports	Ocean renewable energy
Shipbuilding and repair	Marine and seabed mining
Offshore oil and gas (shallow water)	Maritime safety and surveillance
Marine manufacturing and construction	Marine biotechnology
Maritime and coastal tourism	High-tech marine products and services
Marine business services	Others
Marine R&D and education	

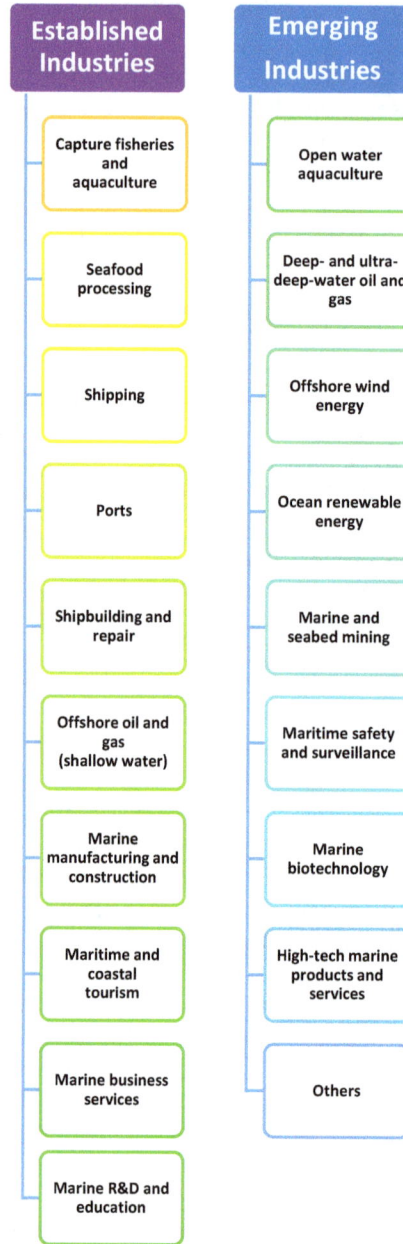

FIGURE 6.1   Established and emerging ocean-based industries (adapted from OECD, 2016).

be driven by aquaculture, offshore wind, fish processing, shipbuilding, and recycling. Australia, Europe, the U.S.A. China, Africa, and the Small Island Developing States (SIDs) make impressive profits from the blue economy (Roberts and Ali, 2016). The whole maritime economy of China would reach 8327 and 8894 billion RMB in 2018 and 2019, respectively, with annual growth rates of 7.3% and 6.8% (To et al. 2018). Despite having resources, only 7% of the oceans of the earth have been investigated (Toropova et al. 2010). UN (2016) set a timeline for SDG-14 targets (Figure 6.3).

In Bangladesh, the blue economy is a new concept although China, Japan, and the Philippines have been earning revenue from the ocean economy for almost 300 years. After the maritime victory

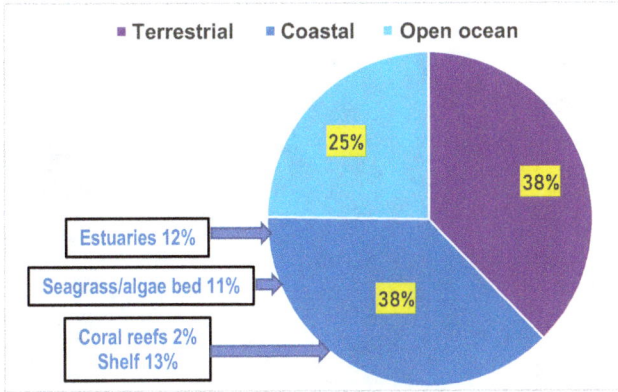

FIGURE 6.2 Contribution of resources from the coast and the sea in global GDP.

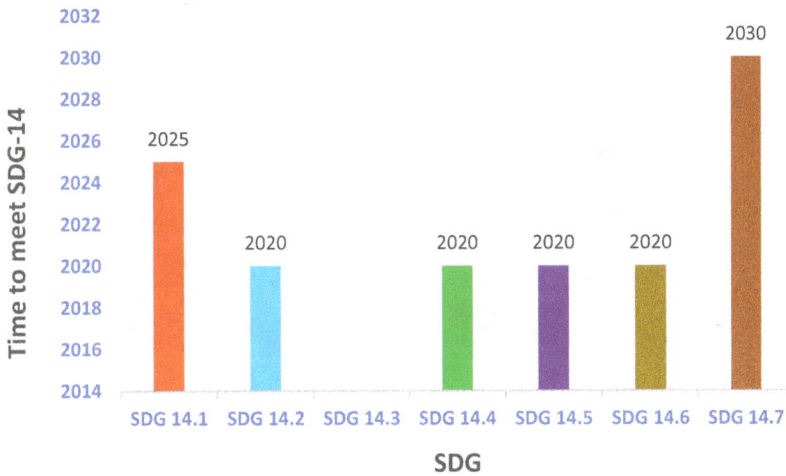

FIGURE 6.3 Time Line for SDG-14 Targets (United Nation, 2016).

over Myanmar (14 March 2012) and India (07 July 2014), the blue economy becomes a hot topic/buzzword. Bangladesh can earn a huge amount of international money by utilizing the Bay of Bengal's resources. About 255 trawlers, 67669 mechanized and non-mechanized boats engaged in fishing. 6.55 lakh MT fish were caught from the ocean in the year 2017-18 (DoF, 2020).

The Bay of Bengal is one of the world's largest bays (Figure 6.4), with 64 bays in total. It is the Indian Ocean's unprotected area, which is 1,300 miles long and 1,000 miles wide, bordered on the west by Sri Lanka and India, on the north by Bangladesh, and on the east by Myanmar and Thailand. The most important feature of the Bay of Bengal is that it is located downstream of one of the most active deltas in the world (Shoeib and Rahman, 2014). The Bay of Bengal (BoB) is also important for politics of influence as well as geopolitical tension. It's considered that the BoB is the biggest reserve source of offshore hydrocarbon including gas, oil, and living resources like fishing resources, herbs, and corals. Capital, both living and non-living is important for the economic development of Bangladesh. Sustainable use and management of fisheries will increase jobs and food security.

Bangladesh is situated in a lucrative geographic location that will grow a significant economy capable of generating US$ 100 trillion by 2030 (Roy, 2017). The contributions from the ocean economy sectors in Bangladesh are shown in Figure 6.5.

FIGURE 6.4   Maritime region (Bay of Bengal) of Bangladesh.

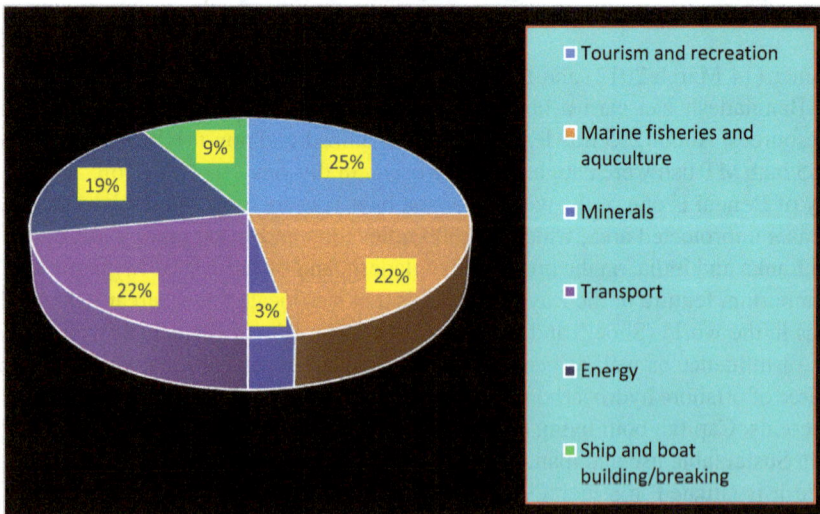

FIGURE 6.5   Contribution of ocean economy sectors in Bangladesh, percentage of GVA (2014–18).

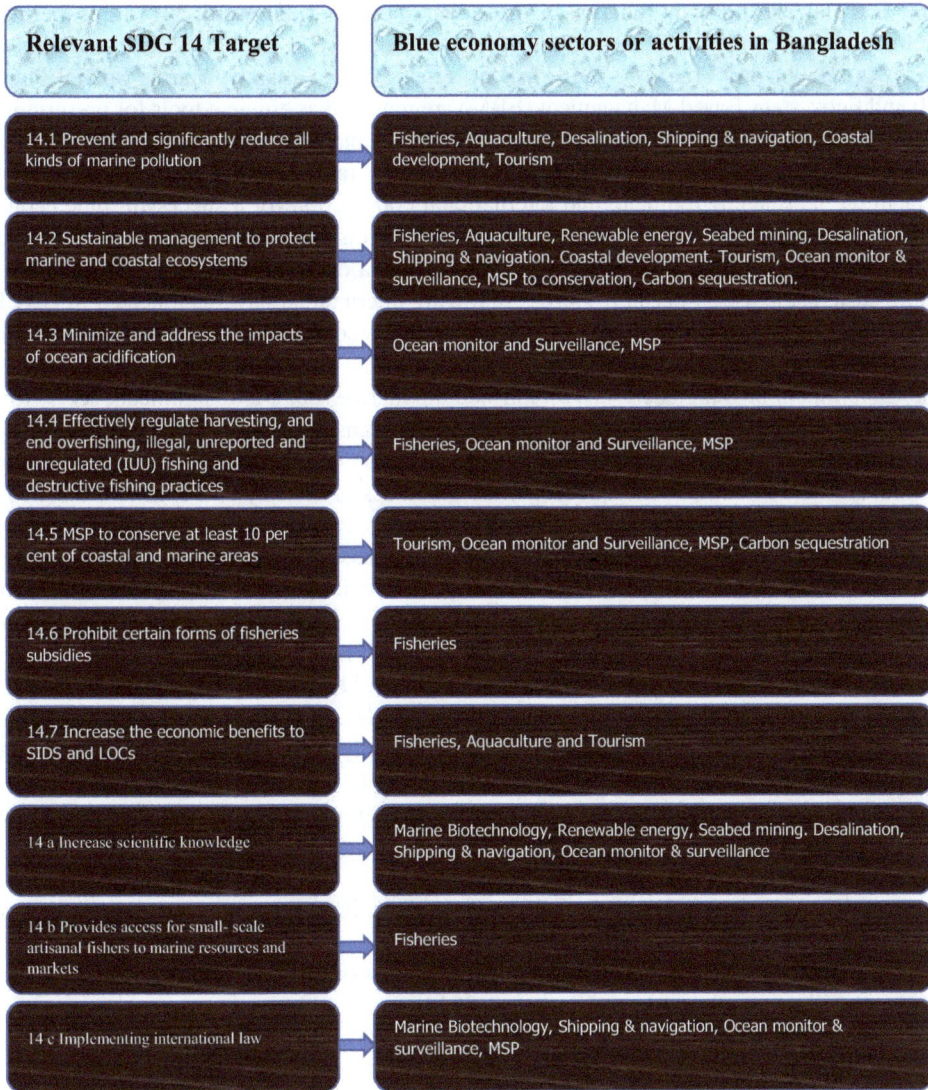

Relevant SDG 14 Target	Blue economy sectors or activities in Bangladesh
14.1 Prevent and significantly reduce all kinds of marine pollution	Fisheries, Aquaculture, Desalination, Shipping & navigation, Coastal development, Tourism
14.2 Sustainable management to protect marine and coastal ecosystems	Fisheries, Aquaculture, Renewable energy, Seabed mining, Desalination, Shipping & navigation. Coastal development. Tourism, Ocean monitor & surveillance, MSP to conservation, Carbon sequestration.
14.3 Minimize and address the impacts of ocean acidification	Ocean monitor and Surveillance, MSP
14.4 Effectively regulate harvesting, and end overfishing, illegal, unreported and unregulated (IUU) fishing and destructive fishing practices	Fisheries, Ocean monitor and Surveillance, MSP
14.5 MSP to conserve at least 10 per cent of coastal and marine areas	Tourism, Ocean monitor and Surveillance, MSP, Carbon sequestration
14.6 Prohibit certain forms of fisheries subsidies	Fisheries
14.7 Increase the economic benefits to SIDS and LOCs	Fisheries, Aquaculture and Tourism
14 a Increase scientific knowledge	Marine Biotechnology, Renewable energy, Seabed mining. Desalination, Shipping & navigation, Ocean monitor & surveillance
14 b Provides access for small- scale artisanal fishers to marine resources and markets	Fisheries
14 c Implementing international law	Marine Biotechnology, Shipping & navigation, Ocean monitor & surveillance, MSP

FIGURE 6.6 Blue economy sectors and activities' importance to SDG 14 targets (Hussain, 2019).

Linkages between the blue economy, sustainable development, and economic growth have been recognized in different national and international conferences, seminars, and forums. Unfortunately, the Bay of Bengal and its marine habitats on the coast are being threatened by the rapid pace of urbanization, population growth, IUU fishing, marine pollution, climate change, ineffective marine regulation, the non-traditional security menace, and other land-based issues (Rahman, 2017; IPCC, 2018). Rising water temperature, rising sea level, ocean acidification, changing ocean circulation patterns, and increasing nutrient input have a great impact on ocean productivity, species composition, biodiversity, and population dynamics (Harley et al. 2006; Doney et al. 2012).

A sustainable blue economy is required to focus on the conservation and justifiable exploitation of aquatic resources (Wenhai et al. 2019). SDG targets and blue economy sectors are presented in Figure 6.6. A good conceptual framework for the blue economy can be used to measure coastal management sustainability and a blue economy management framework promotes blue growth and achieves sustainable development goals (SDGs) (Keen et al. 2018; Sarker et al. 2018). Different stakeholders play an important function in blue economic growth (Howard, 2018). For the convergence of the blue economy and the marine ecosystem, ecosystem accounting is closely linked to blue growth (Häyhä and Franzese, 2014; Lillebø et al. 2017). Globally, Marine Spatial Planning (MSP) and Ecosystem-Based Management (EBM) are regarded as efficient tools for preserving and developing coastal and ocean resources in a sustainable manner (Danovaro et al. 2017).

Bangladesh needs to increase oceanographic awareness, research capacity creation, technology transfer in the marine environment, and international cooperation for a successful blue economy. The 2030 Agenda of SDGs comes when Bangladesh becomes an upper-middle-income country by 2021 and will be a developed country by 2041. To be a developed country, Bangladesh needs to ensure sustainable use of ocean resources following international law (as reflected in UNCLOS III) and formulate a strong ocean policy. For sustainable ecosystem management, marine spatial planning can be implemented.

The Government of Bangladesh has established the marine sector as a significant sector to contribute to increasing food security, alleviating poverty, increasing job opportunities, and lifting the country's trade and industrial profiles in terms of the 'blue economy'. This would have positive economic and environmental consequences if the resources of the Bay of Bengal are handled sustainably (Singh, 2019).

Bangladesh, being a littoral state in the Bay of Bengal is situated in Southeast Asia, facing difficulties with ocean governance, maritime security, and regional/international cooperation to get blue economy benefits from the Bay of Bengal. This paper uncovers the prospect and opportunities of a blue economy for Bangladesh compared to the rest of the ocean-based countries. This article also focuses on the challenges and risks faced and which will be faced by Bangladesh in near future towards the blue economy's effective implementation. It also emphasizes the sustainable development pathways to overcome these challenges and risks. Finally, it provides some overall and sectoral recommendations for sustainable blue economic growth.

The economic contributions from the Oceans at the regional, national, and sub-national levels include:

- Australia: AU$ 47.2 billion contribution to GDP in 2012, or more than 3% of the overall (National Marine Science Committee, 2015);
- China: In 2010, the overall gross value added (GVA) was US$ 239 billion, or 4% of GDP, with over 9 million people working (Zhao et al. 2014).
- European Union: Complete annual GVA of EUR 500 billion with over 5 million employees (EC, 2017).
- Ireland: In 2016, the gross GVA amounted to EUR 3.37 billion, or 1.7% of GDP (Vega and Hynes, 2017).
- Mauritius: For the period from 2012 to 2014, on average, 10% of GDP (Cervigni and Scandizzo, 2017).
- United States: In 2013, the contribution to GDP was US$ 359 billion, or more than 2% of the total, employing 3 million people (Kildow et al. 2016).

Different maritime industries adding value to the global ocean economy are shown in Figure 6.7.

6.2 MARITIME HISTORY AND THE VICTORY OF BANGLADESH

The Father of the Nation, Bangabandhu Sheikh Mujibur Rahman, predicted the need for marine resources for the country's overall development after the War of Liberation in 1971. The Territorial Water and Maritime Zones Act 1974 was implemented long before UNCLOS III came into being in 1982, under his visionary leadership. The Honorable Prime Minister Sheikh Hasina subsequently carried forward Bangabandhu Sheikh Mujibur Rahman's vision and agreed on 25 July 2001 to ratify the UNCLOS, which also granted the right to claim the continental shelf or to claim the continental

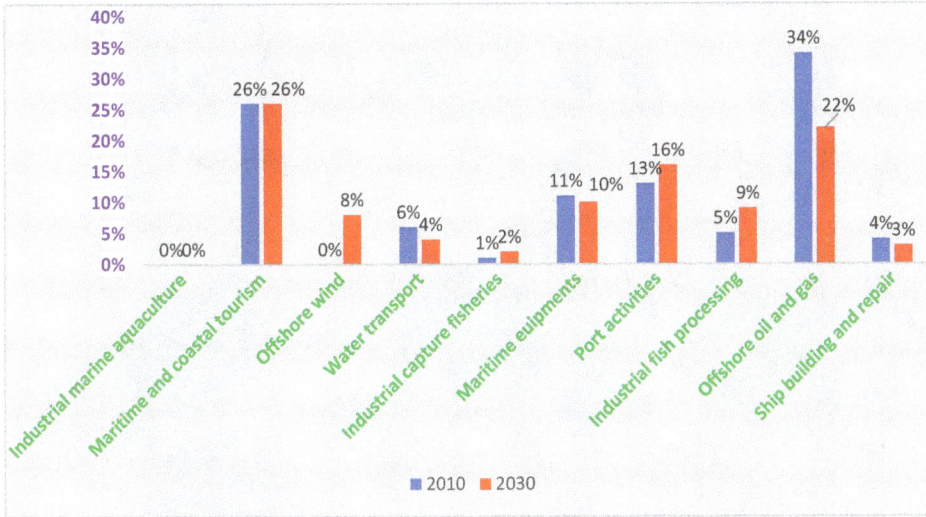

FIGURE 6.7   Ocean industry value-added to double (1.5 to 3 trillion dollars) by 2030 (OECD, 2016).

shelf within 10 years. Afterward, under the sensible leadership of the Honorable Prime Minister Sheikh Hasina, the International Tribunal for the Law of the Sea resolved the maritime boundary demarcation with Myanmar in 2012 and with India at the Permanent Court of Arbitration in 2014. Bangladesh gained sovereignty over the Bay of Bengal's living and nonliving capital. Bangladesh won over 118,813 sq km of the territorial sea, 200 nautical miles (NM) of the Exclusive Economic Zone (EEZ), and the continental shelf up to 354 nautical miles off the coast of Chittagong. It is a large region, with economic and commercial concerns and environmental stakes. It, therefore, guarantees adequate defense and security, and this confirmation is Bangladesh's outstanding achievement (Bhuiyan et al. 2015). The Bay of Bengal has enormous potential as a source of the blue economy. Bangladesh can take advantage of its ocean-borne properties, such as research into oil and gas, fisheries, shipbuilding and shipbreaking, salt collection, tourism development, and the like (Figure 6.8).

6.3   MATERIALS AND METHODS

6.3.1   STUDY SITES

Bangladesh is a maritime country, blessed with the Bay of Bengal, which is around 1.5 times bigger than the area on the land. The coastal zone occupies an area of approximately 36,000 sq. km, almost 25% of the total land surface of the country. The continental shelf is roughly 37000 sq. km and has an area of around 37000 sq. km. The Exclusive Economic Zone (EEZ) has roughly 1,64,000 sq. km (Begum, 2013). Almost all forms of activities in the economy are connected to this maritime field. So, one of the latest and very prospective avenues is the maritime sector for the development of the country.

6.3.2   DATA COLLECTION

Identification of the related research was part of the initial phase. We developed the following requirements for our database to perform a systematic literature review:

- **Searching database:** Scopus, Web of Science, Google Scholar, PubMed, Dimension
- **Searching conditions:** English-language journal articles; Blue economy-related journals, book chapters, conference proceedings; available on the internet (with no time limitation)
- **Searching strings**

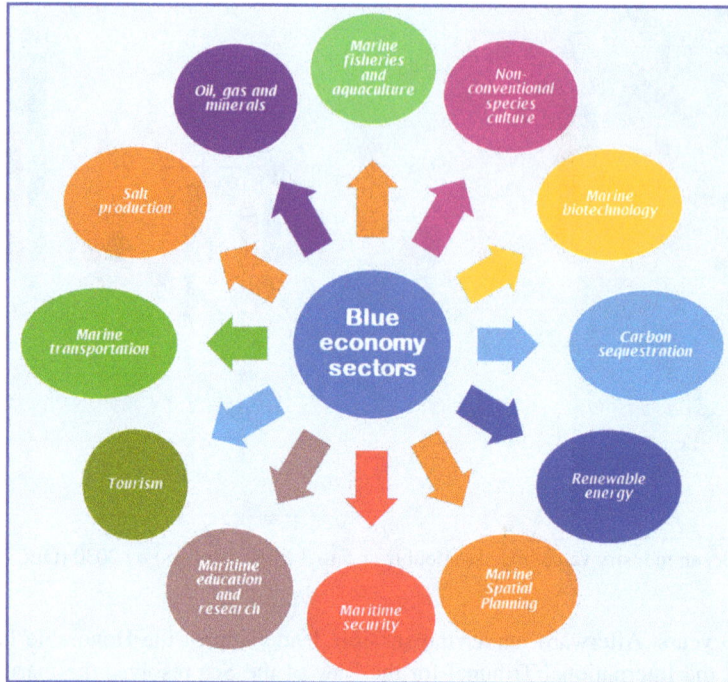

FIGURE 6.8   Sectors of blue economy in Bangladesh.

Data was collected using keywords: blue economy concept; sustainable blue economy; blue economy in Bangladesh; the prospect of the blue economy in Bangladesh; opportunities of the blue economy; opportunities of the blue economy in Bangladesh; risks towards successful blue economy; risks towards the successful blue economy in Bangladesh; threats towards successful Blue Economy; threats towards the successful blue economy in Bangladesh; sustainable development pathways for the blue economy; recommendation for sustainable blue economic growth; recommendation for sustainable blue economic growth in Bangladesh.

6.4   PROSPECT OF A BLUE ECONOMY IN BANGLADESH

Being a maritime country, Bangladesh gained a vast area of the ocean which can play a dynamic role in its national economic development and the well-being of its future generations. Since land resources are depleting rapidly, people will depend on ocean resources for their livelihoods in the future. Bangladesh has a huge prospect of a blue economy, that's why the government is giving priority to the collaboration and formulation of ocean-related policies and plans. Bangladesh can fulfill the blue economic growth with marine fisheries and aquaculture, minerals and mining, oceanic renewable energy, increasing maritime trade, shipping, and transport, tourism, education and research on maritime affairs, marine spatial planning, and so forth. These maritime activities will increase job opportunities, reduce poverty, and develop the national economy.

6.5   OPPORTUNITIES FOR A BLUE ECONOMY IN BANGLADESH

Approximately half of the world's population, most of its major cities and industries, along with vital supply chains, tend to be centered along the coast to ensure access to transport routes and ensure steady flows of energy and goods. Worldwide, different countries make significant income from the blue economy sectors and boost their national GDP (Table 6.1).

TABLE 6.1

Selected Estimates of Value of Ocean-Based Industries, by Country, Region and World

Country	Author	Date of Study	Date of Data	Contribution of Ocean Sectors to GDP or GVA	% of GDP or GVA	Employment (total) FTE
China (People's Republic of)	APEC	2014	2012		9.6% GDP	=
	Jiang et. al.	2014	2000–11		13.83% GDP	=
	CMIEN	2013	2012	CNY 50,087 tn GDP	9.6% GDP	34 0240 000
	Zhao, Hynes and He	2013	2010	CNY 239.09 bn GVA	4.3% GDP	9 000 000
Dubai	Gujarat Maritime Board	2014	2013		4.6% GDP	=
France	Kalaydjian et al.	2009	2007	EUR 28 bn GVA	1.4% GDP	484 548
	Kalaydjian et al.	2011	2009	EUR 26,122 bn GVA	2.5% GDP	460163
	Kalaydjian et al.	2014	2012	EUR 30,252 bn GVA	2.75% GDP	460396
Hong Kong (China)	Gujarat Maritime Board	2014	2013		25% GDP	=
Iceland	Sigfusson and Gestsson	2012	2010		26% GDP	ca. 30 000
Ireland	Vega, Hynes and O'Toole	2015	2012	EUR 1.3 bn GVA	0.7% GDP	17 425
	Vega, Hynes and Coriess	2013	2010	EUR 1.2 bn GVA	0.7% GDP	16614
Japan	Nomura Research Institute	2009	2005	JPY7 863 bn GVA	1.6% GDP	981 234
Korea	APEC	2014	2005		8% GDP	=
	Hwang et al.	2011	2008	KRW13,435 bn GVA		
	4.9% GDP	919314				
Netherlands	Maritime by Holland	2014	2012	EUR 21 bn GVA	3.3% GNP	224 000
New Zealand	Statistics New Zealand	2006	1997–2002	NZD 3.3 bn GVA	2.9% GDP	21 000
Portugal	DGPM	2013	2010		2.5% GVA	=
Singapore	MPA - Maritime Singapore	2014			7% GDP	=
United Kingdom	Pugh	2008	2005–06	GBP 46,041 bn GVA	4.2% GDP	890 416
United States	Kildow et al.	2014	2010	US$ 258 bn GDP	4.4% GDP	2.8 million
Europe	Ecorys	2012	2011	EUR 495 bn GVA		5.6 million
				US$ 2.5 trillion		=
Worldwide	Hoegh-Guldberg et al.	2015	2011–14	'Gross marine product'	3.2% GDP	=

Source: Individual reports by country and region.

Notes: = data not available. The German study focuses only on maritime technology and ocean engineering. FTE = full-time equivalent. The value added of Iceland and the China Marine Statistical Yearbook include also indirect effects on the economy.

In Bangladesh, about 30 million people connect with marine resources and ocean-related work for their livelihoods. In the blue economy sector, Bangladesh has a lot of potential. It aims to be a developed country by the year 2041 and this is possible if we can use ocean resources properly. Few statistics and a birds-eye overview of marine sectors will prove the above statement. The contribution (from 2010-2015) of major blue economic sectors in Bangladesh is tabulated in Table 6.2.

6.6 MARITIME SHIPPING AND TRANSPORTATION

Shipping is the safest, most secure, efficient, and environmentally sound transportation system that is regarded as the lifeblood of the global economy. 90% of global commerce, 80% of global trade by volume, and 70% by value are being transported by sea (Alam, 2018).

The shipping sector is developing very fast in Bangladesh because of its geographic location. The Bay of Bengal is a blessing for Bangladesh (Hung et al. 2010). Bangladesh is earning 80 billion dollars from imports and exports and thus shipping trade is increasing by 10–15% every year (Uddin et al. 2017). These growth percentages are higher than average in terms of the Bangladeshi national economy (Alam, 2014). In 2011, during the global economic crisis, global trade increased by 4%. By 2030, container traffic is expected to triple. The top import and export partners were China, India, Europe, Indonesia, Singapore, and Thailand (World Bank, 2017). Shipping provides facilities to landlocked South Asian countries and connects South, Southeast, and East Asia as the Indo-Pacific Corridor.

Bangladesh earned about US$ 67 billion (2013–14) from exported and imported goods transported by 2500 ships from other countries docked in the country's ports. Moreover, US$ 95 billion was paid by importers, exporters, and buyers who paid shipping firms for freight and associated costs, airlines, and freight operators to carry goods in and out of Bangladesh during the last ten year (Alam, 2014). In 2012, inland and coastal networks transported 231.5 million passengers and 32.6 million Mt of freight (Alam, 2018). Over the last ten years, the annual rate of increase in imports was 15.79% and the export growth rate was 15.43%. According to these estimates, freight value is predicted to be around US$ 435 billion over the next ten years. To keep US$ 400 billion in Bangladesh over the next ten years, the country needs to add more ships to its current fleet and more freight operators.

Maritime transport plays a dynamic role in Bangladeshi social-economic prospects that mostly rely on their ability to connect to the rest of the world and access international markets.

Coastal shipping from India, Sri Lanka, Singapore, Malaysia, Thailand, and Myanmar ports can play a great role in achieving 5–6% long-term annual growth for the coming decade. Transshipment at Singapore, Kelang, Colombo, and other ports could be cost-effective, time-saving, and favorable for employment growth.

6.6.1 Port Facilities

Ports play a strategic role in national trade and economy. For maritime economic activities, ports are critical crystallization sites. Port facilities are needed for cruise ships, coastal shipping, foreign shipping, passenger ferries, fishing, marine mineral mining, oil drilling, and offshore or maritime surveillance (Alam, 2018).

Bangladesh's economy largely depends on international trade. Ports transport 94% of foreign trade. almost 3000 foreign ships visit Bangladeshi ports every year carrying cargo. Bangladeshi ports are not located adjacent to main international shipping lanes and have a draft restriction. Have

TABLE 6.2

Financial Evaluation of Major Blue Economics Sectors in Bangladesh from 2010 to 2015 (million US$)

Economic Sector	2010	2011	2012	2013	2014	2015
Marine fisheries	843.75	949.48	1107.42	1231.06	1384.77	1475.66
Oil	21.90	23.84	26.82	28.77	29.35	34.05
Gas	948.35	956.30	1041.35	1127.73	1158.13	1305.42
Sea salt	119.25	123.48	160.90	206.00	212.35	214.84
Sand, Mineral and Coals	735.18	944.39	1183.79	1452.46	1644.08	1893.14
Water Transport	1215.14	1330.36	1450.21	1606.10	1682.31	1816.67
Trade and Shipping	31390.15	36178.04	41728.94	47156.44	52078.80	58466.90

Source: Data adapted from Bangladesh Bureau of Statistics (BBS, 2017; Hussain et al. 2017).

TABLE 6.3
Annual Gross Value Added from Bangladesh's Blue Economy

Ocean economy industry/service (Nominal USD Millions)	2009–10	2010–11	2011–12	2012–13	2013–14	2014–15
Marine capture fisheries	664.00	777.00	786.23	907.49	1,037.49	1,167.79
Marine aquaculture and shellfish farming (shrimps and crabs)	78.65	92.48	99.76	122.05	144.99	163.20
Sea salt production	123.20	124.11	145.51	184.35	195.45	197.88
Crude petroleum extraction	22.42	23.65	23.69	25.16	26.40	30.55
Natural gas (liquid) extraction	971.13	948.62	919.94	986.25	1,041.87	1,174.58
Maritime freight transportation	307.90	319.55	295.81	300.33	327.15	375.58
Maritime passenger transportation	617.61	659.27	606.66	663.14	720.69	788.35
Port and harbour operations	104.95	103.29	135.57	145.32	172.37	202.17
Shipbuilding and repairing	110.32	114.77	106.68	109.58	108.59	387.06
Ship breaking	127.39	130.80	134.27	136.83	138.31	138.21

Source: World Bank, 2018.

no recommended routes or an international ship traffic separation scheme. Consequently, all these constraints create an adverse situation for foreign ships coming to Bangladesh (Uddin et al. 2017).

Currently, there are only two ports in Bangladesh, including Chittagong Port and Mongla Port to support this huge trade volume. Bangladeshi ports, along with Kolkata and Chennai, serve as hubs within the Bay of Bengal. Facilities and infrastructure development of ports will increase the chances of being a central focus of the country's future ocean economy. Matarbari port, Payra Port, Sonadia Deep Sea Port, Kutubdia FSRU LPG, and the LNG Terminal and Bay terminal would be suitable sites to develop seaports with more capabilities and modern handling equipment (Saha and Alam, 2018). This will open a new window in maritime trade and commerce. The establishment of new seaports will reduce export times and earn a steady flow of revenue for the country.

6.7 SHIPBUILDING

Shipbuilding is regarded as an important sector that is a vital component of the blue economy because of its income generation feature (Table 6.3). Marine engineering consists of the construction and repair of boats, ships, fishing vessels, yachts, and floating structures that need routine maintenance. In Bangladesh, there are currently over 300 shipyards and workshops. These use almost 100% capacity for inland vessels, fast patrol boats, dredging barges, passenger vessels, landing craft, tug, supply barges, deck loading barges, speed boats, cargo coasters, troop-carrying vessels, hydrographic survey vessels, survey boats, pilot boats, water taxis, and pontoons, which are being built by these yards (Alam, 2018). Currently, shipbuilding yards are building 10,000 DWT seagoing ships for exportation and are projected to advance their capabilities to 25,000 DWT. Annually, approximately 15 ships are undergoing repairs in the dry docks of Bangladesh and contribute to earning foreign exchange.

6.8 SHIPBREAKING OR RECYCLING

Bangladesh recycles 300 ships every year, which is 24% of the total scrapped ships in the world (MOI, 2019). Bangladesh ranked 2nd considering the quantity of ships while rated third in terms of gross tonnage. Bangladesh has 105 yards and around 60-70 percent of the yards are in operation

(Alam et al. 2019). As a result, Bangladesh is considered a graveyard of scrap ships of the world that produces continuous capital for Bangladesh (Table 6.3).

Around 70 - 75% of scrap steel used in steel and re-rolling mills comes from the ship recycling industry. As a result, this industry is harnessing a large amount of foreign currency. Furthermore, this industry met the rising demand for furniture, all forms of household fittings, boilers, life-saving vessels, and generators, as well as creating jobs.

Ship recycling industries are the main polluters of the coastal environment. Ship recycling should be promoted with an all-eco-friendly infrastructure and compliance with an international agreement. The Bangladesh government launched steps to control pollution from ship recycling. A ship recycling yard has to obtain clearance from the Environment Department before recycling a ship (Rabbi and Rahman, 2017).

6.9   COASTAL AND MARINE TOURISM

Tourism is a huge global business that creates new jobs and reduces poverty since tourism is human resource intensive. Furthermore, every job created in the core industry generates one and a half jobs in the tourism-related economy. International tourism has increased from 25 million in 1950 to 1.035 million in 2012, with the UNWTO estimating average growth of 3.3% a year, with a projection of 1.8 billion for 2030. In 2016, tourism and leisure contributed just over 10 billion US dollars to the national GDP and more than 2 million direct and indirect jobs (WTTC, 2017).

According to the United Nations World Tourism Organization (UNWTO), approximately one of every two tourists visited the seaside (UNWTO, 2013). Rising incomes and low transport costs make coastal and ocean tourism more popular. Cruise tourism is getting more popular day by day. Between 1970 and 2005 the number of passengers increased 24-fold to 16 million by 2011. Annual passenger growth rates average about 7.5%, with annual passenger expenses expected to be around US$ 18 billion.

Marine and coastal tourism are important to Bangladesh (one of the developing countries) since it offers 5% of world GDP and contributes to 6 - 7% of total employment globally. In 2012, tourism supported 9% of Global GDP, 9% of global jobs were created, and US$ 1.3 trillion, or 6% of global export earnings, was produced (Alam, 2018). Bangladesh can be a major tourist attraction site since it has Cox's Bazar (longest unbroken sea beach), St. Martin's Island (Coral Island), and the Sundarbans (largest single tract mangrove forest). Tourism can create employment for local communities through business facilities (hotels, restaurants, other entertainment industries, and so forth). Tourism will be sustainable if it focuses on the other factors of the environment (Figure 6.9).

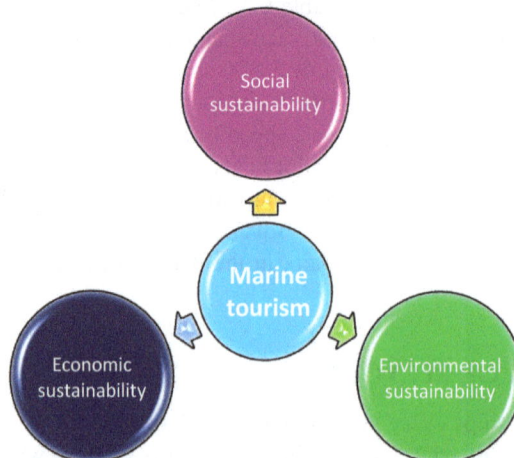

FIGURE 6.9   Marine tourism and three facets of long-term development.

In Bangladesh, coastal and marine tourism is not developed in the same way as in other south Asian countries (for example, India, Sri Lanka, and Maldives). Recreational fishing is very important but the facilities in Bangladesh are limited (Hassan et al. 2014). In the country's coastal waters, there is no recreational fishing (Humayun et al. 2016). Recreational fishing is very small, according to the United Nations, with imports of fishing rods, reels, hooks, and other tackle totaling just US$127,180 in 2013 (UN Comtrade, 2017).

Diverse water sports have been introduced as recreational activities. The construction and maintenance of seaworthy pleasure boats, and the requisite infrastructure support, including marina ports, could promote a rise in coastal tourism. About 100 hotels and motels were recorded in Cox's Bazar where more than 10,000 people work (Bari, 2017).

6.10 CONSTRUCTION OF ARTIFICIAL ISLANDS

Bangladesh should follow an effective strategy for the creation of new islands formed artificially in the territorial sea and the EEZ authorized by the 1982 UNCLOS to minimize demographic pressure on land. By planting salt-tolerant/mangrove plants, the survival of the current 75 marine islands or newly developed islands can be ensured. Agricultural production on sandy soils through the enhancement of existing crops must be adopted. Seawater desalination for the use of freshwater for agriculture, irrigation, commercial use for inhabitants, and marine/offshore island animals may be considered.

6.11 MARINE FISHERIES

Almost 350 million people are working with marine fisheries worldwide (FAO, 2012). Globally, 15.7% of animal protein demand is met by fish. Fish's contribution to the supply of animal-based protein in Bangladesh is 52%, Indonesia 68%, Malaysia 61%, Sri Lanka 65%, and Thailand 52%. The global catch of fish was 4 million tonnes in 1900, 16.7 million tonnes in 1950, 62 million tonnes in 1980, and 86.7 million tonnes in 2000 but then it plateaued. In 2009 marine capture production was 79 million tonnes. Overall catch probabilities declined with 75% of stocks fully exploited or depleted. Almost 90% of fishermen living in developing countries earned US$ 25 billion from fish trading.

Bangladesh is blessed with a Bay of Bengal wide range of shrimp, fish, molluscs, crabs, mammals, seaweed, and other marine creatures. There are about 740 species of fish found in the EEZ of Bangladesh (Habib and Islam, 2020). Total fish production in Bangladesh was reported as 3.68 million tonnes, of which 1.0 million tonnes (28%) come from capture fisheries, 2.2 million tonnes (56%) from aquaculture, and 0.6 million tonnes from marine fisheries (16%). It contributes 3.69% of the GDP of the nation (DoF, 2016). Artisanal small-scale fisheries contribute 86.8%, while trawl fishing (considered a large industrial fishery) accounts for 14.2% of the total marine output (DoF, 2016). The last 5-year fish production from different sources is tabulated in Table 6.4.

The Bay of Bengal's coastal and offshore fishing industry plays a vital role in economic development, social development, and ecological balance. A recent estimate reported that 1.85 million people are engaged in fisheries on a full-time basis covered by the Bay of Bengal Program (BOBP). Other countries catch 8 million tonnes of fish in the Bay of Bengal, while Bangladesh only catches 637379 tonnes (DoF, MoF and LS, 2018).

Fishing activities in Bangladesh can be separated into artisanal fishing and industrial fishing (Islam et al. 2017). 70,000 wooden fishing boats (Lengths 30 to 35 feet) are fishing on the coastline since they cannot go beyond 20 to 30 km from the coastline. Most of the wooden vessels are not registered (DOF, 2018). Over 250 steel body registered trawlers go up to 50 nautical miles for fishing. Bangladesh has a 660 km long sea area from the coastline, and 2,300 m deep sea. In the

TABLE 6.4
Last 5 Years' Fish Production

	Source-wise Production (MT)			
Year	Inland Open	Closed	Marine	Total
2013–2014	1054585	1460769	546333	3061687
2014–2015	1029937	1351979	517282	2899198
2015–2016	1123925	1062801	514644	2701370
2016–2017	1060181	1005542	497573	2563296
2017–2018	10067761	955812	634746	

Source: MoF and LS.

Bay of Bengal, most of the deep-sea fish are dying due to aging because they are not being caught due to a lack of appropriate technology to go deep-sea fishing. Since a high number of fishing boats fish near the shoreline, it creates conflicts with other ocean users (for example, shipping, tourism, and gas exploration) (DOF, 2018). Bangladesh has few fish processing zones or factories in its coastal area to export different types of fish including, Hilsa, tuna, shrimp, and crab (Hussain et al. 2018).

6.12   MARINE AQUACULTURE

Aquaculture is considered to be the world's fastest-growing food market. Fish for human consumption makes up 47% of the total and is provided by the aquaculture sector. Between 1960 and 2009, fish production increased by more than 90 million tonnes (from 27 to 118 million tonnes) which are mostly used as food for human consumption 26. It is projected that aquaculture fisheries production will surpass capture fisheries production soon. In Asia, about 91% of aquaculture is taking place, contributing to over 89% of global production (more than 5%/year).

Bangladesh has huge potential for aquaculture development. Aquaculture offers food and a means of subsistence that ultimately helps in poverty reduction. It will consider the importance of natural resources in its production whilst adhering to ecological guidelines throughout the production cycle, ensuring long-term decent jobs, and supplying high-value export commodities.

Marine aquaculture based on tiger shrimp (*Penaeus monodon*) culture has evolved into a heavily traded export industry. Tiger shrimp is largely cultured in the coastal districts of Satkhira, Khulna, Bagerhat, and Cox's Bazar. Between 1970 and 1990, the culture in these areas grew steadily, covering approximately 183,221 hectares (Belton et al. 2011). Softshell crab (*Scylla serrata*) and finishes are export items also being cultured. In 2015, Bangladesh received US$582 million from shellfish and finfish exports (DoF, 2016). Sea bass (*Lates calcarifer*) is a high-value aquaculture species that can be bred and farmed in the Philippines, Thailand, and Vietnam.

Increasingly, fishmeal is made from fishery by-products that constitute more than 25% of global production (OECD, 2012). Findings show that at least 50% of fishmeal, 50-80% of salmonid oil, 30-80% of fishmeal, and up to 60% of marine fish diet oil can eventually be supplemented by vegetable replacements that greatly expand the potential for industry expansion.

There is a high export value and great potential for marine aquaculture species. However commercial marine aquaculture, as well as marine stock enhancement and sea ranching, are all promising but are still to be established.

6.13 NON-CONVENTIONAL MARINE SPECIES CULTURE

Non-conventional marine species include seaweed, other macroalgae, mussels, oysters, and other shellfish (edible oysters, *Crassostrea* sp. *Saccostrea* sp. pearl oyster, *Anadra* sp. green mussel, *Perna viridis*, clam, *Meretrix* sp., *Marcia opima*, sea snails), sea urchins, sea cucumbers, and so forth. Favorable sites for seaweed, sea urchin, and sea cucumber culture are Cox's Bazar, St. Martin's Island, and the coastal areas. These marine invertebrates are used for valuable cancer-treating pharmaceuticals in many developing countries. Marine pearl culture can be developed in Bangladesh's suitable inshore and coastal areas following Japan and other countries

Crab fattening was carried out on a limited scale, and these were exported to other countries. Over the last 10–15 years, live giant mud crabs (*Scylla serrata*) and estuarine eels (*Muraenesox bagio*) have been exported to East Asian countries. The marginal farmers of Satkhira, Bagerhat, and Cox's Bazar produced less than 20% of the exported live crab. Softshell crab farming is being practised on a limited scale in Sathkhira, Cox's Bazar, and the Moheshkhali areas and this could also be extensively practised in other feasible coastal areas such as Cox's Bazar, Moheshkhali, Kutubdia, Chittagong, Khulna Bhola, and the Barisal regions.

6.14 FOOD SECURITY

In developing countries like Bangladesh, one billion people rely on seafood as their main protein source. Apart from that, many people all over the world enjoy seafood. The oceans can be the largest source of food for all developing countries, and they can also help to solve food security concerns.

6.15 ENERGY

6.15.1 GAS AND OIL

The seabed is a hidden source of treasure that currently provides 32% of the global supply of hydrocarbons which had been 20% in 1980. Gas and oil have long been the key sources of energy in the industrialized world, and they will continue to be so for many decades to come. Offshore fields accounted for 32% of global crude oil output in 2009, and this figure is expected to grow to 34% by 2025, and even higher thereafter. Methane hydrates, a potentially massive source of hydrocarbons, are now being investigated and tapped from the ocean floor.

In Asia, Bangladesh is regarded as the 19th largest producer of natural gas (Alam et al. 2019). Geologists predicted that within its maritime boundaries, Bangladesh might have a huge amount of gas and oil reserves as do India and Myanmar. Adjacent to the Arakan offshore blocks of Myanmar, there is the possibility of finding significant oil and gas reserves.

A total of 20 wells were dug in the Bay of Bengal's offshore locations until 2014. Only two gas reserves were discovered during the exploration (Hossain et al. 2014). Today, Bangladesh discovered 26 blocks in the Bay of Bengal. Among 26 blocks, 15 are deep-sea blocks and 11 are shallow water blocks (Bari, 2017). Bangladesh has been producing and using natural gas from offshore block-16 since 1998. Primary tender and negotiations took place on blocks 4, 9, 11, 12, 16, and 21 (Hussain et al. 2018).

26 TCF (Trillion Cubic Feet) of the gas reserve has been discovered in Bangladesh. About 1 TCF of gas was recorded in offshore gas reserves where the drilling success ratio is also less attractive (9:1) compared to onshore (3:1) (Hossain et al. 2014). 0.8 TCF gas in the Sangu reserves has already depleted. 0.04 TCF gas reserves in the Kutubdia reserves are yet to be established. To generate commercial quantities of hydrocarbons, the Magnama (3.5 TCF) and Hatia (1.0 TCF) fields must be drilled.

Bangladesh needs to explore the gas and oil fields (Alam, 2013). More extensive research, drilling rigs, specialized support ships, platform building, production, and exploitation is required. Downstream activities include refining and distribution to consumer markets.

6.15.2 MINERAL RESOURCES

The world is trying to explore and exploit mineral deposits (for example, cobalt, copper, and zinc) on and beneath the seafloor. The ocean floors may contain 5% of the world's minerals, such as cobalt, copper, and zinc. By 2020, 5% of the world's minerals will have made their way to the ocean floor, with the number increasing to 10% by 2030. In the next ten years, global annual mineral mining is projected to rise from virtually nothing to €5 billion, and then to €10 billion by 2030.

Polymetallic nodules, cobalt crusts, massive sulfide deposits, yttrium, dysprosium, and terbium are important in renewable energy systems and new ICT hardware due to increasing product prices in industries. The International Seabed Authority Initiated Mining Code regulates and licenses bodies to mine in the international sea bed. Coastal countries like Bangladesh need to prepare to explore ocean resources to the best of their abilities from the EEZ.

Bangladesh has the prospect of finding valuable heavy minerals from Patenga to the Teknaf belt (250 km) (Table 6.5). According to the Beach Sand Exploration Center, 17 deposits discovered in this belt enriched with minerals like zircon, rutile, ilmenite, leucoxene, kyanite, garnet, magnetite, and monazite can be useful (Alam, 2004). Extraction in the right way and management of minerals will promote different industries (for example, welding electrodes, glass, paper, and ceramics) and will provide a large number of job prospects for the local population (Alam, 2004).

Mineral extraction and commercialization from beach sand may boost the growth of a variety of industries (such as welding electrodes, glass, paper, and ceramics) and create huge job opportunities for local communities (Alam, 2004).

6.16 MARINE RENEWABLE ENERGY

The demand for renewable energy is expected to increase two and a half times by 2035 all over the world. Renewable energy enjoys an almost 22% share of the global energy mix. There are several

---

TABLE 6.5
Heavy Mineral Reserves are Located in Deposits Along Bangladesh's Coastal Belt

Reserve of minerals (in MT)	Name of the sites						
	Badarmokam	Sabrang	Teknaf	Sikhali	Inani	Cox's bazar	Kuakata
Crude sand	1765000	347556	1939580	2759828	729286	5119000	2872486
Heavy minerals	411000	68582	442291	489714	175476	920000	831668
Zircon	4932	4184	38306	33300	10880	23000	9647
Rutile	3288	1372	13230	10744	4036	6440	3911
Ilmenite	94530	19614	163170	173360	53170	161000	76015
Leucoxene	18002	3470	20124	10970	439	10488	9647
Kyanite	*	727	14728	4407	1404	*	16800
Monazite	4932	206	3045	3918	965	2024	83.2
Magnetite	10275	1001	7209	3085	5545	33214	4325
Garnet	*	3018	22424	39422	12810	50229	52229

Source: Modified after Alam (2004).

*   Not estimated.

forms of marine renewable energy: (1) offshore solar energy, (2) offshore wind energy, (3) wave energy, (4) tidal energy, (5) ocean thermal energy, (6) salinity gradient, (7) ocean current energy, and (8) energy from marine biomass. In both onshore and offshore islands, marine-based renewable energy may provide an alternate source of electric power for homestead houses, small mills, and factories. It also provides a strong source of income for coastal communities that depend on fishing.

Globally speaking, offshore wind energy (electricity generation) is the most advanced ocean-based energy source. Global installed capacity was slightly over 6 GW in 2012; however, this is predicted to quadruple by 2014 and, according to optimistic projections, might reach 175 GW by 2035. The growth from producing 7100 MW of electricity in 2013 to today is astounding at a rate of 40% /year.

Bangladesh has huge potential for renewable energy. In the coastal region of Kutubdia, Bangladesh, a wind generator with a capacity of 2 MW was constructed, but it is still inactive.

6.17 SEA SALT PRODUCTION

The salt reserves in the world's oceans have already been measured at more than 50 million billion tonnes, covering more than half of the world's supply (Mannar, 1982). In Bangladesh, marine salt production has been traditionally produced mostly in Cox's Bazar coastal areas (Alam, 2014). Salt can be generated in onshore areas such as Chakaria, Cox's Bazar, Banshkhali, Teknaf, and offshore islands such as Moheshkhali, Kutubdia, and others. Salt farmers can produce about 20 tonnes/ha during dry seasons. The average production of such crude salt is about 7000–10,000 kg/ha. Production can be increased up to 20,000 kg/ha/season in these locations (Alam, 2014). The coastal region of Cox's Bazar produces 22 metric tonnes of salt per year, while Thailand's Samut Sakhon produces 43 metric tonnes (Hossain et al. 2006).

Traditionally, salt (NaCl) is produced by evaporating marine waters. Most salt farms are small-scale and run with local machinery. The land is rented from local landowners or the government on a yearly basis.

6.18 MARINE BIOTECHNOLOGY

By 2020, the global demand for marine biotechnology goods will have grown from US$ 2.8 billion to US$ 5 billion. Marine biotechnology produces pharmaceutical medicines, chemical materials, enzymes, and other industrial products and processes. Biomaterials, health-care diagnostics, fisheries and aquaculture, seafood conservation, bioremediation, and biofuels all benefit from it (Thakur and Thakur, 2006). Marine bio-resources like fish, algae, bacteria, invertebrates, and other marine organisms are used to produce bio-products needed for the benefit of mankind.

Over 36 marine drugs were discovered in 2011, of which 15 were used for cancer treatment. Yondelis was the first marine drug derived from small soft-bodied animals to fight cancer. Antiviral drugs Zovirax and Acyclovir were discovered in sponges from the Caribbean. Marine bioprospecting detected over 14,000 novel chemicals in 2006, and 300 patents on marine natural products were filed. The European Science Foundation assumes that a production volume of 20–80 thousand liters of oil per hectare per year can be achieved from microalgal cultures, considerably higher than terrestrial biofuel crops. Approximately 20% of all living marine species are screened for chemical and medicinal purposes (Schlosser, 2013).

The future of marine biotechnology in Bangladesh looks bright. In the Bay of Bengal, a large number of marine species remain untapped, with enormous potential for marine-based biotechnology products in the region. Biotechnology can create and produce new processes, products, and services (Hossain et al. 2014). The economy of Bangladesh will be developed if marine bio-resources can be turned into medicines, bioactive compounds, nutrient supplements, and food.

6.19   CARBON SEQUESTRATION AND CLIMATE CHANGE MITIGATION

The world's biggest environmental or emissions trading market is the carbon market (Newell et al. 2013). Carbon sequestration is a process of taking up CO_2 over a long period by seaweeds, mangrove forests, intertidal salt marshes, and seagrass beds. Seaweed is a powerful tool to mitigate carbon, hence reducing climate change (Bhuyan et al. 2022; Bhuyan et al. 2021). This stored carbon is referred to as blue carbon since it offers carbon credit (Trumper et al. 2009; Nellemann et al. 2009; Chowdhury et al. 2015). Carbon is the greenhouse gas responsible for climate change. Some effective measures such as solar panels, the California Climate Action Plan, and the 2015 Paris Climate Agreement, were undertaken to reduce the human carbon footprint. Carbon offsetting (credits for reducing, avoiding, or sequestering carbon) is widely considered to be the most effective method for reducing climate change (van Kooten et al. 2004; Brotto and Pettenella, 2018).

6.20   CHALLENGES TO THE BLUE ECONOMY IN BANGLADESH

Bangladesh should prepare a comprehensive strategy to explore and exploit ocean resources. It is challenging to balance blue growth and conservation of the environment. The ocean's resources are valued at US$ 24 trillion, with a US$ 2.5 trillion annual value addition. It is believed that the ocean will offer more economic benefits to ocean-based countries like Bangladesh. Despite having a huge number of marine resources in the Bay of Bengal, Bangladesh faces numerous obstacles to developing its blue economy.

6.21   RESOLVING THE ISSUES OF CLIMATE CHANGE

Climate change worsens the overall situation. Livelihoods and food security are threatened. Increasing population and less socio-economic development in the exposed zone would result in a high-risk situation (IUCN, 2010). Bangladesh is a low-lying country most vulnerable to climate change. Huge property and crop damages as well as about 718,000 deaths took place because of cyclones over the past 50 years (Haque et al. 2012). Moreover, potential fish production was reduced by 10% in the Bay of Bengal. Sea level rise, ocean acidification, and changes in ecosystem status occur because of changing temperatures. As a result, the ability of marine organisms to shape and sustain shells and skeletons, to survive, grow, proliferate, and for their larvae to develop, was harmed. Unfortunately, the long-term impacts of climate change on ocean systems are not yet fully understood. Currently, no international mechanism to specifically address acidification exists and so appropriate means need to be elaborated to enable coordinated international action.

6.22   MARKET GENERATION FOR BLUE CARBON

Long-term carbon sequestration by mangroves, seaweeds, salt marshes, and seagrass can mitigate climate change and open a new window for carbon trading mechanisms. Carbon credits mechanisms can be a great sector for reducing climate impacts. Coastal blue carbon (seagrass, mangroves, salt marshes) is estimated to be worth about US$180 million in Europe (Luisetti et al. 2013). Other ecosystem services offered by these ecosystems, such as fisheries, can also be protected and capitalized on with such financial incentives for blue carbon. Beyond the immediate carbon sequestration benefits, there would be long-term benefits. For example, hundreds of thousands of dollars per hectare is the value of ecosystem services offered by mangroves and tidal marshes. The blue economy strategy would put policies, regulations, infrastructure, and incentives in place to make the transition to a low-carbon economy as smooth as possible. Andrew et al. (2019) proposed a blue carbon framework for the successful utilization of blue carbon (Figure 6.10).

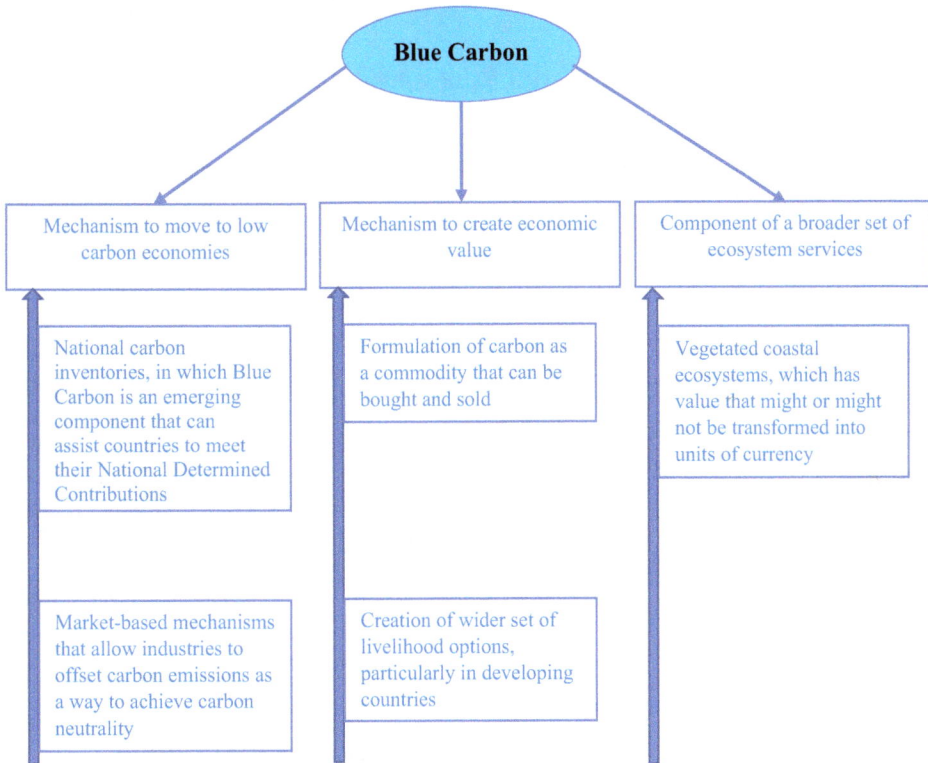

FIGURE 6.10 Function of Blue Carbon (Andrew et al. 2019).

6.23 CREATING AWARENESS

The creation of awareness of the 'blue economy' among the general population is urgent. This issue in Bangladesh has been confined largely to intellectuals and interested people. Media coverage (for example, Television, Radio, Newspaper, and the like) are essential to transforming our people into a more maritime-oriented nation.

6.24 THE GENERATION OF JOBS

Employment generation is a great challenge for Bangladesh. There are jobs linked to the creation of the latest maritime sectors. The generation of new sectors linked to research and innovation (for example, minerals exploration, mariculture, and the like). Some sectors need careful attention if they are to develop (for example, tourism). A successful blue economy will generate a great number of jobs for the coastal people (Table 6.6).

6.25 BIODIVERSITY PROTECTION AND CONSERVATION

The biodiversity of the marine ecosystem is being degraded and this has significant effects on livelihoods. Around 20% of the world's coral reefs have been destroyed, and another 20% are being degraded, according to estimates. 30–50% of mangrove cover has been reduced while 29% of the seagrass ecosystem has been extinct since the late 1800s. In Bangladesh, corals, seaweeds, mangroves, seagrass, salt marshes, and fish are being reduced day by day. To restore biodiversity in the Bay of Bengal, an ecosystem approach is needed.

TABLE 6.6

Overview of Forecasts of Industry-Specific Value Added and Job Growth Rates for 2010 and 2030

Industry	Compound Annual Growth Rate for GVA between 2010 and 2030	Total Change in GVA Between 2010 and 2030	Total change in employment between 2010 and 2030
Industrial marine aquaculture	5.69%	303%	152%
Industrial capture fisheries	4.10%	223%	94%
Fish processing	6.26%	337%	206%
Maritime and coastal tourism	3.51 %	199%	122%
Offshore oil and gas	1.17%	126%	126%
Offshore wind	24.52%	8 037%	1 257%
Port activities	4.58%	245%	245%
Shipbuilding and repair	2.93%	178%	124%
Maritime equipment	2.93%	178%	124%
Shipping	1.80%	143%	130%
Average of the total ocean-based industries	3.45%	197	130
Global economy between 2010 and 2030	3.64%	204	120

Source: OECD, 2016.

6.26   THE UNSUSTAINABLE EXTRACTION OF MARINE RESOURCES

In Bangladesh, the traditional fishing boat can fish up to 60 meters. This causes depletion of fish stock in inland and shallow marine water by overexploitation and unsustainable fishing practices (use of destructive and illegal gear) (BOBLME, 2012). Piracy by domestic and international criminal gangs is also a factor in the decline of fish stocks. Underexploited or moderately exploited marine fish stocks are believed to have decreased from 40% in the mid-1970s to 15% in 2008 whereas stocks that have been exhausted or are recovering have risen from 10% in 1974 to 32% in 2008. The annual loss of overfishing is projected to be in the region of US$ 50 billion. Around 57% of fish populations are completely exploited as a result of technological advancements, while the remaining 30% are overexploited, declining, or recovering (FAO, 2016). The fish stock is also being depleted by IUU fishing, responsible for fish catches ranging from 11 to 26 million tonnes per year.

6.27   ENSURING FOOD SECURITY

The world's population will be 900 cores by 2050 and this huge amount will be fed largely from the sea (Bhuyan et al. 2020). In developing countries, about 1 billion people rely on seafood as a source of protein. Food security is strongly related to the sustainable use of biodiversity. Mariculture can be a good sector of the blue economy since it provides food for the people. In Bangladesh, mariculture is being done on small scale, but it needs to expand industrially.

6.28   THE DESTRUCTION OF MARINE AND COASTAL HABITATS

Cutting down plants along the coast and uprooted submerged plants causes damage to marine and coastal habitats. Erosion is also responsible for the destruction of coastal infrastructure and hence affects coastal livelihoods. Haphazard development along the coastal area increased externalities in different industries, infrastructure placement that isn't ideal, conflicting ground and sea-based uses, the demonization of poor communities, and sensitive ecosystems are being destroyed or degraded. Habitat degradation leads to the destruction of marine biodiversity.

6.29   GREENING OF THE COASTAL BELT

Coastal belt greening is very important to reduce the wind pressure of cyclones. It also solidifies new lands along the coast. Bangladesh has already started working on the coastal belt greening which will help sustainable agriculture, and river courses. It will also help to prevent saline water intrusion.

6.30   MARINE POLLUTION REDUCTION

Anthropogenic causes (the growing human population, intensification of agriculture and the rapid urbanization of coastal areas, sewage, siltation, turbidity, oil spills, maritime shipping, submarine hydrocarbon/mineral exploration and extraction, and so forth.) are responsible for marine pollution. Marine pollutants can be categorized as liquids, metals, gaseous, solids, or harmful microbes (BOBLME, 2011). Marine pollution led to the destruction of marine habitats and damage to marine organisms (Hossain et al. 2014). There are some regulations to stop marine pollution in Bangladesh, but these are outdated and ineffective.

6.31   CONFLICT MINIMIZATION AMONG USERS (COASTAL AND MARINE)

Conflicts between the ocean and the coastal resource user are a common phenomenon. In the past conflict was reported between the agricultural sector and the shrimp farming sector. Saline water enters agricultural land when it is converted to shrimp farming. Chakaria Sundarban contained large Mangrove forests but was converted to shrimp farming because of huge potential profits (Iftekhar, 2006). This created significant environmental damage. Further, agricultural fields were converted for salt production. The Government is planning to build a seaport for blue economy development, and it is also attempting oil and gas exploration, to develop fisheries, tourism, and other ocean-based activities. These will also potentially create conflicts amongst the users.

6.32   MARITIME SPATIAL PLANNING (MSP)

Maritime spatial planning is a very important tool or mechanism for the effective management of ocean resources. MSP recognizes which part would be suitable for mariculture, which part is suitable for port development, and so forth. It would reduce the conflicts among users once developed. MSP identifies that the ocean is the driver of blue economic development with blue growth and innovation. But it is very challenging for countries like Bangladesh. Recently, Bangladesh has been attempting to develop MSP. In-depth knowledge of the marine environment is a prerequisite for successful MSP development. It will give a clear idea of how maritime activities impact each other and the environment.

6.33   MARITIME SECURITY

Maritime security is very important for successful blue economic development. Each sovereign country is liable for its marine resource uses and sustainable development. Without sovereignty over the entire coastal and marine areas, it will be difficult to use marine resources. The activities of pirates will increase in the oceanic region. For Bangladesh, it is challenging to maintain safety over the commercial coastal area of this Asian nation. Bangladesh needs to be more careful and tactful about maritime security.

6.34   SKILLED MANPOWER

Bangladesh faces difficulties with expert personnel in the blue economy sectors. There is a lack of knowledge, expert human resources, and technology that is not yet available to explore and exploit deep-sea fish and seabed resources. A group of experts on blue technology is required for maximum benefits from the blue economy.

6.35   OCEAN GOVERNANCE AND INTERNATIONAL COOPERATION

Structured foreign/international cooperation supports all facets of the blue economy. Updating and advancing governance structures to ensure the sustainable development of waters outside national control (for example, maritime protection, high seas MPAs, sustainable fisheries, and the exploitation of petroleum and minerals) or to support the efficient management and use of national EEZs (for example, technology transfer, technical assistance, and marine spatial planning), capacity building, funding for the promotion of national marine spatial planning and efficient tracking, control, and surveillance. The Economist Intelligence Unit (2015) has prepared a database of ocean-based countries regarding their performance in coastal governance (Figure 6.11). Research is considered to be an essential feature of international cooperation in the blue economy approach. The scientific approach is vital for the sustainable development of the blue economy. The initial assessment and the critical valuation of blue capital will begin. This will provide a foundation for up-to-date decision-making and adaptive governance. In line with changing circumstances, emerging technology, and our growing awareness, this major undertaking must be tackled and constantly refined, and upgraded for the blue economy approach to be developed. This emphasizes the significance of technical support, the transfer of technology, and capacity building in the pursuit of sustainable growth.

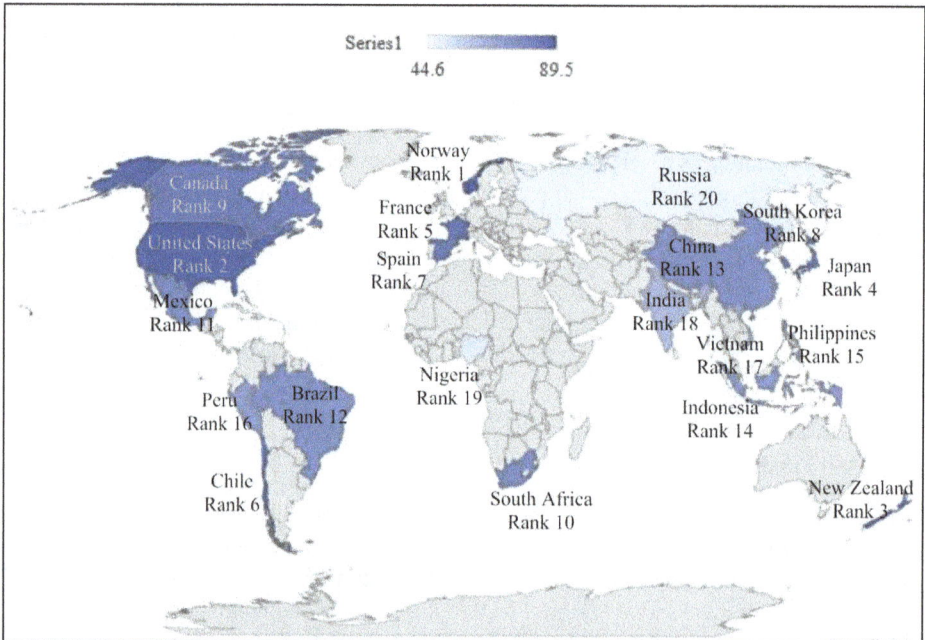

FIGURE 6.11 Countries in the Coastal Governance Index, 2019: overall performance (The Economist Intelligence Unit).

6.36   A STRONG MARINE POLICY FORMULATION

A strong marine policy is key for successful blue economic development. The lack of proper policy and marine experts at policy-making levels, ministerial levels, and national organizations means that the department is regarded as a hindrance to the implementation of many issues related to the development of marine sectors in Bangladesh. In Bangladesh, there is not enough concentration on integrated coastal management and ocean governance policy. Bangladesh is currently following scattered laws, regulations, and policy statements for maritime management. UNCLOS 1982, FAO Code of Conduct for Responsible Fisheries, 1995, United Nations Fish Stock Agreement, Ramsar Agreement on Wetland issues, Convention on Biological Diversity 1992, and Paris Climate Agreement of 2015 are notable (Rahman et al. 2017). Most of them are fit for the international arena but are not suitable for national legislation, laws, and policymaking processes. As a result, several initiatives are slow to materialize. Bangladesh should adopt a strong and comprehensive maritime strategy concentrating on maritime safety, pollution, resource protection, research, and the technology transfer issue. Bangladesh must follow the existing international laws and policies (for example, UNCLOS, 1982) during the drafting of its national marine policy (Bhuiyan, 2014). It is high time for Bangladesh to adopt and implement policies along with maritime technical development and research capacity. Without a strong maritime policy, Bangladesh will not be able to get full benefits from the blue economy.

6.37   RISKS/THREATS TO A BLUE ECONOMY

The blue economy faced lots of threats in its route towards successful execution. These risk factors can hamper the ultimate goals of the blue economy. The overall risk is depicted in Figure 6.12.

6.38   THE TENDENCY FOR ILLEGAL FISHING OR OVERFISHING

Illegal fishing or overfishing is a great threat to the blue economy. It depletes the fish stock from the ocean which is difficult to recover. These types of fishing reduce not only the fish but also destroy the other organisms (by-catch) essential for the marine ecosystem. Consequently, it abolishes a valuable source of food and income for coastal people. About US$ 50 billion is lost per annum due to overfishing. Illegal fishing is a risk to the sustainability of fisheries (FAO, 2014). Increasing

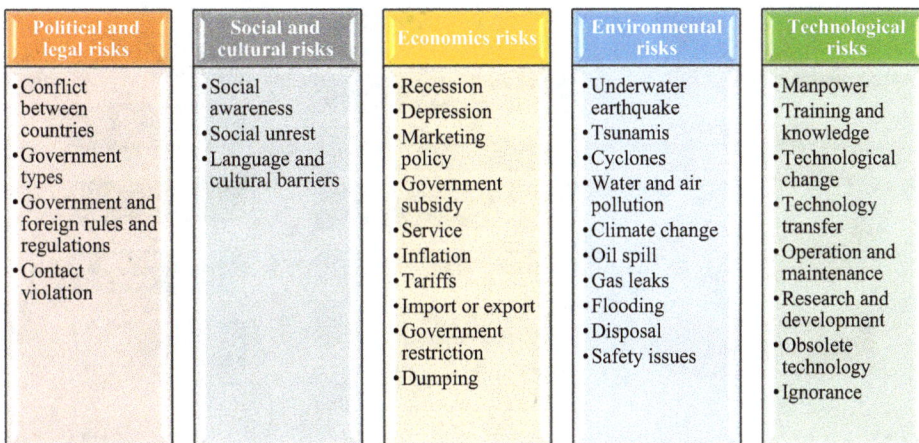

Political and legal risks	Social and cultural risks	Economics risks	Environmental risks	Technological risks
•Conflict between countries •Government types •Government and foreign rules and regulations •Contact violation	•Social awareness •Social unrest •Language and cultural barriers	•Recession •Depression •Marketing policy •Government subsidy •Service •Inflation •Tariffs •Import or export •Government restriction •Dumping	•Underwater earthquake •Tsunamis •Cyclones •Water and air pollution •Climate change •Oil spill •Gas leaks •Flooding •Disposal •Safety issues	•Manpower •Training and knowledge •Technological change •Technology transfer •Operation and maintenance •Research and development •Obsolete technology •Ignorance

FIGURE 6.12   Risk and uncertainty associated with blue economy (Hossain, 2020).

world populations pressurize fish production. Global fish production is growing at a 3.2% annual rate, surpassing the global population growth of 1.6% (FAO, 2014).

6.39   THE INCREASING TREND OF MARINE POLLUTION

Marine pollution originates from land-based and marine sources, creating adverse conditions for marine life and human beings as well (Figure 6.13). Pollution is destroying the marine ecosystem, and, hence threatening marine species. Marine pollution is increasing at an unprecedented rate. This increasing trend is in opposition to sustainable blue economic growth.

6.40   CLIMATE CHANGE IMPACT

Climate is changing now, due mostly to anthropogenic causes, posing a threat to marine animals and humans. The ocean has been warmed during the period from 1971 to 2010 at an average rate of >0.1°C per decade in the upper 75 m of the water column and 0.015°C per decade at a depth of 700 m. As a result, there is glacial melting and rising sea levels (Rhein et al. 2013). The ocean is becoming more acidic due to the increasing pattern of CO_2 emissions from man-made sources (for example, industrialization). Acidification of the ocean is considered to be one of the key drivers of oceanophysical and biological changes.

Ocean acidification can cause decreased calcification, slower repair rates, and damaged calcified systems in most marine species (Kroeker et al. 2013). Moreover, reproductive success, early life-stage survival, feeding rate, and stress-response mechanisms can be affected by the acidification phenomenon (Pörtner et al. 2014).

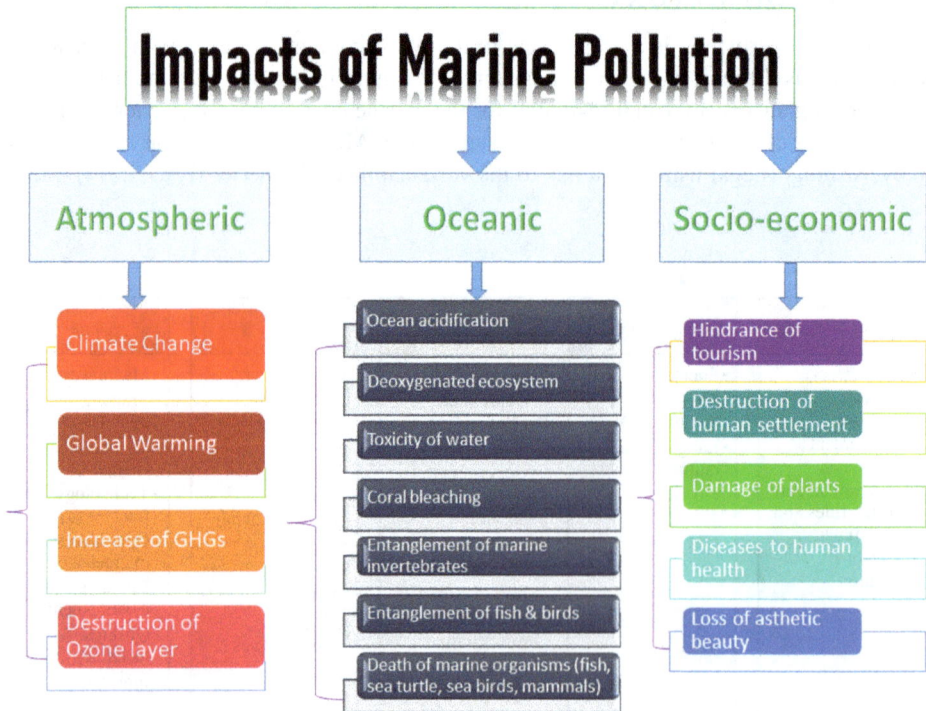

FIGURE 6.13   Possible impacts of marine pollution in the environment.

6.41 THE DESTRUCTION OF NURSERY, BREEDING, AND FEEDING HABITATS

Coastal development, industrialization, trawling, and aquaculture risk the destruction of suitable nursery, breeding, and feeding habitats. As a result, ocean productivity is being reduced gradually. Ocean productivity is also being harmed because of ocean stratification and reduced nutrient mixing in the open seas. The Global Ocean Observing System (GOOS) and LME assessments forecast (2040-2060) a consistent drop in ocean productivity. This poses a threat to the marine environment because animals will perish.

6.42 HYPOXIC OCEAN WATER

Oxygen-depleted water has an impact on pelagic and benthic marine organisms and their physiological performance and distribution are affected by the hypoxic condition (Pörtner, 2010). Large, mobile, and more active fish would face a greater survival challenge than small ones. There are social and economic effects of coastal hypoxia on coastal areas, including loss of tourism due to restrictions on swimming and boating, beach closures, public health issues, and fish and shellfish consumption, all of which have adverse impacts on estuarine and coastal fishery resources.

6.43 THE REDUCTION OF OFFSHORE ENERGY PRODUCTION RISK

Bangladesh needs to produce offshore energy for blue economic development. But it is risky work for the marine environment. Off-shore drilling has some negative environmental impacts. Whale beaching is a result of explosive charges and sounds from seismic operations, which can affect dolphins' and other marine mammals' ears. During tanker transportation, oil spills occur. Pollutants and radioactive elements are brought up from deep earth during drilling and cutting mud. Off-shore drilling requires natural gas and burns off the gas that appears to prevent any hazardous conditions. Offshore drilling operations are expensive and environmentally sensitive processes that require terminals and shipyards which also pose a risk to the marine environment.

6.44 SHIPPING EMITS AND DISCHARGES POLLUTANTS

Shipping has an impact on the environment, whether at sea or in port. The harmful effects on the ecosystem of the oceans have been known for decades. Ships emit SOx about 8% annually, while global NOx emissions accounted for 15% (Islam, 2018). Around 1 billion tonnes of CO_2 emissions per year have been attributed to the global shipping industry. CO_2 alone accounts for 3% of the world's overall Green House Gas (GHG) emissions. Global temperatures have therefore risen by around 0.8°C since 1880, as stated by the Goddard Center for Space Studies of NASA (GISS) (Islam, 2018). Ships discharge oil and chemicals, ballast water, bilge water, sewage and garbage, and antifouling agents in the seawater. Then these pollutants sink and become permanent in the ocean water and are responsible for huge environmental pollution and biodiversity degradation (Saha and Alam, 2018). Ships discharge oil and contaminants into the seawater, ballast water, bilge water, sewage and waste, and antifouling agents (Figure 6.14). These contaminants then sink and are irreversibly in ocean water, responsible for immense environmental contamination and the destruction of biodiversity (Saha and Alam, 2018).

6.45 ENSURING MARITIME SURVEILLANCE

Maritime security is required for the increasing number of both legal and illegal activities at sea. Maritime safety and security must be strict for marine environmental protection, piracy control, fisheries control, trade, and economic interest. In Bangladesh, densely populated coastal cities are

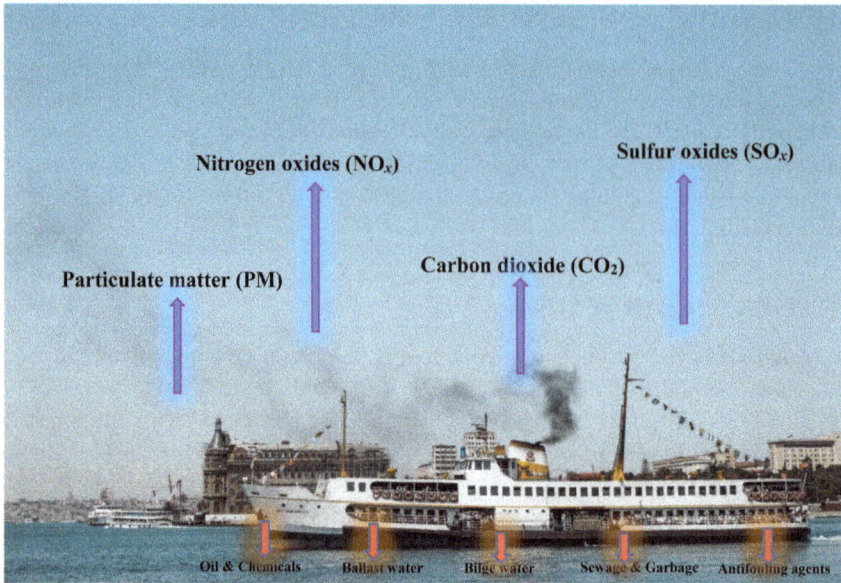

FIGURE 6.14   Pollutants emit and discharge from ships.

working as a hub for terrorism. The networking between organized criminals and terrorists with transnational capabilities poses maritime security threats. Moreover, the Bay of Bengal remains very rough for almost nine months in a year causing a threat to maritime security.

6.46   RISKS AT WORKING SITES

People are suffering in the workplace, sometimes they get injured working in shipping lines and related terminals, yards, and so forth. Lack of protective clothing and training in handling equipment causes accidents. At ship recycling or ship breaking yards, workers become affected by various chronic diseases. Sometimes the workers get injured on the working sites (Onselen, 2018).

6.47   ACCIDENTAL OIL TANKER SPILL

Oil tanker accidents can be disastrously damaging to the surrounding environment. The effects of oil pollution are even more severe when tanks explode along the shore/coast (Khondaker, 1998). Worldwide oil tanker explosions are a common phenomenon. The global accidental oil spill trend is presented in the following Figure 6.15.

In Bangladesh, oil spill risk is associated with oil transportation during regular oil tanker movements. About 20 crude oil carriers and 80 product carriers call at Chittagong Port every year. Chittagong Port handles approximately 1200 international ships on an annual basis and more than 400 ships call at Mongla Port each year.

The causes of oil tanker accidents are mechanical failures and human negligence or errors. Moreover, collision, allision, grounding, fire on board, or explosion resulting from flammable gas are also responsible for oil tanker accidents, and oil spill incidents.

A large oil spill took place back in 1989. MT Filothei, an old Cypriot vessel arrived at the outer anchorage of Chittagong Port fully loaded with crude oil imported for Eastern Refinery Limited. Another oil tanker accident took place in May 1998.

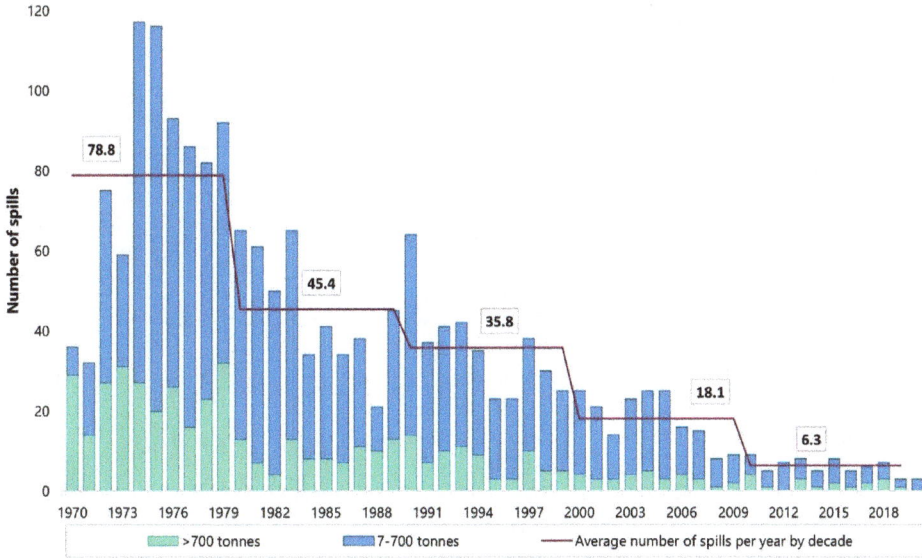

FIGURE 6.15 Global accidental oil spill trend (ITOPF, 2020).

6.48 KARNAPHULI OIL SPILL

An oil tanker leaving for Khulna Desh-1 collided with City-38, a vegetable oil tanker, carrying 1200 tonnes of diesel. On Saturday, October 26th, 2019, it was anchored at Karnaphuli River's Padma Jetty.

6.49 SUNDARBANS OIL SPILL

On December 9, 2014, an oil spill occurred in the Sundarbans (a UNESCO World Heritage site) at the Shela River. An oil tanker, named Southern Star VII that was carrying 350,000 litres (77,000 imp gal; 92,000 US gal) of furnace oil collided with a cargo vessel and sank in the river. The oil had spread out over a 350 km² (140 sq mi) area including a second river, canals, and shoreline.

A clear picture of the top 20 major oil spills around the world is given in Table 6.7.

The spills have devastating effects on trees, plankton, fisheries, agricultural resources, and dolphins. Highly toxic spilt oil is responsible for the death of phytoplankton and zooplankton (primary producers of the marine ecosystem). This results in the destruction of the food chain in that area. Marine mammals and seabirds are particularly vulnerable to spilt oil.

6.50 SUSTAINABLE DEVELOPMENT PATHWAYS

Sustainable development is the guiding principle for achieving the objectives of human development while retaining the capacity of natural systems to provide the natural resources and ecosystem services on which the economy and community depend (Figure 6.16). A state of society is needed in which living standards and the utilization of resources continue to meet human needs whilst considering the marine environment. An integral and vital part of sustainable development is the oceans, seas, and coastal areas. Oceans and seas account for 80% of all life forms and 90% of global trade is sea-borne. More than 3 billion people rely for their livelihoods on marine and coastal resources. Due to the massive facilities, 13 of the world's 20 megacities are along the coastline. Despite having a wide range of valuable uses, ocean resources should be extracted sustainably.

TABLE 6.7
Top 20 Major Oil Spills around the Globe

Position	Ship Name	Year	Location	Spill Size (tonnes)
1	Atlantic Empress	1979	Off Tobago, West Indies	287,000
2	Abt Summer	1991	700 nautical miles off Angola	260,000
3	Castillo De Bellver	1983	Off Saldanha Bay, South Africa	252,000
4	Amoco Cadiz	1978	Off Brittany, France	223,000
5	Haven	1991	Genoa, Italy	144,000
6	Odyssey	1988	700 miles off the coast of Nova Scotia, Canada	132,000
7	Torrey Canyon	1967	Scilly Isles, UK	119,000
8	Sea Star	1972	Gulf of Oman	115,000
9	Sanchi	2018	Off Shanghai, China	113,000
10	Irenes Serenade	1980	Navarino Bay, Greece	100,000
11	Urquiola	1976	La Coruna, Spain	100,000
12	Hawaiian Patriot	1977	300 nautical miles off Honolulu	95,000
13	Independenta	1979	Bosphorus, Turkey	94,000
14	Jakob Maersk	1975	Oporto, Portugal	88,000
15	Braer	1993	Shetland Islands, UK	85,000
16	Aegean Sea	1992	La Coruna, Spain	74,000
17	Sea Empress	1996	Milford Haven, UK	72000
18	Khark 5	1989	120 nautical miles off Morocco's Atlantic coast	70,000
19	Nova	1985	Off Kharg Island, Gulf of Iran	70,000
20	Katina P	1992	Off Maputo, Mozambique	67,000

Source: ITOPF, 2020.

6.51 INITIATIVES ADOPTED BY BANGLADESH FOR SUSTAINABLE MARITIME DEVELOPMENT

It is high time for Bangladesh to achieve unparalleled economic growth and phenomenal growth in regional connectivity, seaports, coastal manufacturing, offshore oil and gas production, special economic zones, and energy clusters. Different factors are responsible for a sustainable blue economy in Bangladesh (Figure 6.17). The Bangladesh Delta Plan 2100 is a big blue economy project focusing on long-term delta management, integrated management of water resources, long-term land reclamation, climate change adaptation, and so forth. Besides, Vision 2041, a long-term strategy for a developed Bangladesh, has established the blue economy as one of the main drivers of sustainable growth. Besides, the goals of SDG 2030, particularly SDG Target 14 are being introduced with various maritime growth agendas in Bangladesh. To ensure that both government and private sector maritime stakeholders in the country are properly organized in their blue economy activities, the government has created an autonomous blue economy cell. Bangladesh established a Maritime Affairs Unit to increase international cooperation for the blue economic growth of Bangladesh. Bangabandhu Sheikh Mujibur Rahman Maritime University, Bangladesh Oceanographic Research Institute, and Marine and Technology Station of Bangladesh Fisheries Research Institute (BFRI) were established for higher ocean education and research. There are various public and private universities, academies/departments, and departments that teach marine and oceanography-related subjects.

Three voluntary commitments were made in New York by Bangladesh at the UN Ocean Conference in June 2017. Firstly, 5% of marine areas should be designated as Marine Protected Areas (MPAs) of around 7,500 sq km by 2020 (Target 14.2). The percentage of marine protected areas as a percentage of the total territorial waters in South Asia is shown in Figure 6.18. The Sundarbans, the world's

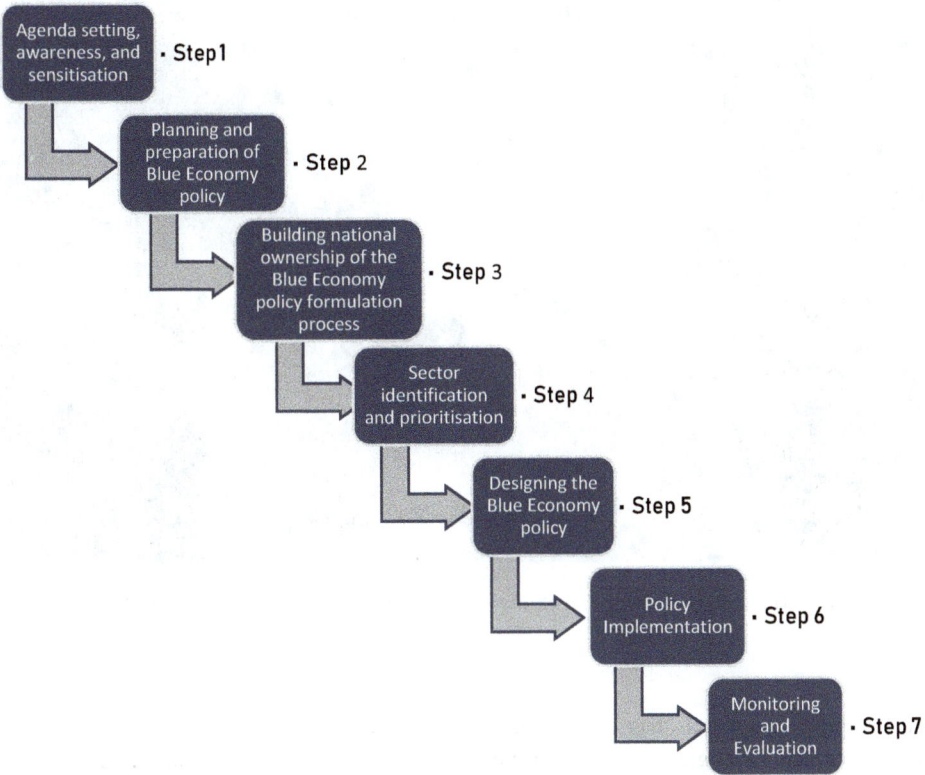

FIGURE 6.16 Flowchart of working process to activate sustainable blue economy.

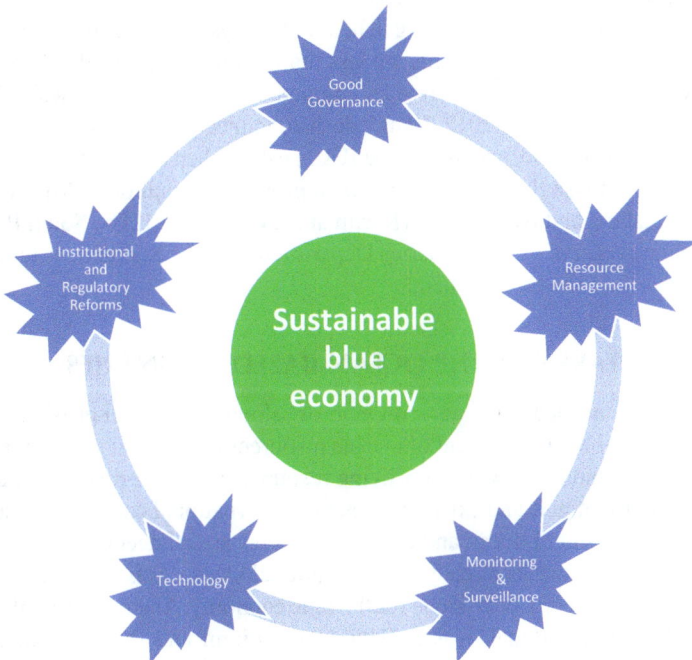

FIGURE 6.17 Factors governing success of blue economy.

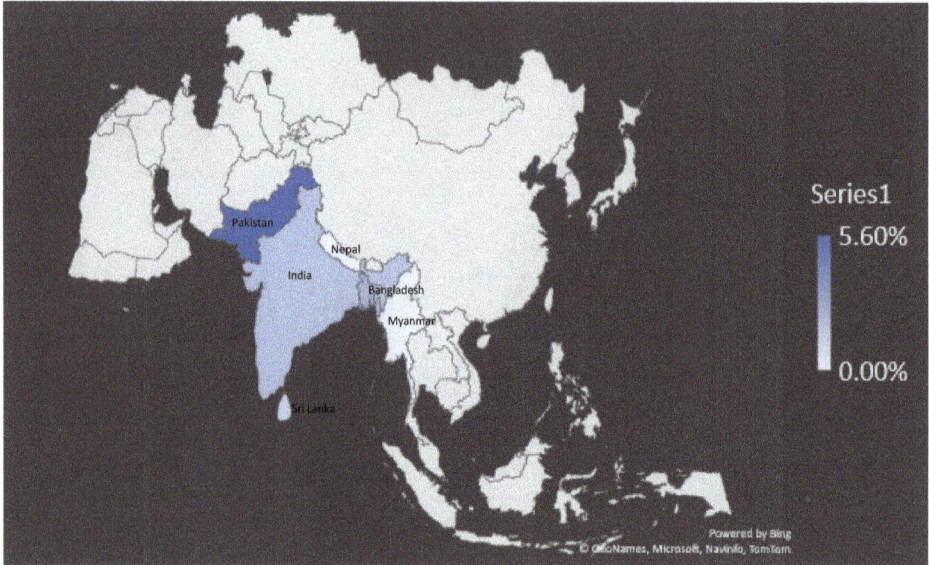

FIGURE 6.18   Marine protected areas (%) of total territorial waters in South Asia (OECD).

largest mangrove ecosystem, has been declared a UNESCO World Heritage Site. Secondly, IUU fishing and harmful fishing activities in marine areas should be effectively controlled (Target 14.4). The third promise is to avoid and significantly reduce all kinds of marine pollution, particularly pollution from land-based sources (Target 14.1). The establishment of collaborations with companies and international institutes, but also at the national level, will promote research in the field of marine aquaculture for the advancement of projects.

Both types of fish are prohibited from being fished for 65 days, and juvenile Hilsa fish is prohibited from being fished for 8 months if specimens are less than 25 cm long, a 22-day fishing ban for large Hilsa fish, and a complete ban on destructive fishing activities in the Bangladesh Maritime Region (Islam and Shamsuddoha, 2018). In 2012, Bangladesh announced the first MPA-SONG in the world under the Wild Conservation Act. In addition, the IUCN established and published a proposal for the establishment of MPAs in Bangladesh in 2015, which proposed a total of 67 MPA declaration sites, including St. Martin Island and Nijhum Dip (Karim and Uddin, 2019). The South Patch has already been declared a marine reserve by the Fisheries Department and the Sundarbans have been declared an ECA by the Ministry of the Environment.

6.52   INITIATIVES TAKEN BY THE OCEAN-BASED COUNTRIES

A scientific approach is needed for blue economic development that emphasizes the appropriate conservation and sustainable management of marine resources (living and non-living). Fisheries and aquatic plants are living resources, while non-living resources are made up of minerals, oil, and gas. The identification and use of new and existing resources, international cooperation, successful ocean governance, and management of ocean and coastal resources are all needed.

Different ocean-based countries already passed National Ocean Acts15 while some of the countries have made special budgetary provisions for the blue economy. Australia, Brazil, Canada, China, Colombia, Japan, Norway, Portugal, Russia, United Kingdom, and the USA already formulated

National Ocean Policy. USA, Canada, and Australia have special legislation and specific provisions for ocean policy in their national budgets. Canada and Australia established a systematic hierarchy of institutions at federal and state levels to plan, coordinate and monitor progress on the various pillars of ocean policy. In Canada, three different layers of institutions comprising the National Ocean Act, the Ocean Strategy, and the Ocean Action Plan govern the development, management, and governance of ocean resources (Mohanty et al. 2017).

At the federal and state levels, Canada and Australia have formed a systematic hierarchy of institutions to prepare, organize and track progress on the different pillars of ocean policy. In Canada, the growth, management, and governance of ocean resources are governed by three distinct layers of institutions comprised of the National Ocean Act, Ocean Policy, and the Ocean Action Plan. Likewise, coastal and marine issues are handled by a multi-layer institutional framework consisting of Ocean Policy, Regional Maritime Strategies, Integrated Ocean Planning and Management, the Minister for the Environment and Heritage, the National Advisory Council, and others (Repetto, 2005).

Seychelles and Mauritius have established separate ministries for the blue economy. For example, the Government of Mauritius formulated a Roadmap for Ocean Economy in 2013 which included diverse objectives, action plans, and mutual-reinforcing sectoral components. In 2009, the Government of Australia launched a Strategic National Framework for Marine Research and Innovation (Mohanty et al. 2017). In Europe, the Netherlands, Denmark, and Norway are quite successful in implementing their blue economy policies. Following their success, Ireland introduced a Marine Knowledge, Research, and Innovation Strategy in 2006 for the period 2007 - 13 which aimed at policy measures to promote blue economy sectors in the Irish economy. The CARIFORUM-EU Economic Partnership Agreement signed in 2008 encourages the importance of fisheries and other living marine resources in the CARICOM member states and the Dominican Republic. The Seychelles and Mauritius have set up separate blue economy ministries (Mohanty et al. 2017). In 2013, for instance, the Government of Mauritius developed a roadmap for the ocean economy that included different goals, action plans, and sectoral components of mutual reinforcement. A Comprehensive National Framework for Marine Exploration and Innovation was initiated by the Government of Australia in 2009. The Netherlands, Denmark, and Norway have been very active in the introduction of blue-economy policies in Europe. Following their performance, in 2006, Ireland launched a Marine Awareness, Research, and Innovation Plan for the period 2007-13, aimed at promoting the blue economy sectors in the Irish economy through policy initiatives. The Economic Partnership Agreement between CARIFORUM and the EU, signed in 2008, promotes the value of fisheries and other living marine resources in the Member States of CARICOM and the Dominican Republic. Similarly, the Temporary Cooperation Agreement between the European Community and the Pacific States includes clauses on trade in fishery products that are derogated directly from the Rules of Origin (Mohanty et al. 2017).

Regional research and networking projects, such as the EU Joint Programming Initiative for Safe and Sustainable Seas and Oceans (JPI-OCEANS), the European Marine Biological Resource Centre (EMBRC), CSA MarineBiotech, the European Research Area Network (ERA-NET), the Association of European Marine Biological Laboratories (ASSEMBLE) and others, are intended to establish popular research.

Trans-regional projects such as the Mediterranean Science Commission (CIESM), the Sustainable Use of Baltic Marine Resources Program (SUBMARINER), and the BioMarine Program are intended to contribute to the promotion of marine biotechnology research and applications. To build capacity and promote innovation in the core industries and sectors of the blue economy, these national and regional initiatives are important. Bangladesh established the Maritime Affairs Unit at MOFA, for exploring international cooperation for maritime development.

TABLE 6.8
Real GDP Growth in South Asia

Country Name	2018 (%)	2019 (e) (%)	2020 (f) (%)	2021(f) (%)
Afghanistan (CY)	1.8	2.5	3.0	3.5
Bangladesh (FY)	7.9	8.1	7.2	7.3
Bhutan (FY)	4.6	5.0	7.4	5.9
India (FY)	6.8	6.0	6.9	7.2
Maldives (CY)	6.7	5.2	5.5	5.6
Nepal (FY)	6.7	7.1	6.4	6.5
Pakistan (FY and factor price)	5.5	3.3	2.4	3.0
Sri Lanka (CY)	3.2	2.7	3.3	3.7

Source: GDP: gross domestic product, CY: calendar year, FY: fiscal year, e: estimate, f: forecast; in Bangladesh, Bhutan, Nepal and Pakistan, 2019 refers to FY2018/2019 and ended in June 2019. For India, 2019 refers to FY2019/2020 and will end in March 2020 (World Bank, 2019).

6.53 PARTNER AGENCIES WORKING ON BLUE ECONOMIC DEVELOPMENT WORLDWIDE

Potential Partner Agencies and Processes include AIMS, AOSIS, CARICOM, Convention on Biological Diversity, Commission on the Limits of the Continental Shelf, Convention on Migratory Species, Indian Ocean Commission, Commonwealth Secretariat, Duke University USA, GLISPA, Global Ocean Acidification Observing Network, Global Ocean Commission (GOC), Global Ocean Forum, the IOC/UNESCO Global Ocean Observing System (GOOS), World Bank (IBRD), International Maritime Organisation, International Seabed Authority, IOR-ARC, National Oceanic and Atmospheric Administration USA, Ocean Acidification International Coordination Centre, Ocean Acidification International Reference User Group, Plymouth Marine Laboratory UK, Regional Seas Programmes, RFMOs, SPREP, UNDESA, UNDOALOS (UNCLOS), UNDP, UNEP, UNESCO's Intergovernmental Oceanographic Commission, UNFAO, UNIDO, UNWTO, World Ocean Council. CBD Sustainable Oceans Initiative, UNEP Green Economy Initiative, European Project on Ocean Acidification.

6.54 GDP IN SOUTH ASIA

South Asian countries are showing growth in their GDPs. In the case of Bangladesh, GDP was 7.9% in 2018 while it reached 7.3% in 2021 (Table 6.8). Bangladesh can make significant growth in GDP using ocean resources. Some recommendations are provided for sustainable use of the blue economy, hence the growth of national GDP.

6.55 RECOMMENDATIONS FOR SUSTAINABLE BLUE ECONOMIC DEVELOPMENT

A successful blue economy needs a collaborative effort between oceanographers, engineers, navigators, merchant mariners, fisheries technologists, biologists, biotechnologists, and so forth. Since Bangladesh acts as a central cohesive source of support for the economic hub connecting inter-Asian states, it is necessary to formulate blue economy strategies carefully. A comprehensive blue economy framework has been proposed for Bangladesh's sustainable extraction of blue resources (Figure 6.19). The blue economy initiative specifically aims to increase job opportunities in the short, medium, and long-term time frames. The government should emphasize coastal poverty reduction in the Bay of Bengal by developing a good policy and action plan. The following policy recommendations are needed for successful blue economy execution.

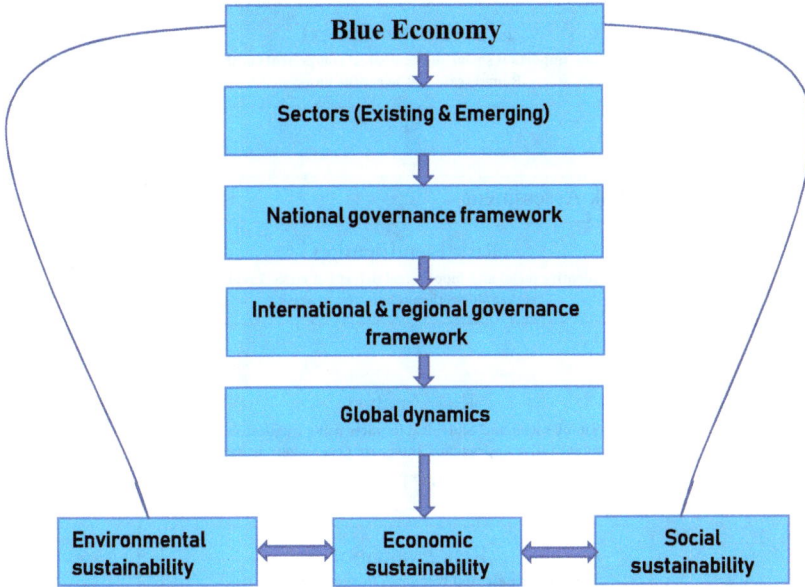

FIGURE 6.19 Proposed blue economy framework for Bangladesh.

6.55.1 Overall Policy Recommendations

Blue economic risk can be reduced/minimized by following the prescribed recommendations. A traditional risk management approach also proves useful in the minimization of threats (Figure 6.20).

6.55.2 Maritime Skills Development

Bangladesh should increase its marine educational institutions. Experts are needed in the field of oceanography, marine dynamics, marine engineering, biotechnology, marine fisheries, marine trade, offshore engineering, naval architectural engineering, marine geological aspects, the marine environment, ecosystem science, and so forth.

6.55.3 Expert Marine Panel

An expert panel from various fields (marine biologists, fisheries and aquaculture specialists, marine trade experts, and economists) can be engaged in tracking and assessing a proposed project. They will provide technical and other assistance. Without the approval of the expert group, no project work should be undertaken.

6.55.4 Marine Research Organizations

Pure marine research-based organizations must be established with up-to-date modern technologies (for example, deep-sea research vessels, underwater cameras, and the like).

6.55.5 Marine Spatial Planning (MSP)

Marine Spatial Planning is an important tool for Bangladesh to manage and use blue economy resources sustainably. An effective MSP is a prerequisite to ensuring a stabilized and sustainable blue economy in any ocean-based country through policy and legal protection, despite the many challenges.

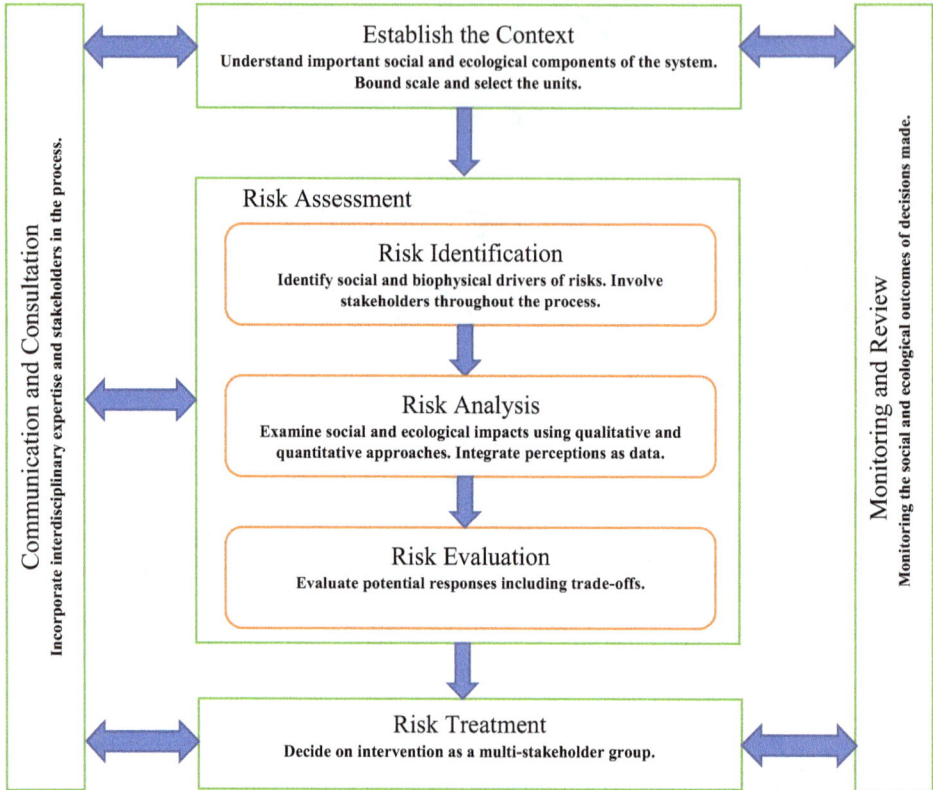

FIGURE 6.20   The traditional risk management approach was adapted from ISO Risk Assessment Principles (ISO, 2009). The italicized text refers to the forms in which this work's ideas are translated into improvements to the conventional method.

6.55.6   COMPREHENSIVE OCEAN POLICY

A comprehensive ocean policy based on a shared regional vision for integrated ocean planning and management is needed. Bangladesh should design its foreign policy in such a way that it can actively participate in regional and global forums or associations. The participation will help to fulfill its vision of the 'blue economy'.

6.55.7   OCEAN GOVERNANCE

Investment is needed for the strong ocean governance that promotes a blue economy.

6.55.8   SCIENTIFIC REVIEW AND ECONOMIC ANALYSIS

Regular systematic scientific review and economic analysis should take place for a successful blue economy.

6.55.9   PUBLIC–PRIVATE PARTNERSHIP

The public-private partnership plays a key role in the sustainable blue economy to enhance capacity building.

6.55.10 Identification of Priority Sectors

Priority sectors must be recognized based on domestic and global needs. Then research work must be carried out on the identified sectors.

6.55.11 International Collaboration

Regional collaboration is needed for technical assistance, technology transfer, and capacity building. Ocean resources management and exploration are highly dependent on this kind of cooperation. The cooperative mechanism among members of the Bay of Bengal Initiative for Multi-Sectoral Technical and Economic Cooperation (BIMSTEC) can be developed to promote collaboration for reaping the maximum benefit of the blue economy.

6.55.12 Maritime Security and Surveillance

Security is essential to protect the EEZ and high seas areas from international smugglers, trafficking of drugs, humans, arms, fish pirates, and narco-terrorism. Maritime security is also needed to reduce ocean-based crimes on the national border.

6.55.13 Environment and Biodiversity Protection

Marine resources exploration and exploitation should be executed in an environmentally eco-friendly way. Marine ecological balanced must be ensured. The increasing temperature is responsible for ocean acidification, coral bleaching, and loss of biodiversity. Mangroves, seaweed, and seagrass have a great role in managing carbon emissions and hence can aid in climate change mitigation. These resources must be protected. There is a need to protect the sea area to keep it free from both marine and land-based pollution. Sustainable development can be achieved by protecting marine biodiversity from possible threats (Figure 6.21).

6.55.14 Monitoring and Evaluation

A strong monitoring and evaluation committee should be formed for regular checking of the ongoing blue economy activities. This committee would report on the progress and failure from time to time to the relevant authorities.

6.55.15 Lessons from Successful Countries

Bangladesh can follow the successful blue economy countries reported by UNEP (2015). Bangladesh can learn a lesson from these countries and can build its own expertise.

6.55.16 Sector-Wise Recommendations

Table 6.9 represent the sector-wise recommendations for the sustainable use of ocean resources.

6.56 NATIONAL POLICIES, ACTS, AND RULES ADOPTED BY BANGLADESH

The Bangladesh government adopted some legislation and regulatory frameworks for maritime and ocean governance (Table 6.10). These are Territorial Water and Maritime Zones Act, Coastal Zone Policy-2005, Coastal Development Strategy-2006, Port Act 2006, Arbitration Act 2016, Merchant Ship Ordinance of 1983, Flag Ordinance Convention 1982, Inland Shipping Ordinance 1976,

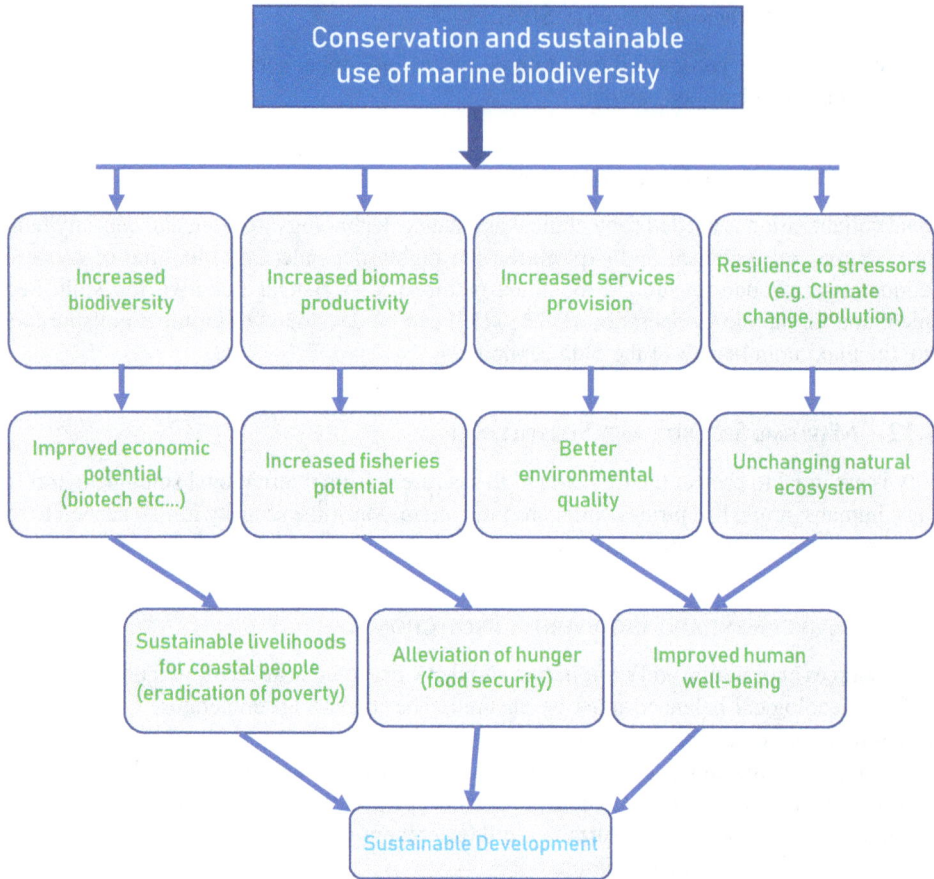

FIGURE 6.21   Sustainable blue economy framework.

Protection and Conservation of Fish Act of 1950, Marine Fisheries Ordinance and Rules 1983, Wild Conservation Act-2002, Forest Act-1927, Provision of Ecologically Critical Area (ECA) Regulation 1995, the Bangladesh Petroleum Act-1974, National Energy Policy-2004, Bangladesh Water Policy-1999, Bangladesh Water Act-2013 (Shamsuzzaman et al. 2017). Most of the legislation and regulations were enacted a long time ago. Some of them have been updated from time to time based on the changing social, economic, cultural, and environmental circumstances, but most of the time the legal framework failed to tackle new challenges (Islam and Wahab, 2005). Updating this outdated and inactive legislation is required. The Department of Shipping (DOS) and the Department of Fisheries (DOF) are trying to create a new shipping and fisheries act (Islam et al. 2017).

6.57 POLICYMAKING AUTHORITY IN BANGLADESH

Blue Economy Cell (BEC) is the sole authorized cell to formulate ocean policy. It was formed in January 2017 as part of the Ministry of Power, Energy, and Mineral Resources' Energy and Mineral Resources Division (Islam and Shamsuddoha, 2018). But the activities of the BEC are confined to holding occasional meetings since the administrative body is equipped with inadequately few officials temporarily. Although the Parliamentary Standing Committee on Power, Energy, and Mineral Resources Ministry recommended upgrading the BEC as an authority with a permanent setup, the recommendation has been ignored (MOFA, 2018). Cooperation between ministries at the regional level is very important for successful blue economic development.

TABLE 6.9
Sector-Wise Recommendations for the Fruitful Use of Blue Economy to be Developed Country

Maritime Sectors	Sub-sectors	Recommended Activities
Fishing	Capture fishery, Aquaculture, and seafood processing.	1) Upcoming Bangabandhu Satellite-based monitoring system 2) Fish capture and protection capability increase 3) Promotion of environment-friendly and sustainable shrimp, crab, and potentially important other species farming systems in the coastal region 4) Mitigation and adaptive measures on the impacts of climate change 5) Declaration of fish sanctuaries 6) Seasonal banning of fishing 7) Advanced deep-sea vessel and technologies 8) Should be a member state of IOTC 9) Encourage Integrated Multi Trophic Aquaculture (IMTA) 10) Maritime security and surveillance to stop piracy and IUU fishing 11) Ensure entrance for native and poor fishermen and ban foreign trawlers 12) Women empowerment
Minerals	Oil and gas, deep-sea mining (exploration of rare earth metals).	1) Determine the location of possible oil and gas fields, as well as their reserves 2) A well-thought-out strategy is needed to carry out a multiline survey 3) Massive exploration and drilling activities 4) Monitoring and assessment 5) Public and private partnerships 6) Foreign support 7) Update the potential evaluation initially that was done in 1994
Salt production	Salt production	1) Land-leasing structures that are focused on the community 2) Sufficient credit facilities 3) Mechanical machinery is used (water pump, leveler, and the like) 4) Reliable weather forecasting 5) Modern techniques even collecting and refining common salt by mining 6) Public-private partnership
Marine Renewable Energy	Solar, wind, wave, and Tidal energy production.	1) Marine Renewable Energy mapping project to determine the true potential 2) Financial support from national and foreign 3) Technology transfer and training 4) Long-term policies
Marine Biotechnology	Pharmaceuticals, chemicals, seaweed harvesting, seaweed products, marine-derived bio-products.	1) Mid and Long-term research on preservation, processing, and quality control of marine products 2) Encouragement of Sea food-based restaurant and sea product based industry
Marine Manufacturing	Boat and ship manufacturing and repairing, marine instrumentation, marine industrial engineering.	1) Loan and project to implement Maine manufacturing

(continued)

TABLE 6.9 (Continued)
Sector-Wise Recommendations for the Fruitful Use of Blue Economy to be Developed Country

Maritime Sectors	Sub-sectors	Recommended Activities
Port, Shipping and Maritime Logistics	Ship building and repairing, ship owners and operators, shipping agents and brokers, ship management, container shipping services, customs clearance, freight forwarders.	1) Sonadia deep (Island) seaport establishment 2) Capability building of three existing ports 3) Need more sophisticated ships 4) Increase modern merchant ships 5) Needs to add more fleets 6) Proper monitoring and management system and training 7) Effective planning and actions 8) Diplomatic efforts to consolidate Bangladesh's international position 9) Need expertise in technical areas 10) Construction of a deep-sea port (DSP)
Marine Tourism	Sunbath, Sailing, and boating at sea, surfing, scuba diving, swimming in the sea, bird watching in coastal areas, whale, dolphin watching, trips to the beach, seaside, and islands.	1) Eco-friendly tourism 2) Safety and security 3) Easy communication system 4) Marine tourism zone 5) National and regional planning 6) Technical cooperation 7) Public financial support and 8) Public-private partnerships
Education and Research	Education and training, Research and Development.	1) Capability enhancement of maritime educational department, institutions, and research centre 2) Making high-quality maritime experts 3) Greater use of economic analysis 4) Better use of innovations in science and technology (for example, drones, unmanned airborne vehicles (UAVs), sensors, mapping, imaging) 5) Government and stakeholder engagement 6) Formulate comprehensive Ocean Policy

Source: Hasan et al. 2018.

TABLE 6.10
Reviewed or Developing Policies Related to the Blue Economy in Bangladesh

Sector	Policies	Laws and Acts	Responsible Institutions
Coastal Protection Climate change resilience and adaptation (including coastal *protection)*	*Bangladesh Climate Change Strategy and Action Plan* (BCCSAP) is to be completed by 2020. National Action Plan for Adaptation (NAPA) is to be completed by 2020.	The 2010 Climate Change Trust Act established the Bangladesh CC Trust, the Bangladesh CC Trust Fund, and the Bangladesh Climate Change Resilience Fund.	Ministry of Environment and Forests Disaster Management Information Centre of Ministry of Food and Disaster Management
Existence of Biodiversity, including mangrove ecosystems ('blue forests') Plan is under review.	*Coastal and Wetland Biodiversity Management Plan is under review.*	Wetland Conservation Act. Environment Conservation Act, 1995, 2000, and 2002. Environment Conservation Rules, 1997, 2000, 2001. National Conservation Strategy, 2005; National River Protection Commission Act, 2013; Forest Act, 1927; Wildlife Protection and Security Act, 2012	The Ministry of Environment and Forests The Bangladesh National Herbarium
Waste Disposal, including addressing externalities from industrial and agriculture pollution creating marine dead zones	*Bangladesh Water Act is under review and revision.*	Integrated Water Resources Management (IWRM), 2005. Participatory Water Management Regulations, 2014	Ministry of Water Resources
Energy (including renewable energy from wave, wind, and solar from ocean areas and explicit gender dimensions)	*Renewable Energy Policy, 2008 and National Energy Policy, 2004 are under review*	The Bangladesh Petroleum Act of 1974 supports planning, organizing, and implementation of exploration, exploitation, development, and production of petroleum wealth from the sea (including all territorial waters, continental shelf, and EEZ).	Ministry of Power, Energy, and Mineral Resources Sustainable and Renewable Energy Development Authority (SREDA) Bangladesh Power Development Board (BPDB) Local Government Engineering Directorate (LGED) Blue Economy Cell
Living Resources: Capture Fisheries, supporting sustainability	*National Marine Fisheries Policy, undergoing consultations and review*	The proposed National Marine Fisheries Policy includes provision for development of new laws in support of sustainable capture fisheries.	Ministry of Fisheries and Livestock, Dept. of Fisheries Bangladesh Fisheries Development Corp.; Bangladesh Coast Guard. Bangladesh Navy

(continued)

TABLE 6.10 (Continued)
Reviewed or Developing Policies Related to the Blue Economy in Bangladesh

Sector	Policies	Laws and Acts	Responsible Institutions
Living Resources: Aquaculture, including mariculture	*National Aquaculture Development Strategy and Action Plan (2013–2020) is reviewed annually. 2014 National Shrimp Policy is under review*	Fish Hatchery Act 2010; Fish Hatchery Rules 2011; Fish Feed and Animal Feed Act 2010; Fish Feed Rules 2011; Fisheries Research Institute Ordinance, 1984	Ministry of Fisheries and Livestock
Tourism, including marine tourism	*National Tourism Policy, 2009 is under review*	Tourism Board Act, 2010; Bangladesh Tourism Protected Areas and Special Tourism Zone Act, 2010; Bangladesh Tourism Protected Areas and Special Tourism Zone Rules, 2011	Ministry of Civil Aviation and Tourism Chambers of Commerce Bangladesh Parjatan Corp. Ministry of Shipping
Shipping and Transport including measures to address marine pollution	*Maritime and Shipping Strategy of Bangladesh*	Clean Air Act; Import Policy Orders; 2012–2015; Payra Port Authority Act, 2013; Chittagong Port Authority (Amendment) Act, 1995; Mongla Port Authority (Amendment) Act, 1995; Navy Ordinance, 1961; Coast Guard Act, 1994	Ministry of Power, Energy and Mineral Resources Infrastructure Financing Facility Inland Water Transport Authority
Ocean-based industry development and growth via access to finance	*Comprehensive Credit Policy for SMEs, including encouraging investment in ocean industries*	Inclusive Digital Financial Systems, 2015	Ministry of Industries Bangladesh Standards and Testing Institution Bangladesh Small and Cottage Industries Corporation Bangladesh Chemical Industries Corporation Bangladesh Bank

Source: Patil et al. 2018.

6.58   CONCLUSION

Blue economic prospects in Bangladesh are notable but challenges and risks are also working against blue economy development. A successful blue economy can change the economic status (developing to developed) of developing countries like Bangladesh. It is now time for data, political will and financial capital to be marshaled to bring Bangladesh onto the blue path and form global and regional partnerships towards ocean governance and sustainable maritime growth. For the blue economy's growth, the government should follow a future policy agenda that can focus on institutional cooperation, translating product science, a holistic approach to the blue economy, and empowering and educating young generations.

REFERENCES

Alam, A. M. K. 2018. *Harvesting Blue Economy-Need for Intertwined Economics for the Bay of Bengal Littorals.* BSMRMU International Seminar Proceedings.

Alam, B. 2013. *Energy Resources of Bangladesh,* University Grants Commission of Bangladesh, 2nd edition. Agargaon, Dhaka.

Alam, I. Barua, S. Ishii, K. Mizutani, S. Hossain, M. M. Rahman, I. M. Hasegawa, and H. 2019. Assessment of health risks associated with potentially toxic element contamination of soil by end-of-life ship dismantling in Bangladesh. *Environmental Science and Pollution Research*, 26: 24162–24175.

Alam, M. K. 2004. *Bangladesh's Maritime Challenges in the 21st Century*. Dhaka: Pathak Shamabesh Book.

Alam, M. K. 2014. Ocean/Blue economy for Bangladesh. In: *Proceedings of International Workshop on Blue Economy, Ministry of Foreign Affairs*. Dhaka, Bangladesh, pp 28–49.

Andrew, D. L. Steven, M. Vanderklift, A. and Bohler-Muller, N. 2019. A new narrative for the Blue Economy and Blue Carbon. *Journal of the Indian Ocean Region,* 15: 123–128.

APEC. 2014. *Marine Sustainable Development Report*. APEC Ocean and Fisheries Working Group, Asia-Pacific Economic Cooperation. http://publications.apec.org/publication-detail.php?pub_id=1552.

BALance Technology Consulting. 2014. Competitive Position and Future Opportunities of the European Marine Supplies Industry. European Commission. http://ec.europa.eu/DocsRoom/documents/4233/atta chments/1/translations/en/renditio ns/native.

Bangladesh Bureau of Statistics (BBS). 2017. Gross Domestic Product. www.bbs.gov.bd/site/page/dc2bc6ce-7080-48b3-9a04-73cec782d0df/Gross-Domestic-Product-(GDP.

Bari, A. 2017. Our oceans and the blue economy: opportunities and challenges. *Procedia Engineering*, 194: 5–11.

Begum, H. 2013. *The Role of Maritime Cluster in Enhancing the Strength and Development of Maritime Sectors of Bangladesh*. Dhaka: Bangladesh Economic Association (BEA), 1–16.

Belton, B. Karim, M. Thilsted, S. Murshed-E-Jahan, K. Collis, W. and Phillips, M. November 2011. *Review of Aquaculture and Fish Consumption in Bangladesh*. Studies and Reviews 2011–53. The WorldFish Center, pp. 72.

Bhuiyan, K. U. 2014. *Integrated Maritime Policy for Blue Economy*, Alam, Md. Jahangir. http://thedailystar. net/integrated-maritimepolicy-for-blue-economy-46456.

Bhuiyan, M. N. U. Ali, M. A. Rahman, M. M. and Selim, S. A. M. 2015. Marine boundary confirmation of Bangladesh: potentials of sea resources and challenges ahead. *The Cost and Management,* 43: 7.

Bhuyan, M. Abid Husain, S. Chowdhury, E. and Bat, L. (2022). Assessment of Carbon Sequestration Capacity of Seaweed in Climate Change Mitigation. Journal of Climate Change, 8(1), 1-8.

Bhuyan, M. Islam, M. Sharif, A. S. M. and Hoq, M. (2021). Seaweed: A Powerful Tool for Climate Change Mitigation That Provides Various Ecological Services. Bangladesh II: Climate Change Impacts, Mitigation and Adaptation in Developing Countries, 159-192.

Bhuyan, M. S. Sharif, A. S. M. Islam, M. M. Mojumder, I. A. Das, M. and Islam, M. S. 2020. Blue economy and the prospect of seaweed in Bangladesh. *Journal of Marine Science Research and Oceanography*, 3: 1–2.

BOBLME. 2011. *Country Report on Pollution in the BOBLME-Bangladesh*. BOBLME-2011-Ecology-01.

BOBLME. 2012. *Managing Advisory for the Bay of Bengal Hilsa Fishery*.

Brotto, L. and Pettenella, D. 2018. *Forest Management Auditing: Certification of Forest Products and Services*. Routledge, UK, 264p.

Cervigni, R. and Scandizzo, P. L. 2017. *The Ocean Economy in Mauritius: Making It Happen, Making It Last*. Washington, DC: The World Bank.

Chowdhury, S. R. Hossain, M. S. Sharifuzzaman, S. M. and Sarker, S. 2015. *Blue Carbon in the Coastal Ecosystems of Bangladesh*. Project Document, Support to Bangladesh on Climate Change Negotiation and Knowledge Management on Various Streams of UNFCCC Process Project, funded by DFID and Danida, implemented by IUCN Bangladesh Country Office.

CMIEN (China Marine Information Economic Network). January 2013. *Statistical Bulletin of China's Ocean Economy 2012*.

Danovaro, R. Aguzzi, J. Fanelli, E. Billett, D. Gjerde, K. Jamieson, A. Van Dover, C. L. 2017. An ecosystem-based deep-ocean strategy. *Science,* 355: 452–454.

DGPM (Directorate General for Maritime Policy) Government of Portugal. 2013. Annex A—The Sea Economy in Portugal. *The National Ocean Strategy 2013–2020*. https://maritime-spatial-planning.ec.europa.eu/ practices/national-ocean-strategy-2013-2020.

DoF. 2016. *National Fish Week 2016 Compendium*. Department of Fisheries, Ministry of Fisheries and Livestock in Bangladesh, pp. 148.

DoF. 2018. *Year Book of Fisheries Statistics.* https://fisheries.portal.gov.bd/sites/default/files/files/fisheries.portal.gov.bd/page/4cfbb3cc_c0c4_4f25_be21_b91f84bdc45c/Fisheries%20Statistical%20Yearbook%202017-18.pdf.

DoF. 2020. http://fisheries.gov.bd/site/page/43ce3767–3981–4248–99bd-d321b6e3a7e5/-.

Doney, S. C. Ruckelshaus, M. Duffy, J. E. Barry, J. P. Chan, F. and English, C. A. 2012. Climate change impacts on marine ecosystems. *Annual Review of Marine Science,* 4: 11–37.

Economist Intelligence Unit. 2015. *The Blue Economy: Growth, Opportunity, and a Sustainable Ocean Economy.* London: EIU. http://perspectives.eiu.com/sustainability/blue-economy/white-paper/blue-economy.

Ecorys. 2012. *Blue Growth: Scenarios and Drivers for Sustainable Growth from the Oceans, Seas and Coasts.* Third Interim Report, Rotterdam/Brussels, March 13. http://ec.europa.eu/maritimeaffairs/documentation/studies/documents/blue_growth_thi rd_interim_report_en.pdf.

European Commission (EC). 2017. *Report on the Blue Growth Strategy: Towards a More Sustainable Growth and Jobs in the Blue Economy.* Brussels: European Commission Report No.: SWD, 128.

European Union. 2018. *The 2018 Annual Economic Report on EU Blue Economy,* 5. https://ec.europa.eu/maritimeaffairs/sites/maritimeaffairs/files/2018-annual-economic-report-onblue-economy_en.pdf.

FAO, 1995. Code of Conduct for Responsible Fisheries. www.fao.org/3/v9878e/v9878e.pdf

FAO. 2012. The State of World Fisheries and Aquaculture 2012. Rome. 209 pp.

FAO. 2014. *The State of World of Fisheries and Aquaculture 2014: Opportunities and Challenges.* Rome: Food and Agriculture Organization. www.fao.org/3/a-i3720e.pdf.

FAO. 2016. *The State of World Fisheries and Aquaculture 2016.* Contributing to food security and nutrition for all. Rome, FAO, pp. 200

Gujarat Maritime Board. 2014. Roadmap of the development of maritime cluster in Gujarat. https://balticcluster.pl/?page_id=949.

Habib, K. A. and Islam, M. J. 2020. An updated checklist of marine fisheries in Bangladesh. *Bangladesh Journal of Fisheries,* 32: 357–367.

Haque, U. Hashizume, M. Kolivras, K. N. Overgaard, H. J. Das, B. and Yamamoto, T. 2012. Reduced death rates from cyclones in Bangladesh: what more needs to be done? *Bulletin of the World Health Organization,* 90: 150–156.

Harley, C. D. Randall Hughes, A. Hultgren, K. M. Miner, B. G. Sorte, C. J. and Thornber, C. S. 2006. The impacts of climate change in coastal marine systems. *Ecology Letters,* 9: 228–241.

Hasan, M. M. Hossain, B. M. S. Alam, M. J. Chowdhury, K. M. A. Al Karim, A. and Chowdhury, N. M. K. 2018. The prospects of blue economy to promote Bangladesh into a middle-income country. *Open Journal of Marine Science,* 8: 355–369.

Hassan, S. R. Hassan, M. K. and Islam, M. S. 2014. Tourist-group consideration in tourism carrying capacity assessment: a new approach for the Saint Martin's Island, Bangladesh. *Journal of Economics and Sustainable Development,* 5: 150–158.

Häyhä, T. and Franzese, P. P. 2014. Ecosystem services assessment: A review under an ecological-economic and systems perspective. *Ecological Modelling,* 289: 124–132.

Hoegh-Guldberg, O. et al. 2015. *Reviving the Ocean Economy: Action Agenda for 2015.* WWF International, Geneva. https://wwf.de/fileadmin/fm-wwf/PublikationenPDF/WWF-Report-Reviving-the-Ocean-Economy-Summary.pdf.

Hossain, F. 2020. Adaptation measures (AMs) and mitigation policies (MPs) to climate change and sustainable blue economy: A global perspective. *Journal of Water and Climate Change,* 12 (5): 1344–1369.

Hossain, M. S. Chowdhury, S. R. Navera, U. K. Hossain, M. A. R. Imam, B. and Sharifuzzaman, S. M. 2014. *Opportunities and Strategies for Ocean and River Resources Management.* Background paper for preparation of the 7th Five Year Plan. Dhaka, Bangladesh: FAO, pp. 61.

Hossain, M. S. Hossain, M. Z. and Chowdhury, S. R. 2006. An analysis of economic and environmental issues associated with sea salt production in Bangladesh and Thailand coast. *International Journal of Ecology and Environmental Sciences,* 32: 159–172.

Howard, B. C. 2018. Blue growth: stakeholder perspectives. *Marine Policy,* 87: 375–377.

Humayun, N. M. Das, A. C. Barua, S. Al Mamun, A. and Singha, N. K. 2016. *Bangladesh National Report to the Scientific Committee of the Indian Ocean Tuna Commission.* (IOTC), 2016. IOTC and FAO Report No.: IOTC–2015–SC18–NR33.

Hung, S. Lu, W. and Wang, T. 2010. Benchmarking the operating efficiency of Asia container ports. *European Journal of Operational Research,* 203: 706–713.

Hussain, G. M. Failler, P. Karim, A. and Alam, M. K. 2017. Review on opportunities, constraints and challenges of blue economy development in Bangladesh. *Journal of Fisheries and Life Sciences,* 2: 45–57.

Hussain, M. G. 2019. *Sustainable Aquaculture and Fisheries Development in Bangladesh: Country Needs and Sectoral Priorities.* Seminar held on BARC, Farm Gate, Dhaka, Bangladesh.

Hussain, M. G. Failler, P. Karim, A. A. and Alam, M. K. 2018. Major opportunities of blue economy development in Bangladesh. *Journal of the Indian Ocean Region,* 14: 88–99.

Hwang, K. H. et al. 2011. *Assessment of Gross Ocean Products in Korea, Korea Institute of Marine Science and Technology Promotion.*

Iftekhar, M. S. 2006. Conservation and management of the Bangladesh coastal ecosystem: overview of an integrated approach. *Natural Resources Forum,* 30: 230–237.

International Tanker Owners Pollution Federation (ITOPF). 2020. http://itopf.com/.

IPCC. 2018. Summary for policymakers. In: *Global Warming of 1.5°C. An IPCC Special Report on the Impacts of Global Warming of 1.5°C Above Pre-Industrial Levels and Related Global Greenhouse Gas Emission Pathways, in the Context of Strengthening the Global Response to the Threat of Climate Change, Sustainable Development, and Efforts to Eradicate Poverty,* V. Masson-Delmotte, et al. (eds). Geneva: IPCC. 24p.

Islam, M. M. and Shamsuddoha, M. 2018. Coastal and marine conservation strategy for Bangladesh in the context of achieving blue growth and sustainable development goals (SDGs). *Environmental Science and Policy,* 87: 45–54.

Islam, M. M. Shamsuzzaman, M. M. Mozumder, M. M. H. Xiangmin, X. Ming, Y. and Jewel, M. A. S. 2017. Exploitation and conservation of coastal and marine fisheries in Bangladesh: do the fishery laws matter? *Marine Policy,* 76: 143–151.

Islam, M. R. 2018. *Energy Efficient Shipping and Smart Port—a Vision for the Future.* BSMRMU International Seminar Proceedings.

Islam, M. S. and Wahab, M. A. 2005. A review on the present status and management of mangrove wetland habitat resources in Bangladesh with emphasis on mangrove fisheries and aquaculture. *Hydrobiologia,* 542: 165–190.

IUCN. 2010. *Integrated Coastal Management (ICM): Best Practices and Lessons Learned.* Workshop Report. Colombo.

Jiang, X. Z. et al. December 2014. China's marine economy and regional development. *Marine Policy,* 50(Part A): 227–237.

Kalaydjian, R. et al. 2009. *French Maritime Economic Data 2009.* Ifremer, Paris: Marine Economics Department.

Kalaydjian, R. et al. 2011. *French Maritime Economic Data 2011.* Ifremer, Paris: Marine Economics Department.

Kalaydjian, R. et al. 2014. *French Maritime Economic Data 2014.* Ifremer, Paris: Marine Economics Department.

Karim, M. S. and Uddin, M. M. 2019. Swatch-of-no-ground marine protected area for sharks, dolphins, porpoises and whales: legal and institutional challenges. *Marine Pollution Bulletin,* 139: 275–281.

Keen, M. R. Schwarz, A. M. and Wini-Simeon, L. 2018. Towards defining the Blue Economy: practical lessons from Pacific Ocean governance. *Marine Policy,* 88: 333–341.

Khondaker, A. M. 1998. *Protection of Bangladesh Waters Against Accidental Oil Pollution from Ships.* World Maritime University Dissertations, 1256.

Kildow, J. et al. 2014. State of the U.S. Ocean and Coastal Economies 2014. *Publications,* Paper 1. http://cbe.miis.edu/noep_publications/1.

Kildow, J. T. Colgan, C. S. Johnston, P. Scorse, J. D. and Farnum, M. G. 2016. *State of the U.S. Ocean and Coastal Economies: 2016 Update.* National Ocean Economics Program. Middlebury Institute. http://midatlanticocean.org/wp-content/uploads/2016/03/NOEP_National_Report_2016.pdf.

Kroeker, K. J. Kordas, R. L. Crim, R. Hendriks, I. E. Ramajo, L. Singh, G. S. and Gattuso, J. P. 2013. Impacts of ocean acidification on marine organisms: quantifying sensitivities and interaction with warming. *Global Change Biology,* 19: 1884–1896.

Lee, K. H. Noh, J. and Khim, J. S. 2020. The Blue Economy and the United Nations' sustainable development goals: challenges and opportunities. *Environment International,* 137: 105528.

Lillebø, A. I. Pita, C. Garcia Rodrigues, J. Ramos, S. and Villasante, S. 2017. How can marine ecosystem services support the blue growth agenda? *Marine Policy,* 81: 132–142.

Luisetti, T. Jackson, E. L. and Turner, R. K. 2013. Valuing the European 'coastal blue carbon' storage benefit. *Marine Pollution Bulletin,* 71: 101–106.

Mannar, M. G. V. 1982. *Guidelines for the Establishment of Solar Salt Facilities from Seawater, Underground Brines and Salted Lakes.* Industrial and Technological Information Bank (INTIB), Industrial Information Section, United Nations Industrial Development Organization (UNIDO), UNIDO Technology Program, pp. 149.

Maritime by Holland. 2014. *The Dutch Maritime Cluster.* High Tech, Hands On, Monitor 2014 summary. www.maritimebyholland.com/download-link.php?file=/wpcontent/uploads/2015/03/150227-NML003-samenvatting-los_UK.pdf.

MOFA. 2018. *Press Statement of the Hon'ble Foreign Minister on the Potential of Blue Economy* (Press Release). www.mofa.gov.bd.

Mohanty, S. K. Dash, P. and Gupta, A. 2017. Unleashing the potential of blue economy. *Policy Brief,* 1.

MOI. 2019. *Press Statement of the Hon'ble Industry Minister on the Ship Recycling* (Press Release), www.moi.gov.bd.

National Marine Science Committee. 2015. National Marine Science Plan 2015–2025: driving the development of Australia's blue economy. www.marinescience.net.au/nationalmarinescienceplan/.

Nellemann, C. Corcoran, E. Duarte, C. Valdés, L. De Young, C. Fonseca, L. and Grimsditch, G. 2009. *Blue Carbon. A Rapid Response Assessment.* United Nations Environment Programme, GRID-Arendal.

Newell, R. G. Pizer, W. A. and Raimi, D. 2013. Carbon markets 15 years after Kyoto: lessons learned, new challenges. *Journal of Economic Perspectives,* 27: 123–146.

Nomura Research Institute. 2009. *The Report on of Japan's Marine Industry.*

OECD. 2012. *Education at a Glance 2012.* OECD Indicators, OECD Publishing. http://dx.doi.org/10.1787/eag-2012-en. ISBN 978–92–64–17715–4 (print).

OECD. 2016. *The Ocean Economy in 2030.* Paris: OECD Publishing.

Onselen, C. V. 2018. *Changing Trends in Shipping and Maritime Business.* BSMRMU International Seminar Proceedings.

Parletta, N. 2019. *Making the Blue Economy Sustainable.* https://forbes.com/sites/natalieparletta/2019/06/21/making-the-blue-economy-a-sustainable-reality/#337b0b9826f8.

Patil, G. Virdin, J. Colgan, C. S. Hussain, M. G. Failler, P. and Vegh, T. 2018. *Toward a Blue Economy: A Pathway for Sustainable Growth in Bangladesh.* Washington, DC: The World Bank Group.

Pörtner, H. O. 2010. Oxygen- and capacity-limitation of thermal tolerance: a matrix for integrating climate-related stressor effects in marine ecosystems. *Journal of Experimental Biology,* 213: 881–893.

Pörtner, H. O. Karl, D. M. Boyd, P. W. Cheung, W. Lluch-Cota, S. E. Nojiri, Y. and Wittmann, A. C. 2014. Ocean systems. In *Climate Change 2014: Impacts, Adaptation, and Vulnerability. Part A: Global and Sectoral Aspects.* Contribution of working group II to the fifth assessment report of the intergovernmental panel on climate change. Cambridge University Press, UK, pp. 411–484.

Pugh, D. 2008. *Socio-economic Indicators of Marine-related Industries in the UK Economy,* London: The Crown Estate. www.thecrownestate.co.uk/media/5774/socio_economic_uk_marine.pdf.

Rabbi, H. R. and Rahman, A. 2017. Ship breaking and recycling industry of Bangladesh: issues and challenges. *Procedia Engineering,* 194: 254–259.

Rahman, M. R. 2017. Blue economy and maritime cooperation in the Bay of Bengal: role of Bangladesh. *Procedia Engineering,* 194: 356–361.

Rayner, R. Jolly, C. and Gouldman, C. 2019. Ocean Observing and the Blue Economy. *Frontier in Marine Science,* 6: 330.

Repetto, M. S. 2005. *Towards an Ocean Governance Framework and National Ocean Policy for Peru.* The United Nations—The Nippon Foundation of Japan Fellow (mimeo).

Rhein, M. Rintoul, S. R. Aoki, S. Campos, E. Chambers, D. Feely, R. A. and Wang, F. 2013. Observations: ocean. Chapter 3 In: *Climate Change 2013: The Physical Science Basis. Contribution of Working Group I to the Fifth Assessment Report of the Intergovernmental Panel on Climate Change,* Stocker, T.F. Qin, D. Plattner, G.K. Tignor, M. Allen, S.K. Boschung, J. Nauels, A. Xia, Y. Bex, V. and Midgley, P.M. (eds). Cambridge University Press, Cambridge, United Kingdom and New York, NY, USA, 1535 pp.

Roberts, J. P. and Ali, A. 2016. *The Blue Economy and Small States.* Commonwealth Blue Economy Series, No. 1. Commonwealth Secretariat, London.

Roy, R. K. 2017. *Blue Economy: Prospects and Challenges*. https://daily-sun.com/magazine/details/256349/Prospects-And-Challenges/2017–09–22.

Saha, K. and Alam, A. 2018. Planning for blue economy: prospects of maritime spatial planning in Bangladesh. *AIUB Journal of Science and Engineering (AJSE)*, 17: 59–66.

Sarker, S. Bhuyan, M. A. H. Rahman, M. M. Islam, M. A. Hossain, M. S. Basak, S. C. and Islam, M. M. 2018. From science to action: exploring the potentials of blue economy for enhancing economic sustainability in Bangladesh. *Ocean and Coastal Management*, 157: 180–192.

Schlosser, P. 2013. Ocean services-what is their value? In *Sustainable Oceans: Reconciling Economic Use and Protection*, P. Pissulla, M. Visbeck, and Schlosser (eds). Kiel: Drager Foundation, p. 15.

Seitz, R. D. Wennhage, H. Bergstr öm, U. Lipcius, R. N. and Ysebaert, T. 2013. Ecological value of coastal habitats for commercially and ecologically important species. *ICES Journal of Marine Sciences*, 71: 648–665.

Shamsuzzaman, M. M. Islam, M. M. Tania, N. J. Al-Mamun, M. A. Barman, P. P. and Xu. X. 2017. Fisheries resources of Bangladesh: present status and future direction. *Aquaculture and Fisheries*, 2: 145–156.

Shoeib, M. J. and Rahman, M. M. 2014. Emerging landscape in the Bay of Bengal and maritime capability building of Bangladesh. *BIISS Journal*, 35: 23–43.

Sigfusson, T. and Gestsson, H. 2012. *Iceland's Ocean Economy*, http://sjavarklasinn.is/e n/wp-content/uploads/2014/11/2.-IcelandsOceanEconomy2011low.pdf.

Singh, R. 2019. *India's Maritime Security and Policy: An Imperative for the Blue Economy*. India in South Asia, Springer Singapore, pp. 269–289.

Statistics New Zealand. 2006. *New Zealand's Marine Economy 1997–2002*, Experimental Series Report, Statistics New Zealand, https://statsnz.contentdm.oclc.org/digital/collection/p20045coll1/id/1405/

Territorial Waters and Maritime Zones Act (No. XXVI) 1974, Section 7.

Thakur, N. L. and Thakur, A. N. 2006. Marine biotechnology: an overview. *Indian Journal of Biotechnology*, 5: 263–268.

To, W. M. and Lee, P. K. (2018). China's maritime economic development: A review, the future trend, and sustainability implications. Sustainability, 10(12), 4844.

Toropova, C. Meliane, I. Laffoley, D. Matthews, E. and Spalding, M. 2010. *Global Ocean Protection: present Status and Future Possibilities*, IUCN.

Trumper, K. Bertzky, M. Dickson, B. Heijden, J. M. and Manning, P. 2009. *The Natural Fix? The Role of Ecosystems in Climate Mitigation*. A UNEP rapid response assessment. Cambridge: UNEPWCMC, pp. 65. https://wedocs.unep.org/handle/20.500.11822/7852.

Uddin, M. I. Islam, M. R. Awal, Z. I. and Newaz, K. M. S. 2017. An analysis of accidents in the inland waterways of Bangladesh: lessons from a decade (2005–2015). *Procedia Engineering*, 194: 291–297.

UNCLOS, 1982. United Nation Convention Law of the Sea. www.un.org/depts/los/convention_agreements/texts/unclos/unclos_e.pdf

UN. 2017. Factsheet: people and oceans. In *The Ocean Conference*. New York, NY: United Nations.

UNEP (United Nations Environment Programme). 2015. *Blue Economy: Sharing Success Stories Inspire Change*, UNEP Regional Seas Report.

UN Comtrade. 2017. International trade statistics database. united nations comtrade database. https://comtrade.un.org/

UNWTO. April 2013. *World Tourism Barometer*. Vol. 11. www.e-unwto.org/doi/epdf/10.18111/wtobarometer eng.2013.11.2.1

van Kooten, G. C. Eagle, A. J. Manley, J. and Smolak, T. 2004. How costly are carbon offsets? A meta-analysis of carbon forest sinks. *Environmental Science and Policy*, 7: 239–251.

Vega, A. and Hynes, S. 2017. *Ireland's Ocean Economy Report* 2017 (No. 1154-2021-688). https://ageconsearch.umn.edu/record/309536/.

Vega, A. Hynes, S. and Corless, R. 2013. *Ireland's Ocean Economy: Reference Year 2010*, NUI Galway, Ireland, www.nuigalway.ie/semru/documents/irelands_ocean_economy_report_series_no2.pdf.

Vega, A. Hynes, S. and O'Toole, E. 2015. *Ireland's Ocean Economy: Reference Year: 2012*, Ryan Institute Research Day, September, Galway, Ireland, https://ageconsearch.umn.edu/record/210705/.

Wenhai, L. Cusack, C. Baker, M. Tao, W. Mingbao, C. Paige, K. Xiaofan, Z. Levin, L. Escobar, E. Amon, D. Yue, Y. Reitz, A. Neves, A. A. S. O'Rourke, E. Mannarini, G. Pearlman, J. Tinker, J. Horsburgh, K. J.

Lehodey, P. Pouliquen, S. Dale, T. Peng, Z. Yufeng, Y. 2019. Successful Blue Economy Examples with an Emphasis on International Perspectives. *Frontier in Marine Science*, 6: 261.

World Bank. 2017. *The Potential of the Blue Economy: Increasing Long-Term Benefits of the Sustainable Use of Marine Resources for Small Island Developing States and Coastal Least Developed Countries.* Washington, DC: World Bank.

World Bank. 2018. *Toward a Blue Economy: A Pathway for Sustainable Growth in Bangladesh.* http://docume nts.worldbank.org/curated/en/857451527590649905/pdf/126654-REPL-PUBLIC-WBG-Blue-Econ omy-Report-Bangladesh-Nov2018.pdf.

World Bank. 2019. *The World Bank in South Asia*, updated on October 11, 2019. https://worldbank.org/en/reg ion/sar/overview.

World Travel and Tourism Council (WTTC). 2017. *Travel and Tourism Economic Impact 2017 Bangladesh.*

Zhao, R. Hynes, S. and He, G. S. 2013. Defining and quantifying China's ocean economy. *Marine Policy*, 43: 164–173.

Zhao, R. Hynes, S. and He, G. S. 2014. Defining and quantifying China's ocean economy. *Marine Policy*, 43: 164–173.

7 Application of Blue Economy for Polymetallic Nodules from the Central Indian Ocean Basin

Ankeeta A. Amonkar,[1]* Niyati Gopinath Kalangutkar,[2] and Sridhar D. Iyer[3]
[1]Dnyanprassarak Mandal's College and Research Centre, Mapusa, Goa, India
[2] School of Earth, Ocean and Atmospheric Science, Goa University, Taleigao Plateau, Goa, India
[3] Formerly with CSIR-National Institute of Oceanography, Dona Paula, Goa, India
*Corresponding author: Ankeeta Amonkar; E-mail: ankeetaamonkar9@gmail.com

CONTENTS

DOI: 10.1201/9781003184287-7

7.1 INTRODUCTION

Under the United Nations' Sustainable Development Goals (SDG-14) emphasis has been given to exploiting in a sustained way the oceans and seas for their energy and marine resources, especially for Small Island Developed States (SIDS) and Least Developed Countries. The background information and importance of the blue economy (BE) and maritime zones is discussed in Chapter 10 where the seafloor massive sulfides (SMS) of the Indian Ocean are discussed. Although the polymetallic nodules and crusts of the Central Indian Ocean Basin (CIOB) (Figure 7.1) have been thoroughly investigated over the four decades or more (Mukhopadhyay et al. 2018), this is the first time that the detailed role of BE as applicable to these deposits is presented.

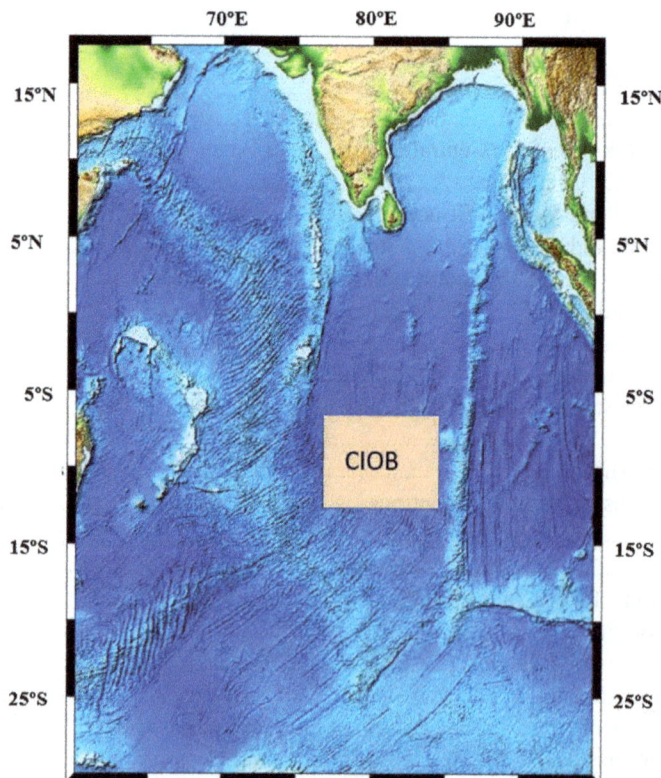

FIGURE 7.1 Contract area in the Central Indian Ocean Basin (CIOB) allotted to India (https://isa.org.jm/map/government-india).

7.2 POLYMETALLIC NODULES OF THE CENTRAL INDIAN OCEAN BASIN

7.2.1 REGIONS OF OCCURRENCE

The knowledge about the Indian Ocean originated from the geophysical studies carried out during the International Indian Ocean Expedition (IIOE) between 1961 and 1965. Subsequently, the international scientific drilling campaigns namely, Deep-Sea Drilling Project (DSDP), Ocean Drilling Program (ODP), and International Ocean Drilling Program (IODP) drilled several sites in the Indian Ocean. The results obtained of the drilled sediments and rocks have provided a large amount of data that have helped to validate the inferences made from initial geophysical and geological investigations.

Polymetallic nodules (also called ferromanganese or manganese nodules; henceforth we will be using the terms 'nodules') were first collected from the Pacific Ocean during the HMS Challenger expedition (Glasby, 1977) but for a few decades, no interest was shown by the scientific community. According to the report entitled by 'The mineral resources of the sea' by Mero (1965) the economic viability of mining the nodules provided an impetus to launch full-scale exploration in the Pacific Ocean in the 1970s. Exploratory programs in the Clarion-Clipperton Zone (CCZ) were initiated by the US, UK, Germany, Russia (then USSR), Japan, and France and followed later by China and South Korea.

In the Pacific and Indian oceans, the nodules generally occur in water depths >5,000 m, are black in colour, and of variable size (mostly 2 - 6 cm diameter) (Figure 7.2). The nodules have variable nuclei such as weathered rock pieces, sharks' teeth, older nodules, and sediment clasts. Over these nuclei, iron (Fe), manganese (Mn), copper (Cu), cobalt (Co), nickel (Ni), zinc (Zn), and several other elements accrete mainly from seawater. These elements could also be contributed through diagenesis and hydrothermal sources or be from mixed sources and hence, the nodules may exhibit one or more accretionary processes. The elements are in the oxide form and ore beneficiation of the nodules would provide metals of economic interest such as Mn, Cu, Co, and Ni. Some nodules occur as large concretions (Figure 7.3) while crusts are common along seamounts and abyssal hills (Iyer and Sharma, 1990) .

FIGURE 7.2 Polymetallic nodules occurring in siliceous pelagic sediments recovered in a box core from the CIOB. Marker pen is used as a scale (Length = 14cm).

FIGURE 7.3 A large manganese nodule from the CIOB collected during 48th Expedition of Research Vessel Sindhu Sadhana in April 2018. Pencil is used as a scale (18cm long).

7.3 GEOMORPHOLOGY AND MORPHO-STRUCTURES

The Indian Ocean has four major spreading ridge systems: Carlsberg Ridge, Central Indian Ridge (CIR), South West Indian Ridge (SWIR), and South East Indian Ridge (SEIR) (Iyer and Ray, 2003). The details of hydrothermal activities along these ridges are discussed in Chapter 10 by Kalangutkar et al. (this volume). The aseismic ridges, Chagos-Laccadive and Ninety East Ridge (NER) were formed from the Reunion and Kerguelen hotspots, respectively. The other ridges are the Broken, Madagascar, Mozambique, Kerguelen-Gaussberg, and Laxmi. The tectonically active regimes are the Andaman back-arc basin and Indonesian trench which are subduction zones in the Indian Ocean. There are also microcontinents such as Socotra with seamounts, and continental islands such as Madagascar, Seychelles, and Sri Lanka. The identified seamounts in the Indian Ocean are Error, Sagar Kanya, Panikkar, Wadia, Alcock, Sewell and the Afanasy-Nikitin Seamount complex (Iyer et al. 2012). In addition, there are several sedimentary basins such as Wharton, Somali, Owen, Madagascar, Laccadive, and Central Indian Ocean Basin (CIOB), the Bay of Bengal, and the Arabian Sea.

Among the sedimentary basins, the CIOB is the largest that extends between 6° S and 20° S and 72° E and 80° E and has an average water depth of 5,000m. The basin is open in the north and bounded by the NER on the east, SEIR in the south, and on the west by the CIR. To the north of the basin, there is a prominent deformed boundary zone which has a smooth seafloor, buried hills, folded and faulted sediments, and the Afanasy-Nikitin seamount complex (Mukhopadhyay et al. 2008). The CIOB formed from the Indian Ocean Triple Junction that has been proposed to be of ~65Ma age and with a half rate of spreading of 67.5mm/year. The initial or 'soft' collision occurred between the north-moving Indian Plate and stable Eurasian Plate at ~60Ma and the first 'hard' collision between ~51 and ~50Ma. During these collisional events the half rate of spreading reduced from 95mm/year to 26mm/year. The CIOB is complex in terms of its tectonic and morphologic nature, vast latitudinal extent, seamounts, abyssal hills, faults, crenulations, lineations, fracture zones (FZ), and with a variety of sediments and rocks. Most importantly the basin has significant and rich economical deposits of nodules that are second to the North Pacific nodule belt (Mukhopadhyay et al. 2018 and references therein).

Seamounts in the CIOB have helped to understand the evolutionary history of the basin. Mukhopadhyay and Khadge (1990) suggested that the seamounts have originated from a hotspot while Kodagali (1991) took the view that the seamounts resulted from mid-plate volcanism. Das et al. (2005, 2007) reported 200 seamounts, the majority of which occur between 10°S and 14°S. Most of these seamounts are along eight chains that trend approximately in an N-S direction while a few are in isolated places. These authors suggested that the seamounts were mostly emplaced along propagative FZ during the northward travel of the Indian Plate. Petrological studies indicated multiple basaltic volcanism episodes during the formation of the CIOB seafloor and seamounts (Iyer et al. 2018).

The basalts occur as outcrops and fragments on the slope and summit of the seamount, abyssal hills, and on the seafloor. Compositionally, the basalts are Normal-Mid-Ocean Ridge Basalts (N-MORB) similar to those from the Mid-Atlantic Ridge (MAR) and East Pacific Rise (EPR) (Mukhopadhyay et al. 2008; Das et al. 2012). Ferrobasalts occur with plagioclase (predominant), olivine (rare), and small euhedral magnetite and hematite grains, and have high Fe (>12 wt.%) and Ti (>2%). These basalts were recovered near topographic highs and high amplitude magnetic zones and perhaps formed from a fractionated melt that was emplaced at shallow crustal depth (Iyer et al. 1996). The basalts along the flank and summit of the seamounts form substrate for the ferromanganese oxides while the weathered fragments at the foothill form nuclei for the nodules.

Spilites that occur near the 79° E FZ resulted from low-temperature alteration of basaltic lava piles (Karisiddaiah and Iyer, 1992). Pumices of variable colours, shapes, sizes, and vesicularities encompass a large field (600,000 sq km) and are trachyandesitic to rhyodacitic in composition. Studies of the pumice provide evidence of their formation from intraplate volcanism (Iyer and Sudhakar, 1993a; Iyer, 1996; Kalangutkar et al. 2011; Kalangutkar, 2011). The possibility of drift pumice from the 1883 eruption of Krakatoa volcano, Indonesia is not ruled out (Iyer and Karisiddaiah, 1988; Mukherjee and Iyer, 1999; Pattan et al. 2008; Kalangutkar et al. 2011).

The volcanogenic-hydrothermal materials (vhm) in the CIOB comprise delicate bread-crust-like magnetic particles, magnetite spherules, spherules of Fe-Ti and Al composition, glass shards (silicic and basaltic), and palagonite grains (Iyer 2005; Amonkar et al. 2020a, Amonkar et al. 2020b, Amonkar and Iyer, 2021). The presence of vhm, at seamounts' base, and near FZ, and their ages from 625 ka (Iyer et al. 1997a; 1999b) to as recent as 100 years (Nath et al. 2008) attest to ongoing volcanic and hydrothermal activities in the basin.

7.4 SEDIMENTS IN THE INDIAN OCEAN

The nature and distribution of seafloor sediments in the Indian Ocean are influenced by five interrelated factors. These are climatic and current patterns, nutrient and organic production in surface waters, relative solubility of calcite and silica, submarine topography, and detrital input (Mukhopadhyay et al. 2008). The wide geographical extent of the CIOB has resulted in a change in the sediment composition and lithofacies. The four types of sediment in the basin are (Udintsev, 1975; Kolla and Kidd 1982).

a) Terrigenous sediments: These sediments are mainly sourced from the Indo-Gangetic plain and Himalaya and debouch into the CIOB at a rate of 1.670 x 10^6 tons/yr from the Ganges-Brahmaputra rivers. Although the northern area of the basin is mainly of terrigenous sediments but their presence is reported up to 15° S (Nath, 2001).

b) Siliceous sediments: Siliceous sediments are composed of diatoms, radiolarians, silico-flagellates, and sponge spicules. In the Indian Ocean, diatoms are the primary producers of silica followed by radiolarians in the CIOB. The region between 5° S and 15° S has a distinct zone of ooze with more than 70% of radiolarians. The sediments are associated with a high biogenic productivity region wherein the production of biogenic silica outweighs its dissolution, resulting in siliceous materials. Siliceous sediment is a favourable host for nodules in the Pacific and Indian oceans (Mukhopadhyay et al. 2018).

c) Calcareous sediments: These sediments composed of foraminifera and pteropods occur above the Carbon Compensation Depth (CCD) and form more than 54% of the surface of the Indian Ocean floor. A small patch of calcareous sediment is present at around 12 - 14° S and 82.5 - 83.5° E in the CIOB. Interestingly, a minor patch of calcareous sediment is present near a seamount summit which is below the CCD (Nath et al. 2012).

d) Red clays: Red clays or pelagic sediments are present between 15° S and 25° S in the CIOB. The red clay sediments occurring below the CCD and outside the zones of biogenic inputs have the lowest sedimentation rate.

The authigenic minerals in the CIOB sediments are zeolites (phillipsite, clinoptilolite, and the like), and hydrogenous Fe-Mn oxy-hydroxides (todorokite, birnessite, vernadite, and so forth) that occur as a coating on existing minerals and rocks and as nodules and crusts. Besides these there are also extra-terrestrial materials such as cosmic spherules and dust, tektites, and microtektites, and most importantly an abundance of volcanic glass shards that may be contributed from nearby terrestrial volcanoes and/or formed in the abyssal depth due to phreatomagmatic activity (Amonkar et al. 2020a and references therein). Evidence in the form of spherules and fragments of iron, titanium, and aluminium, extensive palagonite formation in sediment cores with the simultaneous absence of radiolarians, all vouch for the occurrence of episodic albeit localized hydrothermal activities in the CIOB (Amonkar and Iyer, 2021; Amonkar et al. 2021).

7.5   CHARACTERISTICS OF THE CIOB NODULES

7.5.1   SHAPE AND SIZE

Generally, the nodules are classified based on their shape, size, nucleus, and surface texture. The nodules display variation in shapes with larger size nodules being elongated, discoid, flattened, or irregular while the smaller ones are spheroid and sub-spheroid. The nodules with a typical bulge and knobby band around their equator are termed as 'hamburger-shaped' nodules. The nodule size is an important criterion because the collecting device has to be designed, based on the dominant size present in the mining area. The CIOB nodules vary from <10 mm to a few centimeters but are mostly in the size range of 20 - 60 mm.

7.5.2   NODULE NUCLEUS

A nucleus is vital for the growth of nodules since ferromanganese oxides around the core are deposited as concentric layers from hydrogenous, diagenesis or hydrothermal processes. The CIOB nodules commonly have rock fragments that are derived from weathering of the basaltic hills, seamounts, and seafloor. The other types of nuclei are sharks' teeth, palagonite grains, and rarely, older nodules. There are examples of phillipsite crystals (21 x 10 x 8 mm) that formed a nucleus for the growth of the surface nodules (Ghosh and Mukhopadhyay, 1995) and micronodules (<10 mm size) within sediment cores (Iyer et al. 2012, 2018).

The sizes of the nodule and nucleus are unrelated and are independent of the environment of formation. The CIOB nodules have a large nucleus and less oxide growth and vice-versa, while sometimes the nucleus is fully altered by seawater and replaced by ferromanganese oxides.

The shape of the nucleus often determines the shape of the nodules because the initial growth around it influences the overall layering of the nodules. This is distinctly seen in younger nodules that have a larger nucleus and a thin oxide layer. Nodules with a single nucleus are called mononucleate while those with two or more nuclei are termed as polynucleate. The CIOB mononucleate nodules generally occur in the abyssal plain, whereas polynucleate nodules are commonly on the flanks of seamounts and abyssal hills (Sarkar et al. 2008). A study of the relation between nucleus type and size suggests that the CIOB nodules are of first-generation while the Equatorial North Pacific (ENP) nodules are more matured and of second generation. Martin-Barajas and Lallier-Verges (1993) suggested that the CIOB nodules that formed from the diagenetic process are younger than those formed by the hydrogenous process.

The ratio of thickness of oxide:nucleus (O:N) is an important factor as this would indicate the viability of mining the nodules. It has been noted that the ratio is low for nodules on the upper slope of seamounts close to the summit in contrast to the nodules on the seabed. The former types of nodules have a smooth texture, a larger nucleus with rock fragments and indurated sediments as compared to the seabed nodules with rough texture. Also, the seamount nodules contain high concentrations of Co and Fe as compared to seafloor nodules that are rich in Mn, Ni, and Cu (Mukhopadhyay and Nath, 1988). The nodule size and nucleus affect the bulk composition in that the large nucleus composed of material different from the nodule composition, can dilute the bulk composition of the nodule and in turn lower the metal value of the nodules. Therefore, during mining, areas with nodules containing smaller nuclei need to be identified and these nodules have to be separately treated during the ore beneficiation processes.

7.5.3 Surface Texture

The mammillae on the surface (top and bottom) of the CIOB nodules have variable relief, shape, size, and abundance, and they form due to oxides precipitated from the seawater. Normally, the nodules are rough or smooth textured. The latter results from the fact that the upper parts of the nodules are exposed to the seawater while the undersides are resting on the sediment and are rough-textured. Some nodules with fine granulated smooth surfaces have cracks and pits on the upper. The hamburger-shaped nodules have a very rough bottom side due to the intense biological activities that are visible in the form of broken worm tubes. In such nodules, the bottom surface gets oxides from the underlying sediments' pore water through sub-oxic diagenetic remobilization (Glasby et al. 1983; Mukhopadhyay, 1988).

The overall morphology of a nodule is controlled by the topography of the seafloor. For example, nodules on the flank or near seamounts and hills receive a large amount of nucleus material in the form of rock fragments. Further, these nodules are relatively smooth compared to the seafloor nodules because of the flow of water along the seamounts washes off the sediments sitting over the nodules. In contrast, nodules at the base of topographic highs are smaller in size and have a rough surface texture (Iyer and Sharma, 1990).

7.5.4 Internal Structure and Growth

Sliced and polished sections of nodules depict well-preserved structures in the form of concentric rings that record the complex growth history from the nucleus to the periphery of the nodules. Considering the types of concentric rings and chemical compositions, Sorem and Fewkes (1977) identified five textural patterns in nodules from the world's oceans. The patterns are (1) mottled (discontinuous and chaotic layers), (2) columnar (radial pattern), (3) compact (dense layers with very low Mn/Fe ratio), (4) laminated (short, dense, and columnar) and (5) massive (dense layers, high Mn/Fe ratio).

The CIOB nodules mostly show colloformic (columnar/cuspate patterns) and parallel layers or dendritic features with the former made up of sub-concentric laminae (arcuate cusps). The light colour laminae indicate a high growth rate and are enriched in Mn, Ni and Cu because of diagenetic remobilization of metals. The dark layers with Fe or FeMn-rich oxides with some amounts of Co are precipitated from the water column. The parallel oxide layers indicate slow growths that were perhaps interrupted because of unfavorable conditions (Banerjee et al. 1991).

7.5.5 Age

Based on radiochemical measurements, radiometric dating and empirical equations the CIOB nodules grew at a rate of between 1.2 and 3.2 mm/10^6 years. The study of growth rate along with oxide thickness indicates the initiation of nodule growth. In the Indian Ocean, the growth of nodules commenced between 8 and 3 Ma (Late Miocene to Early Pliocene) (Mukhopadhyay et al. 2018). The CIOB nodules are older than those of the Southwest Pacific nodules that started forming at 3.5 Ma (during Pliocene) but younger than ENP nodules which are ~15 Ma (Lower Miocene) (Glasby et al. 1982; Martin-Barajas et al. 1991).

7.5.6 Composition

The CIOB nodules have quite similar mineralogical and chemical compositions but are different than those of the Pacific Ocean nodules.

a) Mineralogy: The nodules are mainly composed of Fe and Mn minerals with silicate phases as accessories. The nodules with rough texture predominantly have todorokite (10 Å manganite) and these are relatively rich in Mn, Ni, and Cu while the smooth surface nodules have δ-MnO$_2$ (vernadite) and are rich in Fe and Co. But irrespective of the nuceli type, todorokite is the main mineral in the CIOB nodules (Sarkar et al. 2008). Todorokite and birnessite are typical minerals in abyssal nodules on the seabed and are formed in a mildly oxidizing environment while δ-MnO$_2$ is abundant in an oxidizing environment at shallower depth, namely, on flanks and summits of seamounts. Todorokite is associated with high-to-moderate biological productivity in the Indian Ocean. Phase transformations or post-depositional changes within the nodule result in the conversion of δ-MnO$_2$/vernadite to todorokite (Mukhopadhyay et al. 2002).

The accessory components in the CIOB nodules include volcanic glass, rock fragments, diverse silicate minerals, aragonite, zeolitic grains, and fossil tests. The occurrence of clay minerals such as montmorillonite-chlorite and phillipsite crystals indicate alteration of volcanic material (Ghosh and Mukhopadhyay, 1995; Iyer and Sudhakar, 1993a). In contrast, detrital quartz in some of the nodules point to an influx of clastic terrigenous sediments from the Ganges-Brahmaputra rivers. There are also reports of the presence of merlinoite (Mohapatra and Sahoo, 1987) and detrital and authigenic baddeleyite (ZrO_2) in the form of single isolated, subrounded to elliptical grains in the CIOB nodules (Nayak et al. 2011) but these are very rare.

b) Chemical Composition: The CIOB nodules are mainly hosted by siliceous sediment and are rich in Mn, Co, Ni, and Cu, however, there are intrabasinal compositional variations as reflected by the Mn : Fe ratio and overall concentrations of Cu, Co, Ni, and Zn. On average, the nodules have Mn 24.40 wt.%, Fe 7.10 wt.%, Ni 1.10 wt.%, Cu 1.04 wt.%, and Co 0.11 wt.% (Jauhari and Pattan, 2000). The diagenetic nodules in the southern part of the basin are rich in Fe and Co while the hydrogenetic nodules in the central part have Mn, Cu, and Ni. Nodules from the seabed have a high Mn/Fe ratio and high Cu and Ni while those from topographic highs have high Fe and Co and low Mn/Fe ratio (Mukhopadhyay and Nath, 1988). In general, the CIOB nodules are of para-marginal grade with Ni + Cu + Co ≥ 2.0% but the smaller nodules (<4 cm) with smooth surface and present in terrigenous and red clay sediments have 1.21%. The larger nodules (>4 cm) with a rough surface and occurring in siliceous, siliceous-pelagic clay, and calcareous-pelagic clay transition-zone sediments have a grade of 1.8% (Banerjee and Miura, 2001).

In the CIOB nodules, the rare earth elements (REE) occur mainly in the Fe oxyhydroxide and titanium and phosphatic phases. The concentrations of La and Nd are >100 ppm while Ce is very high (1,000 ppm and above) while the rest of the REE have concentrations <100 ppm with Lu being the lowest (<3 ppm). The nodules are rich in Middle REE (MREE) (Sm) compared to Light and Heavy REE (LREE) (La-Lu). The probable intake of these elements could be mainly from the host sediments through diagenesis (Nath et al. 1993). The moderate positive Ce anomaly (~0.25) indicates that the nodules formed under an oxidizing bottom environment which is created by the nutrient-rich, cold Antarctica Bottom Watermass (AABW). In contrast, negative Ce anomaly in the sediments indicates movement of Ce into the nodule phase by Fe-hydroxide colloidal flocks (compare this with Glasby et al. 1987). Alabarède (1995) noted the isotopic composition of Nd in the nodules to be on par with those of the ambient seawater and similar to the average Nd in the deep-water masses over periods of 10^5 - 10^6 years. The study suggests the present-day patterns of deep-ocean circulation to have been predominant throughout the Pleistocene.

7.5.7 PROCESSES OF NODULE FORMATION

The formation, composition, and distribution of nodules show significant intra- and inter-oceanic variations. This is because of the controlling factors such as sedimentation rate, bottom sediment type, availability of nucleating material, the influence of bottom currents, source, and supply of elements, biological productivity, topography, and the physico-chemical environment amongst others, play important roles in the formation of nodules.

Bonatti et al. (1972) and Halbach et al. (1981) identified three types of growth mechanisms for the formation of nodules. These are:

(i)   Early diagenetic in which the elements are transferred from sediment pore water to nodules through remobilization of elements from the sediment column.

(ii)   Hydrogenetic process during which elements are precipitated at a very slow rate from the seawater column onto different nuceli and/or already formed nodules. The elements are mainly in colloid form and occur near the bottom seawater.

(iii) A mixed process consisting of processes (i) and (ii).

Glasby (2000) suggested that enrichment of elements in nodules is highly dependent on diagenesis and that the acoustically transparent sediment layer (ATSL), a thin outermost layer in nodules that forms an interface between the seafloor sediment-bottom water, is involved in elemental uptake. The degree of metal mobilization depends on the sedimentation rate and upward flux of the dissolved elements in the sediment column (Mukhopadhyay and Nath, 1988).

7.6 MICRONODULES

Micronodules of up to 2 mm in diameter, but mostly of <1 mm in size, commonly occur in surface, sub-surface, and tens of centimetres below the seafloor sediments. Micronodules have similar surface texture, mineralogy, chemistry, and internal features like the macro-nodules. In the CIOB the abundance of micronodules decreases with depth in the siliceous but increases in red clay sediments (Mukhopadhyay et al. 1988). Micronodules are associated with biota (radiolarians, diatoms, ichthyoliths, phytoliths, bacterial cells), volcanic glass shards, phillipsite crystals, and palagonite grains (Banerjee and Iyer, 1991; Iyer et al. 2012). Bulk analysis of micronodules, on an average have Mn 35% and Ni+Cu 3% and high Mn/Fe (5-101) and very low Fe 1.8% and total REE (522 ppm) in contrast to the surface nodules. The average REE concentration is around 650 ppm in micronodules, 1,164 ppm in surface nodules, and 210 ppm in the sediments (Mukhopadhyay et al. 2018 and references therein).

7.7 BURIED NODULES

Although most of the nodules (>1 cm) are present at the sediment-water boundary, but there are examples where nodules of different sizes are seen buried at different depths within the sediment column. Buried nodules can be recovered by coring the sediments by using spade, box and gravity cores. Hence, it is difficult to estimate the total nodule reserve in the mining area, determining the actual bulk composition of buried nodules, and to understand as to whether the nodules formed on the seafloor sediments and were subsequently buried or grew within the sediment column.

Nearly 60 spade and gravity cores from water depths of 4,700-5,800 m were studied from different geographical sites in the CIOB. Of the 60 cores, 50 buried nodules were found in 12 cores at different depths within the sediment and mostly in the cores below 8°S in the basin (Banerjee et al. 1991, Pattan and Parthiban, 2007). Most of the buried nodules are ~2 cm in diameter and are irregular, discoidal, polynucleated, and largely elliptical in shape. Their smooth and rough surface textures are quite similar to those of the surface nodules. The buried nodules show alternate recrystallized thick δ-MnO_2-rich layers and todorokite with dendritic texture. Since these nodules are surrounded by sediments, intercalations of clay-rich zones within the nodules are also seen (Banerjee et al. 1991). Pattan and Parthiban (2007) reported the composition of five buried nodules that were found at different depths (166-168, 172-174, 228-230, 328-330, and 418-420 cm) from the siliceous sediment. Nodules in the first three levels with a high Mn/Fe ratio (9.3-15.1) were formed by early diagenesis. The fourth level (328-330 cm) nodule with a very low Mn/Fe ratio (1.6) indicated a hydrogenetic process. The nodule in the fifth level (418-420 cm) with a moderate Mn/Fe ratio (3) again points to diagenesis. The buried nodules have REE between 164 and 497 ppm with an enrichment in HREE and MREE contents.

7.8   THE 5ES, INDIA'S EFFORTS IN THE CIOB

Four economically potential nodule-rich areas have been identified in the world ocean and of these three are in the Pacific Ocean (CCZ, Peru Basin, and Penrhyn Basin) and one in the CIOB (Figure 7.1).

The International Seabed Authority (ISA, previously the United Nations Convention on the Law of the Sea, UNCLOS) has allotted 75,000 km² each to 19 Contractors for exploitation of nodules in the CCZ and CIOB. Of the 18 Contractors 17 are in the CCZ and one in the Western Pacific (China Minmetals Corporation). The 17 Contractors are Interocean Metal Joint Organization (Bulgaria, Czech Republic, Cuba, Poland, Slovakia, and Russia), JSC Yuzmorgeologiya (Russian Federation), Government of the Republic of Korea, China Ocean Mineral Resources Research and Development Association (China), Deep Ocean Resources Development Co. Ltd. (Japan), Institut français de recherche pour l'exploitation de la mer (France), Government of India, Federal Institute for Geosciences and Natural Resources of Germany (Germany), Nauru Ocean Resources Inc. (Nauru), Tonga Offshore Mining Limited (Tonga), Global Sea Mineral Resources NV (Belgium), UK Seabed Resources Ltd (UK of Great Britain and Northern Ireland, 2 sites), Marawa Research and Exploration Ltd. (Kiribati), Ocean Mineral Singapore Pte Ltd. (Singapore), Cook Islands Investment Corporation (Cook Islands), China Minmetals Corporation (China), Beijing Pioneer Hi-Tech Development Corporation (in Western Pacific) (China) and Blue Minerals Jamaica Ltd (Jamaica). India is the sole Contractor for the CIOB nodules.

In the above background we now provide a gist of the information concerning the CIOB nodules in terms of 5Es: Exploration, Environmental studies, Exploitation, Enrichment, and Economics. Later we discuss the applicability of the BE to the nodule deposits.

7.8.1   EXPLORATION

On 26th January 1981 India achieved success by recovering nodules and this led the government to launch a major exploratory programme, "Surveys for Polymetallic Nodules," in the CIOB. This multi-dimensional, multi-disciplinary and multi-institutional programme was funded by the Department of Ocean Development (later re-christened as the Ministry of Earth Sciences, MoES New Delhi) and the nodal laboratory is CSIR-National Institute of Oceanography, Goa. The programme has four components: Survey and Exploration, Environmental Impact Assessment, Technology Development for Mining, and Technology Development for Metallurgy.

Tens of expeditions were made to obtain thousands of kilometers of bathymetric, magnetic and gravity data and to collect nodules and associated crusts, sediments and rocks. A reconnaissance survey of about 4 million sq km area in the CIOB was conducted and this led to identifying 300,000 km² as the Application Area. On 18th December 1987, the erstwhile UNCLOS III recognized India as the first country in the world to be the Pioneer Investor (PI) and allocated a Pioneer Area (PA) of 1,50,000 km² area in the CIOB. Sampling was initially carried out in a grid pattern with a distance of 111 km between the sampling stations. For an accurate assessment of the nodule resources in the PA the sampling distance was halved from 50 km to 25 km to 12.5 km to 6.25 km. By 2002, 50% of the PA was relinquished to ISA in phases and India retained 75,000 km². About 10% of this Retained/Exploration Area has the best nodule resources and forms the First Generation Mine site (FGMS). The nodule resources are estimated to be about 670 million tonnes and with an average abundance of 5 kg/m² on the seafloor. These para-marginal grade nodules have 2-2.4 wt.% nickel + copper + cobalt that constitutes 11 mt of metals (Mukhopadhyay et al. 2018).

Several thousands of samples of nodules, crusts, and rocks were collected through grabs (Pettersson, Okean), dredges (net, box, chain-bag), and also tens of sediment samples by using corers (spade, box, gravity). In addition, geophysical surveys (single and multi-beam bathymetry, gravity, and magnetic) and thousands of underwater photographs were obtained. Various reports have highlighted the morpho-structural, geological, sedimentological, volcanic, and evolutionary history of and hydrothermal episodes in the CIOB (Iyer et al. 2018, Amonkar and Iyer, 2021). Considering these and other parameters, the first geological model to explain the formation and genesis of the nodules in the basin was proposed by Mukhopadhyay et al. (2002). Based on grade and abundance of nodules, seafloor geomorphology and other factors, the geological model helped to identify fours sectors within the basin that had mineable nodules. Another model for the CCZ was reported which, besides grade, abundance and geomorphology, considered a different set of factors (such as chlorophyll content and biota) (ISA, 2010).

7.8.2 ENVIRONMENTAL STUDIES

The Mining Code of the ISA mandates that prior to mining of nodules, it is imperative to carry out an environmental impact assessment (EIA) of the mineable and nearby areas. This would help to chalk out the environmental management plan (EMP) and remedial measures to protect and sustainably exploit the nodules. During an EIA study of the nodule-bearing areas, several broad scientific topics and associated parameters need attention. These are geology (seafloor features, rocks, sediments), biology (flora and fauna that thrive in the seawater and also on the seabed), microbiology (bacteria, fungi in the seafloor sediments), chemistry (physical and chemical properties of seawater column and bottom water), physics of the water (optical, conductivity, temperature), and meteorology (weather and climate conditions in the potential mining area), among others. Once the protocols of the ISA are followed and satisfied then the Contractor is one step closer to be granted permission to mine the nodules.

In 1995, EIA studies were commenced in the CIOB by artificially disturbing the seabed using a hydraulic disturber 'Deep-Sea Sediment Resuspension System.' This multidisciplinary 3-phase study called INDEX (Indian Deep-Sea Environment Experiment) was carried out with technical cooperation from the Central Marine Geological and Geophysical Expedition, Gelendzhik Russia. The INDEX collected baseline environment data from Reference and Test sites in the CIOB. These sites with similar characteristics were selected with an optimum distance from one another to ensure that disturbance at the test sites does not affect the reference site. The data collected were of pore water, sediment geochemistry, sediment size, clay mineralogy, biostratigraphy, geotechnical, macro- and meiobenthos, and microbiology. In addition, moored buoys and sediment traps were also deployed.

The impact of disturbance on the distribution of radiolarians, macrofauna, meiofauna and on microbial and biochemical parameters, was documented. Similarly, the changes in the composition of surface sediments and particles in sediment traps and rosette water sampler were examined. The results indicated the temperature, salinity, potential density, and geostrophic circulation to decrease below 3,500 m water depth. This was ascribed to a southwestward weak flow of abyssal water around 10°S (Ramesh Babu et al. 2001).

Due to the disturbance, the daily flux rate of sediment particles (avg. 50 mg/m²/day) increased by 300% and later fell drastically to 33% within 5 days, because of a rapid settlement of particles. Modeling and kriging estimation indicated that most of the 3,600 tonnes of disturbed sediment discharged from 5 m above the seafloor in the water column did not either spread, laterally nor vertically from the Test sites (Parthiban, 2000). The disturbed sediment samples showed an increase in their water content capacity and a decrease in undrained shear strength in the top 10–15 cm sediment layer (Khadge, 2000). These findings are important to develop an appropriate mining technology to avoid sinking of the collector during nodule recovery. It was noticed that after the disturbance at the Test site, there was decrease in the population of megafauna and benthic biomass but over a period of days there was a prolific growth due to rehabilitation and occurrence of new species of organisms (Rodrigues et al. 2001). The results from the INDEX helped to understand the restoration and recolonization processes of benthic biota in the mineable areas (Sharma, 2015 and references therein).

7.8.3 Exploitation

Prior to mining, the geotechnical properties of the seabed sediments (besides their type and characteristics) need to be investigated. Some of these properties that need to be determined (either in situ or in the laboratory) are water content, porosity, permeability, wet bulk density, void ratio, shear strength, and liquid and plastic limits. These measurements are necessary to understand the bearing capacity of the seabed sediments that would help during the development and deployment of the mining equipment and nodule recovery systems. In addition, the height, aerial extent and slope of topographic features and geology of the mining area have to be mapped. These studies would help to avoid steeply sloping and also rocky areas during mining operations and save time and equipment (Sharma et al. 1994).

Mining of nodules comprises of recovery of nodules from the seabed, lift these to the mother ship, dry and transport the nodules to onshore processing laboratories. In 1970, the first company to test a nodule mining technology was Deep-Sea Ventures Inc. (USA) in the Blake Plateau at a water depth of 750 m and in 1978 nodules were collected from >5,000 m water depth in the North Pacific Ocean (Cronan, 1980). Four mining systems have been tested globally to collect nodules and these are hydraulic lift, air lift, shuttle or modular, and continuous line bucket. The Integrated Mining System (IMS) which many countries prefer, is a remote-controlled ocean-floor miner with a self-propelled collector connected to a free bottom end of the pipe. The miner-to-buffer link, and a buffer at the pipe's other end, completes the system.

Deep-sea mining activities would involve three main phases (Hajkowicz et al. 2011):

(i) Research, exploration, feasibility, and funds
(ii) Development and manufacturing of equipment, operation, and rehabilitation, and
(iii) Long-term monitoring and recycling of new products

The impact of mining on the oceanic environment would be in 3-ways: operational, spatial, and temporal and there are several ways to mitigate these concerns (SPC, 2013).

FIGURE 7.4   Remotely Operable In-situ Soil Tester (ROSIS) (https://niot.res.in).

Presently some mining systems are available in the international market that are used to exploit the beach and nearshore placer minerals. Various equipment has been used to demonstrate recovery of the nodules but since commercial operations are yet to take off there is no large-scale production. The National Institute of Ocean Technology (NIOT, Chennai India), in collaboration with the University of Siegen (Germany), and also the CSIR-Central Mechanical Engineering Research Institute (Durgapur, India), have been working on a mining concept where a crawler-based mining machine collects, crushes, and pumps crushed nodules from 6,000 m water depth to the mother ship. The NIOT has also fabricated systems such as a subsea solid pump system that works at a depth of little more than 1,000 m, a fully electrical remotely operable subsea in-situ soil tester (Figure 7.4) that was operated at 5,462 m water depth, and an acoustic underwater positioning system for 5,400 m water depth.

The NIOT has built a Remotely Operated Submersible (ROSUB 6000) (Figure 7.5) which was deployed at a water depth of 6,000 m in the CIOB to exhibit its capability to pick up the nodules from the seabed. Work is underway to develop a battery-operated submersible that would have an endurance of 12 hours, dive up to 6,000 m, and can be manually operated (Atmanand et al. 2019). In June 2021 the Government of India approved about US$ 550 million to implement the Deep Ocean Mission over a period of 5 years. This multi-institutional mission mode project would support the BE and the Ministry of Earth Sciences would be the nodal agency. As a part of this initiative development for deep-sea and manned submersible (Figure 7.6) would be indigenously developed and deployed in the CIOB and an IMS would be fabricated to mine the nodules. This equipment as components of the BE could significantly help to explore and exploit the CIOB nodules in a sustainable and environmental-friendly manner.

7.8.4   ENRICHMENT

The recovered nodules need to be transported to onshore laboratories or beneficiation plants where the metals from the nodules would be extracted. It was envisaged that extraction of the metals could account for 60-70% of the total cost of the nodule mining project (Kunzendorf, 1986; Padan, 1990).

Since late 1980, the various methods to extract the metals from the CIOB nodules have been undertaken in three organizations: the Institute of Minerals and Materials Technology (IMMT, erstwhile Regional Research Laboratory, Bhubaneswar, using the ammoniacal sulfur dioxide leach route), National Metallurgical Laboratory (Jamshedpur, following the roast reduction ammoniacal leach route), and Hindustan Zinc Limited (Udaipur, using the acid leach-pressure leach route). Over

FIGURE 7.5   ROSUB-6000 (https://niot.res.in).

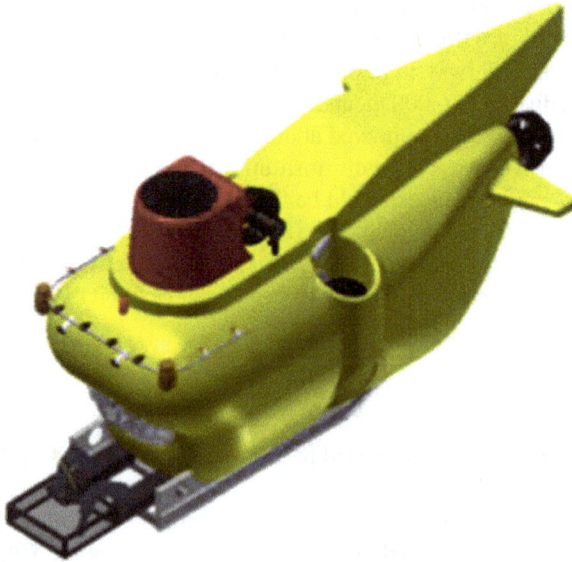

FIGURE 7.6   Design and development of a manned submersible for 6,000 m water depth (https://niot.res.in).

the last four decades more than 50 tonnes of nodules have been processed for three metal (Cu, Ni, Co) extractions in these laboratories and in recent years the process has been refined to obtain Mn (which forms the bulk of the nodules).

The IMMT identified two main extractive methods: pyrometallurgy and hydrometallurgy. In pyrometallurgy, nodules are smelted at high temperature under reducing conditions to produce a crude alloy of Ni, Cu, Co, and Fe. In contrast, hydrometallurgy is a low-temperature aqueous process in which a copious amount of sulfuric acid (H_2SO_4) is used. Both the processes efficiently extract Ni (~90%) and Cu (80%–90%), but Co and Mn are better recovered through pyrometallurgy. Both the processes result in an enriched Mn-bearing slag and this value-added by-product could be mixed with low-grade terrestrial manganese ores for use in industries.

Because of the high-energy cost of pyrometallurgical processing, it was decided to combine hydro- and pyro- metallurgical processes in a cost-effective manner. This helped to enhance pure metal extraction to about 99.8%, comprising of Ni (94.2%), Cu (93.9%), and Co (62.5%; Sen, 1999). Cobalt was recovered up to 79% by using liquid and gaseous reductants (Srikanth et al. 1997). Bioelectro-chemical in the presence of *Thiobacillus ferrooxidans* and *T. thiooxidans* results in recovery of Ni (0.735 gm), Cu (0.709 gm), and Co (0.308 gm) from every 100 gm of nodules (Kumari and Natarajan, 2002). By recycling and use of bulk chemicals, bioelectro-chemical could be congenial for commercial use (Agarwal and Goodrich, 2003). In some processes H_2SO_4 acid can be avoided by using a stoichiometric amount of $FeSO_4$ (Vu et al. 2005), oxidative precipitation (Zhang and Cheng, 2007), and use of ammonia (Sen, 2010). Further research could result in better ways to economically extract the metals from the nodules and also address the environmental concerns.

7.8.5 Economics

According to the ISA (ISA, 2010), a nodule-rich area may be categorized as an individual commercially feasible mine site if 3 million metric tonnes of nodules per year are possible to be recovered continuously for 20 years. The economic interest in nodules has received attention from several mining companies because the metals within the nodules have applications in many hi-tech and green-tech industries such as ferroalloys, dry cell and rechargeable batteries, steel manufacturing, and in critical sectors like energy, military and IT.

It is predicted that by 2030 about 10% of the global minerals could be recovered from the ocean floor (European Commission, 2012). But this is easily said than done since the Contractor has to address the above 4Es, following the stringent guidelines of ISA. An in-depth analysis of the cost-benefit factors could help to realize the feasibility of profits through mining the nodules. Towards this, the factors to be considered are capital costs (mother ship, support vessels), manpower, salaries, daily expenses, hire/ purchase of mining equipment, conveyance of personnel, transport of nodules to land, setting-up an onshore metal beneficiation plant, price variations of the metals in the market, vagaries of nature that could restrict mining operations and other known/ unknown aspects.

The total resource in the 75,000 km^2 PA of India is 365 mt, with Mn of 95.17 mt, Ni 4.508 mt, Cu 4.455 mt, and Co 0.418 mt (all wet nodules), and the approximate value of the four metals were estimated at US$331,840 million (Sharma, 2015). A comprehensive financial feasibility study of a possible nodule mining in the CIOB was made by Mukhopadhyay et al. (2019). Those authors considered the above parameters together with nodule reserve (200 mt), metal grade (Mn + Ni + Cu + Co = 54 mt), and an operating cycle of the mining system for 6 months in a year over a period of 25 years. It was calculated that with an initial start-up cost of US$ 5,000 million, and based on the EBIDT (i.e. earnings before interest, depreciation, and tax), a negative net present value (NPV) and with a profitability index (PI) less than 1, in the first 5 years of the mining operation in the CIOB may not be highly profitable. But by the eighth year, due to a notable increase in NPV and PI, the Contractor could begin to book profits for at least a decade. Though presently far-fetched but if new ore deposits are discovered on land, then any Contractor may have to forego substantial profits by mining the nodules.

Nodule mining is viable provided we have an appropriate ecofriendly mining technology, the demand for the metals in the global market shows an upward trend, availability of an efficient metallurgical process, and that future tax and royalty fee structure remain investor-friendly. Satisfactory completion of the 5Es could help to appreciate the concept of "Minerals to Market" and justification to mine the nodules.

7.9   BLUE ECONOMY OF POLYMETALLIC NODULES

In the background of the above information and considering the need for marine minerals, we now discuss the 5Es of the CIOB nodule deposits with respect to the BE paradigm. To sustainably mine the nodules the various facets of the BE need to be covered. These are optimum use of finance for best results, transport of personnel and nodules, human resources and training, employment and business opportunities, skill development, minimum damage to the environment, collaborations among others.

7.9.1   Exploration

Exploratory activities for the nodules need to be well-planned since huge costs are involved and frequent use of research vessels to sample the nodules could leave behind large footprints of carbon. Further, regular cruises to the exploration area could also severely affect the biota of the ocean. Hence, it could be prudent if more than one Contractor in the Pacific Ocean join hands during the exploratory work so that cost overruns and environmental issues could be minimized. India, the sole Contractor for the CIOB, could conduct her cruises from the nearest country such as Seychelles or Mauritius. This would help to share finance, ship-time, result in scientific collaborations, create jobs in the host countries, and help improve the economy of these countries. Interestingly, both these countries have made significant progress in implementing the BE, and furthermore, Seychelles has a separate Ministry of Finance, Trade, and Blue Economy.

7.9.2   EIA

In 1970, the Deep-Sea Ventures Inc. (USA) carried out the first EIA studies to decipher the impact of mining of deep-sea minerals. It was opined that it is important to understand the nature of bottom sediment, sub-bottom water, and overlying water column during mining. To address these concerns several environmental parameters are required to be measured and monitored before, during, and after mining.

Water samples need to be collected from several desired depths from different locations since the characteristics of flux, colloidal material, and particulate matter, in the water column, influences metal adsorption in nodules. Several sensors are to be lowered along with rosette samplers to measure temperature gradient, conductivity, pH, dissolved oxygen, light transmissivity, and other parameters in the water column.

Based on their body size, deep-sea biota are classified into four classes (Raghukumar et al. 2001; Smith et al. 2008):

   (i)   Megafauna of >3 cm (fish, octopus, squids, starfish, and the like) thriving in the water column and sediment surface,
   (ii)   Macrofauna of 3 cm to 62 μm (polychaete) living within the sediment,
   (iii)   Meiofauna 62 to <500 μm (foraminifera, shrimps, nematodes) within the sediment, and
   (iv)   Microfauna (bacteria, fungi) within sediments and on nodules.

Biological sampling is essential to recover biota of various sizes (mega-, macro-, micro-, and meiofauna). The seabed sediments that would be disturbed by the collector device and suspended in the water would disrupt, displace and bury the biota. During cleaning and crushing of nodules on the seabed prior to their recovery, the resultant slurry could affect the ecology.

A discharge depth of 1,000 m for the slurry was suggested which would reduce the sediment plume to depths greater than the biologically active zone defined by more intensive vertical migration (Thiel, 2001). But we also need to account for the bottom current regime before fixing the final discharge depth of the tailings (Young and Richardson, 1998; Murty et al. 2001; Raghukumar et al. 2006).

Deep-sea mining results in the oceanic environment being subjected to short- and long- time disturbances and the recovery could be slow or fast in the mining area. The seabed would undergo long-term disturbance and a slow rate of recovery due to physical impact and sediment suspension. The water column would show both the effects: reduction in nutrients because of the sediment plume (long term) and on marine life (short term) (Berge et al. 1991; Sharma et al. 2015).

Several environmental experimental programs namely, Metalliferous Sediment in Atlantis II Deep (MESADA, 1977-81), Red Sea; Deep Ocean Mining Environmental Study (DOMES), Disturbance and Recolonization (DISCOL), Benthic Impact Experiment (BIE), and Japan Deep Sea Impact Experiment (JET) were undertaken in the world's oceans to study the impact of mining on the environment. The objective of MESADA (Germany) was to exploit metalliferous mud and sediment from a depth of 10 m below the seabed from an area of 60 km^2. A total of 12,000 m^3 of tailings containing 225 tonnes of particulate matter was discharged at 400 m water depth during this test. The resultant plume was traced to a depth of 1,100 m with lateral extension up to 900 m around the discharge point, which later extended to 5,000 m within 10 days. MESDA's work showed that a mining unit with a capacity of 100,000 tonnes per day would discharge 400,000 m^3 of toxic tailings every day and this could severely damage the marine food chain.

The DOMES, DISCOL, BIE, and JET were undertaken in the Pacific Ocean. The DOMES (NOAA, USA) studied the effect of mining on surface discharge plume, the particulate and dissolved phases, trace metals, particle accumulation at the pycnocline, and light attenuation. The bacterial growth and oxygen demand, the standing stock of phytoplankton, nutrients, trace metal uptake, fish and benthos were also examined (Thiel, 2001). The study suggested regular monitoring in the mining area to determine in-situ settling velocities of particulate phases and their role on the biota.

DISCOL, a long-term large-scale experiment was conducted in the tropical South Pacific by the then West German government to understand the recolonization of the fauna. The study was carried out in an area of 10 km^2 at a water depth of 4,150 m about 600 km south of the Galapagos Islands. During the experiment, about 20% bottom sediment was turned and still and video photographs of the seabed were obtained before, during, and after the disturbance. It was inferred that (1) bottom-dwelling and nodule-crevice fauna have little chance to recolonize, (2) stalked species, penetrating through the sediment blanket, can recover after initial shocks of the sediment plume, (3) tailings should preferably be discharged below the euphotic zone to ensure productivity, and (4) scattering of tailings discharge depends on the nature of the water column and the current regime, which differs from place to place (Thiel, 2001).

The BIE, a joint study by the US and erstwhile Commonwealth of the Independent States (CIS), was made in the North Pacific. The test area was experimentally disturbed in May 1992 and sampled after 4 months. Both continuous line bucket and hydraulic lift systems were used and the impact on the near-surface biological productivity and bottom-dwelling fauna was examined. The Metal Mining Agency of Japan and the NOAA investigated the impact studies of artificial disturbance under the JET, in the North Pacific (off Mexico). The area was monitored to understand pre- and post-disturbance conditions.

Deep scattering layer depths and vertical migration behaviors are proxies for mesopelagic micronekton and zooplankton communities by using acoustic Doppler current profilers. Environmental data are acquired of mean midwater oxygen partial pressure, surface chlorophyll-a, and sea surface height anomaly. Such studies were carried out in the CCZ by Perelman et al. (2021). In the midwater (i.e. the epipelagic, abyssopelagic, meso- and bathypelagic zones) the environmental impacts are expected to be less because no equipment is deployed (Leal Filho et al. 2021). The surface water could be affected by noise and light from ships, hydraulic oil, and waste discharges (Washburn et al. 2019).

Due to the creation of the plumes caused as a result of removal, discharge, and re-deposition of sediments; benthic organisms may get buried, the respiratory surfaces of filter feeders could get clogged reduce the metal content, and also cause depletion in oxygen (Leal Filho et al. 2021). Light and noise pollutions from ships, cameras, machineries, and other instruments could disturb the marine life which would avoid habitats, be temporarily blinded among others (Stanley and Jeffs, 2016; Deep Sea Mining Campaign, 2019).

During sediment, plumes would be generated, and these may have severe ecological effects in deep midwaters that extend from a water depth of 200 m to 5,000 m. The deep midwater hosts more than 90% of the biosphere and has fish biomass 100 times greater than the global annual fish catch. The midwater fauna would suffer (Gillard et al. 2019) and since the midwater is connected with the deep-sea, therefore ecosystems that play key roles in carbon export and regeneration of nutrients could be affected (Drazen et al. 2020). The biodiversity and dynamics of midwater ecosystems are important factors that need to be considered and necessary steps need to be taken to lessen the environmental impact on the midwater regime by the Contractors.

Mining of nodules, as and when it commences, needs to be sustainable and have mitigated measures in place. Any deep-sea mining venture would have three main phases.

Phase I: research, exploration, feasibility, and funding
Phase II: system fabrication, operation, and rehabilitation
Phase III: long-term monitoring and recycling of new products (Hajkowicz et al. 2011).

Mining would affect the environment in three ways: during operation, and on spatial and temporal scales. Some of the operational mitigating measures are: (1) to use a closed lifting mechanism so that the nodules do not fall onto the seabed and result in additional operations, (2) spread the discharge fluid close to the seafloor rather than releasing it in the water column, (3) minimize waste production, (4) have effective sewage treatment plants. The spatial mitigating management plans need to include: (1) identify nearby areas for conservation, (2) delineate areas for total mining, partial mining, fisheries, and tourism, and (3) evaluate the location of waste discharge to ensure minimal impact on ecosystems. The temporal measures suggest that the million-year scale of nodule growth and long time that is required for faunal recovery in the abyssal would make short timescale measures impractical. Yet, as best mining practices environmental it is essential to have in place management plans.

Time-series (few months, a year, few years) information need to be undertaken. The collation and analysis would significantly help to identify the possible threats to the organisms and the vertical and lateral dispersal of the sediment plume. Although several investigations have/are been conducted in the Pacific and Indian oceans yet, these are insufficient to predict the true extent of impact when full-scale deep-sea mining commences.

7.9.3 Exploitation

Mining systems that are under development have to consider the following. (a) The mining equipment should be made of materials that are non-toxic to the seawater environment, be of low weight, able to bear the huge hydrostatic pressure, and be minimally non-corrosive. (b) The system must be able to move freely over the seafloor without either toppling or getting stuck in the sediments. Hence, there is a need to study the microtopography of the seabed and sediment properties. (c) The mining system needs to be easily steered around seafloor mounds and rocks and to be moved from one site to another. (d) Transport the recovered nodules to beneficiation plants on land.

The Blue Nodule project involves 14 industry and research partners from nine European countries. The project aims to: manufacture a state-of-art and industrially viable mining system that can be used between 3,000 and 6,000 m water depths and to develop an in-situ sediment separation and sizing equipment. The nodules would be processed (dewatering and conditioning) onboard and it would be ensured that there is minimum impact on the environment (www.blue-nodules.eu).

Besides the scarce availability of technology, certain related factors limit deep-sea mining viz. availability of land ore and mineral deposits, payment of millions of dollars by Contractors to the ISA to obtain a license and for its annual renewal, huge capital investments either by consortia or governments or through a partnership among them. Once technologies are developed, environmental concerns are accounted for and mining mechanisms are established, then the day is not far off when countries would be retrieving metals from the deep sea.

Though there are proponents for deep-sea mining but Miller et al. (2021) opine that there are ample risks to biodiversity, ecosystem function, and related ecosystem services, and a lack of equitable benefit sharing amongst the global community now and for future generations. They justify a moratorium on deep-sea mining so as to protect the marine ecosystems and to focus on baseline research. These steps could improve governance, preservation, and conservation of ocean biome in the nodule mining areas.

7.9.4 ENRICHMENT

Simultaneous processing of SMS and nodules was reported by Kowalczuk et al. (2019). Those authors leached different ratios of nodules and SMS by using H_2SO_4 and sodium chloride (NaCl). The experiment resulted in high extraction of metals while the residues had silicates (quartz, muscovite, microcline, albite), elemental sulfur, barite, and traces of sulfides (<1wt.%). These products could be purified and used in various applications e.g. electronics, glass, ceramics, oil and gas drilling (baryte), etc. Simultaneous leaching of marine minerals lessens the need for additional oxidizing and reducing agents and is also a less costly, time-saving, and more environmentally friendly process. This method helps to avoid pyrometallurgical pretreatment of nodules and the use of expensive hydrochloric acid.

A suitable ore beneficiation method for nodules must study if the selected process is techno-economically feasible, market demand for the metals, infrastructure for the plant and its location, and significant, at least 30%, return on investments. In addition, environmental consequences of the processing methods on the groundwater and atmosphere and noise pollution need to be considered.

7.9.5 ECONOMICS

There are several sectors of the BE and these need to be made inclusive in the economics of nodule mining. Mining being a capital-intensive venture requires massive funds, machineries, and manpower. These could be undertaken by a Contractor through collaborations within the country or with other Contractors. Investments could be sought from government and corporates and by raising the capital amount through public offerings. The cost and time involved in developing technology and systems for mining could be reduced by outsourcing these to private firms instead of being carried out by one or two entities. The technology developed by a Contractor could be sold or hired to others at a reasonable price. Since different areas have been allotted and not all Contractors would be simultaneously carrying out mining in their respective areas, hence exchange of information and sharing of mining systems could be viable options. And these recommendations are more apt for the several Contractors in the Pacific Ocean.

Mining and shipping of the nodules, and ore beneficiation activities could provide ample jobs both skilled and unskilled, and also help to train hundreds of people in various ways. Small Island Developing States (SIDS) could be used by the Contractors for their ports, maintenance of the mining systems, and if feasible create facilities for metal beneficiation plants. Obviously, during such activities we need to consider the environmental impacts on the SIDS and have precautions in place.

In summary, India's efforts to sustainably mine the nodules could bear fruit for the reasons that since more than four decades copious amount of a spectrum of data has been gathered pertaining to the various aspects of the CIOB, including EIA studies. A few hundred promising mineable blocks have been demarcated in which close grid sampling (nodules and sediments) would be undertaken and microtopography would be examined. These inputs are needed prior to deploying an ROV and the mining system which are being developed. Further, beneficiation routes have been defined which would be able to extract four metals from the nodules with less wastage of chemicals and low impact on the environment. And finally, although the expiration of the Contract of India was on 24th March 2022 and is likely to be renewed; the Indian government has invested large capitals to conduct the above activities in different laboratories.

It is envisioned that the allied aspects of the BE paradigm could get an impetus once deep-sea mining starts. Some of the sectors that could benefit are employment opportunities, maritime activities (ship building and maintenance, upgradation of ports to handle and transport the nodules, etc.), scientific collaborations (within and outside India), indigenous technology, skill development, increase in production of raw materials and chemicals that would be required in beneficiation plants, better ways to dispose of the treated tailings, improvements in infrastructure, IT, among others.

7.10   EPILOGUE

Deep-sea mining involves key challenges but in the near future, we may have to turn our attention to the ocean for mineral deposits due reasons such as depleting terrestrial resources, non-availability or scarcity of some important minerals/metals, or no discoveries of land deposits. Hence, it is prudent that we develop and refine deep-sea mining technologies, create a detailed EIA database, and have in place robust EMP and remedial measures. The work towards these could be hastened and the enormous cost and time overruns could be reduced if Contractors and corporates work in unison instead of in isolation.

After the Goa Declaration in 2015 (Mohanty et al. 2015), the Indian government has taken several steps to recognize and enhance the BE sectors that pertain to marine minerals. The National Institute of Transforming India (NITI) is collaborating with several stakeholders and the Ministry of Earth Sciences to successfully implement a sustainable use of the several blue minerals (placers, SMS, polymetallic nodules). This is in tune with the policy statement which states that, "The blue economy refers to the exploring and optimizing the potential of the oceans and seas which are under India's legal jurisdiction for socio-economic development while preserving the health of the oceans." It is envisaged that the blue minerals could significantly contribute to India's projected economy of US$10 trillion by 2030, despite the recent set-backs due to the ongoing COVID-19 pandemic.

Deep-sea mining aims for a "Green Economy in a Blue World," and to maintain a fine balance between economic, social, and natural gains. The green economy is a low-carbon, resource-efficient, and socially inclusive initiative to help improve human well-being, enhance social security, reduce ecological risk, and limit environmental sacrifices (UNEP et al. 2012). This would result in a greener and more resilient economy, with low-carbon footprints, social inclusiveness, increased social security, decrease in ecological damages and risks, among others. We need to learn from the mistakes made during terrestrial mining that have resulted in environmental damage, demographic change, and low economic returns. The notion of green economy is that no single form of capital grows more than the others (SPC, 2013). National politics and international geopolitics may be hurdles to deep-sea mining activities. Several important trials and tribulations are needed to be overcome in a phased manner to make deep-sea mining an economically viable and environmentally sustainable venture.

ACKNOWLEDGMENTS

We acknowledge Prof. Dr. Md. Nazrul Islam for the invitation to contribute this chapter. We thank the support extended by DMC College and Goa University during the preparation of the manuscript.

REFERENCES

Agarwal, H. P. and Goodrich, J. D. 2003. Extraction of copper, nickel and cobalt from Indian Ocean polymetallic nodules. *The Canadian Journal of Chemical Engineering*, 81, 303–306. https://doi.org/10.1002/cjce.5450810218

Albarède, F. 1995. *Introduction to Geochemical Modeling*. Cambridge University Press, New York, 1–50.

Amonkar, A. and Iyer, S. D. 2021. Influence of low-temperature fluids and post-depositional changes in the siliceous–pelagic sediments of the Central Indian Ocean Basin. *Journal of Sedimentary Environments*, 6, 603-620. https://doi.org/10.1007/s43217-021-00073-4

Amonkar, A. Iyer, S. D. EVSSK, B. and Manju, S. 2020a. Extending the limit of widespread dispersed Toba volcanic glass shards and identification of new in-situ volcanic events in the Central Indian Ocean Basin. *Journal of Earth System Science*, 129(175), 1–24.

Amonkar, A. Iyer, S. D. EVSSK, B. Sardar, A. Shailajha, N. and Manju. S. 2020b. Fluid-driven hydrovolcanic activity along fracture zones and near seamounts: evidence from deep-sea Fe-rich spherules. *Acta Geologica Sinica*, 95 (5), 1591-1603.

Atmanand, M. A. Jalihal, P. Ramanamurthy, M. V. Ramadass, G. A. Ramesh, S. Gopakumar, K. A. Vedachalam, N. and Dharani, G. 2019. Blue economy—opportunities for India. *IEEE India Info*, 14, 106–115.

Banerjee, R. and Iyer, S. D. 1991. Biogenic influence on the growth of ferromanganese micronodules in the Central Indian Basin. *Marine Geology*, 97, 413–421.

Banerjee, R. and Miura, H. 2001. Distribution pattern and morphochemical relationships of manganese nodules from the Central Indian Basin. *Geo-Marine Letters*, 21, 34–41.

Banerjee, R. Iyer, S. D. and Dutta, P. 1991. Buried nodules and associated nodules from the Central Indian Basin. *Geo-Marine Letters*, 11, 103–107.

Berge, S. Markussen, J. M. and Vigerust, G. 1991. *Environmental Consequences of Deep-Seabed Mining. Problem Areas and Regulations.* In: Ocean Mining Series. The Fridtj of Nansen Institute, Oslo, Norway, 135.

Bonatti, E. Kraemer, T. and Rydell, H. 1972. Classification and genesis of submarine iron-manganese deposits. In: *Ferromanganese Deposits on the Ocean Floor*. National Science Foundation, Washington, DC, 149–165.

Cronan, D. S. 1980. *Underwater Minerals*. Academic Press, Oxford, 363.

Das, P. Iyer, S. D. and Hazra, S. 2012. Petrological characteristics and genesis of the Central Indian Ocean Basin basalts. *Acta Geologica Sinica*, 86, 1154–1170.

Das, P. Iyer, S. D. and Kodagali, V. N. 2007. Morphological characteristics and emplacement mechanism of the seamounts in the Central Indian Ocean Basin. *Tectonophysics*, 443, 1–18.

Das, P. Iyer, S. D. Kodagali, V. N. and Krishna. K. S. 2005. Distribution and origin of seamounts in the Central Indian Ocean Basin. *Marine Geodesy*, 28, 259–269.

Drazen, J. C. Smith, C. R. Gjerde, K. M. Haddock, S. H. D. Carter, G. S. Choy, C. A. Clark, M. R. Dutrieux, P. Goetze, E. Hauton, C. Hatta, M. Koslow, J. A. Leitner, A. B. Pacini, A. Perelman, J. N. Peacock, T. Sutton, T. T. Watling, L. and Yamamoto, H. July 2020. *Proceedings of the National Academy of Sciences*, doi: 10.1073/pnas.2011914117.

European Commission, 2012. *Blue Growth*. Opportunities for Marine and Maritime Sustainable Growth: Communication from the Commission to the European Parliament, the Council, the European Economic and Social Committee and the Committee of the Regions. Publications Office of the European Union, Luxembourg, 494 final, 12, doi: 10.2771/43949.

Ghosh, A. K. and Mukhopadhyay, R. 1995. Large phillipsite crystal as ferromanganese nodule nucleus. *Geo-Marine Letters*, 15, 59–62.

Gillard, B. Purkiani, K. Chatzievangelou, D. Vink, A. Iversen, M. H. and Thomsen, L. 2019. *Physical and Hydrodynamic Properties of Deep-Sea Mining-Generated, Abyssal Sediment Plumes in the Clarion-Clipperton Fracture Zone (Eastern-Central Pacific)*. Elementa: Science of the Anthropocene, 7.

Glasby, G. P. 1977. Why manganese nodules remain at the sediment-water interface. N.Z.J. 1988. Hydrothermal manganese deposits in island arcs and related to subduction process: a possible model for genesis. *Ore Geology Review*, 4, 145–153.

Glasby, G. P. 2000. Lesson learned from deep-sea mining. *Science*, 289, 551–553.

Glasby, G. P. and Thijssen, T. 1982. Control of the mineralogy and composition of marine manganese nodules by the supply of divalent transition metal ions. *Neues Jahrbuch für Mineralogie*, 145, 291–307.

Glasby, G. P. Friedrich, G. Thijssen, T. Pluger, W. L. Kunzendorf, H. Ghosh, A. K. and Roonwal, G. S. 1983. Distribution, morphology and geochemistry of manganese nodules from the Valdivia 13/2 area, equatorial north Pacific. *Pacific Science*, 36, 241–263.

Glasby, G. P. Gwozdz, R. Kunzendorf, H. Friedrich, G. and Thijssen, T. 1987. The distribution of rare earth and minor elements in manganese nodules and sediments from the equatorial and S. W. Pacific. *Lithos*, 20, 97.

Hajkowicz, S. A. Heyenga, S. and Moffat, K. 2011. The relationship between mining and socio-conomic well-being in Australia's regions. *Resource Policy*, 36, 30–38.

Halbach, P. Hebisch, U. and Scherhag, C. 1981. Geochemical variations of ferromanganese nodules and crusts from the different provinces of the Pacific Ocean and their genetic control. *Chemical Geology*, 34, 447–453.

ISA, 2010. *Technical Study 6: A Geological Model of Polymetallic Nodule Deposits in the Clarion-Clipperton Fracture Zone*. International Seabed Authority, Jamaica.

Iyer, S. D. 1996. *A Study of the Volcanics of the Central Indian Ocean Basin and Their Relationship to the Ferromanganese Deposits*. Unpubl. PhD thesis, Jadavpur University, India, 222.

Iyer S. D. 2005. Evidence for incipient hydrothermal event(s) in the Central Indian Basin: a review. *Acta Geologica Sinca*, 79, 77–86.

Iyer, S. D. and Karisiddaiah, S. M. 1988. Morphology and petrography of pumice from the Central Indian Ocean Basin. *Indian Journal of Marine Science*, 17, 333–334.

Iyer, S. D. and Ray, D. 2003. Structure, tectonic and petrology of mid-oceanic ridge and the Indian scenario. *Current Science*, 85, 277–289.

Iyer, S. D. and Sharma, R. 1990. Correlation between the occurrence of manganese nodules and rocks in a part of the Central Indian Ocean Basin. *Marine Geology*, 92, 127–138.

Iyer, S. D. and Sudhakar, M. 1993a. Coexistence of pumice and manganese nodule fields—evidence for submarine silicic volcanism in the Central Indian Basin. *Deep-Sea Research*, 40, 1123–1129.

Iyer, S. D. and Sudhakar, M. 1993b. A new report on the occurrence of zeolites in the abyssal depths of the Central Indian Basin. *Sedimentary Geology*, 84, 169–178.

Iyer, S. D. Amonkar, A. and Das, P. 2018. Genesis of Central Indian Ocean basin seamounts: morphological, petrological, and geochemical evidence. *International Journal of Earth Sciences*, 107(7), 2517–2538. https://doi.org/10.1007/s00531-018-1612-z.

Iyer, S. D. Fernandes, G. Q. and Mahender, K. 2012. Coarse fraction components in a red-clay sediment core, Central Indian Ocean Basin: their occurrence and significance. *Journal of Indian Association of Sedimentologists*, 31, 123–135.

Iyer, S. D. Pinto, S. M. and Sardar, A. A. 2018. Characteristics and genesis of phillipsite grains in a sediment core from the Central Indian Ocean Basin. *Indian Journal of Geo-Marine Sciences*, 47, 1121–1131.

Iyer, S. D. Gupta, S. M. Charan, S. N. and Mills, O. P. 1999b. Volcanogenic hydrothermal iron-rich materials from the southern part of the Central Indian Ocean Basin. *Marine Geology*, 158, 15–25.

Iyer, S. D. Prasad, S. M. Gupta, S. M. Charan, S. N. and Mukherjee, A. D. 1997a. Hydrovolcanic activity in the Central Indian Ocean Basin. Does nature mimic laboratory experiments? *Journal of Volcanology and Geothermal Research*, 78, 209–220.

Jauhari, P. and Pattan, J. N. 2000. Ferromanganese deposits in the Indian Ocean. In *Handbook of Marine Mineral Deposits*, CRC Publications, London, 171–195.

Kalangutkar, N. G. Iyer, S. D. and Ilangovan, D. 2011. Physical properties, morphology and petrological characteristics of pumices from the Central Indian Ocean Basin. *Acta Geologica Sinica*, 85, 826–839.

Karisiddaiah, S. M. and Iyer, S. D. 1992. A note on incipient spilitisation of central Indian basin basalts. *Journal of Geological Society of India*, 39, 518–523.

Khadge, N. H. 2000. Geotechnical properties of surface sediments in the INDEX area. *Marine Georesources and Geotechnology*, 18, 251–258.

Kodagali, V. N. 1991. Morphologic investigation of uncharted seamount from Central Indian Basin revisited with multibeam sonar system. *Marine Geodesy*, 15, 47–56.

Kolla, V. and Kidd, R. 1982. Sedimentation and sedimentary processes in the Indian Ocean. In: *The Ocean Basins and Margins*. Vol. 6 (Nairn, A.E.M. and Stehli, F.G. eds). The Indian Ocean. Plenum, New York, 1–45.

Kowalczuk, P. B. Bouzahzah, H. Kleiv, R. A. and Aasly, K. 2019. Simultaneous leaching of seafloor massive sulfides and polymetallic nodules. *Minerals*, 9, 482.

Kumari, A. and Natarajan, K. A. 2002. Development of a clean bioelectro-chemical process for leaching of ocean manganese nodules. *Mineral Engineering*, 15, 103–110.

Kunzendorf, H. 1986. *Marine Mineral Exploration (Edited)*. Elsevier, Amsterdam, 300.

Leal Filho, W. Abubakar, I. R. Nunes, C. Platje, J. Ozuyar, P. G. Will, M. Nagy, G. J. Al-Amin, A. Q. Hunt, J. D. and Li, C. 2021. Deep seabed mining: A note on some potentials and risks to the sustainable mineral extraction from the oceans. *Journal of Marine Science and Engineering*, 9, 521.

Martin-Barajas, A. and Lallieri-Verges, E. 1993. Ash layers and pumice in the Central Indian Basin: relationship from the formation of manganese nodules. *Marine Geology*, 115, 307–329.

Martin-Barajas, A. Lallier-Verges, E. and Leclaire, L. 1991. Characteristics of manganese nodules from the Central Indian Basin: relationship with the sedimentary environment. *Marine Geology*, 101, 249–265.

Mero, J. 1965. *The Mineral Resources of the Sea*. Elsevier Oceanography Series. Elsevier, The Netherlands, 312.

Miller, K. A. Brigden, K. Santillo, D. Currie, D. Johnston, P. and Thompson, K. F. 2021. Challenging the need for deep seabed mining from the perspective of metal demand, biodiversity, ecosystems services, and benefit-sharing. *Frontiers in Marine Science*, 8, 706161. doi: 10.3389/fmars.2021.706161.

Mohanty, S. K. Dash, P. Gupta, A. and Gaur, P. 2015. *Prospects of Blue Economy in the Indian Ocean*. RIS, Research and Information System for Developing Countries, New Delhi, India, 87.

Mohaptra, B. K. and Sahoo, R. K. 1987. Merlinoite in manganese nodules from the Indian Ocean. *Mineralogical Magazine*, 51, 749–750.

Mukherjee, A. D. and Iyer, S. D. 1999. Synthesis of morphotectonics and volcanics of the Central Indian Ocean Basin. *Current Science*, 75, 296–304.

Mukhopadhyay, R. 1988. *Morphological and Geochemical Studies of Polymetallic Nodules from Four Sectors in the Central Indian Ocean Basin between 11° S and 16° S Latitudes*. Unpubl. PhD thesis. The University of Calcutta, 93.

Mukhopadhyay, R. and Khadge, N. H. 1990. Seamounts in the Central Indian Ocean Basin: indicators of the Indian plate movement. *Proceedings of Indian Academy of Sciences*, 99, 357–365.

Mukhopadhyay, R. and Nath, B. N. 1988. Influence of seamount topography on the local facies variation in ferromanganese deposits in the Indian Ocean. *Deep-Sea Research*, 35, 1431–1436.

Mukhopadhyay, R. Ghosh, A. K. and Iyer, S. D. 2008. *The Indian Ocean Nodule Field: Geology and Resource Potential*. Elsevier, Amsterdam, 292.

Mukhopadhyay, R. Ghosh, A. K. and Iyer, S. D. 2018. *The Indian Ocean Nodule Field: Geology and Resource Potential*. 2nd edition, Elsevier, Amsterdam, 413.

Mukhopadhyay, R. Iyer, S. D. and Ghosh, A. K. 2002. The Indian Ocean nodule field: petrotectonic evolution and ferromanganese deposits. *Earth-Science Reviews*, 60, 67–130.

Mukhopadhyay, R. Naik, S. De Souza, S. Dias, O. Iyer, S. D. and Ghosh, A. K. 2019. The economics of mining seabed manganese nodules: a case study of the Indian Ocean nodule field. *Marine Georesources and Geotechnology*, doi: 10.1080/1064119X.2018.1504149.

Murty, V. S. N. Savin, M. Ramesh Babu, V. and Suryanarayana, A. 2001. Seasonal variability in the vertical current structure and kinetic energy in the Central Indian Ocean Basin. *Deep-Sea Research II*, 48, 3309–3326.

Nath, B. N. 2001. Geochemistry of sediments. In: *The Indian Ocean: A perspective*, Vol. 2. Oxford and IBH, New Delhi, India, 2, 645–689.

Nath, B. N. Borole, D. V. Aldahan, A. Patil, S. K. Mascarenhas-Pereira, M. B. L. Possnert, G. Ericsson, T. Ramaswamy, V. and Gupta, S. M. 2008. ^{210}Pb, ^{230}Th, and ^{10}Be in Central Indian Basin seamount sediments: signatures of degassing and hydrothermal alteration of recent origin. *Journal of Geophysical Research*, 35, 1-6. doi:10.1029/2008GL033849.

Nath, B. N. Sijinkumar, A. V. Borole, D. V. Gupta, S. M. Mergulhao, L. P. Mascarenhas- Pereira, M. B. L. Ramaswamy, V. Guptha, M. V. S. Possnert, G. Aldahan, A. Khadge, N. H. and Sharma, R. 2012. Record of carbonate preservation and the Mid-Brunhes climatic shift from a seamount top with low sedimentation rates in the Central Indian Basin. *Boreas*, 42(3), 762–778.

Nath, B. N. 1993. Rare earth element geochemistry of the sedimnets, ferromanganese nodules and crusts from the Indian Ocean.

Nayak, B. Das, S. K. and Bhattacharyya, K. 2011. Detrital and authigenic (?) baddeleyite (ZrO_2) in ferromanganese nodules of Central Indian Ocean Basin. *Geoscience Frontiers*, 2, 571–576.

Padan, J. W. 1990. Commercial recovery of deep seabed manganese nodule: twenty years of accomplishment. *Marine Minerals*, 9, 87–103.

Parthiban, G. 2000. Increased particle fluxes at the INDEX site attributable to simulated benthic disturbance. *Marine Georesources and Geotechnology*, 18, 223–235.

Pattan, J. N. and Parthiban, G. 2007. Do manganese nodules grow or dissolve after burial? Results from the Central Indian Ocean Basin. *Journal of Asian Earth Sciences*, 30, 696–705.

Pattan, J. N. Mudholkar, A. V. Jai Sankar, S. and Ilangovan, D. 2008. Drifted pumice in the central Indian Ocean Basin: geochemical evidence. *Deep-Sea Research I*, 55, 369–378.

Perelman, J. N. Firing, E. van der Grient, J. M. A. Jones, B. A. and Drazen, J. C. 2021. Mesopelagic scattering layer behaviors across the Clarion-Clipperton zone: implications for deep-sea mining. *Frontiers in Marine Science*, doi: 10.3389/fmars.2021.632764.

Raghukumar, C. Loka Bharathi, P. A. Ansari, Z. A. Nair, S. Ingole, B. Sheelu, G. Mohandass, C. Nath, B. N. and Rodrigues, N. 2001. Bacterial standing stock, meiofauna, and sediment-nutrient characteristics: indicator of benthic disturbance in the Central Indian Basin. *Deep-Sea Research- II*, 48, 3381–3399.

Raghukumar, C. Nath, B. N. Sharma, R. Loka Bharathi, P. A. and Dalal, S. G. 2006. Long-term changes in microbial and biochemical parameters in the Central Indian Basin. *Deep-Sea Research*, 53, 1695–1717.

Ramesh Babu, V. Suryanarayana, A. and Murty, V. S. N. 2001. Thermohaline circulation in the Central Indian Ocean Basin (CIB) during austral summer and winter periods of 1997. *Deep-Sea Research-II*, 48, 3327–3342.

Rodrigues, N. Sharma, R. and Nath, B. N. 2001. Impact of benthic disturbance on megafauna in Central Indian Basin. *Deep-Sea Research-II*, 48, 3411–3426.

Sarkar, C. Iyer, S. D. and Hazra, S. 2008. Inter-relationship between nuclei and gross characteristics of manganese nodules, Central Indian Ocean Basin. *Marine Georesources and Geotechnology*, 26, 259–289.

Sen, P. K. 1999. Processing of sea nodules: current status and future needs. *Metals Materials and Processes*, 11, 85–100.

Sen, P. K. 2010. Metals and materials from deep-sea nodules: an outlook for the future. *International Materials Reviews*, 55, 364–391.

Sharma, R. 2015. Environmental issues of deep-sea mining. *Procedia Earth and Planetary Science*, 11, 204–211.

Sharma, R. Sudhakar, M. and Iyer, S. D. 1994. Distribution of manganese nodules, rocks and sediments in the Central Indian Ocean Basin: factors influencing the performance of mining system. In: *Ocean Technology: Perspectives* (Kumar, S. et al. eds). Publication and Information Directorate, CSIR, New Delhi, 817–826.

Smith, C. R. Levin, L. A. Koslow, A. Tyler, P. A. and Glover, A. G. 2008. Near future of the deep-sea floor ecosystems. In: *Aquatic Ecosystem: Trends and Global Prospects* (Polunin, N.V.C. ed.). Cambridge University Press, New York, 334–352.

Sorem, R. K. and Fewkes, R. H. 1977. Internal characteristics. In: *Marine Manganese Deposits* (Glasby, G.P. ed.). Elsevier, Amsterdam, 147–184.

SPC*, 2013. Deep sea mineral. In: *Deep Sea Mineral and the Green Economy*, Vol. 2 (Baker, E. and Beaudoin, Y. eds), 122. Secretariat of the Pacific Community.

Srikanth, S. Alex, T. C. Agrawal, A. and Premchand, A. 1997. Reduction roasting of deep-sea manganese nodules using liquid and gaseous reductants. In: *The Proceedings of the Second ISOPE Ocean Mining Symposium, Seoul, Korea, November 24–26, 1997* (Chung, J.S. and Hong, S. eds). International Society of Offshore and Polar Engineering, Colorado, USA, 177–184.

Stanley, J. A. and Jeffs, A. G. 2016. Ecological impacts of anthropogenic underwater noise. In: *Stressors in the Marine Environment* (Solan, M. and Whiteley, N.M. eds). Oxford University Press: Oxford, UK, 282–297.

Thiel, H. 2001. Emerging studies for the mining of polymetallic nodules from the deep sea. *Deep-Sea Research-II*, 48, 3427–3882.

Udinstev, G. B. 1975. *Geological-Geophysical Atlas of the Indian Ocean*. Pergamon, London, 151.

UNEP, FAO, IMO, UNDP, IUCN, Centre, and Grid-Arendal, W.F. 2012. *Green Economy in a Blue World*. United Nations Environment Program.

Vu, H. Jandova, J. Lisa, K. and Vranka, F. 2005. Leaching of manganese deep ocean nodules in FeSO4-H2SO4-H2O solutions. *Hydrometallurgy*, 77, 147–153.

Washburn, T. W. Turner, P. J. Durden, J. M. Jones, D. O. B. Weaver, P. and Van Dover, C. L. 2019. Ecological risk assessment for deep-sea mining. *Ocean* and *Coastal Management*, 176, 24–39.

Young, D. K. and Richardson, M. D. 1998. Effects of waste disposal on benthic faunal succession on the abyssal seafloor. *Journal of Marine Systems*, 14, 319–336.

Zhang, W. and Cheng, C. Y. 2007. Manganese metallurgy review. Part I: leaching of ores/secondary materials and recovery of electrolytic/chemical manganese dioxide. *Hydrometallurgy*, 89, 137–159.

8 Development and Challenges of Indian Ocean Blue Economy and Opportunities for Sri Lanka

Nawalage S. Cooray,[1] Upul Premarathna,[2] Keerthi Sri Senarathna Atapaththu,[3] and Tilak Priyadarshana[3]*
[1]Graduate School of International Relations, International University of Japan (IUJ), Japan
[2]Department of Oceanography and Marine Geology, Faculty of Fisheries and Marine Sciences & Technology, University of Ruhuna, Sri Lanka
[3]Department of Limnology and Water Technology, Faculty of Fisheries and Marine Sciences & Technology, University of Ruhuna, Sri Lanka
*Corresponding Author: Tilak Priyadarshana

CONTENTS

DOI: 10.1201/9781003184287-8

8.1　INTRODUCTION

Oceans are approximately three-quarters of the earth's surface, accountable for more than 90% of the biosphere, providing an array of goods and services to the global community which includes food, employment, recreation and cultural well-being, minerals, oxygen production, greenhouse gas absorbance, climate change impact mitigation, and serve as highways for seaborne international trade (United Nations, 2017). For the development of economies along with the threats posed by the climate change and global warming the concept of the Blue Economy was first introduced in 1994 by Professor Gunter Pauli of the United Nations University (UNU). Because of this significant role played by the oceans, the importance of implementing sustainable development measures for marine environment was discussed at the United Nations Conference on Sustainable Development (Rio+20) held in Rio de Janeiro, Brazil in June 2012. Consequently, the conference adopted a set of ground-breaking guidelines on green economic policies which are considered as essential tools for achieving sustainable development goals (United Nations, 2012). With strong support from coastal and island nations at the conference, because of the contribution made by the oceans to their economies, the ocean derived green economy was accepted as the blue economy.

8.1.1　Blue Economy as a Concept

It has been nearly a decade since the conference held at Rio de Janeiro which led to the development of the 'Blue Economy' concept. The blue economy is a way of looking at the economic growth of a nation through its contribution from ocean and costal based activities while assuring environmental sustainability and assuring livelihood improvement. The oceans are huge natural resource pools that consist of both renewable and non-renewable resources that can be utilized in many industries to strengthen the economies of nations (Sumaila, 2021). Among the many definitions for the concept of blue economy Kathijotes (2013) identifies it as a system of ocean based green economy that connects to create neo-science and technologies. In other words, strengthening economic systems through ocean-based resources can be referred as 'ocean economy', while using ocean resources sustainably, giving due consideration to the ecological aspects for economic development can be defined as 'blue economy'. Therefore, efficient, and optimal utilization of marine natural resources within ecological capacities are important aspects of blue economy and can be identified as:

i.　a subset of the economy,
ii.　encouraging production from ocean inputs,
iii.　utilizing ocean resources sustainably without degrading the environment,
iv.　aiming to reduce environmental risks and ecological scarcities, and
v.　supporting livelihoods in an equitable manner with benefit-sharing

According to Keen et al. (2018) there are five key components of the blue economy: ecosystem resilience, economic sustainability, community engagement, institutional integration, and technical capacity. Thus, the blue economy can be referred to as the sustainable management of ocean resources to support livelihoods, the fostering of more equitable benefit-sharing, while enhancing ecosystem resilience to climate change, eliminating destructive fishing practices, and managing pressures from external sources to the fisheries sector. According to Wenhai et al. (2019), in all definitions of the blue economy, there are common key areas that includes a strategic framework, a kind of policy, being part of the green economy, a sustainable marine economy, and a marine-based new technology economy, despite the country specific views, because the core of the blue economy is ocean based.

Ocean based economic activities such as fishing, shipping, harnessing offshore wind, maritime and coastal tourism, and marine biotechnology accounts for nearly US$ 1.5 trillion, or 2.5% of the global gross value added in the year 2010. Further, global blue economy sustains 350 million livelihoods, with a continuous increasing contribution over the last few years, and the predicted size of the blue economic contribution accounts to US$ 3.0 trillion by 2030 (Sumaila, 2021). Consequently, in managing economies, the concept of the blue economy is becoming increasingly popular as it is one of the vital tools that can be used by many nations.

8.1.2 Historical Records of the Indian Ocean

The Indian Ocean (Figure 8.1) is unique in terms of its mix of natural resources, cultures, ethnic groups, and shipping routes across its nations and has a long history of human involvements mainly associated with its unique wind pattern for sailing (Cordner, 2011). High intensity temperature oscillations of water masses between the Western Pacific and the Eastern Indian Ocean, attributed to a large scale Tropical Warm Pool, interact with the atmosphere to build a seasonal wind pattern. From May to September (the summer monsoon season) the Indian Ocean faces south-westerly winds and from December to February (the winter monsoon season) north-easterly winds, while from March to April (the pre summer monsoon) and October to November (the post summer monsoon) are considered as inter-monsoon seasons. Sailors used the advantage of south-westerly and north-easterly winds to travel and return between the countries of the region and conduct their international business. Maritime trade records found between Egypt and Somalia (circa 3,000 BC) and Mesopotamia and the Indus Valley (circa 2,500 BC) are the early records and wind pattern might be the reason. The years between 1960s and 1970s and, after the Cold War, it has experienced a politically unstable period.

FIGURE 8.1 Indian Ocean on the World Map.

TABLE 8.1
Indian Ocean Counties and Islands

Continent	Countries	Islands	Population	Population density (per sq km)
African	Djibouti, Egypt, Kenya, Eritrea, Mozambique, Somalia, Somaliland, South Africa, Sudan, Yemen, Tanzania	Mauritius, Seychelles, Comoros, Madagascar, Mayotte, Réunion	1,340,598,147	45
Asian	Bahrain, Iran, Iraq, Israel, Jordan, Kuwait, Oman, Palestine, Qatar, Saudi Arabia, UAE, Pakistan, India, Bangladesh, Myanmar, Indonesia, Malaysia, Singapore, Thailand, Timor-Leste	Chagos and Diego Garcia, Maldives, Sri Lanka, Cocos	4,641,054,775	150
Oceania	Australia	Ashmore and Cartier, Christmas	43,111,704	5
Antarctica		Prince Edward, Heard and McDonald, French Southern and Antarctic Lands, Amsterdam and Saint Paul, Crozet, Kerguelen	0	0

In particular for the Indian Ocean, the concept of the blue economy is vital as is the diversity of the ring countries (Table 8.1), its resources, increasing population and the extent of the ocean. The prevalence of strong southwest monsoon winds induces upwelling which augments the oceanic productivity (Gonaduwage et. al. 2021, Zeng et. al. 2021). Upwelling zones can be identified as major regional components of biological production and are important for the socioeconomics of fishing livelihoods (Tacon and Metian, 2008) which account for about 11% of global marine primary production (Chavez and Toggweiler, 1995). Further, oceanic productivity is largely influenced by the excess freshwater input from monsoon rain and river runoff (Hermes et al. 2019). It is predicted that the Indian Ocean Rim (IOR) countries will be the home for nearly half of the world population by 2050.

8.1.3 Historical Records of Sri Lanka

It is apparent that Sri Lanka's involvement in international maritime trading from ancient times, connecting the east-west maritime route, Greece, Rome and Persia from the west and India, South Asian countries, and China from the east. Sri Lanka had been treated as the meeting point of traders from the east and west. According to the literature, Sri Lankan maritime history dated back to the 6th century B.C, through the well-known seaport of Mahatiththa (Mahatota) that linked to the Indian Ocean trade network for more than thousand years (Kiribamune, 1991; Bohingamuwa, 2017; Sudharmawathie, 2017). The very reason that Sri Lanka became important in maritime history is its geographical location in the Indian Ocean. By the 6th century CE, ancient Mahatiththa had been one of the major seaports and one of the greatest entrepots of Indian Ocean maritime (Indicopleustes, 2010; Bohingamuwa, 2017).

Archaeological findings of megalithic black pottery, red pottery, and several fragments of imported Rouletted Ware which were similar to those found in the South Indian coast, provide strong evidence for the close maritime relations of Sri Lanka with India (Bopearachchi, 2002; Bohingamuwa, 2017). It proves that maritime trade and commerce has played a significant role in Sri Lankan history, specifically with India, Persia, and Ethiopia (Indicopleustes, 2010). Similarly, early Brahmin inscriptions from Andiyagama in Anuradhapura and the Nainativu Tamil Inscription of King Parakramabahu is evidence of maritime voyages and state trading with other countries (Indrapala 1963). According to Murphy (2004), Taprobaneans (Sri Lankans) were also a trading nation whose merchants sailed great distances to India and China similar to competitive Mediterranean traders who were dominant in the Roman world. Besides maritime trade and commerce, Sri Lankans had been engaged in ship manufacturing, repairing, and supplying raw materials for ships (Hall, 2009; Indicopleustes, 2010). Around 100 BC, Sri Lankan maritime trade and commerce was further augmented by the pearl fishery and was famous for natural pearls in the world for more than two millennia, particularly there were high foreign earnings for the country (Katupotha, 2019). These pearls were exported to Rome either by Roman or Greek ships and those were highly valued in Rome. Famous pearl banks were in the proximity of the ancient seaport of Mahatiththa, and thus, this port city was considered to be one of the main pearl processing centres of the region (Bohingamuwa, 2017).

8.2 BLUE ECONOMIC POTENTIALS IN THE REGION AND SRI LANKA

The Indian Ocean covers approximately one fifth of the surface area of the earth's surface which amounts to 73.5 million km^2 extending from longitude 20° E to 147° E and latitudes 30° N to 40° S which includes extensive EEZs of different countries and large seas (Table 8.2). The Indian Ocean is a territory boarded by Asia in the north, Africa in the west, Australia in the east and Antarctica in the south. It has a coastline of 114,172 km boarded by 32 countries and 18 islands. South Asian countries are fortunate to a part of the Bay of Bengal, the largest bay in the world bordered by Bangladesh, India, Myanmar, and Sri Lanka in north, west, east, and southwest respectively (Bari, 2017). Even though the percentage land cover of the IORCs is small, the total share of world's EEZ is nearly 17% (Table 8.2).

The geographical position of the Indian Ocean in terms of industry, international trade, transport, labour, environment, and security together with a considerable portion of the world's population will increasingly influence the global economy in the twenty-first century (Timothey, 2018) by representing approximately a 13% share of the world's GDP (Table 8.2). Further, population of the IORCs has been growing exponentially in the last couple of decades with a population of approximately 2.42 billion in 2020 which represents nearly 30% of the world population. It is expected to be half of the world population by 2050 (Timothey, 2018).

Furthermore, half of the world's trade transport is through this region, due to its geographical significance. The Liner Shipping Connectivity Index (LSCI) captures the level of integration of each country into the existing global liner shipping network by measuring liner shipping connectivity, which is computed by the United Nations Conference on Trade and Development (UNCTAD). The LSCI is a function of the number of ships, their container-carrying capacity, maximum vessel size, number of services, and number of companies that deploy container ships in a country's ports. Sri Lanka ranked fifth among other Indian Ocean Rim Countries (Figure 8.2) indicating the global validity of the Sri Lankan shipping industry. Further, there is an increasing trend in Sri Lankan LSCI over the last two decades, and the increment was approximately two-fold during this period, and we are very close to UAE and France (Figure 8.2).

IORC are rich in both marine and terrestrial natural resources which are essential for the stability of both the built and the natural environment. Blue economy as a concept is well fitting to the Indian Ocean countries and the Indian Ocean region, is defined by a 'maritime regionalism', and the Indian Ocean is considered to be the 'Ocean of the future' (Timothey, 2018, Doyle and Seal, 2015).

Sri Lanka being an island in the Indian Ocean, accounts to 1600 km of coastline and territorial waters extended up to 22 km from the shoreline covering an area of 21,500 km^2. As per the provision set out in the United Nations Convention on the Law of the Sea (UNCLOS), the island has

TABLE 8.2
Potential of the IOR Counties (EEZ, Terrestrial Land Extent, GDP, and Population)

IOR Country	EEZ (km²)	Land Area (km²)	% of EEZ Compared to Total Area	GDP (US$) in 2020	Population as at 2020 (thousand)
Australia	8,505,348	7,692,020	52.51	1,359.33	25,500
Bangladesh	118,813	130,170	47.72	329.12	164,689
Comoros	164,476	1,861	98.88	1.22	870
French Republic	345,240	547,557	38.67	2,598.91	65,274
India	2,305,143	2,973,190	43.67	2,708.77	1,380,004
Indonesia	6,159,032	1,877,519	76.64	1,059.64	273,524
Iran	168,718	1,628,760	9.39	635.72	83,993
Kenya	116,942	569,140	17.04	99.29	53,771
Madagascar	1,225,259	581,800	67.80	13.84	27,691
Malaysia	334,671	328,550	50.46	338.28	32,366
Maldives	923,322	300	99.97	3.76	541
Mauritius	1,284,997	2,030	99.84	11.40	1,272
Mozambique	578,986	786,380	42.41	14.38	31,255
Oman	533,180	309,500	63.27	63.19	5,107
Seychelles	1,336,559	460	99.97	1.13	98
Singapore	1,607	709	69.39	339.98	5,850
South Africa	1,535,538	1,213,090	55.87	302.11	59,309
Sri Lanka	**532,619**	**61,864**	**89.59**	**80.70**	**21,413**
Somalia	50,229	627,340	7.41	4.92	15,893
Tanzania	241,888	885,800	21.45	63.24	59,734
Thailand	299,397	510,890	36.95	501.89	69,800
United Arab Emirates	58,218	71,020	45.05	354.28	9,800
Yemen	552,669	527,970	52.51	20.14	29,826
Share of IORC (%)	**17.1**	**1.5**		**12.9**	**30.0**

Source: EEZ: https://vividmaps.com/exclusive-economic-zones-maps/ and https://marineregions.org/ access on 09.11.2021; Land Area: World Bank data base available at https://data.worldbank.org/indicator/AG.LND.TOTL.K2 access on 09.11.2021.

GDP: World Economic Outlook Database available at https://imf.org/en/Publications/WEO/weo-database/2021/April, access on 08.11.2021; Population: United Nations world population prospects available at https://population.un.org/wpp/ access on 08.11.2021.

been able to claim an offshore area EEZ, which is approximately 517,000 km² and eight times the size of the onshore of the island (Figure 8.3). The outer edge of the EEZ is 200 nautical miles (370 km) from a baseline, which more or less coincides with the coastline of Sri Lanka. Sri Lanka has exclusive rights to the living and non-living resources of the water column, on the seabed and the subsurface under the EEZ. The majority of Sri Lanka's EEZ lies in water depths of more than 3000 m. According to the Annex II of Article 76 of the UNCLOS, coastal states such as Sri Lanka can claim an extension to their continental shelf up to a limit where the sediment thickness is not less than one kilometer. Accordingly, the Sri Lankan Government implemented a project named the 'Delimitation of the Outer Edge of the Continental Margin of Sri Lanka'' (DEOCOM) in 1999 to define an offshore area beyond Sri Lanka's EEZ, where the sediment thickness is not less than one kilometer. Based on the results of the project, the Government submitted its claim on the Limit of the Continental Shelf beyond the EEZ to the United Nation's Commission on Limit of the Continental Shelf (UNCLCS) on 8th May 2009. The area of the proposed extension is approximately 17 times

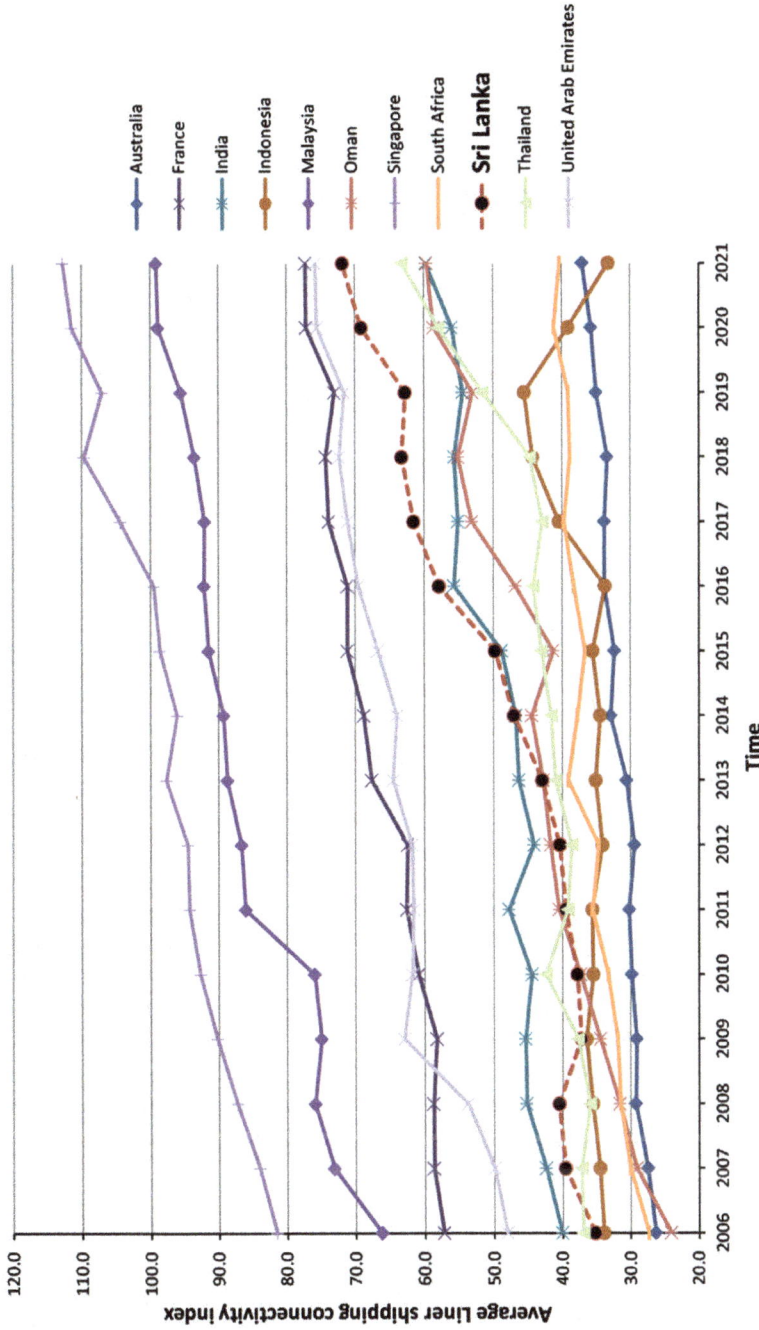

FIGURE 8.2 Average annual linear shipping connectivity index (LSCI) of selected Indian Ocean rim countries. Countries having LSCI below 20 was not include (annual index is the average of four quarters. LCI of 2021 is the average of first three quarters. https://unctadstat. unctad.org/wds/TableViewer/tableView.aspx?ReportId=92).

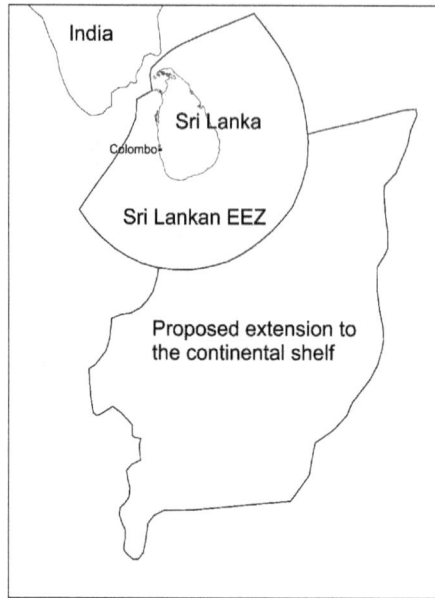

FIGURE 8.3 Sri Lanka's EEZ and the extended 'continental shelf' claimed by Sri Lanka (DEOCOM).

the size of its land area. This proposal is currently under deliberation before the UN. If the proposal is accepted as it is, Sri Lanka will have the right to explore and exploit non-living resources on the sea bed and in the subsurface over the continental shelf extended beyond the EEZ. On the other hand, among other south Asian countries, Sri Lanka has a greater maritime area compared to their land (Bari, 2017).

Non-living resources presently extracted from the world's oceans vary from common constituents to high-tech metals within the water itself. The chemical composition of seawater has demonstrated that it contains about 3.5% dissolved solids, with more than 60 chemical elements (Bardi, 2010; Batapola et al. 2021; Berman et al. 1980; Magazinovic et al. 2004).

The economical extraction of elements dissolved in seawater and offshore mineral resources depends on the available technology and geographic location (ownership and transport distance). The non-living ocean resources that could be profitably exploited from the Sri Lankan offshore include oil, natural gas, gas hydrates, mineral sands (placer deposits), common salt, gypsum, limestone, and metals such as magnesium and lithium dissolved in seawater (Subasinghe, 2021, Batapola et al. 2021).

Moreover, Sri Lanka has the potential to promote recreational opportunities such as surfing, whale and dolphin watching, deep sea diving, sea entertainment, sea sports, and the like. Maritime recreation and sea sports, which remain relatively subtle, could be a substantial foreign exchange earner for the country.

8.2.1 Fisheries and Aquaculture

Sri Lankan coastal waters are rich in diversity of marine biological resources, which consists of 620 species from 137 families, mainly represented by teleost fish having a density of $8.13 \pm 0.86 t/ Nm^2$ followed by fewer elasmobranchs, cephalopods, crustaceans, decapods, echinoderms, gastropods, and reptiles (Athukoorala et al. 2021). There is a gradual increase in total marine fish production over the last couple of decades and it has reached approximately 400,000 Mt in 2019 (Figure 8.4).

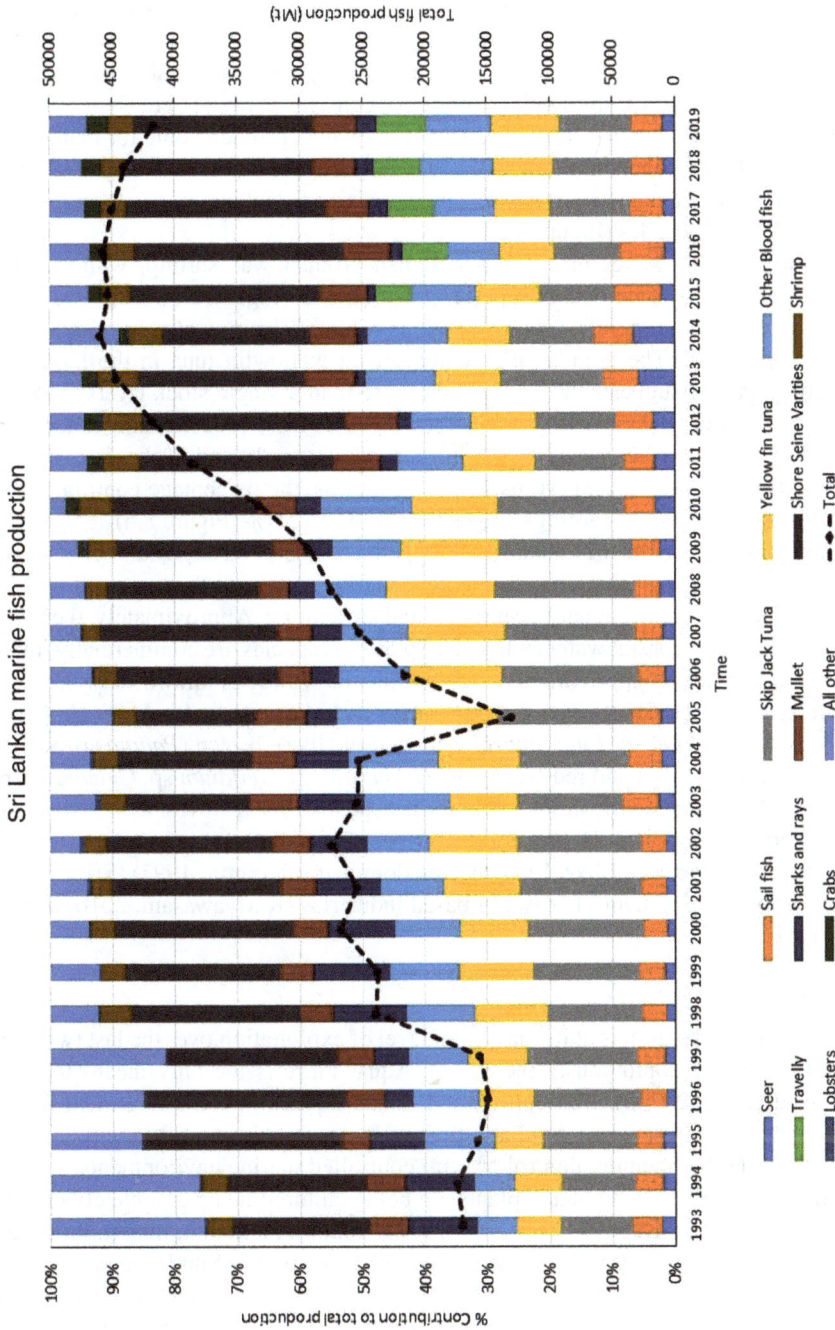

FIGURE 8.4 Sri Lankan marine fish production (Ministry-of-Fisheries Yearbook 2017, 2018, 2019, and 2020 and NARA Fisheries Yearbook 1998, 2000, 2008).

8.2.1.1 Fisheries

A large majority of this production is represented by tuna species (33%) followed by different types of small pelagic shore seine varieties (29%) such as sardines, herrings, and sprats. Besides finfish, Sri Lankan coastal waters are rich in non-finfish living resources. For instance, there are 24 sea cucumbers species including the most expensive *Holuthuria scraba* (approximately US$ 7.00 per fresh animal) (Prasada, 2020). Further, six lobster species are available in Sri Lankan waters including *Panulirus homarus*, *P. ornatus* , *P. versicolor* , *P. longipes* , *P. polyphagus* and *P. penicillatus* . As an island, the entire Sri Lankan coastal belt (~1600 km), consists of several brackish water bodies that provide habitats for an array of bottom fauna and flora which provide home for mangrove mud crab; *Scylla serrate* (How-Cheong and Amandakoon, 1992). All these marine living resources play a significant role in the Sri Lankan export market and there is a gradual increase in foreign exchange earnings over the last two decades (Figure 8.5).

During the early 1990s, the dominant exported fish product was shrimp, while in the last two decades shrimp was to some degree replaced by food fish (Figure 8.5). Exported food fish is mainly represented by tuna varieties such as yellowfin tuna (*Thunnus albacares*) and skipjack tuna (*Katsuwonus pelamis*). The population structure of the yellowfin tuna in the Indian Ocean may be more complex and suggests the presence of more than a single stock (IOTC, 2006). The average annual tuna export over the last two decades was 21403 ± 4605 Mt and it accounts for 121 million US$ in 2020. Besides food fish, shrimps, molluscs, crab, and bêche-de-mer products are also significant seafoods in the export market. Nevertheless, the percentage contribution of the fisheries sector for the total export earnings ranges from 1.6% to 2.5% (Figure 8.6).

Similarly, floral diversity in the Sri Lankan coastal region consists of a large number of seaweeds. Seaweeds are considered one of the important natural marine resources that are used in different industries such as food, cosmetics, pharmaceuticals, and agriculture. Approximately, there are 320 seaweeds in the Sri Lankan coastal waters while nearly 50 - 60 species are commercially important (Kariyawasam, 2016). Among approximately 320 species of seaweeds known in both inter-tidal and deep-water zones of the Sri Lankan coastal waters including green (such as, *Halimeda* sp. *Ulva* sp. *Caulerpa* sp. *Codium* sp. *Enteromorpha* sp. and the like), brown (*Padina* sp. *Sargassum* spp. *Turbinaria* sp. and the like) and red seaweeds (*Gracilaria* sp. *Gelidium* sp. *Gelediella* sp. and the like). Along the southern coast, in the areas most exposed to sea waves with high dissolved oxygen contents brown seaweed such as *Sargassum* species are the most common, whereas in in the shallow depths green seaweed like *Ulva* species dominate (Pernetta, 1993). Sri Lanka has a great potential for the development of seaweed-based industries (Kariyawasam, 2016) because of the high diversity.

8.2.1.2 Mariculture

Global capture fishery production is at its maximum level of exploitation over the last two decades, but the demand for fish is continuously increasing. Aquaculture is the only means to fulfil the increasing demand (FAO, 2020b). Mariculture and inland aquaculture are important fields that can play a significant role in securing food at a time of increasing demand. Mariculture refers to rearing marine flora and fauna under control or semi controlled marine environments. In mariculture systems, finfish, shellfish, and aquatic plants are being cultured using completely or partially artificial structures, near sea (Kapetsky et al. 2013, FAO, 2020b). Although, mariculture practices are common all over the world, such setups are mostly concentrated in South, Southeast and East Asian countries and some Latin American countries, in farming finfish, shellfish and to a lesser extent, seaweed (FAO, 2020b). The use of the open ocean for offshore mariculture is in its infant stage compared to coastal mariculture (Kapetsky et al. 2013). Global total mariculture production was 63.1 million tonnes (Table 8.3) with a value of approximately 120 US$ billion in 2018 (FAO, 2020a).

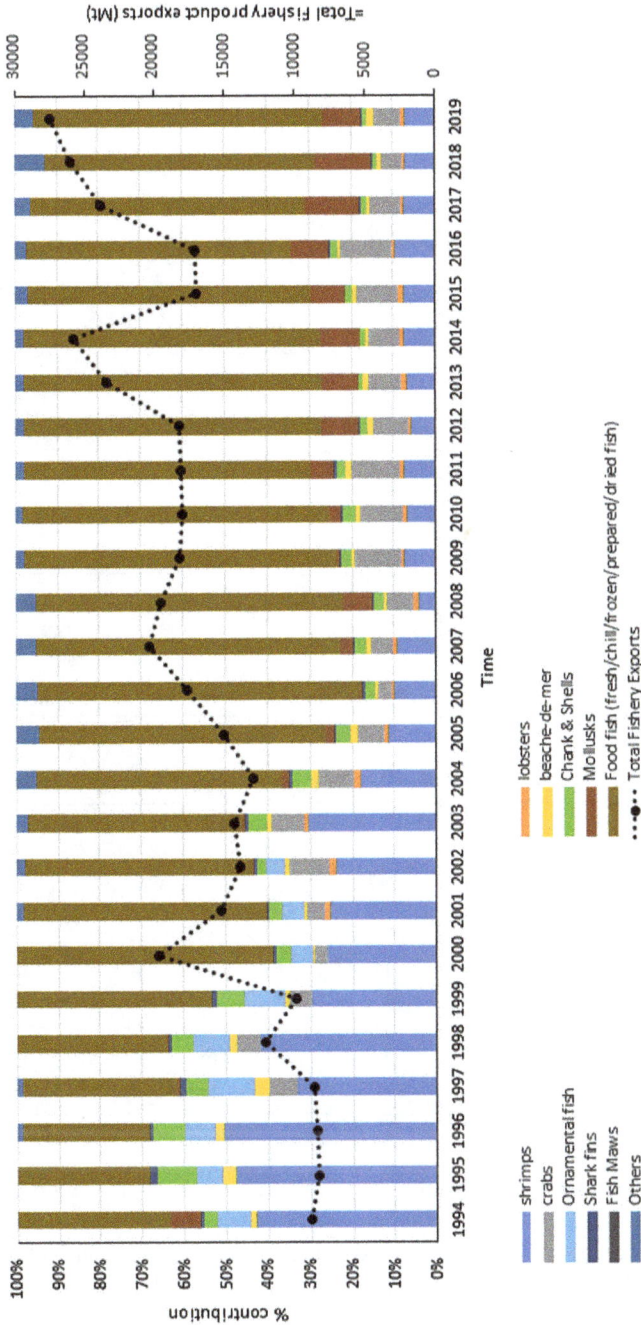

FIGURE 8.5 Exports of fishery products in Sri Lanka (Fisheries Statistics, 2020, Department of Fisheries, Sri Lanka).

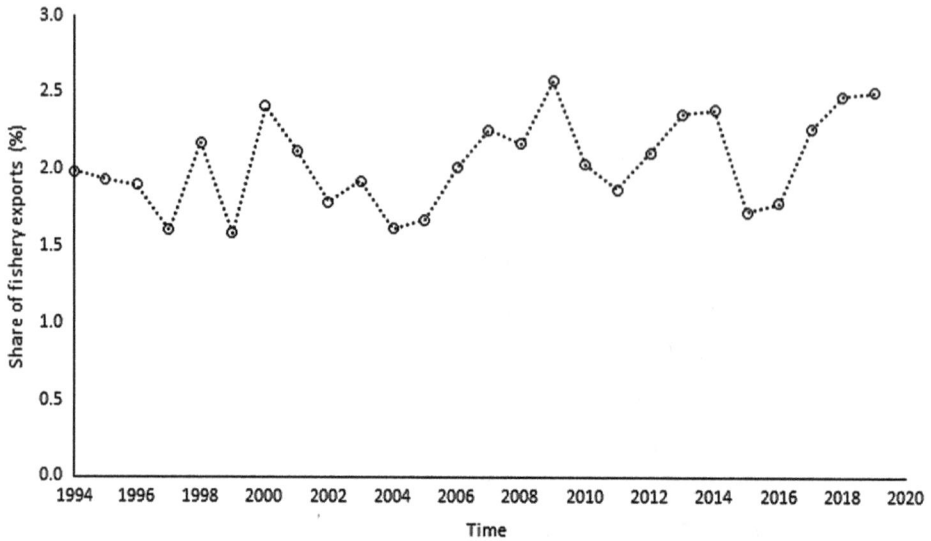

FIGURE 8.6 Percentage contribution of the fisheries sector to the total exports of Sri Lanka (Ministry of Fisheries (MOF), 2020).

TABLE 8.3
Global and Sri Lankan Mariculture Production

	Global Mariculture Production (Mt x 10⁷)	Sri Lankan Production by Different Mariculture Systems (mt)							
		Shrimp	Lagoon Fish	Ponds	Cages	Pens	Rafts	Seaweeds	Sea cucumber
2010	4.2	3550							
2011	4.4	3480							
2012	4.9	4150							
2013	5.3	3310							
2014	5.5	4430							
2015	5.8	6836	548	15	52	5		165	2802
2016	6	6030	3319	32	120	71		200	950
2017	6.3	4630	5920	42	232	19		668	578
2018	6.4	8180	4866	137	352	196	27	320	1171
2019	6.7	6400	5683	587	28	280	11	247	5711

Since Sri Lanka is an island surrounded by the Indian Ocean, there is a big potential to develop mariculture in the island. Along the 1600 km coastline, diverse environments such as lagoons, estuaries and coastal bays can be seen that provide excellent opportunities for mariculture. Furthermore, these coastal waters are rich in economically important finfish, shellfish, and seaweed species. During the last century Sri Lankan mariculture production was dominated by the shrimp (*Penaeus monodon*) which suddenly declined with the spread of global white spot disease (Table 8.3).

Research conducted by the National Aquaculture Development Authority (NAQDA) and other research institutes in Sri Lanka found various possible species for coastal aquaculture including finfish, shellfish, and seaweeds. According to investigations, there are two commercially important seaweeds, *Gracilaria edulis* and *G. verucosa* together with several other commercially important algae (Deepananda, 2011). Further, there is a big potential to develop sea cucumber farming in Sri Lankan coastal waters. Although there is a potential for farming several shellfish species including mussel, oyster, and sea cucumber, it is at very primitive stage in Sri Lanka. Sri Lanka used to have a prominent place in seaweed farming in the 1930s, but it is limited to a small-scale farming at present. Farming seaweed could be viable and profitable if initial challenges are overcome with the support of coastal communities.

At present, coastal aquaculture is being conducted using either floating cages, net enclosures, earthen ponds, or constant water recirculating systems (Jayasinghe, 2019). There is a small scale production of a few other species (Table 8.3), namely, groupers, sea bass, milkfish, oysters, and crabs are also being cultured on different scales in Sri Lankan coastal waters. However, Jayasinghe (2019) identifies that multi-trophic and a mixed culture of seaweed and shellfish farming systems works more effectively to ensure the environmental safety and sustainability of mariculture.

8.2.2 RENEWABLE OCEAN ENERGY

8.2.2.1 Wind Power

The Sri Lankan Government has made a policy decision to increase nonconventional renewable Energy generation (NCRE) by up to 20% in the near future. The contribution of NCRE in 2020 was approximately 12%. Accordingly, the Hambantota and Mannar areas, amongst other places, have been identified as suitable locations to establish large-scale wind farms in Sri Lanka. In 2019 the Ceylon Electricity Board (CEB) awarded an EPC contract to a Danish wind turbine producer to build the country's first large-scale wind farm in Sri Lanka, along the southern coast of Mannar Island which comprises of 30 wind turbines, each having capacity of 3.45 MW. The electricity generating capacity of this wind farm amounts to 103.5 MW. The project is the result of a long-term effort of the country to harness the potential of wind energy on a large scales, exploiting the major monsoonal wind systems across the country. Out of this project, CEB anticipates generating significant amounts of electricity from wind power.

Funding requirements for the project, which is approximately 141 million US$ has been facilitated by the Asian Development Bank. Once completed, it is expected to generate more than 380 million units of clean electricity annually. The estimated cost of generation amounts to 5.0 US Cents/kWh, which is cheap compared to any other source of production and reduces the use of fossil fuels for electricity generation. This is expected to reduce the emission of CO_2 by 285,000 tonnes per year.

8.2.2.2 Wave Energy

There is a huge potential in the ocean for harnessing energy from ocean waves and currents mainly affected by southwest and northeast monsoon patterns in the Indian Ocean. Power plants can be either coastal structures or floating types. They may either use the kinetic energy of ocean waves or the decreasing temperature with depth.

The southwest monsoon that mainly affects the western coast, is the strongest compared to the northeast monsoon that affects the northern and eastern coastal belts. The southern coastal belt, especially from Yala to Thirukkovil is influenced by both monsoons and is suitable for the installation of power plants operating on wave energy (Chamara and Vithana, 2018). Power plants operating on energy from ocean waves can be used as part of the Sri Lankan government's policy to increase nonconventional renewable energy generation. Hence, wave energy remains as a future potential energy source in Sri Lanka.

8.2.3 SEAPORTS AND SHIPPING

Sri Lanka owns four major ports and three minor ports, which are managed by the Sri Lanka Ports Authority (SLPA), a state arm constituted under the Sri Lanka Ports Authority Act, No. 51 of 1979

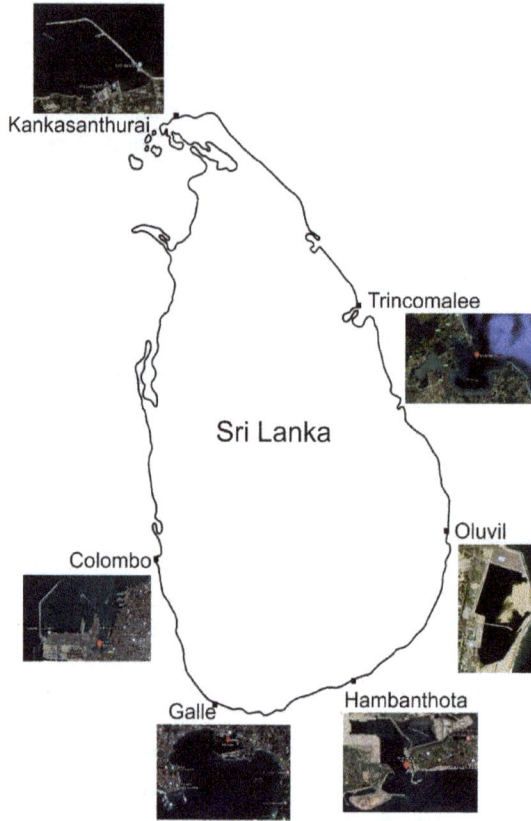

FIGURE 8.7   Seaports of the Island.

and subsequent amendments by Act No. 7 of 1984 and Act No. 35 of 1984. The SLPA is operated by revenue generated by itself and the main ports that it operates are Colombo, Galle, Hambantota and Trincomalee, while minor ports are Oluvil, Kankesanthurai and Point Pedro (Figure 8.7).

Total Vessels arriving in Sri Lanka in 2019 were 4697, out of which 4198 arrived in Colombo harbor (Annual report of the CBSL, 2020). Remaining vessels arrived at Galle, Trincomalee and Hambantota harbors and the numbers were 43, 142, and 314, respectively. Figure 8.8 shows the container handling, transshipment volume and ship arrivals for five consecutive years from 2016 to 2020.

According to the Central Bank of Sri Lanka (annual report 2020), the financial performance of the SLPA indicates progress despite the decline in port operations. Further, the CBSL annual report states that the total revenue of the SLPA declined by 4.5% to Rs. 38.9 billion, while operational cost decreased by 20.6% to Rs. 29.7 billion. In the year 2020 the SLPA recorded a profit before taxes of Rs. 20.3 billion compared to 16.2 billion recorded in 2019. Container handling, transshipment volume and ship arrivals in 2020 have been shown in Figure 8.9.

Colombo Port is primarily a container port, which handled about five million twenty-foot equivalent units (TEU) of containerized cargo in 2020. Colombo Port handles cargo originating from and destined for Europe, East and South Asia, the Persian Gulf, and East Africa. Originally the port had a harbor area of 184 hectares. Then in 2008, about 285 hectares were added through the construction of the South Harbor area, which accommodates deep water berths and the latest generation of mainline vessels. At present, the two-way channel of the harbor has an initial depth of 20 m, and a width of 570 m. Colombo Port can be identified as an emerging maritime hub in the Indian Ocean region.

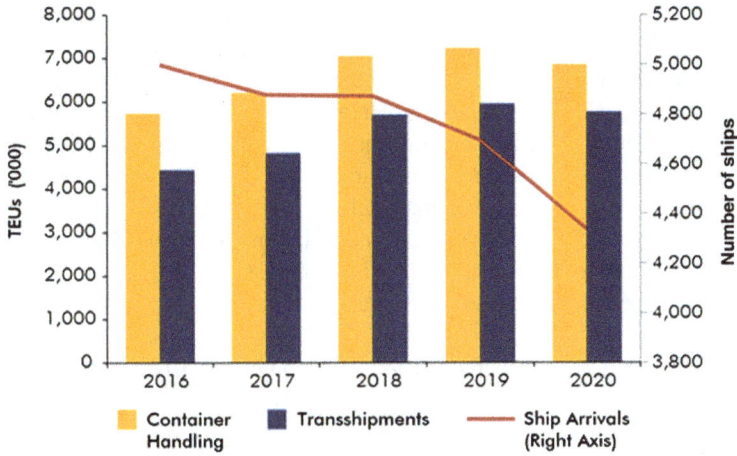

FIGURE 8.8 Container handling, transhipment volume and ship arrivals (annual report of the CBSL, 2020).

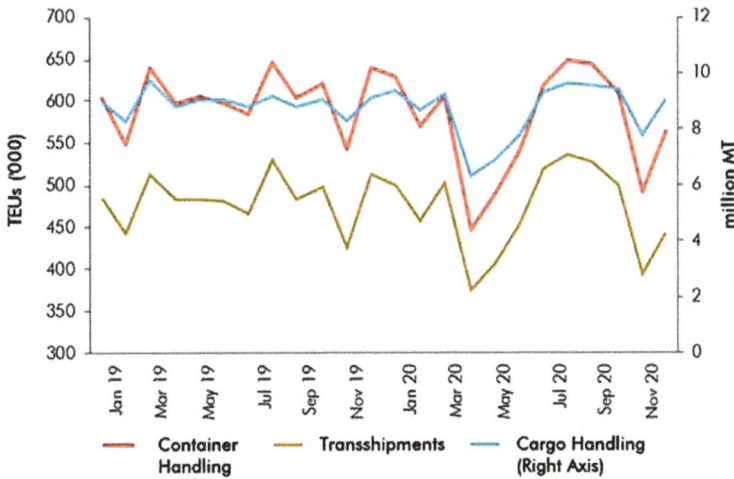

FIGURE 8.9 Performance of port activities (annual report of the CBSL, 2020).

Galle Port, which is a natural harbor, is located on the south-western coast of the island. It is one of the oldest harbors in the region, and has been in operation since the pre-Christian era, but gained prominence after the 12th century A.D. Galle port has been an important service station for boats and ships traveling between Europe and Asia. It remained as the main harbor in Sri Lanka until the construction of Colombo harbor in 1873. Thereafter, much of international shipping was diverted to Colombo from Galle. Since then, Galle port became less important, yet it still handles some ships and boats. It is the only Sri Lankan port that provides facilities for pleasure yachts.

SLPA has a plan to further develop Galle port to handle the increasing demand for freight handling for Sri Lanka. SLPA has plans to construct a deep-water passenger vessel terminal and breakwaters, to increase the depth by the dredging of entrance channel and basin, and other auxiliary facilities at the port of Galle. In addition, upgrading to a fully-fledged yacht marina, which will be beneficial to the economic development of Sri Lanka.

Trincomalee Harbor, is located in Trincomalee, along the eastern coast of the island. It is known as the second-best natural harbor in the world. As far as the land and water areas are considered, it is approximately ten times the size of Colombo Port. Trincomalee has been identified as a potential center to accommodate bulk and break-bulk shipments and port-associated industrial activities. As a result, it can be expected that the Trincomalee harbor will further develop in future.

Hambantota International Port is located in Hambantota on the southern coast of Sri Lanka. It is in a very strategic location close to the major international shipping route between East Asia and the west. Annually, many ships sail along this maritime route passing Sri Lanka and this provides a good opportunity for providing services like fueling, water, vessel staff adjustments, in addition to the usual port operations. The port has been one of the major development projects undertaken by the Sri Lankan Government in recent times. The first phase of the port has been completed and phase II is in the construction process. Hambantota port is connected to Colombo by an expressway constructed parallel with the port construction project. The harbor is protected by 312 m and 988m long break waters. The port access channel is about 210 m wide and 17 m deep allowing it to facilitate vessels up to 100,000 DWT. In addition, Hambantota port will provide fuel bunkering facilities as well as acting as a point for re-exporting goods, especially vehicles. Cargo handling at the Hambantota port increased in 2020 compared to the previous year with the diversion of some vessels from the Colombo Port due to the COVID-19 pandemic. However, the total number of vehicles handled at the Hambantota Port declined by 14.3% to 35,291 with the policy measures taken by the Government to pause importation of motor vehicles in 2020 (Annual report of the CBSL, 2020).

Oluvil Port is on the southeast coast of Sri Lanka. The Government of Sri Lanka constructed this port with the aim of economic expansion in the Eastern region of the island. In terms of access to and from the south-eastern region for goods and cargo originating from the west coast this port can facilitate access effectively.

Kankesanthurai (KKS) Port is in the Jaffna peninsula in the north of Sri Lanka. It connects the north of the island with the rest of the country by sea. The government of Sri Lanka is planning to rehabilitate Kankesanthurai harbor, with the project aiming to repair and rehabilitate the existing breakwaters, piers, and roads including dredging and wreck removal and the construction of a new pier, which has growing economic potential.

In addition to the above-mentioned harbors operated by Sri Lanka Ports Authority, there are 19 fisheries harbors located on the coastal belt from east to west (Figure 8.10). Ten fisheries harbors have been proposed to be constructed mainly in the northern part of the country to tap fishery resources and to promote the economic activities in the northern part of the island, which was hard hit by the 30-year civil war. The construction of a fishery harbor is underway in Suduwela on the southern coast of the country.

Fishery harbors of the country are managed by the Ceylon Fishery Harbor Corporation (CFHC), established in 1972. The CFHC has the mandate to deliver fishery-harbor related services and provide the modern infrastructure and facilities for fishing communities. The corporation has come up with a development plan aimed at better managing itself while developing fishery related businesses in the country. Through this plan it is expected the Sri Lankan fishing industry would be competitive with other prominent fishing-oriented countries in the region. In addition, plans to develop marinas and marine-related sporting and leisure activities, such as diving, whale and dolphin watching.

8.2.4 MARINE AND COASTAL TOURISM

8.2.4.1 Coastal Tourism

Coastal tourism is another important economic sector expanding in the South Asian region. The coral reefs of the shallow coasts of Sri Lanka, Maldives, and India and the dry land mangroves of Pakistan which consist of vast numbers of floral and faunal species (Bari, 2017) are excellent tourist destinations. Coastal regions of South Asia are extraordinarily rich in ecological diversity that can attract tourists to the region, for its climates, biodiversity, clear water, and long sandy beaches.

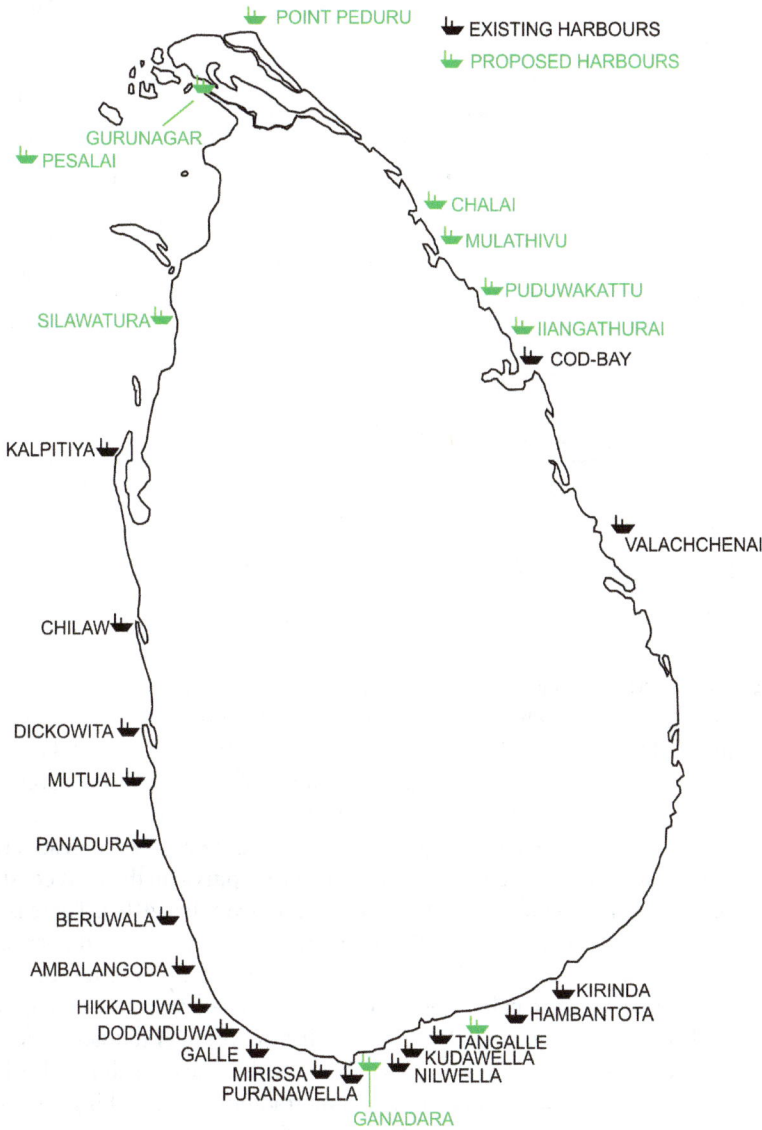

FIGURE 8.10 Locations of existing and proposed fishery harbours in Sri Lanka (Ceylon Fisheries Harbour Corporation 2021).

Sri Lanka is a tourist destination filled with numerous tourist attractions and it has been selected as the number one destination for the year 2019 by leading travel agency, the Lonely Planet. Tourist attractions can be classified as natural or anthropogenic. The coastal belt of approximately 1600 km consists of lagoons, estuaries, mangrove swamps, sandy beaches and sand dunes providing immense opportunities to develop coastal tourism. Among them, tropical sandy beaches are popular natural attractions both for local and foreign visitors. In particular to Sri Lanka, there is a narrow continental shelf, inhabited by abundant cetacean species, nine species of whale and two dolphin species (Buultjens et al. 2016; Ilangakoon, 2012; Sankapala et al. 2021) which is getting extra attention.

Approximately 82% of tourists visit Sri Lanka either for pleasure or for a vacation and their popular tourist destinations include coasts (SLTDA, 2019a; SLTDA, 2019b). Most tourists prefer

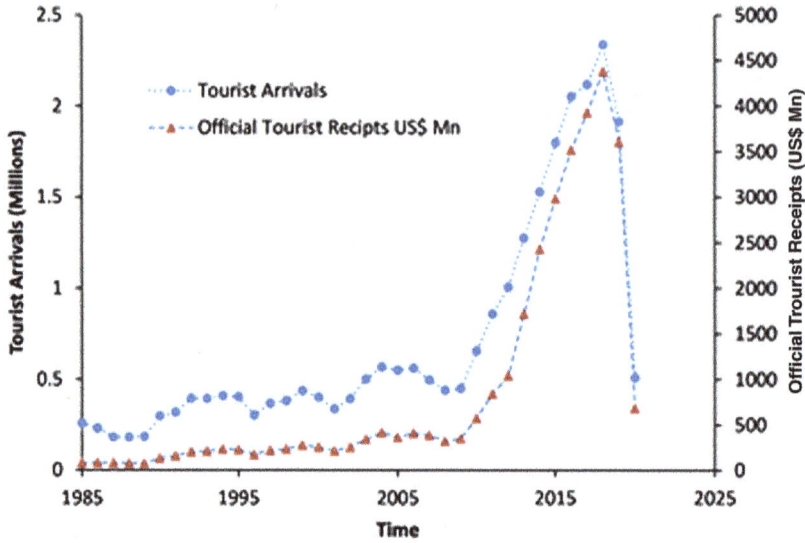

FIGURE 8.11 Tourist arrivals and official tourist receipts from 1985–2020 (Sri Lanka Tourism Development Authority (SLTDA)).

different water-based activities such as lying on the beach, swimming in the sea (72.2%), surfing (18.6%), snorkeling (15.8%), or whale/dolphin watching (9.8%) that are fundamental elements of the blue economy (SLTDA, 2019b). Out of total arrivals in 2019, there were 98,834 tourists who arrived via sea routes (SLTDA, 2019a) which is an indication of blue economic development potential. Whale watching is getting popular worldwide and generates approximately US$2 billion a year, supported by professional operators and guides (Nanayakkara, 2000). Three destinations, namely Mirissa in the south-west, Trincomalee in the north-east and Kalpitiya on the west coast of Sri Lanka offer opportunities for whale watching (Buultjens et al. 2016, De Silva 2019). There is a continuous increase in whale watching tourist records in the Southern coast, where this number in 2014, 2015, 2016, 2017, and 2018 were 76,465, 86,138, 146,150, 166,104 and 117,772 respectively (De Silva 2019). Besides whale watching, the calm beach and coastal waters around the country are preferred for surfing where Mustafa and Majeed (2021) observed that the younger tourists prefer Arugam Bay rather than other tourist spots due to the surfing nature of the beaches which is helped by favorable winds. Further, there is a significant tourist attraction on traditional stilt fishing which is unique to certain locations in the Southern coast of Sri Lanka. Furthermore, a recently opened underwater museum in Galle also provides an extension to develop coastal tourism.

The tourism industry has gradually increased over the last couple of years (Figure 8.11), and the sector is one out of the top six foreign exchange earners for Sri Lanka (Ranasinghe and Sugandhika, 2018). Among other foreign exchange earnings, the tourism industry remains at the fifth place in Sri Lanka, and this sector directly and indirectly influences economic growth via producing tourism revenue, foreign direct investments, new employment opportunities and also increasing the gross domestic product (Ranasinghe and Sugandhika, 2018). The GDP contributions of this sector in 2018 and 2019 were 4.3 and 4.9% respectively.

8.2.5 OFFSHORE HYDROCARBONS AND MINERALS

8.2.5.1 Hydrocarbons

Sri Lanka's energy need is met entirely by the import of crude oil and refined petroleum products. The following graph shows the country's oil bill for the last decade. Fuel oil is used for electricity

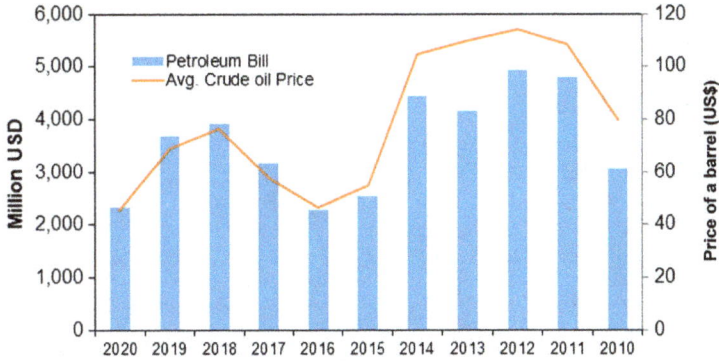

FIGURE 8.12   Sri Lanka's oil bill for the last decade (Central Bank of Sri Lanka).

generation in Sri Lanka, 27% of electricity generation in 2020 has come from fuel oil and 36 % from coal. The same year the Sri Lankan government spent around 217 US$ on coal imports.

Contribution of various sources for electricity generation in Sri Lanka in 2020 includes 25% from hydropower (excluding mini hydro), 12% from non-conventional renewable energy including mini hydropower, 36% from coal and 27% from fuel oil (Central Bank of Sri Lanka, 2020).

Accordingly, Sri Lanka's expenditure on crude oil, refined petroleum products and coal is a significant percentage compared to the country's Gross Domestic Product (GDP) (Figure 8.12). According to the Central Bank of Sri Lanka, Sri Lanka's GDP in 2020 was around 80.71 billion US$. Hence, the economic growth of Sri Lanka is highly dependent on world crude oil and coal prices.

Annually, the Sri Lankan Government spends a significant percentage of its foreign exchange earnings on the import of crude oil, natural gas, and other petroleum products. If a significant percentage of the country's oil and gas demand could be met by a domestic supply, it would be a huge relief to the country's balance of payment and the exchange rate. The best way to free Sri Lanka's dependency on imported petroleum and petroleum products is to explore and produce hydrocarbon resources in the country. Being a metamorphic terrain the hydrocarbon potential on Land Sri Lanka is almost zero. Sri Lanka's offshore basins with a higher petroleum potential include the Cauvery Basin and the Gulf of Mannar basin. The Sri Lankan sector of the Gulf of Mannar, which is commonly named the Mannar Basin, covers approximately 45,000 km² of the western offshore areas of the island (Figure 8.13). The Cauvery Basin is located between the south-eastern region of India and the north-western area of Sri Lanka. In fact, the Sri Lankan sector of the Cauvery Basin has no indication of any hydrocarbon deposits, oil and natural gas have been produced from the Indian sector of the Cauvery Basin.

Sri Lankan history of the upstream petroleum industry can be divided into two phases, the first phase lasted from 1957 to 1984 and the second phase began in 2001 and continues to date. During the first phase, numerous hydrocarbon explorations wells (Pesalai-1, 2 and 3 and, Palk Bay-1, Delft-1, and Pedro-1) were drilled in the Cauvery Basin and Pearl-1 was drilled in the northern part of the Mannar Basin in shallow water (Premarathne, 2015). However, all these wells failed to encounter any economically feasible hydrocarbon deposits and they were plugged and abandoned as dry wells. Further explorations were conducted during the 1984 to 2000 period. The second exploration phase in Sri Lanka began with the acquisition of 1050 km² two-dimensional (2D) marine seismic data over the Mannar Basin by a Norwegian geophysical company (TGS-NOPEC) in 2001. Based on the interpretation of this data, the Mannar basin was identified to have a significant hydrocarbon potential with a rift-basin structure and more than nine-kilometer-thick sediment succession on the crystalline basement (Baillie et al. 2002).

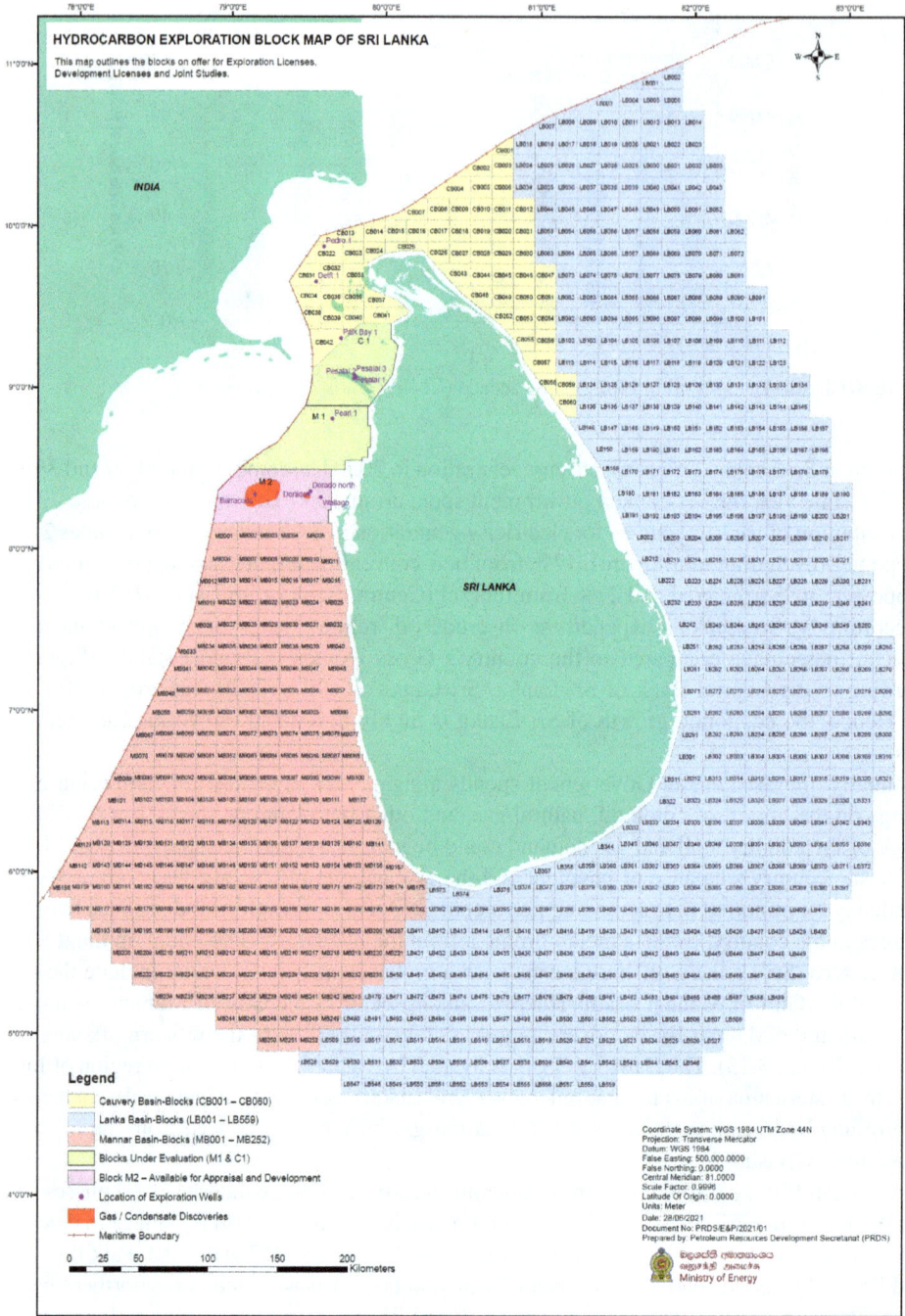

FIGURE 8.13   Hydrocarbon exploration block map of Sri Lanka (PDASL).

Accordingly, the Sri Lankan Government offered an exploration license to Cairn Lanka Private Limited (CLPL), a wholly owned subsidiary of Cairn India Limited in 2008 to explore petroleum in the M2 block identified in the northern part of the Mannar Basin. After preliminary geophysical studies, CLPL drilled three exploration wells in the northern part of the Mannar Basin in 2011. Two of these wells discovered natural gas making them the first hydrocarbon discovery of the island as well as in the whole of the Gulf of Mannar. These discoveries demonstrated the presence of an active petroleum system in the gulf. As a result, the potential of finding oil and natural gas on Sri Lanka's continental shelf is very high. The higher hydrocarbon potential of the Mannar Basin is further supported by the petroleum system modelling studies carried out recently.

In addition to the exploration block that discovered natural gas in the Mannar Basin, the Petroleum Development Authority, Sri Lanka (PDASL), which was earlier called as Petroleum Resources Development Secretariat would offer other exploration blocks in the Mannar Basin and Cauvery Basin to international oil companies (IOCs) at future licensing rounds. Meanwhile, research is being carried out to see the technical and fiscal viability of using natural gas discovered in the Mannar Basin as a fuel for generating electricity and public transportation. To facilitate this move, PDASL, regulator of Sri Lanka's upstream petroleum Industry, gazetted the National Policy on Natural Gas in October 2020.

8.2.5.2   Minerals

Mineral resources that are quantities of rocks enriched with one or more useful materials can be divided into two major categories: metallic and non-metallic. Sri Lanka is endowed with a variety of marine origin mineral resources.

Heavy minerals like ilmenite, rutile, monazite, garnet, zircon, magnetite, and so forth (mineral sand) tend to form onshore, on the beach and on the ocean floor. Prominent deposits occurring along the coastline of Indian Ocean rim countries like India, Sri Lanka, Indonesia, Malaysia, and Australia are among the largest marine non-living resources in the world. The main difference between normal sand and mineral sand is the higher density ($> 2.9\,g/cm^3$) and the hardness of the mineral sand. Most of the mineral sand deposits originate from high quality metamorphic and igneous rocks. Mineral sands produced by weathering and erosion of rocks reach the sea via streams and rivers. The coastal processes give rise to their deposition in the coastal zone. Sediments are modified by waves, tides, longshore currents, and wind, which are effective natural mechanisms for separating the mineral grains based on differences in their size and density.

The coastline of Sri Lanka is approximately 1600 km long and open to the Indian Ocean (Silva et al. 2013). Several mineral sand deposits occur along the coastal zone of Sri Lanka. These occurrences contain minerals such as ilmenite, rutile, garnet, zircon, or monazite. Most of these mineral sands are higher amphibolite to granulite facies metamorphic rocks and they mainly occur in the Highland Complex, this is one of the four lithotectonic units of crystalline rocks in Sri Lanka (Figure 8.14). The coastal line of the Pulmoddai areas in the northeast coast of Sri Lanka has a prominent ilmenite and rutile sand deposit, while a garnet sand exists in the southern coast of Sri Lanka from Hambantota to Yala.

The mineral sand deposit that occurs in the Pulmoddai area on the north-eastern coast of the island (Figure 8.17) has been known for a long time and has been exploited economically since 1918 (Wickremeratne, 1986). The deposit extends over an area of $3.6\,km^2$ on the coastal zone, which is about 6 km long and 600 m wide. It is one of the high-grade mineral sand deposits in the world, containing 60% to 70% of heavy mineral sand. The deposit is a collection of many valuable non-ferrous minerals, including ilmenite, rutile, zircon, monazite, and garnet. Among these mineral sands, the deposit contains around 65% of ilmenite, 10% rutile, 10% zircon, and 15% other minerals. It has been estimated that 12.5 million tons of mineral sand reserves are available in the Pulmoddai deposit.

Mining operations of the Pulmoddai deposit is undertaken by Lanka Mineral Sands Ltd, a company fully owned by the state. It is the successor to Ceylon Mineral Sands Corporation, which was founded in 1957 under the Industrial Corporation act. The Ceylon Mineral Sands Corporation was renamed Lanka Mineral Sands Ltd in 1992. The functions of the company are to manage mining, processing, and the exportation of heavy mineral beach sands. This enterprise is regulated by the Ministry of Industry and Commerce. The company is headquartered in Colombo, while

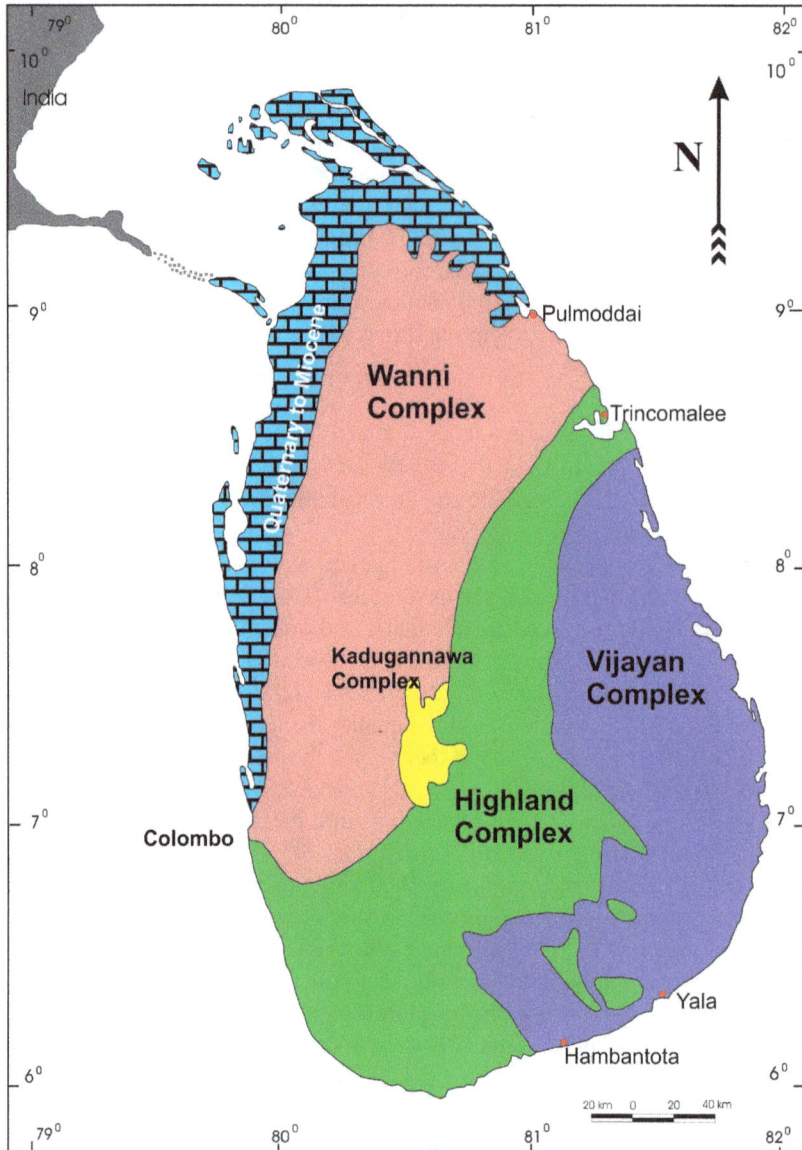

FIGURE 8.14   Outline geological map of Sri Lanka.

the processing plant is in Pulmoddai about 54 km north of Trincomalee (Figure 8.15). 99% of the annual production of mineral sand is exported to countries like Japan, UK, USA, India, Pakistan, China, and Eastern European countries. In 2019, the company exported 34,976 MT of ilmenite and 4,286 MT of rutile and earned 1.839 billion Sri Lankan Rupees (National Output, Expenditure, and Income, 2020). Ilmenite and rutile are mainly used to produce titanium dioxide (TiO_2). Ilmenite and rutile contain 53% and 95% titanium dioxide, respectively. Titanium dioxide is used in many paper, plastic, and ink industries.

In Sri Lanka, there are several garnet rich beach sand deposits. Among them, the most significant deposits in terms of mineral concentration and quantity occur along the south and southeast coast of Sri Lanka, especially in the Palatupana and Yala areas. These deposits consist of natural industrial quality garnets, which can be used as an important raw material in the manufacture of abrasive

FIGURE 8.15 Mineral sand deposit in Pulmoddai (Lankan Mineral Sands Ltd.).

powders and papers due to their abrasive qualities and low material cost. However, commercial mining of these deposits has not been carried out due to the negative environmental impacts that may arise during mining.

Common salt (sodium chloride) is available in seawater at a concentration of about 3.0% and hence constitutes more than 80% of the dissolved chemical elements in the seawater. Like some countries such as the USA, natural rock salt deposits do not occur in Sri Lanka. Therefore, salt in seawater is separated by evaporating water at salterns in the Hambantota and Puttalam areas. The Arid and windy climatic conditions in these areas are ideal for producing salt from seawater. The amount of salt produced in each year varies due to different climatic conditions. Annual demand for salt in Sri Lanka is 160,000 MT (120,000 MT for domestic and 40,000 MT for industrial use), while the combined output is 255,000 MT, which translated into an excess production of 95,000 MT. Salt is harvested largely in the southern, north-western, northern, and eastern parts of Sri Lanka where there's relatively less rainfall (Figure 8.16). The global industrial salt market size in 2018 was around 17 billion US$. Sri Lanka cannot compare itself with the global salt market. Owing to the insufficient infrastructure of the country between 10,000 and 40,000 metric tonnes of salt are imported each year. In 2020 the government spent nearly one billion rupees on salt importation. There are four major salt producers in the country, that are Lanka Salt Limited, Puttalam Salt Limited, Raigam Wayamba Salterns PLC and the government-owned Matai/Elephant Pass Salt Limited, and there are many other salterns run by private organizations. Forty percent of the country's salt requirement is supplied by the Hambantota Salt Company.

The United Nations (UN) has a red alert flag for Sri Lanka because of its high salt consumption, which could pose a huge potential for the prevalence of conditions such as high blood pressure. Therefore, steps should be taken to reduce salt consumption in Sri Lanka. Excessive salt could be exported or used as a by-product of other industries such as the manufacturing of caustic soda and chlorine.

FIGURE 8.16   Harvesting salts at Hambantota saltern.

Gypsum ($CaSO_4.2H_2O$) forms during the evaporation of seawater and can be produced as a by-product in the salterns at Hambanthota and Putlam. Gypsum is used for the manufacturing of wall-board, cement, plaster of Paris, in soil conditioning, and as a hardening retarder in Portland cement. In addition, gypsum can be used as a fertilizer.

The limestone rocks composed of calcium carbonate ($CaCO_3$), are formed extensively in the tropical to subtropical part of the world today because of precipitation by biological organisms ranging from molluscs to corals and plants. There is little exploitation of the modern limestone as it is forming in the oceans. North, north-eastern, and north-western coastal areas of Sri Lanka have thick limestone sequences deposited during the period 5-23 million years ago (Miocene age). During this period, the sea level was much higher than the present sea levels. As a result, a large part of the island's coastal areas was covered by the sea during the Miocene age. This gave rise to the limestone sequences in the north, north-eastern and north-western coastal areas of Sri Lanka. The thickness of this limestone in some areas reaches more than 900 m. Not only the northern part of the country, but also the south such as in the Akurala area, have inland coral deposits, which are thought to have formed due to high sea levels during the Miocene age.

Limestone is used as a raw material for manufacturing Portland cement, quicklime (calcium oxide), and slaked lime (calcium hydroxide). Crushed limestone is used as a soil conditioner to neutralize acidic soils. Limestone required for the Puttalam cement factory is mined at a quarry site located in the Aruwakkalu area in the Puttalam district (Figure 8.17).

Ocean basins are the ultimate depositional sites of sediments that are eroded from the land, while the beaches represent the largest residual deposits of sand (Figure 8.18). Even though beaches and near-shore sediments are locally extracted to be used in civil engineering constructions, they are generally regarded as too valuable as recreational areas and pristine ecosystems to allow mining of sand for construction purposes.

Being an island, the beach is extremely important to Sri Lanka in many ways. Some beaches such as Unawatuna, Nilaweli and Arugam Bay have become world famous tourist destinations.

FIGURE 8.17   Limestone quarrying at Aruwakkalu areas in the north-western province of Sri Lanka.

FIGURE 8.18   Sand dunes occur at Yala in the southeast coast of Sri Lanka.

Some beaches act as very important ecological niches for pristine flora and fauna. Though places with huge sand mounds can be identified in the island's coastal belts as possible sand mining sites, however such practices may not be suitable due to environmental concerns.

8.2.6 Marine Biotechnology, Research and Development

The vast biological and chemical diversity of marine ecosystems is a sink of chemicals that could be use in various human needs such as pharmaceuticals, nutritional supplements, cosmetics, food additives, pesticides, insecticides, agrochemicals, and biopolymers, to name but a few. The use of marine organisms for the production of traditional medicines has a long history in many parts of the world. Pharmaceutical and nutritional supplements from microalgal species, red algae, seaweeds, jellyfish, sponges, and other fish species have a great potential for developing cures for ailments such as diabetes, cancer, viral and bacterial infections, and allergy inflammations (Sarvanan, 2013). Similarly, many natural marine products derived from marine organisms have been reported and patented, leading to the evolving field of marine biotechnology, especially with the manufacture of pharmaceutical products (In economic terms, marine biotechnology has a big role to play as it has key important categories namely, food, energy, health, environment, and industries). The global market share of marine biotechnology was estimated to be US$ 4.8 billion in 2020 and it is expected to reach US$ 6.4 billion in 2025. Countries in the Indian Ocean need to identify the potential and devise a strategy for joint exploration to find sources and tap the marine organisms from the Indian Ocean that are of potential interest in marine biotechnology. Especially, islands such as Sri Lanka could be used as manufacturing centres for drugs from these marine organisms from the extensive costal zones.

8.3 ECONOMIC ANALYSIS

8.3.1 FISHERIES AND AQUACULTURE

8.3.1.1 Fisheries

The contribution of marine fisheries to the overall economy is very significant. Fish products are a vital source of animal protein supply to the population, contributing to 34.5% calorie intake, 50.0% protein intake, and 22.2% fat intake in 2019 (Ministry of Fisheries, 2020). In per capita terms, fish consumption was 16.6 kg in 2019 (Ministry of Fisheries, 2020). Moreover, the sector contributes added value to the national income, employment, and foreign exchange through exports.

The total value added from the fisheries sector, which includes marine and inland fisheries, has gone down to about 1% of GDP in 2019, contributing nominal value to GDP growth. The highest contribution of 2.8 per cent was in 1999 (see Figure 8.3 for long term trends). Sri Lanka produced 415,490 tonnes in 2019, about an eightfold increase from the 51,003 tonnes of 1960. The estimated instantaneous (at a point in time) growth rate, using a linear trend model for total marine production is 3.1% (or 7, 085 tonnes annual increase) between 1960 and 2019. Our econometric analysis shows that the number of fishing boats affects marine fishing production more than any other factor. Following the Cobb-Douglas production function, we estimated an Autoregressive Distributed Lag Model (ARDL), which shows a long-run cointegration (long-run relation) between the two variables. For example, if the number of the boats increases by one, then fish production goes up by 10.26 tonnes in the long run, while the figure for the short term is 4.1 tonnes. Production is also dependent on the number of active fishers (fishermen and women). In 2019, the country had 224,610 active fishers, 1.95 times higher than 115,014 in 2001. On average, if one fisher is added, the total production increases by 3.4 tonnes in the long term and it goes up by 1.3 tonnes in the short term.

The marine fishing sector generates livelihood for 185,390 fishing households while the sector provides direct employment to 224,610 fishers, both men and women, in 2019, which is a 3.6% instantaneous annual growth rate since 2001. The number of fishers in 2001 was 115,014.

The foreign exchange received from the export of fish and fishery products increased to Rs.53, 483 million in 2019 from Rs.854 million in 1990, contributing 1.5% of the country's total exports. The estimated instantaneous annual growth rate of export earning is 12% during 1990–2019. The highest contribution to export earnings of 2.6% was recorded in 2000 and 2009 and, since then, the figure has gone down continuously. It is disappointing to note that with great potential for production and export, Sri Lanka still imports a tremendous amount of fish and fishery products (mainly dried fish and canned fish), creating a negative balance of fish trade in some years (2015 and 2016). On average, Sri Lanka exports only about 5% of the total fish production quantity. One of the biggest challenges that Sri Lanka is facing today is to increase fish production, enabling the country to reduce imports of dried fish and canned fish and then similarly reduce foreign exchange.

8.4 CHALLENGES TO DEVELOPING THE BLUE ECONOMY IN THE REGION AND IN SRI LANKA

8.4.1 DRIVING FORCES IMPACTING THE BLUE ECONOMY

Population of the IOR countries are expected to increase drastically in the future and in consequence, food security and economic activities derived from marine resources would become more important. The Indian Ocean economy is highly diversified with small countries like Comoros and Madagascar, and densely populated countries like Indonesia and Malaysia belonging to this territory. Also, in terms of low income, from Mozambique and Tanzania to high income countries such as Singapore and Australia. According to the World Bank (World Bank Report, 2017; 2018) the combined GDP of the Indian Ocean economy was around 12% in the year 2018 while for South Asia it was 3.9%. If the current trend continued the figure would be 20% of world GDP by 2025 and by that time the proportion of the population who are in abject poverty will decrease by about 50% from its present level.

Many of the countries in the Indian Ocean are developing countries with many of the people relying on marine resources for their food security. An increasing rate of utilization of marine resources builds pressure on the marine systems that can lead to over exploitation, habitat degradation and pollution (Link, 2020, 2021, Larik et. al. 2017). With the current trends it is evident that sustainability of the marine environment needs strengthening of marine polices on marine resources. Among the future challenges of the Indian Ocean and the impact on the region of climate change such as increasing temperature, acidification, sea level rise, changes to distribution of marine species, extreme weather conditions, structural changes to communities as of resettlements, and declined economic productivity are some of the multidimensional challenges faced by the regions (Roy, 2019). With the continuous growth of the human population, need of a governance structure to preserve resources of the Indian Ocean, including areas beyond national territories are significant.

The blue economy captured the attention of all IORA member states at the 14th Ministerial Meeting held in Perth, Australia, in October 2014. Considering its wide range of resources, and growing interest in the blue economy, IORA Member States recognised the establishment of a common vision that would drive for balanced economic development in the Indian Ocean Rim Region. However, economic, and sustainable development concerns of the IOR countries are challenging as is the growing competition over marine resources among the IOR countries. It indicates a need for an enhanced regional framework for harnessing resources in a more substantiable manner. Action needed in this domain involves cooperation between all the stakeholders with an interest of the region. These are the private sector, non-governmental organizations, the scientific community, and local communities, as existing partnerships do not include all the relevant players.

Also, maintaining an effective monitoring control, enhancing surveillance capacity, increasing domain awareness, and information sharing will remain critical in ocean affairs, and efforts should continue to strengthen regional capacities, that is especially for those counties with large exclusive economic zones and weak enforcement capacities.

Non-traditional security threats, for instance, the impacts of climate change and environmental degradation, as well as unlawful activities such as drug smuggling, human trafficking, terror campaigns, and IUU fishing, continue to be critical issues, also many of such activities take place outside the jurisdiction of coastal states, and all are unfavourable to the development of blue economy. In these circumstances, it is crucial to guarantee an appropriate environmental management of the marine ecosystem, supported by an effective law and policy framework (Benzaken, 2017).

8.4.1.1 Fisheries and Aquaculture

The impact of environmental degradation and climate change on global fish stocks seems to be significant, but hard to predict (Rumley et al. 2009). In the case of fisheries, the current status of the Indian Ocean fish stocks indicates an impossibility for further expansion, as some of the species are overexploited (Chang et a. 2020). Over 800 million people around the IOR depends on fisheries as their main source of protein. The region is wealthy in seafood resources, in terms of abundance and varieties and is one of the prominent areas of global fishing. Commercial fisheries are equipped with modern fishing gear mainly targeting tuna or tuna-like species and controlled by distant, mostly foreign vessels that are from Europe and Asia. Artisanal fisheries can be seen throughout the region and mainly by coastal communities accessing the ocean and resources for local livelihoods and subsistence, but these are not managed properly.

In the Indian Ocean many of the living resources are not restricted to one particular area with the fish crossing borders of nearby countries, and sometimes not so nearby countries. Specifically, large pelagic fish species such as tuna and billfish roam over a large ocean space and cross through many boarders of countries inside and outside of the Indian Ocean. A significant number of species at higher trophic levels are exploited in the sea with human consumption itself utilizing 400 species from the seas. In comparison to the past, fisheries technologies are so developed that many marine species are threatened to extinction. Although the Indian Ocean Rim countries have their own laws, regulations, and policies to manage marine resources, specifically, fish stocks, those measures have been ineffective in most cases of illegal, unreported, and unregulated fishing (Stefan et al. 2010). The main four driving issues that can be found related to fisheries are:

i. a decrease in the availability of fish resources
ii. a shift in the species composition of catches
iii. the fraction of immature fish in the catch
iv. a decrease in the marine biodiversity
v. the loss of vulnerable and endangered species.

The rapid growth of human population during the last few decades has led to an overutilization of marine resources, largely to meet the increasing demand for food. To provide fish to meet the increasing demand needs a two to three-fold increase of fishing fleets all over the world and it is not certain whether our ocean can support this sustainably. In the case of the Indian Ocean, many indicators reflect that depleted state of the fisheries resources (Nisar, 2021; Polacheck, 2006):

i. the deflation of production in many of the Indian Ocean countries
ii. the shifting species composition of the catches
iii. the higher percentages of juvenile fish now being taken.
iv. an assessment of major fish stocks reported by Asia-pacific Fishery Commission and Indian Ocean Tuna Commission (IOTC)
v. a decrease in catch-rates in trawler surveys carried out over a long time period
vi. circumstantial evidence from fisherman.

8.4.1.2 Tourism

Among the various tourisms sectors in the world, coastal tourism is getting a higher place because of the varying climate along the coasts. This sector is an important GDP contributor and employment generator for IOR countries. For Sri Lanka, an all-time high of 4,381 million US$ was recorded in 2018 which is 4.9% direct contribution to GDP. Coastal tourism covers a variety of activities such as sunbathing, snorkeling, surfing, scuba diving on coral reefs, or whale watching. The potential of this sector is continuously increasing as more people choose to spend their holidays on this kind of leisure. Tourism related authorities of IOR countries need to strategically plan and design new sites and activities. Such coastal areas need to include other infrastructure facilities such as road networks for easy access, transport, accommodation, restaurants, hotels, and the like. In addition to that, museums, cultural centres, amusement parks can help to increase the added value (Valle et al. 2001).

8.4.1.3 Ports and Shipping

Ports and harbors are crucial components of a country for movement of goods within that country and also in and out of the country. The berthing of cruise ships and marine recreational actives are organised through such infrastructures. Auxiliary activities of shipping are hauling, use of tugs, anchorage at berths, repairs to ships, immigration, customs services, handling of consignments, warehousing of cargoes, and the like. The Indian Ocean is responsible for 11.2% of shipping in DWT terms (Dead Weight Tonnage). The increase in maritime activities and trade through the Indian Ocean is a blessing to IOR countries and an opportunity for countries to develop their ports and harbors to contribute to the shipping networks of the region. Sri Lanka located facing the east-west bound maritime route is one of the major trans-shipment hubs in Asia. The maritime logistic industry is growing but has not yet fully tapped many of the opportunities for trans-shipment and e-commerce. With the initiation of the transshipment trade in the 1980s, Colombo port has witnessed a threefold increase in container traffic, from 1.7 TEU in 2000 to 6 TEU in 2017 and is projected to touch 8 TEU in 2020. However, the quality of Colombo port stays at the regional average (4.5 scales of OECD), compared to the high-end ports in Singapore (6.7) and UAE (6.2), World Economic Forum Report, 2017). The coastal wetlands in Sri Lanka act as a first layer of defense to control floods (IUCN, 2007).

8.4.1.4 Energy Resources

Every nation requires energy for general use by its people, production by its industry and construction activities for its development. Considering the rising demand for energy and the limited supply of fossil fuels, environmental concerns due to extraction and emissions, shifting to alternative sources are unavoidable. A renewable energy source means energy that is extracted sustainably. The most popular renewable energy sources harnessed from oceans are offshore solar energy, offshore wind energy, ocean thermal energy conversion, tidal energy, and salinity gradient energy.

The quantum of energy that can be harnessed from waves depends on the location, wave height, and wave frequency, amongst other factors. According to the distinct wave resource parameters of Sri Lanka, such as mean significant wave height, mean energy period, and mean omni-directional wave power, there is a great potential to generate energy using ocean waves (Lokuliyana et al. 2020). The ocean energy available in this region falls into the category of a moderate wave energy source and its omni-directional wave power density ranges from 10 to 25 kW/m (Lokuliyana et al. 2020).

8.4.1.5 Climate Change

The average sea surface temperature of the ocean has increased by 0.6°C in the past 100 years. This has increased the heat content of the water, leading to number of direct physical and biochemical impacts that include thermal expansion, sea level rise, increased meltwater, reduced salinity, increased storm intensity, and greater stratification of the water column. Strong stratification due to elevated density differences affects nutrient availability and primary production due to changes in the abundance of algal populations, plankton, and fish populations as well as benthic organisms. This may restrict the annual productivity and abundance of individual species, together with the concomitant cascading changes and damage to food webs (Behrenfeld et al. 2006).

Climate change will have a range of effects, leading to reduction in food security and economic distress. In the Indian Ocean the threat that this will pose to humans will be food scarcity, disruption to livelihoods, and extreme meteorological events. Loss of coastal habitats, such as mangrove vegetations and wetlands through erosion and inundation will damage juvenile nursery grounds hampering reproduction and recruitment (Techera, 2018). Ocean acidification due to elevated CO_2 levels is a direct threat to marine organisms, especially reef-forming corals (Scleractinia) but also protozoans, molluscs, crustaceans, echinoderms, and some algae because of their calcium skeletons. Climate change aggravates current issues and contributes to the degradation of fisheries; therefore, it is an important factor when considering blue economy goals. Throughout many Indian Ocean countries, economic reliance on marine resources needs effective management (Techera, 2018). Climate changes also have significant impacts on erosion and inundation that could lead to destruction of coastal habitats, such as mangroves. This would not only be disastrous for the marine organisms, but also for the people that depend on those coastal resources.

8.4.1.6 Pollution

Dumping of industrial and radioactive wastes was banned in 1993 by amending the 1972 London Convention on 'Convention on the Prevention of Marine Pollution by Dumping of Wastes and Other Matter', there were more additions to this in 1996, combined with discharge, emission and losses from land and sea. Though the intentional dumping at sea of industrial and radioactive wastes was banned in 1993 by amendments to the 1972 London Convention, with more limitations added in 1996. The toxins already present, combined with the discharge, emission and losses from land and sea, and ubiquitous plastics pollution, remain major challenges. However, 60 to 80% of ocean pollutants originate from land-based activities that include fertilizers, pesticides, sewage, garbage, plastics, radioactive and other hazardous substances, and oil. For an example, eastern Indonesia which is a highly populated area contributes to more than 90% of the upper shore and strandline (Uneputtey and Evans, 1997). Also, fisheries are being severely affected by marine pollution thus destabilizing ecosystems, and negatively affecting the population of various marine species.

8.4.2 POLICIES AND REGULATIONS

Most of the Indian Ocean countries are parties to major international agreements pertaining to bio-diversity and protection of ecosystems, including, the United Nations Convention on the Law of the Sea, Convention on Biological Diversity, the Cartagena Protocol on Biosafety to the Convention on Biological Diversity. Except for the Maldives, all Bay of Bengal rim countries are parties to the Convention on Wetlands of International Importance especially as Waterfowl Habitat (Ramsar Convention); and the Convention on International Trade in Endangered Species of Wild Fauna and Flora. All Indian Ocean countries are also parties to key international instruments concerning the protection of the atmosphere such as the United Nations Framework Convention on Climate Change and the Kyoto Protocol to the United Nations Framework Convention on Climate.

International environmental law evolves with an integrated legal approach to management and solves environment related issues at regional and global levels. The negotiation of solutions, suggestions or declarations in important international forums often carries normative weight and facilitates their access to customary law. The 'soft approach' of a nonbinding framework or 'umbrella legislations' becomes a step on the way to 'hard law' in the form of conventions, agreements, treaties or protocols. State laws, bilateral agreements and national instruments play a complementary role in developing a sustainable blue economy. There is considerable variance across the Indian Ocean countries regarding their respective political, legislative, and administrative structures. They have all enacted legislation that seeks to regulate activities in the Indian Ocean to ensure that the marine living resources and critical habitats of the Indian Ocean are offered a certain level of protection (Table 8.4) while benefiting sustainably.

The existing legal and policy framework among the Indian Ocean countries dealing with coastal and marine resource management and the sustainable use of the Indian Ocean are, in general, comprehensive in their content and coverage. However, they are fragmented, sectoral in scope, and not effectively implemented. Maritime security of the IOR countries is an issue of common interest, regionally and extra-regionally as that impacts economic, environmental, energy, human, food, and national security. In the maritime domain, the coastal states need regional leadership to cooperate in the face of growing risks to maritime security (Cordner, 201).

Conservation, management, and development of all fisheries, including the capture, processing and trade of fish and fishery products, fishing procedures, aquaculture, fisheries research, and the integration of fisheries into coastal area are important fields to be controlled. The Code of Conduct is supported by the FAO Compliance Agreement and specific International Plans of Action which require the development and implementation of corresponding national plans of action. These binding and non-binding international instruments jointly provide the framework for the implementation of sustainable and responsible fishing practices and sound marine environmental management, including better management of fisheries, protection of migratory and threatened species, ecosystem and biodiversity protection and marine pollution prevention. Key regional fishery bodies relevant to the Indian Ocean are the Indian Ocean Tuna Commission (IOTC), the South Indian Ocean Fisheries Agreement (SIOFA), the Bay of Bengal Programme Inter-Governmental Organization (BOBP-IGO), the Regional Commission for Fisheries (RECOFI) and the Regional

TABLE 8.4
Status of Major Environmental Treaties of Some Indian Ocean Countries

Environmental treaties	Indonesia	Malaysia	Thailand	Myanmar	Bangladesh	Sir Lanka	India	Maldives
Law of the Sea Convention (LOSC)	✓	✓	✓	✓	✓	✓	✓	✓
Convention on Biological Diversity (CBD)	✓	✓	✓	✓	✓	✓	✓	✓
Convention on International Trade in Endangered Species (CITES)	✓	✓	✓	✓	✓	✓	✓	✓
Convention on Migratory Species (Bonn Convention)	✓	✓	✓	✓	✓	✓	✓	✓
Ramsar Convention on Wetlands of International Importance (Ramsar)	✓	✓	✓	✓	✓	✓	✓	✓
Stockholm Convention on Persistent Organic Pollutants	✓	✓	✓	✓	✓	✓	✓	✓
UN Framework Convention on Climate Change (FCCC) and Kyoto Protocol	✓	✓	✓	✓	✓	✓	✓	✓
Basel Convention on the Control of Tran boundary Movements of Hazardous Wastes and Their Disposal	✓	✓	✓	✓	✓	✓	✓	✓
UN Fish Stocks Agreement	✗	✗	✗	✗	✗	✓	✓	✓
FAO Compliance Agreement	✗	✓	✓	✓	✓	✓	✓	✓
Fund Convention	✗	✓	✗	✗	✗	✗	✓	✓
International Convention for the Prevention of Pollution from Ships (MARPOL 73/78)	✓ (Annex I-II)	✓ (Annex I, II, V)	✓ (Annex I and II)	✓ (Annex I andII)	✓ (Annex I - VI)	✓ (Annex I - V)	✓ (Annex I - V)	✓ (Annex I, II, V)

Organization for the Conservation of the Environment of the Red Sea and Gulf of Aden (PERSGA). First, the limited coverage of the Indian Ocean Tuna Commission (IOTC) and the Southern Indian Ocean Fisheries Commission excludes various species unprotected in different geographical areas, notably non-highly migratory, shared, and straddling fisheries resources in the high seas in the northern region of the Indian Ocean (Larik et al. 2017)

Only a few of the Indian Ocean countries have ratified the UN Fish Stocks Agreement and none have accepted the FAO Compliance Agreement. Only a few Indian Ocean countries have developed national action plans to implement the various FAO International Plans of Action, namely on capacity, seabirds, sharks and illegal, unreported, unregulated (IUU) fishing, and the like.

The majority of the IOR states have some fisheries regulation and management practices in operation, but these are not standardized, informed by limited data, and inconsistently enforced, due to capacity constraints (Techera, 2018). Analysis of the laws of individual Indian Ocean countries, in the context of achieving the objectives for the Indian Ocean reveals that many of the laws do not embody modern management concepts reflected in international instruments and sustainable marine environmental management practices. Major gaps exist in relation to ensuring the objectives of long-term sustainable use, the precautionary approach, and ecosystem approach to underpin governmental actions in the marine sector.

There is a complex set of laws and regulations for aquaculture, coastal zone management, environment, capture fisheries, forests, pollution, critical habitats and certain defined commercially attractive and/or endangered species. The domestic legal and administrative structures are largely sectoral, uncoordinated and need to be simplified, streamlined, and complement national and regional efforts in managing the Indian Ocean effectively. Other constraints noticeable are inadequate budgetary commitments, lack of community stakeholder consultation and empowerment.

Some legislations are to protect the Indian Ocean to some extent from the main categories of pollution, mainly in the form of controls on effluent discharges. Even if these controls are rigorously enforced, controlled discharges can still destroy an ecosystem if there are enough of them. The effluent control approaches do not consider the combine effects of pollutants in ecosystems, or even if ecosystems are already polluted, physically damaged, or otherwise stressed. A shortcoming of pollution-specific legislation in the region is the absence of a 'polluter-pays' rule and other fines severe enough to ensure that breaking the law is a serious economic cost of doing business. It needs to be more expensive to break the law than to comply with it.

In contrast to the governing systems of Indian Ocean Rim countries, all the governments are encouraging economic growth and development, including through exploitation of living resources. As a result, all the states have created impressive marine and freshwater production goals, which

---

TABLE 8.5
Some Marine-Related Legislations of Sri Lanka

Aquaculture (Monitoring of Residues) Regulations 2002, 2002.

Aquaculture Management (Disease Control) Regulations 2000, 2000.

National Institute of Fisheries and Nautical Engineering Act (No. 36 of 1999), 1999.

National Aquaculture Development Authority of Sri Lanka Act, No. 53 of 1998, 1998.

Fish Products (Export) Regulations, 1998.

Export and Import of Live Fish Regulations, 1998, 1998.

Fish Processing Establishments Regulations, 1998, 1998.

Aquaculture Management Regulations of 1996, 1996.

Fishing Operations Regulations of 1996, 1996.

Inland Fisheries Management Regulations of 1996, 1996.

Fisheries and Aquatic Resources Act 1996 (No. 2 of 1996), 1996.

Madel (Beach Seine) Fishing Regulations 1984, 1984.

National Aquatic Resources Research and Development Agency Act 1981 (No. 54 of 1981), 1981.

Foreign Fishing Boat Regulations, 1981, 1981.

Sri Lanka Ports Authority Act (No. 51 of 1979), 1979.

Inland Water Fishing Regulations, 1978, 1978.

Proclamation of the President delimiting the breadth of the maritime zones (unofficial title), 1977.

Spiny Lobster and Prawn (Shrimp) Regulations, 1973.

Fisheries Regulations, 1941, 1968.

in many cases do not consider the biological limits of the production of these renewable resources. Also, most countries have well-developed legislative systems and polices in the different sectors, but these policies are often not harmonized across sectors. Examples of these domestic legislations implemented for management of marine related resources, particularly for the fisheries sector are given in the Table 8.5.

8.5 CONCLUSION

This chapter discussed the contribution of the blue economy to the Sri Lankan economy, together with the identification of potentials and challenges. As an island nation, our investigation shows that the country is yet to reap the full possibilities of the blue economy. Geopolitically, Sri Lanka is positioned in a strategically imperative place in the Indian Ocean, one of the vital oceans in the 21st century. Sri Lanka's Ocean is home to essential Sea lines of communications (SLOCs) and maritime chokepoints, and about 100,000 commercial vessels are currently transiting across per year. 50% of world container transportation, 30% of bulk cargo transportation, and 36% of crude oil transportation take place in this area. Energy demand will likely rise by a significant amount over the next 20 years, increasing the importance of Sri Lanka's Ocean in global energy transportation.

The Indian subcontinent has around 20 important ports and 200 minor ports, including the ports of the East Asian nations. The draught of most ports is in the range of 6 m - 10 m, whilst some have up to 14 m, and 4 - 6 major ports can handle deep draught vessels catering up to 18 m. However, Sri Lankan ports have deep water, Colombo with 19 m, Hambantota with 17 m, and Trincomalee with 25 m. These facilities have more potential to contribute to the blue economy of Sri Lanka.

Who dominates Sri Lanka's or Indian Ocean, and shipping routes may dominate the world economy by controlling the flow of oil, close to 90% of global trade? Being the nearest neighbor of Sri Lanka, India will overtake Japan, UK, and Germany by 2050. India is to become World's 3rd largest economy by 2050. The other two largest economies in Asia, Japan, and China, heavily depend on trade traversing Sri Lanka's Ocean and the Strait of Malacca.

Achieving the full benefits of the blue economy in Sri Lanka depends on many factors. Many Indian fishers are crossing to Sri Lanka's Ocean for illegal fishing, and to address this grave issue, Sri Lanka needs a delicate diplomatic effort with India. How to engage with superpowers in the Ocean is a big challenge for the country. Sri Lanka aims to be a 'Wonder of Asia', achieving high economic growth and enhancing the wellbeing of the citizens in the years ahead. Sri Lanka already plans to create five hubs, namely, (a) naval and logistics, (b) knowledge, (c) aviation, (d) commercial, and (e) energy. All these planned hubs reinforce the additional development of the blue economy.

Sri Lanka needs very well-thought-out strategic plans and deals with three big countries in Asia (China, India, and Japan), the US, and other countries as friendly and non-aligned countries. Future actions or cooperation for the country involve the identification of areas and sectors of the blue economy in consultation with stakeholders at different levels: the stakeholder include: (a) the state (central and local governments), (b) markets, and (c) civil Society. The delicate balancing of stakeholders, including local fishers, should be prioritised and

REFERENCES

Aparna, R. 2019. *Blue Economy in the Indian Ocean: Governance Perspectives for Sustainable Development in the Region*. Observer Research Foundation Occasional Paper # 181, 38.

Athukoorala, A. S. H. Bhujel, R. C. Krakstad, J. O. and Matsuishi, T. F. 2021. Regional variation in fish species on the continental shelf of Sri Lanka. *Regional Studies in Marine Science* 44:101755. https://doi.org/10.1016/j.rsma.2021.101755

Baillie, P. W. Shaw, R. D. Liyanaarachchi, D. T. P. and Jayaratne, M. G. 2002. *A New Mesozoic Sedimentary Basin, Offshore Sri Lanka*, Proceedings of 64th Conference and Exhibition of European Association of Geoscientists and Engineers (EAGA). Florence, Italy, 9–12.

Bardi, U. 2010. Extracting minerals from seawater: an energy analysis. *Sustainability* 2(4):980–992. https://doi.org/10.3390/su2040980

Bari, A. 2017. Our oceans and the blue economy: opportunities and challenges. *Procedia Engineering* 194:5–11. https://doi.org/10.1016/j.proeng.2017.08.109

Batapola, N. Dushyantha, N. Ratnayake, N. Premasiri, R. Abeysinghe, B. Dissanayake, O. Rohitha, S. Ilankoon, S. and Dharmaratne, P. 2021. Rare earth element potential in the beach placers along the southwest coast of Sri Lanka. In: *2021 Moratuwa Engineering Research Conference (MERCon). IEEE*, pp. 415–420. https://doi.org/10.1109/MERCon52712.2021.9525678

Behrenfeld, M . J. O'Malley, R. T. Siegel, D. A. McClain, C. R. Sarmiento, J. L. et al. 2006. Climate-driven trends in contemporary ocean productivity. Nature 444:752–55. https://doi.org/10.1038/nature05317.

Benzaken, D. 2017. Blue economy in the Indian Ocean Region: status and opportunities. In S. Bateman, R. Gamage, and J. Chan (Eds), *ASEAN and the Indian Ocean: The Key Maritime Links* (RSIS Monograph No. 33), pp. 80–90. Singapore: S. Rajaratnam School of International Studies.

Berman, S. S. McLaren, J. W. and Willie, S. N. 1980. Simultaneous determination of five trace metals in sea water by inductively coupled plasma atomic emission spectrometry with ultrasonic nebulization. *Analytical Chemistry* 52(3):488–492. https://doi.org/10.1021/ac50053a025

Bohingamuwa, W. 2017. Ancient 'Mahātittha' (Māntai) in Sri Lanka: A Historical Biography. *Journal of the Royal Asiatic Society of Sri Lanka* 62(2):23–50.

Bopearachchi, O. 2002. Archaeological evidence on shipping communities of Sri Lanka. In Barnes. R and Parkin. D (Eds). *Ships and the Development of Maritime Technology in the Indian Ocean*, Taylor and Francis Group, London. pp. 92–127.

Buultjens, J. Ratnayke, I. and Gnanapala, A. 2016. Whale watching in Sri Lanka: perceptions of sustainability. *Tourism Management Perspectives* 18:125–133. https://doi.org/10.1016/j.tmp.2016.02.003

Ceylon Fisheries Harbor Corporation 2021. Locations of Fisheries Harbors of Sri Lanka. http://cfhc.gov.lk/Harbour_Main.php. Visited 28.08.2021.

Central Bank of Sri Lanka. 2020. Annual Report 2020 https://cbsl.gov.lk. Visited 28.08.2021.

Chamara, R. N. and Vithana, H. P. V. 2018. Wave energy resource assessment for the southern coast of Sri Lanka. *06th International Symposium on Advances in Civil and Environmental Engineering Practices for Sustainable Development*, 40–46.

Chang, Y. S, Kim, M. J. and Kim, S. M. 2020. Convergence Analysis on the Fish Stock Status Index for 131 Countries. *Gachon Center of Convergence Research Working paper 2021–02*. Available at SSRN: https://ssrn.com/abstract=3644145. Visited 28.08.2021

Chavez, F. P. and Toggweiler, J. R. 1995. Physical estimates of global new production: the upwelling contribution. In C. P. Summerhayes et al. (Eds), *Upwelling in the Ocean: Modeling Processes and Ancient Records*, pp. 313–320. New Jersey: John Wiley, Hoboken.

Cordner, L. 2011. Progressing maritime security cooperation in the Indian Ocean. *Naval War College Review* 64(4):68–88. http://jstor.org/stable/26397244. Visited 28.08.2021

Deepananda, K. H. M. A. 2011. Mariculture, present trends, and future prospects for Sri Lanka. In B. W. Åmo (Ed.), *Conditions for Entrepreneurship in Sri Lanka: A Handbook*. pp. 285–314. Shaker Verlag, Achen Germany

De Silva, M. I. U. 2019. *Estimating the Economic Values of Sea Whales Based on Coastal and Marine Tourism in Mirissa of Sri Lanka using Individual Travel Cost Method (Master Thesis)*. The Graduate School, Pukyong National University.

Doyle, T. and Seal, G. 2015. Indian Ocean futures: new partnerships, alliances, and academic diplomacy. *Journal of the Indian Ocean Region* 11(1):2–7. https://doi.org/10.1080/19480881.2015.1019994

FAO. 2020a. *FAO Yearbook: Fishery and Aquaculture Statistics*. Food and Agriculture Organization of the United Nations. Rome, Italy

FAO. 2020b. *The State of World Fisheries and Aquaculture 2020*. Sustainability in Action. Food and Agriculture Organization of the United Nations Rome, Italy p. 224.

Flothmann, S. von Kistowski, K. Dolan, E. Lee, E. Meere, F. and Album, G. 2010. Closing loopholes: getting illegal fishing under control. *Science* 328(5983):1235–1236. https://doi.org/10.1126/science.1190245

Gonaduwage, L. P. et al. 2021. Interannual variability of summertime eddy-induced heat transport in the Western South China Sea and its formation mechanism. Clim Dyn 57: 451–468. https://doi.org/10.1007/s00382-021-05719-7

Hall, K. R. 2009. Ports-of-trade, maritime diasporas, and networks of trade and cultural integration in the Bay of Bengal region of the Indian Ocean: c. 1300–1500. *Journal of the Economic and Social History of the Orient* 53:109–145. https://doi.org/10.1163/002249910X12573963244287

Hermes, J. C. G. et al. 2019. A sustained ocean observing system in the Indian Ocean for climate related scientific knowledge and societal needs. *Frontiers in Marine Science* 6(355). https://doi.org/10.3389/fmars.2019.00355

How-Cheong, C. and Amandakoon, H. 1992. *Status, Constraints and Potential of Mud Crab Fishery and Culture in Sri Lanka.* In Angel C.A (Ed.). MUD CRAB. Report of the seminar on the mud crab culture and trade held at Swat Thani, Thailand. Bay of Bengal Programme, Madras, India. pp.231

Ilangakoon, A. D. 2012. Cetacean diversity and mixed-species associations off southern Sri Lanka. In *Proceedings of the 7th International Symposium on SEASTAR2000 and Asian Bio-logging Science (The 11th SEASTAR2000 Workshop).* Graduate School of Informatics, Kyoto University, 23–28.

Indicopleustes, C. 2010. *The Christian Topography of Cosmas, an Egyptian Monk: Translated from the Greek, and Edited with Notes and Introduction.* Cambridge University Press. London

Indrapala, K. 1963. The Nainativu Tamil inscription of Parakramabahu-I. *University of Ceylon Review* XXI: 68–70.

IOTC, 2006. *Report of the Ninth Session of the Scientific Committee*, Victoria, Seychelles, pp. 1–120.

IUCN (International Union for Conservation of Nature). 2007. Counting coastal ecosystems as development assets. *Coastal Ecosystems* 5. https://portals.iucn.org/library/sites/library/files/documents/2006-076.pdf. Last visit 22.07.2022

Jayasinghe, J. M. P. K. 2019. Mariculture development in Sri Lanka: Ensuring sustainability. In S. S. Giri, S. M. Bokhtiar, S. K. Sahoo, B. N. Paul, and S. Mohanty (Eds), *Aquaculture of Commercially Important Finfishes in South Asia*, 178. SAARC Agriculture Centre, Dhaka

Kapetsky, J. M. Aguilar-Manjarrez, J. and Jenness, J. 2013. A global assessment of offshore mariculture potential from a spatial perspective: FAO Fisheries and aquaculture technical paper-549. *Food and Agiriculture Organization of the United Nations*, Rome

Kariyawasam, I. 2016. Seaweed mariculture: a potential multi-million-dollar industry. SATH SAMUDURA Healthy Oceans Healthy Planet ISSN 2279–3208. *Mar. Environ. Prot. Auth.* Sri Lanka 2016:44–49.

Kathijotes, N. 2013. Keynote: Blue economy-environmental and behavioural aspects towards sustainable coastal development. *Procedia-Social and Behavioral Sciences* 101:7–13. https://doi.org/10.1016/j.sbspro.2013.07.173

Katupotha, J. 2019. Pearl fishery industry in Sri Lanka: a review. *Wildlanka* 7:33–49.

Keen, M. R. Schwarz, A.M. and Wini-Simeon, L. 2018. Towards defining the blue economy: practical lessons from Pacific Ocean governance. *Marine Policy* 88:333–341. https://doi.org/10.1016/j.marpol.2017.03.002

Kiribamune, S. 1991. The role of the port city of Mahatittha (Manthota) in the trade networks of the Indian Ocean. *Sri Lanka Journal of Humanities* (17 and 18) 171-192

Larik, J. E. Daniëls, L. Oosterom, J. de Ruiter, L. Smit, L. Vermeij, A. and van Vliet, V. 2017. *Blue Growth and Sustainable Development in Indian Ocean Governance.* The Hague: The Hague Institute for Global Justice.

Link, J. S. Watson, R. A. Pranovi, F. Libralato. S. 2020. Comparative production of fisheries yields and ecosystem overfishing in African Large Marine Ecosystems Environmental Development. *Environmental Development 36* , 100529. https://doi.org/10.1016/j.envdev.2020.100529

Link, J. S. 2021. Evidence of ecosystem overfishing in U.S. large marine ecosystems. *ICES Journal of Marine Science*, fsab185, 78(9) 3176–3201. https://doi.org/10.1093/icesjms/fsab185.

Lokuliyana, R. L. K. Folley, M. Gunawardane, S. D. G. S. P. and Wickramanayake, P. N. 2020. Sri Lankan wave energy resource assessment and characterisation based on IEC standards. *Renewable Energy* 162:1255–1272. https://doi.org/10.1016/j.renene.2020.08.005.

Magazinovic, R. S. Nicholson, B. C. Mulcahy, D. E. and Davey, D. E. 2004. Bromide levels in natural waters: its relationship to levels of both chloride and total dissolved solids and the implications for water treatment. *Chemosphere* 57(4):329–335. https://doi.org/10.1016/j.chemosphere.2004.04.056.

Ministry of Fisheries. 2020. *Annual Report*, Ministry of Fisheries. Colombo.

Murphy, T. 2004. *Pliny the Elder's Natural History: The Empire in the Encyclopedia.* Oxford University Press Inc. New York. pp. 233

Mustafa, M. and Majeed, A. 2021. Impact of seasonal variation on travel and tourism sector: a study of the post-civil unrest in Arugam Bay in Sri Lanka. *Journal of Asian Finance, Economics and Business* 8:431–442.

Nanayakkara, R. P. 2000. To flourish, Sri Lanka's whale-watching industry must operate responsibly (commentary). *Monographbay*. Available at https://news.mongabay.com/2020/08/to-flourish-sri-lankas-whale-watching-industry-must-operate-responsibly-commentary/. Last vist at 21.07.2021

National Output, Expenditure, and Income. 2020. *Central Bank of Sri Lanka*, 45.

Nisar, U. Ali, R. Mu, Y. and Sun, Y. 2021. Assessing five major exploited tuna species in India (Eastern and Western Indian Ocean) using the Monte Carlo Method (CMSY) and the Bayesian Schaefer Model (BSM). *Sustainability* 13(16):8868. https://doi.org/10.3390/su13168868

Pernetta, J. C. (Ed). 1993. Marine Protected Area Needs in the South Asian Seas Region. Volume 5: Sri Lanka. *A Marine Conservation and Development Report*. Gland, Switzerland: IUCN, *vii* pp. 67.

Polacheck, T. 2006. Tuna longline catch rates in the Indian Ocean: did industrial fishing result in a 90% rapid decline in the abundance of large predatory species? *Marine Policy* 30(5):470–482. https://doi.org//10.1016/j.marpol.2005.06.016.

Prasada, D. V. P. 2020. Assessing the potential for closure rules in the Sri Lankan sea cucumber fishery: empirical models of practices and preferences of fishers. *Marine Policy* 120:104130. https://doi.org/10.1016/j.marpol.2020.104130

Premarathne, U. 2015. Petroleum potential of the Cauvery Basin, Sri Lanka: a review. *Journal of Geological Society of Sri Lanka* 17: 41–52.

Ranasinghe, R. and Sugandhika, M. 2018. The contribution of tourism income for the economic growth of Sri Lanka. *Journal of Management and Tourism Research* 1:67–84.

Roy, A. 2019. *Blue Economy in the Indian Ocean: Governance Perspectives for Sustainable Development in the Region*, ORF Occasional. New Delhi: Observer Research Foundation, 181.

Rumley, D. et al. 2009. Fisheries exploitation in the Indian Ocean region. In Rumley et al. (Eds), *Fisheries Exploitation in the Indian Ocean: Threats and Opportunities*, Singapore: ISEAS Publishing. pp. 1–17.

Sankapala, D. M. R. Thilakarathne, E. P. D. N. Lin, W. Thilakanayaka, V. Kumarasinghe, C. P. Liu, M. Lin, M. and Li, S. 2021. Cetacean occurrence and diversity in whale-watching waters off Mirissa, Southern Sri Lanka. *Integrative Zoology* 16:462–476. https://doi.org/10.1111/1749-4877.12540

Saravanan, A. 2013). Patenting Trends in Marine Biodiversity: Issues and Challenges. *Pharma Utility* 7(4): CphI 2013 Special Issue. https://doi.org//10.2139/ssrn.2390067

Silva, E. I. L. Katupotha, J. Amarasinghe, O. and Ariyarathne, R. 2013. *Lagoons of Sri Lanka: From the Origins to the Present*. International Water Management Institute (IWMI), Sri Lanka. p. 122. https://.doi.org//10.5337/2013.215.

SLTDA. 2019a. *Annual Statistical Report*. Sri Lanka Tourism Development Authority, Colombo, Sri Lanka.

SLTDA. 2019b. *Survey of Departing Foreign Tourists from Sri Lanka in 2018/2019*. Sri Lanka Tourism Development Authority, Colombo, Sri Lanka.2021

Subasinghe, H. C. S. Ratnayake, A. S. and Sameera, K. A. G. 2021. State-of-the-art and perspectives in the heavy mineral industry of Sri Lanka. *Miner Econ* 34:427–439. https://doi.org/10.1007/s13563-021-00274-3

Sudharmawathie, J. 2017. Foreign trade relations in Sri Lanka in the ancient period: with special reference to the period from 6th century BC to 16th century AD. *Humanities and Social Sciences Review* 7:191–200.

Sumalia, U. et al. 2021. Financing a sustainable ocean economy. *Nat. Commun.* 12:3259. https://doi.org//10.1038/s41467-021-23168-y

Tacon, A. G. J. and Metian, M. 2008. Global overview on the use of fish meal and fish oil in industrially compounded aquafeeds: trends and future prospects, *Aquaculture* 285:146–156. https://doi.org//10.1016/j.aquaculture.2008.08.015.

Techera, E. J. 2018. Supporting blue economy agenda: fisheries, food security and climate change in the Indian Ocean. *J. Indian Ocean Reg.* 14:7–27. https://doi.org/10.1080/19480881.2017.1420579

Timothy Doyle. 2018. Blue economy and the Indian Ocean Rim. *Journal of the Indian Ocean Region* 14(1):1–6, https://doi.org// 10.1080/19480881.2018.1421450.

Uneputty, P. and Evans, S. M. 1997. The impact of plastic debris on the biota of tidal flats in Ambon Bay (Eastern Indonesia). *Mar. Environ. Res.* 44: 233–242. https://doi.org//10.1016/S0141-1136(97)00002-0.

United Nations. 2012. *The Future We Want*. Outcome document of the United Nations Conference on Sustainable Development.

United Nations. 2017. *The First Global Integrated Marine Assessment: World Ocean Assessment 1*. Cambridge: Cambridge University Press. https://cbsl.gov.lk/sites/default/files/cbslweb_documents/publications/ess_2021_national_output_expenditure_and_income_e.pdf. Visited 18.08.2021.

Valle, P. O. D. Guerreiro, M. Mendes, J. and Silva, J. A. 2001. The cultural offer as a tourist product in coastal destinations: The case of Algarve, Portugal. *Tourism and Hospitality Research.* 11(4), 233–247. https://doi.org/10.1177/1467358411420623

Wenhai, L. et al. 2019. Successful Blue Economy Examples with an Emphasis on International Perspectives. *Frontiers in Marine Science* 6 :261. https://doi.org//10.3389/fmars.2019.00261.

Wickremeratne , W. S. 1986 . Preliminary studies on the offshore occurrences of monazite bearing heavy mineral places, southwestern Sri Lanka . Marine Geology 72: 1 – 9. https://doi.org/10.1016/0025-3227(86)90095-2

World Bank Annual Report. 2017. Washington, DC: World Bank. https://doi.org10.1596/978-1-4648-1119-7.

World Bank Annual Report. 2018. Washington, DC: World Bank

World Economic Forum Report. 2017. *The Global Competitiveness Report.* www.weforum.org/reports/the-glo bal competitiveness-report-2017–2018. Visited July 2018.

Zeng, L. et al. 2021. A Decade of Eastern Tropical Indian Ocean Observation Network (TIOON). *Bulletin of the American Meteorological Society* 102(10): 1–54. https://doi.org/10.1175/BAMS-D-19-0234.1

9 Marine Ecosystem Services
SDGs Targets, Achievement, and Linkages with a Blue Economy Perception

Md. Nazrul Islam,[1]* Sahanaj Tamanna,[2]
S. M. Rashedul Islam,[1] and Md. Shahriar Islam[1]
[1]Department of Geography and Environment, Jahangirnagar University, Savar, Dhaka, Bangladesh
[2] Bangladesh Environmental Modeling Alliance (BEMA), Non-Profit Research and Training Organization, Mirpur, Dhaka, Bangladesh
*Corresponding author: Md. Nazrul Islam. E-mail: nazrul_geo@juniv.edu

CONTENTS

DOI: 10.1201/9781003184287-9

9.1 INTRODUCTION

Marine ecosystems can be defined as the interaction of plants, animals, and the marine environment (Atkins et al. 2011; Liquete et al. 2013). The term 'marine' refers to anything related to or produced by the sea or ocean. The word refers to the Earth's salty oceans, and it's also known as a salt water ecosystem (Adrianov, 2004; Kaiser et al. 2011; Martin, 2019). The marine ecosystem, in general, refers to the oceans, seas, and other salt water environments as a whole; but, deeper examination reveals that it can be separated into smaller, unique ecosystems (Fuhrman et al. 2015; Barbier, 2017; Xu et al. 2021). Salt marshes, estuaries, the ocean bottom, the open ocean, the intertidal zones, coral reefs, lagoons, and mangroves are examples of marine ecosystems (Barbier et al. 2011). Healthy marine ecosystems are important for society since they provide services including food security, feed for livestock and raw materials for medicines (Wilson and Verlis, 2017). Also, it has provided building materials from coral rock and sand, and natural defenses against hazards such as coastal erosion and inundation (Larsen et al. 2016; Smail and Hasson, 2021). The scientific evidence shows that the marine ecosystem includes: marshes, tidal zones, estuaries, the mangrove forest, lagoons, sea grass beds, the sea floor, and coral reefs (Gullström et al. 2002; Martin et al. 2020). Just like every other ecosystem in the world, the aquatic ecosystems rely on each other for maintaining a balanced marine ecosystem (Klain and Chan, 2012; Pendleton et al. 2016). The marine ecosystems are important to the world, because without them, the marine life would not have and protection from predators, which could eventually make the marine life go extinct (Lele et al. 2013; Michael et al. 2018). Although there is some disagreement, several types of marine ecosystems are largely agreed on: estuaries, salt marshes, mangrove forests, coral reefs, the open ocean, and the deep-sea ocean. Recreation, tourism and water transport are familiar services provided by many marine ecosystems (Palumbi et al. 2009; Lange and Jiddawi, 2009; Lillebø et al. 2017).

Some unique estuarine, coastal, and marine habitats are also important stores of genetic material and have educational and scientific research value as well (Beaumont et al. 2008; Martin et al. 2020). Figure 9.1 shows the major maritime activities and marine ecosystem services that generate the blue economy. But ecosystem services are not limited to the terrestrial space (Hattam et al. 2015; Buonocore et al. 2021). On the contrary, the ocean plays a major role in climate regulation. Many studies have shown that the seas absorb almost a third of the carbon dioxide emitted annually.

Moreover, marine and coastal ecosystems are home to numerous plant and animal species, which all produce various useful services for humans (Armstrong et al. 2012; Barbier, 2017). For instance, mangroves help retain friable, or crumbly soil on the coast, and therefore help prevent coastal erosion. Because the friability of soil refers to its crumbly texture, which is somewhere between sand and clay, not so fine and grainy like sand, or so thick and mushy like clay. Friable soil is organically rich soil. Friable soil is crumbly, and very fertile (Konar and Ding, 2020). It may be able to hold up to 15 times its weight in water like a sponge, preventing erosion, and storing these water deposits like a reservoir. They are also natural barriers to water currents, and as such constitute a favored habitat for the birth and development of many species of fish. Mangroves therefore help maintain the available fish stock (Seidensticker and Hai, 1983; Ajonina, 2008). Whale faeces, on the other hand, contains high quantities of iron, which is an essential nutrient for photosynthesis. The level of iron present in the ocean has a direct impact on the development of phytoplankton, a key component of carbon storage (Lovelock, 2008).

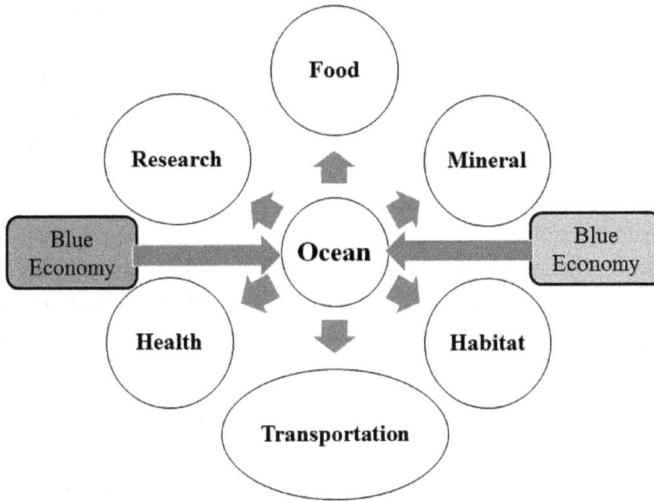

FIGURE 9.1 Major maritime activities and marine ecosystem services to generate the blue economy.

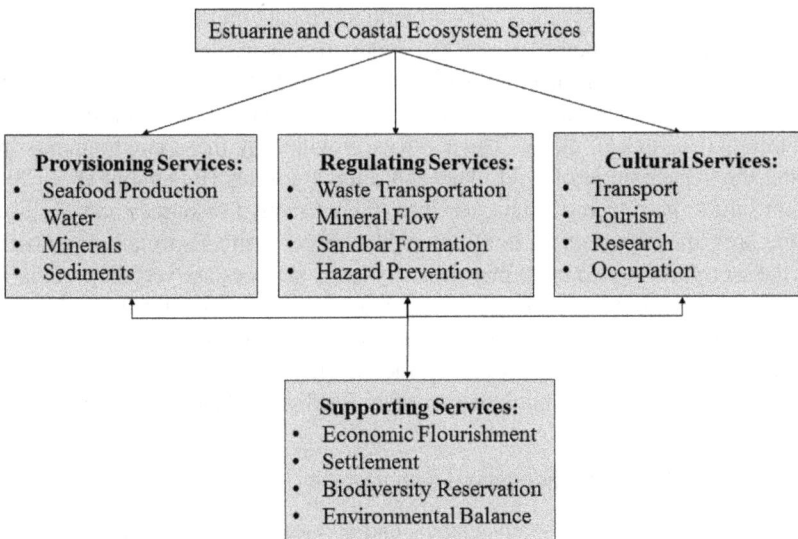

FIGURE 9.2 Marine and coastal ecosystems produce various services, including: provisioning services: fisheries, building materials and supporting services, etc.

9.2 MARINE ECOSYSTEM SERVICES AND THE BLUE ECONOMY PERSPECTIVE

The Millennium Ecosystem Assessment (MEA) defined four types of ecosystem services (Hrabanski, 2017). Marine and coastal ecosystems provide various services, including: provisioning services: fisheries, building materials; supporting services: life-cycle maintenance for both fauna and local, element and nutrient cycling; regulating services (Figure 9.2) (Atkins et al. 2011; IPPC, 2019; Lubchenco et al. 2020), carbon sequestration and storage, erosion prevention, waste-water treatment, moderation of extreme events and they also provide cultural services: tourism, recreational, aesthetic, and spiritual benefits (Van Jaarsveld et al. 2005; Atkins et al. 2011; Lau et al. 2020). Fish and shellfish for food (Lee et al. 2013), seawater for drinking, raw materials (such as algae, salt, sand, and gravel), amber and other bio-items for decoration, space for delivering energy from offshore wind farms, ship traffic routes, cultural services such as recreational opportunities for local visitors and tourists, and biodiversity are all provided by marine ecosystems to the benefit of society (Ashie et al. 1996; Guldner et al. 2007; Vincent, 2012; Zanoli et al. 2015).

Other services provided by the maritime environment include maintenance and regulatory services, such as nutrient and hazardous material control, carbon sequestration, and climate regulation (Ajonina, 2008; Ross et al. 2019; IPBES, 2019). The economic value of intermediate services is enormous, despite the fact that they are more difficult to price and evaluate than other more tangible services (Devassy and Nair, 1987; Beaumont et al. 2008). A few examples of final services that have a direct impact on human wellbeing are food, raw materials, energy distribution, and recreational activities (Liquete et al. 2013). Maintenance and regulatory services, on the other hand, are examples of intermediate services that have an impact on wellbeing indirectly (Hasler, 2016).

9.3 CLASSIFICATION OF MARINE ECOSYSTEM SERVICES

Components of marine ecosystem services benefit human health and provide beneficial commodities and services (Summers et al. 2012; Abrantes et al. 2015). These services contribute to the economic system's growth. Because of the way marine ecosystems are structured and operate, they are able to provide ecological production of ecosystem services (Benilli et al. 2008; Guerry et al. 2012; Edward, 2017). Some of these commodities, services, and cultural advantages have a direct influence on human wellbeing, whilst others have an indirect impact on human wellbeing by sustaining or maintaining significant economic assets and production activities (Ayompe et al. 2021). The following Table 9.1 summarizes some of the services supplied by marine ecosystems. They include a range of products and services, as well as cultural and other advantages. Marine habitats generate a variety of things (Nash et al. 2020). The services provided by these environments include fish harvesting and the conservation of wild plant and animal species (Beaumont et al. 2008; Obura, 20220). Among these marine ecosystem services, the extractive resources and the uses of large bodies of water are quite important for developing the blue economy. Especially for most of the poor countries in the world, these two main marine ecosystems services are very important (Buonocore et al. 2020). The primary functions of marine life ecosystems often include both. Commercial and 'informal' gatherings on a modest scale help to support the livelihoods of indigenous peoples, for example, via fishing, hunting, and fuel wood extraction (Biswas et al. 2014; Barbier, 2017). These services can help to build a strong economic structure for Small Island Developing States (SIDS).

9.3.1 ESTUARINE AND COASTAL ECOSYSTEM SERVICES

Estuaries give a variety of resources, advantages, and services to us. Recreational activities, scientific research, and aesthetic pleasure can all be found in estuaries (Martin et al. 2020; Stuchtey et al. (2020).). Estuaries are a unique natural resource that must be carefully managed for the mutual benefit of all those who use them and rely on them (Friedman, 1977; Martin, 2019). Thousands of birds, animal, fish, and other wildlife species rely on estuarine ecosystems to survive, breed, and feed (Friedman, 1977; MoFa (2014). Moreover, many marine organisms, including the majority of commercially important fish species, rely on estuaries at some stage during their lifecycle (Lenanton and Potter, 1987). Estuaries are good places for migratory birds to stop and recharge throughout their lengthy trips since they are biologically productive.

Estuaries are known as the 'nurseries of the sea' because many kinds of fish and wildlife rely on their protected breeding grounds in the sheltered waters of estuaries (Abrantes et al. 2015). Estuaries are beneficial to the economy. Estuaries have significant commercial value, and their resources support tourism, fishing, and leisure activities (Barbier, 2012; Brown et al. 2018). Estuaries maintain significant public infrastructure by serving as harbors and ports, which are essential for shipping and transportation (Potter and Hyndes, 1999; Ross et al. 2019). The natural beauty and wealth of

TABLE 9.1
Marine Ecosystems Services and Enhancing Components of Blue Economy

Type of Activity/ goods	Marine Ecosystem Services	Enhancing Components of Blue Economy		
		Industry	Divers of Growth	Source
Harvest of Living Resources	Seafood	Fisheries	Food Security	Barbier (2017)
		Aquaculture	Demand for Protein	Hrabanski (2017)
	Marne biotechnology	Pharmaceuticals, chemicals	Healthcare and industry	Ross et al. (2019)
Non Living Resources	Minerals	Seafood Mining	Demand for minerals	Lele et al. (2013)
	Energy	Oil and Gas	Demand for Alternative Energy Sources	Hattam et al. (2015)
	Freshwater	Desalination	Demand for Freshwater	Barbier (2017)
Commerce and Trade in and around the Oceans	Transport and trade	Shipping	Growth in seaborne trade. International regulations	Ross et al. (2019)
		Port infrastructure and services		Hrabanski (2017)
	Tourism and recreation	Tourism	Growth of Global tourism	Barbier (2017)
		Coastal Development	Coastal urbanization Domestic regulations	Wang and Tang (2010)
Response to ocean Health challenges	Ocean monitoring and surveillance	Technology and RandO	RandD in ocean technologies	Lele et al. (2013)
	carbon Sequestration	Blue carbon	Growth in Coastal and ocean protection and conservation activities	Hrabanski (2017)
Water	Coastal Protection	Habitat protection and restoration		
	Waste Disposal/ Pollution Protection	Assimilation of nutrients and wastes		Ross et al. (2019)
	Storm Protection		Habitat regeneration	Liquete et al. (2013)
	Breeding and nursery habitats	Bequest for future generations		

Source: After modified from Barbier, 2017; Ross et al. 2019 and above others mentioned sources.

estuaries are vital to the economies of many coastal communities. The livelihoods of persons who live and work in estuary basins are jeopardized when natural resources are depleted (Lenanton and Potter, 1987; Fox, 2012).

Coastal ecosystems contain unique and recognizable landforms like beaches, cliffs, and coral reefs, all of which are extremely vulnerable to disturbances (Venkataraman and Raghunathan, 2015). Coastal locations are home to some of the world's most diverse ecosystems. In the Indian Ocean, the biodiversity hotspot of the Andaman and Nicobar Islands can be found (Rajalingam et al. 2016). As many distinct species of marine life, as in a tropical rainforest, can be found on the coral reefs there. Sadly, coastal erosion is destroying habitat and causing lasting damage to coastal communities.

Marine communities in coastal habitats are very biodiverse and vary based on topography and climate. Bays, estuaries, mangroves, salt marshes, and wetlands are a few examples of coastal habitats (Ranjitkumar et al. 2020). Because of the abundance of food and the fact that they are protected from some of the perils of the deep ocean, many fish, turtles, and migratory birds breed in coastal areas. Disturbances have a high impact on these populations (Tittensor et al. 2009). The availability of sunlight and a consistent supply of nutrients allows creatures that live in coastal environments to thrive. The shallow seas of coastal habitats allow sunlight to reach the ocean floor, where nutrients from dead species can gather and nourish life (Chen et al. 2020). Because sunlight can only reach a depth of 50 to 100 m in the ocean, this type of nourishing environment does not exist in the deep ocean, where nutrients sink to depths where most living species cannot survive (Duarte et al. 2008).

9.3.2 ROCKY SHORE ECOSYSTEM SERVICES

A marine ecosystem is any that occurs in or near salt water, which means that marine ecosystems can be found all over the world, from sandy beaches to the deepest parts of the ocean (Hall, 2001; Lenanton and Potter, 1987). An example of a marine ecosystem is a coral reef, with its associated marine life including fish and sea turtles and the rocks and sand found in the area. The ocean covers 71% of the planet, so marine ecosystems make up most of the Earth (Roberts et al. 2003). Along a rocky shore, you might find rock cliffs, boulders, small and large rocks, and tidal pools (puddles of water that can contain a surprising array of marine life) (Kandasamy et al. 2022). You will also find the intertidal zone, which is the area between low and high tide (Wilson and Verlis, 2017). Rocky shores can be extreme places for marine animals and plants to live. At low tide, marine animals have an increased threat of predation (Peterson, 1991; Lee et al. 2013). There may be pounding waves and lots of wind action, in addition to the rising and falling of the tides. Together, this activity has the ability to affect water availability, temperature, and salinity (Powell and Domack, 2002). Specific types of marine life vary with location, but in general, some types of marine life you'll find at the rocky shore include (Ewing, 2012): marine algae, lichens, birds, invertebrates such as crabs, lobsters, sea stars, urchins, mussels, barnacles, snails, limpets, sea squirts (tunicates), and sea anemones, plus fish, seals and sea lions, and so forth. (Jackson and Sala, 2001).

9.3.3 SANDY BEACH ECOSYSTEM SERVICES

Sandy beaches may seem lifeless compared to other ecosystems, at least when it comes to marine life. However, these ecosystems have a surprising amount of biodiversity (Griffiths and Waller, 2016). Like the rocky shore, animals in a sandy beach ecosystem have had to adapt to the constantly changing environment. Marine life in a sandy beach ecosystem may burrow in the sand or need to move quickly out of reach of the waves (Adrianov, 2004; Palumbi et al. 2009). They must contend with tides, wave action, and water currents, all of which may sweep marine animals off the beach (Lee et al. 2020). This activity can also move sand and rocks to different locations (Shields et al. 2011). Within a sandy beach ecosystem, you'll also find an intertidal zone, although the landscape isn't as dramatic as that of the rocky shore (Palumbi et al. 2009). Sand generally is pushed onto the beach during summer months, and pulled off the beach in the winter months, making the beach more gravelly and rocky at those times (Mauzey et al. 1968). Tidal pools may be left behind when the ocean recedes at low tide (Wright and Mella, 1963). Marine life that is occasional to be found on sandy beaches includes: sea turtles, who might nest on the beach, or pinnipeds, such as seals and sea lions, who might rest on the beach (Eckert et al. 1986). Regular sandy beach inhabitants are: algae, plankton, invertebrates (such as amphipods, isopods, sand dollars, crabs, clams, worms, snails, flies, and plankton), fish (including rays, skates, sharks, and flounder) can be found in shallow waters along the beach, birds (such as plover, sanderlings, willet, godwits, herons, gulls, terns, ruddy turnstones, and curlews, and the like. (Fertl and Fulling, 2007; Schlacher et al. 2014).

9.3.4 Salt Marsh Ecosystem Services

Salt marshes are areas that flood at high tide and are composed of salt-tolerant plants and animals (Roman et al. 1984). Salt marshes are important in many ways. They provide habitats for marine life, birds and migratory birds, they're important nursery areas for fish and invertebrates (Power et al. 1988). They also protect the rest of the coastline by buffering wave action and absorbing water during high tides and storms. Examples of salt marsh marine ecosystem life are algae, plankton, birds, fish, and occasionally marine mammals, such as dolphins and seals (Gibson et al. 2011).

9.3.5 Marine Deep Sea Ecosystem Services

The term 'deep sea' refers to parts of the ocean that are over 1,000 m (3,281 ft). One challenge for marine life in this ecosystem is light and many animals have adapted so that they can see in low light conditions, or don't need to see at all (Jones et al. 2020). Another challenge is pressure. Many deep-sea animals have soft bodies so they aren't crushed under the high pressure that is found at extreme depths (Palumbi et al. 2009). The deepest parts of the ocean are more than 30,000 ft deep, so we're still learning about the types of marine life that live there (Marcus, 2004; Da Ros et al. 2019). Here are some examples of general types of marine life that inhabit these ecosystems: invertebrates such as crabs, worms, jellyfish, squid, and octopus, corals, and fish, such as anglerfish and some sharks, and marine mammals, including some types of deep-diving marine mammals, such as sperm whales and elephant seals (Balcombe, 2016).

9.3.6 Open Sea Ecosystem Services

The open ocean is by international convention the largest transboundary space, with ocean areas beyond national jurisdictions covering about half of the surface of planet Earth (ocean areas under national jurisdiction cover a further 20%), under the ultimate governance of the UN General Assembly (Lee et al. 2013). Governance of the open ocean is mediated mainly through global international treaties based on particular themes (climate change, fisheries, pollution, biodiversity), as well as some regional conventions.

9.3.7 Services from Marine Mangrove Ecosystem

Mangrove trees are salt-tolerant plant species with roots that dangle into the water. Forests of these plants provide shelter for a variety of marine life and are important nursery areas for young marine animals (Webber et al. 2014). These ecosystems are generally found in warmer areas between the latitudes of 32° N and 38° S (Smardon, 2009). Species that may be found in mangrove ecosystems include algae, birds, invertebrates such as crabs, shrimp, oysters, tunicates, sponges, snails, and insects, fish, dolphins, manatees, reptiles such as sea turtles, land turtles, alligators, crocodiles, caimans, snakes, and lizards (MoFa, 2014; Sheaves, 2017).

9.3.8 Services from Coral Reef Ecosystems

Healthy coral reef ecosystems are filled with an amazing amount of diversity, including hard and soft corals, invertebrates of many sizes, and even large animals, such as sharks and dolphins (Friedlander et al. 2008; Tittensor et al. 2009). The reef-builders are the hard (stony) corals. The basic part of a reef is the skeleton of the coral, which is made of limestone (calcium carbonate) and supports tiny organisms called polyps (Tittensor et al. 2009). Eventually, the polyps die, leaving the skeleton behind. There are several major marine species that might live there. The invertebrates

FIGURE 9.3 Mangrove Sundarbans Forest at the Coast of Northern Bay of Bengal in Bangladesh (BSS, 2017; https://farmsandfarmer24.com/worlds-largest-mangrove-forest-sundarban/).

include hundreds of species of coral, sponges, crabs, shrimp, lobsters, anemones, worms, bryozoans, sea stars, urchins, nudibranchs, octopuses, squid, and snails. Vertebrate inhabitants might include a wide variety of fish, sea turtles, and marine mammals (such as seals and dolphins) (Shields et al. 2011).

9.3.9 Services from Mangrove Forests

Mangrove forests make up one of the most productive and biologically diverse ecosystems on the planet (Figure 9.3) (Smardon, 2009). They grow in a variety of depths of salt water, their roots sticking up out of the mud, with fish, crustaceans and a host of other species living between tree trunks. These mangroves are a unique, diverse and complex ecosystem providing lot of natural resources (Shamsuzzaman et al. 2017). They act as a coastal stabilizer as well as tempest defenders. Almost 1.5 million people are protected by these mangroves against the effect of water interruption, water contamination and siltation

9.3.10 Services from Marine Biological Habitats

Many ecosystems in coastal and open ocean areas benefit from the marine environment, which supports biodiversity (Sink et al. 2012). Marine ecosystems produce various resources that benefit humanity, and the seas and coasts are vital to the survival and well-being of a large section of the world's population (Raudsepp-Hearne et al. 2010). As challenges such as changes in land-use, overfishing, climate change, the invasion of non-native species, and other anthropogenic activities have an impact on biodiversity, the demand on marine ecosystems and the resources they provide is increasing (Atkins et al. 2011; Liquete et al. 2013). Species must evolve and adapt to changing environmental conditions as the world around them changes. Food security, animal feed, raw materials for medications, building materials from coral rock and sand, and natural defenses against threats such as coastal erosion and inundation are all benefits provided by healthy marine ecosystems (Figure 9.4) (Hossain et al. 2021). Policymakers can respond to, protect, and manage threatened ecosystems by using ocean observations to monitor biodiversity and estimate species distribution and density in marine environments (Atkins et al. 2011). Marine ecosystems are inextricably related to global climate, and scientists can better predict the impact of climate change on biodiversity and human populations by monitoring and researching them (US EPA, 2015).

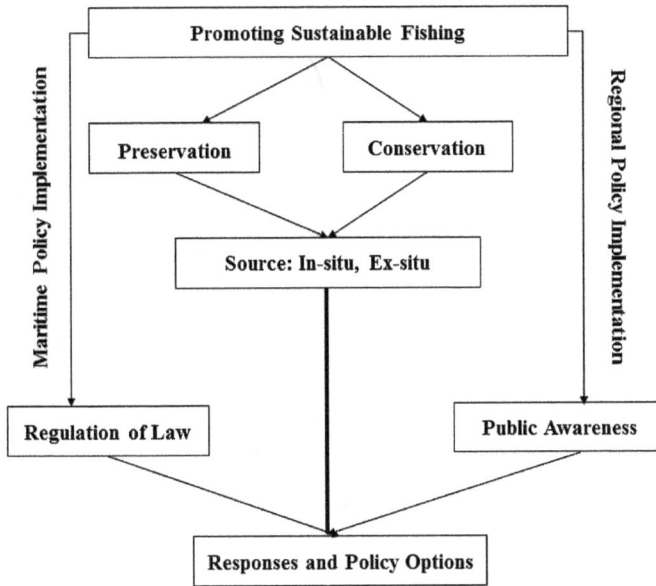

FIGURE 9.4 Promoting responsible and sustainable fisheries resources to enhancing the global blue economy.

9.4 OTHER MAJOR MARINE ECOSYSTEM SERVICES

9.4.1 FISHERIES RESOURCES

Most of the coastal people of these areas live their lives with fishing providing the primary source of food and nutrition to the teeming millions (Kurien, 2007). The coastal areas are over-crowded by people, resulting in an average density of 80 people/km^2, which is almost double the world's average (Moriarty, 2007). At least 41,500 large and small fishing craft are plying this area, directly operated by approximately 3.7 million fishermen (Manson et al. 2005). These fishermen catch 6 million tonnes of fish per year worth US$ 4 billion which is 7% of the world's total catch (Harper et al. 2020). Around 185 million people live in the coastal areas of this region, with almost 3% of these directly involved with fishing. This is approximately 4.4 million (BOBLME, 2015) people are directly involved with fishing in the Bay of Bengal regions.

International fisheries play a critical part in ensuring food and economic security around the world. Still, many fish populations are not managed according to scientific advice and best practices, a reality that is taking a toll on many important marine species (Salomidi et al. 2012; Wilson, 2016). Sustainably managing the world's shared fish stocks, and securing the well-being of the habitats and ecosystems on which they depend, requires effective governance at every level (Shamsuzzaman et al. 2017). Policies should be implemented by the best available science and technology of the countries and each government will have to commitments to enforce these compliance (Wilson, 2016). According to the suggestions of the United Nations Food and Agriculture Organization (FAO) Committee on Fisheries (COFI) it is important to ensure the policymakers of the small island countries take steps to reduce overfishing, protect sharks, combat illegal fishing, and safeguard marine habitats and ecosystems. Here are 10 ways they can have a positive impact (Wilson, 2016):

9.4.2 SERVICES FROM SEAWEEDS AND REEFS

Seaweeds and other reefs, along with tropical rain forests, are one of the most ecologically diverse ecosystems on the planet (Guldner et al. 2007). They are made up of sponges, crabs, mollusks, fish, sea turtles, sharks, dolphins, and much more, in addition to hard and soft corals (Shamsuzzaman

FIGURE 9.5 Scenarios of marine ecosystem services from coastal tourisms at Cox's Bazar Sea beach in Bangladesh (Islam et al. 2017; https://c.mi.com/thread-1665290-1-0.html).

et al. 2017; Sarker et al. 2018). The amount and diversity of creatures on a reef are mostly determined by competition for resources such as food, space, and sunlight. Countless other plants, animals, and organisms are dependent on and associated with each component of a coral reef. Less than 1% of the earth's surface is covered by these reefs providing shelter to 25 % of all marine fish species (Michael et al. 2018). Approximately 0.7 million people are protected by coral reefs. The Bay of Bengal large marine ecosystem region which represents almost 8% of the world's coral reef with an area of 22,602 km² (Hossain, 2001; Shamsuzzaman et al. 2017), the Maldives, the Andaman and the Nicobar Islands, Myanmar and the Andaman Sea area of Thailand are the main areas where corals reefs are seen (Figure 9.5).

Moreover, coral reefs supply a wide range of ecosystem services and goods, such as:

- **Coastal Protection**
  Coral reefs acts as natural wave barriers that protect coastal communities and beaches from storm damage, help to control erosion, and help is sand formation.
- **Food and Fishing**
  Seaweeds and coral reefs sustain the fish and shellfish population that provide protein for 1 billion people. Reefs are nurseries for many commercially valuable species.
- **Medicine**
  Seaweeds and coral reef species provide new medical compounds and technologies to treat serious diseases. More than half of all new cancer drug research is focusing on marine organisms.
- **Tourism and Recreation**
  Coral reefs attract millions of tourists every year, bringing important income to coral reef communities. Some countries derive more than half of their gross national product from coral reef industries.

9.4.3 SERVICES FROM OCEAN ENERGY

Marine renewable energy (MRE), also known as ocean energy or marine and hydrokinetic energy (MHK), refers to the various ways of generating electricity from the world's oceans, seas, and rivers (Duarte et al. 2008). Movement of water occurs naturally in these bodies of water in the form of waves, tides, and currents (Cassotta, 2021). All coastal areas consistently experience two high and low tides over a period of slightly greater than 24 hours. For those tidal differences to be harnessed into electricity, the difference between the high and low tides must be at least 5 meters, or more than 16 feet. There are only about 40 sites on the earth with tidal ranges of this magnitude (Liguo et al. 2022). The Pacific Northwest and the Atlantic Northeast are locations with tidal potential in the United States.

9.4.4 Services from Coastal and Marine Tourism

Tourism is one of the main sources of earnings for islands like the Maldives, Andaman and the Nicobar Islands. There are a lot of fantastic beaches in these countries which are the obvious tourist destinations and which provide a lot of revenue to these counties (Hall, 2001). Over the past decades the number of tourists increased dramatically in this region providing ass important source of wealth. But these extra tourists are creating extra pressure on the marine ecosystems by discharging garbage, plastics, and the like, in these zones, leading to environmental degradation (Islam, 2003; Shamsuzzaman et al. 2017). Cox's bazaar, the longest continuous sea beach of the planet is heavily polluted by plastics left by tourists. Because tourism is one of the pillars of various economies, governments very often ignore the pollution aspect (Islam et al. 2017).

9.4.5 Services from Shipping and Transportation

Ships have to pass through BOBLME (Bay of Bengal Large Marine Ecosystem) before entering the Malacca straits which remain busy throughout the year as more than 50,000 merchant ships cross this strait which connects the two big oceans, the Indian Ocean and the Pacific Ocean. These ships carry almost 25% of the goods traded by the world. Major ports like Colombo, Mumbai, Chittagong, and Bangkok have become the shipping hub of the world.

9.5 VALUATION OF DEEP-SEA GOODS AND SERVICES

The deep sea, defined as water and seafloor areas below 200 m, accounts for 90% of the biosphere, although humans understood little about it until recently (Van et al. 2005; Folkersen et al. 2018). Because little or no light penetrates these depths, it was supposed that deep-sea life was limited (Nellemann and Corcoran, 2009; George, 2013). However, life is rich and diverse, and the deep marine habitats differ greatly from the uniform and desert-like plains recorded by pioneer voyages (Salomidi et al. 2012; Folkersen et al. 2018). Though deep water is still a largely unexplored region, national and international research initiatives are rapidly expanding our understanding (Van Dover et al. 2014). Deep-sea ecosystems are critical to global biogeochemical cycles, on which much terrestrial life and human civilization rely (Colwell and D'Hondt, 2013). However, given its remoteness, the deep sea is far less clean and undisturbed than one might imagine. Deep-sea ecosystems are under increasing pressures and threats. And our understanding of the deep-water is still patchy and incomplete (Roberts et al. 2006).

There are still significant gaps in our understanding of the presence and operation of deep-sea ecosystems, as well as their specific roles in global biogeochemical cycles (Frades et al. 2020). The relationships of biochemical processes, habitats, ecosystems, and species are mostly unknown (Van Dover et al. 2014). As a result, we know little about the resilience and vulnerabilities of deep-sea commodities and services systems (Armstrong et al. 2012). As a result, understanding how these deep-sea habitats work and the risks they face is critical. We also need a deeper understanding of how deep-sea ecosystems work and how this contributes to the provision of ecosystem goods and services to humans (Figure 9.6) (Tittensor et al. 2009). Measuring and valuing these functions and marine ecosystem services is difficult since our understanding of products and services, as well as the trade-offs between them, are so limited (Armstrong et al. 2012).

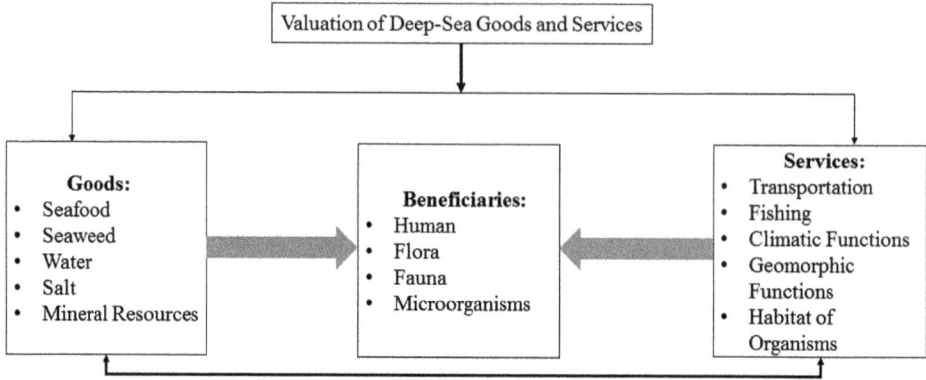

FIGURE 9.6 Classification and valuation of deep-sea goods, services and their functional characteristics.

9.6 MAJOR MARITIME ACTIVITIES AND SERVICES TO GENERATE THE BLUE ECONOMY

Maritime activities and services play a vital role in generating the blue economy. The most important sectors are the following (Hussain et al. 2017).

- **Marine biotechnology:** which is the use of biotechnology, molecular and cell biology, and bioinformatics to create goods and processes from marine creatures.
- **Carbon sequestration:** which is a method of trapping and storing atmospheric carbon dioxide. It is one way of decreasing carbon dioxide levels in the atmosphere with the purpose of slowing global climate change.
- **Oil, gas, and mineral mining** concerns all operations involved in the exploration, evaluation, and extraction of minerals, metals, petroleum, and fossil fuels from the earth and these are included in mining, oil, and gas research.
- **Marine renewable energy:** which is renewable energy that is installed and operated at sea and necessitates access to offshore grid and distribution infrastructure. Offshore wind, tidal stream, tidal range, and wave energy technologies are examples of this.
- **Sea salt production:** in the form of solar salt manufacturing is typically accomplished by collecting salt water in shallow ponds where the sun evaporates the majority of the water.
- **Marine trade, shipping, and transport:** where the transportation of commodities (cargo) through waterways is known as maritime transport.
- **Marine tourism:** encompasses a diverse range of activities that take place in the deep oceans.
- **Maritime education and research:** are distinct but vital components of the supply of skilled and competent human resources to the connected businesses.

9.7 MARINE SERVICES TO SMALL ISLAND DEVELOPING STATE (SIDS)

The Small Island Developing States (SIDS) seem to be the world's most vulnerable group of nations, both in terms of their ecologies and their economies (Ghina, 2003). Their problems and limitations, which are tied to their ecological fragility and environmental sensitivity, are many and complex. They must overcome these obstacles in order to achieve sustainable development goals and their respective countries vision and missions (Ghina, 2003). Marine services with sustainable blue economy would be the best solution to achieve these goals and vision.

9.7.1 ECONOMIC GROWTH FROM ECOSYSTEM SERVICES

Although it is quite difficult to place a monetary figure on all of the ways in which the natural world enhances our lives, there are several concrete advantages to living in a world where ecosystems are robust and healthy (Frades et al. 2020). As a consequence of wildlife and natural ecosystems, we have a stronger economy, a greater variety of food goods, and improvements in medical research to our credit (Folkersen et al. 2018). Despite the fact that the benefits of nature have long been known, it has only been in recent years that the idea of ecosystem services has been devised to represent these many advantages (Luisetti et al. 2011). A good advantage that animals or ecosystems give to humans is referred to as an 'ecosystem service' (Hattam et al. 2020). The advantages might be either direct or indirect, and they can be either minor or enormous (Van Dover et al. 2014). There are hundreds of advantages associated with marine ecosystem service that benefit not only human-kind but also the economic structure. Economic development will be tremendous if SIDS take use of the appropriate services (Liquete et al. 2013). In many cases, small island developing states (SIDS) suffer from distinct disadvantages as a result of their size, insularity, isolation, remoteness, and vulnerability to natural catastrophes (Ghina, 2003; Keen, et al. 2018). The combined effect of these variables makes the economies of these countries very sensitive to forces outside their control a situation that might occasionally jeopardize their long-term economic survival (Briguglio, 1995; Clark Howard et al. 2018).

9.7.2 POSSIBILITIES FOR SEAWEED CULTIVATION, BLUE CARBON AND MANAGEMENT

Despite the fact that the great bulk of seaweed production continues to take place in nations such as China, Japan, and Indonesia, seaweed is gradually becoming more popular in the Western world (Wenhai et al. 2019). In the United States and Canada, there are an increasing number of seaweed farms (Heisler, 2008). However, the most difficult obstacle to overcome is the growing demand for fresh, unprocessed seaweed as a dietary source (IUCN, 2017). If farming is ever to become a main-stream type of aquaculture for SIDS, people will need to acquire a preference for seaweed as an ingredient in their diet. Seaweed farming is also beneficial to the environment since it urges people to protect and respect their ocean (Troell et al. 2022). Seaweed farming is a vital source of employment and local food supply in developing nations, particularly in the Caribbean. This alleviates strain on local fish supplies and has the potential to minimize overfishing by lessening the economic need for fishing for survival.

9.7.2.1 Blue Carbon Management

Blue carbon restoration and protection activities may potentially be used to provide market-based carbon offsets (also called carbon credits). Coastal blue carbon accounting stock value is about $180 million (Luisetti et al. 2013). Understanding carbon stocks and fluxes will speed up the construction of solid non-market estimates. Blue carbon incentives may also help conserve and benefit other ecosystem services offered by these areas, such as fisheries (Lau et al. 2020). This will have advantages beyond carbon sequestration. For example, mangroves and tidal marshes offer ecological services worth millions of dollars per hectare (Steven et al. 2019).

The Commonwealth Scientific and Industrial Research Organisation (CSIRO) is an Australian Government agency responsible for scientific research, the Blue Carbon cooperation cluster has provided the most complete blue carbon sequestration estimations (Hossain, 2001; Islam et al. 2017). This research has influenced national reporting and emissions policies. The Blue Carbon Initiative and the Blueprint for Ocean and Coastal Sustainability are multinational projects that encourage blue carbon habitat conservation and restoration (Vanderklift et al. 2019). Quantifying coastal carbon reserves in mangroves, saltmarsh, and sea grass is critical, but there are still challenges (Shamsuzzaman et al. 2017). For example, the seasonal and geographical variability of certain sea grass meadows makes estimating carbon reserves difficult. Thus, knowing blue carbon habitat limits and carbon flows is critical for policy and market development (Steven et al. 2019).

9.8   KEY BUSINESS INDICATORS IN THE CONTEXT OF A BLUE ECONOMY

The Global Reporting Initiative (GRI) is an organization which exists to ensure that corporate blue economy sector firms and companies disclose their environmental, social and governance performance (Waddock, 2008). Companies need to follow these guidelines and report their performance. The latest version of the guideline is G4. The GRI provides a framework by which companies provide reports publicly on sustainable issues (Koseoglu et al. 2021).

CDP (previously known as Carbon Disclosure Project) is an organization which provides companies, states, cities, and regions a system to measure and manage environmental impacts and globally disclose the reports. These reports are important to inform decisions taken in business and to enable countries to plan future steps to ensure a safe environment for everyone. The business indicators in the context of the Blue Economy follow the GRI guidelines and CDP guidelines to measure their performance and progress of their goals. Below are some business indicators and the related guidelines used to measure this performance.

TABLE 9.2
Business Indicators with Corresponding Guidelines

Business indicators	Guidelines to measure performance	References
Total Water Discharge: the total quantity and quality is measured in every location	GRI G4 Sustainability Reporting Guidelines, G4-EN22 CDP's Water Questionnaire, W1.2b	Koseoglu et al. (2021) Cisneros-Montemayor et al. (2021)
Effects of Organization's water discharge: The impacts of the discharged water from an organization on the biodiversity and surrounding habitats of the water body is noted	GRI G4 Sustainability Reporting Guidelines, G4-EN26	Cisneros-Montemayor et al. (2021) Waddock (2008)
Mining and Metal Sector Land: The total amount of land used for mining and metal sector is calculated	GRI G4 Mining and Metals Sector Disclosures, MM1	Cisneros-Montemayor et al. (2021)
Electric utilities sector impacts: The impact of the electric utility sector on the biodiversity of affected areas is compared to the biodiversity of the adjacent non-affected areas.	GRI G4 Electric Utilities Sector Disclosures, EU13	Cisneros-Montemayor et al. (2021) Frades et al. (2020)
Climate Change Information: Total number of projects is noted, and which projects are at what stage of implementation is noted. Then CO_2 saving is calculated.	CDP's Climate Change Information Request, CC3.3a	Cisneros-Montemayor et al. (2021)

Source: After modified adopted from Cisneros-Montemayor et al. 2021 and Koseoglu et al. 2021.

9.9 SDGs TARGETS, ACHIEVEMENT AND LINKAGES WITH A BLUE ECONOMY PERSPECTIVE

According to the World Bank, the blue economy is the 'sustainable use of ocean resources for economic growth, improved livelihoods, and jobs while preserving the health of the ocean ecosystem' (Cicin-Sain, 2015). The European Commission defines it by saying that all economic activities related to oceans, seas and coasts would be considered as a blue economy ((Binelli et al. 2008; Biswas et al. 2014; Hossain et al. 2021). The blue economy is not just about market opportunities but also provides for the protection and development of more intangible 'blue' resources such as traditional ways of life, carbon sequestration, and coastal resilience to help vulnerable states mitigate the often devastating effects of climate change (Islam and Shamsuddoha, 2018). To ensure that the SDG can achieve the goals of the blue economy, some targets were set with certain timelines (Lee et al. 2020). It was decided that by 2020 the target was to achieve healthy and productive oceans. The blue economy comprises the economic activities and supporting institutions, relationships, and choices that create sustainable wealth from the world's oceans and coasts (Le Blanc et al. 2017; Nash et al. 2020). The center examines ways that ocean and coastal resources can support economic development that provides decent livelihoods and maintains, restores, and enhances the critical coastal and marine ecosystems that provide the foundations for human wellness and prosperity (Frades et al. 2020).

As per the article published, SDG-14 is one of the sustainable development goals that focuses highly on the sustainable use of ocean, sea, and marine resources to achieve SDG (United Nations. Economic Commission for Africa (2018-11); Obura, 2020).The global blue economy developing countries include a large number of economic activities which include a total of 26 marine economic functions (Hossain et al. 2021). Fishery, maritime trade, and shipping, energy, tourism, coastal protection, maritime monitoring, and surveillance are some of the notable economic activities to contribute to the blue economy of the country helping to achieve the SDG and Society 5.0 factors (Binelli et al. 2008; Biswas et al. 2014). In order to achieve a better and happy society, Society 5.0 focuses on technological and industrial advancement and makes use of potential digital transformation for economic growth as well as giving a solution to the social issues with the coexistence of nature (Luisetti et al. 2013). In achieving Sustainable Development Goals (SDG's) and Society 5.0 the holds a major significance (World Resources Institute, 2020). Many developing countries like Bangladesh has set goals to reach the SDG and the concept of Society 5.0 to help in the economic and structural growth of the country (Islam et al. 2018). In relation to the development process of SDG the blue economies of developing nations have a crucial role to play as well (Cumming and von Cramon-Taubadel, 2018). The appropriate use of the ocean, sea, and other marine resources will contribute largely to the growth of the economies in both developed and developing countries hence, moving a step forward to achieve SDGs mission and vision.

Ocean growth is one of 17 Sustainable Development Goals (SDGs) and 169 objectives set by the United Nations to be achieved by 2030 (Ferreira et al. 2021). Each target is significant in and of itself, and each one is regarded as indivisible SDG 14, entitled 'Life below water,' consisting of ten goals primarily focused on the health of the oceans and the living resources found within them (Lee et al. 2020). Other SGD components such as poverty, food, economic development, cities, production and consumption, and environment have strong links to this component. Seas, oceans, and coastal areas are an integral and necessary part of the Earth's environment, critical to long-term sustainability. Oceans are crucial for global food stability and human health. As a result, ocean resources are vital to society and the economy (Hasan et al. 2018). The blue economy is the use of ocean resources to increase food security, improve nutrition and health, alleviate poverty and create employment, and so forth (Lau et al. 2020). The notion of the blue economy has been supported to generate alternative energy, improve seaborne trade and industrial profiles.

TABLE 9.3

SDGs Targets, Achievement and Linkages with the Blue Economy Perceptions

Blue economy sector or activity	Relevant SDG 14 target (in addition to 14.7)	Rationale
Fisheries	*Target 14.1 By 2025, prevent and significantly reduce marine pollution of all kinds, in particular from land-based activities, including marine debris and nutrient pollution*	Improved fisheries management will contribute to a reduction in sea-based pollution from fishing vessels, including in the form of discarded fishing gear, which will help reduce marine debris and ghost fishing
	Target 14.2 By 2020, sustainably manage and protect marine and coastal ecosystems to avoid significant adverse impacts, including by strengthening their resilience, and take action for their restoration in order to achieve healthy and productive oceans	Improved fisheries management will build resilience of ocean ecosystems as a whole
	Target 14.4 By 2020, effectively regulate harvesting and end overfishing, illegal, unreported and unregulated fishing and destructive fishing practices and implement science-based management plans, in order to restore fish stocks in the shortest time feasible, at least to levels that can produce maximum sustainable yield as determined by their biological characteristics	Achievement of targets 14.7 and 14.4 depend on each other
	Target 14.6 By 2020, prohibit certain forms of fisheries subsidies which contribute to overcapacity and overfishing, eliminate subsidies that contribute to illegal, unreported and unregulated fishing and refrain from introducing new such subsidies, recognizing that appropriate and effective special and differential treatment for developing and large developed countries should be an integral part of the World Trade Organization fisheries subsidies negotiation	Achievement of targets 14.7 and 14.6 depend on each other
	Target 14.9 Provide access for small-scale artisanal fishers to marine resources and markets	Access to markets will allow artisanal fishers to benefit from the blue economy
Aquaculture	**Target 14.1**	Sustainable aquaculture causes minimal pollution and in the case of seaweed and mollusk culture is a net remover of nutrients from the aquatic environment
	Target 14.2	Sustainable, climate-smart aquaculture can help build resilience by increasing incomes and diversifying livelihoods

Bioprospecting and biotechnology	**Target 14.8** Increase scientific knowledge, develop research capacity and transfer marine technology, taking into account the Intergovernmental Oceanographic Commission Criteria and Guidelines on the Transfer of Marine Technology, in order to improve ocean health and to enhance the contribution of marine biodiversity to the development of developing countries, in particular Small Island Developing States and Least Developed Countries	Capacity building and technology transfer are required for SIDS and developing countries to participate in marine bioprospecting and biodiscovery activities
	Target 14.10 Enhance the conservation and sustainable use of oceans and their resources by implementing international law as reflected in UNCLOS, which provides the legal framework for the conservation and sustainable use of oceans and their resources, as recalled in paragraph 158 of The Future We Want	Benefit sharing from the use of marine genetic resources is tied to the implementation of international law, including the Nagoya Protocol for areas under national jurisdiction; discussions are ongoing on a new international legally binding instrument under UNCLOS on the conservation and sustainable use of marine biodiversity of areas beyond national jurisdiction.
Extractive industries	**Target 14.2**	Deep-sea mining can undermine the resilience of marine ecosystems and species and should thus be preceded by effective social and environmental impact procedures
	Target 14.8	Capacity building and technology transfer are required for SIDS and developing countries to participate in extractive activities
Renewable (offshore) energy	**Target 14.2**	Ocean energy helps build self-sufficiency and reduce pollution, thus increasing resilience of SIDS and coastal countries
	Target 14.8	Capacity building and technology transfer are required for SIDS and developing countries to benefit from ocean energy and other renewables
Desalination (fresh water generation)	**Target 14.1**	Desalination technologies may cause pollution in the form of brine and CO_2 emissions, which will need to be reduced through appropriate technologies, including renewable sources of energy
	Target 14.2	Desalination, together with water conservation and good water governance, can help build self-sufficiency
	Target 14.8	Desalination plants are expensive: financing, capacity building, and technology transfer are required for SIDS and developing countries to benefit from desalination

(continued)

TABLE 9.3 (Continued)
SDGs Targets, Achievement and Linkages with the Blue Economy Perceptions

Blue economy sector or activity	Relevant SDG 14 target (in addition to 14.7)	Rationale
Maritime transport, ports and related services, shipping and shipbuilding	Target 14.1	Improved implementation of shipping regulations will reduce sea-based pollution
	Target 14.2	Improvement in management of ballast water, biofouling, and other transportation-related vectors of invasive species will improve overall resilience of marine and coastal ecosystems
	Target 14.8	Implementation of more-sustainable and low-carbon transportation systems globally will require both capacity building and technology transfer
	Target 14.10	Implementation of international law pertaining to the conservation and sustainable use of oceans and their resources, including, e.g. shipping
Coastal development	Target 14.1	Coastal development can increase in increased sedimentation and pollution, which will need to be reduced through sustainable operations
	Target 14.2	Sustainable coastal development and integrating climate change considerations into planning and development can enhance economic, social, and environmental resilience
Coastal and maritime tourism	Target 14.1	Sustainable tourism reduces marine pollution both from land-based and ship based sources
	Target 14.2	Sustainable tourism can help build ecosystem and human resilience
	Target 14.5 *By 2020, conserve at least 10 per cent of coastal and marine areas, consistent with national and international law and based on the best available scientific information*	Sustainable tourism can provide financing for marine protected areas
Ocean monitoring and surveillance	Target 14.2	Ocean monitoring provides better data for sustainable management and protection
	Target 14.3 *Minimize and address the impacts of ocean acidification, including through enhanced scientific cooperation at all levels*	Monitoring ocean acidification is an important component of gaining better scientific understanding about acidification and its impacts
	Target 14.4	Monitoring and surveillance are important components of sustainable fisheries
	Target 14.5	Monitoring and surveillance are important for marine protected area management

	Target 14.8	Capacity building and technology transfer are required for SIDS and developing countries to benefit from ocean surveillance technologies
	Target 14.10	Ocean monitoring and surveillance will assist in implementing international law, including UNCLOS
Coastal and marine area management, protection, and restoration activities	Target 14.2	Coastal and marine area management, protection, and restoration are key components of Target 14.2
	Target 14.3	While there are scientific uncertainties, marine protection may help provide marine ecosystems and species a better chance to adapt to the impacts of ocean acidification
	Target 14.4	IMCAM, MPAs, and restoration activities help achieve more-sustainable fisheries
	Target 14.5	Marine protection will help achieve Target 14.5
	Target 14.10	Implementing IMCAM, MSP, and MPAs is part of a number of existing international agreements; area-based management tools, including MPAs, are also being considered as part of United Nations discussions on an international legally binding instrument under UNCLOS on the conservation and sustainable use of marine biodiversity of areas beyond national jurisdiction
Activities supporting carbon sequestration (blue carbon)	Target 14.2	Management of blue carbon ecosystems will not only maintain their capacity to store carbon and provide possible economic benefits, but will also strengthen their resilience
	Target 14.5	Where blue carbon ecosystems are conserved via marine protected areas or other effective means, they would also contribute to achievement of Target 14.5.
Waste disposal management	Target 14.1	Waste disposal management is a key activity for reducing pollution of the coastal and marine environment
	Target 14.2	Waste disposal management contributes to sustainable management of marine ecosystems and builds resilience

Source: After modification and adopted from World Bank and United Nations Department of Economic and Social Affairs. 2017. The Potential of the Blue Economy: Increasing Long-term Benefits of the Sustainable Use of Marine Resources for Small Island Developing States and Coastal Least Developed Countries. World Bank, Washington DC. https://sustainabledevelopment. un.org/content/documents/15434Blue_Economy Jun1; Pisano et al. 2015; Nash et al., 2020; Thomson, P., 2018; https://impakter.com/ocean-future-global-action-achieve-sdg-14/; Ritchie et al. 2018.

9.10   CONCLUSION

Marine ecosystems are vital to the survival and well-being of a large section of the world's population. The benefits provided by healthy marine ecosystems are food security, animal feed, raw materials for medications, building materials from coral rock and sand, natural defenses against threats such as coastal erosion and inundation. The ocean, on the other hand, plays a critical role in climate regulation to maintain the global carbon balance. Estuaries are known as the 'nurseries of the sea' because of many kinds of wildlife such as fishes, birds, marine organisms, plants, and animals. Other services we get from marine biological habitats are erosion prevention, waste-water treatment, moderation of extreme events, for example, the impacts of tornados, and cyclones is reduced, and also cultural services are provided, such as tourism, recreational, aesthetic, and spiritual benefits. It is needed to protect the marine ecosystem by achieving a blue economy because it acts as carbon sequestration, provides us with oil, gas and mineral, it is a source of renewable energy, salt production, a route for transportation and trade, it facilitates tourism, and it is a source of education and research. The export of Hilsha fish and shrimp, seafood, sea plants such algae/seaweeds, and the like, adds to the economy of a country.

ACKNOWLEDGEMENTS

I would like to express my gratitude to all the anonymous authors and contributors whose articles I reviewed on many occasions to produce this scientific chapter. I have also used many websites, open access domains, blogs and other sources for reviewing the literature, concepts and ideas in order to build the scenarios of marine pollution and challenges to developing sustainable blue economy. Also, I would like to show my gratitude to the Ministry of Science and Technology, Government of Bangladesh that have provided the funding support to the corresponding author of this chapter for continuing his research on the global blue economy and seafood production and policies.

REFERENCES

Abrantes, K. G. Barnett, A. Baker, R. and Sheaves, M. (2015). Habitat-specific food webs and trophic interactions supporting coastal-dependent fishery species: an Australian case study. *Reviews in Fish Biology and Fisheries*, 25(2), 337–363.

Adrianov, A. V. (2004). Current problems in marine biodiversity studies. *Russian Journal of Marine Biology*, 30(1), S1–S16.

Ajonina, G. N. (2008). Inventory and Modelling Mangrove Forest Stand Dynamics Following Different Levels of Wood Exploitation Pressures in the Douala-Edea Atlantic Coast of Cameroon, Central Africa. Mitteilungen der Abteilungen für Forstliche Biometrie, Albert-Ludwigs-Universität Freiburg, 2.

Armstrong, C. W. Foley, N. S. Tinch, R. and van den Hove, S. (2012). Services from the deep: steps towards valuation of deep sea goods and services. Ecosystem Services, 2, 2–13.

Ashie, I. N. A. Smith, J. P. Simpson, B. K. and Haard, N. F. (1996). Spoilage and shelf-life extension of fresh fish and shellfish. *Critical Reviews in Food Science and Nutrition*, 36(1–2), 87–121.

Atkins, J. P. Burdon, D. Elliott, M. and Gregory, A. J. (2011). Management of the marine environment: integrating ecosystem services and societal benefits with the DPSIR framework in a systems approach. *Marine Pollution Bulletin*, 62(2), 215–226.

Ayompe, Lacour M. Schaafsma, Marije, and Egoh, Benis N. (2021).Towards sustainable palm oil production: the positive and negative impacts on ecosystem services and human wellbeing. *Journal of Cleaner Production*, 278, 123914.

Balcombe, J. (2016). *What a Fish Knows: The Inner Lives of Our Underwater Cousins*. Scientific American/ Farrar, Straus and Giroux.

Barbier, E. B. (2012). Progress and challenges in valuing coastal and marine ecosystem services. *Review of Environmental Economics and Policy*, 6(1), 1–19.

Barbier, E. B. (2017). Marine ecosystem services. Current Biology, R507–R510. 10.1016/j.cub.2017.03.020. Cell Press, 50 Hampshire St. 5th Floor, Cambridge, MA 02139, USA

Barbier, E. B. Hacker, S. D. Kennedy, C. Koch, E. W. Stier, A. C. and Silliman, B. R. (2011). The value of estuarine and coastal ecosystem services. *Ecological Monographs*, 81(2), 169–193.

Beaumont, N. J. Austen, M. C. Mangi, S. C. and Townsend, M. (2008). Economic valuation for the conservation of marine biodiversity. *Marine Pollution Bulletin*, 56(3), 386–396.

Binelli, A. Sarkar, S. K. Chatterjee, M. Riva, C. Parolini, M. deb Bhattacharya, B. ... and Satpathy, K. K. (2008). A comparison of sediment quality guidelines for toxicity assessment in the Sunderban wetlands (Bay of Bengal, India). Chemosphere, 73(7), 1129–1137.

Biswas, S. N. Rakshit, D. Sarkar, S. K. Sarangi, R. K. and Satpathy, K. K. (2014). Impact of multispecies diatom bloom on plankton community structure in Sundarban mangrove wetland, India. *Marine Pollution Bulletin*, 85(1), 306–311.

BOBLME (2015) An ecosystem characterization of the Bay of Bengal. Compiled by Brewer, D. Hayes, D. Lyne, V. Donovan, A. Skewes, T. Milton, D. and Murphy, N. BOBLME-2015-Ecology-13. xvii + 287p.

Briguglio, L. (1995). *Small island developing states and their economic vulnerabilities. World Development*, 23, 1615–1632.

Brown, E. J. Vasconcelos, R. P. Wennhage, H. Bergström, U. Støttrup, J. G. van de Wolfshaar, K. ... and Le Pape, O. (2018). Conflicts in the coastal zone: human impacts on commercially important fish species utilizing coastal habitat. *ICES Journal of Marine Science*, 75(4), 1203–1213.

Buonocore, E. Donnarumma, L. Appolloni, L. Miccio, A. Russo, G. F. Franzese, P. P. (2020). Marine natural capital and ecosystem services: an environmental accounting model. *Ecol. Model.* 424, 109029.

Buonocore, E. Grande, U. Franzese, P. P. and Russo, G. F. (2021). Trends and evolution in the concept of marine ecosystem services: an overview. *Water*, 13(15), 2060.

Cassotta, S. (2021). Ocean acidification in the Arctic in a multi-regulatory, climate justice perspective. *Frontiers in Climate*, 3: 3:713644. doi: 10.3389/fclim.2021.713644

Chen, H. Wang, X. and Wang, Q. (2020). Microalgal biofuels in China: The past, progress and prospects. *GCB Bioenergy*, 12(12), 1044–1065.

Cicin-Sain, B. (2015). Conserve and sustainably use the oceans, seas and marine resources for sustainable development. *UN Chronicle*, 51(4), 32–33.

Cisneros-Montemayor, A. M. et al. (2021). Enabling conditions for an equitable and sustainable blue economy. *Nature*, 591(7850), 396–401.

Clark Howard, B. (2018). Blue growth: stakeholder perspectives. *Marine Policy*, 87, 375–377.

Colwell, F. and D'Hondt, S. 2013. Nature and extent of the deep biosphere. Rev. Mineral. Geochem. 75: 547–574.

Cumming, G. S. and von Cramon-Taubadel, S. (2018). Linking economic growth pathways and environmental sustainability by understanding development as alternate social–ecological regimes. *Proceedings of the National Academy of Sciences*, 115(38), 9533–9538.

Da Ros, Z. Dell'Anno, A. Morato, T. Sweetman, A. K. Carreiro-Silva, M. Smith, C. J. and Danovaro, R. (2019). The deep sea: the new frontier for ecological restoration. *Marine Policy*, 108, 103642.

Devassy, V. P. and Nair, S. R. S. (1987). Discolouration of water and its effect on fisheries along the Goa coast. *Mahasagar*, 20(2), 121–128.

Duarte, C. M. Dennison, W. C. Orth, R. J. and Carruthers, T. J. B. (2008). The charisma of coastal ecosystems: addressing the imbalance. *Estuaries and Coasts*, 31(2), 223–238.

Eckert, S. A. Nellis, D. W. Eckert, K. L. and Kooyman, G. L. (1986). Diving patterns of two leatherback sea turtles (Dermochelys coriacea) during internesting intervals at Sandy Point, St. Croix, US Virgin Islands. *Herpetologica*, 42, 381–388.

Edward, B. B. (2017). Marine ecosystem services, *Current Biology*, 27(11), 507–510, https://doi.org/10.1016/j.cub.2017.03.020; ISSN 0960–9822.

Ewing, S. (2012). *The Great Alaska Nature Factbook: A Guide to the State's Remarkable Animals, Plants, and Natural Features*. Graphic Arts Books.

Ferreira, J. C. Vasconcelos, L. Monteiro, R. Silva, F. Z. Duarte, C. M. and Ferreira, F. (2021). Ocean literacy to promote sustainable development goals and agenda 2030 in coastal communities. *Education Sciences*, 11(2), 62.

Fertl, D. and Fulling, G. L. (2007). Interactions between marine mammals and turtles. *Marine Turtle Newsletter*, 115, 4–8.

Folkersen, M. V. Fleming, C. M. and Hasan, S. (2018). The economic value of the deep sea: a systematic review and meta-analysis. *Marine Policy*, 94, 71–80.

Fox, H. E. et al. (2012). Explaining global patterns and trends in marine protected area (MPA) development. *Marine Policy*, 36(5), 1131–1138.

Frades, J. L. Barba, J. G. Negro, V. Martín-Antón, M. and Soriano, J. (2020). Blue economy: compatibility between the increasing offshore wind technology and the achievement of the SDG. *Journal of Coastal Research*, 95(SI), 1490–1494.

Friedlander, A. Aeby, G. Brainard, R. Brown, E. Chaston, K. Clark, A. ... and Wiltse, W. (2008). The state of coral reef ecosystems of the main Hawaiian Islands. *The State of Coral Reef Ecosystems of the United States and Pacific Freely Associated States*, 17, 222–269.

Friedman, J. M. (1977). *The Growth of Economic Values in Preservation: An Estuarine Case Study.*

Fuhrman, J. A. Cram, J. A. and Needham, D. M. (2015). Marine microbial community dynamics and their ecological interpretation. *Nature Reviews Microbiology*, 13(3), 133–146.

George, R. Y. (2013). Deep-sea organisms. *Functional Adaptations of Marine Organisms*, Editors: F. John Vernberg, Winona B. Vernberg; eBook ISBN: 9781483266541, Elsevier Publication, 279.

Ghina, F. (2003). Sustainable Development in small Island Developing States. Environment, Development and Sustainability, 5, 139–165.

Gibson, R. N. Atkinson, R. J. A. Gordon, J. D. M. Smith, I. P. and Hughes, D. J. (2011). Bioengineering effects of burrowing thalassinidean shrimps on marine soft-bottom ecosystems. *Oceanography and Marine Biology: An Annual Review*, 49, 137–192.

Griffiths, H. J. and Waller, C. L. (2016). The first comprehensive description of the biodiversity and biogeography of Antarctic and Sub-Antarctic intertidal communities. *Journal of biogeography*, 43(6), 1143–1155.

Guerry, A. D. Ruckelshaus, M. H. Arkema, K. K. Bernhardt, J. R. Guannel, G. Kim, C. K. ... and Spencer, J. (2012). Modeling benefits from nature: using ecosystem services to inform coastal and marine spatial planning. *International Journal of Biodiversity Science, Ecosystem Services and Management*, 8(1–2), 107–121.

Guldner, L. Monfort, C. Rouget, F. Garlantezec, R. and Cordier, S. (2007). Maternal fish and shellfish intake and pregnancy outcomes: a prospective cohort study in Brittany, France. *Environmental Health*, 6(1), 1–8.

Gullström, M. de la Torre Castro, M. Bandeira, S. O. Björk, M. Dahlberg, M. Kautsky, N. ... and Öhman, M. C. (2002). Seagrass ecosystems in the western Indian Ocean. *Ambio*, 31, 588–596.

Hall, C. M. (2001). Trends in ocean and coastal tourism: the end of the last frontier?. *Ocean and Coastal Management*, 44(9–10), 601–618.

Harper, S. Adshade, M. Lam, V. W. Pauly, D. and Sumaila, U. R. (2020). Valuing invisible catches: estimating the global contribution by women to small-scale marine capture fisheries production. *PLoS One*, 15(3), e0228912.

Hasler, B. (2016). *Marine ecosystem services: Marine ecosystem services in Nordic marine waters and the Baltic sea-possibilities for valuation*. Nordic Council of Ministers.

Hasan, M. M. Hossain, B. S. Alam, M. J. Chowdhury, K. A. Al Karim, A. and Chowdhury, N. M. K. (2018). The prospects of blue economy to promote Bangladesh into a middle-income country. *Open Journal of Marine Science*, 8(03), 355.

Hattam, C. Atkins, J. P. Beaumont, N. Börger, T. Böhnke-Henrichs, A. Burdon, D. De Groot, R. Hoefnagel, E. Nunes, P. A. Piwowarczyk, J. and Sastre, S. (2015). Marine ecosystem services: linking indicators to their classification. *Ecological Indicators*, 49, 61–75.

Hattam, C. Evans, L. Morrissey, K. Hooper, T. Young, K. Khalid, F. Bryant, M. Thani, A. Slade, L. Perry, C. and Turrall, S. (2020). Building resilience in practice to support coral communities in the Western Indian Ocean. *Environmental Science and Policy*, 106, 182–190.

Heisler, J. Glibert, P. M. Burkholder, J. M. Anderson, D. M. Cochlan, W. Dennison, W. C. Dortch, Q. Gobler, C. J. Heil, C. A. Humphries, E. and Lewitus, A. (2008). Eutrophication and harmful algal blooms: a scientific consensus. *Harmful Algae*, 8(1), 3–13.

Hossain, M. (2001). Biological aspects of the coastal and marine environment of Bangladesh. *Ocean and Coastal Management*, 44(3–4), 261–282.

Hossain, M. S. Sharifuzzaman, S. M. Nobi, M. N. Chowdhury, M. S. N. Sarker, S. Alamgir, M. ... and Chowdhury, S. (2021). Seaweeds farming for sustainable development goals and blue economy in Bangladesh. *Marine Policy*, 128, 104469.

Hrabanski, M. (2017). Private Sector Involvement in the Millennium Ecosystem Assessment: using a UN platform to promote market-based instruments for ecosystem services. *Environmental Policy and Governance*, 27(6), 605–618.

Hussain, M. G. Failler, P. Karim, A. A. and Alam, M. K. (2017). Major opportunities of blue economy development in Bangladesh. *Journal of the Indian Ocean Region*, 5(19), 2020.

Intergovernmental Panel on Climate Change (IPCC) (2019). *Special Report on the Ocean and Cryosphere in a Changing Climate.*

Intergovernmental Science-Policy Platform on Biodiversity and Ecosystem Services (IPBES) (2019). *Summary for Policymakers of the Global Assessment Report on Biodiversity and Ecosystem Services.*

Islam, M. (2003). Perspectives of the coastal and marine fisheries of the Bay of Bengal, Bangladesh. *Ocean and Coastal Management*, 46(8), 763–796.

Islam, M. M. and Shamsuddoha, M. D. (2018). Coastal and marine conservation strategy for Bangladesh in the context of achieving blue growth and sustainable development goals (SDGs). *Environmental Science and Policy*, 87, 45–54.

Islam, M. M. Shamsuzzaman, M. M. Mozumder, M. M. H. Xiangmin, X. Ming, Y. and Jewel, M. A. S. (2017). (2017). Exploitation and conservation of coastal and marine fisheries in Bangladesh: do the fishery laws matter? *Marine Policy*, 76, 143–151.

Jackson, J. B. and Sala, E. (2001). Unnatural oceans. *Scientia Marina*, 65(S2), 273–281.

IUCN, S. (2017). Antelope Specialist Group (2017). *Tetracerus quadricornis*, 2017-2.

Jones, D. O. Ardron, J. A. Colaço, A. and Durden, J. M. (2020). Environmental considerations for impact and preservation reference zones for deep-sea polymetallic nodule mining. *Marine Policy*, 118:103312

Kaiser, Michel J. Attrill, Martin J. Jennings, Simon, Thomas, David N. and Barnes, David K. A. (2011). *Marine Ecology: Processes, Systems, and Impacts.* Oxford University Press.

Kandasamy, S. Zhang, B. He, Z. Bhuvanendran, N. EL-Seesy, A. I. Wang, Q. ... and Dar, M. A. (2022). Microalgae as a multipotential role in commercial applications: current scenario and future perspectives. *Fuel*, 308, 122053.

Keen, M. R. Schwarz, A.-M. and Wini-Simeon, L. (2018). Towards defining the blue economy: practical lessons from pacific ocean governance. *Marine Policy*, 88, 333–341.

Klain, S. C. and Chan, K. M. A. (2012). Navigating coastal values: participatory mapping of ecosystem services for spatial planning. *Ecol. Econ.* 82, 104–113. doi: 10.1016/j.ecolecon.2012.07.008

Konar, M. and Ding, H. A. (2020). *Sustainable Ocean Economy for 2050: Approximating Its Benefits and Costs.* World Resources Institute.

Koseoglu, M. A. Uyar, A. Kilic, M. Kuzey, C. and Karaman, A. S. (2021). Exploring the connections among CSR performance, reporting, and external assurance: evidence from the hospitality and tourism industry. *International Journal of Hospitality Management*, 94, 102819.

Kurien, J. (2007). The blessing of the commons: small-scale fisheries, community property rights, and coastal natural assets. *Reclaiming Nature: Environmental Justice and Ecological Restoration*, 1, 23.

Lange, G. M. Jiddawi, N. (2009) Economic value of marine ecosystem services in Zanzibar: implications for marine conservation and sustainable development. *Ocean and Coastal Management*, 52(10), 521–532.

Larsen, P. S. Andrup, P. Tang, B. Riisgård, H. U. (2016). Biomixing in stagnant water above population of blue mussels (mytilus edulis). *J Oceanogr Mar Res*, 4, 147. doi: 10.4172/2572–3103.1000147.

Lau, J. D. Cinner, J. E. Fabinyi, M. Gurney, G. G. and Hicks, C. C. (2020). Access to marine ecosystem services: examining entanglement and legitimacy in customary institutions. *World Development*, 126, 104730.

Le Blanc, D. Freire, C. and Vierros, M. (2017). *Mapping the Linkages Between Oceans and Other Sustainable Development Goals (SDGs): A Preliminary Exploration.*

Lee, H. J. Yu, H. Oh, E. G. Shin, S. B. Park, K. and Kim, J. H. (2013). Germicidal effect of electrolyzed seawater on live fish and shellfish. *Korean Journal of Fisheries and Aquatic Sciences*, 46(5), 534–539.

Lee, K. H. Noh, J. and Khim, J. S. (2020). The blue economy and the United Nations' sustainable development goals: challenges and opportunities. *Environment International*, 137, 105528.

Lele, S. Springate-Baginski, O. Lakerveld, R. Deb, D. and Dash, P. (2013). Ecosystem services: origins, contributions, pitfalls and alternatives. *Conserv. Soc.* 11, 343–358. doi: 10.4103/0972–4923.125752

Lenanton, R. C. and Potter, I. C. (1987). Contribution of estuaries to commercial fisheries in temperate Western Australia and the concept of estuarine dependence. *Estuaries*, 10(1), 28–35.

Liguo, X. Ahmad, M. Khattak, S. (2022). Impact of innovation in marine energy generation, distribution, or transmission-related technologies on carbon dioxide emissions in the United States. *Renewable and Sustainable Energy Reviews*, 159, 112225.

Lillebø, A. I. Pita, C. Rodrigues, J. G. Ramos, S. and Villasante, S. (2017). How can marine ecosystem services support the Blue Growth agenda?. *Marine Policy*, 81, 132–142.

Lovelock, C. E. (2008). Soil respiration and belowground carbon allocation in mangrove forests. *Ecosystems,* 11, 342–354. doi: 10.1007/s10021–008–9125–4.

Liquete, C. Piroddi, C. Drakou, E. G. Gurney, L. Katsanevakis, S. Charef, A. and Egoh, B. (2013). Current status and future prospects for the assessment of marine and coastal ecosystem services: a systematic review. *PloS one,* 8(7), e67737.

Lubchenco, Jane, Haugan, Peter M. and Pangestu, Mari Elka (2020). Five priorities for a sustainable ocean economy. *Nature,* 588, 30–32.

Luisetti, T. Jackson, E. and Turner, R. (2013). Valuing the European 'coastal blue carbon' storage benefit. *Marine Pollution Bulletin,* 71(1–2), 101–106.

Luisetti, T. Turner, R. K. Bateman, I. J. Morse-Jones, S. Adams, C. and Fonseca, L. (2011). Coastal and marine ecosystem services valuation for policy and management: managed realignment case studies in England. *Ocean and Coastal Management,* 54(3), 212–224.

Manson, F. J. Loneragan, N. R. Skilleter, G. A. and Phinn, S. R. (2005). An evaluation of the evidence for linkages between mangroves and fisheries: a synthesis of the literature and identification of research directions. *Oceanography and Marine Biology,* 43, 483.

Marcus, G. F. (2004). *The Birth of the Mind: How a Tiny Number of Genes Creates the Complexities of Human Thought.* Basic Civitas Books.

Martin, C. L. (2019). *Cultural Ecosystem Services Derived from Estuaries in New South Wales,* Australia (Doctoral dissertation, The University of Newcastle).

Martin, C. L. Momtaz, S. Gaston, T. and Moltschaniwskyj, N. A. (2020). Estuarine cultural ecosystem services valued by local people in New South Wales, Australia, and attributes important for continued supply. *Ocean and Coastal Management,* 190, 105160.

Mauzey, K. P. Birkeland, C. and Dayton, P. K. (1968). Feeding behavior of asteroids and escape responses of their prey in the Puget Sound region. *Ecology,* 49(4), 603–619.

Michael, T. Kate, D. Nicholas, H. Judi, E. H. Carolyn, J. L. Andrew, M. L. (2018). The challenge of implementing the marine ecosystem service concept. *Frontiers in Marine Science,* 5, 359. doi: 10.3389/fmars.2018.00359; ISSN: 2296–7745.

MoFa (2014). Ministry of Foreign Affairs, Press Release: Press Statement of the Hon'ble Foreign Minister on the Verdict of the Arbitral Tribunal/PCA. Dhaka: MoFa.

Moriarty, P. (2007). Environmental and resource constraints on Asian urban travel. *International Journal of Environment and Pollution,* 30(1), 8–26.

Nash, K. L. Blythe, J. L. Cvitanovic, C. Fulton, E. A. Halpern, B. S. Milner-Gulland, E. J. Addison, P. F. Pecl, G. T. Watson, R. A. and Blanchard, J. L. (2020). To achieve a sustainable blue future, progress assessments must include interdependencies between the sustainable development goals. *One Earth,* 2(2), 161–173.

Nellemann, C. and Corcoran, E. (Eds) (2009). *Blue Carbon: The Role of Healthy Oceans in Binding Carbon.* A Rapid Response *Assessment.* UNEP/Earthprint.

Obura, D. O. (2020). Getting to 2030-Scaling effort to ambition through a narrative model of the SDGs. *Marine Policy,* 117, 103973.

Palumbi, S. R. Sandifer, P. A. Allan, J. D. Beck, M. W. Fautin, D. G. Fogarty, M. J. Halpern, B. S. Incze, L. S. Leong, J. A. Norse, E. and Stachowicz, J. J. (2009). Managing for ocean biodiversity to sustain marine ecosystem services. *Frontiers in Ecology and the Environment,* 7(4), 204–211.

Pendleton, L. H. Thebaud, O. Mongruel, R. C. and Levrel, H. (2016). Has the value of global marine and coastal ecosystem services changed? *Marine Policy,* 64, 156–158. doi: 10.1016/j.marpol.2015.11.018.

Peterson, C. H. (1991). Intertidal zonation of marine invertebrates in sand and mud. *American Scientist,* 79(3), 236–249.

Pisano, U. Lange, L. Berger, G. and Hametner, M. (2015). The Sustainable Development Goals (SDGs) and their impact on the European SD governance framework. *ESDN Quarterly Report,* (35), 6.

Potter, I. C. and Hyndes, G. A. (1999). Characteristics of the ichthyofaunas of southwestern Australian estuaries, including comparisons with holarctic estuaries and estuaries elsewhere in temperate Australia: a review. *Australian Journal of Ecology,* 24(4), 395–421.

Powell, R. and Domack, G. W. (2002). Modern glaciomarine environments. In *Modern and Past Glacial Environments* (pp. 361–389). Editor: John Menzies, eBook ISBN: 9780080497327, Elsevier Publication.

Power, M. E. Stout, R. J. Cushing, C. E. Harper, P. P. Hauer, F. R. Matthews, W. J. ... and De Badgen, W. (1988). Biotic and abiotic controls in river and stream communities. *Journal of the North American Benthological Society,* 7(4), 456–479.

Rajalingam, A. Jani, S. Kumar, A. and Khan, M. (2016). Production methods of biodiesel. *Journal of Chemical and Pharmaceutical Research*, 8(3), 170–173. Available online https://www.jocpr.com/articles/product ion-methods-of-biodiesel.pdf

Ranjithkumar, S. Suthan, C. Earnestpremkumar, B. Revand, R. and Rajesh, M. (2020). Effect of methanol fumigation on performance and emission characteristics in a Jatropha bio-diesel fueled single cylinder constant speed CI engine. *International Journal of Scientific and Technology Research*, 9(4), 332–337.

Raudsepp-Hearne, C. Peterson, G. D. Tengö, M. Bennett, E. M. Holland, T. Benessaiah, K. ... and Pfeifer, L. (2010). Untangling the environmentalist's paradox: why is human well-being increasing as ecosystem services degrade? *BioScience,* 60(8), 576–589.

Roberts, C. M. Andelman, S. Branch, G. Bustamante, R. H. Carlos Castilla, J. Dugan, J. ... and Warner, R. R. (2003). Ecological criteria for evaluating candidate sites for marine reserves. *Ecological Applications*, 13(sp1), 199–214.

Ritchie, Roser, Mispy, Ortiz-Ospina.c(2018). *Measuring progress towards the Sustainable Development Goals.* SDG-Tracker.org, website: https://sdg-tracker.org/oceans; License: the SDG Tracker is a project of the Global Change Data Lab, a registered charity in England and Wales (Charity Number 1186433); Website link: https://indicators.report/targets/

Roberts, J. M. Wheeler, A. J. and Freiwald, A. (2006). Reefs of the deep: the biology and geology of cold-water coral ecosystems. *Science*, 312(5773), 543–547.

Roman, C. T. Niering, W. A. and Warren, R. S. (1984). Salt marsh vegetation change in response to tidal restriction. *Environmental Management*, 8(2), 141–149.

Ross, H. Adhuri, D. S. Abdurrahim, A. Y. and Phelan, A. (2019). Opportunities in community-government cooperation to maintain marine ecosystem services in the Asia-Pacific and Oceania. *Ecosystem Services,* 38, 100969.

Sarker, S. Bhuyan, M. A. H. Rahman, M. M. Islam, M. A. Hossain, M. S. Basak, S. C. and Islam, M. M. (2018). From science to action: exploring the potentials of blue economy for enhancing economic sustainability in Bangladesh. *Ocean and Coastal Management*, 157, 180–192.

Salomidi, M. Katsanevakis, S. Borja, A. Braeckman, U. Damalas, D. Galparsoro, I. ... and Fernández, T. V. (2012). Assessment of goods and services, vulnerability, and conservation status of European seabed biotopes: a stepping stone towards ecosystem-based marine. *Mediterranean Marine Science*, 13(1), 49–88.

Schlacher, T. A. Jones, A. R. Dugan, J. E. Weston, M. A. Harris, L. Schoeman, D. S. ... and Peterson, C. H. (2014). Open-coast sandy beaches and coastal dunes. *Coastal Conservation,* 19, 37–92.

Seidensticker, J. and Hai, M. A. (1983) *The Sundarbans Wildlife Management Plan: Conservation in the Bangladesh Coastal Zone*. Gland, Switzerland: International Union for the Conservation of Nature and Natural Resources.

Shamsuzzaman, M. M. Xiangmin, X. Ming, Y. and Tania, N. J. (2017). Towards sustainable development of coastal fisheries resources in Bangladesh: an analysis of the legal and institutional framework. *Turkish Journal of Fisheries and Aquatic Sciences*, 17(4): 831–839; DOI: 10.4194/1303-2712-v17_4_19

Sheaves, M. (2017). How many fish use mangroves? The 75% rule an ill-defined and poorly validated concept. *Fish and Fisheries*, 18(4), 778–789.

Shields, M. A. Woolf, D. K. Grist, E. P. Kerr, S. A. Jackson, A. C. Harris, R. E. ... and Side, J. (2011). Marine renewable energy: the ecological implications of altering the hydrodynamics of the marine environment. *Ocean and Coastal Management*, 54(1), 2–9.

Sink, K. J. Holness, S. Harris, L. Majiedt, P. A. Atkinson, L. Robinson, T. ... and Awad, A. (2012). *National Biodiversity Assessment 2011: Technical Report*. Volume 4: Marine and Coastal Component. South African National Biodiversity Institute, Pretoria.

Smardon, R. (2009). The Mankote mangrove: microcosm of the Caribbean. *In Sustaining the World's Wetlands* (pp. 267–300). Springer.

Smail, E. and Hasson, A. (2021). *Earth System Science Interdisciplinary Center*. University of Maryland, USA, Satellite Oceanography and Climatology Division, NOAA Center for Satellite Applications and Research, https://www.star.nesdis.noaa.gov/star/socd_index.php

Steven, A. D. L. Vanderklift, M. A. and Bohler-Muller, N. (2019). A new narrative for the blue economy and blue carbon. *Journal of the Indian Ocean Region*, 15(2), 123–128.

Stuchtey, M. Vincent, A. Merkl, A. Bucher, M. Haugan, P. M. Lubchenco, J. and Pangestu, M. E. (2020). *Ocean Solutions that Benefit People, Nature and the Economy*. Report. World Resources Institute.

Summers, J. K. Smith, L. M. Case, J. L. and Linthurst, R. A. (2012). A review of the elements of human well-being with an emphasis on the contribution of ecosystem services. *Ambio*, 41(4), 327–340.

Tittensor, D.P. Baco, A. R. Brewin, P. E. Clark, M. R. Consalvey, M. Hall-Spencer, J. Rowden, A. A. Schlacher, T. Stocks, K. I. and Rogers, A. D. (2009). Predicting global habitat suitability for stony corals on seamounts. *Journal of Biogeography*, 36(6), 1111–1128.

Troell, M. Henriksson, P. J. Buschmann, A. H. Chopin, T. and Quahe, S. (2022). Farming the Ocean–Seaweeds as a Quick Fix for the Climate? *Reviews in Fisheries Science and Aquaculture*, 1–11.

U.S. EPA. Report on the 2015 U.S. Environmental Protection Agency (EPA) International Decontamination Research and Development Conference. U.S. Environmental Protection Agency, Washington, DC, EPA/600/R-15/283, 2015.

Vincent, J. R. (2012). *"Vincent, Jeffrey R. 2012. Ecosystem Services and Green Growth. Policy Research Working Paper; No. 6233. World Bank. © World Bank.* https://openknowledge.worldbank.org/handle/10986/12084 *License: CC BY 3.0 IGO."*

Van Dover, C. L. Aronson, J. Pendleton, L. Smith, S. Arnaud-Haond, S. Moreno-Mateos, D. ... and Warner, R. (2014). Ecological restoration in the deep sea: Desiderata. *Marine Policy*, 44, 98–106.

Van Jaarsveld, A. S. Biggs, R. Scholes, R. J. Bohensky, E. Reyers, B. Lynam, T. ..and Fabricius, C. (2005). Measuring conditions and trends in ecosystem services at multiple scales: the Southern African Millennium Ecosystem Assessment (SAfMA) experience. *Philosophical Transactions of the Royal Society B: Biological Sciences*, 360(1454), 425–441.

Vanderklift, M. A. Marcos-Martinez, R. Butler, J. R. Coleman, M. Lawrence, A. Prislan, H. Steven, A. D. and Thomas, S. (2019). Constraints and opportunities for market-based finance for the restoration and protection of blue carbon ecosystems. *Marine Policy*, 107, 103–429.

Venkataraman, K. and Raghunathan, C. (2015). Coastal and marine biodiversity of India. In *Marine Faunal Diversity in India* (pp. 303–348). Academic Press. Waddock, S. (2008). Building a new institutional infrastructure for corporate responsibility. *Academy of Management perspectives*, 22(3), 87–108.

Wang, Q. X. and Tang, X. X. (2010). Connotation and classification of marine ecosystem services. *Marine Environmental Science*, 29(1).1-24.

Webber, M. Webber, D. and Trench, C. (2014). Agro-ecology for sustainable coastal ecosystems. Agroecology, Ecosystems, and Sustainability, 239. Edited: Noureddine Benkeblia, Imprint, CRC Press, 26, eBook ISBN9780429159374

Wenhai, L. Cusack, C. Baker, M. Tao, W. Mingbao, C. Paige, K. Xiaofan, Z. Levin, L. Escobar, E. Amon, D. and Yufeng, Y. (2019). Successful blue economy examples with an emphasis on international perspectives. *Frontiers in Marine Science*, 6; 261. doi: 10.3389/fmars.2019.00261.

Wilson, E. (2016). *Ways World Leaders Can Improve Fishery Management*. Recommendations for the 32nd UN Committee on Fisheries meeting, Pew's International Ocean Policy Work.

Wilson, S. P. and Verlis, K. M. (2017). The ugly face of tourism: Marine debris pollution linked to visitation in the southern Great Barrier Reef, Australia. *Marine Pollution Bulletin*, 117(1–2), 239–246.

World Bank and United Nations Department of Economic and Social Affairs. 2017. *The Potential of the Blue Economy: Increasing Long-term Benefits of the Sustainable Use of Marine Resources for Small Island Developing States and Coastal Least Developed Countries*. World Bank, Washington DC. https://openknowledge.worldbank.org/handle/10986/26843 *License: CC BY 3.0 IGO*

Wright, C. and Mella, A. (1963). Modifications to the soil pattern of South-Central Chile resulting from seismic and associated phenomena during the period May to August 1960. *Bulletin of the Seismological Society of America*, 53(6), 1367–1402.

World Resources Institute (2020). *The High Level Panel for a Sustainable Ocean Economy*. Transformations for a Sustainable Ocean Economy: A Vision for Protection, Production and Prosperity. Report. World Resources Institute.

Xu, Y. Wei, J. Li, Z. Zhao, Y. Lei, X. Sui, P. and Chen, Y. (2021). Linking ecosystem services and economic development for optimizing land use change in the poverty areas. *Ecosystem Health and Sustainability*, 7(1), 1877571.

Zanoli, R. Carlesi, L. Danovaro, R. Mandolesi, S. and Naspetti, S. (2015). Valuing unfamiliar Mediterranean deep-sea ecosystems using visual Q-methodology. *Marine Policy*, 61, 227–236.

10 The Blue Economy Paradigm and Seafloor Massive Sulfides along the Indian Ocean Ridge Systems

Niyati Gopinath Kalangutkar,[1] Ankeeta A. Amonkar,[2] and Sridhar D. Iyer[3]*
[1] School of Earth, Ocean and Atmospheric Science, Goa University, Taleigao Plateau, Goa, India
[2] Dnyanprassarak Mandal's College and Research Centre, Mapusa, Goa, India
[3] Formerly with CSIR-National Institute of Oceanography, Dona Paula, Goa, India
*Corresponding author: Niyati Gopinath Kalangutkar
E-mail: niyati@unigoa.ac.in; ankeetaamonkar9@gmail.com; sdiyer2001@gmail.com

CONTENTS

10.1 INTRODUCTION

The 70,000 km long global system of mid-ocean ridges (MOR) manifest in the Indian Ocean as four major ridge systems which are collectively called, the Indian Ocean Ridge System (IORS). The ridge systems are the Carlsberg Ridge (CR) which trends in an NW direction and protrudes into the Red Sea through the Gulf of Aden. The CR snakes towards the equator to form the Central Indian

Ridge (CIR) which bifurcates at the Rodriguez Triple Junction (RTJ, 25°S, 70°E) into the South West Indian Ridge (SWIR) and the South East Indian Ridge (SEIR) (Iyer and Ray, 2003). This inverted 'Y' IORS is less investigated relative to the Mid-Atlantic Ridge (MAR) which is also a slow to medium spreading ridge and has a comparable geology and tectonic architecture. Yet, the MAR has tens of hydrothermal vent sites of variable dimensions with abundant seafloor massive sulfides (SMS), also known as the volcanogenic massive sulfides (VMS) that are hosted by lava flow, basalt outcrops and serpentinites.

The IORS was believed to be less favourable for hydrothermal metallogenesis until the discovery of hot brine and metalliferous sediments in the Red Sea (Degens and Ross, 1969). In the Indian Ocean, a number of low and high intensity hydrothermal sites have been reported. Among the low intensity sites are some segments along the CR, regions near the Vityaz fracture zone and areas between latitudes 24° and 37°S and longitudes 49° and 60°E; and along the SEIR. Under the bilateral India-Germany collaborative programme *GEMINO*, several low intensity sites were found (Herzig and Plüger, 1988). A few high intensity sites such as the Red Sea spreading centre, the Sonne hydrothermal plume site (24°00.3'S and 69°39.6'E) and Geodyn plume site (19°29'S, 65°44'E) were located. The seafloor at the slow spreading Red Sea rift, representing an early stage in the opening of an ocean basin, contains one of the largest deep sea mineral deposits. The Atlantis II Deep (21°24'N and 38°03'E) is the most significant active hydrothermal site in the Red Sea, consisting of a stratified pool of high temperature (~ 56°C) brine, about 10 times more saline than the seawater. The metalliferous sediments have high concentrations of zinc (Zn 1.7%), copper (Cu 0.43%), silver (Ag 0.18%), and cobalt (Co 0.14%). The best estimate suggests that Atlantis II Deep deposits contain about 200 million tones (mt) of ore, including 3.2 mt of Zn and 0.8 mt of Cu (Swallow and Crease, 1965; Scholten et al. 2000).

The United Nations Convention on the Law of the Sea (UNCLOS) proposed a detailed legal framework for rights and obligations of countries to access, use and reclaim marine resources from territorial waters and open oceans. The UNCLOS document (Article 76) was signed on 10th Dec 1982 in Jamaica and implemented on 16 Nov 1994. The ocean space under the jurisdiction of a country can be classified into several maritime zones. A coastal nation has full rights over resources that can be derived from the air, the water column, the seabed, and sub-surface from its respective coastal waters (5.55 km into the sea from the coast), its territorial sea (5.55 to 22.2 km) and its contiguous zone (22.4 to 44.4 km). A nation can access resources from the water column, seabed and sub-surface that occur within its Exclusive Economic Zone (EEZ) (44.4 to 370 km), while resources only from the seabed and sub-surface strata can be exploited from the Extended Continental Shelf (ECS/CS) 370 to 647.5 km).

The global coasts and oceans are repositories of placer minerals (coastal and nearshore), phosphorites, fossil fuels (oil, gas, methane), SMS along the MOR, cobalt-rich crusts over seamounts and polymetallic manganese nodules in the abyssal depth. The exploration, mining and allied activities for these resources can be sustainably carried out by applying the various features of the blue economy (Mukhopadhyay et al. 2020).

The United Nations Conference on Sustainable Development (UNCSD) at the Rio+20 Conference (Rio de Janeiro, June 20–22, 2012) emphasized the concept of the 'Blue Economy' (BE) as it pertains to oceans and seas. The major sectors of the BE are food security, harnessing energy-minerals-pharma products, climate change, increasing trade and investments, improving maritime activities, tourism (leisure, recreation), employment opportunities, and socio-economic growth (Pauli, 2010). It has been suggested that the BE could support sustained fiscal growth, enhance social integration, and improve human welfare (UNCSD, 2018). During exploration and exploitation of marine minerals there are opportunities to develop and utilize innovative technology and further, there would be ample scope for skilled and unskilled workers, onboard and on land.

During the first IORA (Indian Ocean Rim Association) Blue Economy Dialogue that was held on 17th and 18th Aug 2015 in Goa (India) the sectors that were discussed were fisheries and aquaculture, renewable marine energy, accounting frameworks, ports, shipping and related activities, and explorations for marine minerals. The Dialogue was followed by the First Ministerial Blue Economy Conference (Mauritius, Sep 2–3 2015) and the Second Indian Ocean Dialogue (Perth, Australia, Sep 2015). In Mauritius, the Blue Economy Declaration was adopted and this sought to use ocean resources to boost a country's economy, create jobs, progress technologically, amongst others, while simultaneously protecting the environment *(www.iora.int)*. During the second ministerial BE conference (Indonesia, May 8–10 2017) the IORA Secretariat identified major sectors: Fisheries and Aquaculture; Renewable Ocean Energy; Seaports and Shipping; Minerals and Hydrocarbons; Tourism; Marine Biotechnology; and Research and Development.

The BE paradigm envisages mining resources in the above-mentioned maritime zones in best, efficient, responsible and workable ways. This is along the line of the UN's Sustainable Development Goal (SDG-14) that is concerned with conserving and a justifiable use of the oceans and seas. The resources available beyond the ECS are reserved for the common heritage of mankind, and cannot be mined by any country unless permitted by the UNCLOS (presently it is the International Seabed Authority, ISA based in Jamaica).

Deep sea minerals have been recognised as principal sources of base metals that are useful in high- and green-technology industries (Hein et al. 2013). In this chapter, firstly we synthesize the studies made of the IORS hydrothermal vents in terms of their geology, mineralogy, composition and other parametres. Secondly, this is followed by a discussion of the application of the BE to recover the SMS.

10.2 HYDROTHERMAL MINERALIZATION AND MORPHO-TECTONIC CONTROLS

According to Veizer et al. (1989) the modern seafloor hydrothermal ore deposits that are related to the mineralization of base metals reflect the ~100 Ma geological history of the Earth. Sea water-rock interaction leads to the leaching of metals and the sformation of hydrothermal convection systems in areas of rifting, subsidence and thinning of the crust. Initially, hot mafic-ultramafic magma acts as a heat source and initiates convective circulation of hydrothermal fluids that ascend within the serpentinized mantle peridotite and deposits SMS (Franklin et al. 2005; Garuti et al. 2008). The SMS is precipitated from the hot solution at ~600°C from aqueous solutions within the upper crust (Barnes and Rose, 1998). The congenial sites for a variety of mineral deposits are active magmatic arcs, continental margins, MOR, fore-arcs and back-arcs (Bierlein et al. 2009).

The formation and deposition of SMS are influenced by morpho-structural features such as MOR, fracture zones with deep roots into the upper mantle (Kutina, 1983), syn-volcanic structures, folds, faults, unconformities and shear zones, fault-bounded axial rifts, and seamount calderas adjacent to extensional structures submerged island arcs (Scott, 1992; Fouquet, 1997). Because these structures are the pathways for the ascending hydrothermal solution and control the geometry of the ore deposits, it is important to locate the economic mineral deposits through geological, geophysical, and geochemical approaches.

10.3 MORPHO-TECTONICS AND HYDROTHERMAL SITES OF THE IORS

We provide a gist of the studies of the work carried out along the IORS, by the international community and by India. Table 10.1 is a compilation of the hydrothermal areas that occur along the IORS while some of the hydrothermal sites are shown in Figure 10.1.

TABLE 10.1

Hydrothermal Vent Location and Work Carried out by Various Researchers

Sr no.	Name of the Site	Latitude	Longitude	Authors	Studies Carried out
	South West Indian Ridge				
	Western part	40°–60°S	10°–25°E	Suo et al. (2017)	Analysed spreading rate, bathymetry, gravity and geochemical data.
	Eastern part	20°–45°S	49°–70°E		
	–	37°47′S	49°39′E	Tao et al. (2011)	Mineralogy and geochemistry of sulfide chimneys.
	Segment 27	37°80′–37°50′S	50°80′–50°40′E	Yue et al. (2019)	Identified turbidity anomalies and oxidation reduction potential values.
	–	38°–38.4°S	48.1–48.7°E	Chen et al. (2021)	Bathymetry, normal faults, anomalous turbidity values and oxidation reduction potential.
	–	28°50′–26°50′S	63°–68°E	Agarwal et al. (2019)	Major, Trace and REE of RTJ and Mt Jourdanne samples.
	Tiancheng	27°51′S	63°55′E	Chen et al. (2018)	Discovered 2 hydrothermal fields.
	Tianzuo	27°57′S	63°32′E		
	Longqi field	37°47′S	49.6°E	Ji et al. (2017)	H_2, CH_4, and other chemical data of hydrothermal fluids.
	Longqi	37°47′S	49.6°E	Zhou et al. (2018)	Investigated biodiversity and biogeographical relationship.
	Tiancheng	27°51′S	63°55′E		
	Duanqiao	37°39′S	50°24′E		
	Kairei	25°S	70°E	Han et al. (2018)	Mineralogy and geochemistry of hydrothermal precipitates.
	Pelagia	26°S	71°E		16S rRNA tags from different sites were analysed and compared with other marine environments.
	• SWIR from 49°E to 53°E	38°S	49°E	Tao et al. (2014)	REE, XRD of pyrite, silica, opal and sulfide deposits.
		36°S	53°E		
	• Longqi	37°47′S	49.6°E		
	• 50°24′E hydrothermal field	37°39′S	50°24′E		
	• 50°56′E carbonate field	37°37′S	50°56′E		
	• SWIR 63° field	27°57′S	63°32′E		
	–	43°S	40°E	Ren et al. (2016)	Analysis of topographic, geology, geophysics and metallogenic data. Proposed a prospecting prediction model.
	Longqi	37°47′S	49.6°E	Wang et al. (2018)	He-Ar-S isotopes
	Tiancheng	27°51′S	63°55′E		
	Duanqiao	37°39′S	50°24′E		
	Yuhuang	36°–38°S	49°–52°E		
	Kairei	25°S	70°E		
	Edmond	23.8778°S	69.5973°E		

Site	Latitude	Longitude	Reference	Description
Different sites	27°–38°S	46°–63°E	Chen et al. (2021)	Geochemistry of surface sediments and hydrothermal deposits.
Yuhuang-1	36°–38°N	49°–52°E	Liao et al. (2019)	Zn isotope compositions, element ratios of Zn, Fe, Cu, and Cd in sulfides.
Yuhuang-1	36°–38°N	49°–52°E	Liao et al. (2019)	Major and trace elements, sulfur isotopes analysis.
Duanqiao	37°39′ S	50°24′ E	Zhu et al. (2020)	Elemental concentration and Hg isotope analysis.
Yuhuang	36°–38°S	49°–52°E		
–			Kalangutkar et al. (2021)	Hydrothermal signatures in FeMn from SWIR

Central Indian Ridge

Site	Latitude	Longitude	Reference	Description
Onnuri Vent Field	8°10.1′S	68°08.2′E	Kim et al. (2020)	Plume sample, water column, fauna of vent samples.
Segment 2	9°47′S	66°41.9′E		
Segment 3	11°20′S	66°26′E		
Edmond vent field	23°52.68′S	69°35.80E	Wu et al. (2018)	Mineralogy and geochemistry of sphalerite to identify different textures and micro-environments.
A1A	20°20′S	68°E	Briais (1995)	Analysis of segments between 20°30'S and 25°30'S (Rodriguez Triple Junction).
A1B	26°S	70°30′E		
A2A	23°23.56′S	69°14.53E	Halbach et al. (1995)	Sonne Field – First SMS in the Indian Ocean.
Yokoniwa	25°16′S	70°05′E	Fujii and Okino (2018)	Magnetization of hydrothermally altered zone and host lava flows.
Kairei	25°21′S	70°03′E		
Solitaire field	19°33.410S	65°50.89′E	Kawagucci et al. (2016)	Fluid chemistry and microbial communities in chimney habitats.
Dodo hydrothermal field	18°20.190S	65°17.99′E		
Between 10°18'S and 10°57'S	10°47.5′S	66°38.6′E	Ray et al.(2020)	Analysis of dissolved Mn and He
Yokoniwa	25°16′S	70°04′E	Fujii et al.(2016)	Magnetic studies using an AUV and manned vehicle.
Dodo	18°20′S,	65°17E	Nakamura et al. (2012)	Measured chlorine, dissolved gases (H_2, CH_4, CO_2, and so forth.), pH, fauna and flora, rRNA gene sequencing was done.
Solitaire	19°33′S	65°50E		
A1A	20°20′S	68°E	Briais (1995)	Bathymetry, fracture zones, tectonics.
A1B	26°S	70°30′E		
A2A	23°23.56′S	69°14.53′E	Halbach et al. (1995)	Massive sulfide mineralogy and chemistry.
MESO	23°23.56′S	69°14.53′E	Halbach et al. (1998)	Geology, mineral zonation, different sulfide types, stages of formation and decay of a modern SMS.
MESO	23°23.63′–38°3.38′S	69°14.43′– 69°14.48′E	Halbach et al. (2002)	Hydrothermal sulfide impregnated and pure silica precipitates, sulfides chimney.

(continued)

TABLE 10.1 (Continued)
Hydrothermal Vent Location and Work Carried out by Various Researchers

Sr no.	Name of the Site	Latitude	Longitude	Authors	Studies Carried out
	MESO	23°23.56'S	69°14.53'E	Halbach and Münch (1997)	Study of sulfide chimneys.
	MESO	21.5°–23° S	68.5°–69.25° E	Herzig and Plüger (1988)	Mapping, photography, sampling to locate fossil/recent hydrothermal activity. Geochemistry of basalts, sediments, and water.
	MESO	27°–28°S	65°20'–66°40'E	Muller et al. (1999)	Variation of oceanic crustal thickness using seismic velocity model.
	MESO	23°23.63'S	69°14.43'E	Münch et al. (1999)	Hydrothermal mineralization, structural control, mineralogy, and geochemistry of sulfide chimneys.
	MESO	23°23.63' - 38°3.38'S	69°14.43' - 69°14.48'E	Lalou et al. (1998)	Radiochronological investigation of hydrothermal deposits.
	MESO	23°23.56'S	69°14.53'E	Plüger et al. (1990)	Geology
	—	23°52.68'S	69°35.80'E	Gallant and Von Damm (2006)	Chemical composition of hydrothermal fluids.
	—	23.88°S	69.60°E	Kumagai et al. (2008)	Geology and tectonics.
	Carlsberg Ridge				
	Wocan	6°22'N	60°31'E	Wang et al. (2020)	Sulfur and iron isotope analysis.
	Wocan 1	6°21'40'– 6°21'50'N	60°31'45'–60°31'30'E	Qiu et al. (2021)	Mineralogy, chemistry, Pb–Sr isotopes.
	Wocan 2	6°22'30'– 6°23'N	60°30'45'–69°30'15E		
	Carlsberg ridge	3°42'–3°41.5'N	63°40'–63°50'E	Ray et al. (2012)	Temperature anomaly, oxidation-reduction potential, dissolved Mn and ^{3}He were analysed.
	Carlsberg	6°21.796'N	60°31.534'E	Popoola and Akintoye (2019)	Geochemistry of sediments and sulfides.
	Wocan-1	6°21.866'N	60°30.372'E		
	Wocan-2	6°35.675'N	60°13.190'E		
	Ridge flank	4° 07.52'N	69°20.201'E		
	Core sediments				

Tianxiu	3.67°N	63.83°E	Chen et al. (2020)	Precipitation of calcite veins in serpentinized, Carbon and Oxygen isotopes.
Wocan	6°22'N	60°31'E	Popoola et al. (2019)	Morphology, mineralogy and geochemistry of Fe-Si-Mn oxyhydroxides, sulfur isotopes.
Daxi	6°48'N	60°10'E	Wang et al. (2020)	Mineralogy, chemistry, and bathymetry studies.
Along ridge segment	10°N	66°E	Yu et al. (2016)	Major element and REE of 30 sediments from 24 sites.
Wocan	6°22'N	60°31'E	Wang et al. (2020)	Mineralogy and chemistry of Cu - rich chimneys and massive sulfides.

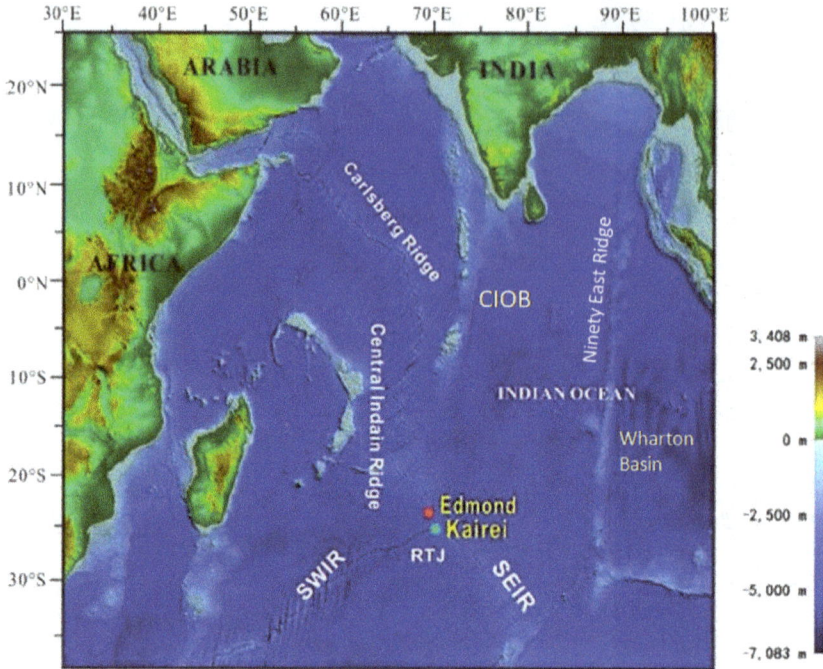

FIGURE 10.1   India's exploration area for SMS along the SWIR (Modified after https://isa.org.jm/index.php/map/government-india-0).

10.3.1   THE CARLSBERG RIDGE (CR)

The Carlsberg Ridge is a slow spreading ridge with half-spreading rates between 11 and 16 mm/yr (henceforth half-rate will be used). The CR is devoid of major transform faults and is segmented by dextral, non-transform, and second-order discontinuities. Indications of weak hydrothermal activity were earlier detected in the CR (Kempe and Easton, 1974) and this was confirmed by iron-rich (28%) basal sediments from the DSDP Site 236 (1°40'S, 57°38'E) (Baturin and Rozanova, 1975).

During the maiden voyage in 1983 of *ORV Sagar Kanya* (India) from Germany to Goa, a segment of the CR was dredged and the basalts (Banerjee and Iyer, 1991; 1993; 2003; Iyer and Banerjee, 1993) and geophysical aspects were reported (Ramana et al. 1993). Subsequently, the ridge section between 2°30'S and 4°30'S and 62°30' to 66°15'E was mapped and basalts and upper mantle rocks were recovered (Mudholkar et al. 2002). Studies reported event plumes (Murton et al. 2006) and identification of hydrothermal activities along the various segments of the CR. Ray et al. (2012) reported hydrothermal plumes from unknown active vent(s) near 3°42'N/63°40E and 3°41.5'N/63°50'E. The magmatic/hydrothermal chalcopyrite, pyrite, and magnetite in the basalts at 3°37'S/64°07N (Banerjee and Iyer, 1993; 2003) are similar to sulfide - oxide minerals in the basalts at 5°23'N (Baturin and Rozanova, 1975).

During the 26th Chinese COMRA (China Ocean Mineral Resources Research and Development Association) a hydrothermal activity field with SMS was located along the CR at 3.5° - 3.8°N. Evidence for two separate vent fields were identified, one near 3°42'N, 63°40'E (Wocan) and another at 3°41.5'N, 63°50'E (Daxi). Prominent optical backscatter and thermal anomalies coupled with chemical (for example, helium ^{3}He, manganese Mn) signatures in seawater demonstrated the existence of hydrothermal sources on off-axis highs on the south wall of the CR. Although ultramafic rocks have been recovered near these sites, the light-scattering and dissolved Mn anomalies indicate that the plumes do not arise from a system driven solely by exothermic serpentinization (Ray et al. 2012). It was suggested that the source fluids for these two active sites may be a product of both ultramafic and basaltic/gabbroic fluid-rock interaction, similar to the Rainbow and Logatchev fields, MAR.

i) Wocan Hydrothermal Field: During the Chinese DY28[th] cruise along the CR in 2013, the basalt-hosted Wocan Hydrothermal Field (WHF) was found on the NW slope of an axial volcanic ridge at a water depth of ~3,000 m. The hydrothermal precipitates that were recovered were classified into four groups: (i) Cu-rich chimneys; (ii) Cu-rich massive sulfides; (iii) Fe-rich massive sulfides; and (iv) silicified massive sulfides (Wang et al. 2017). The mineralogy and geochemistry of metalliferous sediment were studied at the Wocan hydrothermal field active site (Wocan-1) and an inactive site (Wocan-2). Based on the mineralogy and morphology of sulfide and non-sulfide grains, bulk composition, and sulfur isotopes it was concluded that at Wocan-1 there is an intermediate - high temperature hydrothermal discharge; while Wocan-2 shows a moderate - extensive oxidation and secondary alterations by seawater in a low - intermediate environment (Popoola et al. 2019).

ii) The Daxi Vent Field: The Daxi Vent Field (DVF) is a basalt-hosted hydrothermal field located on a rifted volcanic ridge along a non-transform offset between two second-order ridge segments. At the DVF there are three hydrothermal sites, namely Central mound, NE mound, and South mound. Eight black smokers were observed in the Central mound which hosts the largest sulfide chimney 'Baochu Pagoda' of ~24 m height. Another inactive silica-rich chimney was observed in the NE mound. The sulfide chimneys are dominated by sphalerite and pyrrhotite with high Sn, Co and Ag; and silica-rich chimneys have high SiO_2 and Ba contents (Wang et al. 2020).

10.3.2 Central Indian Ridge (CIR)

The CIR with a half spreading rate of 20 - 30 mm/yr has structures, spreading kinematics and isotope geochemistry of erupted lava that are remarkably different from the other MOR (Drolia et al. 2003). Exploration related activities in the Indian Ocean commenced about four decades ago sometime in early 1983. The initial results were encouraging with the finding of characteristics He and Mn anomalies that indicated hydrothermal plume activities along segments of the CIR (Herzig and Plüger, 1988). The discovery of two fossil hydrothermal vent fields, the Sonne Field, and Mount Jourdanne Field, led to several new exploration programs in this region (Halbach et al. 1998; Munch et al. 2001). These were followed by the detection of the active hydrothermal fields, Kairei, and Edmond, where the first direct observations were made of active hydrothermal discharge, vent biota, and shimmering water (Hashimoto et al. 2001; Gamo et al. 2001; Van Dover et al. 2001). Some details of the hydrothermal sites found along the CIR are provided below.

(i) Sonne: Herzig and Plüger (1988) and Plüger et al. (1990) respectively reported the existence of Sonne (an inactive hydrothermal field, named after the famous German research vessel *FS Sonne*), and a first indication of a hydrothermal plume site (24°00.3'S) along the CIR. The Sonne field at 23°23.6'S and ~200 km NW of the RTJ, consists of hydrothermally influenced basalts and sediments, layered FeMn precipitates, and blocks of massive sulfides. The Edmond and Kairei hydrothermal fields were first recognized in 1993 and reported by Gamo et al. (1996; 2001) and Hashimoto et al. (2001), whereas the Dodo and Solitaire hydrothermal fields were discovered later (Nakamura et al. 2012).

(ii) Meso zone: The Meso zone is named after the *RV Meteor* and *RV Sonne* zone and is located at 23.3927°S and 69.2422°E. The Meso zone is at a distance of 270 km N of the RTJ on a neo-volcanic intra-rift ridge and covers an area of ~0.6 km². Three sites were identified with evidence of hydrothermal activity (Halbach et al. 1998). The sites are the Talus-Tips-Site (TTS) in the northern part, the Sonne Field (SF) in the central part and the Smooth Ground (SG) in the southern part of the mineralized zone *(www.interridge.org)*. Hydrothermal mineralization and structural control in the Meso zone region were detailed by Munch et al. (1999) and sulfide-impregnated and pure silica precipitates of hydrothermal origin were reported by Halbach et al. (2002).

(iii) Kairei and Edmond: The Kairei and Edmond hydrothermal fields are located ~6 km to the east of the spreading axis on the eastern wall of the axial valley (Wilson, 1993). The Kairei field is developed on shoulder of the west-facing slope of the abyssal hill of CIR-1 (Hakuho Knoll) with flat or lobate lava flows, whereas the Edmond field has flat, partly wrinkled lava flows. Kairei field is along a linear ridge which is perhaps an abandoned ridge axis formed during ridge jump and has dunite and troctolite and a regional seafloor morphology that is distinctly heterogeneous within 30 km of the current ridge axis while regular ridge-parallel abyssal hills occur along the Edmond field (Kumagai et al. 2008; Van Dover et al. 2001). Both the fields have large and complex chimney structures with large massive sulfide mounds at their bases (Nakamura et al. 2012). But the morphological contrast between the two fields might have influenced the pathway of the recharged vent fluid, as evident from the composition of the fluids (Gallant and Von Damm, 2006).

Copper-rich chimney edifices and fragments rich in chalcopyrite, with pyrite, marcasite, wurtzite, and sphalerite occur in Kairei. Granular chalcopyrite decreases in amount and grain size towards the outer parts of the chimneys, while disseminated sphalerite and pyrite increase in the outer parts. This fact indicates a fall in temperatures towards the outer parts of the chimney. Active chimneys are nearly fresh and weathered products are present on the outer wall in contact with seawater or in inactive vents (Han et al. 2018). At the Edmond site are native Cu and Cu-sulfides (covellite, digenite and chalcocite), altered chalcopyrite, and outer walls have plentiful abundant sub-microscopic Au-Ag alloys (Wu et al. 2018).

(iv) Dodo and Solitaire: The Dodo hydrothermal field with active vents (18°20.1′S, 65°17.9′E; water depth 2,745 m) is located in the Dodo Great Lava Plain on the spreading axis of CIR segment 16 (Nakamura et al. 2012). The hydrothermal field is 10 km with smooth sheet flow lavas along the axis that indicate high production rates of basaltic melt, a feature similar to the East Pacific Rise (EPR). Potsunen, Tsukushi-1, and Tsukushi-2 are the three main chimneys. Black smoker discharges occur at Tsukushi-1 whereas, active chimneys and several inactive chimneys are near Tsukushi-2 (Nakamura et al. 2012). Extensive plume surveys using vertical and tow-yo hydrocasts and an autonomous underwater vehicle (AUV) led to identify anomalous concentrations of methane (CH_4), Mn, and ^3He (Kawagucci et al. 2008).

The Solitaire field (19°33.413'S, 65°50.888′E; at a depth of 2,606 m) is located on the Roger Plateau on the western ridge flank of CIR segment 15. Plume signatures of hydrothermally derived CH_4, Mn and ^3He abundance and a light transmission signal anomaly were evident (Kawagucci et al. 2008). In this field, three major chimney sites (Toukon-3, Tenkoji, and Liger) were identified with chimneys <5 m in height. At the Toukon-3 chimneys the emissions are clear fluids and a few black smoker discharges (Nakamura et al. 2012).

(v) OCC 1-1, OCC 2-1, OCC 3-1, OCC-3-2, OCC-4-1, and OCC-4-2: Strong hydrothermal plume signals were measured over the Oceanic Core Complexes (OCCs) along long-lived detachment faults that formed because of tectonic extension in the middle part of the CIR (8°S to 17°S) which has a morphology typical of slow spreading ridges (Pak et al. 2017; Kim et al. 2020). The OOCs are conduits for hydrothermal fluids which rise at off-axis regions. Pak et al. (2017) felt that the serpentinization and latent/cooling heat of the underlying mantle and magma supply heat for hydrothermal circulation, resulting in high-CH_3 concentration in the plumes. The Onnuri Vent Field (OVF) is located at the summit of OCC-3-2, and vents clear, low-temperature fluids, located on the ridge flanks of typical abyssal hill structures of a symmetrical ridge section. Hydrothermal mineralization is primarily silica-rich and disseminated sulfide with secondary Cu minerals, associated with hydrothermal precipitates (Kim et al. 2020)

India commenced the investigations of the CIR (initially funded by the Office of Naval Research and NSF, USA and later by the Indian government under the InRidge programme) and undertook studies between 3°S and 11°S, and between 66°E and 69°E. The areas included the transform faults (TF) Sealark, Vityaz, and Vema and the intervening ridge segments (Drolia et al. 2003). Later the morphotectonic features and petrological variations between 20°30′S and 23°07′S were detailed

(Mukhopadhyay et al. 2015). The possibility of hydrothermal activity in certain segments of the CIR was postulated by Banerjee and Ray (2013, 2015 and references therein). The magmatic and tectonic processes that resulted because of the interaction between the Reunion plume and CIR at the Vema Trench and along the Vema Fracture Zone was detailed (Dhawaskar et al. 2020). The InRidge programme also included studies of the CR and Andaman Back Arc Basin (ABAB) which are separately discussed.

10.3.3 SOUTH-EAST INDIAN RIDGE (SEIR)

The SEIR is an intermediate spreading (30–35 mm/yr) and this is the fastest spreading rate of all the IORS (DeMets et al. 1990).

(i) Antarctic Australian Ridge (AAR): The AAR with a series of ridge segments and transform faults extending from 140°E to 180°E, has an intermediate spreading (~39-30 mm/yr) and its axial depth is relatively shallow (~2,200 m) (Choi et al. 2013). The KR1 and KR2 are first-order segments and bounded by transform faults. Hydrothermal activity has been noted at two first-order segments of the AAR: KR1 and KR2. Optical and oxidation-reduction-potential anomalies indicate multiple active sites on both segments (Hahm et al. 2015).

The KR1 segment (139.5°E, 122°W) shows large variations in its axial morphology that point to a variable magma supply. Alkalic to tholeiitic magmatism along KR1 may be potential source materials for alkaline basalts and are considered to be ancient, recycled oceanic crust (namely, eclogite) as well as sub-KR1 depleted MOR basalt mantle (DMM). Whereas the main source materials for the KR1 tholeiites are presumed to be the DMM-dominant lithology with minor recycled material (Yi et al. 2021). An off-axis seamount chain intersects the ridge at 158.33°E, where the ridge morphology changes from axial rift to axial high. Seventeen sites were identified along the KR1 with the Mujin hydrothermal site, near the centre, having ^{3}HeA of up to 3.8 fmol/kg in water samples (Hahm et al. 2015).

KR2 is a 180 km long segment that progressively deepens from 2,200 m in the west to 2,500 m in the east. An offset divides KR2 into two segments, an eastern rift valley, and a western axial high. The variability of the magma supply is apparently lower than at KR1(Hahm et al. 2015).

(ii) Boomerang Seamount: This active seamount was discovered during a bathymetric survey in 1996. This basaltic seamount lies along the SEIR axis, 18 km NE of Amsterdam Island and marks the site of the Amsterdam - St. Paul hotspot. The seamount rises to within 650 m of the ocean surface and has a 2 km wide summit caldera that is 200 m deep. Rift zones that extend to the SE and N give the seamount its arcuate shape. Water column temperature anomalies above the seamount suggest the presence of hydrothermal activity within the caldera (Johnson et al. 2000).

(iii) Pelagia vent: Pelagia hydrothermal field (26°09.40'S, 71°26.26'E) is located within the neovolcanic zone of the SEIR and near the RTJ at water depth of 3,690 m. Active smoking vents in this site were found to be up to 20 m high on top of a mound of sulfide talus (Noowong et al. 2021). The chimneys have intricate intergrowth of different minerals, while weathered products occur on the outer wall of the vent. The vent fluid flows towards the chimney walls because of abundant pore spaces that have resulted due to aggregates of collomorphic pyrite/marcasite, and sphalerite surrounded by chalcopyrite, lath-shaped pyrrhotite, and amorphous silica, lined with traces of sulfides. All these are evidence of the high-temperature environment prevalent in the area. An inactive chimney depicts replacement of chalcopyrite-isocubanite by secondary copper minerals (Han et al. 2018).

10.3.4 SOUTH-WEST INDIAN RIDGE (SWIR)

The ultraslow-spreading SWIR represents one of the important end-member MOR types because of its very slow and oblique spreading rate of 7-9 mm/yr. The first evidence of high-temperature hydrothermal activity was identified by German et al. (1998). Later a survey was carried out using submersible Shinkai 6500 and temperature anomalies of ~0.1°C were recorded at 31°05'S, 59°00'E and 27°54'S, 64°29'E (Sohrin and Gamo, 1999). The hydrothermal structures are related with E-W trending graben and smaller fissures and cracks (Munch et al. 2000). Geophysical, optical back-scatter and deep-tow side-scan sonar surveys of the rift-valley floor (54° - 67°E) helped to detect six sets of plume signals (German, 2003). A recent morphological and compositional study of the FeMn crusts from a segment of the SWIR indicated distinct hydrothermal signatures from their formation (Kalangutkar et al. 2021). Information about four hydrothermal sites that occur along the SWIR are given below.

(i) Mount Jourdanne: The Melville fracture zone acts as a dividing line for two distinct morphological characteristics. The western side of Melville fracture zone is associated with the highest number of volcanoes per segment indicating a shallower spreading centre (4,400 m) (Mendel et al. 1997). Abyssal tholeiites occur up to the Atlantis II fracture zone while to its east and until the RTJ, the number of volcanoes are less, the spreading centres are deeper (4,800 m) and host sodic and titaniferous glasses (Natland, 1991).

Hydrothermal precipitates in water depths of about 2,960 m close to the top of a neovolcanic ridge (Mount Jourdanne) and weathered reddish-brown SMS of about 5 m are present as small mounds along with small tube-like chimneys. The strongest temperature anomalies of ~0.1°C were recorded at Mt. Jourdanne (27°50.97'S, 63°56.15'E) (Fujimoto et al. 1999). Due to a volcanic heat source and conduits for fluid convection, several extinct hydrothermal sites occur within an area of approximately 0.5 km² at a water depth of about 2,941 m within graben or smaller fissures. The chimney edifices rise for approximately 40 to 50 cm from the seafloor and are about 10 cm in diameter. No hydrothermal activity, shimmering waters, chemical anomalies, or biological features were recorded. The summit of Mt. Jourdanne is characterized by E - W trending graben and by basaltic pillows and lava tubes whereas the shallower slopes are dominated by sheet flows (lobate, folded) that are often covered by a thin sediment layer (Munch et al. 2000; Munch et al. 2001).

(ii) Tiancheng and Tianzuo: In Tianzuo hydrothermal field, two inactive, ultramafic-hosted vents (Tiancheng and Tianzuo) occur in the ridge section 63° - 64°E between the Melville fracture zone and RTJ and southwest of the relict Mt. Jourdanne field (Tao et al. 2012). Hydrothermal signatures in sediments reported from 63°E to 68°E (Agarwal et al. 2020)

(iii) Duanqiao and Yuhuang: The Duanqiao and Yuhuang hydrothermal fields are between the Indomed and Gallieni fracture zones at the central volcano along the SWIR (Zhu et al. 2020). Bouguer gravity results indicate the crustal thickness to be between 3 and 10 km (average: 7.5 km) with the maximum crustal thickness of 10 km in the Duanqiao field. This is the thickest crust discovered along the SWIR (Sun et al. 2018). The Duanqiao (inactive) (50.5°E) field lies on an axial highland with a shallow depth of ~1,700 m and relatively flat surrounding terrain (Sun et al. 2018). This field has relict chimneys, massive sulfides, opals, basalts, and metalliferous sediments (Tao et al. 2012). As compared with other areas of the SWIR, abundant siliceous samples such as opals have been recovered that are evident of low-temperature hydrothermal activity (Yang et al. 2019). The Yuhuang (49.2°E) inactive field is located on the south rift wall of segment 29 of the SWIR, approximately 7.5 km from the ridge axis and at water depth ranging from 1,400 to 1,600 m.

(iv) Dragon Horn: The Dragon Horn field with sulfide-bearing vent was identified along an OCC and is located on the south flank of the SWIR segment 28 (~49.7°E) and it exhibits high-temperature hydrothermal vents that are associated with a major detachment fault system. Twin detachment faults penetrate to a depth of 13±2 km below the seafloor. Dragon Horn is a basalt-hosted active vent field at water depths of 2,700 - 2,900 m comprising of two sulfide-bearing vents: Longqi-1 and Longqi-3 (Tao et al. 2012). The Longqi-1 field is located at segment 28 (~49.7°E) along the Dragon Horn region on the southern flank of the ultra-slow spreading SWIR. Three hydrothermal vents, namely S, M, and N have been confirmed at the Longqi-1 field (Tao et al. 2012). The inactive Longqi-3 field is a hydrothermal plume anomaly site with a possible linear mineralized zone along the detachment fault 2 in serpentinized peridotite along with carbonate sediments. The evidence points to the presence of low-temperature to the east side of the OCC. At the hydrothermal field the calculated Bouguer gravity shows a crustal thickness of ~3 km (Tao et al. 2012).

Considering the above reports it has been shown that areas where SMS occur in the Indian Ocean can be classified into two types: (1) at or near the neovolcanic ridges of the rift valley floor, for example, Meso (Halbach et al. 1998), Mount Jourdanne (Münch et al. 2001), Dodo (Nakamura et al. 2012), Solitaire (Nakamura et al. 2012), Wocan (Wang et al. 2017), Duanqiao (Yang et al. 2017), and Pelagia (Han et al. 2018), and (2) elevated off-axis deposits on the rift valley wall, for example, Edmond (Van Dover et al. 2001), Kairei (Van Dover et al. 2001), Longqi (Tao et al. 2012), 3.69°N (Tao et al. 2013), Yuhuang (Liao et al. 2018), and several sites related with OCCs on the CIR (OCC-3-2, OCC-4-1, and OCC-4-2; Pak et al. 2017).

10.4   THE BLUE ECONOMY OF SEAFLOOR MASSIVE SULFIDES

Although several countries have investigated and even discovered tens of hydrothermal vents in the ocean, surprisingly, only a handful of countries are registered contractors with the ISA (erstwhile UNCLOS). In contrast to the 19 contractors for polymetallic nodules in the Pacific and Indian oceans, there are only seven contractors for exploration of the SMS. This is even though the SMS deposits occur at relatively shallower water depths than the nodules (>5,000 m). The contractors for SMS in the Indian Ocean are one each in the SWIR and CIR, and five in the MAR. These contractors are India, China, Germany, and Korea. The three contractors along the MAR are Poland, France and Russia.

The National Centre for Polar and Ocean Research (NCPOR, Goa India) (erstwhile the National Centre for Antarctica and Ocean Research) under the support of the Ministry of Earth Sciences initiated a mission-mode multi-disciplinary program on exploration of the SMS along the SWIR and CIR. In 2014, India obtained a 15-year licence from the ISA to explore 10,000 km of the CIR and SWIR for SMS and in 2016 India signed a 15-year exploration contract which would expire on 25th March 2031. The first cruise was undertaken on 12th Jan 2017 along segments of the CIR and SWIR. Multi-disciplinary efforts were made to locate potential SMS deposits. Seamounts are easier to sample since these may occur at shallow water depth (500 m) and host ferromanganese (FeMn) oxides with significant contents of cobalt (~1%). Such seamounts have been targeted for mining by five contractors: Korea, Japan and Russia in the Pacific Ocean, Brazil in the South Atlantic Ocean and China in the Western Pacific Ocean (Mukhopadhyay et al. 2018). In the Indian Ocean, the Afanisy-Nikitin Seamount (ANS) has significant cobalt (Co 0.3-0.9%, average: 0.65%), rare earth elements and platinum (200 - 900 ppb) (Rajani et al. 2005; Balaram et al. 2012). But beyond this preliminary work, the ANS has so far not been earmarked for exploitation by India.

Sustainable and profitable mining, either on land or from the deep-seas, involve one or more of the 5Es: Exploration, Environmental studies, Exploitation, Enrichment and Economics. A successful completion of all these 5Es could result in 'Minerals to Market.' We discuss the importance of the BE for India (as a contractor) in her endeavour to comprehend the potential economic viability and related issues of the SMS deposits.

10.4.1   EXPLORATION

In the late eighties and early nineties, ridge research in India was mostly individual driven (for example, Mukhopadhyay and Iyer, 1993) and there was an absence of integration. Hence, to have a synergy at a national level, in 1997 the Council of Scientific and Industrial Research-National Institute of Oceanography (CSIR-NIO), Goa initiated a major programme, 'Tectonic and Oceanic Processes Along the Indian Ridge System and Backarc Basins.' Simultaneously, the InRidge (India's Ridge Research initiative) was formed and India became an Associate Member in the global

InterRidge body. InRidge provided opportunities to individuals and institutions to collaborate, save funds and resources, help avoid duplication of research efforts, share samples, ship time and train researchers and students. The other reasons were lack of data pertaining to the IORS and their easy accessibility from Indian shores.

The areas of studies chosen under the InRidge were the CR, CIR and ABAB, to understand tectonic architecture, transform faults (TF), ridge-transform interaction (RTI), incipient triple junction formation, interaction of the wide deformation zone, seamounts, petrological variations, and most importantly, to search for hydrothermal vents. The Geological Survey of India (GSI) also conducts its studies in the CR and ABAB. Although India started her ridge studies much later than other countries who have extensively studied the MAR, SWIR, EPR and other areas in the world's oceans, nevertheless much impact and many findings have resulted from the InRidge. Currently, investigations are underway along segments of the CR, CIR and SWIR and plans are afoot to examine the SEIR.

The hydrothermal fields along the SWIR, SEIR, CIR and CR are generally sited within axial rift valleys or rift flanks on segment centers, hence these need to be targeted to locate hydrothermal vents. In the exploratory work, India could consider collaborating with the IORA countries such as Sri Lanka, Maldives, Mauritius, Seychelles, and Madagascar who would assist to expedite the expeditions to the IORS in a faster, easier and less expensive manner. This effort could lead to training of personnel, job opportunities and development in scientific and port infrastructure and economic growth in the collaborating countries.

10.4.2 ENVIRONMENT

Extracting the SMS deposits could be a challenging task. This is because mining will inevitably affect the environment, but several metals from the SMS are required in technologies that are vital to society to have a low-carbon future and to achieve the global sustainable development goals (Lusty and Murton, 2018). It is mandated that under Regulation 32 the Contractors need to undertake environmental baseline studies (ISA, 2013) as recommended and outlined by ISA's Legal and Technical Commission (LTC). The LTC stipulates that 'the best available technology and methodology for sampling should be used in establishing baseline data for environmental impact assessments.' Besides following the protocols, plausible solutions and mitigatory measures also have to be outlined by the contractor.

Before, during and after mining, an array of complex environmental impacts need to be assessed. These would include physical oceanography, sediment characteristics (physical and chemical), the sinking rate of particles, the aggregation of particles, toxic discharges, biological studies (microbes to mammals), biodiversity, ecosystem functioning, hydrodynamic plume modelling, noise and light hazards, amongst others (Billett et al. 2019). Noowong et al. (2021) reported the molecular composition of dissolved organic matter (DOM) of Kairei (CIR) and Pelagia (SEIR) vents. The vent fluids (>330°C) were extremely rich in dissolved Fe, Si, K, Li, Mn and Zn compared to the seawater. The DOM from these fluids was different than that from diffuse fluids and plumes, which had a predominant signature of the seawater DOM.

Hydrothermal vents shelter a variety of biota that rely on microbial chemosynthesis by using hydrogen sulfide and methane in the hot vent fluid as sources of energy (Van Dover, 2000). Globally, about 600 of such sites have been located (Beaulieu and Szafrański, 2020) and most of these are in the Pacific and Atlantic Oceans (Thaler and Amon, 2019). The active vent ecosystem is a rare habitat and comprises an estimated 50 km^2, that is < 0.00001% of the Earth's surface area (Van Dover et al. 2018). Therefore, mining could potentially harm the biota that are endemic around the vents.

The deep-sea environment could also change in other ways during mining operations. For example, a deep drilling operation at the Iheya North hydrothermal field (Okinawa Trough, Pacific Ocean) revealed that the vent-clam/soft sediment habitat transforms into a crust with higher temperature flow and the presence of bacterial mat and squat lobsters (Nakajima et al. 2015). On the other hand, when drilling took place at ODP Leg 158 in the TAG area of the MAR there was hardly any change in the nature of the shrimp-dominated vents (Copley et al. 1999, 2016). Under such circumstances it is difficult to foresee the damage, or lack of it, at hydrothermal vents.

Both terrestrial and deep-sea mining influence the environment in different ways and more so when sea-minerals are mined. This is because during mining the water column, bottom water, biota (bottom-dwelling, surface and water column inhabitants) and bottom sediments would be altered. Therefore, after exploration and prior to exploitation, an in-depth study is required concerning the Environment Impact Assessment (EIA) and Environment Monitoring and Planning (EMP). The EIA and EMP are multidisciplinary approaches to address the concerns that the stakeholders and people would have once mining starts. Baseline data need to be obtained, namely, observations and measurements of several parameters of geological, biological, physical, and chemical nature, of the water, biota, and sediments. The data, samples, and observations must be validated, compared, and checked for variations in reference (control) and test (experimental) areas. The investigations could range from a few days to 2 - 3 years and be a continuing process, even after mining.

Filho et al. (2021) reviewed the potentials and risks of deep seabed mining by considering the legal aspects and environmental impact. During mining, the seabed could be significantly disturbed, the creation of sediment plumes, and these together with light and noise pollution would affect the surface, benthic, meso, and bathypelagic zones. A systems approach to adaptive management was proposed by Hyman et al. (2021) that could help to guide and better manage the environmental aspects deep-sea mineral extraction.

Except for routine geological, biological and seawater sampling at the licensed site of the SWIR, India has not reached the stage where EIA/EMP tasks are being carried out. But the data so far collected would help to undertake the detailed EIA work in the future.

10.4.3 Exploitation

During the mid-1970s successful pre-pilot mining and metallurgical testing operations were carried out at the Atlantis II deep site. It was then presumed that very soon there would be a clamour to mine for deep-sea minerals from the world's ocean. To-date, this dream of the scientific community is unrealised due to factors other than technological advancements in mining. Further we need to allay the growing global fear for the marine environment, both by people and regulatory bodies such as the ISA. Despite such concerns, there is now a renewed interest ini the exploitation of the SMS deposits given the ever-growing global population, industrialisation, an enhanced demand for metals, and geopolitical issues, amongst others.

Deep-sea mining is an expensive proposition as it is essential not only to have high resolution mapping techniques but also to fabricate equipment that can withstand the erosive effect of the sea-water and hydrostatic pressures. In addition, low-cost autonomous or remotely operated vehicles (AUV/ROV) would be necessary to locate and evaluate SMS deposits. The mined materials that are recovered by the mother ship must be transported to onshore processing laboratories using supply vessels. Presently some limited mining systems exist but not for commercial operations as is the case for nearshore placer deposits that are mined by a few countries, including India. Once the mining lease and environmental clearances are obtained then the SMS deposits of the Solwara 1 site in the Bismarck Sea (Papua New Guinea) could be the first to be commercially mined. The SMS deposits are 50 km from land and at a water depth of 1,600 km. The inferred total mineral resource is ~1.54 million tonnes with a grade of 6% of gold grams/tonne and 8% copper (Lipton, 2008).

The other constraints to seabed mining are the availability of ore and mineral deposits on land that are similar to the marine deposits, procurement of a license and its yearly renewal at a huge price from the ISA to explore and exploit marine resources, and a large capital investment which is usually only possible either by consortia or by governments or through partnerships. India has developed a nodule mining system, a soil tester and an ROV (ROSUB 6000), all of which are operable at 6,000 m water depth. Work is underway to develop a battery-operated manned submersible that would have an endurance of 12 hours and dive at least 6,000 m (www.niot.org). Recently the Indian government has approved the 'Deep Ocean Mission,' in which there are plans to develop suitable mining systems for the SMS and polymetallic manganese nodules.

India could collaborate with some of the IORA or advanced countries that may have expertise to jointly develop and produce deep-sea mining and other ancillary systems. The efforts would mutually benefit the countries in terms of exchange of scientific and technological ideas, the creation of employment opportunities, a shared and reduced cost of manufacturing of machinery, and the hastening of the process of mining the SMS deposits, amongst others.

10.4.4 ENRICHMENT

The hydrothermal deposits are a concoction of various metals, oxides, sulfides and sulfates of iron, copper, nickel, cobalt, zinc, lead, gold, silver, barium, silica amongst others. The extraction of a particular metal from such a conglomeration is difficult unlike the processes that help to separate 3 (copper, cobalt nickel) or 4 metals (copper, cobalt, nickel, and manganese) from polymetallic nodules. Several metallurgical techniques must be developed or existing ones need to be refined to separate as many of the metals as possible from the SMS. This could involve the setting-up of large metallurgical plants that would require extensive treatment and efficient disposal of their chemical effluents so as not to contaminate the subsurface and groundwaters. The gangue that would be produced during metal extraction needs to be either examinedto see if it could be put to some use or disposed of. Noise and air pollution produced from the beneficiation plants must be alleviated as much as possible.

Once the flow charts to recover the metals from the SMS are finalized then as a part of the BE initiative, India could establish ore beneficiation plants in one or more of the IORA countries that are near the area. This step would help to save the cost of transporting the SMS from the SWIR which is located tens of kilometres from India, reduction in the carbon footprint, and help to boost metal production. The host country could benefit by way of improved or new infrastructure, creation of jobs, advances in science and technology, and financial gain through foreign investments.

10.4.5 ECONOMICS

The overall accumulation of the SMS in the MOR is estimated to be ~6×10^8 tonnes and of this, 86% is accounted for by deposits present along slow- and ultraslow-spreading ridges (Hannington et al. 2011). It has been reported that a majority of hydrothermal fields with more than 1 million tonnes are found along such ridges that have spreading rates between 20 and 55 mm/yr and <20 mm/yr, respectively (Dick et al. 2003). So far, a little more than 20 fields have been found and confirmed on ultraslow-spreading ridges *(www.vents-data.interridge.org/)* and among these, only for the Mount Jourdanne deposit (SWIR), was the size reported. The estimated SMS is <3,000 tonnes, using the area versus tonnage relationship for the Solwara-1 deposit as a reference, and this is much smaller than expected (Hannington et al. 2011). Therefore, we need more data to demarcate the distribution and content of SMS on ultraslow-ridges.

The Yuhuang-1 hydrothermal field (YHF), situated on the SWIR, has two SMS deposits that are ~500 m apart, one in the SW and other in NE. Calculations reveal that the total volume of SMS in the YHF is ~10.6×10^6 tonnes, with at least ~7.5×10^5 tonnes of Cu and Zn and ~18 tonnes of Au. Accounting for the coverage of layered hydrothermal sediment together with sulfide-rich breccias and underlying massive sulfide deposits, the maximum total mass was estimated at ~45.1×10^6 tonnes. Apparently, the YHF is one of the largest SMS deposits worldwide and reaffirms that ultraslow-spreading ridges have the greatest potential to form large-scale SMS deposits (Yu et al. 2021).

It is tempting for a country or consortia to follow the above 4Es in the hope of extracting several metals simultaneously from SMS deposits and make a handsome profit in the long term. Reportedly, the metals in the SMS deposits may sometimes exceed terrestrial reserves that are now economically mined (Hein et al. 2013). After having been granted the license and strictly adhering to the ISA guidelines to mine the SMS deposits, a critical analysis of the cost-benefit and the multiple issues involved with it need to be worked out by the investors. The investors must account for factors such as capital costs (mother ship, supply vessels), manpower, salaries, daily expenses, hiring/purchase/replacement of mining equipment, transport of personnel and mined ores to land, establishing an onshore facility for metal extraction by following all protocols of EIA and EMP, fluctuating market values of the metals, import-export of the metals, and various other known/unknown factors. Importantly, the profits could fluctuate depending on distribution and resource estimates of the SMS deposits and whether active or inactive vents would be mined.

The investors must consider the fact that future discovery of new ore deposits on land could substantially reduce their profits. In the above constraints we have not accounted for the vagaries of nature that could hamper mining operations. Although it is predicted that by 2030 about 10% of global minerals could be recovered from the ocean floor (European Commission, 2012), we still have a long way to go to make deep-sea mining a profitable venture.

Considering a host of parametres (abundance, grade, topography, metal prices, infrastructure, capital costs, and the like), it was estimated that by mining the Central Indian Ocean Basin (CIOB) polymetallic nodules (water depth 5,000 m) that profits could be made from the 8th year of a 25 year life of the mine (Mukhopadhyay et al. 2019). In the case of shallow-seated SMS deposits perhaps the profits could be realized much earlier provided that the other 4Es have been properly addressed.

10.5 CONCLUSION

Currently, the world is managing with its available land resources but a time could come when it needs to turn to the oceans for mining useful metals and minerals. But the day is not far off when countries will be able to recover marine minerals when technologies are developed, environmental concerns are addressed and mechanisms for sustainable marine mining are set in place. These factors could be accelerated, and the cost reduced, if countries and global corporates worked in unison instead of working in isolation. Regarding seabed mineral exploration, India is quite ahead in the game with respect to its neighbouring countries and, indeed most IORA countries. This is because the Indian Ocean is easily accessible, its availability of large quantities of human resources (scientific, skilled, unskilled), technological advancements for exploration, environmental studies, exploitation, and its ore beneficiation plants, amongst other favourable factors.

After the Goa Declaration (2015), the Indian government seriously took up several initiatives to identify and boost blue economy sectors, with emphasis on marine minerals. The National Institute of Transforming India (NITI) is working hand-in hand with different stakeholders and the Ministry of Earth Sciences to successfully implement a sustainable use of several blue minerals (placers, SMS, polymetallic nodules). This is in tune with the Government's policy statement which is, 'The blue economy refers to the exploring and optimizing of the potential of the oceans and seas which are under India's legal jurisdiction for socio-economic development while preserving the health of the oceans.'

Several recommendations were made by Nayak (2019) to the government of India. Some of these are to expedite technology development for exploration, and the like, the setting up of a national placer mission, financial and human resources, exploration rights for cobalt, a comprehensive study of the Andaman and Nicobar Islands, protecting marine biodiversity, and to establish an appropriate institutional framework to implement the BE activities. It was estimated that the size of the BE in India, measured in Gross Value Added (GVA), in 2016–17 was US$81.8 billion. Currently, the magnitude of the BE is akin to several coastal nations although in some countries (like Malaysia and

Mauritius), the input of BE to Gross Domestic Product (GDP) is quite significant. In India the GVA steadily rose from 3% 2012 - 13 to 10.5% in 2015 - 16. The present contribution of BE to the GDP is ~4% but this could surge if we consider outputs from all the marine sectors or marine-related activities (Rajeevan, 2019).

We suggest that India could establish an independent Ministry of Blue Economics that would chalk out programmes, outline work plans, take policy decisions and successfully implement the BE by networking with the other sectors of the BE. By doing so, the blue minerals could be sustainably resourced in an environmentally-friendly manner and thus substantially contribute to India's projected economy of US$10 trillion by 2030, despite the recent set-backs caused by the ongoing COVID-19.

ACKNOWLEDGEMENTS

We acknowledge Prof. Dr. Md. Nazrul Islam for the invitation to contribute this chapter. We thank Ms. Mansi Shinde during the preparation of Table 10.1. We also acknowledge the support of Goa University and DM college during the preparation of the chapter.

REFERENCES

Agarwal, D. K. Roy, P. Prakash, L. S. and Kurian, P.J. 2020. Hydrothermal signatures in sediments from eastern Southwest Indian Ridge 63 E to 68 E. *Marine Chemistry*, 218: 103732.

Balaram, V. Banakar, V. K. Subramanyam, K. S. V. Roy, P. Satyanarayan, M. Ram Mohan, M. and Sawant, S. S. 2012. Yttrium and rare earth element contents in seamount cobalt crusts in the Indian Ocean. *Current Science*, 103: 1334–1338.

Banerjee, R. and Iyer, S. D. 1991. Petrography and chemistry of basalts from the Carlsberg Ridge. *Journal of Geological Society of India*, 38: 369–386.

Banerjee, R. and Iyer, S. D. 1993. A note on sulfide-oxide mineralisation in Carlsberg Ridge basalts. *Journal of Geological Society of India*, 42: 579–584.

Banerjee, R. and Iyer, S. D. 2003. Genetic aspects of basalts from the Carlsberg Ridge. *Current Science* (Special Section on Mid-Oceanic Ridges), 85: 299–305.

Banerjee, R. and Ray, D. 2013. Metallogenesis along the Indian Ocean Ridge System. *Current Science* (Special Section on Mid-Oceanic Ridges), 85: 321–327.

Barnes, H. L. and Rose, A. W. 1998. Origins of hydrothermal ores. *Science*. 279: 2064–2065.

Baturin, G. N. and Rozanova, T. V. 1975. Ore mineralization in the rift zone of the Indian Ocean. In: Vinogradov, A.P. and Udintsev, G.B. (eds), *Rift Zones of the World Oceans*. Wiley, New York and Israel Prog Sci Trans, pp. 431–441.

Beaulieu, S. E. and Szafrański, K. M. 2020. InterRidge global database of active submarine hydrothermal vent fields version 3.4. *PANGAEA*. doi:10.1594/PANGAEA.917894 Chapman, A.S.A. Beaulieu, S.E. Colaço, A. Gebruk, A.V. Hilario A. Kihara.

Bierlein, F. P. Groves, D. I. and Cawood, P. 2009. Metallogeny of accretionary orogens—the connection between lithospheric processes and metal endowment. *Ore Geology Reviews*, 36(4): 282–292.

Billett, D. S. M, Jones, D. O. B. and Weaver, P. P. E. 2019. Improving environmental management practices in deep-sea mining. *Environmental Issues of Deep-Sea Mining*, 403–446.

Chen, D. Tao, C. Wang, Y. Chen, S. Liang, J. Liao, S. and Ding, T. 2021. Seafloor Hydrothermal activity around a large non-transform discontinuity along ultraslow-spreading Southwest Indian Ridge (48.1–48.7°E). *Journal of Marine Science and Engineering*, 9(8): 825.

Chen, J. Tao, C. Liang, J. Liao, S. Dong, C. Li, H. Li, W. Wang, Y. Yue, X. and He, Y. 2018. Newly discovered hydrothermal fields along the ultraslow-spreading Southwest Indian Ridge around 63°E. *Acta Oceanologica Sinica*, 37(11): 61–67.

Chen, X. Sun, X. Wu, Z. Wang, Y. Lin, X. and Chen, H. 2021. Mineralogy and geochemistry of deep-sea sediments from the ultraslow-spreading Southwest Indian ridge: implications for hydrothermal input and igneous host rock. *Minerals*, 11(2): 138.

Chen, Y. Han, X. Wang, Y. and Lu, J. 2020. Precipitation of calcite veins in serpentinized harzburgite at Tianxiu hydrothermal field on Carlsberg Ridge (3.67° N), Northwest Indian Ocean: implications for fluid circulation. *Journal of Earth Science*, 31(1): 91–101.

Choi, H. Kim, S. S. and Park, S. H. 2013. Interpretation of bathymetric and magnetic data from the eastern-most segment of Australian-Antarctic Ridge, 156°–161°E. *American Geophysical Union Fall Meeting Abstracts*, 2013: T13A–T2498.

Copley, J. Tyler, P. VanDover, C. Schultz, A. Dickson, P. Singh, S. and Sulanowska, M. 1999. Effects of ODP drilling on the TAG hydrothermal vent community, 26 N Mid-Atlantic Ridge. *Marine Ecology*, 20: 291–306.

Copley, J. T. Marsh, L. Glover, A. G. Hühnerbach, V. Nye, V. E. Reid, W .D. K. Sweeting, C. J. Wigham, B. D. Wiklund, H. 2016. Ecology and biogeography of megafauna and macrofauna at the first known deep-sea hydrothermal vents on the ultraslow-spreading Southwest Indian Ridge. *Scientific Reports*, 6(1): 39158. doi: 10.1038/srep39158.

Degens, E. T. and Ross, D. A. 1969. *Hot Brines and Recent Heavy Metal Deposits in the Red Sea: Geochemical and Geophysical Account*. Springer-Verlag, Berlin, Heidelberg, p. 600. doi: 10.1007/978–3-662–28603-6.

DeMets, C. Gordon, R. G. Argus, D. F. and Stein, S. 1990. Current plate motions. *Geophys. J. Int.* 101: 425–478.

Dhawaskar, P. Ganguly, S. Mukhopadhyay, R. Manikyamba, C. Iyer, S. D. Karisiddaiah, S. M. and Mahender, K. 2020. Geochemical heterogeneity along the Vema Fracture Zone, Indian Ocean: mixing of melts from the reunion plume and the Central Indian Ridge. *Geological Journal*, 55: 330–343. https://doi.org/10.1002/gj.3395.

Dick, H. J. B. Lin, J. and Schouten, H. 2003. An ultraslow-spreading class of ocean ridge. *Nature*, 426(6965): 405–412.

Drolia, R. K. Iyer, S. D. Chakraborty, B. Kodagali, V. N. Ray, D. Misra, S. Andrade, R. Sarma, K. V. L. N. S. Rajasekhar, R. P. and Mukhopadhyay, R. 2003. The Northern Central Indian Ridge: geology and tectonics of fracture-zones dominated spreading ridge segments. *Current Science* (Special Section on Mid-Oceanic Ridges), 85: 290–298.

European Commission. 2012. *Blue Growth: Opportunity for Marine and Maritime Sustainable Growth*. Communication from the Commission to the European Parliament, the Council. The European Economic and Social Committee of the Regions. European Union, Luxembourg, 494 final, pp. 12. doi: 10.2771/43949.

Filho, W. L. Abubakar, I. R. Nunes, C. Platje, J. Ozuyar, P. G. Will, M. Nagy, G. J. Al-Amin, A. Q. Hunt, J. D. and Li, C. 2021. Deep seabed mining: a note on some potentials and risks to the sustainable mineral extraction from the oceans. *Journal of Marine Science and Engineering*, 9: 521. https://doi.org/10.3390/jmse9050521.

Franklin, J. M. Gibson, H. L. Jonasson, I. R. and Galley, A. G. 2005. Volcanogenic massive sulfide deposits. In Hedenquist, J. W. Thompson, J. F. H. Goldfarb, R. J. and Richards, J. P. (eds), *Economic Geology One Hundredth Anniversary Volume: Society of Economic Geologists*, pp. 523–560.

Fouquet, Y. 1997. Where are the large hydrothermal sulfide deposits in the oceans? *Proceedings of the Royal Society A*, 355(1723): 427–441.

Fujii, M. and Okino, K. 2018. Near-seafloor magnetic mapping of off-axis lava flows near the Kairei and Yokoniwa hydrothermal vent fields in the Central Indian Ridge. *Earth, Planets and Space*, 70(1): 1–17.

Fujii, M. Okino, K. Sato, T. Sato, H. and Nakamura, K. 2016. Origin of magnetic highs at ultramafic hosted hydrothermal systems: insights from the Yokoniwa site of Central Indian Ridge. *Earth and Planetary Science Letters*, 441: 26–37.

Fujimoto, H. Cannat, M. Fujioka, K. Gamo, T. German, C. Mével, C. Münch, U. Ohta, S. Oyaizu, M. Parson, L. Searle, R. Sohrin, Y. and Yama-ashi, T. 1999. First submersible investigations of mid-ocean ridges in the Indian Ocean. *InterRidge*, 8: 22–24.

Gallant, R. M. and Von Damm, K. L. 2006. Geochemical controls on hydrothermal fluids from the Kairei and Edmond Vent Fields, 23°–25°S, Central Indian Ridge. *Geochemistry, Geophysics, Geosystems*, 7: 6, Q06018, doi: 10.1029/2005GC001067.

Gamo, T. Chiba, H. Yamanaka, T. Okudaira, T. Hashimoto, J. Tsuchida, S. Ishibashi, J. Kataoka, S. Tsunogai, U. Okamura, K. Sano, Y. and Shinjo, R. 2001. *Earth and Planetary Science Letters,* 193: 371–379.

Gamo, T. Nakayama, E. Shitashima, K. Isshiki, K. Obata, H. Okamura, K. Kanayama, S. Oomori, T. Koizumi, T. Matsumoto, S. and Hasumoto H. 1996. Hydrothermal plumes at the Rodriguez triple junction, Indian ridge. *Earth and Planetary Science Letters*, 142: 261–270.

Garuti, G. Bartoli, O. Scacchetti, M. and Zaccarini, F. 2008. Geological setting and structural styles of Volcanic Massive Sulfide deposits in the Northern Apennines (Italy): evidence for seafloor and sub-seafloor hydro-thermal activity in unconventional ophiolites of the Mesozoic Tethys. *Boletin de la Sociedad Geologica Mexicana*, 60/1: 121–145.

German, C. R. 2003. Hydrothermal activity on the eastern SWIR (50°–70°E): Evidence from core-top geochemistry, 1887 and 1998. *Geochemistry, Geophysics, Geosystems*, 4: 7, 9102. doi: 10.1029/2003GC000522.

German, C. R. Baker, E. T. Mevel C. Tamaki, K. and the FUJI Science Team. 1998. Hydrothermal activity along the southwest Indian ridge. *Nature*, 395: 490–493.

Hahm, D. Baker, E. T. Rhee, T.S. Won, Y. J. Resing, J. A. Lupton, J. E. Lee, W. K. Kim, M. Park, S. Y. 2015. First hydrothermal discoveries on the Australian-Antarctic Ridge: Discharge sites, plume chemistry, and vent organisms. *Geochemistry, Geophysics, Geosystems*, 16: 3061–3075.

Halbach, M. Halbach, P. and Luders, V. 2002. Sulfide impregnated and pure silica precipitates of hydrothermal origin from the Central Indian Ocean. *Chemical Geology*, 182: 357–375.

Halbach, P. Blum N. Munch U. Pluger W. Garbe-Schonberg D. and Zimmer, M. 1998. Formation and decay of a modern massive sulfide deposit in the Indian Ocean. *Mineralium Deposita*, 33: 302–309.

Han, Y. Gonnella, G. Adam, N. Schippers, A. Burkhardt, L. Kurtz, S. Schwarz-Schampera, U. Franke, H. and Perner, M. 2018. Hydrothermal chimneys host habitat-specific microbial communities: analogues for studying the possible impact of mining seafloor massive sulfide deposits. *Scientific Reports*, 8(1): 1–12.

Hannington, M. Jamieson, J. Monecke, T. Petersen, S. and Beaulieu, S. 2011. The abundance of seafloor massive sulfide deposits. *Geology*, 39(12): 1155–1158.

Hein, J. R. Mizell, K. Koschinsky, A. and Conrad, T. A. 2013. Deep-ocean mineral deposits as a source of critical metals for high-and green-technology applications: comparison with land-based resources. *Ore Geology Reviews*, 51: 1–14.

Herzig, P. M. and Plüger, W. L. 1988. Exploration for hydrothermal activity near Rodriguez Triple Junction, Indian Ocean. *Canadian Mineralogist*, 26: 721–736.

Hyman, J. Stewart, R. A. and Sahin, O. 2021. Adaptive management of deep-seabed mining projects: a systems approach. *Integrated Environmental Assessment and Management*, doi: https://doi.org/10.1002/ieam.4395.

ISA. 2013. Recommendations for the guidance of contractors for the assessment of the possible environmental impacts arising from exploration for marine minerals in the Area. ISBA/LTC/19/8. https://isa.org.jm/sites/default/ files/ files/documents/isba-19ltc8_0.pd.

Iyer, S. D. and Banerjee, R. 1993. Mineral chemistry of Carlsberg Ridge basalts at 3° 35'–3° 41' N. *Geo-Marine Letters*, 13: 153–158.

Iyer, S. D. and Ray, D. 2003. Structure, tectonic and petrology of mid-oceanic ridges and the Indian scenario. *Current Science* (Special Section on Mid-Oceanic Ridges), 85: 277–289.

Ji, F. Zhou, H. Yang, Q. Gao, H. Wang, H. and Lilley, M. D. 2017. Geochemistry of hydrothermal vent fluids and its implications for subsurface processes at the active Longqi hydrothermal field, Southwest Indian Ridge. *Deep Sea Research Part I: Oceanographic Research Papers*, 122: 41–47.

Johnson, K. T. M. Graham, D. W. Rubin, K. H. Nicolaysen, K. Scheirer, D. S. Forsyth, D. W. Baker, E. T. and Douglas-Priebe, L. M. 2000. Boomerang Seamount: the active expression of the Amsterdam–St. Paul hotspot, Southeast Indian Ridge. *Earth and Planetary Science Letters*, 183: 245–259.

Kalangutkar, N. G. Kurian, P. J. and Iyer, S. D. 2021. Characterization of ferromanganese crusts from the Central and South West Indian ridges: evidence for hydrothermal activity, *Marine Georesources and Geotechnology*, doi: 10.1080/1064119X.2021.1886205.

Kawagucci, S. Miyazaki, J. Noguchi, T. Okamura, K. Shibuya, T. Watsuji, T. Nishizawa, H. Watanabe, K. Okino, N. Takahata, Y. Sano, K. Nakamura, A. Shuto, M. Abe, Y. Takaki, T. Nunoura, M. Koonjul, M. Singh, G. Beedessee, M. Khishma, V. Bhoyroo, D. Bissessur, L.S. Kumar, D. Marie, K. Tamaki, K. and Takai, K. 2016. Fluid chemistry in the Solitaire and Dodo hydrothermal fields of the Central Indian Ridge. *Geofluids*, 16(5): 988–1005.

Kempe, D. R .C. and Easton, A. J. 1974. Metasomatic garnets in calcite (micarb) chalk at SITE 251, Southwest Indian Ocean. *Deep Sea Drilling Project, Initial Reports DSDP*, 26: 593–599. doi: 10.2973/dsdp.proc.26.1974.

Kim, J. Son, S. K. Kim, D. Pak, S. J. Yu, O. H. Walker, S.L. Oh, J. Choi, S. K. Ra, K. Ko, Y. Kim, K. H. Lee, J. H. and Son, J. 2020. Discovery of active hydrothermal vent fields along the Central Indian Ridge, 8–12 S. *Geochemistry, Geophysics, Geosystems*, 21(8): e2020GC009058.

Kumagai, H. Nakamura, K. Toki, T. Morishita, T. and Okino, K. 2008. Geological background of the Kairei and Edmond hydrothermal fields along the Central Indian Ridge: implications of their vent fluids' distinct chemistry. *Geofluids*, 8: 239–251.

Kutina, J. 1983. Global tectonics and metallogeny: deep roots of some ore-controlling fracture zones. A possible relation to small-scale convective cells at the base of the lithosphere? *Advances in Space Research*, 3(2): 201–214.

Liao, S. Tao, C. Li, H. Barriga, F. J. A. S. Liang, J. Yang, W. Yu, J. and Zhu, C. 2018. Bulk geochemistry, sulfur isotope characteristics of the Yuhuang-1 hydrothermal field on the ultraslow-spreading Southwest Indian Ridge. *Ore Geology Reviews*, 96: 13–27.

Liao, S. Tao, C. Zhu, C. Li, H. Li, X. Liang, J. Yang, W. and Wang, Y. 2019. Two episodes of sulfide mineralization at the Yuhuang-1 hydrothermal field on the Southwest Indian Ridge: insight from Zn isotopes. *Chemical Geology*, 507: 54–63.

Lipton, I. 2008. Mineral resource estimate, Solwara 1 project, Bismarck Sea, Papua New Guinea: NI43–101 Technical Report for Nautilus Minerals Inc. http://nautilusminerals.com/i/pdf/2008–02–01_Solwara1_43–101.pdf (July 2011).

Lusty, P. A. J. and Murton, B. J. 2018. Deep-ocean mineral deposits: metal resources and windows into earth processes. *Elements*, 301–306, doi: 10.2138/gelements.14.5.301.

Mendel, V. and Sauter, D. 1997. Seamount volcanism at the super slow-spreading Southwest Indian Ridge between 57° and 70°. *Geology*, 25(2): 99–102.

Mudholkar, A.V. Kodagali, V. N. Kamesh Raju, K. A. Valsangkar, A. B. Ranade, G. H. and Ambre, N. V. 2002. Geomorphological and petrological observations along a segment of slow-spreading Carlsberg Ridge. *Current Science*, 82: 982–989.

Murton, B. J. Baker, E. T. Sands, C. M. and German, C. R. 2006. Detection of an unusually large hydrothermal event plume above the slow-spreading Carlsberg Ridge: NW Indian Ocean. *Geophysical research letters*, 33(10) L10608, doi:10.1029/2006GL026048.

Mukhopadhyay, R. and Iyer, S. D. 1993. Petrology of tectonically segmented Central Indian Ridge. *Current Science*, 65: 623–628.

Mukhopadhyay, R. Ghosh, A. K. and Iyer, S. D. 2018. *The Indian Ocean Nodule Field: Geology and Resource Potential*. 2nd edition, Elsevier, Amsterdam, pp. 413.

Mukhopadhyay, R. Iyer, S. D. Ray, D. Karisiddaiah, S. M. and Drolia, R. K. 2015. Morphotectonic and petrological variations along the southern Central Indian Ridge. *International Journal Earth Sciences* (Geol Rundsch), 105: 905–920. doi: 10.1007/ s00531–015–1193-z.

Mukhopadhyay, R. Loveson, V. J. Iyer, S. D. and Sudarsan, P. K. 2020. *Blue Economy of the Indian Ocean: Resource Economics, Strategic Vision, and Ethical Governance*. CRC Press, Boca Raton, Florida, USA, pp. 297.

Mukhopadhyay, R. Naik, S. De Souza, S. Dias, O. Iyer, S. D. and Ghosh, A. K. 2019. The economics of mining seabed manganese nodules: a case study of the Indian Ocean nodule field. *Marine Georesources and Geotechnology*, doi: 10.1080/1064119X.2018. 1504149.

Munch, U. Blum, N. and Halbach, P. 1999. Mineralogical and geochemical features of sulfide chimneys from the MESO zone, Central Indian Ridge. *Chemical Geology*, 155: 29–44.

Münch U. Halbach P. and Fujimoto H. 2000. Sea-floor hydrothermal mineralization from the Mt. Jourdanne, Southwest Indian Ridge. *JAMSTEC. Journal of Deep Sea Research*, 16: 125–132.

Munch, U. Lalou, C. Halbach, P. and Fujimoto, H. 2001. Relict hydrothermal events along the super-slow Southwest Indian spreading ridge near 63°56'E—Mineralogy, chemistry and chronology of sulfide samples. *Chemical Geology*, 177: 341–349.

Nakajima, R. Yamamoto, H. Kawagucci, S. Takaya, Y. Nozaki, T. Chen, C. Fujikura, K. Miwa, T. and Takai, K. 2015. Post-drilling changes in seabed landscape and megabenthos in a deep-sea hydrothermal system, the Iheya North field, Okinawa trough. *PLoS One*, 10: e0123095. http://dx.doi.org/10.1371/journal.pone.0123095.

Nakamura, K. Watanabe, H. Miyazaki, J. Takai, K. Kawagucci, S. Noguchi, T. Nemoto, S. Watsuji, T. Matsuzaki, T. Shibuya, T. Okamura, K. Mochizuki, M. Orihashi, Y. Ura, T. Asada, A. Marie, D. Koonjul, M. Singh, M. Beedessee, G. Bhikajee, M. and Tamaki, K. 2012. Discovery of new hydrothermal activity and chemosynthetic fauna on the Central Indian Ridge at 18–20 S. *PLoS One*, 7(3): e32965.

Natland, J. 1991. Indian Ocean crust. In: Floyd, P.A. (eds), *Oceanic Basalts*. Springer, Boston, MA. https://doi.org/10.1007/978–1-4615–3540–9_12.

Nayak, S. 2019. *Report of Blue Economy Working Group on Coastal and Deep-Sea Mining and Offshore Energy*. Ministry of Earth Sciences, New Delhi, India, p. 80 .

Noowong, A. Gomez-Saez, G. V. Hansen, C. T. Schwarz-Schampera, U. Koschinsky, A. and Dittmar, T. 2021. Imprint of Kairei and Pelagia deep-seahydrothermal systems (Indian Ocean) on marine dissolved organic matter. *Organic Geochemistry*, 152. https://doi.org/10.1016/j.orggeochem.2020.104141.

Pak, S. J. Moon, J. W. Kim, J. Chandler, M. T. Kim, H. S. Son, J. Son, S. K. Choi, S. K. and Baker, E. T. 2017. Widespread tectonic extension at the Central Indian Ridge between 8°S and 18°S. *Gondwana Research*, 45:163–179.

Plüger, W. L. Herzig, P. M. Becker, K. P. Deissmann, G. Schops, D. Lange, L. Jenisch, A. Ladage, S. Richnow, H. H. Schulz, T. and Michaelis, W. 1990. Discovery of hydrothermal fields at the Central Indian Ridge. *Marine Mining*, 9: 73–86.

Popoola, S. O. and Akintoye, A. E. 2021. Integrated geochemical investigations on Fe-Mn nodules, polymetallic sulfides and Fe-Mn oxides recovered from marine sediments of Carlsberg Ridge, Northwest Indian Ocean. *Advances in Environmental Studies*, 5(1): 394–403.

Popoola, S. O. Han, X. Wang, Y. Qiu, Z. Ye, Y. and Cai, Y. 2019. Mineralogical and geochemical signatures of metalliferous sediments in Wocan-1 and Wocan-2 hydrothermal sites on the Carlsberg Ridge, Indian Ocean. *Minerals*, 9(1): 26.

Qiu, Z. Han, X. G. Li, M. Wang, Y. Chen, X. Fan, W. Zhou, Y. Cui, R. and Wang, L. S. 2021. The temporal variability of hydrothermal activity of Wocan hydrothermal field, Carlsberg Ridge, northwest Indian Ocean. *Ore Geology Reviews*, 132: 103999.

Rajani, R. P. Banakar, V. K. Parthiban, G. Mudholkar, A. V. and Chodankar, A. R. 2005. Compositional variation and genesis of the ferromanganese crust of the Afanasiy-Nikitin Seamount, Equatorial Indian Ocean. *Journal of Earth System Sciences*, 114: 51–61.

Rajeevan, M. 2019. *Report of Blue Economy Working Group on National Accounting Framework and Ocean Governance*. Ministry of Earth Sciences, New Delhi India, p. 75 .

Ramana, M. V. Ramaprasad, T. Kamesh Raju, K. A. and Desa, M. 1993. Geophysical studies over a segment of the Carlsberg Ridge, Indian Ocean. *Marine Geology*, 115: 21–28.

Ray, D. Kamesh Raju, K. A. Baker, E. T. Srinivas Rao, A. Mudholkar, A. V. Lupton, J. E. Surya Prakash, L. Gawas, R. G. and Vijaya Kumar, T. 2012. Hydrothermal plumes over the Carlsberg Ridge, Indian Ocean. *Geochemistry, Geophysics, Geosystems*, 13(1).

Ray, D. Kamesh Raju, K. A. Srinivas Rao, A. Surya Prakash, L. Mudholkar, A. V. Yatheesh, V. Samudrakala, K. and Kota, D. 2020. Elevated turbidity and dissolved manganese in deep water column near 10° 47'S Central Indian Ridge: studies on hydrothermal activities. *Geo-Marine Letters*, 40: 619–628.

Ren, M. Chen, J. Shao, K. and Zhang, S. 2016. Metallogenic information extraction and quantitative prediction process of seafloor massive sulfide resources in the Southwest Indian Ocean. *Ore Geology Reviews*, 76: 108–121.

Scholten, J. C. Stoffers, P. Garbe-Schönberg, D. and Moammar, M. 2000. Hydrothermal mineralization in the Red Sea. In Cronan, D.S. (ed.), *Handbook of Marine Mineral Deposits*. CRC-Press, Boca-Raton, FL, pp. 369–395.

Scott, S. D. 1992. Olymetallic sulfide riches from the deep: Fact or fallacy? In Hsu, K.J. and Thiede, J. (eds), *Use and Misuse of the Seafloor*. New York, Wiley-Interscience, pp. 87–115.

Sun, C. Wu, Z. Tao, C. R. A. Zhang, G. Guo, Z. and Huang, E. 2018. The deep structure of the Duanqiao hydrothermal field at the Southwest Indian Ridge. *Acta Oceanologica Sinica*, 37: 73–79.

Swallow, J. C. and Crease, J. 1965. Hot salty water at the bottom of the Red Sea. *Nature*, 205: 165–166.

Tao, C. Li, H. Jin, X. Zhou, J. Wu, T. He, Y. Deng, X. Gu, C. Zhang, G. and Liu, W. 2014. Seafloor hydrothermal activity and polymetallic sulfide exploration on the southwest Indian Ridge. *Chinese Science Bulletin*, 59(19): 2266–2276.

Tao, C. Wu, G. Deng, X. Qui, Z. Han, C. and Long, Y. 2013. New discovery of seafloor hydrothermal activity on the Indian Ocean Carlsberg Ridge and Southern North Atlantic Ridge—progress during the 26th Chinese COMRA cruise. *Acta Oceanologica Sinica*, 32: 85–88. https://doi.org/10.1007/s13131–013–0345-x.

Tao, C. H. Li, H. M. Huang, W. Han, X. Q. Wu, G. H. Su, X. Zhou, N. Lin, J. He, Y. H. and Zhou, J. P. 2011. Mineralogical and geochemical features of sulfide chimneys from the 49° 39′ E hydrothermal field on the Southwest Indian Ridge and their geological inferences. *Chinese Science Bulletin*, 56(26): 2828–2838.

Tao, C. H. Lin, J. Guo, S. Q. Chen, Y. S. J. Wu, G. H. Han, X. Q. German, C. R. Yoerger, D. R. Zhou, N. Li, H. M. Su, X. and Zhu, J. 2012. First active hydrothermal vents on an ultraslow spreading center: Southwest Indian Ridge. *Geology*, 40(1): 47–50. doi: 10.1130/G32389.1.

Thaler, A. D. and Amon, D. 2019. 262 Voyages Beneath the Sea: a globalassessment of macro- and mega-faunal biodiversity and research effort atdeep-sea hydrothermal vents. *PeerJ*, 7: e7397. doi: 10.7717/peerj.7397.

UN-DESA. World Population Prospects. 2019. Highlights. ST/ESA/SER.A/423. United Nations: New York, NY, USA.

Van Dover, C.L. 2000. *The Ecology of Deep-Sea Hydrothermal Vents*. Princeton University Press, Princeton, p. 412.

Van Dover, C. L. Arnaud-Haond, S. Gianni, M. Helmreich, S. Huber, J. A. Jaeckel, A. L. Metaxas, A. Pendleton, L. H. Petersen, S. Ramirez-Llodra, E. Steinberg, P. E. Tunnicliffe, V. and Yamamoto, H. 2018. Scientific rationale and international obligations for protection of active hydrothermal vent ecosystems from deep-sea mining. *Marine Policy*, 90: 20–28. doi: 10.1016/j.marpol.2018.01.020.

Van Dover, C. L. Humphris, S. E. Fornari, D. Cavanaugh, C. M. Collier, R. Goffredi, S. K. Hashimoto, J. Lilley, M. D. Reysenbach, A. L. Shank, T. M. Von Damm, K. L. Banta, A. Gallant, R. M. Götz D. Green, D. Hall, J. Harmer, T. L. Hurtado, L. A. Johnson, P. McKiness, Z. P. Meredith, C. Olson, E. Pan, I. L. Turnipseed, M. Won, Y. Young III, C. R. and Vrijenhoek, R.C. 2001. Biogeography and ecological setting of Indian Ocean hydrothermal vents. *Science*, 294(5543): 818–823. doi: 10.1126/science.1064574.

Veizer, J. Laznicka, P. and Jansen, S. L. 1989. Mineralization through geologic time: recycling perspective. *American Journal of Science*, 289(4): 484–524. doi: https://doi.org/10.2475/ajs.289.4.484.

Wang, S. Sun, W. Huang, J. and Zhai, S. 2021. Iron and sulfur isotopes of sulfides from the Wocan hydrothermal field, on the Carlsberg Ridge, Indian Ocean. *Ore Geology Reviews*, 103971.

Wang, Y. Han, X. Petersen, S. Frische, M. Qiu, Z. Li, H. Wu, R. and Cui, R. 2017. Mineralogy and trace element geochemistry of sulfide minerals from the Wocan Hydrothermal Field on the slow-spreading Carlsberg Ridge, Indian Ocean. *Ore Geology Reviews*, 84: 1–19.

Wang, Y. Han, X. Zhou, Y. Qiu, Z. Yu, X. Petersen, S. Li, H. Yang, M. Chen, Y. Liu, J. Wu, X. and Luo, H. 2020. The Daxi Vent Field: an active mafic-hosted hydrothermal system at a non-transform offset on the slow-spreading Carlsberg Ridge, 6°48′N. *Ore Geology Reviews*, 103888. doi: 10.1016/j.oregeorev.2020.1038.

Wang, Y. Wu, Z. Sun, X. Deng, X. Guan, Y. Xu, L. Huang, Y. and Cao, K. 2018. He–Ar–S isotopic compositions of polymetallic sulfides from hydrothermal vent fields along the ultraslow-spreading Southwest Indian Ridge and their geological implications. *Minerals*, 8(11): 512.

Wilson, D. S. 1993. Confidence intervals for motion and deformation of the Juan de Fuca plate. *Journal of Geophysical Research*, 98: 16053–16071.

Wu, Z. Sun, X. Xu, H. Konishi, H. Wang, Y. Lu, Y. Cao, K. Wang, C. and Zhou, H. 2018. Microstructural characterization and in-situ sulfur isotopic analysis of silver-bearing sphalerite from the Edmond hydrothermal field, Central Indian Ridge. *Ore Geology Reviews*, 92: 318–347.

Yang, W. Tao, C. Shili, L. Liang, L. Liu, J. and Li, W. 2019. Geological fate of seafloor polymetallic sulfides at the Duanqiao hydrothermal field (Southwest Indian Ridge). In *AGU Fall Meeting Abstracts* (Vol. 2019, pp. OS33C-1821).

Yang, W. Tao, C. Li, H. Liang, J. Liao, S. Long, J. Ma, Z. and Wang, L. 2017. 230Th/238U dating of hydrothermal sulfides from Duanqiao hydrothermal field, Southwest Indian Ridge. *Marine Geophyshysical Research*, 38: 71–83. https://doi.org/10.1007/s11001–016–9279-y.

Yi, S. B. Lee, M. J. Park, S. H. Nagao, K. Han, S. Yang, S. Y. Choi, S. H. Baek, J. and Sumino, H. 2021. Alkalic to tholeiitic magmatism near a mid-ocean ridge: petrogenesis of the KR1 Seamount Trail adjacent to the Australian-Antarctic Ridge. *International Geology Review*, 63(10): 1215–1235, doi: 10.1080/00206814.2020.1756002.

Yu, J. Tao, C. Liao, S. Alveirinho Dias, A. Liang, J. Yang, W. and Zhu, C. 2021. Resource estimation of the sulfide-rich deposits of the Yuhuang-1 hydrothermal field on the ultraslow-spreading Southwest Indian Ridge. *Ore Geology Reviews*. doi: https://doi.org/10.1016/j.oregeorev.2021.104169.

Yu, Z. Li, H. Li, M. and Zhai, S. 2018. Hydrothermal signature in the axial-sediments from the Carlsberg Ridge in the northwest Indian Ocean. *Journal of Marine Systems*, 180: 173–181.

Yue, X. Li, H. Ren, J. Tao, C. Zhou, J. Wang, Y. and Lü, X. 2019. Seafloor hydrothermal activity along mid-ocean ridge with strong melt supply: study from segment 27, southwest Indian ridge. *Scientific Reports*, 9(1): 1–10.

Zhou, Y. Zhang, D. Zhang, R. Liu, Z. Tao, C. Lu, B. Sun, D. Xu, P. Lin, R. Wang, J. and Wang, C. 2018. Characterization of vent fauna at three hydrothermal vent fields on the Southwest Indian Ridge: implications for biogeography and interannual dynamics on ultraslow-spreading ridges. *Deep Sea Res Part I: Oceanographic Research Papers*, 137: 1–12. doi: 10.1016/j.dsr.2018.05.0018.

Zhu, C. Tao, C. Yin, R. Liao, S. Yang, W. Liu, J. and Barriga, F.J. 2020. Seawater versus mantle sources of mercury in sulfide-rich seafloor hydrothermal systems, Southwest Indian Ridge. *Geochimica et Cosmochimica Acta*, 281: 91–101.

11 Global Scenarios of Seaweed Cultivation

Science-Policy Nexus for Enhancing the Seaweeds and Algae Farming

Md. Nazrul Islam,[1]* Sahanaj Tamanna,[2] Md. Shahriar Islam,[1] and Md. Noman[1]

[1] Department of Geography and Environment, Jahangirnagar University, Savar, Dhaka, Bangladesh

[2] Bangladesh Environmental Modeling Alliance (BEMA), Non-Profit Research and Training Organization, Mirpur, Dhaka, Bangladesh

* Corresponding author: Md. Nazrul Islam. E-mail: nazrul_geo@juniv.edu

CONTENTS

DOI: 10.1201/9781003184287-11

11.1 INTRODUCTION

Recent climate change impacts on the agricultural sectors as well as threatening traditional livelihoods, and thus increases global concern for ensuring food security in the forthcoming days after Covid-19 impacts (Islam and Amstel, 2021). Seaweed holds enormous potential to transform food systems for improved sustainability, equity, and nutrition (Benemann, et al. 2008; Fiorella et al. 2021). However, seaweed's current contributions and future potential are often overlooked by the global food systems agenda (Short et al. 2021; Simmance et al. 2022). Civil society, researchers and academia must work in collaboration to utilize seaweed to improve food and nutrition security and foster nature-positive solutions to overcome the food system challenges. Global seaweed output increased from roughly 21 million tonnes in 2010 to 33 million tonnes in 2018. (Amosu et al. 2013; Hussain et al. 2020; Karningsih et al. 2021). On average, farming accounts for 97% of total production. Production grows at a rate of roughly 4% each year on average. The top three species are Japanese kelp, Eucheuma, and Gracilaria (Munoz et al. 2004; Ferdouse et al. 2018; Cai et al. 2021). They account for around 70% of total production in the contexts of global scenarios (Munoz et al. 2004; Roohinejad et al. 2017). Asia's largest producers include China, Indonesia, and the Republic of Korea (Muthayya et al. 2014). Because of growing consumer awareness of the benefits of seaweed extracts, which are used as supplements to nutrient food (as a staple diet), as flavor enhancers, in beauty enhancement, in diabetes control, and as fertilizers, amongst other things. Consumption of seaweed has increased by 125% in the UK solely (Midmore and Rank, 2002; Gupta and Abu-Ghannam, 2011). In 2017, 0.48 metric tonnes of seaweed worth US$ 880 million were shipped internationally. Indonesia dominated the global export market with a 21% share of the market, followed by Chile (9%), and Ireland (7%) (Wurmann, 2022). Their contribution accounts for around 87% of world output. Growing seaweed requires no land, fresh water, fertilizers, or pesticides, making it a resource-free process. It sequesters CO_2, so mitigating the negative consequences of global warming (Michalak et al. 2017). Marine macroalgae, sometimes known as seaweeds, are plant-like creatures that thrive in coastal locations adhering to rock and perhaps other hard substrates (Manzelat et al. 2018; Sudhakar et al. 2018). Bioprospecting operations between 1965 and 2012 yielded a total of 3129 marine natural products (MNPs) or bioactive compounds from seaweeds, according to scholarly publishing metrics. However, the transition from discovery to development has been gradual (EI-Moslamy et al. 2017; Falkenberg et al. 2019).

There are around 1000 different types of seaweeds accessible all over the world. From 14.70 million metric tonnes in 2005, to 30.40 million metric tonnes in 2015, production has increased (Region, 2014). The worldwide seaweed sector is valued around US$ 6 billion per year. The increasing commercial market demand for seaweed extract is primarily driven by its use in nutritional supplements, in food service, pharmaceutical, medical and healthcare-related businesses, and other industries (Seaweed: An alternative protein source, Hossain et al. 2020). Producers, Fast Moving Consumer Goods (FMCG) firms, research organizations, pharmaceutical and cosmetics makers, fertilizer producers, and so on are key participants (Agarwal and Agrawalla, 2017).

11.2 FUNDAMENTALS OF MARINE ALGAE AND THE VALUE OF SEAWEEDS

Algae are farmed commercially in the open air to take advantage of the ample sunshine (Bajhaiya et al. 2010). As a result, information learned from algae activity in research facility labs must be applied to efficiency in larger types of reactors that must cope with fluctuations in light and temperature at the manufacturing location (Acién et al. 2017; Wang et al 2021). It is possible to obtain lipids, proteins, and carbohydrates from microalgae, which may then be refined into a variety of goods (Chung et al. 2013; Malvis et al. 2019). Microalgae have lately aroused the interest of academics all over the world due to their wide range of applications in the renewable power, pharmaceutical, and nutritional industries (Muthayya 2t al. 2014). Seaweeds are a renewable, long-lasting, and low-cost source of biofuel, bioactive medicines, and food ingredients (Khan et al. 2018; Bajpai, 2019). Similarly, numerous microalgae varieties have been recognized as having promising possibilities as value-added products due to their outstanding pharmacological and biological capabilities (Khan et al. 2018; Sharma et al. 2022). Biofuels are an excellent replacement for liquid fossil fuels in terms of price, renewability, and environmental concerns (Khan et al. 2018). On the other hand, the ,microalgae may convert atmospheric CO_2 into useful substances including polysaccharides, lipids, and other therapeutic molecules (Chung et al. 2011; Saeed et al. 2021).

Although microalgae are feasible sources of energy and biological products in general, considerable limits and impediments must be addressed in order to progress the innovation from trial to full scale implementation (Figure 11.1) (Nazir et al. 2020; Geremia et al. 2021). The most challenging and crucial considerations include increasing algal growth rate and chemical synthesis, drying the algae colony for biomass production, pretreating fodder, and enhancing the brewing process in the example of algal production of bioethanol (Khan et al. 2018; Saad et al. 2019; Samsul et al. 2020). The current scientific literature review highlights the benefits of microalgae for the generation of biofuels and other bioactive chemicals, as well as reviewing the factors affecting its culturing (Thanigaival et al. 2022). Cultivation, harvesting, extraction, and conversion are the four primary stages of the production process (Abdelaziz et al. 2013). To begin, enormous amounts of biomass must be grown under optimal conditions, where nutrients must be given. The biomass must then be extracted from the medium. To extract the lipids, the cells in the recovered algae slurry must be disturbed (Marrone et al. 2018).

Microalgae culture is intertwined with bivalve production and is greatly dependent on water quality used for their production (Eljaddi et al. 2021). To avoid the introduction of contamination (parasites that eat the microalgae, such as copepods, or are harmful for bivalves, such as pathogenic

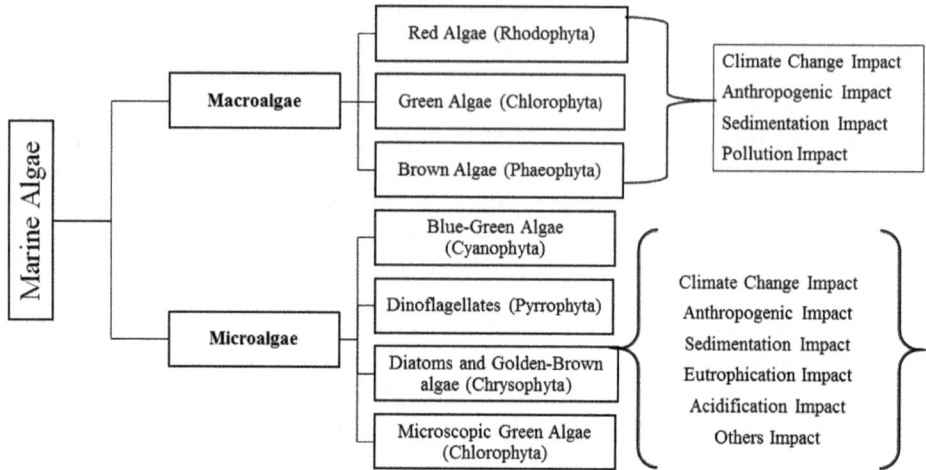

FIGURE 11.1 Basic classification of marine algae and major challenges for harvesting the algae and seaweeds.

bacteria and viruses) in microalgae culture and consequently within aquaculture production during feeding, the water used for phytoplankton growth must be treated (National Geographic Soceiety, n.d.; Cordier et al. 2021). Seaweed is a phrase used to describe multicellular, sea algae that are large enough to be seen with the naked eye. Some species can grow to be 60m long. Seaweeds are made up of red, brown, and green algae. They are not plants because they relate to the kingdom Protista (Kandasamy, et al. 2022). They lack a plant's circulatory tissue (internal transport system), as well as roots, stalks, leaves, flowers, and cone. They use the pigment, chlorophyll for photosynthesis, just like plants, but they also have other carotenoids that can be red, blue, brown, or gold (Chen et al. 2020).

Seaweeds are luminous, hardy, and adaptable, surviving in some of the most severe environmental places on the planet. They are found in the intertidal zone, between the near shore and the foreshore (Meneur et al. 2015). There are four tidal zones: low, mid, high, and spray, with different seaweed species flourishing in each zone. When looking at low tide, you can use the growth of specific seaweed species as a reference to discern where one tidal zone stops and the next begins (Chen et al. 2020).

11.2.1 Brown Algae

Brown algae, which belongs to the class Phaeophyta, is the most prevalent form of seaweed (meaning 'dusky plants'). Brown algae, which can look brown or yellow-brown, is found in both temperate and polar seas (Kandale et al. 2011). While brown algae lack actual roots, they do develop root-like projections known as 'holdfasts' that are utilized to bind the algae to a contact.

Seaweeds can develop in both saltwater and freshwater, but kelp, a brown alga, can only grow in saltwater, typically near rocky coastlines. Kelp comes in over 30 distinct varieties (Kandale et al. 2011; Radulovich et al. 2015). One source is the massive kelp beds off the California coast, while another source is the drifting kelp beds in the North Atlantic Ocean's Sargasso Sea. Kelp, one of the most popular seaweeds, contains a variety of vitamins and minerals (Kandale et al. 2021).

11.2.2 Red Algae

Red algae are usually found all over the world, growing at the bottom of bodies of water or clinging to hard surfaces. Herbivores such as fish, crabs, worms, and gastropods feed on them (Koss et al.

2011; Nagelkerken et al. 2018). Red algae are often slower growing than green algae and reproduce both asexually and sexually. They have the most complex sexual cycles of any organism. Most red algae store their carbohydrates as glycogen (Ball et al. 2011). Glycogen is a lengthy chain of glucose molecules with numerous branching points. It is not the same as the starch stored by brown or green algae (Ball et al. 2011). Rhodophyta feature three essential compounds in their cell walls in addition to the standard cellulose cell walls seen in most algae. These substances are agar, carrageenan, and mucusy sugars. Calcium carbonate is deposited in the cell walls of some red algae, such as coralline algae (Mathew et al. 2019; New guidelines for sustainable European seaweed, 2021). This protects them from being devoured while also providing them with strength and support. These algae play a crucial role in the formation of coral reefs. Some red algae species can build a thin mat over rocks and other hard surfaces and are referred to as crustose when they do so. Red algae, both upright and crustose, link and infill coral skeletons to build vast sedimentary structures that can withstand wave action and erosion (Kennedy, 2019).

11.2.3 GREEN ALGAE

There are around 4,000 distinct types of green algae on the planet. Green algae may be found in both saltwater and freshwater environments, and such species can grow in wet soils (Milano et al 2016). These algae are classified as unicellular, colonial, or multicellular. Oceanic lettuce is a kind of algal species that grows in tidal pools. Codium is another type of green algae enjoyed by some sea slugs, and the variant Codium fragile is also known as Dead Man's Fingers (Kennedy, 2019).

11.2.4 ALGAE IN AQUARIUMS

Even though the stubby-tailed blue-green algae, called Cyanobacteria is not one of the fundamental forms of algae, it is often classified as a sort of seaweed (Cai et al. 2021). These algae (also known as Slime or Smear algae) are frequently found in household aquariums (Kennedy, 2019). While certain algae are a natural part of a healthy aquarium environment, if left unchecked, it will colonize practically every surface in an astonishingly short period of time. Although some aquarium owners employ chemicals to control algae, most prefer to introduce one or more algae-eating catfish also known as suckerfish or snails into the tank (Kennedy, 2019).

11.3   GLOBAL IMPORTANCE OF MARINE ALGAE/SEAWEEDS

Microalgae are components of plankton and benthic communities and can live anywhere, in marine, fresh or transitional waters. Seaweeds, which are non-flowering, primordial sea algae with no foundation, stem, or leaves, play a vital role in marine habitats. This organism's thousands of species, which vary greatly in size, shape, and color, provide habitats for marine life and defend them from dangers (Islam et al. 2021). Figure 11.2 shows the globally significant importance of seaweeds its main sectors (Source: After modified from Anderson, 2001). Giant seaweeds generate kelp forests, which are thick underwater forests that function as incubators for fish, snails, and sea urchins. Its thallus is also consumed by herbivorous marine creatures. Seaweeds, for their part, obtain sustenance through the photosynthesis of sunlight and nutrients found in seawater (Cai et al. 2021).

The algae and sraweeds expel oxygen from all parts of their bodies (Sundararaju, 2021). Some nutrients found in huge amounts of water are hazardous to marine life and can even kill it. Seaweeds, which are usually found in the intertidal zone, shallow and deep-sea waters, estuaries, and backwaters, absorb excess nutrients and help to balance the ecosystem (Islam et al. 2012; Islam et al. 2013). They are also used as bio-indicators. When trash from agriculture, industries, aquaculture, and households is dumped into the ocean, it generates nutrient imbalances that result in algal blooming, which is a symptom of marine chemical harm (Mohan et al. 2022).

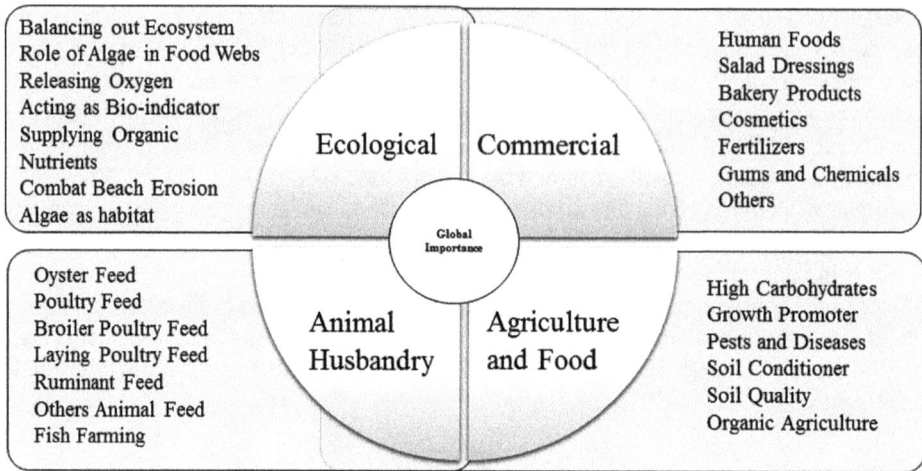

FIGURE 11.2 Global significant importance of seaweeds in main sectors (after and modified from Anderson, 2001).

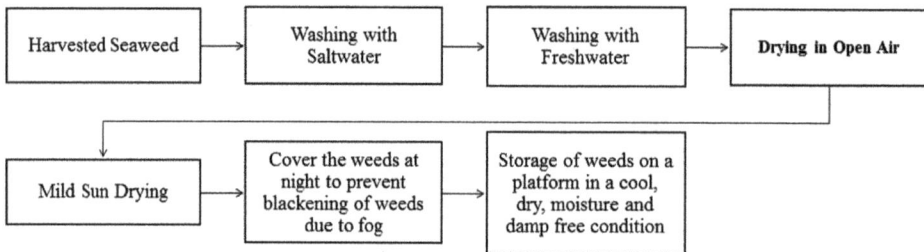

FIGURE 11.3 Processing techniques and post-harvest management of seaweeds.

For photosynthesis, these aquatic creatures rely considerably on iron. When the concentration of this mineral exceeds healthy levels and becomes harmful to marine life, seaweeds catch it and protect it (Calvin, 1974; FPJ Bureau, 2021). Similarly, seaweeds trap and eliminate many heavy metals found in marine ecosystems. Some nutrients found in huge amounts of water are hazardous to marine life and can kill it (Boyd, 1970). The Figure 11.3 shows the cultivating and processing techniques and post-harvest management of seaweeds and algae. The seaweeds, which are usually found in the intertidal zone, shallow and deep-sea waters, estuaries, and backwaters, absorb excess nutrients and help to balance the ecosystem (Islam et al. 2012). They are also used as bio-indicators. When trash from agriculture, industries, aquaculture, and households is dumped into the ocean, it generates nutrient imbalances that result in algal blooming, which is a symptom of marine chemical harm (Newton et al. 2014).

For photosynthesis, these aquatic creatures rely considerably on iron. When the concentration of this mineral exceeds healthy levels and becomes harmful to marine life, seaweeds catch it and protect it (Khandake et al. 2021). Similarly, seaweeds trap and eliminate the majority of heavy metals found in marine ecosystems (Andersen, 2001; Stengel et al. 2011).).

The leftovers of antiquity algae's photosynthetic products, which were later converted by bacteria, are oil and natural gas (Andersen, 2004). As a potential alternative for fossil fuels, many companies have obtained oil from oil-producing microalgae cultivated in high-salinity ponds. As a processed and uncooked food, algae has an economic worth of many billion US$ (Vedantu, 2022). Algal preparations are widely used in the manufacture of meals and other goods, and direct intake of algae has long been a part of the meals of East Asian and Western Pacific populations (Cai et al. 2021). The red seaweed, Nori, often referred as laver, is by far the most substantial business food alga (Porphyra). In Tokyo exclusively, almost 100,000 ha of shallower sea and ocean is planted. The life cycle of Porphyra is separated into two sections. This cycle is a tiny, coral stage that may be deliberately replicated by growing on oyster shells attached to cables or nets and put in certain ocean beds for growth (Andersen, 2001; Cai et al. 2021).

11.4 BLUE GREEN ALGAE BIOFUELS

With the help of light, cyanobacteria, often known as blue-green algae, can create oil from water and carbon dioxide (Vu et al. 20220). A recent study from the University of Bonn demonstrates this. The outcome is surprising: until today, it was thought that this ability was only reserved for plants (Temperton et al. 2019). Blue-green algae may now be of interest as a source of feed or fuel, especially because they do not require arable land (Aizouq et al. 2020). Researchers at UC-Davis and Sandia National Laboratories are investigating the use of cyanobacteria in the production of biofuels (Anderson, 2015; Stanley, 2018). Cyanobacteria are also fuelled by sunlight, the energy source is already plentiful (during daylight hours of course) (Richardson, 2013). Furthermore, they create the material used to build fuel outside the cell, allowing it to be collected without destroying the cell. Whereas in eukaryotic algae, the pre-fuel material must be taken from cells, which damages them and prevents them from producing it again (Banhart, 2001; Smith, 2007). As a result, a fresh generation of algae must be cultivated. Cyanobacteria can be used again. However, present yields are insufficient to be commercially viable, therefore ruffing is being explored to increase them (Motwalli et al. 2017). Figure 11.4 shows their conceptual biochemical and technological framework for converting algae and seaweeds to biofuel and their business policy.

Microalgae are sustainable biofuel resources that may be grown on unsuitable land using salty or salty water (Islam et al. 2012). One big benefit about using microalgae for fuels is that it eliminates the need for farms to cultivate food supplies (Siddiki et al. 2022). As per the Energy Department, microalgae have the capacity to offer at least 30 times the energy of existing land-based crops used to make biofuels. Algae are also capable of successfully recycling carbon from the air (Singh and Gu, 2010). Although algae account for less than 2% of global plant carbon, they capture and fix up to 50 % of CO_2 in the atmosphere, transforming it to hydrocarbons (Cheah et al. 2016).

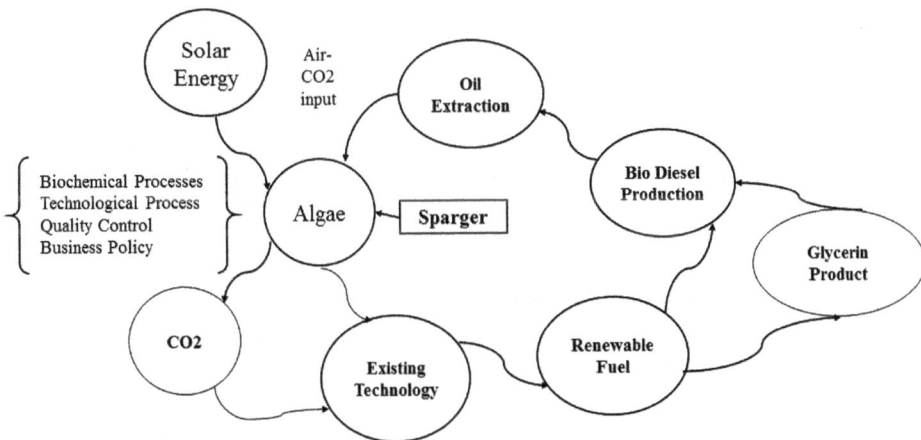

FIGURE 11.4 Conceptual biochemical and technological framework for converting algae and seaweeds to biofuel and their business policy.

11.4.1 Major Compositions of Microalgae Biomass

Algae are eukaryotic organisms, which have cells with a nucleus and other components (organelles) contained by membranes (Sagan, 1967). They dwell in damp, generally watery settings and contain chlorophyll. Algae are not terrestrial plants since they have some attributes lacking. These are: i) genuine roots, stems, and leaves, ii) vascular (conducting) tissues like xylem and phloem, and iii) non-reproductive cells in reproductive structures (Shanks, 2014). Cyanobacteria are prokaryotes with a single circular chromosome and no membrane-bound organelles. Algae is made up of half carbon, ten% nitrogen, and two% phosphorus (Behera et al. 2019).

11.4.2 Biological Importance of Marine Algae

Marine organisms have the chance of being rich sources of essential bioactive metabolites, which might lead to the development of novel pharmacological therapies (Zotchev, 2012). Algae are classified into two types: microalgae (blue green algae), dinoflagellates, bacillariophyta (dinoflagellates), and so forth, and macroalgae (seaweeds), which include greenish, brown, and red microorganisms (Marine Algae (n.d); Suhanya et al. 2016). Microalgae phyla have been acknowledged for their chemical and pharmacological variety and creativity (Hallmann, 2015). Microalgae are also known to be the principal producers of certain highly bioactive compounds found in aquatic resources (Jalilian et al. 2020). Red algae are regarded to become the main component of several physiologically active chemicals when compared to other algal types (Samarakoon and Jeon, 2012).

11.5 FUNDAMENTALS OF SEAWEED FARMING: MECHANISMS AND STRATEGIES

There is hardly any mention of the world's second largest aquaculture business, namely seaweed, in the various aquaculture conversations that actually happen on Seafood Source (Engle et al. 2016). With a worldwide production of 17.3 million metric tonnes, seaweed farming is second only to freshwater fish farming in terms of volume (Verdegem, 2013). The vast majority is cultivated in Asia, which amounts for 17.1 million metric tonnes, or 98.8% of total production, with China being the main producer (Evans et al. 2020). Seaweeds of various species are well recognized in Asian cuisine, and seaweed extracts such as carrageenans, alginates, and agars are utilized in a wide variety of processed foods (Abbott, 1996; Holmyard, 2011; Mac Monagail et al. 2013). Each year, over 114,1000 tonnes of dry red algae are produced worldwide, the majority of which is used in the production of industrial agar (Phi, 2014). However, many of these are harvested in the wild, and there is an immediate need to supplement dwindling native populations with farmed ones (Fletcher and Fletcher, 2022).

   According to a World Bank study from 2016, annual global seaweed output might exceed 1.5 billion dried tonnes by 2050 if the industry can increase its collection by 14% per year (Hossain et al. 2021). Increasing to 500 million would enhance the global food supply by 10% above present levels, provide 50 million direct employments, and substitute around 1% of the fossil fuels used to power automobiles as biofuels (De Fraiture et al. 2008; Glaros et al. 2021). The position of the Worldwide Seaweed Aquaculture Industry (Net value) in thousand tonnes, 2018 and the positioning rank of nation data base are shown in Table 11.1.

   Large-scale seaweed production for dietary purposes, for animal feed, and for biofuels, according to the research (Soleymani and Rosentrater, 2017), *'may represent a fundamental transformation in the global development formula as well as the way we see and utilize the seas.'* However, there is a large initial outlay for starting a farm (Blikra et al. 2021). Producers could get up and operating and start laying seaweed lines in the ocean for around $20,000 plus the cost of a boat (which can vary greatly) (Watson and Dring, 2011). A farmer may make around $37,000 per year by

TABLE 11.1

The Scenarios of Global Seaweed Aquaculture Production ((Net Weight) in Thousand Tonnes) 2018 and the Positional Rank of Country Data Base

| | Production of Seaweed Aquaculture (2018) | |
Country/area	(Net Weight) Expressed in Thousand Tonnes	% of Global Total
World	32,386.2	100
Asia	32,226.3	99.51
China	18,575.7	57.36
Indonesia	9,320.3	28.78
South Korea	1,710.5	5.28
Philippines	1,478.3	4.56
Democratic People's Republic of Korea	553	1.71
Japan	389.8	1.20
Malaysia	174.1	0.54
Viet Nam	19.3	0.06
India	5.3	0.02
Africa	108.5	0.34
Tanzania	103.2	0.32
Madagascar	5.3	0.02
Americas	20.7	0.06
Chile	20.7	0.06
Oceania	14.04	0.04
Solomon Islands	5.5	0.02
Papua New Guinea	4.3	0.01
Kiribati	3.65	0.01
Europe	4.5	0.01
Russian Federation	4.5	0.01
Other producers	21	0.06

Source: FAO-SOFA, 2020; Mantri et al. 2022.

cultivating ten times the amount of kelp and 150,000 mussels per acre, according to GreenWave (Fehrenbacher, 2021).

Because of expanding demand and a strong growth rate, seaweed output has increased in recent years. Seaweed accounts for at least 27% of overall marine aquaculture output (Langton et al. 2019)). The revenue from brown seaweed in 1984 was US$ 737,400.9, whereas it was US$ 5,944,093 in 2017. (Bhuyan et al. 2021). In the case of red seaweed, US$ 751614.6 were made, which is equivalent to US$ 5272332 in 2017 (Smith, 2017; Bhuyan et al. 2021). Red seaweed has recently emerged as a promising candidate for the extraction of essential chemicals (for example, agar, carrageenan, and the like). Consequently, red seaweed production increased over the previous year, while brown seaweed production decreased (Khalil et al. 2018).

11.6 VARIOUS SEAWEED FARMING TECHNIQUES AND PROCESSES

The type of farming approach and its demonstration are critical to the success of seaweed farming. These methods focus solely on agricultural systems that have been shown to be cost-effective, simple to build, operate, and maintain (Sievanen et al. 2005).

11.6.1 Horizontally Net Hauling Method

It was the first commercially successful Eucheuma technique. The sowing component is a 2.5 x 5 m square net with diagonal meshwork and a bar length of 25 cm (Ford, 2021). The net is made of braided nylon or conventional polypropylene lines for the margin (110–150 lb. test) and 30–100 pounds. test for the crystalline lattice (Leisner, 1999). The nets are horizontally strung. Their looping corners are attached to poles or wire strung between stakes (Doty, 1973). Each net unit has 127 mesh intersections. At these areas, soft plastic products should be used to tie Eucheuma shoots together (tie–tie) (Alvarez, 1977).

Net farming gives the benefit of intense production since more crops may be cultivated in a given space. It is divided into three types: the floating bamboo method, the mangrove posts and net method, and the tubular net method (Juanich, 1988).

11.6.1.1   The Floating Bamboo Method

The basic concepts of the floating bamboo methods for seaweeds cultivation are the mainframe of floating bamboo raft is of .5 x 5 m mesh. Usually, the four bamboo poles (each of 4' length) are tied diagonally in four corners of mainframe. Nearly 20 polypropylene-twisted ropes along with seed materials are tied in the raft (Johnson et al. 2017). Around 150-200 g of seaweed fragments are tied at a spacing of 15 cm along the length of the rope. A total of 20 seaweed fragments can be tied in single rope. The total seed requirement per raft is 60-80 kg. Fish net of 4x4 m size is tied at the bottom of the raft to avoid grazing (Johnson et al. 2017).

- Using a cable, tie each corner of a 2.5 x 5 m mesh to a big coral, ensuring that the net is securely stretched.
- Cut one meter of bamboo and knot one to each side netting.
- Add additional net to the one you've previously constructed (Johnson et al. 2017)

11.6.1.2   Mangrove Posts and Net Method

Seaweed farming is an attractive livelihood for fishermen and a high yielding investment. Mangroves provide numerous ecological functions and are instrumental in providing socioeconomic support for the community.

- Methods for building mangrove posts and net method: Install mangrove stake bipod and tripod 6 meters apart in rows with 11 bipods or tripods in each row. The rows should be 6 meters apart (11 rows can hold 20 nets) (Juanich, 1988; Nagelkerken et al. 2008).
- Attach 2.5 × 5 meters net to the bipods and tripods. Make sure all nets are stretched tightly and are at least 2 feet above the bottom but below the lowest tide level (Juanich, 1988; Tengku et al. 2020).

11.6.2 Method of the Bottom Monoline

This strategy is less costly to build, easier to handle, and less subject to ambient weather conditions than the raft approach (Table 11.2). This method is comprised of units, which are hectare-scale planting units. A module is made up of 28 monolines (single wires), each of which is 30 ft (9.8 m) long. Up to 36 stems can be supported by a monoline. As a consequence, a hectare of 35 units has 35000 stems, with around 1000 plants in each module (Juanich, 1988).

Methods for building a monoline:

- Use a hammer to drive fence posts towards the bottom, 1 m apart in lines and 10 m across rows.
- Tie nylon monolines parallel to one another at both ends of the posts, 20–25 cm (8–10 in) from the bottom.

TABLE 11.2
The Sort of Farming Approach and Its Demonstration are Critical to the Success of Seaweed Farming and World Production of Seaweed

Country/Area	Seaweed (Fresh Weight Million Tonnes)	World Production (%)
China	4.093	59
Korea	0.771	11
Japan	0.737	10
Philippines	0.404	6
Far East counties (total)	6.263	90
Norway	0.185	2.6
Chile	0.182	2.6
USA	0.116	1.6
France	0.079	1.1
European countries	0.302	4.3
Total	6.941	100

Source: Forster and Radulovich, 2015; Algo Rhythme, No. 31, CEVA, Pluebian, France, 2022.

While most seaweed collection inside the UK is still done manually at low tide, other nations gather wild seaweed using boats and gear such as rollers or fishing vessels (Juanich, 1988; Morrison et al. 2019). This is significantly faster than hand gathering, but if done excessively, it can harm the ecosystem by removing other seaweeds and altering sea animal habitats (Rahman, 2015). Seaweed harvesters, on the other hand, have given this subject a lot of care and consideration, which is a positive thing. In Norway, for instance, the rake approach only eliminates the top drifting canopy of seaweed, enabling it to regrow throughout the next two years while inflicting minimum damage to the seafloor (Flora, 2019).

11.6.3 SEAWEED CULTIVATION AND COASTAL COMMUNITIES

Seaweeds are often grown in lagoons, coastal waters, and bays. Seaweeds get their nutrients straight from the ocean, thus it's critical to have waves that sweep the area where the seaweeds are growing. Seaweed, as a term, can also include marine aquatic plants or spinach (Rahman, 2015; Harb and Chow, 2022) and is a term used to describe marine macrophages. Seaweed is a major world marine asset that is utilized as a raw material in a variety of nations for diet and commerce (Choudhary, 2021). Seaweed is widely consumed in the East, particularly in Japan, China, and South Korea, and its consumption is increasing in South America, Africa, and Europe (Bixler and Porse, 2011). It was initially called brown algae (66%), red algae (33%), and green algae (1%) of food , describing three types of seaweed. It is commonly utilized as a raw material in manufacturing, in addition to having been used as a food for humans (Buschmann et al. 2017). It has been utilized as jelly in the dairy, pharmaceutical, textile, and paper sectors across the world. Furthermore, seaweed is utilized in the manufacturing of fertilizer, animal feed, and salt in the field. Because of the abundance of minerals in seaweed, it is used as an alternative cuisine (Kim and Venkatesan, 2015).

The requirement for seaweeds to produce phycocolloids for diverse purposes in the food, medical, textile, paper, and other sectors has steadily increased during the last fifty years (Periyasamy et al. 2018). Because of increased demand for previously established industrial seaweed areas, particularly in the last 20 years, new stations are being researched and inquiries are being made in 'various countries of the Eastern and Western continents' (Rao et al. 2019). Cultivation of microalgae is a good potential process for making sustainable energy hydrocarbon feedstock for biodiesel because: i) the chosen species of algae can generate about twice as much oil for every acre than soybeans, ii) the cultivation of microalgae does not necessitate arable land, and iii) the cultivation of microalgae can use minimal water sources that are not fit for drinking or irrigation. Several factors influence algae development and consequently biomass productivity (Gatamaneni et al. 2018). Excessive light, excessive oxygen levels, and inadequate temperature all have a detrimental effect on development (Melis, 1999; Diab, 2018). The extent to which these elements have an influence varies dependent on the particular algae species with some developing well at relatively low temperature and light intensity (for example, Chlamydomonas nivalis), whereas others are adapted to greater irradiation (for example, Chlorella sorokiniana). For algae production, numerous emitter types and layouts are available (Zheng, 2020).

At Bamfield, a seaside community in British Columbia, Ontario, Louis Druehl sails his boat, the Kelp Train, a mile along the rugged beach. For the last 51 years, this boat has taken Druehl to the serendipitously named Kelp Bay, where threads of kelp that Druehl has meticulously gathered for years hang in the chilly Pacific water at the bottom the surface of the ocean (Godin, 2020).

Because of the lack of fishes, Blue Ventures aided people of Belo-sur-Mer and the outlying areas in locating a business activity apart from fishing that would enable people to live off the sea in a responsible way (Gardner et al. 2017). The community chose red seaweed farming. Durable cloth, a red kelp, is an important tactile ingredient in the food industry (Ventures, 2021).

11.7   MAJOR FACTORS THAT AFFECT THE PRODUCTION OF SEAWEED

Seaweed is the foundation of life in the ocean and provides most of the oxygen on Earth. Understanding how seaweed survives and flourishes is critical for the protection of Earth's ecosystems (Mouritesen, 2018). The phrase 'seaweed' refers to a broad category of non-vascular aquatic plants, sometimes known as algae. Seaweed comes in a variety of colors and sizes, ranging from microscopic plants to enormous plants with lengthy fronds (Ecology of Seaweed and Its Environmental Significance | CCBER, n.d.).

11.7.1   Factors Affecting Production

The many environmental conditions are required for cultivating different species of seaweed are variable (Kerrison et al. 2015). In general, seaweed production requires areas with sufficient nutrients and light for growth and salinity and temperatures that are not limiting to the species being cultivated (Campbell et al. 2019)

11.7.1.1   Nutrition

For sustainable growth, seaweeds require essential nutrients (C, N, P) in species-specific ratios (Duarte 1992). According to study by Duarte (1992 found that the percentage tissue carbon (per unit tissue dry weight) is 10%–50% with a median value of 25%; tissue nitrogen range is 0.2%–4.2% with a median range of 0.6%–2.2%, and phosphorous from 0.1% to 0.5% with a median of 0.1%. The ratios of these elements (C:N:P) are often used to infer nutrient limitation (Roleda and Hurd, 2019). All types of seaweed, like terrestrial plants, employ sunshine, carbon dioxide, and water to produce nourishment. As a result, to thrive, seaweed must grow at the ocean's surface, within reach of sunlight, and there must be an excess of carbon dioxide in the water (Martinez, 2019).

11.7.1.2   Hydration

Seaweed, like all living things, requires water to survive. Seaweeds absorb water through the surface of their leaf and stem-like structures since they lack the genuine leaves, stems, roots, and internal vascular systems that most other plants employ to take in water (Jessica, 2019). As a result, seaweed must always be partially or completely submerged (Gambrel, 2019).

11.7.1.3 Limiting Factors

Nitrogen availability is a significant limiting element in seaweed growth, particularly for green algae (Doty, 1973). The rising discharge of fertilizer-related nitrogen from farms and streams into the oceans has produced favorable circumstances for algae growth, particularly during the summer when temperatures are high and the days are long (Mouritsen, 2018).

11.7.1.4 Salinity

The scientific evidences shows that the density influences salinity. This can harm organisms such as fish and kelp due to the volume of salt ions. Even during the evaporation process, salt is left in the water, and the salt may reach the organisms and destroy them (Dawes et al. 1998). All the extra seaweed hovering on the water surface is algae that has been damaged by even more salt. This simply helps the customer by providing them with more excellent meals.

11.7.1.5 Constraints

Pressure is a density-related quantity. Many sea species can be killed by this element because they are unable to withstand the pressure of diving so deep into the water (Dawson, 2006). Many sharks utilize this to sink a large number of fish in order to kill them before devouring them (Shaikh et al. 2019). This increases the shark population as a benefit and decreases the fish population as a disadvantage, but that's how life works.

11.7.1.6 Temperature

Temperature is an important variable of density. This component, like pressure, eliminates microorganisms when the temperature is lower (Buckow et al. 2009; National Geographic Society, n.d.). Squids, for example, dwell extremely deep and there aren't many species down there with them as a benefit, but fish that can't survive in very cold temperatures will die down and even become feed, just as if the pressure isn't right.

11.7.1.7 Light

Light, like that visible in the sun, is an efficiency component. The producers of the maritime biome rely on sunlight to obtain nourishment (Townsend et al. 2011). Phytoplankton absorbs energy from the sun through photosynthesis and subsequently feeds the rest of the marine biome (Halsey and Jones, 2015). Light has the advantage of feeding the primary producer of the marine environment (Riebesell, 2004). One downside of light is that if phytoplankton don't really perform photosynthesis, the entire marine ecosystem may suffer.

11.7.1.8 Tidal Forces

Tides are indeed a variable that changes with density. Tides are classified into two types: high and low. Small creatures, such as fish, might become stranded during high tide, making it easier for fisherman or predators to catch them (Langdon, 2006). This would be detrimental to the fish species while benefiting humans and the shark population. Low tides would benefit fish by allowing them additional time to hide, but predators would be helped to identify them. (*Limiting Factors*, n.d.).

11.7.2 Monitoring Protocols for Seaweed Farming

With all the world's population predicted to exceed 9 billion by 2050, macroalgae can provide an alternate food source, feed, fuel, and income for an ever-growing population if farmed sustainably (Hossain et al. 2021). They also have a vital ecological function in coastal ecosystems, such as supporting the food chain, avoiding shoreline erosion, and extracting nutrients from the environment (Barbier et al. 2011). Worldwide kelp production has almost doubled in the last 10 years.

The principal source of seaweed production in Europe is wild stock collection (Naylor et al. 2000). Increasing human activity, on the other hand, is placing a strain on these communities, with economically valuable species and critical coastal habitats falling in abundance in certain areas. (Mineur et al. 2015).

Several varieties of kelp, for instance, which are used for food, livestock feed, and compost, are disappearing from the most southerly areas where they typically grow (Morrissey et al. 2001; Pereira and Cotas, 2019). Aquaculture (growing) is becoming more popular in order to protect these wild seaweed reserves and fulfill consumer needs for seaweed mass (Buck et al. 2017). However, there may be environmental and ecological implications, such as various ecosystem dynamics caused by the introduction of non-native species or altered species interactions (Van den Burg et al. 2021). Overcoming technological, commercial, and regulatory barriers, such as scaling up production and streamlining legal procedures, is also crucial to the sector's growth (Karltorp, 2014; Karningsih et al. 2021). Because it is a new industry in the EU, there is presently no system in place to oversee the expansion of European seaweed aquaculture that takes all of these variables into consideration (Barbier et al. 2020; Van den Burg et al. 2021).

11.8   GLOBAL SEAWEED CULTIVATION SCENARIOS: COUNTRY CASE STUDIES

More than 300 years ago, China, Japan, the Philippines, Australia, Indonesia, and many other countries focused on the marine business (Ferdouse et al. 2018). Now, coastal fish and plants provide 15–16% of the protein ingested by the globe's 4.3 billion people (Henchion et al. 2017). There are many countries has started to cultivate the seaweeds commercially in developing nations but it is very low in developing and least developed countries. Currently, in this chapter we discuss seaweed cultivation scenarios in some developing countries with comparing other countries. Such as:

11.8.1   African Countries: Kenya and Senegal

11.8.1.1   Seaweeds Cultivation Scenarios in Kenya
Kenya has a US$ 2.4 billion portion in the Western Indian Marine sector. Coastal tourism accounts for the majority of the revenue, amounting to around US$ 1.5 billion each year (Figure 11.5) (Potgieter, 2018). In comparison to the country's yearly GDP of around US$ 60 billion, this is

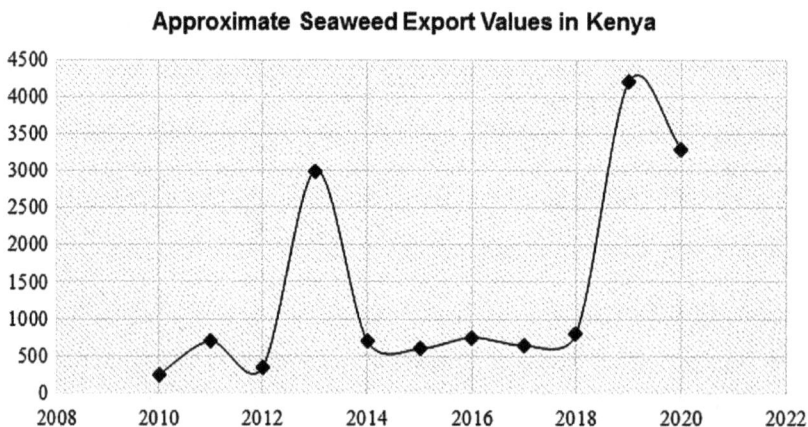

FIGURE 11.5 Approximate seaweed export values in Kenya from 2008 to 2020 and the company primarily exports to the United States and Asia, where demand is increasing.

Approximate Senegal Seaweed Export Values

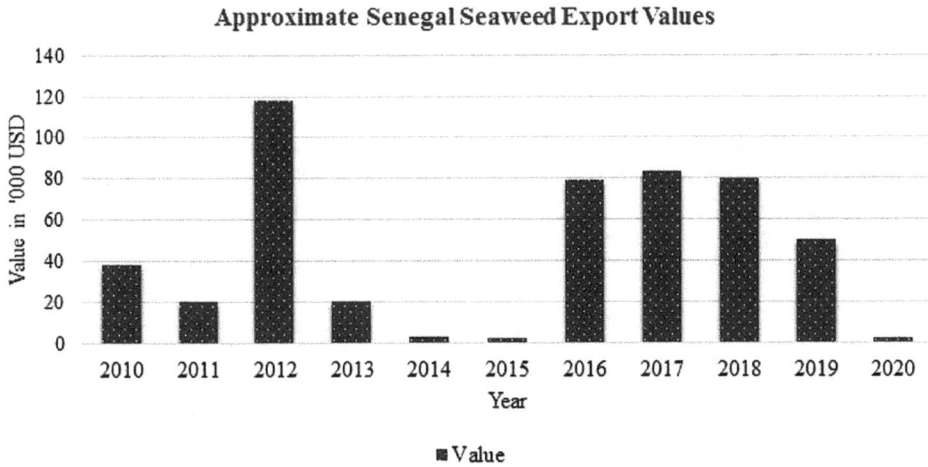

FIGURE 11.6 Approximate seaweed export values in Senegal.

merely a 4% contribution to GDP. Kenya is in a profitable tuna region. Kenya's wide 200,000 nautical mile Special Economic Zone is expected to have 150,000–300,000 metric tonnes of fish (Kennedy, 2019). As part of the Kenyan government's blue economy vision, more efforts should be devoted toward encouraging investment development and management of offshore tuna resources.

In Kenya's coastal communities, such income diversification is critically required. Many of them are the most impoverished in the country and depend on fishing for a living (Geheb and Binns, 1997). According to statistics from Kenya's Central Bureau of Statistics, Kwale County, which is situated just south of Mombasa, has a gap between the poor (which measures the depth and level of poverty) of 41.8%, compared to a national average of 12.2% (County, 2014). The organization offers raw seaweed and makes money by providing value to goods such as soaps, beverages, lunches, and beauty (Mouritsen et al. 2013). Because of its high fiber and vitamin content, seaweed may also double as a natural fertilizer and nutrient. Most of the company's seaweed exports go to the United States, where demand is growing. As an outcome of these activities, the living standards of the communities have improved (Msuya, 2006).

Kenya Seaweed is expected to cost between US$ 0.40 and US$ 0.21 per kg or between US$ 0.18 and US$ 0.10 per lb. in 2022. The price per kilogram is KES 43.57 in Kenyan Shillings. In Mombasa and Nairobi, the average price for a tonne is US$ 400 (Selina Wamucii, 2020). Farmers can currently only manage the growing of between 100 and 200 ropes of seaweed. Only a few of them have been able to reach the desired 300 ropes for optimum yield and earnings (Mirera et al. 2020). 'This is because farmers need a lot of personnel to raise 300 ropes of seaweed,' Mirera explains, implying that a bigger staff is required. Prices for a kilogram of dry seaweed range from US$0.22 to US$0.25, implying that a farmer may profit approximately US$250 from a tonne of seaweed. Growing seaweed also has another significant advantage. It serves as a nursery for fish (Kamadi, 2020).

11.8.1.2 Seaweed Cultivation Scenarios in Senegal

A survey of the literature, as well as specimens collected in December and February 1993–94, were utilized to describe the shallow coastal algae vegetation of northern Senegal (Figure 11.6) (Harper and Garbary, 1994). Northern Senegal now has 242 species categorized, 80 of which were identified in this research (16 *Chlorophyta*, 12 *Phaeophyta* and 52 *Rhodophyta*). This includes 29 previously unknown Senegal seaweed species as well as 8 completely undiscovered West African species (Marquez et al. 2014). Senegal's physiographic affinities are mostly with warm floras to the north, with a significant difference in species diversity between northern Senegal and locations towards the south (Bianchi et al. 2012). The lack of tropical species is defined by the existence of huge stretches of brackish water and mangrove ecosystems to the south, as well as southward-flowing ocean currents (Harper and Garbary, 1997). Senegal exported 0.05 million US$ of seaweed in 2019, a 36.71% reduction from the total seaweed export of 0.079 million US$ in 2018. The yearly change in exchange rate of Senegal seaweed was -11.236% from 2017 and 2018. The yearly change in the amount of Senegal's seaweed trade was -43.33% between 2017 and 2019, comparing to a fluctuation of -32% throughout 2018 and 2019 (Selina Wamucii, 2020b).

11.8.2 ASIAN COUNTRIES: CHINA, INDONESIA, VIETNAM AND BANGLADESH

11.8.2.1 Seaweed Cultivation Scenarios in China

Laminaria is the most significant commercial seaweed in China. Laminaria fisheries on artificial movable rafts started in 1952, and output steadily increased until 1980, when 252,907 tonnes of dry sample was reported as the highest production (Yang et al. 2015; Zhang et al. 2021). Laminaria cultivation areas and total output have dropped in recent years because shellfish growth has evolved so quickly that farmers have begun to prefer farming mussels over Laminaria (Troell et al. 2006). The yearly production of Laminaria is approximately just over 200,000 tonnes. Undaria is farmed in Qingdao and Dalian using the same raft technique as Laminaria and is commonly co-planted on a same flowing raft (Limiting Factors (n.d.). Each year, just a few metric tonnes are produced. It is being used to nourish abalones, but some is delivered to Japan. Porphyra is largely grown in the provinces of Jiangsu, Zhejiang, and Fujian in China. It is used in the manufacture of food and also the distillation of agar. Gracilaria and Eucheuma are grown in Guangdong Province and are used for phycocolloid separation (Feijiu, 2019).

Seafood is a major industry in this seaside Chinese city on the Yellow Sea, where seaweed is the prize of each day. Outside of a small group of caterers, horticulturists, and pharmacists, few are fascinated in the flora whose all-too-familiar pungent 'rotten egg' stink repels people on the beach and surfers, and even less are conscious of its economic value (Figure 11.7). As per market research company World Market Analytics, per capita seaweed demand in China is up to 2.5 kilograms annually, ranking it one of the world's highest (Wijsman et al. 2019).

The appetite suppressor for the seaweed has propelled China to the forefront of a multibillion-dollar business, backed by a boom in the world market as more health-conscious individuals include seaweed in their meals (Behera and Varma, 2017). Seaweed's versatility and contribution to a range

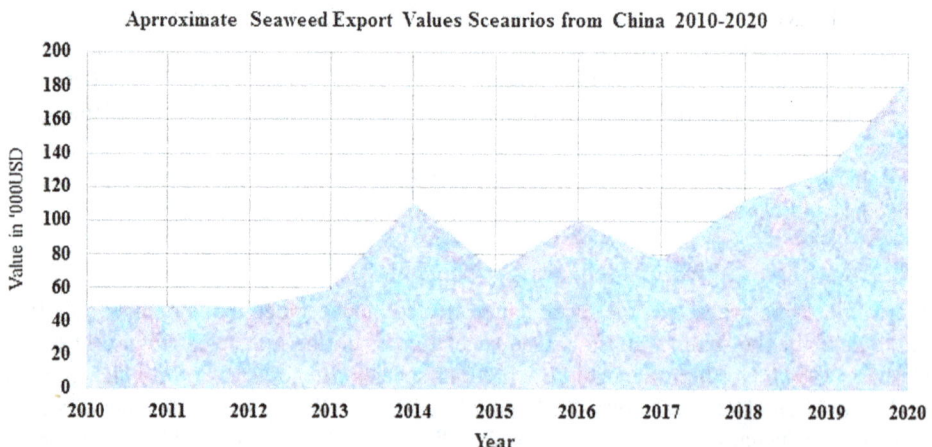

Aprroximate Seaweed Export Values Sceaurios from China 2010-2020

FIGURE 11.7 Seaweed exported from China in 2009–2020.

of sectors are expected to increase future demands. Grand View Research estimates that the global trading market for packaged algae was valued at US$ 10.31 billion in 2015 (Insights, 2021). China, whose long-standing need for seaweed has nurtured superior scientific knowledge in its production and marketing, has been following the trend of the algae industry's positive prognosis (Klinkhamer et al. 2020). Marine industries accounted for more than half of Qingdao's GDP in 2016, the year that the city entered the elite 'trillion-yuan club,' a growth of more than 220% from 2012 (Diab, Wang, 2018).

11.8.2.2   Seaweed Cultivation Scenarios in Indonesia

After China, Indonesia is the approximately world's second largest producer of seaweed, accounting for 38% of the worldwide seaweed market (Ferdouse et al. 2018). In Indonesia, seaweed cultivation is primarily concerned with the manufacturing of carrageenan. This by-product is extracted from edible seaweeds and is commonly utilized as a natural gelling agent in the food and cosmetic sectors. Contrary to the ever-increasing prospects and demand for seaweed, production in Indonesia has been dropping over the last ten years. Rapid tourism expansion, the destruction of seaweed wetland ecosystems, and pollution issues all pose threats to the national market (Vu et al. 2020). To secure a steady income for seaweed growers, seaweed production must be restored (Seaweed/ Rikolto in Indonesia, 2021).

Nevertheless, seaweed production is not really a common source of income in Indonesia. Farmers usually have difficulty drying their seaweed, particularly when it rains. Because seaweed condition (and price) is so closely linked to water content and the presence of contaminants, insufficient drying equipment and methods can severely reduce seaweed grade (Langford and Waldron, 2020). Furthermore, weather patterns influence seaweed development, and disease can destroy productivity. Farmers are also struggling to maintain the growing power of their seaweed crops after a few harvests. Even if farmers can produce high quality seaweed, they seldom receive a similarly high price. The current crude seafood marketing system commonly combines higher and poorer grade seaweeds, which are marketed at a range of average prices (Figure 11.8). This, in turn, diminishes incentives for growers to produce the higher-quality seaweed sought by manufacturers. Despite these challenges, the seaweed industry has immense promise (Langford et al. 2020). In 2019, Indonesia traded around US$ 324.85 million in seaweed. This is the largest amount for Indonesia in recent years.

11.8.2.3   Seaweed Cultivation Scenarios in Vietnam

Because of their nutritional and therapeutic value, Vietnamese edible marine macroalgae are of curiosity (Tanna and Mishra, 2018). Seaweed species have long been utilized as a dietary supplement

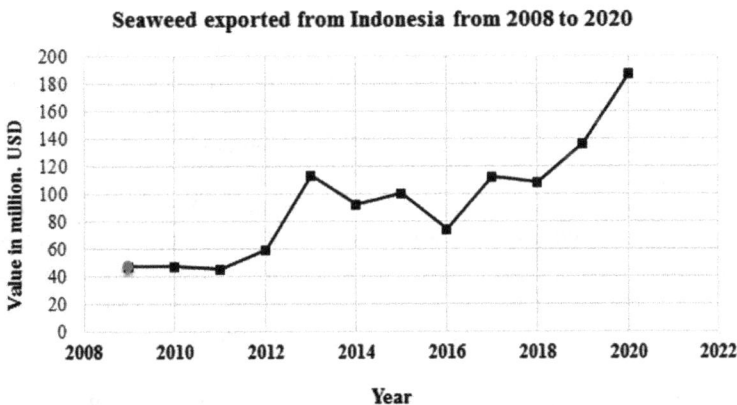

FIGURE 11.8   Seaweed exported scenarios from China in 2008–2020.

and herbal remedies by Vietnamese coastal inhabitants. They ate seaweed uncooked as salads and vegetable, fermented with vinegar, pickle or honey flavored jellies, and cooked for vegetable soup (Hickling, 1971). Seaweed is often used as an herbal medication to cure coughs, asthma, hemorrhoids, sores, goiters, gastrointestinal discomfort, urinary infections, and to reduce the occurrence of tumors, ulcers, and headaches (Hu et al. 2021). Although Vietnam does have a wealth of algal blossoms, the total number of species is expected to be close to 1000 spp. with 638 marine algae species identified (Seaweed harvesting and cultivation | Scotland's Marine Assessment, 2018). However, there has been no comprehensive investigation into alterations in the chemistry of marine algae. This study looked at fifteen edible seaweed species, including green, brown, and red algae (Hong and Hien, 2004). Vietnam now has 10,000 ha of seaweed, with an annual production of much more than 101,000 tonnes of natural seaweed. The majority of seaweed is grown around the north Vietnam coastline (6,600 ha), the north central long coastline (2,000 acres of forest), and the south-central south coast (2,000 ha). Sea grapes, a species of seaweed, has indeed been cultivated in central province Khanh Hoa state since 2004 (Nguyen Van, 2014), and the plant's native territory has been expanded to generate enough for exports. Sea grape has been delivered to Japan and the EU for 110,000 VND per unit of mass of fresh sea grape, but dried sea grape gets three times the cost of fresh sea grape (Algaeworldnews, 2017).

11.8.2.4 Seaweed Cultivation Scenarios in Bangladesh

The term 'seaweed' refers to the popular name for primarily microscopic and gregarious algal species that lack roots, flowers, leaves, stalks, fruits, and seeds and grow and live by adhering to stones or other hard sedimentary layer underneath a high point or drifting in the oceans (Ahmed and Taparhudee, 2005; Towhidy and Alfasane, 2018).) Many seaweeds and plants are also said to have medicinal properties. Taking all these considerations into account, the current approach has been carefully crafted. Until now, fish and other sea creatures have been mostly exploited (Hossain et al. 2019). Around 230,000 marine plant and animal species have been scientifically described all around the world (Sun, 2019).

The majority of seaweeds are present in the tidal and interbreeding zones of Bangladesh's Sathkhira, Bagerhat, and Pathuakhali areas, as well as along the Teknaf and Cox's Bazar coasts of the Bay of Bengal (Rahman, 2013; Islam et al. 2018)). There are some areas where seaweeds can be grown. In general, seaweeds, salad, and chattinis have been eaten in Cox's Bazar, Rakhain, and other tribal tribes in salty locations. Seaweed is referred as 'Hijala' in the local language (Islam et al. 2021). Although there was no information about commercially produced seaweed in the country, locals in coastal areas assumed that the waste would be given to Myanmar dealers for a nominal price (Hossain et al. 2020). It has a high international demand; thus export will make thousands of crores every year. It will strengthen the country's economy. The Seaweed Research Team found 84 seaweeds, 50 of which are marketed at high prices in various nations throughout the world (Reporter, 2020). Due to increased demand in the local and worldwide markets, the coastal population in Bangladesh is gradually adopting seaweed cultivation as a new source of income (Success through Seaweed in Bangladesh, 2020).

Presently, minority groups in Cox's Bazar and the Chattogram Hill Tract are the nation's major consumers of seaweed (Mamun, 2022). Bangladesh, according to scientists, experts, and farm owners, might be amongst the world's top seaweed growers in the future owing to inexpensive labor and a wide coastal area with 480 km of shoreline and 25,000 square km of coastal land (Khan et al. 2016). Seaweed production is one of the fastest developments in the world's aquaculture industries, according to Food and the Food and Agricultural Organization, with a yearly output of around 33 billion metric tonnes, worth US$ 11.8 billion. In the following two years, this is likely to triple (Mohamed et al. 2012; Mamun, 2022).

In search of a new revenue source, fishermen have begun cultivating seaweed in the developing nations as well as Bangladesh. Growing seaweed is less difficult than fishing since the methods are smoother, the technology is less costly, and the seaweed varieties require no feed, grow fast, absorb carbon, and are simple to collect (Alaswad et al. 2015).

11.8.3 LATIN AMERICAN COUNTRIES: ARGENTINA, MEXICO, AND VENEZUELA

11.8.3.1 Seaweed Cultivation Scenarios in Argentina

Subantarctic seaweeds grow in Argentina's coldest regions, near the coasts of Patagonia and Tierra del Fuego (Mystikou et al. 2016). Only in the southern provinces of Patagonia (Chubut and Santa Cruz), along the parallel 42°S, where seaweed is collected commercially, have environmental conditions and seaweed taxonomy been researched systematically (De Zaixsoa et al. 1998; Selina, 2020b). Gracilaria plants are picked by hand from stranded algae, whereas Macrocystis pyrifera is typically harvested from boats (Castro et al. 2022). Private firms collect and trade in specified areas, which are granted permission by provincial governments. Chubut, a province in Argentina's Patagonia noted for southern whale sightings, sheep breeding, and ubiquitous winds, is home to the world's first hamlet dedicated solely to collecting marine algae (Ciancia et al.2020). Top-quality seaweed, located in Bustamante's clear waters or on the beach, was traditionally picked by algueros (seaweed harvesters), who were frequently ex-convicts just out of prison (Selina, 2020). After being boiled and rinsed, the seaweed was sun-dried to help retain minerals such as vitamins A and C, calcium, iron, and iodine. Bustamante had 400 employees at its peak in the 1960s and 1970s, sourcing 5,000 tonnes per year (Rebours et al. 2014). An astounding 220lb. of fresh seaweed are required to generate just over two lb. of the sought-after agar (Moseley-Williams, 2014). Algueros first collected algae species specific to the Argentine Sea, such as Ulva (also known as sea lettuce), a bright green, edible kind; Gracilaria, which resembles a mane of auburn hair and contains a large amount of agar; and kelp (Hayashi et al. 2014). However, the varietals gathered have shifted over time. Japanese species such as nori and wakame, which arrived in Argentina on the bottoms of foreign fishing boats, are now part of Argentina's repertoire (Hayashi et al. 2014; Croce et al. 2015).

Seaweed and its derivatives may be found in anything from Japanese rolls and stews to desserts gelatin and chocolate pudding. La Proveedura, Bustamante's lone restaurant, goes a step further, incorporating raw ulva, nori, and wakame throughout all its cuisine (Perez et al. 2011; Kilinc et al 2013). The prices of seaweed in Argentina per tonne for the years 2016, 2017 and 2018 were US$ 5,368.16, US$ 4,041.67 and US$ 3,947.37 respectively (Selina, 2020). The total values in export for seaweed in Argentina were US$ 1,079,000, US$ 679,000 and US$ 675,000 for the years 2016, 2017 and 2018 respectively. Unfortunately, the seaweed exports from Argentina have declined remarkably between 2017 and 2019 (Selina Wamucii, 2020c).

11.8.3.2 Seaweed Cultivation Scenarios in Mexico

The Mexican beaches feature a diverse marine macroalgae community, including numerous indigenous species (Aguilar-Rosas et al. 2013). The north Coastal region and the Mexican Caribbean coastline have received the most interest in terms of seaweed research in Mexico, with less attention dedicated to other places. The country's utilization of seaweed supplies has primarily been focused on artisan collection of wildlife species, primarily as polysaccharide removal, cleansers, natural resources for livestock feed, and cosmetic items (Vázquez-Delfn et al. 2019). Most exploitation has occurred around the Baja California peninsula's beaches, with a concentration on a few varieties. Harvesting has increased to around 11,500 moist tonnes annually (2013–2016) due to the proper policies and government management rules in Mexico. Nevertheless, seaweed trade volume has increased (5302 tonnes per year typically) in Mexico in recent years (2014–2016), outweighing export volume (2156 tonnes per year on average). As a results, day by it is expanding seaweed recultivation and many commercial companies are interested to cultivate the seaweeds (Iqbal, 2022). Numerous initial studies on seaweed cultivation in Mexico imply that sustainable and responsible techniques might replace natural population harvests while also providing long-term homogeneous production of high quality (Vázquez-Delfn et al. 2019).

11.8.3.3 Seaweed Cultivation Scenarios in Venezuela

Because of the concerns and risks to the ecology and biodiversity, the introduction of exotic species has become a contentious topic amongst environmentalists and aqua culturists. As a result, we have implemented a responsible Seaweed Mariculture Program in Venezuela (Ask et al. 2003). There is a scarcity of data on the detrimental effects of intentional seaweed invasions for aquaculture reasons. Venezuela sold garbage and waste to Trinidad for US$ 99.82 thousands in 2013, as per the United Nations COMTRADE (Common format for Transient Data Exchange for power systems) statistics on international trade. Various clinker and debris (including seaweed ashes) shipments from Venezuela to Barbados and Trinidad – data, historical graph, and stats – were last modified in January 2022 (Trading Economics, 2022).

11.8.4 European Countries: Case Studies from Norway and Scotland

11.8.4.1 Seaweed Cultivation Scenarios in Norway

Norway does have the capacity to develop seaweed harvesting into a new and massive industry. Commercial production of macroalgae opens opportunities for the conversion of biofuel, which may be used to make a number of goods and help Norway become much more prominent in food, meal components, and biofuels (Shudakar et al. 2018). Norway has vast stretches of coastline that are ideal for seaweed farming. Seaweeds are principal sources that may be cultivated without the need for arable land, manure, clean water, pesticides, or medicines in the sea (Skjermo, 2020). Because kelp aquaculture is a young sector in Norway and many other parts of Europe, there are numerous obstacles to overcome before output can rise and become commercially sustainable (Araujo et al. 2021). Norway's active cultivation locations are limited and confined to the southern and mid-coastal regions; however, kelp farming has significant promise in northern Norway (Broch et al. 2019). As a result, Nofima experts are currently working on creating long-term value from kelp production in northern Norway (James et al. 2021).

11.8.4.2 Seaweed Cultivation Scenarios in Scotland

There is a significant seaweed supply, which is especially abundant in three geographic locations: west of the Trobriand Islands, the Isle of Wight, and the northern coast of Orkney (Sanderson et al. 2008; Hiatt, 2016). The Outer Hebrides have the best commercial wild seaweed collection in Scotland, focusing on egg or knotted wrack. The wild seaweed harvesting sector is mostly small-scale, with a wide range of brown, red, and greenish macroalgae gathered (Al-Dulaimi et al. 2021). Seaweed is utilized all over the world as a supply of nutrition, animal feed, and fertilizer, as well as in several sectors such as skincare, cosmeceuticals, and medications (Polat et al. 2021). As the properties of various kinds and potential applications in various goods are understood, there is substantial economic interest in Scotland's seaweed potential. The Scottish Government is expanding its scientific basis, and in 2019 began an algal review to collect evidence on the sustainable growth of current and future seaweed farming activities, as well as to appropriate amount for larger business growth (Al-Dulaimi et al. 2021)

11.8.4.3 Prospects, Challenges and Limitations of Global Seaweed Cultivation

Many south-east Asian countries have advanced their seaweed production. However, it is still in its early stages in many countries. People in developing countries are largely unaware of the benefits of seaweed growing (Ahmed and Taparhudee, 2005). In general , seaweed culture in areas appropriate for cultivation by familiarizing poor farmers with cost-effective technologies could open up a new channel for the country's seaweed business to flourish. The production of seaweeds utilizes natural materials such as bamboo and rope (Shomrat, 2021). The primary culture methods involve either vegetative propagation utilizing pieces from mother plants or spore propagation using various types of spores such as zoospores, monospores, tetraspores, and carpospores, amongst others (Msuya and Hurtado, 2017). Adult plant fragments, juvenile plant fragments, and spores are sown onto ropes or other substrata, and the plants grow to maturity in the sea. Seaweeds should be prioritized as part of an integrated coastal and national development strategy. Seaweed polyculture in conjunction with molluscs, shrimp, mud crabs, and fish appears to have promising potential for increasing harvest and earnings (Ahmed and Taparhudee, 2005). As there is no culture-based production and no processing is done, the limitations at this stage are not identifiable. In coastal areas, seaweeds used in shrimp farming can help detoxify effluent water (Phang et al. 2015). Markets and marketing organizations must be built near agriculture regions to maximize resource use and profit.

11.9 APPLICATION OF THE DPSIR MODEL FOR IDENTIFYING THE CHALLENGES OF SEAWEED CULTIVATION

The proposed DPSIR framework for seaweed cultivation in many developing countries would make it possible to overcome the many hurdles to producing seaweed. This model uses: Driver, Pressure, State, Impact, and Response (Figure 11.9). Where, driver contains alternative food, bioenergy production, Carbon sequestration, climate benefits of circular, nutrients management. Pressure mainly portraits dumping, waste disposal, pesticides, change in ecosystem, emissions of chemical, loss of species due to diversity, increased sedimentation, and destruction of coral reef. State contains degradation of water quality, utilization of coastal resources, greater abundance of fish, and higher diversity of herbivores. Impact of this model implies water pollution, increased turbulence, coastal erosion, loss of underwater species, health hazards, and acidification. Response in this model requires simple farming, ensuring workers safety, marine resource management, application of modern technology, arrangement of subsidies, provision of funding for workers, and the development of a cost-efficient farming system, especially in developing nations.

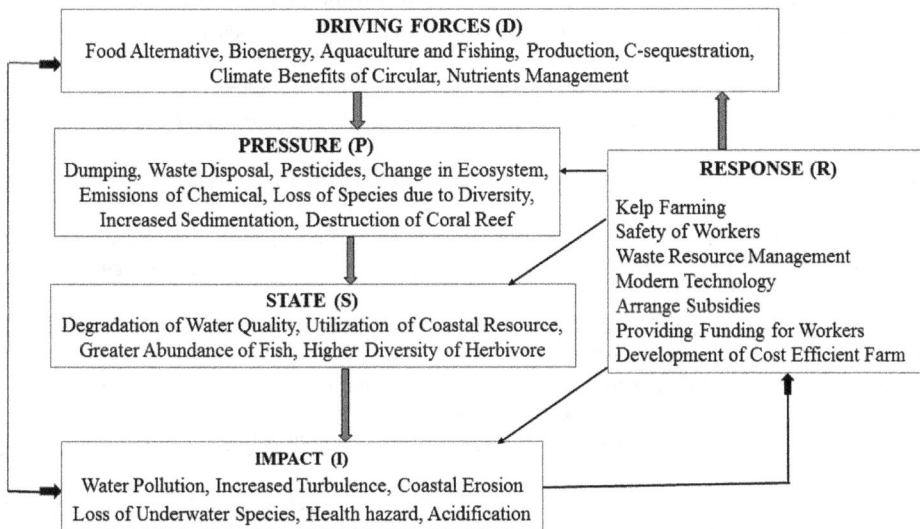

FIGURE 11.9 Identifying the major Driving Forces, Pressure, State, Impact and Response and Challenges to take the future policies and targets.

11.10 CONCLUSION

The businesses of the seaweed sector in developing countries are currently thriving, however the current status of output from local areas is falling significantly. However, numerous studies have revealed that the resources to produce seaweed by residents along the coast, are vast. There are also plenty of seaweed reserves in the intertidal and subtidal zones. Such tools offer a high potential for advancing seaweed-based enterprises.

ACKNOWLEDGEMENTS

I would like to express my gratitude to all the anonymous authors and contributors whose articles I reviewed on many occasions to produce this scientific chapter. I have also used many websites, free domains, blogs and other sources for reviewing the literature, concepts and ideas to build the scenarios of marine pollution and the challenges to developing sustainable blue economy. Also, I would like to show my gratitude to the Ministry of Science and Technology, Government of Bangladesh that have provided the funding support to the corresponding author of this chapter for continuing his research on the global blue economy and seafood production and policies. Also, I would like to show my gratitude to the *Sumitomo Foundation* in Japan who have provided the funding support to the author of this chapter for continuing his research on the global blue economy and seafood production techniques emanating from the Japanese tradition.

REFERENCES

Abbott, I. A. (1996). Ethnobotany of seaweeds: clues to uses of seaweeds. In: Lindstrom, S.C, Chapman, D.J. (eds) Fifteenth International Seaweed Symposium. Developments in Hydrobiology, vol 116. Springer, Dordrecht. https://doi.org/10.1007/978-94-009-1659-3_2

Abdelaziz, A. Leite, G. and Hallenbeck, P. (2013). Addressing the challenges for sustainable production of algal biofuels: II. Harvesting and conversion to biofuels. *Environmental Technology,* 34, 1807–1836. doi: 10.1080/09593330.2013.831487.

Acién, F. Molina, E. Reis, A. Torzillo, G. Zittelli, G. Sepúlveda, C. and Masojídek, J. (2017). Photobioreactors for the production of microalgae. Edited by Cristina Gonzalez-Fernandez and Raúl Muñoz *Microalgae-Based Biofuels and Bioproducts*, 1-44. https://doi.org/10.1016/b978-0-08-101023-5.00001-7

Agarwal, V. and Agrawalla, S. (2017). *Patanjali's Marketing Mix: The Monk's New Ferrari*. Emerald Emerging Markets Case Studies.

Aguilar-Rosas, L. E. Núñez-Cebrero, F. and Aguilar-Rosas, C. (2013). Introduced marine macroalgae in the Port of Ensenada, Baja California, Mexico: biological contamination. *Procedia Environmental Sciences,* 18, 836–843.

Ahmed, N. and Taparhudee, W. (2005). Seaweed cultivation in Bangladesh: problems and potentials. *Journal of Fisheries and Environment,* 28, 13–21.

Aizouq, M. Peisker, H. Gutbrod, K. Melzer, M. Hölzl, G. and Dörmann, P. (2020). Triacylglycerol and phytyl ester synthesis inSynechocystissp. *PCC6803. Proceedings of the National Academy of Sciences,* 117(11), 6216–6222.

Al-Dulaimi, O, Rateb, M. E, Hursthouse, A. S, Thomson, G, and Yaseen, M. (2021). The Brown Seaweeds of Scotland, Their Importance and Applications. *Environments*, 8(6), 59.21

algaeworldnews (March 21, 2017). Seaweed farming—a promising industry and solution to pollution. *Algae World News*. www.facebook.com/algaeworldnews//

Algo Rhythme, No. 31, CEVA, Pluebian, France, (2022) Retrieval on 28 July 2022 from www.ceva-algues.com/en/

Alvarez, V. B. (January 17–19, 1977). Site selection specification for seaweeds (eucheuma) farming. *A Paper Presented at the National Seaweeds Symposium,* Manila, p. 5.

Amosu, A. O. Maneveldt, G. W. Robert, J. and Bolton, J. J. (2013). South African seaweed aquaculture: a sustainable development example for other African coastal countries. *African Journal of Agricultural Research*, 8(43), 5268–5279.

Anderson, L. A. (2015). Chemical Stimulation of Lipid Production in Microalgae and Analysis by NMR Spectroscopy for Biofuel Applications. University of California, Davis.

Andersen, R. (2001). Algae Toxicity. *Encyclopedia Britannica*. Retrieved January 15, 2022, from https://britannica.com/science/algae/Toxicity.

Andersen, R. A. (2004). Biology and systematics of heterokont and haptophyte algae. *American Journal of Botany*, 91(10), 1508–1522.

Araújo, R. Vázquez Calderón, F. Sánchez López, J. Azevedo, I..C. Bruhn, A. Fluch, S. Garcia Tasende, M. Ghaderiardakani, F. Ilmjärv, T. Laurans, M. and Mac Monagail, M. 2021. Current status of the algae production industry in Europe: an emerging sector of the blue bioeconomy. *Frontiers in Marine Science*, 7, 626389.

Ask, E. I. Batibasaga, A. Zertuche-Gonzalez, J. A. and De San, M. (2003). Three decades of *Kappaphycus alvarezii* (Rhodophyta) introduction to non-endemic locations. In *Proc Int Seaweed Symp*, 17, pp. 49–57.

Bajhaiya, A. K. Mandotra, S. K. Suseela, M. R. Toppo, K. and Ranade, S. (2010). Algal biodiesel: the next generation biofuel for India. *Asian J. Exp. Biol. Sci*, 4, 728–739.

Banhart, J. (2001). Manufacture, characterisation and application of cellular metals and metal foams. *Progress in Materials Science*, 46(6), 559–632.

Bajpai, P. (2019). Production of Biofuel from Microalgae. Edited by Patima Bajpai In: Third Generation Biofuels. SpringerBriefs in Energy. Springer, Singapore. pp. 45–66; https://doi.org/10.1007/978-981-13-2378-2_7

Ball, S. Colleoni, C. Cenci, U. Raj, J. N. and Tirtiaux, C. (2011). The evolution of glycogen and starch metabolism in eukaryotes gives molecular clues to understand the establishment of plastid endosymbiosis. *Journal of Experimental Botany*, 62(6), 1775–1801.

Barbier, E. B. Hacker, S. D. Kennedy, C. Koch, E. W. Stier, A. C. and Silliman, B. R. (2011). The value of estuarine and coastal ecosystem services. *Ecological Monographs*, 81(2), 169–193.

Barbier, M. Araújo, R. Rebours, C. Jacquemin, B. Holdt, S. L. and Charrier, B. (2020). Development and objectives of the PHYCOMORPH European Guidelines for the Sustainable Aquaculture of Seaweeds (PEGASUS). *Botanica Marina*, 63(1), 5–16.

Behera, B. K. and Varma, A. (2017). *Microbial Biomass Process Technologies and Management*. Springer International Publishing, Basel, Switzerland.

Behera, S. S. Ray, R. C. Das, U. Panda, S. K. Saranraj, P. (2019). Microorganisms in Fermentation. In: Berenjian, A. (eds) Essentials in Fermentation Technology. Learning Materials in Biosciences. Springer, Cham. https://doi.org/10.1007/978-3-030-16230-6_1

Benemann, J. R. Woertz, I. and Lundquist, T. (2018). Autotrophic microalgae biomass production: from niche markets to commodities. *Industrial Biotechnology*, 14(1), 3–10.

Bianchi, C. N. Morri, C. Chiantore, M. Montefalcone, M. Parravicini, V. and Rovere, A. (2012). Mediterranean Sea biodiversity between the legacy from the past and a future of change. *Life in the Mediterranean Sea: A Look at Habitat Changes*, 1, 55.

Bixler, H. J. and Porse, H. (2011). A decade of change in the seaweed hydrocolloids industry. *Journal of Applied Phycology*, 23(3), 321–335.

Blikra, M. J. Altintzoglou, T. Løvdal, T. Rognså, G. Skipnes, D. Skåra, T. Sivertsvik, M. and Fernández, E. N. (2021). Seaweed products for the future: using current tools to develop a sustainable food industry. *Trends in Food Science and Technology*, 118, 765–776.

Boyd, C. E. (1970). Vascular aquatic plants for mineral nutrient removal from polluted waters. *Economic Botany*, 24(1), 95–103.

Broch, O. J. et al (2019). The kelp cultivation potential in coastal and offshore regions of Norway. *Frontiers in Marine Science*, 5, 529.

Buck, B. H. Nevejan, N. Wille, M. Chambers, M. D. Chopin, T. (2017). Offshore and Multi-Use Aquaculture with Extractive Species: Seaweeds and Bivalves. In: Buck, B, Langan, R. (eds) Aquaculture Perspective of Multi-Use Sites in the Open Ocean. Springer, Cham. https://doi.org/10.1007/978-3-319-51159-7_2

Buckow, R. Weiss, U. and Knorr, D. (2009). Inactivation kinetics of apple polyphenol oxidase in different pressure–temperature domains. *Innovative Food Science and Emerging Technologies*, 10(4), 441–448.

Buschmann, A. H. Camus, C. Infante, J. Neori, A. Israel, A. Hernández-González, M. C. Pereda, S. V. Gomez-Pinchetti, J. L. Golberg, A. Tadmor-Shalev, N. and Critchley, A. T. (2017). Seaweed production: overview of the global state of exploitation, farming and emerging research activity. *European Journal of Phycology*, 52(4), 391–406. https://doi.org/10.1080/09670262.2017.1365175

Cai, J. Lovatelli, A. Aguilar-Manjarrez, J. Cornish, L. Dabbadie, L. Desrochers, A. Diffey, S. Garrido Gamarro, E. Geehan, J. Hurtado, A. Lucente, D. Mair, G. Miao, W. Potin, P. Przybyla, C. Reantaso, M. Roubach, R. Tauati, M. and Yuan, X. 2021. Seaweeds and microalgae: an overview for unlocking their potential in global aquaculture development. FAO Fisheries and Aquaculture Circular No. 1229. Rome, FAO. https://doi.org/10.4060/cb5670en

Calvin, M. (1974). Solar energy by photosynthesis. *Science*, 184(4134), 375–381.

Campbell, I. Macleod, A. Sahlmann, C. Neves, L. Funderud, J. Øverland, M. Hughes, A. D. and Stanley, M. (2019). The environmental risks associated with the development of seaweed farming in Europe-prioritizing key knowledge gaps. Frontiers in Marine Science, 6, 107.

Castro, K. L. Epherra, L. Raffo, M. P. Morsan, E. and Rubilar, T. (2022). Changes in the diet of the native sea urchin Arbacia dufresnii at different scenarios of the Undaria pinnatifida invasion (Patagonia, Argentina). *Food Webs*, *31*, e00221.

Cheah, W. Y. Ling, T. C. Juan, J. C. Lee, D. J. Chang, J. S. and Show, P. L. (2016). Biorefineries of carbon dioxide: From carbon capture and storage (CCS) to bioenergies production. *Bioresource Technology*, 215, 346–356.

Chen, J. L, Hsu, K, and Chuang, C. T. (2020). How do fishery resources enhance the development of coastal fishing communities: Lessons learned from a community-based sea farming project in Taiwan. *Ocean and Coastal Management*, *184*, 105015.

Choudhary, B. (2021). Edible seaweeds: a potential novel source of bioactive metabolites and nutraceuticals with human health benefits. *Frontiers*. Retrieved February 26, 2022, from https://frontiersin.org/articles/10.3389/fmars.2021.740054/full.

Chung, I. K. Beardall, J. Mehta, S. Sahoo, D. and Stojkovic, S. (2011). Using marine macroalgae for carbon sequestration: a critical appraisal. *J Appl Phycol*, 23, 877–886.

Chung, I. K. Oak, J. H. Lee, J. A. Shin, J. A. Kim, J. G. and Park, K. S. (2013). Installing kelp forest/seaweed beds for mitigation and adaptation against global warming: Korean Project overview. *ICES J Mar Sci*, 70, 1038–1044.

Cordier, C. Moulin, P. Stavrakakis, C. Lange, A. and Chenier, F. (2021). Culture of Microalgae with Ultrafiltered Seawater: From a Feasibility Study to an Industrial Development. *FAO Aquaculture Newsletter*, 64, 12–13.

County, N. C. (2014). The project on integrated urban development master plan for the City of Nairobi in the Republic of Kenya. *Nairobi City County Nairobi: Nairobi City County*.

Croce, M. E. Villar, M. A. and Parodi, E. R. (2015). Assessment of alternative sources of seaweed polysaccharides in Argentina: potentials of the agarophyte Gelidium crinale (Hare ex Turner) Gaillon (Rhodophyta, Gelidiales). *Journal of applied phycology*, 27(5), 2099–2110.

Dawes, C. J. Orduna-Rojas, J. and Robledo, D. (1998). Response of the tropical red seaweed Gracilaria cornea to temperature, salinity and irradiance. *Journal of Applied Phycology*, 10(5), 419–425.

Dawson, K. (2006). Enslaved swimmers and divers in the Atlantic world. *The Journal of American History*, 92(4), 1327–1355.

De Zaixsoa, A. B. Ciancia, M. and Cerezo, A. S. (1998). Seaweed resources of Argentina. In *Seaweed Resources of the World* (pp. 372–384). Japan International Cooperation Agency.

Diab, Wang, N. A. (June 4, 2018). Seaweed helps Qingdao's marine economy stay afloat. *CGTN*. Retrieval date: 2018-06-04; https://news.cgtn.com/news/3d3d774d3559444f77457a6333566d54/share_p.html

Doty, M. S. (1973). Seaweeds farms: a new approach for U.S. industry. Reprints from *Marine Technology Society. Proc.* 701–708.

Duarte, C. M. 1992. Nutrient concentration of aquatic plants: patterns across species. *Limnology and Oceanography* 37: 882–889. DOI: 10.4319/lo.1992.37.4.0882

El Gamal, A. A. (2010). Biological importance of marine algae. *Saudi Pharmaceutical Journal*, 18(1), 1–25.

Eljaddi, T. Ragueneau, S. Cordier, C. Lange, A. Rabiller, M. Stavrakakis, C. and Moulin, P. (2021). Ultrafiltration to secure shellfish industrial activities: culture of microalgae and oyster fertilization. *Aquacultural Engineering*, 95, p.102204. ISSN 0144-8609, https://doi.org/10.1016/j.aquaeng.2021.102204.

Engle, C. R. Quagrainie, K. K. and Dey, M. M. (2016). *Seafood and Aquaculture Marketing Handbook*. John Wiley: Hoboken, NJ.

Evans, E. A. Ballen, F. H. and Siddiq, M. (2020). Banana production, global trade, consumption trends, postharvest handling, and processing. Editor(s):Muhammad Siddiq,Jasim Ahmed and Maria Gloria Lobo. In *Handbook of Banana Production, Postharvest Science, Processing Technology, and Nutrition* (pp. 1–18). Willey Inline Library, Australia

Falkenberg, M. Nakano, E. Zambotti-Villela, L. Zatelli, G. A. Philippus, A. C. Imamura, K. B. Velasquez, A. M. A. Freitas, R.P. de Freitas Tallarico, L. Colepicolo, P. and Graminha, M. A. (2019). Bioactive compounds against neglected diseases isolated from macroalgae: a review. *Journal of Applied Phycology*, 31(2), 797–823.

Fehrenbacher, K. (2021). Meet the new US entrepreneurs farming seaweed for food and fuel. *The Guardian*. Retrieved January 12, 2022, from https://theguardian.com/sustainable-business/2017/jun/29/seaweed-farms-us-california-food-fuel

Feijiu, W. (2019). *Seminar Report on the Status of Seaweed Culture in China, India, Indonesia, ROK, Malaysia, Philippines and Thailand*. www.fao.org. Retrieved July 26, 2022, from https://fao.org/3/ab719e/AB719E02.htm

Ferdouse, F. Holdt, S. L. Smith, R. Murúa, P. and Yang, Z. (2018). *The Global Status of Seaweed Production, Trade and Utilization*. Globefish Research Programme, 124, I.

Fiorella, K. J. Okronipa, H. Baker, K. and Heilpern, S. (2021). Contemporary aquaculture: implications for human nutrition. *Current Opinion in Biotechnology*, 70, 83–90.

Fletcher, R. and Fletcher, R. (2022). Indonesia aims for creation of 'aquaculture village' network. *Thefishsite.com*. Retrieved January 12, 2022, from https://thefishsite.com/articles/study-makes-strong-commercial-case-for-seaweed-farming

Flora, K. (June 3, 2019). Seaweed | Growing and Harvesting Farms. Retrieved January 12, 2022, from https://foodunfolded.com/article/seaweed-growing-harvesting-farms

Ford, J. (February 7, 2021). *What Type of Organism Is Seaweed?—Answers to All*.

Forster, J. Radulovich, R. (2015) Chapter 11 – Seaweed and food security, Editor(s): Brijesh K. Tiwari, D, Seaweed Sustainability, Academic Press, 2015, Pages 289–313, ISBN 9780124186972, https://doi.org/10.1016/B978-0-12-418697-2.00011-8.

FPJ Bureau (January 15, 2021). Why its high time we conserve seaweeds urgently? *Free Press Journal*. Retrieved July 26, 2022, from https://freepressjournal.in/science/why-its-high-time-we-conserve-seaweeds-urgently

De Fraiture, C. Giordano, M. and Liao, Y. (2008). Biofuels and implications for agricultural water use: blue impacts of green energy. *Water policy*, 10(S1), 67–81.

Gambrel, E. (2019). What are the characteristics of the Protista Kingdom? *Sciencing*. Retrieved January 13, 2022, from https://sciencing.com/characteristics-protista-kingdom-8576710.html

Gardner, C. J. Rocliffe, S. Gough, C. Levrel, A. Singleton, R. L. Vincke, X. and Harris, A. (2017). Value chain challenges in two community-managed fisheries in western Madagascar: insights for the small-scale fisheries guidelines. In *The Small-Scale Fisheries Guidelines* (pp. 335–354). Springer, Cham.

Gatamaneni, L. Bhalamurugan, O. V. and Lefsrud, M. (2018). Factors affecting growth of various microalgal species. *Environmental Engineering Science*, 35: 1037–1048. doi: 10.1089/ees.2017.0521.

Geheb, K. I. M. and Binns, T. (1997). 'Fishing farmers' or 'farming fishermen'? The quest for household income and nutritional security on the Kenyan shores of lake Victoria. *African Affairs*, 96(382), 73–93.

Geremia, E, Ripa, M, Catone, C. M, and Ulgiati, S. (2021). A Review about Microalgae Wastewater Treatment for Bioremediation and Biomass Production—A New Challenge for Europe. *Environments*, 8(12), 136.

Glaros, A. Marquis, S. Major, C. Quarshie, P. Ashton, L. Green, A. G. Kc, K. B. Newman, L. Newell, R. Yada, R. Y. and Fraser, E. D. (2021). *Horizon Scanning and Review of the Impact of Five Food and Food Production Models for the Global Food System in 2050*. Trends in Food Science and Technology.

Godin, M. (2020). The ocean farmers trying to save the world with seaweed. *Time*. Retrieved January 12, 2022, from https://time.com/5848994/seaweed-climate-change-solution/

Gupta, S. and Abu-Ghannam, N. (2011). Bioactive potential and possible health effects of edible brown seaweeds. *Trends in Food Science and Technology*, 22(6), 315–326.

Hallmann, A. (2015). Algae biotechnology–green cell-factories on the rise. *Current Biotechnology*, 4(4), 389–415.

Halsey, K. H. and Jones, B. M. (2015). Phytoplankton strategies for photosynthetic energy allocation. *Annual Review of Marine Science*, 7, 265–297.

Harb, T. B. and Chow, F. (2022). An overview of beach-cast seaweeds: Potential and opportunities for the valorization of underused waste biomass. *Algal Research*, 62, 102643.

Harper, J. T. and Garbary, D. J. (1994). Host specificity of Podocystis adriatica on the red alga *Heterosiphonia crispella* from Senegal. *Diatom Research*, 9(2), 329–333.

Harper, J. T. and Garbary, D. J. (1997). Marine algae of Northern Senegal: The flora and its biogeography. *Botanica Marina*, 40(1–6). https://doi.org/10.1515/botm.1997.40.1-6.129

Hashim, T. M. Z. T. and Suratman, M. N. Mangrove Biomass: review on estimation Methods and the applications of remote sensing approach. *Mangroves*, 52.

Hayashi, L. Bulboa, C. Kradolfer, P. Soriano, G. and Robledo, D. (2014). Cultivation of red seaweeds: a Latin American perspective. *Journal of Applied Phycology*, 26(2), 719–727.

Henchion, M. Hayes, M. Mullen, A. M. Fenelon, M. and Tiwari, B. (2017). Future protein supply and demand: strategies and factors influencing a sustainable equilibrium. *Foods*, 6(7), 53.

Hickling, C. F. (1971). Estuarine fish farming. Editor(s): Frederick S. Russell, Maurice Yonge, In *Advances in Marine Biology* (Vol. 8, pp. 119–213). Academic Press. Cambridge, MA.

Holmyard, N. (2011). Seaweed farming for profit. www.seafoodsource.com. Retrieved August 17, 2011, from https://seafoodsource.com/features/seaweed-farming-for-profit

Hong, D. D. and Hien, H. T. M. (2004). Nutritional analysis of Vietnamese seaweeds for food and medicine. *BioFactors*, 22(1–4), 323–325. https://doi.org/10.1002/biof.5520220164

Hossain, Delwar, and Islam, Md. Shariful (2019). Unfolding Bangladesh-India maritime connectivity in the Bay of Bengal region: a Bangladesh perspective. *Journal of the Indian Ocean Region*, 15(3), 346–355.

Hossain, M. S. Alamgir, M. Uddin, S. Chowdhury, M. (2020). *Seaweeds for Blue Economy in Bangladesh*.

Hossain, M. S. Sharifuzzaman, S. M. Nobi, M. N. Chowdhury, M. S. N. Sarker, S. Alamgir, M. Uddin, S. A. Chowdhury, S. R. Rahman, M. M. Rahman, M.S. and Sobhan, F. (2021). Seaweeds farming for sustainable development goals and blue economy in Bangladesh. *Marine Policy*, 128, 104469. Elesevier Publication, Amsterdam, The Netherlands.

Hu, Z. M. Shan, T. F. Zhang, J. Zhang, Q. S. Critchley, A. T. Choi, H. G. Yotsukura, N. Liu, F. L. and Duan, D. L. (2021). Kelp aquaculture in China: a retrospective and future prospects. *Reviews in Aquaculture*, 13(3), 1324–1351.

Insights, F. B. (2021). Commercial Seaweed Market Size, Share and COVID-19 Impact Analysis, by Type (Red Seaweed, Brown Seaweed, and Green Seaweed), form (Flakes, Powder, and Liquid), End-Uses (Food and Beverages, Agricultural Fertilizer, Animal Feed Additives, Pharmaceutical, and Cosmetics and Personal Care), and Regional Forecast. 2021–2028.

Islam, M. N. Kitazawa, D. Kokuryo, N. Tabeta, S. Honma, T. and Komatsu, N. (2012). Numerical modeling on transition of dominant algae in Lake Kitaura, Japan, *Ecological Modelling*, 242, 146–163. doi: 10.1016/j.ecolmodel.2012.05.013

Islam, M. N. Al Amin, M. and Noman, M. (2018). Management challenges of sinking of oil tanker at Shela Coastal River in Sundarbans Mangroves in Bangladesh. *Environmental Management of Marine Ecosystems*, 9, 323.

Islam, M. N. Kitazawa, D. Hamill, T. and Park, H. D. (2013). Modeling mitigation strategies for toxic cyanobacteria blooms in shallow and eutrophic Lake Kasumigaura, Japan. *Mitigation and Adaptation Strategies for Global Change*, 18(4), 449–470. doi:10.1007/s11027–012–9396–0

Islam, M. Tamanna, S. Amstel, A.V. Noman, M. Ali, M. Saadat, S. Aparajita, D. M. Roy, P. Tanha, S. R. Sarker, N. and Ashiquzzaman, M. (2021). Climate Change Impact and Comprehensive Disaster Management Approach in Bangladesh: A Review. Edited by Islam and Amstel In: Bangladesh II: Climate Change Impacts, Mitigation and Adaptation in Developing Countries. Springer Climate. Springer, Cham. https://doi.org/10.1007/978-3-030-71950-0_1

Islam, M. M. Hasan, J. Ali, M. Z. and Hoq, M. E. (2021). Culture of three seaweed species in Cox's Bazar Coast, Bangladesh. *Bangladesh Journal of Zoology*, 49(1), 47–56. https://doi.org/10.3329/bjz.v49i1.53681

Jalilian, N. Najafpour, G. D. and Khajouei, M. (2020). Macro and micro algae in pollution control and biofuel production–a review. *ChemBioEng Reviews*, 7(1), 18–33.

James, P. Wang, X. James, P. and Wang, X. (November 25, 2021). Advancing selective breeding in sea bass and sea bream. *Thefishsite.com*.Retrieval Decmber 23, 2021 from https://thefishsite.com/articles/kelp-farming-a-great-opportunity-for-northern-norway-and-the-world

Johnson, B. Narayanakumar, Ramani, Nazar, A. K. Abdul, Kaladharan, P. and Gopakumar, G. (2017). Economic analysis of farming and wild collection of seaweeds in Ramanathapuram District, Tamil Nadu. *Indian Journal of Fisheries*, 64, 94–99. doi: 10.21077/ijf.2017.64.4.61828–13.

Juanich, L. G. (April 1988). Manual on seaweed farming. www.fao.org. Retrieved January 12, 2022, from https://fao.org/3/ac416e/ac416e00.htm.

Kamadi, G. (2020). An ocean of opportunity. *NextBlue*. Retrieval on September 15, 2021, from https://next.blue/articles/an-ocean-of-opportunity

Kandale, A. Meena, A. K. Rao, M. M. Panda, P. Mangal, A. K. Reddy, G. and Babu, R. (2011). Marine algae: an introduction, food value and medicinal uses. *Journal of Pharmacy Research*, 4(1), 219–221.

Kandasamy, S. Zhang, B. He, Z. Bhuvanendran, N. EL-Seesy, A.I. Wang, Q. Narayanan, M. Thangavel, P. and Dar, M.A. (2022). Microalgae as a multipotential role in commercial applications: Current scenario and future perspectives. *Fuel*, 308, 122053.

Karltorp, K. (2014). *Scaling up Renewable Energy Technologies—The Role of Resource Mobilisation in the Growth of Technological Innovation Systems*. Chalmers University of Technology.

Karningsih, P. D. Kusumawardani, R. Syahroni, N. Mulyadi, Y. and Saad, M. S. B. M. (2021). Automated fish feeding system for an offshore aquaculture unit. Edited Manik Mahachandra In *IOP Conference Series: Materials Science and Engineering* (Vol. 1072, No. 1, p. 012073). doi:10.1088/1757-899X/1072/1/012073; IOP Publishing, Temple Circus, Temple Way, Bristol BS1 6HG, UK

Kennedy, J. (November 29, 2019). *What Are the 3 Types of Sea Weed (Marine Algae)?*

Kerrison, P. D. Stanley, M. S. Edwards, M. D. Black, K. D. and Hughes, A. D. (2015). The cultivation of European kelp for bioenergy: site and species selection. *Biomass Bioenergy* 80, 229–242. doi: 10.1016/j.biombioe.2015.04.035

Khalil, H. P. S. Lai, T. K. Tye, Y. Y. Rizal, S. Chong, E. W. N. Yap, S. W. Hamzah, A. A. Fazita, M. R. and Paridah, M.T. (2018). A review of extractions of seaweed hydrocolloids: Properties and applications. *Express Polymer Letters*, 12(4).

Khan, M. I. Shin, J. H. and Kim, J. D. (2018). The promising future of microalgae: current status, challenges, and optimization of a sustainable and renewable industry for biofuels, feed, and other products. *Microbial Cell Factories*, 17(1), 1–21.

Khan, M. S. K. Hoq, M. E. Haque, M. A. Islam, M. M. and Hoque, M. M. (2016). Nutritional evaluation of some seaweeds from the Bay of Bengal in contrast to inland fishes of Bangladesh. *IOSR J. Environ. Sci. Toxicol. Food Technol*, 10(11), 59–65.

Khandaker, M. U. Chijioke, N. O. Heffny, N. A. B. Bradley, D. A. Alsubaie, A. Sulieman, A. Faruque, M. R. I. Sayyed, M. I. and Al-Mugren, K.S. (2021). Elevated concentrations of metal (Loids) in seaweed and the concomitant exposure to humans. *Foods*, 10(2), 381.

Kılınç, B, Cirik, S. Turan, G. Tekogul, H. and Koru, E. (2013). Seaweeds for food and industrial applications. Edited by Innocenzo Muzzalupo In *Food Industry*. IntechOpen. https://doi.org/10.5772/53172, Vieana, Austria.

Kim, S. K. and Venkatesan, J. (2015). Introduction to seafood science. *Europe*, 16(19.8), 2–3.

Klinkhamer, A. .J. Vann, A. D. Swain, E. Madden, J. Luce, J. Rosen, J. Schuler, J. Zalesny, K. Kamman, K. Lael, L. and Minnick, M. (2020) EfficienSea. Retrieved March 16, 2022, from www.nama.org/uploads/1/2/6/6/126666192/purdue_univ_exec_summary.pdf

Koss, R. Bellgrove, A. Lerodiaconou, D. Gilmour, P. and Bunce, A. (2005). Sea search: community-based monitoring of Victoria's Marine National Parks and Marine Sanctuaries: subtidal reef monitoring. *Parks Victoria Technical Series*, 1(8).

Langdon, S. J. (2006). Selective traditional Tlingit salmon fishing techniques on the west coast of the Prince of Wales Archipelago. *Traditional Ecological Knowledge and Natural Resource Management,* 21.

Langford, Z. Waldron, S. and Sulfahri (November 21, 2020). Seaweed farmers' flexibility makes Indonesia a major player in global markets, but there is more work to be done. *The Conversation*. Retrieved January 15, 2022, from https://theconversation.com/seaweed-farmers-flexibility-makes-indonesia-a-major-player-in-global-markets-but-there-is-more-work-to-be-done-150371

Langford, A. and Waldron, S. (2020). Seaweed farmers' flexibility makes Indonesia a major player in global markets, but there is more work to be done. *The Conversation*.

Langton, R. Augyte, S. Price, N. Forster, J. Noji, T. Grebe, G. St Gelais, A. and Byron, C. J. (2019). An ecosystem approach to the culture of seaweed. NOAA Tech. Memo. NMFS-F/SPO-195, 24 p.

Leisner, T. M. (1999). *Bidirectional Transmembrane Modulation of Platelet Glycoprotein IIb-IIIa Conformations* (Doctoral dissertation, University of Illinois at Chicago, Health Sciences Center).

Limiting Factors (n.d.). *The Marine Biome*. Retrieved January 13, 2022, from https://marinebiomescience.wee bly.com/limiting-factors.html

Mac Monagail, M. Cornish, L. Morrison, L. Araujo, R. Critchley, A. T. (2017). Sustainable harvesting of wild seaweed resources. *European Journal of Phycology*, *52*(4), 371–390.

Malvis, A. Hodaifa, G. Halioui, M. Seyedsalehi, M. and Sánchez, S. (2019). Integrated process for olive oil mill wastewater treatment and its revalorization through the generation of high added value algal biomass. *Water Research*, 151, 332–342.

Mamun, S. (2022). Immense potential of seaweed. https://dhakatribune.com/. Retrieved January 13, 2022, from https://dhakatribune.com/bangladesh/2022/01/05/immense-potential-of-seaweed.

Mantri, V. Ghosh, A. and Meenakshisundaram, G. (2022). A sea of opportunities for seaweed farming and processing in India. Retrieval on April 22, 2021, from file:///C:/Users/USER/Downloads/AquaPost_ 2022%20(4).pdf

Manzelat, S. F. Mufarrah, A. M. Hasan, B. A. and Hussain, N. A. (2018). Macro algae of the Red Sea from Jizan, Saudi Arabia. *Phykos*, *48*(1), 88–108.

Marine Algae (n.d.). www.mesa.edu.au. Retrieved January 14, 2022, from http://mesa.edu.au/marine_algae/

Marquez, G. P. B. Santiañez, W. J. E. Trono Jr, G. C. Montaño, M. N. E. Araki, H. Takeuchi, H. and Hasegawa, T. (2014). Seaweed biomass of the Philippines: sustainable feedstock for biogas production. *Renewable and Sustainable Energy Reviews*, *38*, 1056–1068.

Marrone, B. L. Lacey, R. E. Anderson, D. B. Bonner, J. Coons, J. Dale, T. Downes, C. M. Fernando, S. Fuller, C. Goodall, B. Holladay, J. E. Kadam, K. Kalb, D. Liu, W. Mott, J. B. Nikolov, Z. Ogden, K. L. Sayre, R. T. Trewyn, B. G. and Olivares, J. A. (2018). Review of the harvesting and extraction program within the National Alliance for Advanced Biofuels and Bioproducts. *Algal Research*, 33, 470–485. https://doi. org/10.1016/j.algal.2017.07.015

Melis, A. (1999). Photosystem-II damage and repair cycle in chloroplasts: what modulates the rate of photodamage in vivo? *Trends in Plant Science*, 4(4), 130–135.

Martinez, J. (2019) What Does Seaweed Need to Live? Retrieved on 28 July 2022 from https://sciencing.com/ seaweed-need-live-7426003.html

Mathew, S. Raman, M. Kalarikkathara Parameswaran, M. and Rajan, D. P. (2019). Bioactive Compounds from Marine Sources. In *Fish and Fishery Products Analysis* (pp. 379–443). Springer, Singapore.

Michalak, I. Chojnacka, K. and Saeid, A. (2017). Plant growth biostimulants, dietary feed supplements and cosmetics formulated with supercritical CO2 algal extracts. *Molecules*, 22(1), 66.

Midmore, D. and Rank, A. (2002). A New rural industry-Stevia-to replace imported chemical sweeteners: a report for the Rural Industries Research and Development Corporation.

Milano, J. Ong, H. C. Masjuki, H. H. Chong, W. T. Lam, M. K. Loh, P. K. and Vellayan, V. (2016). Microalgae biofuels as an alternative to fossil fuel for power generation. *Renewable and Sustainable Energy Reviews*, *58*, 180–197.

Mineur, F. Arenas, F. Assis, J. Davies, A. J. Engelen, A. H. Fernandes, F. Malta, E. J. Thibaut, T. Van Nguyen, T. U. Vaz-Pinto, F. and Vranken, S. (2015). European seaweeds under pressure: consequences for communities and ecosystem functioning. *Journal of Sea Research*, 98, 91–108.

Mirera, D. O. Kimathi, A. Ngarari, M. M. Magondu, E. W. Wainaina, M. and Ototo, A. (2020). Societal and environmental impacts of seaweed farming in relation to rural development: the case of Kibuyuni village, south coast, Kenya. *Ocean and Coastal Management*, 194, 105253.

Morrissey, J. Kraan, S. and Guiry, M. D. (2001). *A Guide to Commercially Important Seaweeds on the Irish coast*. Irish Bord Iascaigh Mhara/Irish Sea Fisheries Board.

Moseley-Williams, S. (2014). The one dish to eat in Argentine Patagonia is seaweed. *Cond Nast Traveler*. Retrieved January 15, 2022, from https://cntraveler.com/stories/2014–12–06/the-one-dish-to-eat-in-argentine-patagonia-is-seaweed

Moslamy, S. H. Elkady, M. F. Rezk, A. H. and Abdel-Fattah, Y. R. (2017). Applying Taguchi design and large-scale strategy for mycosynthesis of nano-silver from endophytic Trichoderma harzianum SYA. F4 and its application against phytopathogens. *Scientific Reports*, 7(1), 1–22.

Motwalli, O. Essack, M. Jankovic, B. R. Ji, B. Liu, X. Ansari, H. R. Hoehndorf, R. Gao, X. Arold, S. T. Mineta, K. and Archer, J. A. (2017). In silico screening for candidate chassis strains of free fatty acid-producing cyanobacteria. *BMC Genomics*, 18(1), 1–21.

Morrison, T. H. Hughes, T. P. Adger, W. N. Brown, K. Barnett, J. and Lemos, M. C. (2019). Save reefs to rescue all ecosystems, 1, 333–336.

Mouritsen, O. G. Dawczynski, C. Duelund, L. Jahreis, G. Vetter, W. and Schröder, M. (2013). On the human consumption of the red seaweed dulse (Palmaria palmata (L.) Weber and Mohr). *Journal of Applied Phycology*, 25(6), 1777–1791.

Mouritsen, O. L. E. (2018). The science of seaweeds. *American Scientist*. Retrieved January 13, 2022, from https://americanscientist.org/article/the-science-of-seaweeds

Msuya, F. E. and Hurtado, A. Q. (2017). The role of women in seaweed aquaculture in the Western Indian Ocean and South-East Asia. *European Journal of Phycology*, 52(4), 482–494.

Munoz, J. Freile-Pelegrín, Y. and Robledo, D. (2004). *Mariculture of Kappaphycus alvarezii (Rhodophyta, Solieriaceae) color strains in tropical waters of Yucatán, México. Aquaculture*, 239(1–4), 161–177.

Muthayya, S. Sugimoto, J. D. Montgomery, S. and Maberly, G. F. (2014). An overview of global rice production, supply, trade, and consumption. *Annals of the New York Academy of Sciences*, 1324(1), 7–14.

Mystikou, A. Asensi, A. O. DeClerck, O. Müller, D. G. Peters, A. F. Tsiamis, K. Fletcher, K. I. Westermeier, R. Brickle, P. Van West, P. and Küpper, F. C. (2016). New records and observations of macroalgae and associated pathogens from the Falkland Islands, Patagonia and Tierra del Fuego. *Botanica Marina*, 59(2–3), 105–121.

Muthayya, S. Sugimoto, J. D. Montgomery, S. and Maberly, G. F. (2014). An overview of global rice production, supply, trade, and consumption. *Annals of the New York Academy of Sciences*, 1324(1), 7–14.

Nagelkerken, I. S. J. M. Blaber, S. J. M. Bouillon, S. Green, P. Haywood, M. Kirton, L. G. Meynecke, J. O. Pawlik, J. Penrose, H. M. Sasekumar, A. and Somerfield, P. J. (2008). The habitat function of mangroves for terrestrial and marine fauna: a review. *Aquatic Botany*, 89(2), 155–185.

Nagelkerken, I., Goldenberg, S. U., Coni, E. O., & Connell, S. D. (2018). Microhabitat change alters abundances of competing species and decreases species richness under ocean acidification. *Science of the Total Environment*, 645, 615–622.

National Geographic Society (n.d.). Limiting Factors. (C) *National Geographic Society*. Retrieval on July 10, 2021 from https://nationalgeographic.org/topics/limiting-factors/

Naylor, R. L. Goldburg, R. J. Primavera, J. H. Kautsky, N. Beveridge, M. Clay, J. ... and Troell, M. (2000). Effect of aquaculture on world fish supplies. *Nature*, 405(6790), 1017–1024.

Nazir, Y. Halim, H. Prabhakaran, P. Hamid, A. A. and Song, Y. (2020). Microalgae single cell oil. *Handbook of Microalgae-Based Processes and Products,* 419–444.

Nagelkerken, I. Goldenberg, S. U. Coni, E. O. and Connell, S. D. (2018). Microhabitat change alters abundances of competing species and decreases species richness under ocean acidification. *Science of the Total Environment*, 645, 615–622.

New guidelines for sustainable European seaweed (November 17, 2021). EU Science Hub—European Commission. Retrieved January 12, 2022, from https://ec.europa.eu/jrc/en/news/new-guidelines-sustainable-european-seaweed

Newton, A. Icely, J. Cristina, S. Brito, A. Cardoso, A. C. Colijn, F. Dalla Riva, S. Gertz, F. Hansen, J. W. Holmer, M. and Ivanova, K. (2014). An overview of ecological status, vulnerability and future perspectives of European large shallow, semi-enclosed coastal systems, lagoons and transitional waters. *Estuarine, Coastal and Shelf Science*, 140, 95–122.

Pereira, L. and Cotas, J. (2019). Historical use of seaweed as an agricultural fertilizer in the European Atlantic area. In Seaweeds as plant fertilizer, agricultural biostimulants and animal fodder (pp. 1–22). CRC Press.

Perez, A. A. Fajardo, M. A. Farias, S. S. Perez, L. B. Strobl, A. and Roses, O. (2011). Human dietary exposure to lead and cadmium via the consumption of mussels and seaweeds from San Jorge Gulf, Patagonia Argentina. *International Journal of Environment and Health*, 5(3), 163–185.

Periyasamy, S. and Viswanathan, N. (2018). Hydrothermal synthesis of hydrocalumite assisted biopolymeric hybrid composites for efficient Cr (VI) removal from water. *New Journal of Chemistry*, 42(5), 3371–3382.

Phang, S. M. Chu, W. L. and Rabiei, R. (2015). Phycoremediation. In *The algae world* (pp. 357–389). Springer, Dordrecht.

Phi, T. N. (2014). *Cultivation Characteristics and Biological Responses of Agarophytic Seaweed, Gracilaria fisheri (Rhodophyta), in Southern Thailand* (Doctoral dissertation, Prince of Songkla University, Pattani Campus).

Polat, S, Trif, M, Rusu, A, Šimat, V, Čagalj, M, Alak, G, ... and Özogul, F. (2021). Recent advances in industrial applications of seaweeds. *Critical Reviews in Food Science and Nutrition*, 1–30.

Potgieter, T. (2018). Oceans economy, blue economy, and security: notes on the South African potential and developments. *Journal of the Indian Ocean Region*, 14(1), 49–70.

Radulovich, R. Neori, A. Valderrama, D. Reddy, C. R. K. Cronin, H. and Forster, J. (2015). Farming of seaweeds. In *Seaweed sustainability* (pp. 27–59). Academic Press.

Rahman, M. R. (2015). Causes of biodiversity depletion in Bangladesh and their consequences on ecosystem services. *American Journal of Environmental Protection*, 4(5), 214–236.

Rao, P. S. Periyasamy, C. Kumar, K. S. Rao, A. S. and Anantharaman, P. (2019). Seaweeds: distribution, production and uses. *Bioprospecting of Algae*. Society for Plant Research India, Meerut, India, 59–78.

Rebours, C. Marinho-Soriano, E. Zertuche-González, J.A. Hayashi, L. Vásquez, J. A. Kradolfer, P. Soriano, G. Ugarte, R. Abreu, M. H. Bay-Larsen, I. and Hovelsrud, G. (2014). Seaweeds: an opportunity for wealth and sustainable livelihood for coastal communities. *Journal of Applied Phycology*, 26(5), 1939–1951.

Region, M. (2014). Review of selected California fisheries for 2013: coastal pelagic finfish, market squid, groundfish, highly migratory species, dungeness crab, basses, surfperch, abalone, kelp and edible algae, and marine aquaculture.

Reporter, D. (2020). Three new species of seaweed discovered in the Bay of Bengal. *The Green Page*. Retrieved January 13, 2022, from https://thegreenpagebd.com/seaweed-discovered/

Richardson, J. (2013). Blue-green algae biofuel research continues. *CleanTechnica*. Retrieved January 15, 2022, from https://cleantechnica.com/2013/01/15/blue-green-algae-biofuel-research-continues/

Riebesell, U. (2004). Effects of CO2 enrichment on marine phytoplankton. *Journal of Oceanography*, 60(4), 719–729.

Roleda, M. Y. and Hurd, C. L. (2019). Seaweed nutrient physiology: application of concepts to aquaculture and bioremediation. *Phycologia*, 58(5), 552–562.

Roohinejad, S. Koubaa, M. Barba, F. J. Saljoughian, S. Amid, M. and Greiner, R. (2017). Application of seaweeds to develop new food products with enhanced shelf-life, quality and health-related beneficial properties. *Food Research International,* 99, 1066–1083.

Saad, M. G. Dosoky, N. S. Zoromba, M. S. and Shafik, H. M. (2019). Algal biofuels: current status and key challenges. *Energies*, 12(10), 1920.

Saeed, M. U. Hussain, N. Shahbaz, A. Hameed, T. Iqbal, H. M. and Bilal, M. (2021). Bioprospecting microalgae and cyanobacteria for biopharmaceutical applications. *Journal of Basic Microbiology*. doi: 10.1002/jobm.202100445. Epub ahead of print. PMID: 34914840.

Sagan, L. (1967). On the origin of mitosing cells. *Journal of Theoretical Biology*, 14(3), 225–276.

Samarakoon, K. and Jeon, Y. J. (2012). Bio-functionalities of proteins derived from marine algae—A review. *Food Research International*, 48(2), 948–960.

Samsul M. Henrik N. Tafsir J. and Clive, S. (2020). Enabling stakeholder participation in marine spatial planning: the Bangladesh experience. *Journal of the Indian Ocean Region,* 16(3), 268–291.

Sanderson, J. C. Cromey, C. J. Dring, M. J. and Kelly, M. S. (2008). Distribution of nutrients for seaweed cultivation around salmon cages at farm sites in north–west Scotland. *Aquaculture*, 278(1–4), 60–68.

Seaweed: An alternative protein source (2012). *ScienceDaily*. Retrieved January 13, 2022, from https://sciencedaily.com/releases/2012/10/121012074659.htm

Seaweed harvesting and cultivation | Scotland's Marine Assessment 2020 (2018). *marine.gov.scot*. http://marine.gov.scot/sma/assessment/seaweed-harvesting-and-cultivation

Seaweed | Rikolto in Indonesia (July 28, 2021). indonesia.rikolto.org | Rikolto in Indonesia. Retrieved January 15, 2022, from https://indonesia.rikolto.org/en/crop/seaweed

Selina Wamucii (February 18, 2020). *Page Analysis*. Retrieved January 15, 2022, from https://selinawamucii.com/insights/prices/kenya/seaweed/

Selina Wamucii (February 18, 2020b). *Page Analysis*. Retrieval on May 17, 2021, from https://selinawamucii.com/insights/market/senegal/seaweed/

Selina Wamucii (2020c). *Page Analysis*. Retrieved January 15, 2022, from https://selinawamucii.com/insights/market/argentina/seaweed/

Shaikh, S. F. Mazo-Mantilla, H. F. Qaiser, N. Khan, S. M. Nassar, J. M. Geraldi, N. R. Duarte, C. M. and Hussain, M. M. (2019). Noninvasive featherlight wearable compliant 'Marine Skin': standalone multisensory system for deep-sea environmental monitoring. *Small*, 15(10), 1804385.

Shanks, A. (2014). The Effects of Jasmonic Acid and Chemicals in the JA Pathway on the Defense Systems and Gene Expression in Moss, *Physcomitrella patens* and *Amblystegium serpens*.

Sharma, R. Mishra, A. Pant, D. and Malaviya, P. (2022). Recent advances in microalgae-based remediation of industrial and non-industrial wastewaters with simultaneous recovery of value-added products. *Bioresource Technology*, *344*, 126129.

Shomrat, A. (2021). Seaweed cultivation and algal industry. *Plantlet*. Retrieved January 14, 2022, from https://plantlet.org/seaweed-cultivation-algal-industry/

Siddiki, S. Y. A. Mofijur, M. Kumar, P. S. Ahmed, S. F. Inayat, A. Kusumo, F. Badruddin, I. A. Khan, T. Y. Nghiem, L. D. Ong, H. C. and Mahlia, T. M. I. (2022). Microalgae biomass as a sustainable source for biofuel, biochemical and biobased value-added products: an integrated biorefinery concept. *Fuel,* 307, 121782.

Short, R. E. Stefan, G. David, C. Little, F. M., Edward, H. Allison, X. B. Ben Belton et al. (2021). Harnessing the diversity of small-scale actors is key to the future of aquatic food systems. *Nature Food* 2(9), 733–741.

Sievanen, L. Crawford, B. Pollnac, R. and Lowe, C. (2005). Weeding through assumptions of livelihood approaches in ICM: seaweed farming in the Philippines and Indonesia. *Ocean and Coastal Management*, 48(3–6), 297–313.

Simmance, F. A, Cohen, P. J, Huchery, C, Sutcliffe, S, Suri, S. K, Tezzo, X, ... and Phillips, M. J. (2022). Nudging fisheries and aquaculture research towards food systems. *Fish and Fisheries*, 23(1), 34–53.

Singh, J. and Gu, S. (2010). Commercialization potential of microalgae for biofuels production. *Renewable and Sustainable Energy Reviews*, 14(9), 2596–2610.

Skjermo, J. (2020). Norwegian Seaweed Technology Center. *SINTEF*. Retrieval on November 21, 2021, from https://sintef.no/en/ocean/initiatives/norwegian-seaweed-technology-center/

Smith, A. (2007). Translating sustainabilities between green niches and socio-technical regimes. *Technology Analysis and Strategic Management*, 19(4), 427–450.

SOFA, FAO (2020). Overcoming water challenges in agriculture. Retrieval on May 27, 2021. Available from www.fao.org/publications/sofa/sofa-2021/en/

Soleymani, M. and Rosentrater, K. A. (2017). Techno-economic analysis of biofuel production from macroalgae (seaweed). *Bioengineering*, 4(4), 92.

Stanley, Michele (2018). Textile substrate seeding of *Saccharina latissima* sporophytes using a binder: an effective method for the aquaculture of kelp. *Algal Research,* 33, 352–357. doi: 10.1016/j.algal.2018.06.005.

Stengel, D. B. Connan, S. and Popper, Z. A. (2011). Algal chemodiversity and bioactivity: sources of natural variability and implications for commercial application. *Biotechnology Advances*, 29(5), 483–501.

Success through Seaweed in Bangladesh (2020). *Feed the Future*. Retrieved January 13, 2022, from https://feedthefuture.gov/article/success-through-seaweed-in-bangladesh/

Sudhakar, K. Mamat, R. Samykano, M. Azmi, W. H. Ishak, W. F. W. and Yusaf, T. (2018). An overview of marine macroalgae as bioresource. *Renewable and Sustainable Energy Reviews,* 91, 165–179.

Suganya, T. Varman, M. Masjuki, H. H. and Renganathan, S. (2016). Macroalgae and microalgae as a potential source for commercial applications along with biofuels production: a biorefinery approach. *Renewable and Sustainable Energy Reviews,* 55, 909–941.

Sun, D. (2019). Potentials of seaweed cultivation. *Daily Sun*. Retreival on 25 August, 2021 from https://daily-sun.com/printversion/details/430909/Potentials-of-seaweed-cultivation

Sundararaju, V. (2021). Why seaweeds need to be conserved urgently. www.downtoearth.org.in. Retrieved January 15, 2022, from https://downtoearth.org.in/blog/environment/why-seaweeds-need-to-be-conserved-urgently-75070

Tanna, B. and Mishra, A. (2018). Metabolites unravel nutraceutical potential of edible seaweeds: an emerging source of functional food. *Comprehensive Reviews in Food Science and Food Safety*, 17(6), 1613–1624.

Temperton, V. M. et al. (2019). Step back from the forest and step up to the Bonn Challenge: how a broad ecological perspective can promote successful landscape restoration. *Restoration Ecology*, 27(4), 705–719.

Thanigaivel, S, Priya, A. K, Dutta, K, Rajendran, S, and Vasseghian, Y. (2022). Engineering strategies and opportunities of next generation biofuel from microalgae: A perspective review on the potential bioenergy feedstock. *Fuel*, *312*, 122827.

Towhidy, A. A. S. and Alfasane, M. A. (2018). Sublittoral seaweed flora of the St. Martin's Island, Bangladesh. *Bangladesh Journal of Botany*, 44(2), 223–236. https://doi.org/10.3329/bjb.v44i2.38511

Townsend, S. A. Webster, I. T. and Schult, J. H. (2011). Metabolism in a groundwater-fed river system in the Australian wet/dry tropics: tight coupling of photosynthesis and respiration. *Journal of the North American Benthological Society*, 30(3), 603–620.

Trading Economics (2022). Venezuela Exports of other slag and ash (including seaweed ash) to Trinidad And Tobago—2022 Data 2023 Forecast 2013 Historical. https://tradingeconomics.com/venezuela/exports/trinidad-tobago/other-slag-ash

Troell, M. Robertson-Andersson, D. Anderson, R. J. Bolton, J. J. Maneveldt, G. Halling, C. and Probyn, T. (2006). Abalone farming in South Africa: an overview with perspectives on kelp resources, abalone feed, potential for on-farm seaweed production and socio-economic importance. *Aquaculture*, 257(1–4), 266–281.

Van den Burg, S. W. K. Dagevos, H. and Helmes, R. J. K. (2021). Towards sustainable European seaweed value chains: a triple P perspective. *ICES Journal of Marine Science*, 78(1), 443–450.

Vázquez-Delfín, E. Freile-Pelegrín, Y. Pliego-Cortés, H. and Robledo, D. (2019). Seaweed resources of Mexico: current knowledge and future perspectives. *Botanica Marina,* 62(3), 275–289. https://doi.org/10.1515/bot-2018–0070

Vedantu (February 22, 2022). *Algae.*Retrieval on 20 October 24, 2021, from https://vedantu.com/biology/algae

Ventures, B. (2021). Sustainable living from the sea through community-led seaweed farming. *Blue Ventures-Beyond Conservation Blog.* Retrieved January 12, 2022, from https://blog.blueventures.org/en/sustainable-living-from-the-sea-through-community-led-seaweed-farming/

Verdegem, M. C. (2013). Nutrient discharge from aquaculture operations in function of system design and production environment. *Reviews in Aquaculture,* 5(3), 158–171.

Vu, H. P. Nguyen, L. N. Zdarta, J. Nga, T. T. and Nghiem, L. D. (2020). Blue-green algae in surface water: problems and opportunities. *Current Pollution Reports*, 6(2), 105–122.

Wang, K. Khoo, K. S. Leong, H. Y. Nagarajan, D. Chew, K. W. Ting, H. Y. Selvarajoo, A. Chang, J. S. and Show, P. L. (2021). How does the Internet of Things (IoT) help in microalgae biorefinery? *Biotechnology Advances*, 107819.

Watson, L. and Dring, M. (2011). *Business Plan for the Establishment of a Seaweed Hatchery and Grow-out Farm (Part 2).* Irish Sea Fisheries Board.

Wijsman, J. W. M. Troost, K. Fang, J. and Roncarati, A. (2019). Global production of marine bivalves. Trends and challenges. *Goods and Services of Marine Bivalves*, 7–26.

Wurmann, C. S. (2022). *Regional Review on Status and Trends in Aquaculture Development in Latin America and the Caribbean–2020.*

Yang, Y. Chai, Z. Wang, Q. Chen, W. He, Z, and Jiang, S. (2015). Cultivation of seaweed Gracilaria in Chinese coastal waters and its contribution to environmental improvements. *Algal Research*, 9, 236–244.

Zhang, H. Marjerison, R. K. and Zhao, Y. (2021). The Blue Economy and Its Long-Term Competitive Advantage: An Examination of China's Coastal Tourism. In *Quality Management for Competitive Advantage in Global Markets* (pp. 136–158). IGI Global.

Zheng, Y. (2020). Low-temperature adaptation of the snow alga *Chlamydomonas nivalis* is associated with the photosynthetic system regulatory process. *Frontiers.* Retrieved February 26, 2022, from.https://www.frontiersin.org/articles/10.3389/fmicb.2020.01233/full

Zotchev, S. B. (2012). Marine actinomycetes as an emerging resource for the drug development pipelines. *Journal of Biotechnology*, 158(4), 168–175.

12 Deep-Sea Mining and Potential Risks, Opportunities, and Challenges

Nezha Mejjad*[1] and Marzia Rovere[2]
[1]Faculty of Sciences Ben M'sik, University Hassan II Casablanca, Morocco
[2]Institute of Marine Sciences, National Research Council, Bologna, Italy
*Corresponding author: Nezha Mejjad.
Email: mejjadnezha@gmail.com; marzia.rovere@bo.ismar.cnr.it

CONTENTS

12.1 INTRODUCTION

Deep-sea mining or seabed mining in areas beyond national jurisdictions refers to the extraction of three types of seafloor mineral deposits: polymetallic or manganese nodules (Mn nodules), polymetallic/hydrothermal/seafloor massive sulfides (SMS deposits) and ferromanganese/cobalt-rich crusts (Figure 12.1). These deposits are supposed to compensate the depletion of terrestrial deposits of high economic interest. Deep-sea mining is expected to become operational in the whole world's oceans within the next 25 years (Wedding et al. 2013; 2015). The commercially-viable deposits are located on the international seabed, named 'The Area', while all operations related to these mineral deposits are regulated by the International Seabed Authority (ISA). The ISA was established in 1994 under the 1982 United Nations Convention on the Law of the Sea (UNCLOS) and the 1994 Agreement relating to implementing Part XI of the UNCLOS. As at March 2022, the ISA has finalized initial 15-year contracts, nineteen of which are for the exploration of Mn nodules in the Pacific Clarion-Clipperton Fracture Zone (CCZ) and one in the Central Indian Basin (ISA, 2021). Seven contracts entered into force to explore polymetallic sulfides in the South West Indian Ridge, Central Indian Ridge and the Mid-Atlantic Ridge and five contracts for cobalt-rich crust exploration in the Western Pacific Ocean. Many contracts for exploring polymetallic nodules have been awarded further temporal extension, exceeding five years from the original duration.

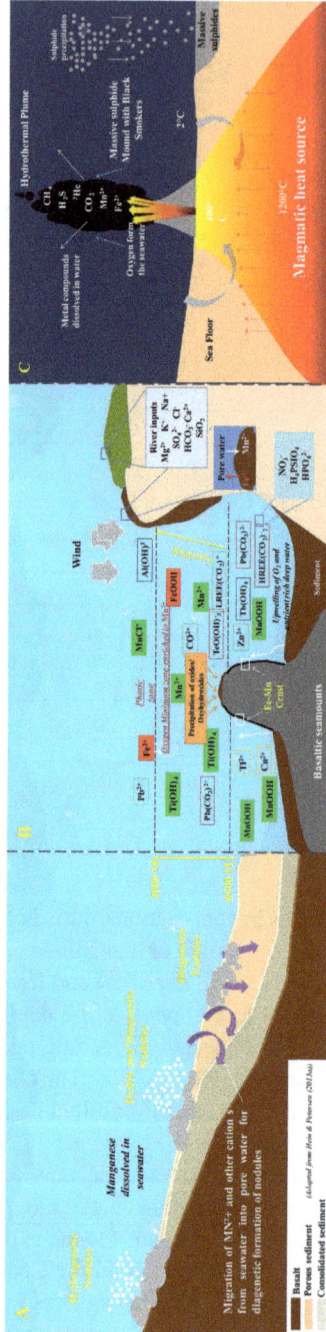

FIGURE 12.1 Deep-sea mineral deposits formation. (A) Polymetallic or manganese nodules (Mn nodules), (B) ferromanganese/cobalt-rich crusts, and (C) polymetallic/hydrothermal/seafloor massive sulfides (SMS deposits).

The growing demand for raw materials such as metals and Rare Earth Elements (REEs) used for electronic devices, construction materials, and renewable energy technologies, together with the growing depletion of these resources on land, have led to the exploration of the seafloor so that such minerals remain available to industry. The path to responsible and sustainable sourcing of minerals has highlighted the need to strengthen the governance of global mineral supply (Ali et al. 2017), envisaging regulating frameworks, such as those implemented by the ISA, as a pre-requisite for future operations in the deep sea. Therefore, to meet the unstoppable increase of mineral demand and the rapid development of technology requiring such raw materials, the exploitation of seabed minerals is seen as a possible alternative to land sources (Koschinsky et al. 2018). This is especially true given that the production of electronic items and green technology relating to many important areas such as energy, transport, digitalization and agriculture depend on metals like cobalt, copper, lithium, platinum and REEs (Takaya et al. 2018; Herrington et al. 2021; IEA, 2021), which are relatively abundant in marine mineral deposits. Furthermore, the comparative analysis showed a substantially lower footprint when processing polymetallic nodules than land ores (Paulikas et al. 2020), although this remains highly questionable in terms of carbon footprint and pollution for an actual scale production scenario (Heinrich et al. 2020).

However, deep-sea mining will have adverse impacts on the environment, different from land mining impacts or even worse (Ramirez-Llodra et al. 2010). Several studies carried out to define the impacts of deep-sea mining have reported a potential impact of deep-sea mining operations on biodiversity and associated ecosystem services (Mejjad and Rovere 2021; Armstrong et al. 2012). A lack of knowledge regarding deep-sea ecosystem functions and components (Danovaro et al. 2017) is among the major challenges and principles against the awarding of exploitation contracts in the deep-sea, as the technology is not sufficiently developed to extract these minerals harmlessly and sustainably (Ribeiro et al. 2018; Niner et al. 2018; Smith et al. 2020; Levin et al. 2020).

Furthermore, it is equally important to recognise that environmental conditions can also impact the progress of exploration and exploitation activities in the deep seabed. Environmental conditions such as seafloor topography, atmospheric and hydrographic conditions, and mineral characteristics do influence the mining system design, performance, and operation. Thus, such data would help better assess the overall deep-sea mining impact and allow for a more accurate design of the mining system and reliable planning of the mining operations (Sharma, 2011).

This chapter analyses the main opportunities and challenges facing deep-sea mining. For this reason, we applied a Political-P, Economic-E, Social-S, Technological-T, Legal-L, and Environmental-E (PESTLE) analysis tool combined with SWOT analysis (the Strengths, Weaknesses, Opportunities, and Threats) powered by an accurate literature review of available data and information related to deep-sea mining.

12.2 DEEP-SEA MINING: PROBLEM CONTEXT

The development of technology and the global transition to a sustainable and green economy in response to climatic change challenges has grown the demand for metals that serve as raw materials for high technology. At the same time, the depletion of land mineral resources is driving the mining sector towards new frontiers. Terrestrial-based mining industries have caused much damage to biodiversity, natural habitat, and the environment with toxic waste and polluting emissions (Farjana et al. 2018; Agboola et al. 2020). Since the discovery of polymetallic nodules, deep-sea mineral exploration has been conducted to investigate and value the associated resources (Petersen et al. 2016), mainly because the deep-sea industry has been sometimes considered less harmful to the environment compared to land-based mining in a scenario of transition from fossil to non-fossil energy resources (Cathles, 2014). The deep-sea is the vastest ecosystem on earth, and little knowledge is available regarding both its biodiversity (Danovaro et al. 2010) and its ecosystem as a whole. Additionally, the lack of information concerning the extent of the deep-sea mineral resources, their composition, global distribution and effects of their extraction on the deep-sea ecosystem has made the exploitation of deep-sea mineral resources uncertain in terms of cost and benefit analysis (Beaulieu et al. 2017).

12.3　PESTLE AND SWOT ANALYSES (BRIEF OVERVIEW)

A PESTLE analysis is a framework to assess the key external factors and forces influencing a sector of activity (Political-Economic-Sociological-Technological-Legal and Environmental). This analysis allows defining the 'business environment' and relevant factors and forces that could impact a project (for example, the biofuels energy industry; Achinas et al. 2019). The PESTLE analysis is often used in combination with the SWOT Analysis (Strength, Weakness, Opportunities and Threats) because it enables the determination of internal parameters related to a project and its organization into the different PESTLE categories. SWOT analysis has been widely used since the 1960s as an important basis for business management (Porter, 1979). This approach is used to evaluate internal (Weaknesses and Strengths) and external (Opportunities and Threats) factors to identify and guide the present and future industry potential. Numerous studies have applied the combined approach PESTLE/SWOT to identify external and internal factors that have an impact, positive or negative, on a project (Vardopoulos et al. 2021; Mostafa et al. 2020; Christodoulou and Cullinane, 2019; Koshesh and Jafari, 2019; Ahmadzai and McKinna, 2018; Islam et Mamun, 2017; Srdjevic et al. 2012).

In the present study, we applied the combined SWOT/PESTLE analysis to identify the key internal and external factors which may influence the deep-sea mining industry and define the main opportunities and challenges that deep-sea mining could present.

The PESTLE framework analysis has been widely used for data collection and decision-making while covering various business and related external issues. In the present chapter, PESTLE analysis is applied to provide data and produce information related to deep-sea mining and the opportunities and challenges of this industry sector. Every component of PESTLE are described below (Figure 12.2):

- *Political factors* comprise decision-makers, stakeholders and government interventions in economic issues. They also include geographical location, international relationships, and corruption.
- *Economic factors* include the gross domestic product (GDP), economic stability, inflation rate, exchange rate, and local and foreign investment.
- *Social factors* include demographic growth and increasing demand for new devices (for example, electronic devices), which depend mainly on nickel, copper, REEs, silver, gold, and others. Additionally, the demand for clean energy and the adaptation to a green economy depends mainly on metals and minerals that could be mined from the deep sea and serve as raw materials for new technologies (for example, tellurium, that is currently used as the main conductivity element in solar panels; USGS, 2015).
- *Technological factors* are mainly related to the advancement and development of sustainable deep-sea technology, which is one of the main issues that hampers the commencement of full commercial-scale exploitation of deep-sea minerals. Thus, we mainly focused on the aspects of technology related to monitoring, control, and research facilities for exploring the deep sea.
- *Legal factors* are related to laws and policies governing the deep-sea mining sector. In this case, we will investigate the laws provided by the International Seabed Authority (ISA) that have an impact on economic and social factors.
- *Environmental factors* are related to the environmental impacts of the exploitation activities on the deep-sea ecosystem and blue economy.

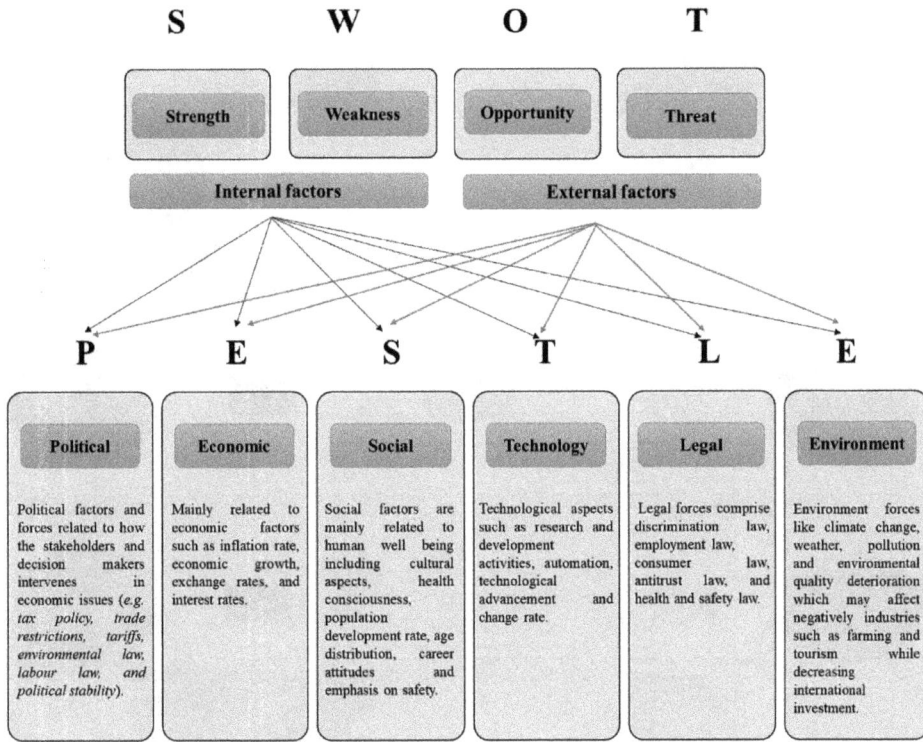

FIGURE 12.2 PESTLE analysis and SWOT analysis.

12.3.1 PESTLE ANALYSIS RESULTS

The global PESTLE analysis is shown in Figure 12.3, where the PESTLE analysis factors are divided into sub-factors for deep-sea mining, highlighting the main opportunities and challenges this sector is facing. These factors are also explained in detail in the following sections.

12.3.1.1 Political Factors

In response to the depletion of natural resources on land, referring in this case to metals and minerals, the deep sea is seen as an alternative source for the supply of these resources that play an important role as raw materials for the new and green technologies that are promising to meet human needs and ensure economic growth in a climate change scenario. However, many stakeholders, including environmentalists, governments, and NGOs, argue against developing this industry in the deep sea as the environmental impacts are still largely unknown. Although the ISA has already granted thirty-one contracts to twenty-two contractors, there are repeated calls from the scientific community for a deep-sea mining regulatory framework to become more stringent in terms of environmental protection and mitigation strategies (for example, Van Dover et al. 2017), placing deep-sea mining at a threshold of law, where the legal governance of the deep ocean space is considered to be an ongoing project (Hannigan, 2016; Ali et al. 2017).

168 countries are Member States of the ISA, whose governance is organized by different bodies, the most important of which are the Council and the Assembly. In these member states are organized into regional groups following United Nations (U.N.) criteria. Members of the ISA have different views on deep-seabed mining implementation, opportunities, and threats. Broadly speaking, several countries have strong environmental concerns, especially those facing the eastern Pacific coast, which will be potentially most seriously affected by the adverse effects and impacts of deep-seabed

FIGURE 12.3 The results of the PESTLE and SWOT analysis applied to deep-sea mining are summarized per subcategory.

mining in the CCZ. These states include Argentina, the Bahamas, Chile, Costa Rica, Cuba, the Dominican Republic, Guyana, Jamaica, Panama and Trinidad and Tobago who collectively submitted a document in late 2021 (ISBA/26/C/47) concerning the letter sent by the Government of Nauru on June 25 2021, through which it notified the Council that Nauru Ocean Resources Inc. (NORI), a Nauruan entity, sponsored by Nauru, intended to apply for approval of a plan of work for exploitation (www.isa.org.jm/news/nauru-requests-president-isa-council-complete-adoption-rules-regulations-and-procedures). The document expressed concern about the crucial pending issues that need to be discussed and agreed upon before any plan of work should be considered by the Council of the ISA. These members of the Latin-American and Caribbean Group to the ISA agreed that issues such as the establishment of regional environmental management plans, rules for inspection, compliance and enforcement, and adoption of standards and guidelines relating to environmental protection must be resolved in advance of the commencement of any exploitation activity in the "Area".

Indeed, the above-mentioned request made by Nauru triggered section 1, paragraph 15, of the annex to the 1994 Agreement, which binds the ISA to complete the adoption of rules, regulations, and procedures necessary to facilitate the approval of plans of work for exploitation in the area within a two-years' period. Pacific Small Island Developing States (PSIDS) have enthusiastically adhered to the seabed mining prospects for polymetallic nodules since the mid-2010s. Many small island states are now sponsoring exploration contracts in the CCZ, notably the Cook Islands (2016), Kiribati (2015), Tonga (2012), and Nauru (2011). Nauru, in particular, a 12,000 population country, adopted the International Seabed Minerals Act in 2015, a national legislation that enables effective control over the activities undertaken by their sponsored contractor, NORI. Nauru, in its letter, citing the IPCC's Special Report on Global Warming of 1.5°C (IPCC, 2018), claimed there is no hope of combating climate change without an industrial revolution based on renewable energies. Nauru complained about the impact of rising seas and salt-water intrusion, frequent droughts, and land mining done by colonial powers making over 80% of the island uninhabitable. On this premise, Nauru intended to promote the responsible collection of polymetallic nodules from the seafloor to deliver a carbon-neutral future to the next generations. On the other hand, other small island states have opposing visions on deep seabed mining. In particular, in June 2022 at the United Nations Ocean Conference in Lisbon, the Government of Palau and Fiji joined the Deep Sea Conservation Coalition and the World Wildlife Fund in officially launching the Alliance of Countries Supporting a Deep-Sea Mining Moratorium. In July 2022 also the Federated States of Micronesia joined the Alliance. The Asian States are one of the largest groups of investors, contractors and sponsoring states (China, Republic of South Korea, Japan, Singapore, India), totaling 12 exploration contracts over the three categories of deep seabed mineral resources. The Western European and Others Group are the second-largest investor group, with France, Belgium, U.K. and Germany granting sponsorship to national contractors. Overall, the WEOG have a balanced approach between industrial development and marine environmental protection. In addition, several WEOGs belong to the European Union. The E.U. Parliament, on June 9 2021, recalling its resolution of January 16 2018, on international ocean governance, called on the Commission and the Member States to promote a moratorium on deep-seabed mining until its effects on the marine environment, biodiversity and human activities at sea have been researched in depth (European Parliament, 2021). Furthermore, the European Parliament has emphasized the need for the Commission to cease funding for the development of seabed mining technology in line with a circular economy based on minimizing, reusing, and recycling minerals and metals and to work through the ISA to ensure transparency in its working methods as well as the effective protection and preservation of the marine environment from harmful effects. During 2021, at the IUCN (The International Union for the Conservation of Nature) World Conservation Congress, over 81% of government and over 94% of civil society/NGO membership approved a motion calling for a moratorium on deep-sea mining.

Civil society and NGOs are collectively against the initiation of deep seabed mining in areas beyond national jurisdiction. For example, in a letter sent to the Secretary-General of the ISA, dated November 9 2021 (http://savethehighseas.org/wp-content/uploads/2021/11/20211109_DSCC-let ter-re-NORI-EISEIA-1.pdf. Accessed November 22, 2021), the Deep Sea Conservation Coalition expressed concern in respect of the NORI environmental impact assessment (EIA) and statement (EIS) that are preliminary steps toward the request for a plan of work for exploitation in the area, citing the absence of environmental baseline data in the study and a general lack of transparency in the decision-making process.

Another important stakeholder is the Enterprise, an organisation of the ISA, which can carry out activities in the area, pursuant to article 153, paragraph 2(a) of the UNCLOS, together with the shipping, processing, and marketing of minerals recovered from the area. As it recalls the Special Representative of the Secretary-General of the International Seabed Authority for the Enterprise (ISBA/26/C/15), there is general support for the operationalization of the Enterprise, despite the many challenges and issues behind it. The African group, reacting to Nauru's letter (ISBA/26/C/40), emphasized the need to operationalize the enterprise to fully implement the common heritage of mankind principle in the area because the Enterprise is the only means of participation in activities in the area for most developing states. In the same document, the African group expressed their concerns regarding the mechanism for equitably sharing benefits derived from seabed mining and the financial regime that properly compensates humanity for its resources and land-based miners for their losses and, therefore, the impact on terrestrial mining economies.

12.3.1.2 Economic Factors

The growing consciousness about the depletion of land-based mineral resources and the economic potential of deep-sea minerals such as manganese nodules and ferromanganese crusts (Rona, 2003; SPC, 2013) increased the interest in exploring and exploiting deep-sea metals. Furthermore, the new economies are highly dependent on metals such as copper, nickel, and REEs, meaning that mining deep-sea minerals and metals appear like an attractive commercial opportunity. Nevertheless, the main economic and environmental challenges facing this emerging industry are the high costs for the initial investment in pilot mining tests, marine robotics and new technologies for the exploration and sustainable collection of minerals (Rovere, 2018). Although uncertainties related to deep-sea resource estimation and mining profitability prevail (Petersen et al. 2018), other studies (ISA, 2008a; ISA, 2008b; Sharma, 2011; Hein et al. 2013; Cathles, 2014) highlight the economic viability of deep-sea mining compared to terrestrial mining operations, especially in the processing of manganese nodules, notwithstanding their high phosphorous content compared to terrestrial ores. Lehnen et al. (2019) investigated the economic feasibility of deep-sea mining as an emerging industry that could provide the E.U. with a new strategy for supplying raw materials of economic importance. In particular, the study highlighted a positive economic assessment of SMS deposits due to their high metal grades and the lack of overlying material. Similarly, manganese nodules exhibit high economic potential, mainly due to the production and marketing of their concentrated metals (Mn, Ni, Cu and Co) (Friedmann et al. 2017).

Numerous studies have analyzed and assessed the deep-sea mining economic value (Cameroun et al. 1981; Folkersen et al. 2018; Volkmann et al. 2018; Volkmann et al. 2019; Dacey, 2020). All the Authors agreed that deep-sea mining resources are important for providing rare metals for technologies such as solar panels, wind turbines, and electric vehicle batteries. Therefore, the deep sea can offer novel opportunities for industrial development, particularly for the green economy. However, the limitations related to the knowledge gaps regarding the deep-sea ecosystem result in uncertainties about this industry's economic value and feasibility (Mejjad and Rovere, 2021).

12.3.1.3 Social Factors

World population growth leads to rising demand for resources, life quality, and enhancement of living standards (Ali et al. 2017). Currently, access to raw materials is crucial for societal wealth and human well-being (Blue mining, 2018). The current global improvement of living standards, in terms of access to health care, education and social mobility, can be facilitated by the transition to carbon-neutral energy and innovative and digital technologies that all require metals such as copper and cobalt. Thus, in response to a human need for well-being and sustainable economic growth, governmental agencies and private companies are investing in the exploration, research and development of technology, and knowledge to explore, assess and mine deep-sea mineral resources (ISA, 2017; Blue nodule, 2019; JPI oceans, 2019).

Social factors also include cultural values. Advancement in education and scientific understanding of the ocean ecosystem is essential for the sustainable use of deep-sea mineral resources for future generations. According to the U.N. ocean science accounts for only between 0.04 and 4% of total research and development funding worldwide (UNESCO, 2019). The United Nations Decade of Ocean Science for Sustainable Development (2021–2030) intends to push the world to come up with a common agenda on how ocean science can help countries achieve the U.N. sustainable development goals (https://sdgs.un.org/goals).

Deep-sea mining exploration opened opportunities for overcoming the gaps in deep-sea ocean knowledge. For example, the Joint Programming Initiative, Healthy and Productive Seas and Oceans (JPI Oceans), which was established in 2011 as an intergovernmental platform, open to all E.U. Member States and Associated Countries who invest in marine and maritime research, supported an action (2013–2022) to study the ecological aspects and risks of deep-sea mining through the realization of two projects (Mining Impact 1 and 2). During the development of the projects, scientists from academia and contractors to the ISA (namely GSR of the Belgian DEME group and BGR of the German Federal Institute for Geosciences and Natural Resources) joined forces to investigate the impact of sediment plumes and dredge trails on the biodiversity of the CCZ.

Furthermore, deep-sea mining exploration and exploitation can also impact cultural traditions and norms (Wakefield and Myers, 2016). These authors identified four main potential cultural changes: i) concession of traditional land ownership, ii) modifications in societal norms, iii) changes in employment patterns and iv) loss of access to subsistence fisheries. On the other hand, they concluded that deep-sea mining could improve the well-being of the host country's people. The revenue from the exploitation of deep-sea minerals could contribute to developing countries' infrastructures (for example, roads, schools, and hospitals) and allow the creation of new jobs (SPC, 2013), which positively affects human well-being.

12.3.1.4 Technological Factors

The deep-sea is the largest ecosystem on earth; however, knowledge about this environment is still in its early stages as the technology is not well advanced to allow a better understanding, and its study requires a large financial capability and human resource capacity. Indeed deep-sea knowledge needs remarkable investment, such as the construction of oceanic vessels and large ship-time availability, remote monitoring vehicles and tools, and an extremely specialized workforce that not all states can sponsor (Ribeiro et al. 2018). Nevertheless, the development of deep-sea and ultra-deep-sea technology is well underway. Various techniques are being employed to explore deep-sea minerals, including ship-based swath sonar bathymetric mapping, autonomous underwater vehicles, geophysical surveying, and remotely operated vehicles. Exploring deep-sea mineral resources involves reconnaissance, mapping, sampling, and drilling (Figure 12.4). The exploration phase, its value chain, and techniques have been described in detail by Ecorys (2014). As shown in Figure 12.4, the equipment and techniques strongly depend on the type of mineral deposit.

12.3.1.4.1 Exploration Phase

12.3.1.4.1.1 Reconnaissance and Mapping   A range of techniques is used to locate, assess, and map the deep-seabed mineral deposits, including geochemical and geophysical surveys. These techniques are all operated from vessels with oceanic capacity and equipped with dynamic positioning systems

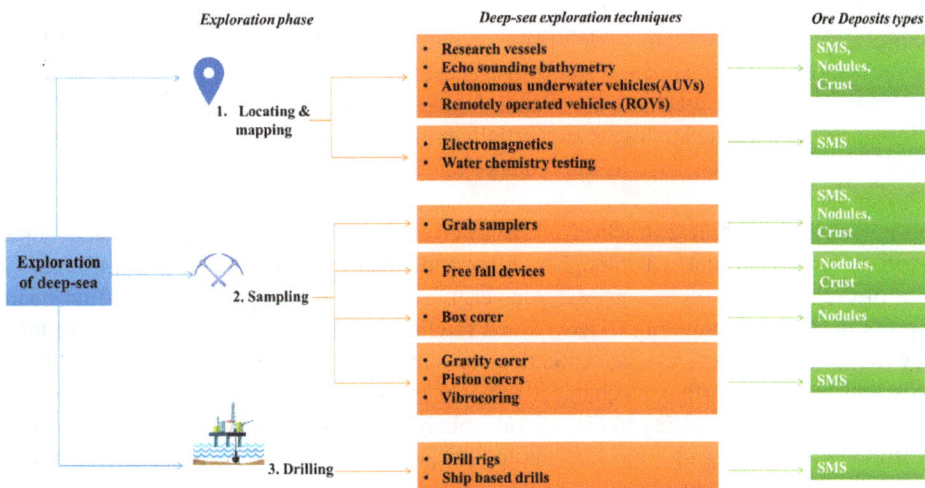

FIGURE 12.4   Used equipment and techniques for deep-sea minerals exploration.

to ensure reliable and efficient deployment and positioning of various geophysical and geochemical devices and probes. These instruments and techniques include, but are not limited to: sonar technology (single beam, multibeam, side scan), electromagnetic methods, water-chemistry sampling and testing, possibly operated from Remotely-Operated Vehicles (ROVs) and Autonomous Underwater Vehicles (AUVs) (Ecorys, 2014). Indeed, the advance in autonomous underwater technology, comprising of improved autonomy, enhanced hovering capabilities, and novel geophysical and geochemical sensing tools is believed to be crucial to more efficiently locating and mapping seabed resources (Wynn et al. 2014).

12.3.1.4.1.2 Sᴀᴍᴘʟɪɴɢ   The available technology for sediment and nodule sampling includes freefall devices. They have been used since the 1970s for sampling manganese nodules in the deep-sea and consist of various devices, including box-corers, grabs, and dredges. These devices can be used where the sea bottom is covered by sediment, which they reach by gravity falling from the ship's stern and are designed to collect sediment samples and nodules. They may be further equipped with cameras and be cable-operated during the sampling operations (ISA, polymetallic nodules).

12.3.1.4.1.3 Drilling   The technique used for drilling includes drill rigs and ship-based drills but is used only in the case of SMS deposits, which can be firmly attached to the rock substrate, may consist of moderately hard deposits and usually are thicker compared to the nodules, which are large pebbles lying on the deep-seabed.

12.3.1.4.2   Estimation and Assessment Phase

This phase consists of assessing the viability and economic feasibility of a potential mining project based on several assumptions and evaluations, comprising technological and metallurgical, economic, marketing, social, legal, governmental, and environmental factors. The data can be input to a numerical 3D resource model, and its uncertainty is evaluated (Ecorys, 2014). Potential modelling techniques include the 3D Geometallurgical Modelling technique for polymetallic sulfide deposits and the 2D Multivariate Modelling technique for nodules and cobalt-rich crust deposits (Ecorys, 2014).

12.3.1.4.2.1 Extraction Phase   Extraction of seabed mineral deposits is conceived in a similar fashion to land mining: scrapping the mineral deposit from the surface or digging and tunnelling to an under seabed deposit or drilling from the seabed. The extracted resource must always be shipped to shore to be processed into a saleable final product.

Seabed collector systems are cable-operated robots that will detach deposits from the seabed through pressurized or mechanical water drills. These operations face many challenges due to possible communication and mechanical failures, high-pressure conditions, and weather constraints. Therefore, new deep-sea technologies are partially borrowed from more mature sectors operating in the deep and ultra-deep sea, such as the offshore oil and gas industry and the dredging and ocean-cable laying sectors.

The polymetallic nodule extraction is projected to be performed through remotely operated collectors or harvesters that will plough, scrape, and vacuum the seafloor over vast areas (300 - 800 km² of deep-sea per mining process per year). The nodules will be transported on board the mining vessel using a pipe string (riser), which might be several kilometers long due to the great average depth of abyssal hills where nodules concentrate (4 - 6,000 m) (Figure 12.5). The SMS deposits will require a further mechanical removal from the substrate compared to nodules and sophisticated mineral processing to extract valuable metals, while the recovery of ferromanganese crusts is the most challenging from a technological point of view and the reliability of prototyped

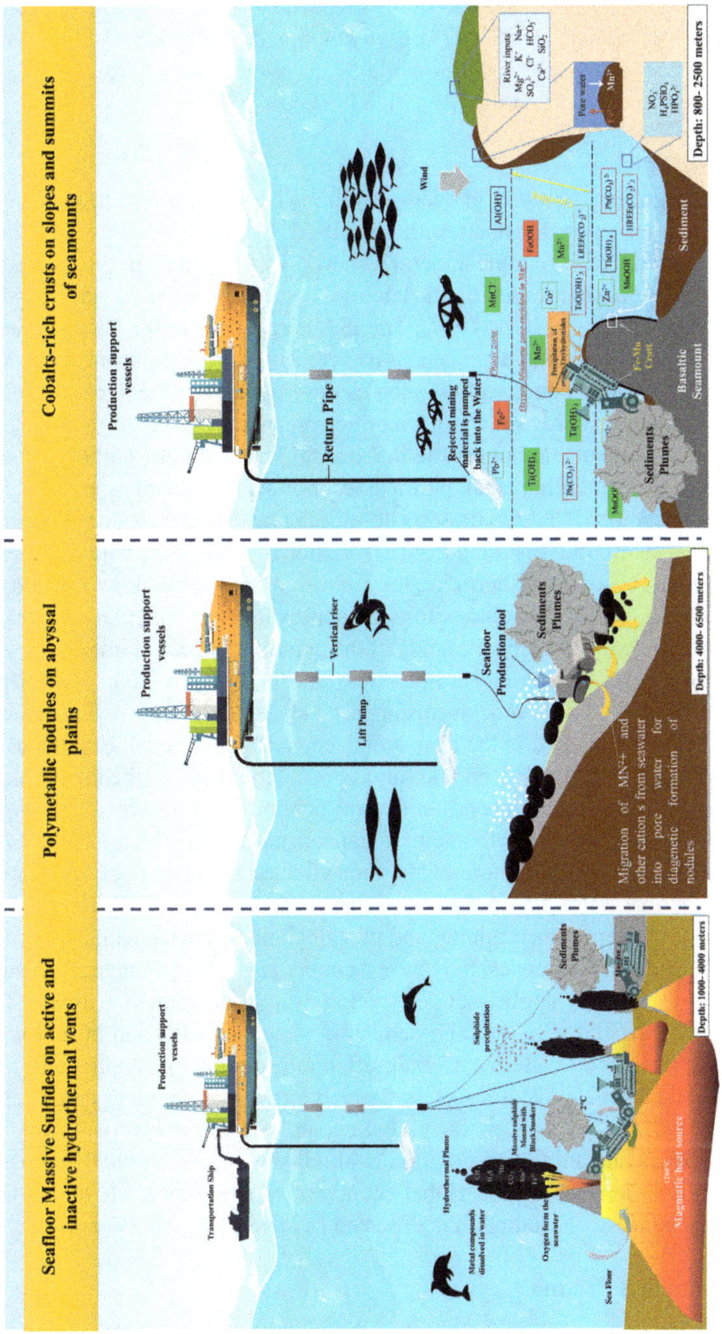

FIGURE 12.5 A schematic indicating the processes involved in mining in the deep-sea of the three types of ore deposits (representation not to scale).

equipment in ~ 5,000 m water depths still is far from being tested in the field (Paul et al. 2018). In summary, removing cemented crusts from hard substrates is technologically complex, very expensive, and environmentally destructive (Hein et al. 2009).

Up to date, none of the designed and prototyped harvesting systems has reached industrial practicability. For this purpose, projects like the EU Horizon2020 research and innovation programme Blue Nodules (Grant Agreement no. 688975) have attempted to develop and improve techniques used for sustainably harvesting polymetallic nodules, to reduce the environmental impact while maintaining and ensuring efficient production (Blue Nodules, 2020).

12.3.1.4.3   Legal Factors

Laws and policies that govern the sector of deep-sea mining beyond national jurisdiction are collectively provided by the International Seabed Authority (ISA), which, through its many mandates, has the overarching goal of administering the mineral resources of the seabed of the area as the common heritage of humankind, as first proposed by the late Ambassador of Malta Arvid Pardo in his speech to the U.N. General Assembly (UNGA, 1967). The ISA is responsible for the activities carried out in the area to promote the economic and social advancement of all peoples in the world, with due regard for the most vulnerable communities, such as developing and landlocked countries. At the same time, the ISA must encourage marine scientific research, coordinate its dissemination, and protect the marine environment and human life in the area. Therefore, the authority is at the core of a sophisticated architecture entailing legal, political, economic, and social aspects.

Notably, Article 150 of the UNCLOS protects developing land-based producers from the possible adverse effects on their economies resulting from a reduction in price or the volume of exports caused by minerals extracted in the area. Furthermore, the 1994 Agreement, relating to the Implementation of Part XI of the UNCLOS, provides that before the approval of the first plan of work for exploitation, the authority must concentrate on the study of the potential impact of mineral production from the area on the economies of developing land-based producers. The first study of such, published by the ISA in 2020 (www.isa.org.jm/files/documents/impactstudy.pdf), concerns polymetallic nodules, the most likely resource to be exploited in the area, which carry copper, nickel, cobalt and manganese.

The market position of polymetallic nodules and how much impact seabed minerals mining will produce on land-based mining will depend on several factors, including the costs and viability of their processing and the projected growth rate of metal consumption in the future years. At the same time, a decrease in prices for one or more of the four affected metals would automatically reduce the profitability of the market of polymetallic nodules. Therefore, it is difficult to strike a balance between the economy of deep-seabed mining and its global market placement.

To give some figures, the ISA impact study gave a copper projection of annual growth of 1.8 - 4% due to several green-technology applications. Nickel consumption growth estimates range from 4 to 14 times increase by 2030 due to the high demand for the electrification of motor vehicles and the production of high-capacity batteries. Cobalt, often a by-product of copper and nickel extraction, is linked to the lithium-ion batteries sector, which aims at a complete metal replacement in the coming years; therefore, projections for global cobalt consumption growth vary significantly from around 5% to 9%. Manganese consumption is highly dependent on the dynamics of the steel industry, which, in turn, depends on the global economic situation as a whole and China in particular. Therefore, manganese consumption is expected to increase by 1 - 3% per year.

12.3.1.5   Environmental Factors

Deep-sea mining of minerals is considered to be a possible mitigation factor for the climate crisis that the world is experiencing. The global transition to clean and green energy implies a huge influx of raw materials such as cobalt, copper and REEs used for batteries and generators. The land-based resource depletion increased the interest in deep-sea minerals exploitation. In contrast, the lack of knowledge about deep-sea ecology and the possible impact of mining operations on the seabed environment encouraged some governments and non-governmental organizations to call for delaying deep-sea mining activities until the environmental impacts are entirely understood (e.g. European Parliament, 2021). Although the technology related to the environmental assessment and monitoring of seabed mining has been largely developed, the lack of data in the deep sea (e.g. habitat distribution, species identities, ecosystem function, and the like) made the prediction of the potential impact of seabed mining on the environment and biodiversity of the area unachievable (SPC, 2013).

Several studies have been developed in the framework of international projects such as JPI Oceans related to studying the environmental impacts and associated risks of mining the seabed. These studies have reported the need for the development of strategies, standards, and regulations associated with the environmental management of mining in the area and to ensure adequate protection of marine life (Amos and Roels, 1977; Boschen et al. 2013; Levin et al. 2016; Durden et al. 2017; Kaikkonen et al. 2018; Weaver and Billett, 2019; Tunnicliffe et al. 2020; Jaeckel et al. 2020; Christiansen et al. 2020).

Furthermore, scientific studies on the biological effects of dredge trails left by exploration carried out in the mid-1990s have shown adverse biological effects characterized by a significant decrease of suspension-feeding populations in areas directly disturbed by polymetallic nodule dredging. Seabed mining for nodules is, therefore, likely to cause a loss of the main ecosystem functions (Simon-Lledó et al. 2019). Similarly, Vonnahme et al. (2020) showed that exploration and dredging of the deep-sea floor had caused a reduction of microbial activity even after 26 years had passed. They also found that the number of microbial cells was reduced by about 30% ~ 50%. Significant negative alterations in the diversity and density of benthic communities were also found in eleven areas of the nodule mining test simulation using meta-analysis techniques (Jones et al. 2017).

The nature and timescale of the impacts of seabed mineral exploitation remain unclear and depend on the type of resource, its related ecosystems, and the employed technology for mineral recovery. According to current knowledge, the main impacts of the operations on the deep-sea floor, which may harm biodiversity and ecology, are as follows (Figure 12.6):

- Seafloor disturbance: The scraping of the seafloor using mining equipment can cause habitat alteration and fragmentation, leading to species loss and ecological function destruction or loss. The machines used for harvesting nodules can wipe out entire species (Cuyvers et al. 2018).
- Sediment plumes: These are identified as the greatest ecological threat deep-sea mining poses to marine life. It is not fully understood how far these plumes can disperse from the mined area and to what extent they may influence species and the whole ecosystem or injure filter-feeding species that depend on clean and clear waters to feed (for example, krill and some marine mammals) (Cuyvers et al. 2018).
- Pollution: Vibration, noise, and light resulting from mining operations in the deep sea can harm species such as tuna, whales and sharks if these operations scale up. Additionally, these activities in the deep sea can lead to toxic products, fuel spills, and smothering of benthic communities (Cuyvers et al. 2018).

Therefore, the impact review of deep-sea mining should consider the nature and the short and long-term environmental consequences of the whole process on marine biodiversity and ecosystems.

12.3.2 SWOT ANALYSIS RESULTS

The purpose of applying a SWOT analysis system is to define deep-sea mining minerals' positive and negative aspects. This analysis describes and assesses external factors (S-strengths and W-weaknesses) and internal factors (O-opportunities and T-threats). The SWOT analysis was used for evaluating the factors connected with the deep-sea mining value chain of polymetallic nodules,

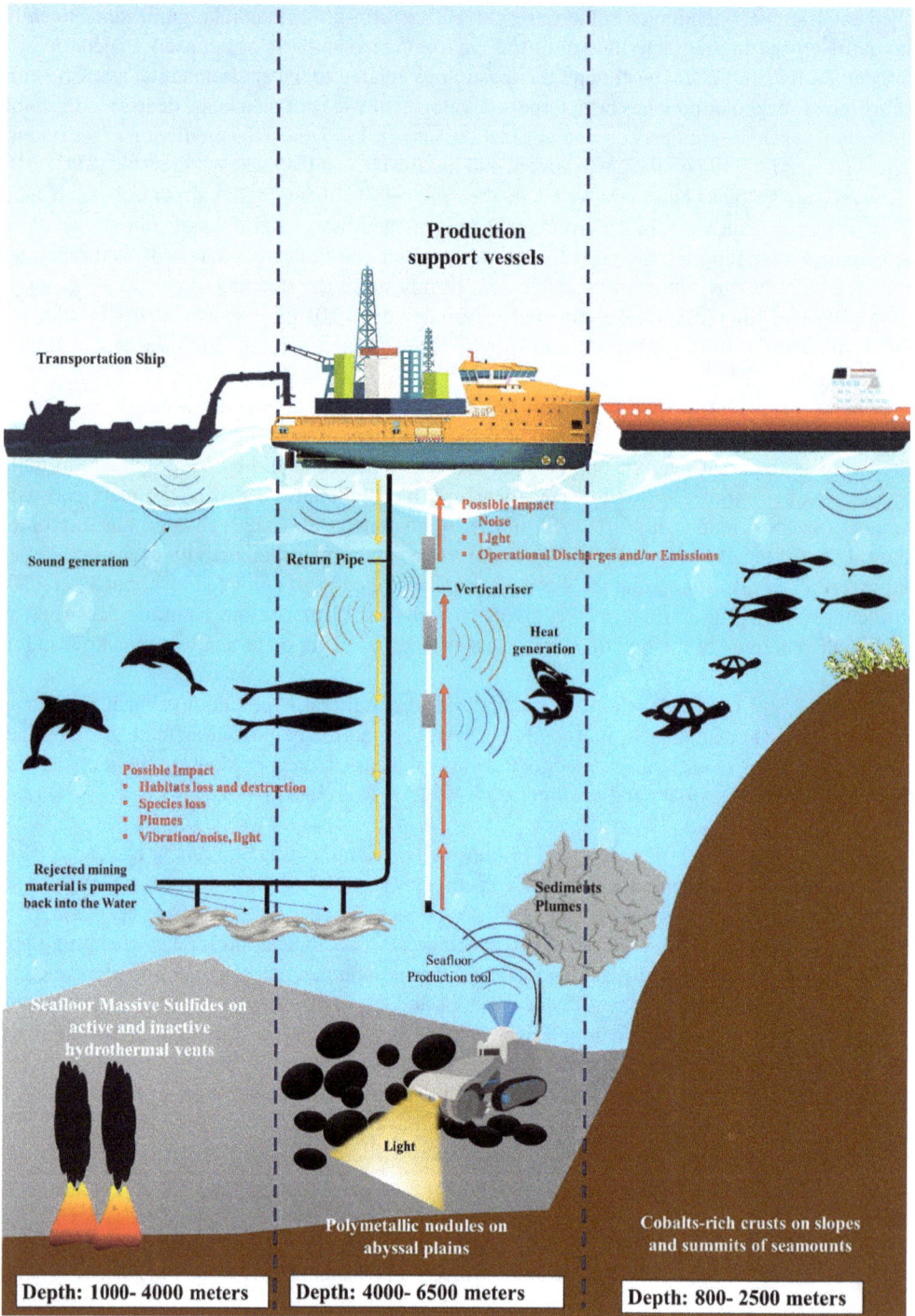

FIGURE 12.6   Potential impacts from deep-sea mining (adapted from Secretariat of the Pacific Community (SPC, 2013)).

as presented in Abramowski and Balaz (2017). The results showed common strengths, weaknesses, opportunities, and threats from resource, technological, economic, and environmental viewpoints.

As presented in Figure 12.3, we highlighted the main strengths, weaknesses, opportunities, and threats. Deep-sea mining of ore deposits may present some strengths and many opportunities for combating climate change and the natural resources depletion that the world is experiencing. However, the analysis also underlines common weaknesses and threats at all levels, for all analyzed factors and also during the whole process of the mining operations.

12.4 CONCLUSIONS

The deep-sea mining sector has attracted attention because of the discovery of deposits presenting significant amounts of metals of economic interest, such as nickel, copper, cobalt, manganese and rare earth elements. The transition toward clean energy and a low to zero-carbon economy increased the demand for such metals, which serve as raw materials for renewable energy infrastructure and green technology.

However, knowledge gaps about life in the deep sea and uncertainty regarding the methods and techniques of deep-sea mining and their potential risks and impacts on marine ecosystems are the main challenges facing the commercial activities of these minerals.

This chapter analyzed the main opportunities and challenges facing deep-sea mining. For this reason, we used a Political-P, Economic-E, Social-S, Technological-T, Legal-L, and Environmental-E (PESTLE) analysis tool combined with SWOT analysis (the Strengths, Weaknesses, Opportunities, and Threats) powered by a literature review of available data and information related to deep-sea mining.

Every pillar of the PESTLE analysis showed opportunities and threats/weaknesses, strengths and threats/weaknesses indicating that this blue sector presents more challenges than opportunities. The uncertainties around economic feasibility, technological constraints and the scientific data gaps of the deep sea call us to focus on the challenges facing the growth of this emerging industry and to prioritize the assessment of the nature and scale of the possible impacts on deep-sea ecosystem services from seabed mining.

Future investigations and studies must consider combining all analyzed factors to provide a solid basis for effective and informed decision-making.

REFERENCES

Abramowski, T. and Balaz, P. 2017. *Structural Aggregation of Feasibility Factors for the Assessment of the Polymetallic Nodules Deep Sea Mining Value Chain.* In The 27th International Ocean and Polar Engineering Conference. OnePetro.

Achinas, S. Horjus, J. Achinas, V. and Euverink, G. J. W. 2019. A PESTLE analysis of biofuels energy industry in Europe. *Sustainability,* 11, 5981. https://doi.org/10.3390/su11215981

Agboola, O. Babatunde, D. E. Isaac Fayomi, O. S. Sadiku, E. R. Popoola, P. Moropeng, L. Yahaya, A. and Mamudu, O. A. 2020. A review on the impact of mining operation: Monitoring, assessment and management. *Results in Engineering*, 8, 100181. Elsevier B.V. https://doi.org/10.1016/j.rineng.2020.100182

Ahmadzai, S. and McKinna, A. 2018. Afghanistan electrical energy and trans-boundary water systems analyses: challenges and opportunities. *Energy Reports*, 4, 435–469.

Ali, S. H. Giurco, D. Arndt, N. Nickless, E. Brown, G. Demetriades, A. Durrheim, R. Enriquez, M. A. Kinnaird, J. Littleboy, A. Meinert, L. D. Oberhänsli, R. Salem, J. Schodde, R. Schneider, G. Vidal, O. and Yakovleva, N. 2017. Mineral supply for sustainable development requires resource governance. In *Nature,* 543(7645), 367–372. Nature Publishing Group. https://doi.org/10.1038/nature21359

Amos, A. F. and Roels, O. A. 1977. Environmental aspects of manganese nodule mining. *Marine Policy*, 1(2), 156–163.

Armstrong, C. W. Foley, N. S. Tinch, R. and van den Hove, S. 2012. Services from the deep: steps towards valuation of deep sea goods and services. *Ecosyst. Serv.* 2, 2–13.

Beaulieu, S. E. Graedel, T. E. and Hannington, M. D. 2017. Should we mine the deep seafloor? *Earth's Future,* 5(7), 655–658. John Wiley https://doi.org/10.1002/2017EF000605

Blue Mining. 2018. Public report—Blue Mining: Breakthrough Solutions for Mineral Extraction and Processing in Extreme Environments. Available at: https://cordis.europa.eu/docs/results/604/604500/final1-blue-mining-public-report-2018.pdf

Blue Nodules. 2019. Blue Nodules: Breakthrough Solutions for the Sustainable Harvesting and Processing of Deep Sea Polymetallic Nodules. Available at: http://blue-nodules.eu/. Accessed date 28 October 2021.

Blue Nodules. 2020. CORDIS Results Pack on mineral extraction. Available via https://blue-nodules.eu/cordis-results-pack-2020/. Accessed December 1, 2021.

Boschen, R. E. Rowden, A. A. Clark, M. R. and Gardner, J. P. A. 2013. Mining of deep-sea seafloor massive sulfides: a review of the deposits, their benthic communities, impacts from mining, regulatory frameworks and management strategies. *Ocean and Coast. Manage.* 84, 54–67. doi: 10.1016/j.ocecoaman.2013.07.005

Cameron, H. Georghiou, L. Perry, J. G. and Wiley, P. 1981. The economic feasibility of deep-sea mining. *Engineering Costs and Production Economics*, 5(3–4), 279–287. doi: 10.1016/0167–188x(81)90019–7

Cathles, L. M. 2014. Future Rx: optimism, preparation, acceptance of risk. *Geological Society Special Publication,* 393, 303–324. https://doi.org/10.1144/SP393.6

Christiansen, B. Denda, A. and Christiansen, S. 2020. Potential effects of deep seabed mining on pelagic and benthopelagic biota. *Marine Policy,* 114, 103442. https://doi.org/10.1016/j.marpol.2019.02.014.

Christodoulou, A. and Cullinane, K. 2019. Identifying the main opportunities and challenges from the implementation of a port energy management system: a SWOT/PESTLE analysis. *Sustainability,* 11, 6046. https://doi.org/10.3390/su11216046.

Cuyvers, L. Berry, W. Gjerde, K. Thiele, T. and Wilhem, C. 2018. *Deep Seabed Mining: A Rising Environmental Challenge.* Gland, Switzerland, IUCN and Gallifrey Foundation. x + pp. 74. https://doi.org/10.2305/IUCN.CH.2018.16.en.

Dacey, J. August 3, 2020. Deep-sea mining may have deep economic, environmental impacts, Eos, 101. https://doi.org/10.1029/2020EO147683.

Danovaro, R. et al. 2010. Deep-sea biodiversity in the mediterranean sea: the known, the unknown, and the unknowable. *PLoS ONE,* 5(8), e11832. https://doi.org/10.1371/journal.pone.0011832

Danovaro, R. Corinaldesi, C. Dell'Anno, A. and Snelgrove, P. V. 2017. The deep-sea under global change. *Current Biology,* 27(11), R461–R465.

Durden, J. M. et al. 2017. A procedural framework for robust environmental management of deep-sea mining projects using a conceptual model. *Marine Policy,* 84, 193–201.

Ecorys, 2014. Study to investigate state of knowledge of deep-sea mining: Final Report under FWC MARE/2012/06—S.C. E1/2013/04. https://webgate.ec.europa.eu/maritimeforum/en/printpdf/3732 Accessed November 26, 2021.

European Parliament. 2021. European Parliament resolution of June 9, 2021 on the E.U. Biodiversity Strategy for 2030: Bringing nature back into our lives (2020/2273(INI)). P9_TA(2021)0277.

Farjana, S. H. Huda, N. and Mahmud, M. P. 2018. Life-cycle environmental impact assessment of mineral industries. *IOP Conference Series: Materials Science and Engineering*, 351(1), 012016. IOP Publishing.

Folkersen, M. V. Fleming, Ch. M. and Hasan, S. 2018. The economic value of the deep sea: a systematic review and meta-analysis. *Marine Policy,* 94, 71–80. doi:10.1016/j.marpol.2018.05.003

Friedmann, D. Friedrich, B. Kuhn, T. and Rühlemann, C. 2017. Optimized, zero waste pyrometallurgical processing of polymetallic nodules from the German CCZ license area. *Proc. of the 46th Underwater Mining Conference*, Berlin.

Hannigan, John. 2016. *The Geopolitics of Deep Oceans.* John Wiley and Sons.

Hein, J. R. Mizell, K. Koschinsky, A, and Conrad, T. A. 2013. Deep-ocean mineral deposits as a source of critical metals for high- and green-technology applications: comparison with land-based resources. *Ore Geology Reviews,* 51, 1–14.

Hein, J. R. Tracey, A. C. and Dunham, R. E. 2009. Seamount characteristics and mine-site model applied to exploration-and mining-lease-block selection for cobalt-rich ferromanganese crusts. *Marine Georesources and Geotechnology*, 27(2), 160–176.

Heinrich, L. Koschinsky, A. Markus, T. and Singh, P. 2020. Quantifying the fuel consumption, greenhouse gas emissions and air pollution of a potential commercial manganese nodule mining operation. Marine Policy, 114. p.103678 https://doi.org/10.1016/j.marpol.2019.103678.

Herrington, R. 2021. Mining our green future. *Nat Rev Mater,* 6, 456–458 https://doi.org/10.1038/s41 578–021–00325–9.

International Energy Agency (IEA). 2021. The Role of Critical World Energy Outlook Special Report Minerals in Clean Energy Transitions. Available via https://iea.blob.core.windows.net/assets/24d5dfbb-a77a-4647-abcc-667867207f74/TheRoleofCriticalMineralsinCleanEnergyTransitions.pdf. Accessed October 10, 2021.

International Seabed Authority. 2008a. *Biodiversity, Species Ranges and Gene Flow in the Abyssal Pacific Nodule Province: Predicting and Managing the Impacts of Deep Seabed Mining.* ISA Technical Study: No. 3. eBook. ISBN 978–976–95217–2-8.

International Seabed Authority. 2008b. Rationale and recommendations for the establishment of preservation reference areas for nodule mining in the Clarion Clipperton Zone. ISBA/14/LTC/2*. pp. 12.

ISA. 2017. *Status of Contracts For Exploration And Related Matters.* ISBA/23/C/7. Twenty-third session. Kingston, Jamaica, International Seabed Authority, pp. 7.

ISA. 2021. Exploration Contracts. Available via https://isa.org.jm/exploration-contracts. Accessed in November 28, 2021.

Islam, F. R. and Mamun, K. A. 2017. Possibilities and challenges of implementing renewable energy in the light of PESTLE and SWOT analyses for island countries. In: F.M. Rabiul Islam, Kabir Al Mamun, Maung Than Oo Amanullah (eds) Smart Energy Grid Design for Island Countries. Cham, Springer, pp. 1–19.

Jaeckel, A. 2020. Strategic environmental planning for deep seabed mining in the area. *Marine Policy,* 114, 103423.

Jones, D. O. et al. February 8, 2017. Biological responses to disturbance from simulated deep-sea polymetallic nodule mining. *PLoS One.* 12(2), e0171750. doi: 10.1371/journal.pone.0171750. PMID: 28178346; PMCID: PMC5298332.

JPI Oceans. 2019. Ecological Aspects of Deep-Sea Mining. Available at: http://jpi-oceans.eu/ecological-aspe cts-deep-sea-mining. Accessed January 28, 2019.

Kaikkonen, L. Venesjärvi, R. Nygård, H. and Kuikka, S. 2018. Assessing the impacts of seabed mineral extraction in the deep sea and coastal marine environments: current methods and recommendations for environmental risk assessment. *Marine Pollution Bulletin,* 135, 1183–1197.

Koschinsky, A. et al. 2018. Deep-sea mining: interdisciplinary research on potential environmental, legal, economic, and societal implications. *Integrated Environmental Assessment and Management,* 14(6), 672–691.

Koshesh, O. S. and Jafari, H. R. 2019. The environmental strategic analysis of oil and gas industries in the Kurdistan Region using PESTLE, SWOT and FDEMATEL. *Pollution,* 5(3), 537–554.

Lehnen, F. Rahn, M. Volkmann, S. E. Peter A. Kukla, P. A. Lottermoser, B. G. 2019. *Economic Evaluation of Deep-Sea Mining.* Mining Report Glückauf, 155, 2.

Lempinen, H. 2016. The geopolitics of deep oceans. John Hannigan. 2015. Cambridge: Polity. 200 p, illustrated, softcover. ISBN 987-0-7456-8019-4.£ 14.99. *Polar Record,* 52(5), 616–617.

Levin, L. A. Amon, D. J. and Lily, H. 2020. Challenges to the sustainability of deep-seabed mining. *Nature Sustainability,* 3(10), 784–794. https://doi.org/10.1038/s41893-020-0558-x

Levin, L.A. Kathryn Mengerink, K. Gjerde, K. M. Rowden, A. A. Van Dover, C. L. Clark, M. R. Ramirez-Llodra, E. Currie, B. Smith, C. R. Sato, K. N. Gallo, N. Sweetman, A. K. Lily, H. Armstrong, C. W. and Brider. J. 2016. Marine Policy Defining 'serious harm' to the marine environment in the context of deep seabed mining. *Marine Policy,* 74 , 245–259. http://dx.doi.org/10.1016/j.marpol.2016.09.032.

Masson-Delmotte, V. Zhai, P. Pörtner, H. -O. Roberts, D. Skea, J. Shukla, P. R. Pirani, A. Moufouma-Okia, W. Péan, C. Pidcock, R. Connors, S. Matthews, J. B. R. Chen, Y. Zhou, X. Gomis, M. I. Lonnoy, E. Maycock, T. Tignor, M. and Waterfield, T. (eds). Cambridge University Press, Cambridge, UK and New York, NY, USA, pp. 3–24. https://doi.org/10.1017/9781009157940.001.Summary for PolicymakersSPM

Mejjad, N. and Rovere, M. 2021. Understanding the impacts of blue economy growth on deep-sea ecosystem services. *Sustainability*, 13(22), 12478. https://doi.org/10.3390/su132212478

Mostafa, A. A. Youssef, K. and Abdelrahman, M. 2020. Analysis of photovoltaics in Egypt using SWOT and PESTLE. *International Journal of Applied Energy Systems,* 2(1), 11–14.

Niner, H. J. Ardron, J. A. Escobar, E. G. Gianni, M. Jaeckel, A. Jones, D. O. B. Levin, L. A. Smith, C. R. Thiele, T. Turner, P. J. van Dover, C. L. Watling, L. and Gjerde, K. M. 2018. Deep-sea mining with no net loss of biodiversity-an impossible aim. *Frontiers in Marine Science*, 5(MAR). p. 53. https://doi.org/10.3389/fmars.2018.00053.

Paul, A. J. Lusty, L. and Bramley J. Murton. 2018. Deep-ocean mineral deposits: metal resources and windows into earth processes. *Elements*, 14, 301–306. 1811–5209/18/0014–0301$2.50. doi: 10.2138/gselements.14.5.301.

Paulikas, D. Katona,, S. Ilves, E. Stone, G. and O'Sullivan, A. (2020). *Where Should Metals for the Green Transition Come From?* Vancouver, DeepGreen Metals Inc.

Petersen, S. Krätschell, A. Augustin, N. Jamieson, J. Hein, J. R. and Hannington, M. D. (2016). News from the seabed—geological characteristics and resource potential of deep-sea mineral resources. *Marine Policy*, 70, 175–187. https://doi.org/10.1016/j.marpol.2016.03.012

Petersen, S. Lehrmann, B. and Murton, B. J. 2018. Modern seafloor hydrothermal systems: new perspectives on ancient ore-forming processes. *Elements*, 14, 307–312.

Porter, Michael E. 1979. HBR. *Harvard Business Review*. Reprint 79208: https://strategygurus.com/wp-content/uploads/2020/05/porter_5competitive_forces.pdf

Ramirez-Llodra, E. et al. 2010. Deep, diverse and definitely different: unique attributes of the world's largest ecosystem. *Biogeosciences*, 7, 2851–2899. doi: 10.5194/bg-7-2851-2010.

Ribeiro, M. C. Ferreira, R. Pereira, E. and Soares, J. 2018. Scientific, technical and legal challenges of deep sea mining. A vision for Portugal—conference report. *Marine Policy*, 114, 103338. S0308597X18307346. doi: 10.1016/j.marpol.2018.11.001

Rona, Peter A. 2003. Resources of the sea floor. *Science*, 299(5607), 673–674.

Rovere, M. 2018. The Common Heritage applied to the resources of the seabed. Lessons learnt from the exploration of deep sea minerals and comparison to marine genetic resources. *Marine Safety and Security Law Journal*, 5(2018–19), 78–98. ISSN 2464–9724.

Sharma, R. 2011. Deep-sea mining: economic, technical, technological, and environmental considerations for sustainable development. *Marine Technology Society Journal*, 45(5), 28–41.

Simon-Lledó, E. et al. 2019. Biological effects 26 years after simulated deep-sea mining. *Science Rep*, 9, 8040. https://doi.org/10.1038/s41598-019-44492-w

Smith, C. R. et al. 2020. Deep-sea misconceptions cause underestimation of seabed-mining impacts. *Trends in Ecology and Evolution*, 35(10), 853–857.

SPC. 2013. *Deep Sea Minerals: Deep Sea Minerals and the Green Economy*. Baker, E. and Beaudoin, Y. (eds), Vol. 2, Secretariat of the Pacific Community. https://dsm.gsd.spc.int/public/files/meetings/TrainingWorkshop4/UNEP_vol2.pdf

Srdjevic, Z. Bajcetic, R. and Srdjevic, B. 2012. Identifying the criteria set for multicriteria decision making based on SWOT/PESTLE analysis: a case study of reconstructing a water intake structure. *Water Resources Management*, 26(12), 3379–3393.

Takaya, Y. Yasukawa, K. Kawasaki, T. Fujinaga, K. Ohta, J. Usui, Y. Nakamura, K. Kimura, J. I. Chang, Q. Hamada, M. Dodbiba, G. Nozaki, T. Iijima, K. Morisawa, T. Kuwahara, T. Ishida, Y. Ichimura, T. Kitazume, M. Fujita, T. and Kato, Y. 2018. The tremendous potential of deep-sea mud as a source of rare-earth elements. *Scientific Reports*, 8(1), 1–8. https://doi.org/10.1038/s41598-018-23948-5

Tunnicliffe, V. Metaxas, A. Le, J. Ramirez-Llodra, E. and Levin, L. A. 2020. Strategic environmental goals and objectives: setting the basis for environmental regulation of deep seabed mining. *Marine Policy*, 114, 103347.

UNESCO. November–December 2019. Why we need a united nations decade of ocean science for sustainable development. *EcoMagazine*, 44–45.

United Nations. November 1, 1967. General Assembly (Doc. A/C.l/PV. 1515, 1516).

USGS, 2015. *Tellurium—The Bright Future of Solar Energy*. Fact Sheet 2014–3077 ISSN 2327–6932 (online). http://dx.doi.org/10.3133/fs20143077

van Dover, C. L. Ardron, J. A. Escobar, E. Gianni, M. Gjerde, K. M. Jaeckel, A. B Jones, D. O. Levin, L. A. Niner, H. J. Pendleton, L. Smith, C. R. Thiele, T. Turner, P. J. Watling, L. and Weaver, P. P. 2017. *Biodiversity Loss from Deep-Sea Mining*. https://doi.org/10.2771/43949

Vardopoulos, I. Tsilika, E. Sarantakou, E. Zorpas, A. A. Salvati, L. and Tsartas, P. 2021. An integrated SWOT-PESTLE-AHP model assessing sustainability in adaptive reuse projects. *Applied Sciences*, 11(15), 7134.

Volkmann, S. E. Kuhn, T. and Lehnen, F. 2018. A comprehensive approach for a techno-economic assessment of nodule mining in the deep-sea. *Miner Econ*, 32(5), 32–336. https://doi.org/10.1007/s13563-018-0143-1

Volkmann, S. E. Lehnen, F. and Kukla, P. A. 2019. Estimating the economics of a mining project on seafloor manganese nodules. *Miner Econ,* 32, 287–306 https://doi.org/10.1007/s13563-019-00169-4

Vonnahme, T. R. Molari, M. Janssen, F. Wenzhöfer, F. Haeckel, M. Titschack, J. and Boetius, A. 2020. Effects of a deep-sea mining experiment on seafloor microbial communities and functions after 26 years. *Sci Adv.* 6(18), eaaz5922. doi: 10.1126/sciadv.aaz5922. PMID: 32426478; PMCID: PMC7190355.

Wakefield, J. R. and Myers, K. 2016. Social cost benefit analysis for deep sea minerals mining. *Marine Policy,* S0308597X16303694–. doi: 10.1016/j.marpol.2016.06.018

Weaver, P. P. and Billett, D. 2019. Environmental impacts of nodule, crust and sulfide mining: an overview. In: Sharma, R. (ed) (Environmental Issues of Deep-Sea Mining. Springer, Cham. https://doi.org/10.1007/978-3-030-12696-4_3

Wedding, L. M. et al. 2013. From principles to practice: a spatial approach to systematic conservation planning in the deep sea. *Proceedings of the Royal Society B: Biological Sciences,* 280(1773), 20131684.

Wedding, L. M. et al. 2015. Managing mining of the deep seabed. *Science,* 349(6244), 144–145.

Wynn, R. B. et al. 2014. Autonomous Underwater Vehicles (AUVs): Their past, present and future contributions to the advancement of marine geoscience. *Marine Geology,* 352, 451–468.

13 Modern Seafood Production to Enhance the Blue Economy

A Proposed Sustainable Model for Bangladesh

Md. Nazrul Islam,[1]* Sahanaj Tamanna,[2] and Khaled Mahamud Khan[1]

[1]* Department of Geography and Environment, Jahangirnagar University, Savar, Dhaka, Bangladesh
[2] Bangladesh Environmental Modeling Alliance (BEMA), Non-Profit Research and Training Organization, Mirpur, Dhaka, Bangladesh
Corresponding author: Md. Nazrul Islam. E-mail: nazrul_geo@juniv.edu

CONTENTS

DOI: 10.1201/9781003184287-13

13.1   INTRODUCTION

Seafood production is very important for now-a-days to meet the additional food demand of the increasing population worldwide. The major kinds of seafood are shellfish like lobsters, mussels, and crabs, and now and then other ocean animals that you can eat (Liu and Ralston, 2021). The seafood shops serve plates of fish including mussels, crabs, and calamari (Hicks, 2016; Niiler, 2018). Particularly, in developed countries, seafood serving restaurants spend significant time selling fish and shellfish but currently many small island countries are also serving seafood in many restaurants (Szuwalski et al. 2020; Gómez and Maynou, 2021). The ingredients in the seafoods soup are shrimp and crab (Nguyen et al. 2019; Adeli et al. 2021). It contains every hard fish and the more crude sharks, skate, beam, sawfish, sturgeon, and lampreys; scavengers like lobsters, crabs, shrimps, prawns, and crawfish; molluscs, including shellfish (Pauli, 2010), clams, cockles, mussels, periwinkles, whelks, snails, abalones, scallops, and limpets; the cephalopod molluscs and so on (Sornaraj, 2014; Venugopal and Gopakumar, 2017; Klarin et al. 2019) also, sea-growing plants, generally green growth, utilizing or growing in fresh water, ocean water, harsh water and inland saline water (Oranusi et al. 2018; FAO, 2020). The worldwide production of seafood has expanded over recent years. Seafood accessible in the ocean in both farm raised and wild varieties of fisheries and others food resources (Funge-Smith and Bennett, 2019). Fish like catfish, tilapia, and salmon are regularly farm raised while swordfish, fish, and sharks are gathered from nature as a seafood (Lin et al. 2016). Fish *is the* most significant food, after cereals, providing around 15% of the total population's protein. The fish has been compared to meat or poultry yet is a lot lower in calories (West WQB, 1973; Mishra, 2020). In fish, one gram of protein is available for 4 to 10 calories, as opposed to 10 - 20 calories for each protein gram for lean meats and up to 30 for greasy meats (Comerford and Pasin, 2016). In excess of 50% of all fish created for human consumption are farm raised, and this number is simply expected to expand currently (Villicana et al. 2019).

By 2030, the World Bank assesses that almost 66% of fish will be farm raised. It is not just that the total population has dramatically increased over this period, but that individuals currently eat twice as much seafood as 50 years ago (Ritchie and Roser, 2021). So, the seafood is overexploited to meet our needs. Because of the increased population and the increased amount of seafood consumption it is needed to develop some innovative ideas to cultivate seafood rather than depending on nature to make it sustainable (Venugopal and Sasidharan, 2021). One innovation has assisted with mitigating a portion of the strain on wild fish catch: aquaculture, the act of cultivating seafood and fish (Neori, 2008). The difference between cultivated fish and wild catch is like the distinction between raising animals rather than hunting wild creatures (Ritchie and Roser, 2021). FAO suggested that the percentage of wild fish catches has been almost stable in the last 50 years with

+14%   +527%

Rise in global capture wild fisheries   Rise in global aquaculture
production from 1990 to 2018   production from 1990 to 2018

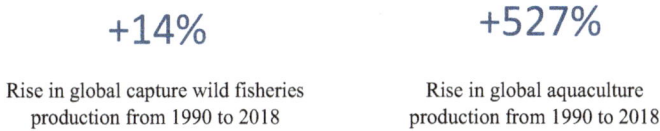

FIGURE 13.1   The percentage of wild fish catches has been almost stable in the last 50 years with a rise of just 14% whereas, the percentage of aquaculture has seen a massive expansion of 527% (approximately) within the same period (FAO, 2020).

a rise of just 14% whereas, percentages coming from aquaculture has seen a massive expansion of 527% within the same period (FAO, 2020) (Figure 13.1).

American Samoa, Guam, Hawaii, the Northern Mariana Islands, and the other Pacific Islands are surrounded by a diverse range of marine life that is essential to the stability of our culture and economy (Cheer et al. 2018). We aim for sustainable seafood. It is the key to our health and well-being (Kleisner et al. 2013). It also benefits from recreational and commercial fishing, bringing nearly US$ 1 billion in revenue and approximately 10,000 jobs to the economy (Blamey and Bolton, 2018; NOAA, 2022). The US fishing department or sector is a very powerful for generating blue economic and to contribute to GDP also. Nationally, it created jobs for almost more than 1.2 million people in 2017 and increased GDP by US$ 69.2 billion (Penn, 1983; Akinyoade, 2019). Commercial harvesting, agriculture, processing and retail are also important factors for both the local and national economies (Farmery et al. 2021). But ultimately, as consumers of fish and shellfish products, we and the retailers determine the future of fishery (Ziegler et al. 2013; Johansen et al. 2019). The consumers need to make responsible purchase choices and retailers need to operate with a responsible sustainable seafood portfolio (Roheim et al. 2018). The greater demand for sustainable products and the diversity of supplies in stores and fish markets will support the transition to more responsible and sustainable fishing and aquaculture (Haby et al. 1993; Vega et al. 2014). The sustainable fish production supports the supply of sufficient fish to the growing world population and secures the long-term livelihoods of millions of people in nations that are developing and least developing. The processing of seafood involves the preparation of seafood for delivery to consumers after harvesting. This includes tasks such as gutting, freezing, canning, and product packaging (Willer et al. 2021). The sustainable seafood processing system helps to reduce food loss and waste, reduces pressure on fishery resources, and promotes sustainable blue economy the country. Processing often produces large amounts of by-products such as heads, bones, internal organs, and shells (Islam and Peñarubia, 2021).

13.2   THE GLOBAL DEMAND FOR SEAFOOD PRODUCTION

For seafood exporters in Europe which is the significant market that provides promise. According to recent data, it is seen that due to the spread of coronavirus, the European fish market and imports from emerging nations fell marginally (Abbott et al. 2021). In general, interest in seafood has remained stable all through 2020. With high demand for seafood, the southern part of Europe presents the greatest chance for trading your seafood products (Figure 13.2) (Lin et al. 2016). Total European seafood imports in 2020 was US$ 54.8 billion. Seafood imports have been on the rise since around 2015, exceeding US$ 58 billion in 2018. In 2019, earnings were down 4% (Feher et al. 2021). However, this top-down pattern continued in 2020, with imports declining only 1% in 2020 (Seafood TIP, 2021). Imports from emerging countries reached US$ 14.4 billion in 2020. It also peaked at US$ 16.6 billion in 2018, declining 6% in 2019 and 8% in 2020 (Seafood TIP, 2021).

Europe is the world's second largest seafood importer. In 2020, Europe imported US$ 19.8 billion from outside Europe, following the US with US$ 22.5 billion. Japan came in third with US$ 12.8

Sceanarios of Global Seafood Export (in billion dollars) 2007-2020

FIGURE 13.2 Scenarios of EU's seafood export 2007–2020 (Seafood TIP, 2021); EUMOFA, based on EUROSTAT (online data code: fish_ld_main) and national sources' data. More details on the sources used can be found in the methodological background.

billion (Moore et al. 2021). China imported US$ 12.7 billion. Imports from these top four seafood markets declined from their 2018 highs, with the exception of China, where imports peaked in 2019 (Seafood TIP, 2021). In 2020, Europe was the largest importer, with US$ 14.4 billion in imports from emerging nations. The United States imported US$ 13.1 billion (Tomascik, 1997). Japan and China imported US$ 7.2 billion and US$ 6.2 billion, respectively. This indicates that Europe is the most important market for fish from developing countries (Seafood TIP, 2021).

13.2.1 EXPORTING SEAFOOD TO ENHANCE THE NATIONAL ECONOMY

If the ocean is healthy then it gives occupations and food, supports monetary development, controls the environment, and provides prosperity for waterfront societies (Fleming et al. 2019). Globally, billions of individuals particularly the world's most unfortunate depend on sound seas for the provision of occupations and food, highlighting the pressing need to prudently utilize, oversee and safeguard this natural asset (Cisneros-Montemayor et al. 2019). If seafoods and other fisheries resources (as a resource item for fishing families) provides food security (Francisco, 2018), then its provision of livelihoods can also support food security (Howson, 1995; Francisco, 2018). As example, the U.S. seafood fish industry is a strong financial driver in the world. Broadly, it upheld 1.2 million positions of employment and added US$ 69.2 billion to the GDP in 2017 (Asusa, 2017). Businesses collecting, cultivating, handling, and retailing are significant resources for nearby and provincial economies. Likewise, in Bangladesh there are 162 fish handling plants in the country. Out of 162 plants European Commission has endorsed 74 plants (Honma and Hayami, 2009). The Hazard Analysis and Critical Control Point (HACCP) guidelines have effectively been adopted in fish handling organizations (Reilly and Käferstein, 1997). Significant adopting nations are European nations, USA and Japan. Around 98% of complete fish items are sent out to those nations (Clarke, 2004). The remaining are traded to the nations in Southeast Asia and the Middle East (Visbeck et al. 2014).

13.2.2 FROZEN FISH SUBSECTORS AND THE BLUE ECONOMY

The worldwide frozen fish market is enjoying significant development and is expecting to see impressive development over the next couple of years as it is becoming popular among consumers. (Nguyen et al. 2019; Adeli et al. 2021). The key factor contributing towards the development of the frozen fish market is the increase in consumption of frozen oceanic items in domestic households (Potts et al. 2016). The rise in consumption of frozen fish is because of specific advantages like value, comfort, and taste. Frozen fish is less expensive than fresh fish (Lin et al. 2016; Le Manach et al. 2020; Blandon and Ishihara, 2021). As the pandemic happens, the offer of frozen fish has declined because of the limitation of access of the general population to their neighborhood markets and grocery stores (Hobbs, 2020). It became important for the providers to handle the unsold supply of fish in the fridges so the fish stays firm and fresh till it is sold (Harmsen and Traill, 1997). The European frozen fish market arrived at a value of US$ 22.4 billion at the end of 2020. Looking forward, IMARC Group anticipates that the market should arrive at a value of US$ 27.8 billion by 2026 (Abila, 2003; Market, 2021). The value of frozen fish items under HS0303 imported from non-industrial nations has expanded by 2% per year. These items incorporate a wide scope of fish, sardines, anchovies, and other species. The overall imports from agricultural nations is around US$ 900 million (Lin et al. 2016; Le Manach et al. 2020; Blandon and Ishihara, 2021).

13.2.3 SHRIMP AND CRAB SUBSECTORS

In 2020, ready and canned shrimp accounted for around 18% of Europe's net imports of warm water shrimp, with an absolute worth of US$ 750 million from developing nations (Suresh et al. 2018). This product group has grown in imports from developing countries by 2.5% per year (Lin et al. 2016; Fletcher, 2021). Worldwide shrimp imports declined imperceptibly in 2020 contrasted with 2019. Total imports from the best four business market sectors, the European Union, United States of America, China and Japan, were 2.4% lower in 2020 from 2019 at 2.485 million tonnes, with an 80 - 82% portion of the global shrimp exchange (FAO, 2020;Lin et al. 2016; Le Manach et al. 2020; Blandon and Ishihara, 2021).

Worldwide shrimp production levels in 2021 are set to be a minimum of 8.9% higher than they were in 2020, while an increase in value of 5% is suggested for 2022 painting an exceptionally positive story for shrimp cultivating areas (Fletcher, 2021).

Shrimp is one of the biggest fish products, addressing around US$ 19 billion or 15% of the absolute worth of globally exchanged fishery items 2012 (Bjørndal and Guillen, 2016). Fundamentally created in shrimp growing districts, most shrimp is bound for global business sectors (FAO, 2020).

The production of Salmon, which has been developing throughout the last ten years because of the extension of aquaculture creation in northern Europe and North and South America, represented US$ 18 billion, a 14% portion of absolute worldwide exchange (FAO, 2020; Pilleron et al. 2019).

Ground fish such as cod, hake, pollock, tilapia and pangasius represented US$ 13 billion in 2012, accounting for 10% of international fish trade. Cod holds the position of being the most expensive bottom fish (Nguyen et al. 2019; Green, 2011). An important source of cheaply traded fish, pangasius entered the global market relatively recently and is mainly produced in Vietnam for the international market. The main suppliers of tilapia are countries in Asia and Central America, mainly targeting the US market. Production of tilapia is increasing in Asia, South America and Africa (FAO, 2020). In 2012, tuna represented US$ 10 billion or 8% of total fish exports (FAO, 2020).

The tuna market has been volatile over the past 3 years due to fluctuations in catches, sustainability matters and the introduction of environmental labels. Japan is a major importer of sashimi grade tuna, and canned tuna is normally directed to the US, European and Asian markets.

Fish dinner (generally produced using little pelagic fish) represented US$ 4 billion or 3% of the world fish exchange (Yao et al. 2019; FAO, 2020). Peru is the world's biggest maker of fish supper, having the biggest fishery delivering the highest yielding species on the planet, the Peruvian anchoveta. The Peruvian fish supper commodity to China is probably the biggest exchange fish and records for roughly US$ 500 billion per year (Abbott et al. 2021), serving generally to help the Chinese aquaculture industry.

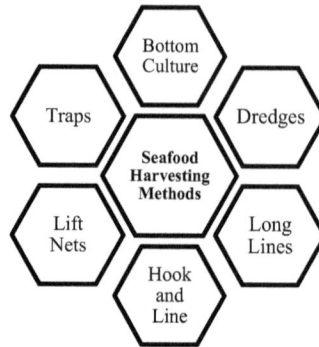

FIGURE 13.3 Several methods of harvesting seafood in the world such as Traps, Bottom culture, Large Metal-Framed Baskets or Dredges, hook and line fishing are some of the most used and popular methods of seafood harvesting (Nguyen et al. 2019).

13.3 GLOBAL SEAFOOD HARVESTING METHODS AND CHALLENGES

General dredges have large iron teeth to dig the seabed to move the catch into a basket (Cisnerosmontemayor et al. 2019). Hydraulic dredges shoot jets of water to disturb the seabed and release their catch. Traditional fish catching techniques include: Ring Seine, Steak Net, China Mesh, Cast Network, Seine River, Trammel Net, Mini Traws, Hook and Line, Trap and Pot (Gudmundsson and Wessels, 2000; Gopal et al. 2017). The world's collected seafood is captured using a variety of fishing gear (Nguyen et al. 2019). Some types of equipment are known to be detrimental to marine health because they are indiscriminate, indiscriminately killing species and damaging habitats (Misund et al. 2002). Up to 10.3 million tonnes of marine animals are unintentionally caught by nets, lines, and other equipment worldwide each year. Fishing is a devastating and wasteful practice that harms many species, including endangered species such as seabirds, sharks, sea turtles, whales, and fish.

There are numerous methods of harvesting seafood (Figure 13.3). Some of the methods are mentioned below along with their procedures and impact on the environment. Traps, Bottom culture, Large Metal-Framed Baskets or Dredges, hook and line fishing are some of the most used and popular methods of seafood harvesting (Nguyen et al. 2019). These methods are used, based on the area and the kind of oceanic patterns of each individual location.

13.3.1 TRAPS

The name of this method itself explains what it is. It is basically a method where fish are trapped using different trapping systems using the flow of water (Figure 13.4).

A variety of materials using barriers, stakes, mesh, stakes, and the like are used to trap fish but with mesh sizes that allow undersized fish to escape. These techniques are mainly used in the marine intertidal zone or in waterways or streams (Nguyen et al. 2019; Le Manach et al. 2020; Blandon and Ishihara, 2021).

As the tide determines the fish movement that's why it is usually used in the intertidal zones to get the best advantage of the fish movement in a particular direction. Salmon, eel, perch, whitebait are some of the most harvested fish using this method of seafood harvesting.

13.3.2 BOTTOM CULTURE AND SEAFOOD CULTIVATION

'Bottom' indicates the sea floor. Thus, a bottom culture technique implies the seafood is cultivated on the seafloor (Connie Lu, 2015; Le Manach et al. 2020; Blandon and Ishihara, 2021).

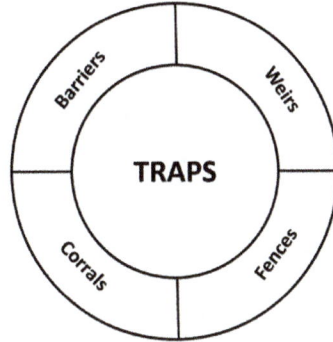

FIGURE 13.4 Popular seafood production methods TRAPS and its characteristics.

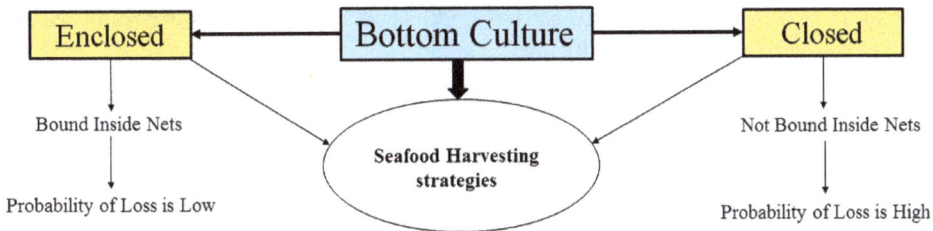

FIGURE 13.5 Traditional techniques of enclosed, bottom, and closed methods for seafood harvesting.

In other words, it is a strategy for aquaculture wherein bivalves (for example, clams or oysters) or seaweeds are grown on the seabed. Bottom culturing is the closest technique for developing seafood as it would grow in the wild (Figure 13.5). In spite of the fact that cultivated shellfish do not set on a surface like their wild counterparts, and assuming that they are spread out on a base, they are similar to their local cousins in that they are sifting the water and growing on the base that influences the color of their shell (Connie Lu, 2015). Bottom culture is of two types. One is called enclosed bottom culture and the other is known as open bottom culture.

In the enclosed bottom culture technique, living specimens are enclosed inside or underneath nets or some other sorts of enclosure set on the ocean floor that can be controlled.

In the open bottom culture technique, the specimens are not enclosed in nets or underneath nets or in any form of enclosure set on the ocean floor that can be controlled.

Every technique has its pros and cons. The primary advantage of bottom culture is the capacity to create strong and generous shells (Le Manach et al. 2020; Blandon and Ishihara, 2021).

There are various theories for bottom culture for aquaculture and fisheries in the oceans. Some consider that it is due to the fact that they retain minerals from the mud, whilst others consider that it is due to the fact that they receive maximum wave activity during the periods of change of the tide and during bad weather (Kaiser et al. 2011). In any case, the main disadvantage is that producers can lose many products to the compelling force of nature (Connie Lu, 2015).

13.3.3 BOTTOM TRAWL AND SEAFOOD HARVESTING

The term of bottom trawl is a method of fishing that involves gathering and catching specific organisms, like catching crabs through towing a net on the seabed (Martin et al. 2014). In this harvesting method floats are joined to a head rope, with the top part of the net open, while weights are appended to the footrope to hold the net down as it travels along the sea-floor (Office of Protected Resources, 2021). Red hake, Dogfish, Crab and Shrimp are mainly harvested using this method.

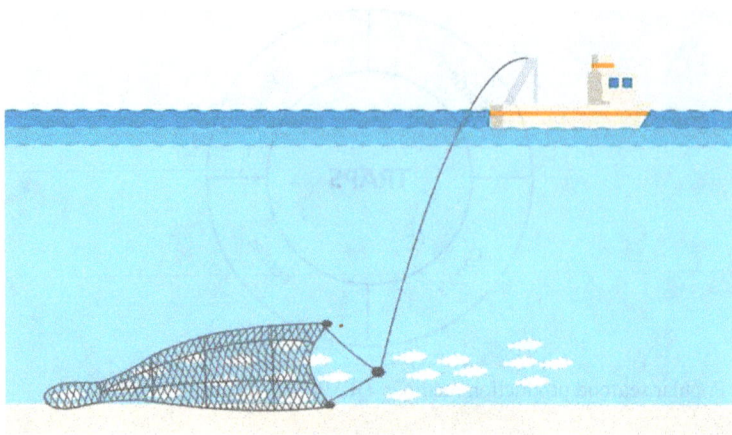

FIGURE 13.6 Bottom trawl and seafood harvesting techniques and its impact to the marine biodiversity (Marine Stewardship Council (MSC), 2020).

a) Fish Trawls: A single net is towed at the back of the boat. Some trawl boats have ramps at the stern of the boat to set up and transport nets, while others have separate boats (Figure 13.6) (Dickison, 1973; AFMA, 2022). Fishing trawls have long metal cables called trawls that connect the trawl gates to the net, allowing the gates to spread much wider than the full width of the net. The fishers 'swarm' the fish until they are exhausted and fall back into the nets, where they are finally caught (Kirmani et al. 2018). Trawl nets are equipped with reels or rollers on ground equipment so that the nets can travel across the seabed without jamming and so that contact with the bottom is minimized (AFMA, 2022). The size of the trawl network varies. Nonetheless, there is a minimum limit to the grid size in the nets.

b) B) Shrimp Trawls: The shrimp trawl has two, three to four otter trawls, towed by outriggers. The net has one large hole and larger species, such as sea turtles are guided to this hole so that they can escape. Trawling shrimp nets use ground chains, so that they can penetrate the seabed and encourage shrimps living on the sea floor into the mouth of the trawl (AFMA, 2022). The trawling mesh for shrimp has no long sweep, and the grid has a smaller mesh than the fish trawl.

Like all other harvesting methods it also has some flaws (Yao et al. 2019). For instance, when we use bottom trawls over a certain type of habitat, this technique can prompt the annihilation of actual constructions like corals and wipes. The effect of trawls can be limited by adjusting nets (FAO, 2020). Numerous ocean turtle species rest, forage on the bottom, and are in danger of being caught in bottom trawls (Office of Protected Resources, 2021). The seafood harvesting techniques bottom trawling, a fishing training that hauls weighty nets along the ocean bottom to collect species like shrimp and plaice, is a disastrous and indiscriminate type of fishing, the effects of which are very detrimental to ocean marine conservation and biological systems (Victorero et al. 2018).

13.3.4 LARGE METAL-FRAMED BASKETS OR DREDGES

The dredging is a fishing method that uses a dredge (Figure 13.7) which is a shovel shaped metal frame for capturing bivalve molluscs such as, oysters, clams and scallops from the seabed, and which is also used for collecting sea urchins (Elias et al. 2011). Previously, we used rakes and tongs to collect large quantities of mussels and extend the collection work to deeper bodies of water (Singh, 2021; Poisson et al. 2022). A dredge is a seafloor scraping or suction device used for digging

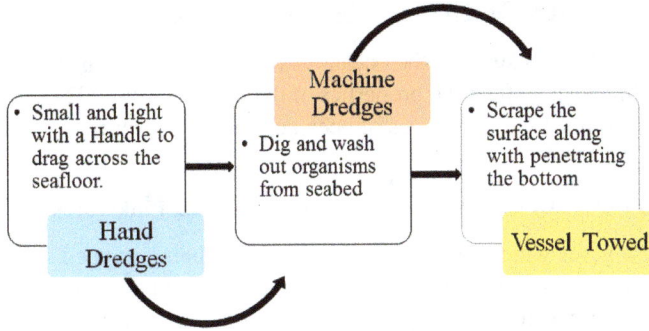

FIGURE 13.7 Dredging is a fishing method that use dredge, which is a shovel shaped metal frame for harvesting seafood.

(Singh, 2021). A dredger is a boat equipped with a dredge. The seafood harvesting tools excavators are typically developed from heavy steel containers such as buckets (Stachowitsch, 2019). The brim is covered with a chain-net with a front that can be pulled open by the chain. Early dredgers had teeth on the bottom called tynes. They have large iron teeth that dig into the ocean floor to lift the mud and sand-covered shells. Dredges are of three types. They are:

i) **Hand dredges:** which are small, lightweight, handheld, with metal contours or mouth contours connected to a retaining sac consisting of a metal ring or net that is dragged across the seafloor (Pokines and Higgs, 2022). They have teeth that are made of iron and large in size that bite into the seafloor to get mud and sand-covered shells.

ii) **Mechanical dredgers**: which are devices specially designed to detect and clean the organisms attached to the seabed. The harvester separates the organisms from the water by limited filtration or pumping (Pokines and Higgs, 2022).

iii) **Vessel towed dredges:** which are mainly of two main types. One which is known for scratching the outer layer of the bottom and other one that digs into the ocean floor to a 30 cm or more to collect specimens (Gaspar and Chícharo, 2007).

These methods of harvesting can cause huge damage to delicate natural surroundings and, as it is an unselective type of collection that can have a large bycatch. Restricting the regions where dredges are allowed and focusing on sandy, rather than hard bottoms can limit the harmful effects (Office of Protected Resources, 2021).

13.3.5 THE SUITABILITY OF HOOK AND LINE FISHING

The seafood cultivation and production tools and techniques namely, hook and line fishing involves placing a fishing line in water, normally with a baited hook, by hand or with a rod and reel or longline (Gabriel et al. 2008; Aneesh Kumar, 2013). In large commercial and marine fisheries, it primarily involves long line fishing. Long line fishing consists of long fishing lines (up to a few kilometers) with various guy lines with baited hooks. Longline materials can be placed in the water or on the seabed to target pelagic or demersal fish species (Poisson et al. 2022). Baited hook fishing is viewed as being relevant to a greater number of animal types and is more size-specific than different sorts of fishing, for example, fishing thus might catch less quantities of non-target species and undesirable smaller fish (Swimmer et al. 2020) Furthermore, the survival rates of undesirable fish caught might likewise be higher after having been caught by hook and line and this strategy for fishing might not affect the seabed and fish territories (Taylor et al. 2021).

13.4　THE CHALLENGES OF GLOBAL SEAFOOD PRODUCTION

Covering about 70% of Earth's surface, the oceans have a strong relationship with the weather and climate. Changing climate can affect the properties of the oceans (EPA, 2021). In this section of the chapter, we are going to discuss the changes of oceanic properties over time due to climate change.

13.4.1　Global Warming, Climate Changes and Seafood Harvesting

When the greenhouse gases which cause global warming are being trapped within the ozone layer then the layer absorbs more heat from the sun, making the earth, and its oceans, warmer (Heuzé et al. 2015). These changes in ocean temperature will make further changes to the major properties of ocean resulting in damage to world climatic patterns (Wilson and Coles, 2005). The ocean chemistry is being changed by the increased amount of dissolved carbon dioxide and this results in damage the life of the organisms living in the ocean. Storms become more frequent and are stronger in the tropical regions (Tabata, 1989). Ocean heat, sea levels rising, ocean acidity, and so forth, are some of the major properties of the ocean that are being changed due to global warming (Figure 13.8).

The irreparable damage to marine life, presented by environmental change, are expanding temperatures of water and sea fermentation. The sea absorbs the heat from the air and ozone harming substances like carbon dioxide (CO_2) (Khadka, 2020). Over the last decade, our seas have retained 90% of the extra warming because of the initial impact and one quarter of human CO_2 output (Khadka, 2020).

13.4.2　Climate Change and Ocean Biological Productivity

The marine ecosystem and its functioning depend on climatic conditions. When these conditions are changed, it creates major changes in the ecosystem structure and function (Sattar et al. 2021). The most noticeable effect is seen in primary production, sea temperature and species dispersion (Singh, 2021).

More than two hundred million individuals from all over the world depend on the ocean for proteins, yet in-spite of this the sector is flimsy (Heuzé et al. 2015). Aquaculture is progressively filling the hole, and this might further develop water quality and decrease the overabundance of nutrients which can cause destructive algal blooms (Islam et al. 2012; Kumar et al. 2021). Nonetheless, as water turns out to be more acidic and as the warming water adjusts microscopic fish development, hydroponics and mollusc creation are undermined. Risk assessment predicts that

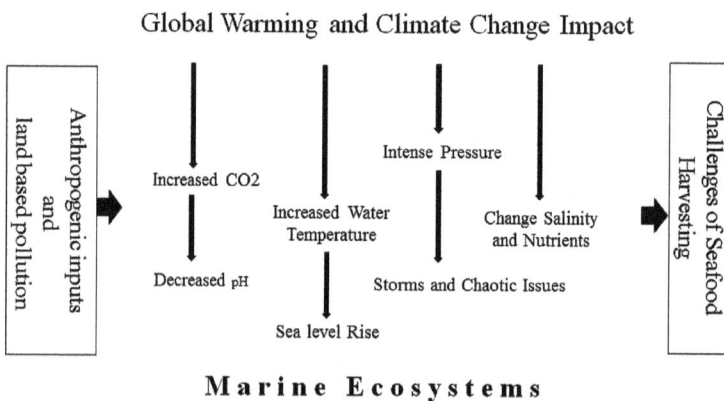

FIGURE 13.8　Global warming, climate change and anthropogenic inputs and land based pollution in the marine ecosystems and challenges of seafood harvesting.

mollusc aquaculture will start to decrease around 2060, with certain nations affected significantly sooner, especially developing and least developed countries (Riskas, 2020).

13.4.3   OVERFISHING AND HABITAT DESTRUCTION

In terms of defining overfishing it can be said that when we capture or harvest fish at a rate that is greater than its ability to reproduce then as a result the total amount of that particular species starts to decline (wikipedia, 2021b). As a development of worldwide fishing undertakings after the 1950s, serious methods of fishing began only from a couple of highlighted regions. Moreover, the different methods of harvesting seafood around the world have created a significant destruction of the seafloor and the physical-chemical properties of the ocean (Figure 13.9). This obliteration changes the working of the environment and can modify species' arrangement and conservation permanently (wikipedia, 2021b).

13.4.4   COASTAL DEVELOPMENT AND MARKET EXPANSION

In the name of the development of the coast is characterized as the human-initiated change in what can be seen on the shoreline (Sattar et al. 2021). This incorporates building structures that are on or close to the coast overall for assurance, trade, correspondence, or entertainment. These designs support financial and social exercises that can contribute, with positive or adverse consequences, to the seaside climate (Sevilla et al. 2019)

At the Conference on Environment and Development organized be the UN, it is clearly stated in section 17 of 'Agenda 21' that seas, oceans, coasts, and marine assets should be appropriately managed for generations to come. Which is reinforced recently in the recent SDGs proposed by the UNDP (Sevilla et al. 2019). The coast is the boundary between landmass and sea that includes around 4% of the earth's overall land mass, which has given jobs to a large number of people for millennia (Raghunathan et al. 2019).

Around the world, 2.5 billion individuals (40% of the total population) at present live within 100 km of the coast, adding increased strain to seaside biological systems (FAO, 2020). Beachfront

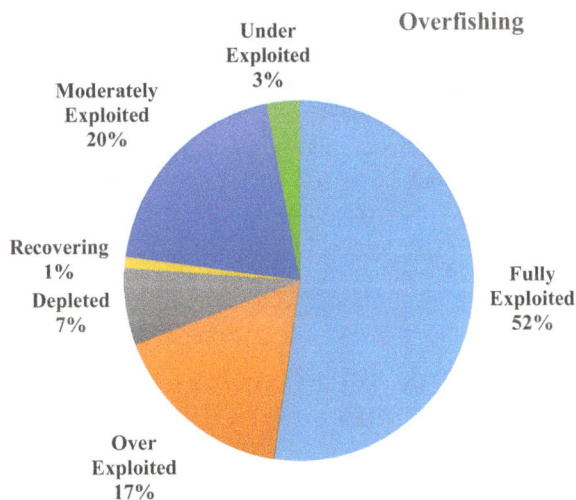

FIGURE 13.9   Sector-wise global overfishing scenarios and capture or harvest fish at a rate in which it cannot cope up through reproducing and as a result the total amount of those particular species starts to decline (Wikipedia, 2021b).

improvements connected to human settlements, industry, hydroponics, or various structures can have serious effects on biological systems that are close to the shore, especially coral reefs (Chauhan, 2019). Waterfront advancement effects might be immediate, for example, land filling, digging, and coral and sand digging for development, or backhanded, like expanded spillover of silt and poisons (Sridhar et al. 2019). Fishing is common among individuals living in the coastal regions and on islands since there is a lot of water in their areas and the majority of what they eat is fish. It is simple for them to get fish by fishing and some people sell fish to bring in cash so it is a significant occupation (Chauhan, 2019).

13.4.5 INTERNATIONAL FRAMEWORK FOR OCEAN GOVERNANCE

Ocean governance is defined as the coordinated conduct of the approaches, acts and affairs regarding the world's oceans to protect the environment of the ocean, its sustainable use and the conservation of coastal and marine resources as well as biodiversity (Haas et al. 2021). The course of ocean governance ought to be coordinated on a level plane since it requires the support of administrative foundations, the private area, NGOs, scholastics, researchers, and so on, just as in an upward direction across all of levels of administration inside an incorporated framework with proportional cooperation and coordination.

However, it is complex. The ocean is not 'owned' by anyone or any one nation thus causing a difficulty in ocean governance (Haas et al. 2021).

We can address three major risks from ocean governance:

- The effect of overfishing of marine assets
- Out of line marine environment administration
- Lacking or unseemly responses to changing ocean conditions

Successful ocean governance requires strong global agreements. To put it plainly, there is a requirement for some type of administration to keep up with the ocean and its different uses, ideally in a manageable way. In the long term, various global settlements have been endorsed to control worldwide ocean administration (wikipedia, 2021a).

13.4.6 SEAFOOD MARKETING POLICIES AND BUSINESS TECHNIQUES

Fish customers are a little contrasted to people who purchase hamburger, pork and chicken (Yapanto and Musa, 2018). Nonetheless, they are likewise a worthwhile group that spends more cash on regular food items every week and are in this way worth seeking, as indicated by a new Food Marketing Institute report (Aday and Aday, 2020). Fish is famous everywhere. With present day transportation and refrigeration frameworks, individuals can appreciate fish that comes from any place on the planet directly in their own localities (Grema et al. 2020). How could you showcase fish to your clients? This shifts to some degree contingent upon your area and your chosen item (Table 13.1).

13.5   SEAFOOD CULTIVATION, PROCESSING AND MARKETING SCENARIOS IN DEVELOPING COUNTRIES

13.5.1 CASE STUDY: BANGLADESH

Amongst all of the leading fish producing countries Bangladesh plays a major role. Besides, Bangladesh has about 12,88,222 ponds which provide a total water area of about 1,46,890 hectares (BFRSS 1986) and the average pond size is 0.011 hectares (Paul and Vogl, 2011; Ghose, 2014;

TABLE 13.1

Global Seafood Marketing Policies and Business Techniques

	Marketing and Business Techniques		
Pre-marketing Policies	**Marketing Policies**		**Post-marketing Policies**
Know your customers: Male/female, local/foreign,	Optimize with Search Engine Optimization (SEO) to increase chances.	Do not forget about local search.	Check your emails daily and remind your customers of your service.
Find out your uniqueness that makes you different from others.	Engage in all kinds of social media to reach out to more people.	Always keep a good relationship with your existing customers.	Analyze your marketing polices using the digital data.
How your product is beneficial for their health.	Create interesting content.	Sponsor different social programs.	Pay attention to which ad or promotion is working best
Make your own website and other social media accounts.	Promote with ads	Reach out to people with campaigns.	Modify the things that are not working for you

Source: Flower, J (2020) 15 Tips For Marketing Your Seafood Business Online, Retrieved on August 8, 2022, Available from https://oscwebdesign.biz/15-tips-for-marketing-your-seafood-business-online/; Bender and Fish, 2000; Jacquet and Pauly, 2008.

Babu et al. 2020). Bangladesh is bounded by the Bay of Bengal on its southern limit. The coastline of the country is about 480 km in length. The area of the sea that makes up the Bangladesh Exclusive Economic Zone (EEZ) is estimated to be about 1,25,000 sq km (Hussain et al. 2017). The area is vast, but the resources are limited and need proper exploration, exploitation, conservation and management for sustainable yields, a huge marine fishery wealth is growing. Traditionally, these inshore, coastal, and ocean waters have been the origin of fish (DoF, 2018). The blue economy debate is a fairly recently developed issue in Bangladesh, beginning very shortly after the maritime demarcation dispute between Myanmar and India (2014) was resolved (Hussain et al. 2017).

Only after this agreement did the authority work with stakeholders in terms of taking steps to introduce the blue economy, along with sustainability policies and plans. (Hasan et al. 2018). The goal is to take advantage of undiscovered capability of the marine climate involving helpful arrangements and advancements for expanding food security, mitigating destitution, further developing sustenance and wellbeing, creating jobs, lifting exchange and modern techniques, and so forth. (Hassan et al. 2014; Hossain et al. 2021). Marine-related issues such as expanding world trade, utilizing resources of ocean mineral for long-term energy certainty, proper management of marine fish resources, and protecting marine weathers and habitat are the future developments for Bangladesh (Table 13.2). There is no doubt that it will determine economic growth. (Hussain et al. 2014; Babu et al. 2020).

The Bay of Bengal's (BoB) fish inventories and other mineral assets can make a significant contribution to the country's economy (Department of Fisheries, 2018). We have to ensure that this is administered by the standards of the policies of the seas, including biodiversity, biological capacity and supporting ecological administrations (Mustafa and Halls, 2007;Ghose, 2014). Nations like Bangladesh, which as of now have immature blue economies, are located in order to foster certain areas (Hasan et al. 2020). For instance, fisheries and seaside hydroponics offer gigantic potential for the arrangement of food and livelihoods, regarding biological boundaries, making maintainable work and delivering high value species for the worldwide product markets (Figure 13.10) (Hussain

TABLE 13.2
Fisheries Sectors, Their Area and Production of Bangladesh

Fisheries Sectors, their Area and Production of Bangladesh					
Inland open water (captured)		Inland closed water (cultured)		Marine fisheries	
Sector	Area (HA)	Sector	Area (HA)	Sector	Area (HA)
River and Estuary	853863	Pond	397775	Industrial	107236
Beels	114161	Seasonal Cultured	144217	Artisanal	552675
Floodplain	2675758	Baor	5671		
Sundarbans	177700	Shrimp/Prawn	258553		
Kaptai Lake	68800	Crab	9377		
Production = 1,235,709 MT		Production = 2,488,601 MT		Production = 659,911 MT	
Total Fish Production = 4,384,221 MT					

Source: Islam et al. 2016; Shamsuzzaman et al. 2017; DoF, 2019; Hasan et al. 2021.

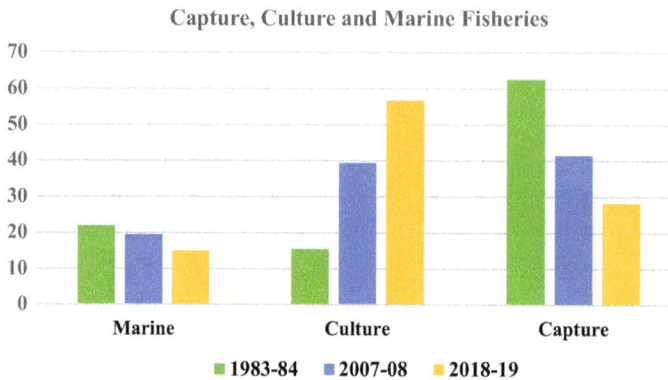

FIGURE 13.10 Ratio of catch, aquaculture and sea fishing to total fish production in Bangladesh (after and modified from FAO, 2018; Shamsuzzaman et al. 2020; Rashid and Sarkar, 2020).

et al. 2017; Parven et al. 2021). In any case, an essential methodology, upheld by an administration structure is basic to fostering the blue economy. Fisheries and waterfront hydroponics improvement could easily lead to an abuse of assets, such as damaging biodiversity and eventually loss of environmental capacity (Hassan et al. 2014; Babu et al. 2020).

The target of this survey article is to feature the significant chances of monetary areas connected with the improvement of the oceanic economy in Bangladesh and addresses the limitations and difficulties in accomplishing this objective (Islam et al. 2017). Two new decisions on the maritime boundaries between Bangladesh and Myanmar and India have allowed Bangladesh to exercise sovereignty over 118,813 km² of water, 12 nautical miles from the region's seas (Mostofa et al. 2018). A more exclusive economic zone (EEZ) of 200 nautical miles to the sea and continental shelf, extending from the Chittagong coast up to 354 nm (Mostofa et al. 2018). It is clearly a huge region with business and financial interests along with ecological stakes (UNCTAD, 2014). Henceforth, it warrants an appropriate assurance and security and this affirmation is a remarkable accomplishment for Bangladesh (Babu et al. 2020).

For ensuring the earning sources from blue economy in Bangladesh, it is needed to expand the awareness and build the programs for sustainable fish and shrimp aquaculture technology, protection

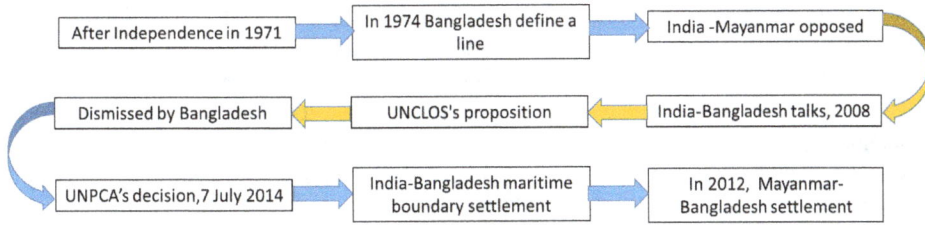

FIGURE 13.11 History of maritime boundary trajectory of Bangladesh and its characteristics for enhancing the blue economy in Bangladesh.

and organization of all sorts of fisheries in the open ocean in northern Bay of Bengal. The fishery sector was a coveted goal (Department of Fisheries, 2018).

13.5.2 THE HISTORY OF THE MARITIME BOUNDARY OF BANGLADESH

Bangladesh has a uniquely inward shoreline. Its two chief stretches of coast meet nearly at right angles. Bangladesh, in this way, risks having the outside location of its oceanic zones reduced by its neighbors, India and Myanmar (Mitra and Zamman, 2015). The number one discussion on ocean restrictions between Bangladesh and Myanmar were held in 1974, shortly after Bangladesh's independence (Bissinger, 2010).

In 1974, Bangladesh described a boundary from the Hariabhanga River with the enclosed west area and the Naaf River with the enclosed easy area, for two hundred nautical miles, bringing the whole internal area within its Exclusive Economic Zone (EEZ) (Figure 13.11) (Haque, 2016). This was questioned by the India and Myanmar, who drew their own lines that incorporated an enormous ocean region guaranteed by Bangladesh. After a wait of more than twenty years, talks at last continued in November 2007 and proceeded all through 2008 (Haque et al. 2021).

In 2008, India and Bangladesh held much awaited sea limits conversations, which remained by and by uncertain as the two players maintained their prior claims (Bissinger, 2010). The UNCLOS's proposition to apply the equidistance rule with the end goal of sea limit boundary is upheld by India, yet dismissed by Bangladesh because of the size of the associated losses (Hoque, 2006). On 7 July 2014, the United Nations Permanent Court of Arbitration (UNPCA) conveyed a decision on this long standing oceanic question of 43 years between Bangladesh and India (Rosen, 2014). In 2012, the settlement of the oceanic limit delimitation question with Myanmar was likewise finished.

13.5.3 IMPORTANCE OF SEAFOOD SECTOR FOR BANGLADESH

As being one of more important agricultural products in Bangladesh, fish plays an important role in the jobs and occupations of millions of people (Islam, 2011). Thus, the lifestyle and use of fish has a significant impact on social income and food security. Bangladeshis often say, 'Mache Bhate Bangali' or 'Make Bengali with fish and rice' (Department of Fisheries, 2018). The fishing sector is an important part of the national economy as it has become home to a large fishing and aquaculture sector. According to the FAO World State of Fishing and Aquaculture Report 2018, Bangladesh is ranked 3rd in the creation of vast inland waters and 5th in the world in hydroponics development (FAO, 2020). Bangladesh is currently the fourth largest tilapia-growing country in the world and the third-largest tilapia-growing country in Asia (Islam, 2003). The State Hillsa fish (Tenualosa ilisha) is the sole species and has the highest participation rate (12.15%) in terms of absolute fish production in the country (Department of Fisheries, 2018).

The fisheries and livestock sector plays a vital role in the socio-economic development of Bangladesh. This sector also has high potential from the perspective of economic development of the country (Ghose, 2014). The contribution of the agricultural sector to gross domestic product (GDP) is very important (Paul and Vogl, 2011). The fisheries and livestock sub-sector alone accounts for 35 - 40% of the total agricultural sector. This contribution is about 78% of total GDP, of which about 3.57% is the fishing sector and 1.53% is the livestock sector (Azim et al. 2017; DoF, 2018). Additionally, this sub-sector provides over 90% of animal protein. According to FAO of the United Nations report, Bangladesh ranked third in inland open water capture production and fifth in inland aquaculture production in the world (FAO, 2020). Already, Bangladesh has achieved self-sufficiency in fish and meat production.

Following 51 years of autonomy, Bangladesh is emerging as an independent fish producer with a target of 62.58 g per capita fish consumption of 60 g/day (Shikder, 2020). In the fiscal year 2017, Bangladesh supplied a total of 4,277,000 tonnes of fish and aquaculture accounted for 56.25% of total fish production (DoF, 2019). Aquaculture shows a solid and steady development, with a normal development rate of just about 10% during the equivalent time span. The Government is attempting to support this development, which ultimately guarantees to accomplish the projected target of producing 4.55 million MT by 2020 - 21 (Depattment of Fisheries, 2018).

13.5.4 FROZEN FISH SUBSECTOR

In Bangladesh frozen seafood will acta as one of the important sub-divisions if the government and policymakers have taken some sustainable measures to ensure the production of good quality seafood (Asaduzzaman, 2015). The frozen seafood products account for about 23% of the total exports of agricultural and industrial complexes, and shrimp account for about 90% of the absolute exports of frozen products, which account for 23% (Table 13.3) (Rahman, 2014; Robayet Ferdous and Syed, 2015).According to Bangladesh Frozen Fish Export Association, among frozen fish the following items are exported all around the world.

In terms of exported seafood products, frozen fish has been playing a most important role in Bangladesh as well as South Asian countries, however along with this fresh fish exporting is also going hand to hand with it by sharing rapid popularity and with the help of advancements in technology (Suresh, 2016). As referenced, the complete worth of frozen fish trades in 2011 was over US$ 80 million. For frozen fish, the main business sectors are the UK, Saudi Arabia, the US, and to some degree, Italy and China (Rothuis et al. 2013; Shamsuzzaman et al. 2017). For new fish, the main business sectors of seafood cultivation and production are India, China, Germany, and Oman. The major seafood production sectors incorporate entire fish (40%), filets (5%), cuts (20%) and head

TABLE 13.3
Exported Frozen Fish Items from Bangladesh to Other Countries

Si No.	Items
01	Frozen Shrimp and Prawn
02	Frozen Fish
03	Fresh and Chilled Fish
04	Frozen Fillets and Steaks of Fish, Sharks Shells Skates and Rays
05	Shark Fins and Fish Maws
06	Salted and dehydrated Fish
07	Dry Fish
08	Live Crabs and Tortoises
09	Fish meals and Crushed Fish
10	Value Added Shrimp and Fish Products

Source: DoF, 2019; Hasan et al. 2021.

and tailless filets (35%). This demonstrates that Bangladeshi frozen and fresh fish is at present primarily heading towards the low-end of market (Van der Pijl and van Duijn, 2012).

13.5.5 SHRIMP SUBSECTOR

13.5.5.1 Production

Though there are very little state policy obligates the land owners to culture shrimp themselves or lease their land on a long-term basis, which forces the lease holders to undertake land development work for improved farming in the coastal areas in Bangladesh (Rahman and Hossain, 2011; Saha, 2017). Shrimp development in Bangladesh has sped up rapidly following immense interest, since the mid-1980s on the global market. Shrimp have been an important part of Bangladesh's economy (Didar-Ul Islam and Bhuiyan, 2016). This is due to great climatic conditions that are helpful for aquaculture, adequate water assets, modest work, and worldwide benefactor organizations. Countless areas where there are shoreline domains have been included in shrimp development (Shahriar and Dhrubo, 2019).

Cultivating shrimp in the seaside regions has become a popular aquaculture in regions like the Asia-Pacific, offering more than 85% of the world's developed shrimp, where Bangladesh is the fifth greatest producer (Hossain et al. 2014; Shameem et al. 2015). After jute and the garment sector it is the third remote exchange winning product. Nowadays, in Bangladesh shrimp is the second most exported item in this sector.

Important shrimp producers are Bagerhat, Sathira, Pirojpur, Khulan, Cox's Bazar and Chittagong. Besides Bangladesh, countries like China, India, Pakistan, Japan, Vietnam, Germany, and the Netherlands are involved in shrimp exports to the world market. India is the world's largest exporter of shrimp, accounting for about 25 percent of the world's demand, and Ecuador is the second largest exporter at 21.1 percent (Hossain, 2021). As an Asian country, Vietnam exports 11.1 percent, Indonesia 8.3 percent and Thailand about 4 percent. On the other hand, Bangladesh ranks 9th in the world with 2.1 percent shrimp exports (Hossain, 2021). A lot of people directly or indirectly depend on shrimp and their development (Shahriar and Dhrubo, 2019).

13.5.5.2 Export

Bangladeshi fish and fishery items are traded to over 50 nations counting the European Union (EU) as a whole and other advanced countries like USA, Russia, Japan, China and so forth (Baldwin and Lopez-Gonzalez, 2015). EU nations are significant merchants of fish exported from this developing country. Among the most important fisheries items the marine and non-marine fish are vital. Different types of shrimp play a most important role here (Department of Fisheries, 2019).

13.5.6 CHALLENGES TO SEAFOOD PRODUCTION IN BANGLADESH

Seafood cultivation, production, processing and marketing will significantly profitable in Bangladesh due to exclusive economic zones in Bay of Bengal and cheap labour forces. But due to the lack of modern technology for cultivating and processing the seafood in Bangladesh this sector is facing many challenges now-a-days. Figure 13.12 shows the bottlenecks mentioned that suggest that the lack of raw materials along with a lack of skilled labor results in the inability to supply high quality fish.

Coordination is especially crucial to every one of the formative issues connected with the blue economy. A part of the Ministry of Foreign Affairs known as the Sea Affairs Unit would assume an essential role for creating this harmony as quickly as time permits (Hossain et al. 2014;DoF, 2019). Whilst India, Myanmar and other financially balanced countries are taking monetary benefits through the BE idea, at the same time Bangladesh is falling behind (Figure 13.13).

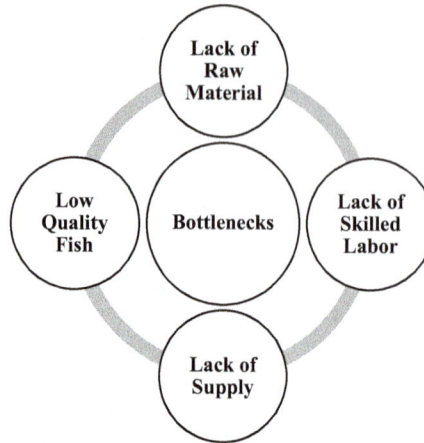

FIGURE 13.12 Some of the bottlenecks mentioned in the above figure that suggest that the lack of raw materials along with lack of skilled labor result in the inability to supply high quality fish.

Steps To Be Taken

Resist illegal use of animal and mineral resources in the exclusive economic zone

01

Besides national efforts, regional or global cooperation is also necessary

02

Navy's coordination with other maritime forces including coast guards and other government agencies

03

Bangladesh ports, shipping, fishing, off-shore oil and gas facilities and shipping lanes in Bangladesh water

04

FIGURE 13.13 Major steps to be taken to enhance the blue economy in Bangladesh.

13.6 SEAFOOD FOR SUSTAINABLE DEVELOPMENT (SSD) MODEL PARAMETERS

Bangladesh a great potential to cultivate and marketing the seafood considering the large number of marine resources in Bay of Bengal. But lack of proper management and shortness of building the social awareness still the seafood industry is not flourished at all. A suitable set of guidelines and model could be the pathways to enhnace the seafood production in the country. For considering these, in this chapter a Seafood for Sustainable Development (SSD) Model has been proposed which will build an awareness to the citizens, countrymen and policymakers for strengthening the seafood production processes (Mulligan and Wainwright, 2004; Perrin et al. (2001). For developing a suitable model it is important to explain modeling approaches to build the process based model. In the SSD model, the parameters are the challenges and prospects in production and marketing of seafood in Bangladesh as well as the possible outcome if we can utilize the prospects and overcome the challenges.

13.6.1 Definitions of Seafood for Sustainable Development (SSD) Model Parameters

13.6.1.1 Sea Food

The US Department of Agriculture (USDA) defined seafood on 28th January 1995 as follows, saying that the term 'seafood' equals 'fish'. Seafood tends to refer to animals (rather than plants such as seaweed) which are considered edible (Oktariani et al. 2022). Mammals are also specifically excluded because no aquatic mammals are processed or marketed commercially.

13.6.1.2 Food Security

According to FAO, Food security exists when all people, at all times, have physical and economic access to sufficient safe and nutritious food that meets their dietary needs and food preferences for an active and healthy life (Figure 13.14) (de Wit and Pebesma, 2001).

13.6.2 Boundary Conditions and Model Assumptions

A model is a combination of some assumptions which can be right or wrong. But it is important to know which assumptions more appropriate and which ones are are wrong. Knocking out or giving less priority to the invalid assumptions and working along the appropriate assumptions will result in a very successful model (Galloul et al. 2020). Easy and understandable assumptions will determine how far your model will be accepted by others. The following figure explains very clearly how model assumptions and complexities are interrelated with each other (Figure 13.15). The easy equation between these two terms is that simple equals complex (Ahmad et al. 2017). If the model is simple, then the assumptions must be complex and if it's the model that is complex then the assumptions must be simple. So, it can be said that they are vice-versa. Simple models are both cost and time effective.

The first step in modeling is to set the boundary of the space in time so that the we can have a clear idea of our data in terms of time and space. As these data are found after the settlement of boundaries, they work to present a spatial domain (Mulligan and Wainwright, 2004). The proposed boundary conditions and assumptions of the SSD model: of the SSD Model:

- Here food security only indicates production of protein type food or availability of food. The accessibility and utilization of food cannot be ensured by this model.
- Two elements of sustainable development (environmental and economic) are considered. The other element, social development is not included in the model.

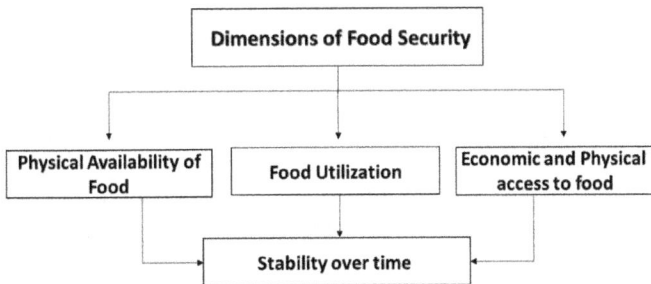

FIGURE 13.14 Physical availability of food, economic and physical access to food, food utilization and stability of the other three dimensions over time.

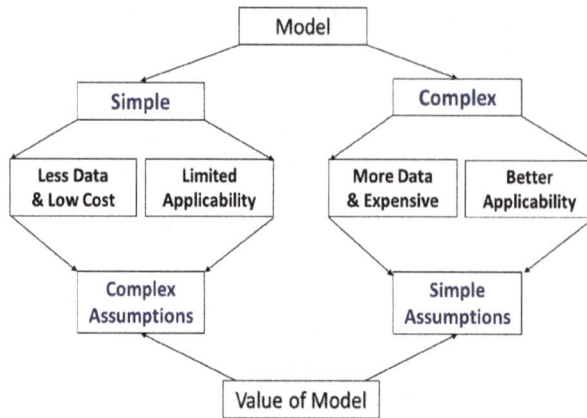

FIGURE 13.15 The selected parameters are the challenges and prospects in production and marketing of Seafood in Bangladesh as well as the possible outcomes of Seafood for Sustainable Development (SSD) Model.

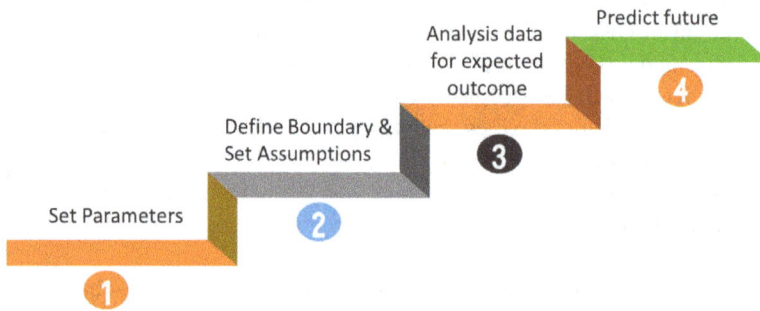

FIGURE 13.16 Steps to Seafood for Sustainable Development (SSD) model build-up, explanation, and future prediction.

As the aim of this task is to develop a model to see the blue economy insights of seafood production the primary parameters are the seafood production, its prospects, challenges, and the outcome, as mentioned in the model name, sustainable development. To comply with the primary parameters some other secondary parameters also come to light under four sub-sectors, namely, the economic sub-sector, the social sub-sector, the development sub-sector and the environmental sub-sector (Figure 13.16). Then the activities under each of these sub-sectors are pointed out in the model to reach the goal.

13.6.3 SEAFOOD FOR SUSTAINABLE DEVELOPMENT (SSD) MODEL BUILD-UP AND EXPLANATION

A conceptual model of the 'Blue Economy Insight of Seafood Production Prospects and Challenges in Bangladesh' has been developed under the task. The model has been named as an SSD Model where SSD stands for 'Seafood for Sustainable Development'. The model has two primary regimes: (i) Prospects for seafood production and (ii) Challenges of seafood production. Both the regimes have three/four sub-sets listed sector wise (economic, social, development and environmental). Some sub-sets have one or more outcomes and some further issues to handle. For example, increased productivity under the economic sub-sector will lead to a market expansion primarily (filled line arrow →) and foreign investment ultimately (when the market will be expanded). But this expanded market may lead to over exploitation of seafood production thus threatening the ecosystem

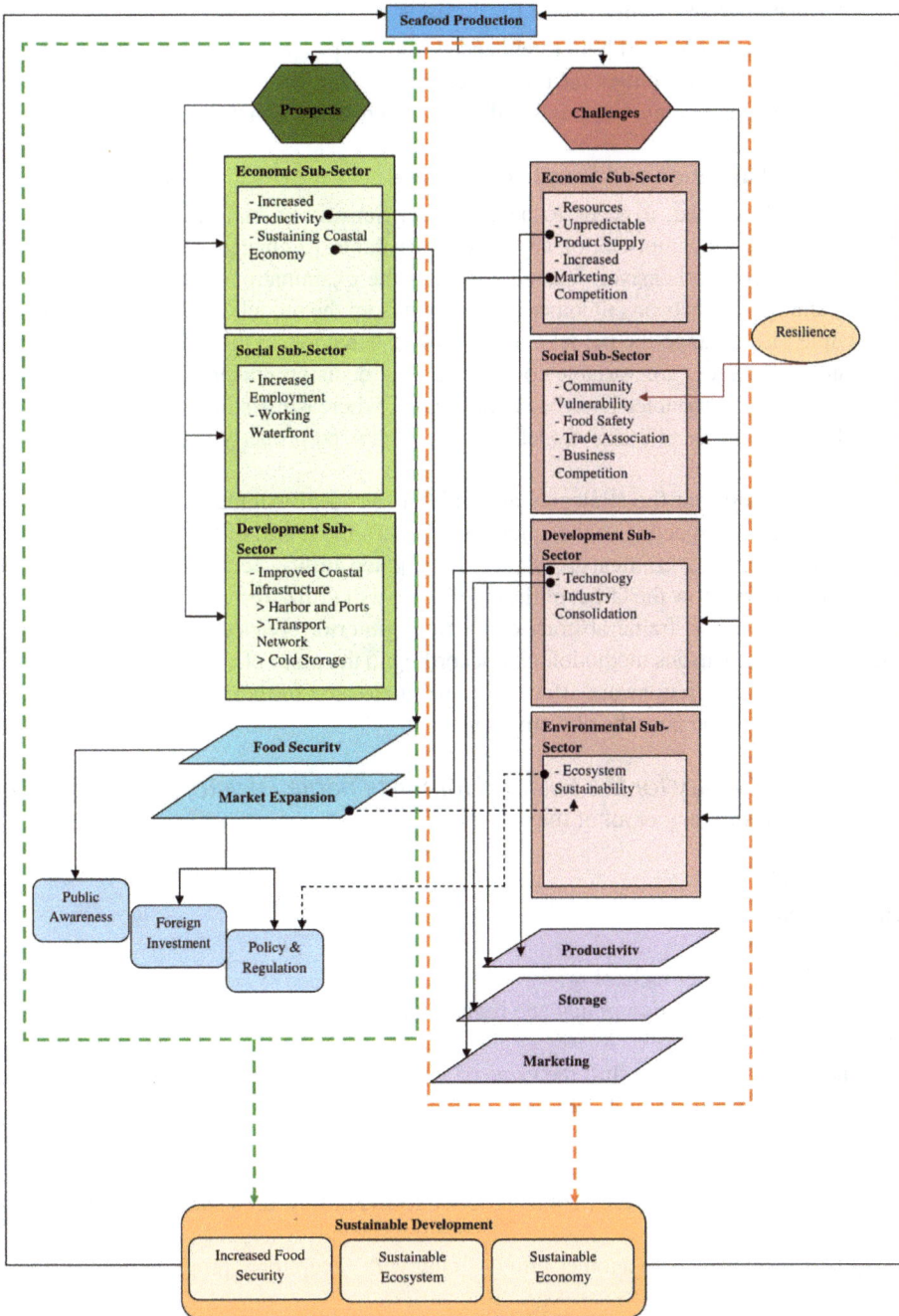

FIGURE 13.17 Proposed Seafood for Sustainable Development (SSD) Model.

sustainability (dashed line arrow - - - >). To address this issue further development of policy and regulations will be required. Considering all the parameters analytically, as in the example, the further possible effects of each of the parameters have been identified and sighposted in the model (Figure 13.17). Then the final model has been developed showing the inter-relationships between each of the parameters and their effects.

13.6.3.1 Future Scenarios and Model Predictions

It is expected that the model will help to enhanced seafood production options, which in turn will create new livelihood opportunities for the coastal people. This in turn will attract investment and strengthen the coastal economy. Further detail is discussed in Section 4.2.

13.6.3.2 Model Validation and Sensitivity Analysis

Once a model is developed, it needs to be validated. Sensitivity analysis is also necessary. The boundary values and presumptions of any model are liable to change and mistakes. Awareness examination (SA), comprehensively characterized, is the examination of these expected changes and mistakes and their effects on outcomes to be drawn from the model. In principle, sensitivity analysis is a simple idea: change the model and observe its behavior. In this model, decision variables (prospects and challenges) are variables over which the decision maker has control as almost all the possible supporting variables have been considered, whereas the strategies (alluding to a set of values for all the choice factors of a model) are also organized properly keeping in mind the goals of the model.

Thus, it can be said that the strategies are optimal [an ideal technique is the system which is best according to the perspective of the chief - it enhances the worth of the leader's goal work (for example benefit, social government assistance, anticipated utility, ecological outcome)]. For this situation, the modelers know the target of the chiefs who will utilize the data produced by the model. The modelers will want to frame abstract convictions (interior convictions, hunches or surmises) about the exhibition of various methodologies according to the point of view of the chiefs. Yet, as the modelers' emotional convictions are affected by the model and furthermore by different variables; these convictions might be near the objective truth.

13.6.3.2.1 What to Vary for the Seafood for Sustainable Development (SSD) Model

One could decide to vary any or all of the following:

- productivity
- employment
- market development
- introduction of proper technology
- absence of implementation policy and regulations.

Normally, the methodology is to shift the value of a mathematical boundary through a few levels. In different cases there is vulnerability around a situation with just two potential results; either a specific occurrence will happen, or it will not. Models include:

- Imagine a scenario where the public authority enacts to boycott a specific innovation for natural reasons
- Imagine a scenario where another info or fixing with exceptional properties opens up.

Regularly this kind of inquiry requires a few underlying changes to the model. When these progressions are made, yield from the overhauled model can measure up to the first arrangement, or the modified model can be utilized in an awareness examination of unsure boundaries to research more extensive ramifications of the change.

13.6.3.2.2 Recommendation for Integrating the SSD Model with other Issues in Bangladesh

Whichever things the modeler decides to shift, there are a wide range of parts of a model outcome to which consideration may be paid:

i)   Food security and market expansion might slow down
ii) Impact on sustainable livelihood
iii) Market expansion might not take place at a rate that was expected
iv) Impact on rate of production, storage, and marketing
v)   Ecosystem threatened.

This model is a conceptual model developed during classwork and thus there was no scope to have a real simulation of the model. Moreover, to develop the model and to consider some more complex parameters related to the study site could only happen if there was scope to visit the relevant communities and chalk out their livelihood activities. The final expected model output of this model is sustainable development achieved through i) increased food security; ii) sustainable economy; and iii) sustainable ecosystem. To achieve the output, the strategies need to be integrated with all possible stakeholders. It will require the involvement of the local community to build their awareness of ecosystem sustainability and avoid over exploitation of seafood produce. It will need to involve local and foreign investors and legislative authorities to enhance investments with a proper legislative guideline to avoid possible economic collapse.

As observed from the model, the variables or parameters of the model are broadly categorized into two separate sections: prospects and challenges. Then the impacting parameters are organized within four subsectors (Hossain, 2001). The model predicts that the economic prospect will lead to food security and market expansion. This market expansion will lead to further domestic foreign investment and then will have impact on social sub-sectors increasing employment opportunities. On the other hand, with the increase in production, an expanded market development of physical infrastructure will be inevitable.

Nevertheless, all these outcomes may lead to food security through increased seafood production and increased livelihood options. But it will require public awareness to accept seafood over traditional food supplies (E-Jahan et al. 2010).

On the other side, if we look at the challenges of seafood production, supply of the produce is unpredictable. It is not at all aware about the total stock of the seafood within our territory (Crona et al. 2020). It will require new technology to estimate the stock; improved industrial facilities to capture adequate amounts of seafood, and cold storage to keep stock and to process seafood for greater market value and increased profit. Once all these challenges can be addressed, it is expected that the economy of the coastal community can be made sustainable.

Nevertheless, with an increased market, there will be competition among the people of a coastal community to maximize their profit. This will lead to trade associations as well as a chaotic social system in the end. Moreover, the profit maximizing drive of the people may increase the threat to ecosystem sustainability.

To address all these issues policy and regulations will need to be developed and implemented so that ecosystem sustainability can be restored (Figure 13.18). People living in the coastal area of Bangladesh are at risk to natural calamities (cyclones) at present and with the changing climate (increased frequency and intensity of cyclones) in future and thus need to build resilience (Saha, 2015). The model also has its focus on this issue under social challenges. With all these parameters and strategies, the model predicts sustainable development as the final outcome through seafood production achieved by:

• increased food security
• sustainable ecosystem
• sustainable economy.

The recent verdict over the dispute resolution on the maritime boundary by the International Tribunal for the Law of the Sea (ITLOS) and Permanent Court of Arbitration increased the

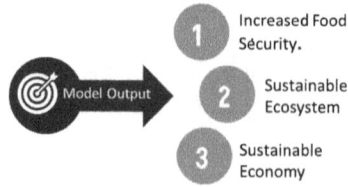

FIGURE 13.18 The SSD model also has its focus on this issue under social challenges. With all these parameters and strategies, the model predicts Sustainable Development as the final model out come through seafood production achieved in the above figure.

horizon and hopes to develop its fisheries within the now undisputed area in future (Alam, 2021). Bangladesh, in the meantime, has instituted several conservation and management measures within the artisanal and industrial fisheries considering the welfare of the small-scale fishers, comprising of gear, temporal and spatial restrictions as well as compensation during the closed season (Mozumder et al. 2019). To explore enhanced economic return from ocean based natural resource, fisheries could be a priority sector where multifaceted and confronting issues as well as opportunities exist which are addressed for future times from its own resources and foreign assistance.

Digital Marine Fisheries Resource Mapping (DMFRM) is a fundamental device for productive and reasonable collection of marine assets. Notwithstanding, in all SAARC nations this is either missing or not in a working condition (Rahman, 2013). Assuming a typical situation, a DMFRM can be produced for the SAARC nations, which will be extremely helpful for every one of the nations and simultaneously will gain the expected interest in this regard.

13.6.3.2.2.1 Managing trans-boundary fisheries resources   There are numerous trans-limit significant fisheries assets that should be overseen on territorial premise. This might be done under the UN settlement on the riding and exceptionally transient fish stocks, which give the two nations to work under the sponsorship of the UN to oversee and preserve these fisheries for ideal advantage (FAO, 2020).

13.6.3.2.2.2 Long-term information generation on ocean dynamics and climate change   Sea elements extraordinarily impact the conduct and life procedure of marine species and add to the environmental change (Wells et al. 2015). Thus, a long term continuous review ought to be directed to produce data on sea elements and changes in environmental factors.

13.6.3.2.2.3 Long-term study on fish behavior and fishing technology   Long haul study on fishing conduct, location of producing, nursery and fishing grounds and fishing innovation to create new climate amicable and asset agreeable collecting advances (Reddy et al. 2022).

13.6.3.2.2.4 Information sharing and database management   Data sharing between the territorial bodies and researchers is vital for feasible innovation and all data should be shared and overseen through an online data set framework.

13.6.3.2.2.5 Skilled manpower development   Marine areas in Bangladesh are amazingly short in skilled labor supply and this deficiency should be satisfied through preparation, higher review and new enlistments (Kabir, 2014).

13.7 CONCLUSION

In Bangladesh, marine catch addresses around 22% to 15% of total fish during 1983 to 2019. Alongside floodplain fisheries the marine culture or seafood creation is additionally in danger of increasing salinity and overfishing, overexploitation, water contamination and so forth. As per FAO, universally, around a quarter of all fish stocks are overexploited and a big portion of them are completely taken advantage of (FAO, 2020). Scientists have shown that ineffective administration of prawn and shrimp culture is e affecting the Sundarbans (the biggest mangrove forests on the planet) where an expected 9700 ha of the backwoods mass has been lost because of extraordinary shrimp cultivating. Changes of numerous regular wetlands to prawn farms has brought about hindrance of water streams and furthermore diminished the extent of movement for some fish species. In Bangladesh because of the absence of execution and authorization of the board measures, numerous open doors in marine assets improvement stay undiscovered. In this context, the proposed *Seafood for Sustainable Development (SSD) Model* could be a pathway to enhance the blue economy in Bangladesh and other developing countries as well.

ACKNOWLEDGEMENTS

We have reviewed many books, articles, blogs, and websites, I would like to express my gratitude to all the anonymous authors and contributors. I would like to show my gratitude to the Ministry of Science and Technology, Government of Bangladesh that have provided the funding support Research Allocation Project Year 2019–2020 to the first author of this chapter for continuing his research on the global blue economy and seafood production and policies. Also, I would like to thank the Faculty of Social Science, Jahangirnagar University, Savar, Dhaka-1342, and University Grants Commission (UGC) of Bangladesh for their financial support to continue my research on the blue seafood cultivation and enhancing blue economy in Bangladesh.

REFERENCES

Aalm, M. K. (2021) Dispute settlement mechanism under the UNCLOS 1982, Maritime Affairs Unit), Ministry of Foreign Affairs, Dhaka, Bangladesh. Retrieved on September 20, 2022. Available from www.theda ilystar.net/views/in-focus/news/dispute-settlement-mechanism-under-the-unclos-1982-2169151

Abbott, J. K. Willard, D. and Xu, J. (2021). Feeding the dragon: The evolution of China's fishery imports. *Marine Policy*, 133, 104733.

Abdullah, S. and Barua, D. Hossain, M. S. (2019). Environmental impacts of commercial shrimp farming in coastal zone of Bangladesh and approaches for sustainable management. *Int J Environ Sci Nat Res.* 20(3), 556038. https://doi.org/10.19080/IJESNR.2019.20.556038

Abila, R. O. (2003). *Fish Trade and Food Security: Are They Reconcilable in Lake Victoria*. Kenya Marine and Fisheries, 31.

Abbott, J. K. Willard, D. and Xu, J. (2021). Feeding the dragon: The evolution of China's fishery imports. *Marine Policy*, 133, 104733.

Aday, S. and Aday, M. S. (2020). Impact of COVID-19 on the food supply chain. *Food Quality and Safety*, 4(4), 167–180.

Adeli, A. Hassannejad, M. and Saadatfar, M. (2021). Factors affecting the purchase of processed and packaged seafood among Gorgan' youth. *Utilization and Cultivation of Aquatics*, 10(1), 71–80.

AFMA (2022). Australlian Fishieries Management Authorities, Australian Government, CRIMFISH: 1800 274 634, Level 3, 15 Lancaster Place, MAJURA PARK ACT 2609, wAustralia.

Ahmed, N. Troell, M. Allison, E. H. and Muir, J. F. (2010). Prawn post larvae fishing in coastal Bangladesh: challenges for sustainable livelihoods. *Marine Policy,* 34, 218–227.

Asaduzzaman, M. (2015). Agriculture: future challenges, special supplement. *The Daily Star*, p. 42.

Ahmad, M. Aftab, S. Muhammad, S. S. and Ahmad, S. (2017). Machine learning techniques for sentiment analysis: A review. *Int. J. Multidiscip. Sci. Eng*, 8(3), 27.

Akinyoade, A. (2019). Nigeria: education, labour market, migration. Annex A to" Dutch labour market shortages and potential labour supply from Africa and the Middle East"(SEO report no. 2019-24). *Nigeria: education, labour market, migration. Annex A to" Dutch labour market shortages and potential labour supply from Africa and the Middle East" (SEO report no. 2019-24).*

Aneesh Kumar, K. V. Meenakumari, B. and Remesan, M. P. (2013). *Investigations on Longline Fishing in Lakshadweep Sea with special reference to Bycatch Issues Bait and Hook Selectivity* (Doctoral dissertation, Cochin University of Science And Technology).

Asusa, P. O. (2017). *US Foreign Policy toward Africa: A contextual analysis of interventions, partnerships, and the role of Congress, 1989 to 2015* (Doctoral dissertation, United States International University-Africa).

Azim, M. S. Tarannum, L. and Patwary, A. K. (2017). The effects of leadership style into fisheries business sector in Bangladesh. *International Journal of Business and Technopreneurship*, 7(1), 13–22.

Babu, M. M. Dey, B. L. Rahman, M. Roy, S. K. Alwi, S. F. S. and Kamal, M. M. (2020). Value co-creation through social innovation: a study of sustainable strategic alliance in telecommunication and financial services sectors in Bangladesh. *Industrial Marketing Management,* 89, 13–27.

Baldwin, R. and Lopez-Gonzalez, J. (2015). Supply-chain trade: A portrait of global patterns and several testable hypotheses. *The World Economy*, 38(11), 1682–1721.

Belton, B. Karim, M. Thilsted, S. Murshed-E-Jahan, K. Collis, W. and Phillips, M. (November 2011). *Review of Aquaculture and Fish Consumption in Bangladesh. Studies and Reviews 2011–53*. The WorldFish Center.

Bender, S. and Fish, A. (2000). The transfer of knowledge and the retention of expertise: the continuing need for global assignments. *Journal of Knowledge Management*.

BFFEA. (2013). *Export Product*. https://bffea.net/product.php

Bissinger, J. (2010). The maritime boundary dispute between Bangladesh and Myanmar: motivations, potential solutions, and implications. *Asia Policy,* 10, 103–142. www.jstor.org/stable/24905004

Bjørndal, T. and Guillen, J. (2016). *Market Competition Between Farmed and Wild Fish: A Literature Survey*. FAO Fisheries and Aquaculture Circular (C1114), I.

Blamey, L. K. and Bolton, J. J. (2018). The economic value of South African kelp forests and temperate reefs: past, present and future. *Journal of Marine Systems*, 188, 172–181.

Blandon, A. and Ishihara, H. (2021). Seafood certification schemes in Japan: Examples of challenges and opportunities from three Marine Stewardship Council (MSC) applicants. *Marine Policy*, 123, 104279.

Chauhan, N. (2019). *Why Are Farming and Fishing Common Occupation of the People in All the Coastal States and Islands?* https://edurev.in/question/1548509/Why-are-farming-and-fishing-common-occupation-of-t

Cheer, J. M. Pratt, S. Tolkach, D. Bailey, A. Taumoepeau, S. and Movono, A. (2018). Tourism in Pacific island countries: A status quo round-up. *Asia and the Pacific Policy Studies*, 5(3), 442–461.

Cisneros-Montemayor, A. M. et al. (2019). Social equity and benefits as the nexus of a transformative Blue Economy: a sectoral review of implications. *Marine Policy,* 109, 103702.

Clarke, S. (2004). Understanding pressures on fishery resources through trade statistics: a pilot study of four products in the Chinese dried seafood market. *Fish and Fisheries*, 5(1), 53–74.

Comerford, K. B. and Pasin, G. (2016). Emerging evidence for the importance of dietary protein source on glucoregulatory markers and type 2 diabetes: different effects of dairy, meat, fish, egg, and plant protein foods. *Nutrients*, 8(8), 446.

Connie, L. U. (2015). *The Different Methods of Growing Oysters*. https://pangeashellfish.com/blog/the-different-methods-of-growing-oysters

De Wit, M. J. M. and Pebesma, E. J. (2001). Nutrient fluxes at the river basin scale. II: the Balance between data availability and model complexity, *Hydrological Processes,* 15, 761–775.

Department of Fisheries (2018). DoF. 2019 Annual Report.

Crona, B. Wassénius, E. Troell, M. Barclay, K. Mallory, T. Fabinyi, M. ... and Eriksson, H. (2020). China at a crossroads: An analysis of China's changing seafood production and consumption. *One Earth*, 3(1), 32–44.

Department of Fisheries (2019). *Yearbook of Fisheries Statistics of Bangladesh 2018–19*.

Dickinson, W. R. (1973). Japanese fishing vessels off Alaska. *Marine Fisheries Review*, 35(1–2), 6–18.

Didar-Ul Islam, S. M. and Bhuiyan, M. A. H. (2016). Impact scenarios of shrimp farming in coastal region of Bangladesh: an approach of an ecological model for sustainable management. *Aquaculture International*, 24(4), 1163–1190.

DoF (2014). *National Fish Week 2014 Compendium (In Bengali)*. Bangladesh: Department of Fisheries, Ministry of Fisheries and Livestock, p. 144.

DoF. (2019) Yearbook of Fisheries Statistics of Bangladesh, 2018–19. Fisheries Resources Survey System (FRSS), Department of Fisheries, Bangladesh: Ministry of Fisheries and Livestock, 36, 135.

E-Jahan, K. M. Ahmed, M. and Belton, B. (2010). The impacts of aquaculture development on food security: lessons from Bangladesh. *Aquaculture research*, 41(4), 481–495.

Elias, I. Carozza, C. R. Di Giácomo, E. E. Isla, M. S. Orensanz, Jose Maria; et al. (2011) Coastal fisheries of Argentina; Food and Agriculture Organization of the United Nations. 13–48; http://hdl.handle.net/11336/111040

EPA (2021). *Climate Change Indicators: Oceans*. https://epa.gov/climate-indicators/oceans

FAO (2020). *Aquaculture*. https://fao.org/state-of-fisheries-aquaculture

FAO (2020) The State of World Fisheries and Aquaculture 2020. Sustainability in action. Sustainability in Action, Food and Agriculture Organization of the United Nations, Rome, Italy, Retrieved on August 9, 2022, from https://doi.org/10.4060/ca9229en

Feher, A. Iancu, T. Raicov, M. and Banes, A. (2021). Food Consumption and Ways to Ensure Food Security in Romania. In *Shifting Patterns of Agricultural Trade* (pp. 305–333). Springer, Singapore.

Farmery, A. K. et al. (2021). Food for all: designing sustainable and secure future seafood systems. *Reviews in Fish Biology and Fisheries*, 32, 1–21.

Fleming, L. E. Maycock, B. White, M. P. and Depledge, M. H. (2019). Fostering human health through ocean sustainability in the 21st century. *People and Nature*, 1(3), 276–283.

Fletcher, R. (2021). *Global Shrimp Production Sees Significant Growth in 2021*. https://thefishsite.com/articles/global-shrimp-production-sees-significant-growth-in-2021-gorjan-nikolik-rabobank

Francisco, D. R. (2018). Change and Resiliency of Traditional Labor Institutions in Response to the Effects of Special Economic Zones: *The Case of Mariveles, Bataan, The Philippines*.

Funge-Smith, S. and Bennett, A. (2019). A fresh look at inland fisheries and their role in food security and livelihoods. *Fish and Fisheries*, 20(6), 1176–1195.

Ghose, B. (2014). Fisheries and aquaculture in Bangladesh: challenges and opportunities. *Ann Aquac Res*, 1(1), 1001.

Gaaloul, K. Menghi, C. Nejati, S. Briand, L. C. and Wolfe, D. (2020, November). Mining assumptions for software components using machine learning. In *Proceedings of the 28th ACM Joint Meeting on European Software Engineering Conference and Symposium on the Foundations of Software Engineering* (pp. 159–171).

Gabriel, O. Lange, K. Dahm, E. and Wendt, T. (Eds.). (2008). *Von Brandt's fish catching methods of the world*. John Wiley and Sons, Australia

Gaspar, M. B. and Chícharo, L. M. (2007). Modifying dredges to reduce by-catch and impacts on the benthos. In *By-catch Reduction in the World's Fisheries* (pp. 95–140). Springer, Dordrecht.

Gómez, S. and Maynou, F. (2021). Alternative seafood marketing systems foster transformative processes in Mediterranean fisheries. *Marine Policy*, 127, 104432.

Gopal, N. M.J. Williams, S. Gerrard, S. Siar, K. Kusakabe, L. Lebel, H. Hapke, M. Porter, A. Coles, N. Stacey and R. Bhujel. 2017. Gender in Aquaculture and Fisheries: Engendering Security in Fisheries and Aquaculture. *Asian Fisheries Science* (Special Issue) 30S. 423 pp.

Green, K. (2011). Annual review of the status of the feed grade fishes stocks used to produce fishmeal and fish oil for the UK market.-2011/12 Band 2011/12 (2012). *Annual review of the status of the feed grade fishes stocks used to produce fishmeal and fish oil for the UK market*, 12.

Grema, H. A. Kwaga, J. K. P. Bello, M. and Umaru, O. H. (2020). Understanding fish production and marketing systems in North-western Nigeria and identification of potential food safety risks using value chain framework. *Preventive Veterinary Medicine*, 181, 105038.

Gudmundsson, E. and Wessells, C. R. (2000). Ecolabeling seafood for sustainable production: implications for fisheries management. *Marine Resource Economics*, 15(2), 97–113.

Haas, B. et al. (2021). The future of ocean governance. *Reviews in Fish Biology and Fisheries*. DOI:10.22541/au.160193487.70124607/v1

Haby, M. G. Edwards, R. A. Reisinger, E. A. Tillman, R. E. and Younger, W. R. (1993). *Importance of Seafood-Linked Employment and Payroll in Texas*. Sea Grant Program, Texas A and M University. Url: http://hdl.handle.net/1834/30024; Texas, USA

Haque, A. E. (2016). *United Nations convention on the law of the sea (UNCLOS) and delimitation of maritime boundaries: A Bangladesh perspective* (Doctoral dissertation, Western Sydney University (Australia)).

Haque, A. B. D'Costa, N. G. Washim, M. Baroi, A. R. Hossain, N. Hafiz, M. ... and Biswas, K. F. (2021). Fishing and trade of devil rays (Mobula spp.) in the Bay of Bengal, Bangladesh: Insights from fishers' knowledge. *Aquatic Conservation: Marine and Freshwater Ecosystems*, 31(6), 1392–1409.

Harmsen, H. and Traill, B. (1997). Royal Greenland A/S: from fish to food. In: Traill, B. Grunert, K.G. (eds) *Products and Process Innovation in the Food Industry* (pp. 147–162). Boston, MA: Springer.

Hasan, M. M. Hossain, B. M. S. Alam, M. J. Chowdhury, K. M. A. Al Karim, A. and Chowdhury, N. M. K. (2018). The prospects of blue economy to promote Bangladesh into a middle-income country. *Open Journal of Marine Science*, 8, 355–369.

Hasan, J. Lima, R. A. and Shaha, D. C. (2021). Fisheries resources of Bangladesh: A review. *Int. J. Fish. Aquat. Stud*, 9, 131–138.

Hassan, S. R. Hassan, M. K. and Islam, M. S. (2014). Tourist-group consideration in tourism carrying capacity assessment: A new approach for the Saint Martin's island, Bangladesh. *Journal of Economics and Sustainable Development*, 5, 150–158.

Heuzé, C. Heywood, K. J. Stevens, D. P. and Ridley, J. K. (2015). Changes in global ocean bottom properties and volume transports in CMIP5 models under climate change scenarios. *Journal of Climate*, 28(8), 2917–2944.

Hicks, D. T. (2016). Seafood safety and quality: The consumer's role. *Foods*, 5(4), 71.

Hobbs, J. E. (2020). Food supply chains during the COVID-19 pandemic. *Canadian Journal of Agricultural Economics/Revue canadienne d'agroeconomie*, 68(2), 171–176.

Honma, M. Hayami, Y. 2008. Distortions to Agricultural Incentives in Japan, Korea, and Taiwan. Agricultural Distortions Working Paper; 35. World Bank, Washington, DC: orld Bank. https://openknowledge.worldbank.org/handle/10986/28189 License: CC BY 3.0 IGO.", pp. 67–114.

Hoque, M. N. (2006). *The Legal and Scientific Assessment of Bangladesh's Baseline in the Context of Article 76 of the United Nations Convention on the Law of the Sea*. United Nations–The Nippon Foundation, National University of Ireland Galway, Republic of Ireland, pp. 107.

Hossain, M. S. Chowdhury, S. R. Navera, U. K. Hossain, M. A. R. Imam, B. and Sharifuzzaman, S. M. (2014). *Opportunities and strategies for ocean and river resources management. Dhaka: Background paper for preparation of the 7th Five Year Plan. Planning Commission, Ministry of Planning, Bangladesh*; Dhaka

Hossain, F. (2021). Adaptation measures (AMs) and mitigation policies (MPs) to climate change and sustainable blue economy: a global perspective. *Journal of Water and Climate Change*, 12(5), 1344–1369.

Hossain, M. S. (2001). Biological aspects of the coastal and marine environment of Bangladesh. *Ocean and Coastal Management*, 44(3–4), 261–282.

Hossain, M. S. Chowdhury, S. R. Navera, U. K. Hossain, M. A. R. Imam, B. Sharifuzzaman, S. M. (2014). Opportunities and Strategies for Ocean and River Resources Management. Background paper for preparation of the 7th Five Year Plan. Dhaka, Bangladesh: FAO, pp. 61.

Howson, J. (1995). Colonial Goods and the Plantation Village: Consumption and the Internal Economy in Monserrat from Slavery to Freedom (Doctoral dissertation, New York University).

Hossain, S. (2021) Business Inspection: Shrimp Farming in Bangladesh: Why Shrimp Sector is Lagging Behind? Retrieved on August 9, 2022. Available from https://businessinspection.com.bd/why-the-shrimp-sector-of-bangladesh-is-lagging-behind/

Hussain, G. M. Failler, P. Karim, A. A. and Alam, M. K. (2017). Review on opportunities, constraints and challenges of blue economy development in Bangladesh. *Journal of Fisheries and Life Sciences*, 2(1), 45–57.

Hussain, Y. Ullah, S. F. Akhter, G. and Aslam, A. Q. (2017). Groundwater quality evaluation by electrical resistivity method for optimized tubewell site selection in an ago-stressed Thal Doab Aquifer in Pakistan. *Modeling Earth Systems and Environment*, 3(1), 1–9.

Islam, M. M. Shamsuzzaman, M. M. Mozumder, M. M. H. Xiangmin, X. Ming, Y. and Jewel, M. A. S. (2017). Exploitation and conservation of coastal and marine fisheries in Bangladesh: do the fishery laws matter? *Marine Policy*, 76, 143–151.

Islam, M. S. (2003). Perspectives of the coastal and marine fisheries of the Bay of Bengal, Bangladesh. *Ocean and Coastal Management*, 46(8), 763–796.

Islam, M. M. (2011). Living on the margin: the poverty-vulnerability nexus in the small-scale fisheries of Bangladesh. In *Poverty mosaics: Realities and prospects in small-scale fisheries* (pp. 71–95). Springer, Dordrecht.

Islam, Md. N. Kitazawa, D. Kokuryo, N. Tabeta, S. Honma, T. Komatsu, N (2012) Numerical modeling on transition of dominant algae in Lake Kitaura, Japan, *Ecological Modelling*, 242: 146–163, Elsevier publication, DOI: 10.1016/j.ecolmodel.2012.05.013

Islam, M. S. Jahan, H. and Al-Amin, A. A. (2016). Fisheries and aquaculture sectors in Bangladesh: an overview of the present status, challenges and future potential. *Journal of Fisheries and Aquaculture Research*, 1(1), 002-009.

Islam, M. J. and Peñarubia, O. R. (2021). Seafood Waste Management Status in Bangladesh and Potential for Silage Production. *Sustainability*, 13(4), 2372.

Jacquet, J. L. and Pauly, D. (2008). Trade secrets: renaming and mislabeling of seafood. *Marine Policy*, 32(3), 309–318.

Johansen, U. Bull-Berg, H. Vik, L. H. Stokka, A. M. Richardsen, R. and Winther, U. (2019). The Norwegian seafood industry–importance for the national economy. *Marine Policy*, 110, 103561.

Klarin, B. Garafulić, E. Vučetić, N. and Jakšić, T. (2019). New and smart approach to aeroponic and seafood production. *Journal of Cleaner Production*, 239, 117665.

Kabir, M. D. (2014). Enhancement of seafarers' employability through capacity building in maritime education and training (MET): a case study of Bangladesh. World Maritime University Dissertations. 465. https://commons.wmu.se/all_dissertations/465

Khadka, Y. (2020). CARBON COMPOUNDS: Pollution Aspects. *Patan Pragya*, 6(1), 127–135.

Kimani, E. Okemwa, G. and Aura, C. (2018). The status of Kenya Fisheries: Towards sustainability exploitation of fisheries resources for food security and economic development.pp135, Kenya Marine and Fisheries Research Institute (KMFRI), Kenya, Retrievd on August 9, 2022, from http://hdl.handle.net/1834/16123

Kaiser, M. J. Attrill, M. J. Jennings, S. Thomas, D. N. and Barnes, D. K.A. *Marine ecology: processes, systems, and impacts*. Oxford University Press, 2011.

Kleisner, K. M. et al. (2013). Exploring patterns of seafood provision revealed in the global Ocean Health Index. *Ambio*, 42(8), 910–922.

Kumar, N. Kumar, A. Marwein, B. M. Verma, D. K. Jayabalan, I. Kumar, A. and Ramamoorthy, D. (2021). Agricultural Activities Causing Water Pollution And Its Mitigation: A review. *International Journal of Modern Agriculture*, 10(1), 590–609.

Le Manach, F. Jacquet, J. L. Bailey, M. Jouanneau, C. and Nouvian, C. (2020). Small is beautiful, but large is certified: A comparison between fisheries the Marine Stewardship Council (MSC) features in its promotional materials and MSC-certified fisheries. *PloS One*, 15(5), e0231073.

Lin, Z. Yuegang, L. Jing, K. Zhanqun, S. and Andrew, B. (2016). Development of transportation supervision information system of frozen seafood processing enterprises based on ERP. *Journal of Applied Science and Engineering Innovation*, 3(1), 18–21.

Liu, C. and Ralston, N. V. (2021). Seafood and health: What you need to know? In *Advances in food and nutrition research* (Vol. 97, pp. 275–318). Academic Press.

Market, B. (2021). *Global Industry Trends, Share, Size, Growth, Opportunity and Forecast 2021–2026*. IMARC Group. https://imarcgroup.com/induction-motor-market

Martín, J. Puig, P. Palanques, A. and Giamportone, A. (2014). Commercial bottom trawling as a driver of sediment dynamics and deep seascape evolution in the Anthropocene. *Anthropocene*, 7, 1–15.

Mishra, S. P. (2020). Significance of fish nutrients for human health. *Int. J. Fish. Aqua. Res*, 5(3), 47–49.

Misund, O. A. Kolding, J. and Fréon, P. (2002). Fish capture devices in industrial and artisanal fisheries and their influence on management. *Handbook of fish biology and fisheries*, 2, 13–36.

Mitra, A. and Zaman, S. (2015). Biodiversity of the Blue Zone. In *Blue Carbon Reservoir of the Blue Planet* (pp. 37–92). Springer, New Delhi.

Moore, C. Morley, J. W. Morrison, B. Kolian, M. Horsch, E. Frölicher, T. ... and Griffis, R. (2021). Estimating the economic impacts of climate change on 16 major US fisheries. *Climate Change Economics*, 12(01), 2150002.

Mostofa, M. Al-Amin, D. M. and Bint-E-Basar, K. T. (2018). Delimitation of Maritime Boundary with India and Bangladesh's Rights over the Sea. *American International Journal of Social Science Research*, 2(1), 108–113.

Mozumder, M. M. H. Pyhälä, A. Wahab, M. Sarkki, S. Schneider, P. and Islam, M. M. (2019). Understanding Social-Ecological Challenges of a Small-Scale Hilsa (Tenualosa ilisha) Fishery in Bangladesh. International journal of environmental research and public health, 16(23), 4814. https://doi.org/10.3390/ijerph16234814

Mulligan, M. and Wainwright, J. (Eds) (2004). *Environmental Modelling Finding Simplicity in Complexity*. London: John Wiley.

Mustafa, M. G. Halls, A. S. (2007). Impact of the Community-Based Fisheries Management on sustainable use of inland fisheries in Bangladesh. In Dickson, M. and A. Brooks (eds.) Proceedings of the CBFM-2 International Conference on Community Based Approaches to Fisheries Management, Dhaka, Bangladesh, 6-7 March. www.worldfishcenter.org/publication/impact-community-based-fisheries-management-sustainable-use-inland-fisheries-bangladesh#:~:text=LINK%20TO%20PDF,WF_778.pdf%3Fsequence1%3D

Neori, A. (2008). Essential role of seaweed cultivation in integrated multi-trophic aquaculture farms for global expansion of mariculture: an analysis. *Journal of Applied Phycology*, 20(5), 567–570.

Nguyen, T. D. P. Tran, T. N. T. Le, T. V. A. Phan, T. X. N. Show, P. L. and Chia, S. R. (2019). Auto-flocculation through cultivation of Chlorella vulgaris in seafood wastewater discharge: Influence of culture conditions on microalgae growth and nutrient removal. *Journal of Bioscience and Bioengineering*, 127(4), 492–498.

Niiler, E. (2018). More Sharks Ditching Annual Migration as Ocean Warms. National Geographic.

Office of Protected Resources. (2021). *Fishing Gear: Bottom Trawls*. https://fisheries.noaa.gov/national/bycatch/fishing-gear-bottom-trawls

Oktariani, A. F. Ramona, Y. Sudaryatma, P. E. Dewi, I. A. M. M. and Shetty, K. (2022). Role of Marine Bacterial Contaminants in Histamine Formation in Seafood Products: A Review. *Microorganisms*, 10(6), 1197.

Oranusi, S. Effiong, E. D. and Duru, N. U. (2018). Comparative study of microbial, proximate and heavy metal compositions of some gastropods, bivalve and crustacean seafood. *African Journal of Clinical and Experimental Microbiology*, 19(4), 291–302.

Parven, A. Pal, I. and Hasan, M. S. (2021). Ecosystem for disaster risk reduction in Bangladesh: a case study after the Cyclone 'Aila.' In *Disaster Resilience and Sustainability* (pp. 277–300). Elsevier Publication: The Netherlands.

Paul, B. G. and Vogl, C. R. (2011). Impacts of shrimp farming in Bangladesh: challenges and alternatives. *Ocean and Coastal Management*, 54, 201–211

Pauli, D. G. (2010), *Blue Economy—10 Years, 100 Innovations, 100 Million Jobs*. Brookline, MA: Paradigm Publications.

Penn, J. W. (1983). An assessment of potential yield from the offshore demersal shrimp and fish stock in Bangladesh water (including comments on the trawl fishery 1981–1982). E:DP/BGD/81/034, Field Doc. p. 22. Fishery Advisory Service (Phase II) Project, Rome, Italy: FAO.

Perrin, C. Michel, C. and Andreassian, V. (2001). Does a large number of parameters enhance model performance? Comparative assessment of common catchment model structures on 429 catchments. *Journal of Hydrology*, 242, 275–301.

Pilleron, S. Sarfati, D. Janssen-Heijnen, M. Vignat, J. Ferlay, J. Bray, F. and Soerjomataram, I. (2019). Global cancer incidence in older adults, 2012 and 2035: a population-based study. *International Journal of Cancer*, 144(1), 49–58.

Poisson, F. Budan, P. Coudray, S. Gilman, E. Kojima, T. Musyl, M. and Takagi, T. (2022). New technologies to improve bycatch mitigation in industrial tuna fisheries. *Fish and Fisheries*, 23(3), 545–563.

Pokines, J. T. and Higgs, N. D. (2022). Marine environmental alterations to bone. In *Manual of Forensic Taphonomy* (pp. 193–250). CRC Press: New York.

Potts, et al. (2016). *State of Sustainability Initiatives Review: Standards and the Blue Economy*. Retrived on July 2, 2022, from www.iisd.org/system/files/publications/ssi-blue-economy-2016.pdf

Raghunathan, C. Raghuraman, R. and Choudhury, S. (2019). Coastal and marine biodiversity of India: challenges for conservation. *Coastal Management: Global Challenges and Innovations*, 201–250. Academic Press: Cambridge, MA. www.sciencedirect.com/science/article/pii/B9780128104736000121

Rahman, M. M. and Hossain, M. M. (2009). Production and export of shrimp of Bangladesh: problems and prospects. *Progressive Agriculture*, 20(1–2), 163–171.

Rahman, M. J. (2013). Coastal and Marine Fisheries Management in. *Coastal and marine fisheries management in SAARC countries*, 127.

Rahman, M. (2014). *Trade benefits for least developed countries: the Bangladesh case market access initiatives, limitations and policy recommendations.* Unitited Nations, Department of Economic and Social Affairs UN Secretariat, 405 East 42nd Street New York, N.Y. 10017, USA, www.un.org/en/development/desa/papers/

Rahman, F. M. M. (2014). Country's first integrated maritime border map introduced. *Dhaka Tribune.* http://archive.dhakatribune.com/bangladesh/2014/jul/17/country%E2%80%99s-first-integrated-maritime-border-map-introduced

Rashid, M. M. O. and Sarkar, M. S. K. (2020). Post-harvest losses of culture, capture and marine fisheries of Bangladesh. *International Journal of Business and Economy*, 2(2), 11–20.

Reddy, R. V. S. K. Omprasad, J. and Janakiram, T. (2022). Technological innovations in commercial high tech horticulture, vertical farming and landscaping. *International Journal of Innovative Horticulture*, 11(1), 78–91.

Reilly, A. and Käferstein, F. J. A. R. (1997). Food safety hazards and the application of the principles of the hazard analysis and critical control point (HACCP) system for their control in aquaculture production. *Aquaculture Research*, 28(10), 735–752.

Riskas, K. (2020). *Farmed Shellfish Is Not Immune to Climate Change.* Coastal Science and Societies Hakai Magazine. https://oceanfdn.org/ocean-and-climate-change/

Ritchie, H. and Roser, M. (2021). *Fish-and-Overfishing @ ourworldindata.org.* Our World in Data. https://ourworldindata.org/seafood-production?utm_source=newsletterandutm_medium=emailandutm_campaign=foodandutm_content=2020–12–03

Robayet Ferdous, S. and Syed, D. (2015). Prospect and challenge of Bangladesh frozen food: a way to overcome. *Online International Interdisciplinary Research Journal*, V, 7–23, ISSN 2249–9598.

Rosen, M. E. (2014). *Philippine Claims in the South China Sea: A Legal Analysis.* CNA Corporation. Arlington County: Virginia.

Roheim, C. A. Bush, S. R. Asche, F. Sanchirico, J. N. and Uchida, H. (2018). Evolution and future of the sustainable seafood market. *Nature Sustainability*, 1(8), 392–398.

Rothuis, A. J. van Duijn, A. P. Roem, A. J. Ouwehand, A. van der Pijl, W. and Rurangwa, E. (2013). *Aquaculture business opportunities in Egypt* (No. 2013-039). Wageningen UR.

Saha, C. K. (2015). Dynamics of disaster-induced risk in southwestern coastal Bangladesh: an analysis on tropical Cyclone Aila 2009. *Natural Hazards*, 75(1), 727–754.

Sattar, Q. Maqbool, M. E. Ehsan, R. and Akhtar, S. (2021). Review on climate change and its effect on wildlife and ecosystem. *Open Journal of Environmental Biology*, 6(1), 008–014.

Seafood TIP. (2021). *What Is the Demand for Fish and Seafood on the European Market?* CBI. https://cbi.eu/market-information/fish-seafood/what-demand

Sevilla, N. P. M. Adeath, I. A. Le Bail, M. and Ruiz, A. C. (2019). Coastal development: Construction of a public policy for the shores and seas of Mexico. In *Coastal Management* (pp. 21–38). Academic Press: Cambridge, MA.

Shameem, M. I. M. Momtaz, S. and Kiem, A. S. (2015). Local perceptions of and adaptation to climate variability and change: the case of shrimp farming communities in the coastal region of Bangladesh. *Climatic Change*, 133(2), 253–266.

Shamsuzzaman, M. M. Islam, M. M. Tania, N. J. Al-Mamun, M. A. Barman, P. P. and Xu, X. (2017). Fisheries resources of Bangladesh: Present status and future direction. *Aquaculture and Fisheries*, 2(4), 145–156.

Shamsuzzaman, M. M. Mozumder, M. M. H. Mitu, S. J. Ahamad, A. F. and Bhyuian, M. S. (2020). The economic contribution of fish and fish trade in Bangladesh. *Aquaculture and Fisheries,* 5(4), 174–181.

Singh, E. S. and A. (2021). Dredging. *Aquafind.* http://aquafind.com/articles/Dredging.php

Shikder, M. S. (2020). Adoption of Catfish Farming at Jashore Sarad Upazila Under Jassore District (Doctoral dissertation, Department of Agricultural Extension and Information System, Sher-E-Bangla Agricultural University, Dhaka, Bangladesh. Retrieved on 9 August 2022 from http://archive.saulibrary.edu.bd:8080/xmlui/bitstream/handle/123456789/3698/final%20correction%20SHOHEL.pdf?sequence=1

Sornaraj, R. (2014). Utilization of seafood processing wastes for cultivation of the edible mushroom Pleurotus flabellatus. *African Journal of Biotechnology*, 13(17): 1779–1785.

Sridhar, M. K. C. Ana, G. R. E. E. and Laniyan, T. A. (2019). Impact of sand mining and sea reclamation on the environment and socioeconomic activities of Ikate and Ilubirin coastal low income communities in Lagos Metropolis, Southwestern Nigeria. *Journal of Geoscience and Environment Protection*, 7(02), 190.

Stachowitsch, M. (2019). Plastic. In *The beachcomber's guide to marine debris* (pp. 87–158). Springer, Cham.

Suresh, M. (2016). *Export Potential of Marine Products and its Impact on Eradication of Poverty in Andaman Islands A Study with Special Reference to Tuna Fish Variety* (Doctoral dissertation).

Suresh, A. Sajeev, M.V. Rejula, K. and Mohanty, A.K. (2018) E-manual on Extension Management Techniques for Upscaling technology Dissemination in Fisheries. Central Institute of Fisheries Technology, Cochin, India. P. 298. Retrieved in August 2022 from http://krishi.icar.gov.in/jspui/handle/123456789/20337

Swimmer, Y. Zollett, E. A. and Gutierrez, A. (2020). Bycatch mitigation of protected and threatened species in tuna purse seine and longline fisheries. *Endangered Species Research*, 43, 517–542.

Szuwalski, C. Jin, X. Shan, X. and Clavelle, T. (2020). Marine seafood production via intense exploitation and cultivation in China: costs, benefits, and risks. *PloS One*, 15(1), e0227106.

Tabata, S. (1989). Trends and long-term variability of ocean properties at Ocean Station P in the northeast Pacific Ocean. *Aspects of Climate Variability in the Pacific and the Western Americas*, 55, 113–132.

Taylor, N. Clarke, L. J. Alliji, K. Barrett, C. McIntyre, R. Smith, R. K. and Sutherland, W. J. (2021). *Marine Fish Conservation: Global Evidence for the Effects of Selected Interventions. Synopses of Conservation Evidence Series.* University of Cambridge.

Tomascik, T. 1997 Management plan for coral resources of Narilel Jinjira (St. Martin's Island). *Draft Final Report Submitted to NCSIP-1,* Ministry of Environment and Forest, Government of Bangladesh. pp. 125.

UNCTAD, 2014. *The Oceans Economy: Opportunities and Challenges for Small Island Developing States.* UNCTAD - Palais des Nations, 8-14, Av. de la Paix, 1211 Geneva 10, Switzerland.

van der Pijl, W. and van Duijn, A. P. (2012). *The Bangladeshi seafood sector : a value chain analysis.* LEI Wageningen UR. https://edepot.wur.nl/238011

Vega, A. Miller, A. C. and O'Donoghue, C. (2014). Economic impacts of seafood production growth targets in Ireland. *Marine Policy,* 47, 39–45.

Venugopal, V. and Gopakumar, K. (2017). Shellfish: Nutritive value, health benefits, and consumer safety. *Comprehensive Reviews in Food Science and Food Safety*, 16(6), 1219–1242.

Venugopal, V. and Sasidharan, A. (2021). Seafood industry effluents: environmental hazards, treatment and resource recovery. *Journal of Environmental Chemical Engineering*, 9(2), 104758.

Villicana, C. Amarillas, L. Soto-Castro, L. Gomez-Gil, B. Lizárraga-Partida, M. L. and Leon-Felix, J. (2019). Occurrence and abundance of pathogenic Vibrio species in raw oysters at retail seafood markets in northwestern Mexico. *Journal of Food Protection*, 82(12), 2094–2099.

Victorero, L. Watling, L. Deng Palomares, M. L. and Nouvian, C. (2018). Out of sight, but within reach: A global history of bottom-trawled deep-sea fisheries from> 400 m depth. *Frontiers in Marine Science*, 5, 98.

Visbeck, M. et al. (2014). Securing blue wealth: the need for a special sustainable development goal for the ocean and coasts. *Marine Policy,* 48, 184–191.

Wells, M. L. Trainer, V. L. Smayda, T. J. Karlson, B. S. Trick, C. G. Kudela, R. M. ... and Cochlan, W. P. (2015). Harmful algal blooms and climate change: Learning from the past and present to forecast the future. *Harmful Algae*, 49, 68–93.

West WQB (1973). *Fishery Resources of the Upper Bay of Bengal.* Indian Ocean Programme, Indian Ocean Fisheries Commission, Rome, FAO, IOFC/DEV/73/28. pp. 28.

Wikipedia. (2021b). *Overfishing.* https://en.wikipedia.org/wiki/Overfishing

Willer, D. F. Nicholls, R. J. and Aldridge, D. C. (2021). Opportunities and challenges for upscaled global bivalve seafood production. *Nature Food*, 2(12), 935–943.

Wilson, C. and Coles, V. J. (2005). Global climatological relationships between satellite biological and physical observations and upper ocean properties. *Journal of Geophysical Research: Oceans*, 110(C10001): doi:10.1029/2004JC002724

Yao, Z. Meng, Y. Le, H. G. Kim, J. A. and Kim, J. H. (2019). Isolation of Bacillus subtilis SJ4 from Saeu (Shrimp) jeotgal, a Korean fermented seafood, and its fibrinolytic activity. *Microbiology and Biotechnology Letters*, 47(4), 522–529.

Yapanto, L. M. and Musa, F. T. (2018). Distribution of Seafood Production in Bajo Sector of Gorontalo Province Indonesia. *International Journal of Innovative Science and Research Technology*, 3(8), 521–523.

Ziegler, F. Winther, U. Hognes, E. S. Emanuelsson, A. Sund, V. and Ellingsen, H. (2013). The carbon footprint of Norwegian seafood products on the global seafood market. *Journal of Industrial Ecology*, 17(1), 103–116.

14 New Challenges for Sustainable Plastic Recycling in Japan

Jeongsoo Yu,[1]* Shiori Osanai,[2] Kosuke Toshiki,[3] Xiaoyue Liu,[4] Tadao Tanabe,[5] Gaku Manago,[6] Shuoyao Wang,[7] Kevin Roy B. Serrona,[8] Kazuaki Okubo,[9] and Ryo Ikeda[10]

[1] Tohoku University, Japan, jeongsoo.yu.d7@tohoku.ac.jp
[2] Tohoku University, Japan, shiori.osanai.p6@dc.tohoku.ac.jp
[3] University of Miyazaki, Japan, toshiki.k@cc.miyazaki-u.ac.jp
[4] Tohoku University, Japan, liu.xiaoyue.p4@dc.tohoku.ac.jp
[5] Shibaura Institute of Technology, Japan, tanabet@shibaura-it.ac.jp
[6] Tohoku University, Japan, gaku.manago.e4@tohoku.ac.jp
[7] Shanghai SUS Environment Co. Ltd, China, 18616676619@163.com
[8] Prince George's County Government, USA. krbserrona@co.pg.md.us
[9] Tohoku University, Japan, kazuaki.okubo.d5@tohoku.ac.jp
[10] Tohoku University, Japan, ryo.ikeda.d2@tohoku.ac.jp
* Corresponding author: Jeongsoo Yu. Email: jeongsoo.yu.d7@tohoku.ac.jp

CONTENTS

DOI: 10.1201/9781003184287-14

14.1   INTRODUCTION

In recent years, the ocean plastic waste problem has become a worldwide concern, especially after more and more people saw the picture that shows the straws stabbed into sea turtles' noses or the dead whales with a massive amount of plastic packaging in their stomachs.

The USA and Japan consume massive amounts of disposable plastic products every year, and most of the waste plastic products are exported to China instead of being recycled domestically. From January 2018, affected by China's ban on waste plastic importation, the USA and Japan had to learn how to deal with waste plastic products domestically.

During the G7 summit meeting held in Canada in 2018, participants adopted the 'Blueprint for Healthy Oceans, Seas and Resilient Communities that outlines commitments related to resilient coasts and coastal communities' to solve ocean plastic waste problems. Moreover, England, France, Germany, Italy, Canada, and the EU also signed on the 'Ocean Plastic Charter' to further strengthen plastic management. Meanwhile, it is noticeable that neither the USA nor Japan signed the charter.

On the other hand, most ocean plastic waste was generated from China and Southeast Asian counties such as Indonesia, the Philippines, and Vietnam. Only with the advanced countries cooperating with developing countries can the ocean plastic waste problem be solved on a global scale. Moreover, since the microplastic waste problem has not been investigated thoroughly, the situation has not been revealed yet. Although Japan has tried to solve these problems by charging consumers for plastic bags and disposable containers, such issues cannot be solved completely by mere laws or regulations.

This chapter will introduce residents' waste discharging characteristics, their cooperation behavior towards waste plastic recycling, issues related to ocean plastic waste, advanced waste plastic selection technology, and the importance of environmental education. In fact, these are essential in achieving effective plastic recycling, and the lack of any single element of these would seriously damage its effectiveness.

14.1.1   Marine Waste in Japan

The Ministry of Environment has been researching the volume and type of marine waste on beaches since FY 2010. According to its report, they had researched it at Awaji in the Fiscal Year (FY) 2014 while Wakkanai, Nemuro, Hakodate, Yuza, and Goto were done in FY 2016. The focus on the volume and proportion of artifact material was higher than natural material except for Wakkanai and Hakodate. The types of artifact materials were different form place to place, and plastic materials such as PET, Polystyrene foam, and fishing gear, were higher in many regions. The weight and volume of waste artifact materials was higher at Nemuro, Shiriya, Yuza, Hachijojima, Matsue, and Goto while natural materials were higher in Wakanai, Hakodate, Awaji and Nichinan. As a result, artificial materials were not different in the region. However, Polystyrene, vinyl bags, containers and packaging of foods and plastic fishing gear were higher in many regions. Then the quantity of artifact waste materials accounted for more than 80% in all regions. Moreover, plastics accounted for the largest number in the whole region. Used PET bottles were the highest at 6 out of the 10 regions, and they were found to have foreign labels (Japanese Ministry of the Environment, 2019a).

As for marine plastics, research conducted by the Ministry of the Environment showed an abundance of distribution around the Japan sea. Especially, it tended to show higher density in the

Pacific Ocean and Japan ocean side of Tohoku, and around the pacific coast of Shikoku and Kyushu (Japanese Ministry of the Environment, 2019b).

14.1.2 Plastic Containers and Packaging Recycling Law in Japan

Plastic products have contributed to solving many industrial problems and promoting social development due to their high functionality. However, the mass consumption of plastic, especially disposable plastic products, causes serious environmental issues. Approximately 9,400 tonnes of plastic products will be discharged as waste in Japan annually, of which 4,260 tonnes were plastic containers and packaging (Consumer Affairs Agency of Japan, 2020).

It is vital to recycle waste plastic containers and packaging to avoid pollution problems while ensuring a stable resource supply in the future. As a resource-dependent country, Japan has various recycling policies aimed at reduceing, as well as using the waste effectively. Before 1995, it was entirely local municipalities' responsibility to collect waste plastic containers and packaging in Japan. In 1995, to further prompt the separate collection and recycling of waste plastic containers and packaging (such as PET bottles, plastic, and paper packaging), the Japanese government published the 'Containers and Packaging Recycling Law', which is also the first law mentioned which promotes the Extended Producer Responsibility (EPR) principle in Japan, and requires plastic products' makers, consumers and also `retailers to take the responsibility in recycling.

14.2 COOPERATIVE BEHAVIOUR FOR COLLECTING CONTAINERS AND PACKAGING WASTE

14.2.1 Stakeholders' Cooperative Behaviour in Recycling Plastic Containers and Packaging

According to Containers and Packaging Recycling Law, consumers should discharge each type of waste plastic container and packaging separately according to their recycling identification marks (as shown in Figure 14.1) and local municipalities' rules. PET bottles are generally collected and recycled separately from other plastic containers and packaging. Theoretically, consumers should rinse PET bottles, remove the label and caps before discharging them. The municipalities will then separate each type of bottle and remove foreign objects before regularly giving them to authorized recyclers. The authorized recyclers will then recycle waste PET bottles.

Some retailers, such as supermarkets, also collect waste PET bottles, food trays, and egg packs from their consumers for free by setting up collecting boxes based on the EPR principle (Figure 14.2). The supermarkets will transport collected plastic wastes to authorized recycling corporations and pay for the recycling fee. Some recycling companies are also doing partnership with supermarkets by setting up waste PET collection machines at supermarkets. The PET collection machine will provide consumers one (1) point, which can be used in the supermarket, for every 5 waste bottles they

FIGURE 14.1 Recycling identification marks on plastic containers and packaging.

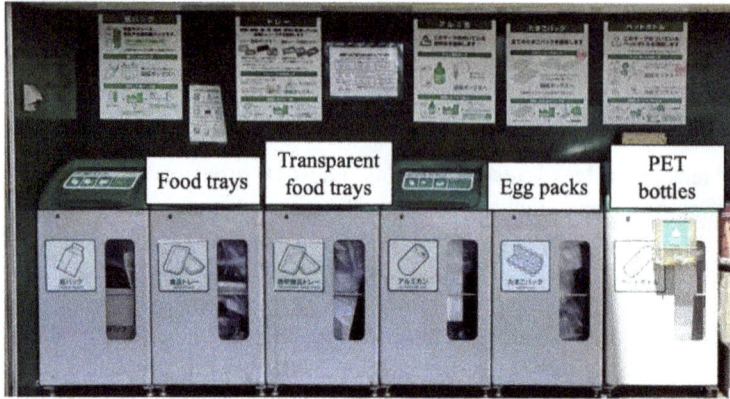

FIGURE 14.2 Waste collection boxes in supermarkets.

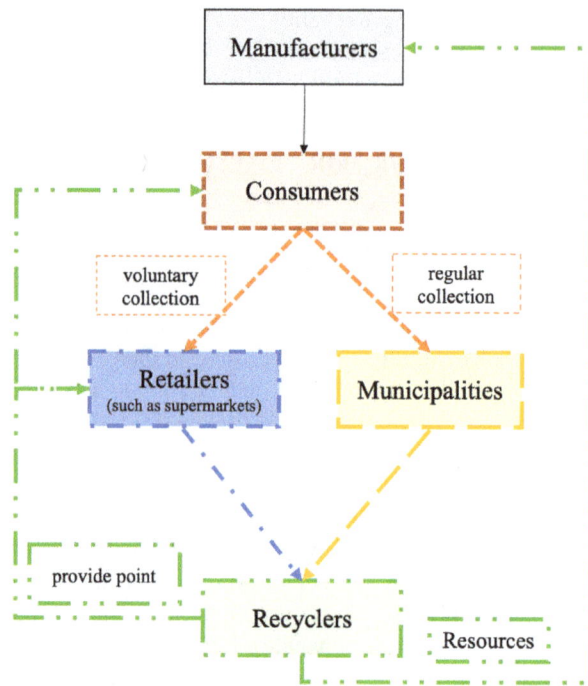

FIGURE 14.3 Roles of each stakeholder in the Containers and Packaging Recycling Law.

throw. The roles of each stakeholder in the Containers and Packaging Recycling Law are shown in Figure 14.3.

Currently, 76.7% of municipalities in Japan collect waste plastic containers and packaging from 85.1% of the total population in Japan (Minister of the Environment of Japan, 2021).[1] According to a survey performed by the National Supermarket Association of Japan, 86.2% of supermarkets collect food trays, and 60.6% of supermarkets collect PET bottles from their consumers (National Supermarket Association of Japan, 2020). With the cooperation of each stakeholder, the generation of waste plastic packaging was reduced by 17.6%, waste PET bottles reduced by 24.8% from 2014 to 2019. Furthermore, the waste PET bottles' collection rate in Japan reached 85.8%, which is also the highest globally (The Japan Containers and Packaging Recycling Association, 2021a).

FIGURE 14.4 PET bottles and food trays collected in supermarkets.

14.2.2 CHALLENGES IN COLLECTING CONTAINERS AND PACKAGING WASTE

The consumption of PET bottles keeps increasing fast, especially in the beverage production industry, by over 10% each year (The Japan Containers and Packaging Recycling Association, 2021b). Influenced by COVID-19, people are discharging more used PET bottles and plastic packaging as they change their lifestyle and spend more time at home. Consequently, municipalities are taking a heavier economic burden to collect waste containers and packaging currently.

Also, the although waste PET bottles' collection rate was high in Japan, most of the PET bottles collected by municipalities were dirty and can only be thermally recycled or be exported abroad. However, affected by increasing stringent waste import policies in China and other east north Asian countries, Japan needs to consider recycling waste plastic containers and packaging with high efficiency domestically in the future.

On the other hand, most of the waste plastic products collected by supermarkets can be recycled as resources since they are comparatively clean and well separated (Figure 14.4).

In general, it is easier for a supermarket resource recycling station to collect recyclable materials than the public collection system (Liu et al. 2021)). Therefore, to improve the recycling of waste plastic products in Japan, stakeholders could consider a new cooperation system, under which retailers (such as supermarkets) instead of the municipalities could take more responsibility.

14.3 DISCUSSION ON OCEAN PLASTIC POLLUTION

14.3.1 BACKGROUND OF PROBLEMS

Marine litter is a growing menace that is affecting economies on global and regional scales. Globally, about 4.8 to 12.7 million tonnes of plastic find their way into oceans yearly, where Asia is the major contributor at 80% (World Bank, 2021). Current production and consumption patterns are leading causes of material pollution in marine environments. As waste infrastructures continue to be overwhelmed by different discarded materials, natural environments are becoming inundated with plastic waste, causing enormous harm to human and animal health. Countries in East Asia and Southeast Asia have been at the receiving end of this occurrence, and it has not spared developed and developing economies. Japan, a highly industrialized country with advanced waste management technologies, has been battling marine litter, not from domestic origins but waste materials swept onto its coastal shores from other countries.

Similarly, the Philippines is facing the same problem by virtue of its geographic location and the challenges brought about by indiscriminate disposal of domestic waste coupled with limited waste management infrastructures. Based on recent figures, Japan generates about 7.99 tonnes of plastic waste and more than 143,000 tonnes of plastic litter annually, and it has 29,751 km of coastline. On the other hand, the Philippines has an annual plastic generation rate of 2,566,766 tonnes and its coastline is 36,289 km (World Population Review, 2021). The Philippines is the third-largest contributor of plastic leakage at about 0.75 million tonnes every year.

Japan and the Philippines are countries which contrast in terms of geographical and economic attributes but are equally faced with worsening marine plastic pollution. Tackling the problem, therefore, requires international/regional collaboration and the introduction of circular economy approaches where material production is closely linked to design, prolonged reusability, and recyclability. Addressing marine litter, specifically plastic waste, is a regional problem requiring macro planning but extensive research at the country level in terms of sources, types, occurrence, and volume to identify sustainable solutions and pathways.

Plastics do not decompose into inorganic materials in nature. They are destroyed by wind, rain, and ultraviolet light. However, they continue to exist as plastics, albeit in smaller pieces. If marine organisms accidentally ingest plastics, they may suffer physical damage such as injury and suffocation. In addition, plastics may contain additives to give them various properties such as flame-retardancy and flexibility. Plastics can also absorb Persistent Organic Pollutants (POPs) in the ocean. If marine organisms accidentally ingest plastics that contain these chemicals, they may suffer chemical damage. Smaller marine organisms such as small fish and zooplankton mistakenly eat the smaller pieces of plastic as food, and then larger predators eat them in the food chain, leading to bioaccumulation of toxic substances in larger marine organisms. Of these plastics, those smaller than 5 mm in size are called microplastics (MPs). Since they are easily contacted by marine organisms, they have been actively studied in recent years. Microscopic particles larger than 5 mm are called mesoplastics, which will eventually become MPs, and their dynamics are also of interest. Of course, plastics of larger sizes are also a problem. When they are scattered along the coast, they spoil the scenery and are a great loss to the tourism industry.

Recent studies on marine plastic pollution point to the multifaceted nature of the problem. One paper looked at the presence of MPs in marine sediments and in commercial fish in select coastal areas in the Southern Philippines. Research confirmed that MPs are ingested by fish which are then consumed by local residents (Bucol et al. 2019). Another study delved into the degradation of wetlands and their biodiversity in the Philippines caused by indiscriminate disposal of solid waste. The recommendation was to develop interventions that use indigenous materials built from local knowledge and practices (Lecciones et al. 2021).

In Japan, some studies have been conducted on the dynamics of MPs in nearshore water (Isobe et al. 2014), with results suggesting that the closer to shore, the more mesoplastics are present, and the finer they are, the more they become MPs, and the more they spread offshore. Another study has been conducted on the number of MPs in feeding fish in the seas around Japan (Ushijima et al. 2018). In this study, MPs larger than 100 μm were recorded in seven fish species collected from five bays and Lake Biwa in Japan. A total of 197 fish were detected with these in the digestive tracts of 37.6 % of all fish species.

14.3.2   Policy and Regulatory Frameworks

Marine plastic pollution places a huge burden on national and local economies in terms of the degradation of aquatic life, loss of biodiversity, and subsequent impacts on human health. It also drains the budgets of local governments to remove waste plastic from marine environments and manage them ecologically. In the case of Japan, a combination of strategies to address it, such as waste separation, 3Rs, incineration, and final disposal are in place. Its strategy for solving marine litter is elucidated in its National Action Plan for Marine Plastic Litter, where it focuses on preventing illegal dumping and unintentional leakage of waste into the oceans, land-based collection, development of alternative materials, local and international collaboration, the sharing of best practices, and the promotion of scientific knowledge.

In the Philippines, some local governments either have limited or lacking in waste infrastructures such as recycling and residual disposal facilities. Republic Act 9003 (R.A. 9003) or the Philippine Ecological Solid Waste Management Act provides guidelines on solid waste avoidance and minimization through reduce, reuse, recycling, and composting. It requires local government units to provide the necessary infrastructure to collect, transport and process recycling and trash streams in an ecological manner. Specific to marine litter, it developed the National Plan of Action for the prevention, Reduction, and Management of Marine Litter where the main objective is to improve current efforts to address marine litter issues and address leakage into water bodies.

14.3.3 Survey of Microplastics Found on the Coast of Miyazaki City, Japan

Miyazaki prefecture is in the southern part of Japan, a prosperous agricultural region. It is also famous for its beaches and surfing spots, which attract many tourists. The climate is mild throughout the year, with more hours of sunshine and more days of clear skies than any other area in Japan. However, typhoons sometimes hit the area in September and October. Miyazaki City, the capital of Miyazaki prefecture, is a core city with a population of approximately 400,000 and faces the Pacific Ocean.

For years, MPs have been found to be drifting in the oceans all over the world. Some of them are carried by waves and drift to the land, and MPs are carried by water currents to the coast of Miyazaki City and can be collected. Those washed up on the sandy beach are particularly easy to collect. However, they are not uniformly scattered on the beach. On closer look, MPs amass from the high tide line to the supratidal zone where the pieces of wood and other debris are gathered. If the number of MPs drifting on multiple beaches is compared, it is better to collect at the points where these MPs are gathered (Ikegai et al. 2017).

In this survey, the number of MPs (1 - 5 mm in size) on seven beaches in Miyazaki City, as shown in Figure 14.5, was investigated. The collection process was carried out from mid-October to early January. To collect MPs, we first defined a 30 cm square plot and sieved all the sand sediment up to a depth of 7.5 cm through a 1 mm mesh sieve. Recently, coastal MPs have often been collected at a

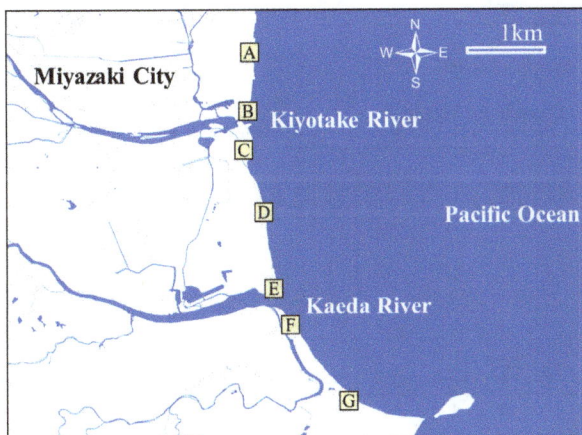

FIGURE 14.5 Coasts in Miyazaki City where MPs were collected.

depth of 5 cm. The size of the plot is usually 30 - 60 square cm. The larger the plot, the more MPs can be collected, but higher labor is required to count them.

Sand and other particles smaller than 1 mm were removed. MPs smaller than 1 mm were similarly removed. The remaining material in the 1 mm mesh sieve is then passed through a 6 mm mesh sieve to remove large pieces of wood and other materials. Samples of 1 to 6 mm in size were collected from five sites on each beach and brought back to the laboratory for analysis.

The sample contained many pieces of wood, pebbles, and shells. A visual examination showed the MPs, and they were collected with tweezers. Then, to find other MPs, the remaining samples were placed in water. In this way, pebbles and shells sank into the water, while MPs made of resin, which is lighter than water, floated on the water, making them easier to find (if sodium iodide solution, which has higher specific gravity, is used instead of water, the MPs, which are heavier than water, will also float, making easier to find. If MPs smaller than 1 mm are to be collected, this solution should be used). Then, the collected MPs were sieved again into 5 mm, 4 mm, 3 mm, 2 mm, and 1 mm mesh sieves and divided into different sizes.

MPs can be divided into two types: primary plastics (primary MPs) and secondary plastics. Primary MPs include pellets, which are raw materials for plastics, and microbeads, which are used in cosmetics. Nylon fiber is also included in primary MPs. However, MPs smaller than 1 mm in size were not included in this survey, so all primary MPs collected were pellets. Secondary plastics are plastic products that have been scraped or broken into fragments. In this survey, Styrofoam and plastic capsules for fertilizer were especially abundant. Plastic capsules are resin-coated fertilizer ingredients. Fertilizer capsules are coated with a resin that slowly dissolves the fertilizer ingredients. Therefore, the MPs collected in this survey were classified into four types: primary MPs, Styrofoam, plastic capsule for fertilizer, and other secondary MPs.

Figure 14.6 shows the number of MPs for each coast, and Figure 14.7 shows the median number of MPs for each coast and its breakdown by type. The abundance of MPs on the coast varies even

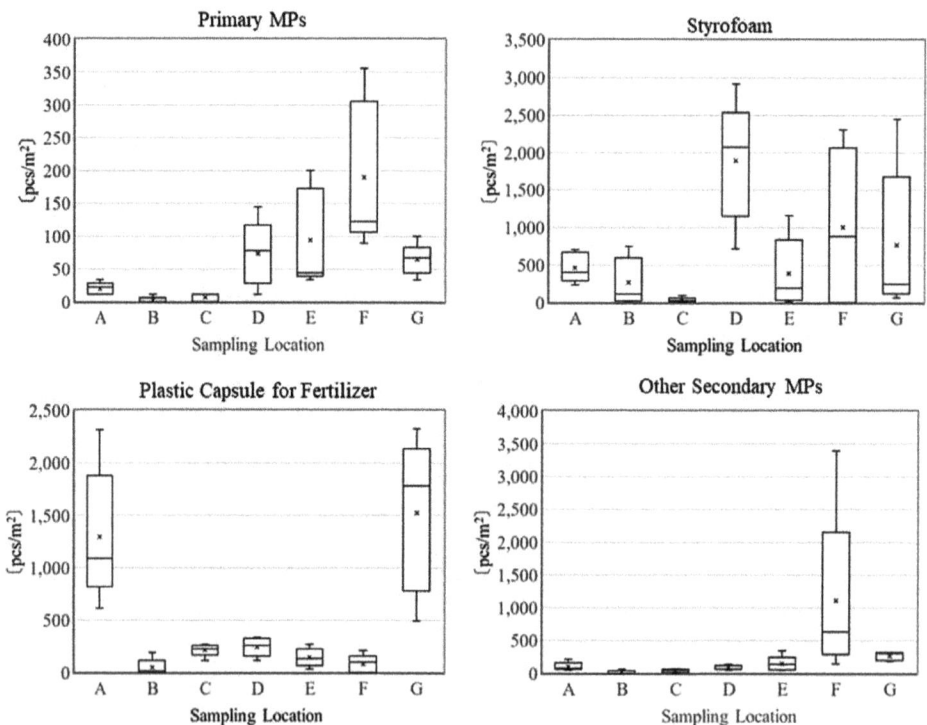

FIGURE 14.6 Distribution of various microplastics on the coast of Miyazaki City, Japan.

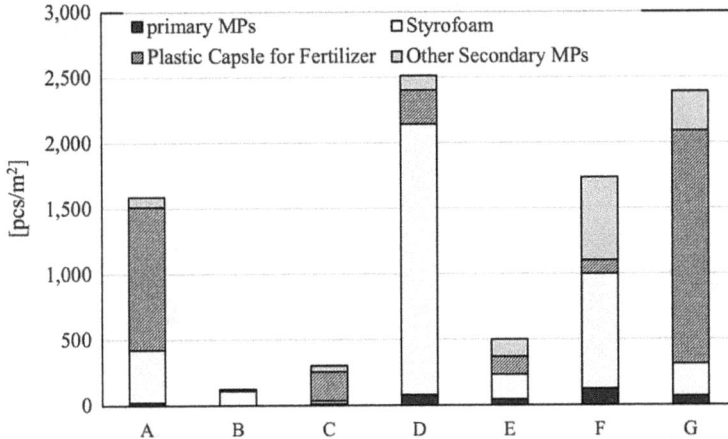

FIGURE 14.7   Median number of MPs for each coast and its breakdown by type.

between the high tidal line and the upper tidal zone. Some coasts have more MPs than others depending on the types of MPs. Characteristically, plastic capsules for fertilizer are abundant in A and G. The other types of MPs are abundant in D, E, and F. Overall, MPs are more abundant in D and G, because D has more Styrofoam and G has more plastic capsules for fertilizer. Other Secondary MPs are especially abundant in F.

Despite this, the number and composition of MPs on each coast are very different. This suggests that they are influenced by inland areas, especially river basins.

Figure 14.8 shows histograms of the MPs collected in this survey according to the size of each type. The size of primary MPs and plastic capsules for fertilizer is with some exceptions, mostly in the range of 2 - 4 mm, because they are all of this size when they are used. On the other hand, the smaller size of Styrofoam and other secondary MPs, the more numerous they are. This is probably because the number increases when one piece of plastic is broken. This implies that the number of Styrofoam and other secondary MPs with a size of less than 1 mm would be very large.

Finally, what kinds of plastics are discharged into the ocean through rivers in Miyazaki City is discussed. A regular check of the riverbeds of Miyazaki city shows that a variety of plastics are left behind. This is the plastic left behind when rivers rise and recede due to heavy rains. Some of them are shown in Figure 14.9 below. In the riverbed, it is difficult to find plastic resin materials and plastic capsules, which are small in nature. However, it can be confirmed that many plastic products are being washed down the river.

In Japan, it is rare to see littering on the street. However, some traces of plastic shopping bags, lunch containers, cigarette filters, and the like, may be found. Another possible cause is the separation of plastics. In Japan, many cities (including Miyazaki City), towns, and villages separate plastics for containers and packaging as a resource. Since these plastics are very light, they are often blown away by the wind when they are left outside on a windy day. Outdoor trash boxes that contain empty PET bottles are sometimes littered with their contents during typhoons. In agriculture, agricultural materials are supposed to be kept indoors, in principle, and it is rare to see plastic bags of fertilizer left outside. With careful investigation, there are many things that can be found as sources of secondary MPs, and it appears to be a challenge to manage them. On the other hand, plastic capsules for fertilizer are spread on the farmland, particularly on paddy fields and whose residual materials drain into rivers and oceans through irrigation channels. Now that marine plastic pollution has become an issue, alternative methods such as the use of biodegradable and compostable plastics are gaining popularity.

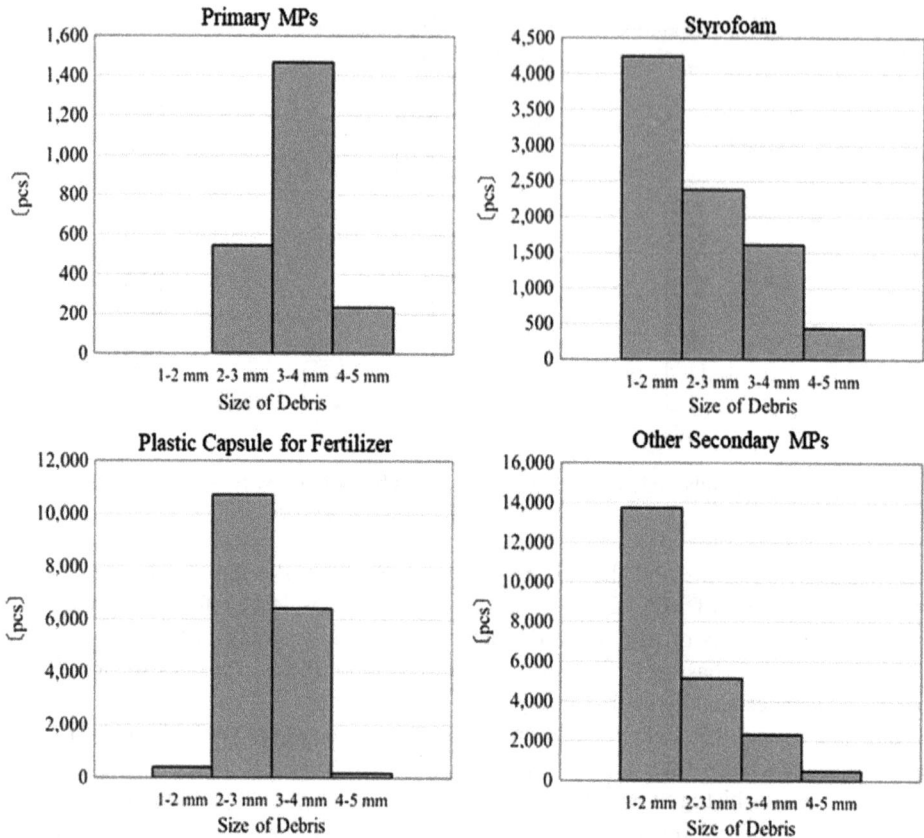

FIGURE 14.8   Size of various microplastics on the coast of Miyazaki city, Japan.

14.3.4   Sources of Marine Plastics

Plastics found in marine environments originate from both land and sea-based sources. Some economic activities such as fishing, agriculture, maritime transport, industrial processes, tourism, and household consumption are triggers for the increase of marine litter (Newman S. et al. 2015). Managing municipal waste in the Philippines is a challenge for local government units (LGUs) because of its complexity and the huge resources (logistics and finance) needed to address it. For example, some LGUs lack or do not have enough resources to have a consistent waste collection system in communities and the ability to establish engineered disposal facilities. In addition, some recycling centers called Material Recovery Facilities (MRFs) are lacking in equipment and systems to process recyclable materials. Thus, leakage into streets and water channels is inevitable due to these challenges. The National Solid Waste Management Commission (NSWMC), the agency tasked to oversee the implementation of R.A. 9003 has issued some guidelines on having a cluster approach to managing a landfill by having multiple LGUs co-locating such facilities to save on costs and resources.

14.3.5   Policy Recommendations

Plastics that have been miniaturized in the ocean can no longer be removed. This is because it is practically impossible. Therefore, we have no choice but to focus on stopping them at the source so that plastics do not increase any further in the ocean. Tackling marine plastic pollution requires a

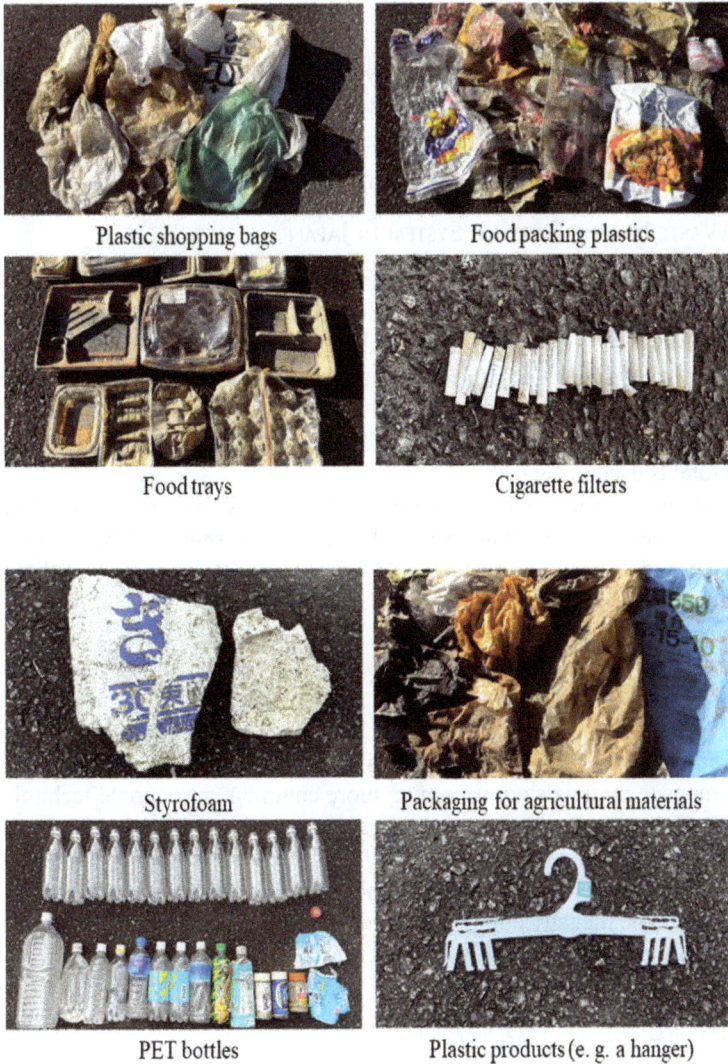

FIGURE 14.9 Plastic litter washed up on a riverbed in Miyazaki city.

mix of policy and regulatory interventions. International and regional initiatives on addressing the proliferation of plastic waste in water bodies exist. Japan, for example, is supporting the ASEAN+ 3 Marine Plastic Debris Cooperative Action Initiative, which aims to assist member countries in constructing waste treatment facilities as well as to build local capacities for marine litter management and monitoring. The Philippines, on the other hand, is involved in the Glolitter Partnership Project, which is being backed by the United Nations to tackle marine litter, clean up the oceans and encourage industries to limit the use of plastics. As the global momentum continues to build on suppressing plastic waste, some policies could be introduced/reiterated:

1. Incentive mechanisms for industries and local government units promoting Extended Producer Responsibility (EPR).
2. Strengthening of Research and Development (RandD) through ties with academic institutions to identify approaches, alternatives, and technologies in managing marine litter with a community participation component.

3. Development of common tools/platforms to qualify and quantify marine plastic waste to aid in the scientific understanding/analysis of waste materials.

4. Establishment of an entity that will unify certification of biodegradable and compostable plastic materials for adoption by manufacturing industries and recognition by governments.

14.4 NEW TECHNOLOGY FOR THE SORTING OF PLASTIC

14.4.1 THE WASTE PLASTIC SORTING SYSTEM IN JAPAN

There are many plastic wastes found in both land marine environments in Japan. Waste plastic must be managed and recycled appropriately locally. In Japan, the amount of used plastic products discharged annually is 8.5 million tonnes (including 680,000 tons of loss during production), of which 1.86 million tonnes (22% of the amount discharged) are recycled materials. The total amount of effective utilization is 7.26 million tons, which means that 85% of the total is recycled, 270,000 tons is used for chemical recycling and 5.13 million tonnes for thermal recycling, indicating that 64% of the total is used as fuel.

Material recycling rate remains low and one of the reasons for this is the economic and technical problems associated with sorting. Waste plastic must be separated by material (PP, PS, PE, and so on) for appropriate material recycling. Material recycling requires that the qualities of the waste plastics must be kept high for a good quality new product, but it is a lower priority because of cost-effectiveness.

As for the technical problems, the near-infrared (NIR) identification system is being used for sorting plastic waste. However, it cannot measure black plastic, and it is difficult to measure composite materials. One of the companies in Japan is disposing of un-recyclable plastics. Its cost is estimated at several hundred million Japanese yen a year.

As plastic composite materials are becoming more common, appropriate technology for sorting must be created. If the appropriate technology is not provided, illegal dumping increases, and the environment deteriorates. After COVID-19, the plastic screen which was used for preventing COVID-19 will be disposed of. If countries do not have strategies and technical solutions, waste plastic will continue to be illegally dumped into the environment and oceans. To solve this problem, the Terahertz (THz) sorting machine is being developed.

14.4.2 THE TECHNOLOGY OF THz SORTING

THz is one frequency of electromagnetic wave that is located between the mm waves and infrared light. Therefore, a THz wave has both the transparency of a mm wave and the direct propagation of infrared light, respectively. Since the dielectric constant of plastics at THz frequencies differs among materials, the material of the plastic can be estimated from its transmittance and reflectance of THz waves, as shown in Figure 14.10. Figure 14.11 shows the relationship between the THz refractive index and the specific gravity of various plastics. The material of the plastic can be identified by THz waves even if it is coloured. THz waves are being studied for non-destructive evaluation applications on non-polar materials such as concrete infrastructure structures (J. Nishizawa, 2004; M. Tonouchi, 2007; M. Naftaly et al. 2019). Further, the energy of the THz wave is so small that it does not ionize biological tissue. Prototype apparatus has been constructed to detect the shape of metal inside a concrete and cable-coated insulation (Y. Oyama et al. 2009; S. Takahashi et al. 2014).

As basic scientific data, there are instruments that measure spectra in the THz frequency region, such as THz time-domain spectroscopy (THz-TDS) and GaP THz Spectrometer (R. E. Miles et al. 2001; K. Suto et al. 2015), respectively. The THz spectrum has information of reflectivity and transmittance, which is effective for practical THz applications with several single-frequency

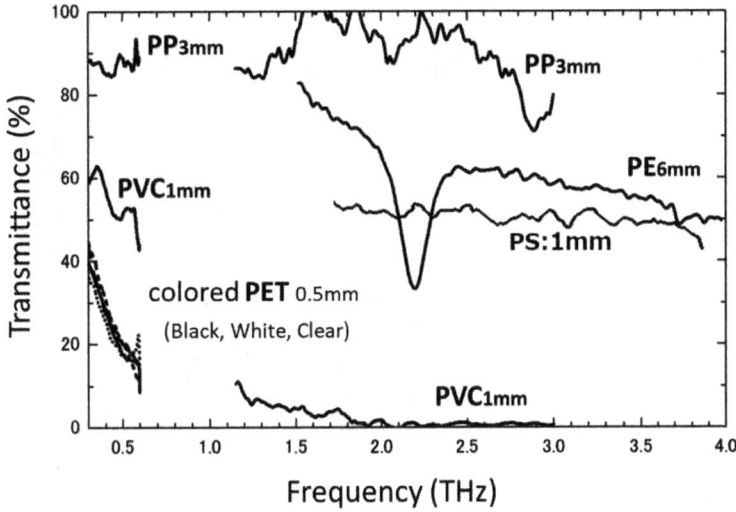

FIGURE 14.10 Transmittance spectra of plastic samples.

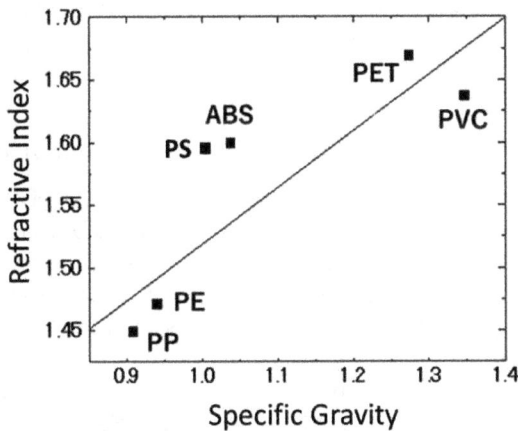

FIGURE 14.11 Relationship between THz refractive index and specific gravity of plastics samples.

THz emitters. The optical configuration of THz-TDS is as follows. A femto-second pulse from a Ti:Sapphire laser is split into pump light and probe light by a beam splitter and the pump light is introduced to the THz generating device and probe light guided to the THz detection device, where each device is a low-temperature grown GaAs photoconductive antenna. The time delay of the THz pulse through the plastic sample can be detected by sweeping the delay. By Fourier transforming the transmitted THz pulse shape and phase delay, it is possible to obtain the refractive index of plastics in the THz region. The refractive index is measured in the frequency range of 0.2 to 0.7 THz. The photoexcitation of lattices vibrating at THz frequencies in GaP crystals can emit THz waves outside the crystals, where the frequency of the THz wave can be tuned from 1 THz to 7 THz. THz emitters, especially at lower THz frequencies, include the Tunnel-injection Transit Time effect diode (TUNNETT) (J. Nishizawa et al. 1958; J. Nishizawa et al. 2008), the Resonant Tunnelling Diode (RTD), and the Impact ionization Avalanche Transit-Time diode (IMPATT). These are compact but fixed frequency (T. Maekawa et al. 2016).

14.5　POTENTIAL OF EDUCATION ON SDGS

14.5.1　SDGs Education Towards Solving Environmental Problems

With the introduction of the SDGs in 2016, people began to feel the importance of maintaining a sustainable society more than ever. Consequently, waste problems became one of the important topics since they may threaten the sustainability of the environment and society. At the Davos Conference in 2016, it was estimated that the amount of plastic in the ocean would exceed the number of fish by 2050. Therefore, to fulfill the 14th goal of the SDGs ('LIFE BELOW WATER'), it is necessary to solve the waste and marine plastic problem.

There are various measures to solve environmental problems, and 'education' is one of the most important solutions. In the 1950s, there was a growing voice in Japan that pollution problems should be solved, and at the same time, pollution education and nature maintenance education was conducted in Japan. Furthermore, as environmental issues are becoming more and more complex every year, it is important to introduce SDGs education into school curriculums.

As a matter of fact, the latest curriculum guidelines published in 2020 have already demonstrated the importance of SDGs education. Clearly, the Japanese government takes the view that SDGs education is important for achieving a sustainable society.

This section introduces the efforts of establishing SDGs education at three elementary schools in Higashi Matsushima City as a case study. These three elementary schools are located near the ocean and were damaged by the tsunami caused by the Great East Japan Earthquake in 2011. The research would test whether SDGs education could lead to the solution of waste problems (such as marine pollution problems) in such areas.

14.5.2　SDGs Education at Schools Stricken by the Great East Japan Earthquake

When referring to sustainability in Japan, one inevitable topic is the reconstruction from the Great East Japan Earthquake. In the aftermath of this earthquake, Tohoku University has provided 'delivery lectures' at elementary schools in the affected areas continuously (Figure 14.12). In this section, the research will introduce the effect of SDGs education under the industry-academia collaboration system performed in Higashi Matsushima City, which was damaged by the great tsunami from the Great East Japan Earthquake.

The topic of visiting classes for Yamoto Nishi Elementary School, Akai Elementary School, and Oshio Elementary School's fourth-grade students was 'plastic recycling' in 2020. The lecture was given in collaboration with Dow Chemical, the world's largest plastic raw material manufacturer, and SEINAN Corporation, the largest recycling company in the Tohoku region. During this kind of experience-based class, students get to learn about plastics while touching real plastic bottles, containers, overseas beverage containers, and recycled eco-bags and pellets. And so, the word 'recycling' for students has changed from a distant concept to a more familiar experience.

Figure 14.13 shows Yamoto Nishi Elementary School 56 students' answers for solutions to reduce plastic waste following the class. This survey was performed at Yamoto Nishi Elementary School on November 17, 2020. The most common idea was to separate the garbage, which accounted for about 35% of the total answer. The next most common idea was 'recycling', which accounted for about 28% of the total answer. In addition, some students believe that plastic products should be carefully repaired and used instead of being discharged (see Figure 14.13).

In addition, based on the survey of the teachers after the class, the teachers felt that the 'delivery lecture' would decrease their burden on preparing lessons and provide their students the opportunity to learn the latest and diverse topics at the same time.

FIGURE 14.12 Delivery lecture in Yamotonishi Elementary School.

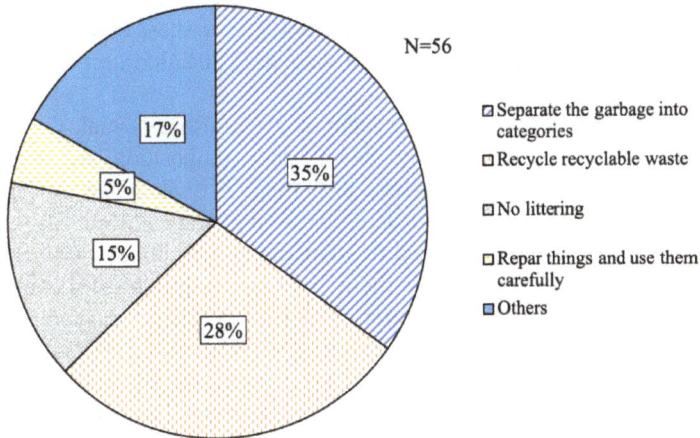

FIGURE 14.13 Breakdown of the answers for the solutions to reduce plastic waste at Yamotonishi Elementary School.

14.5.3 EFFECTS OF THE INDUSTRY-ACADEMIA COLLABORATION SYSTEM

In the previous section, the effects of SDGs education under the industry-academia collaboration system were described from the elementary schools' perspectives. The visiting class also provided an opportunity for raw material manufacturers (stakeholders in the arterial industry) and recyclers (stakeholders in the vein industry), who were rarely involved, to learn about each other's efforts and issues. Furthermore, such kind of lesson provided a chance for the university to return the latest research results to the local community. With the cooperation of the artery industry, venous industry, local educational institutions and the private sector, SDGs activities could make significant social contribution.

14.5.4   SDGs Education's Role in Solving the Waste Problem

Higashi Matsushima is a city that has been selected as one of the 'SDGs future cities' in 2018. It is also a city damaged by the tsunami in the Great East Japan Earthquake.

The people who live in Higashi Matsushima City need to learn how to coexist with the sea.

Therefore, it is important to find clues to solve marine plastic pollution and create a sustainable society through reconstruction.

Through SDGs education, the young generations who will lead society in the future will learn the importance of recycling and the necessity for waste separation. Thus, as a long-term effect, environmental problems could be solved in their generation.

14.6   CONCLUSION

As the waste plastic problem became a social issue, the Japanese government has decided to enact new plastic recycling laws from 2022 to reduce the amount of waste plastic.

However, impacted by COVID-19, residents in Japan tend to discharge more waste plastic products and put less effort into supporting recycling activities.

Consequently, it is hard to collect and recycle waste plastics through municipal systems only. Therefore, it is expected that private waste collection systems (such as the waste collection box in supermarkets) can be installed and play an important part in improving residents' recycling consciousness and reducing the Japanese government's burden.

As for the ocean plastic wastes, countries such as the Philippines cannot fundamentally solve such problems properly by merely instituting domestic laws since they do not have the necessary technologies and financial resources. Therefore, it is necessary to investigate the amount of waste plastic and the waste collection/recycling technologies in these countries first and then provide suitable waste selection technology or environmental education according to their features, individually.

Moreover, advanced waste plastic sorting technology is also essential. Current near-infrared (NIR) identification systems to verify the color, impurity, composition, and degradation level of plastic waste can be improved when THz sorting technology is implemented in the recycling field.

The 17 SDGs brought up by the UN focus on waste problems as well. The experience-based visiting classes (SDGs education) performed in cooperation with Tohoku University, earthquake-stricken elementary schools, stakeholders in the arterial and vein industries could be an effective way for the future generation to solve waste plastic problems and achieve a more sustainable society eventually. In fact, most teachers do not have enough knowledge of the SDGs and feel uncomfortable conducting SDGs education at present (Okubo et al. 2021). Therefore, a variety of stakeholders support SDGs education in each region through industry, academia, and government collaboration.

In conclusion, to solve ocean plastic waste problems, the discharge amount of waste plastic in advanced countries and developing countries should be clarified, and the most appropriate recycling technology chosen accordingly, and SDGs education should be provided for students in school. Without these data and efforts, there would not be an effective environmental policy or international cooperation.

Residents' cooperation in recycling activities, retailers' efforts to develop environment-friendly products, advanced recycling technology development in the education department, and quality management in the waste recycling industry are all indispensable for the sustainable resource circulation network and for a solution to the ocean waste problem.

NOTE

1   This work was supported by the program for Creating Start-ups from Advanced Research and Technology (2021) by the Japan Science and Technology Agency (JST)

REFERENCES

Bucol, L. Romano, E. Cabcaban, S. Siplon, L. Madrid, G. Bucol, A. and Polidoro, B (2019), 'Microplastic in marine sediments and rabbitfish (Siganus fuscescens) from selected coastal areas of Negros Oriental, Philippines,' Marine Pollution Bulletin, 150, 110685, https://doi.org/10.1016/j.marpolbul.2019.110685

Consumer Affairs Agency of Japan (2020), White Paper on Consumer Affairs 2020, www.caa.go.jp/policies/policy/consumer_research/white_paper/2020/, accessed on August 8, 2021.

Ikegai, T. Hasebe, Y. Mishima, S. and Kobayashi, Y. (2017), 'The sampling method of drifted microplastics for evaluation of abundance on the coast,' Journal of Environmental Laboratories Association, 42, 4 (in Japanese).

Isobe, A. Kubo, K. Tamura, Y. Kako, S. Nakashima, E. and Fujii, N. (2014), 'Selective transport of microplastics and mesoplastics by drifting coastal waters,' Marine Pollution Bulletin, 89, 1–2. https://doi.org/10.1016/j.marpolbul.2014.09.041

Japanese Ministry of the Environment (2019b), Marine Litter Survey Result in 2017, www.env.go.jp/press/H29%20MarineLitterSurveyResults_1.pdf

Lecionnes, A. J. Serrona, K. R. Devandadera, C. Leciciones, A. and Yu, J. (2021), 'Creative approaches in engaging the community towards ecological waste management and wetland conservation,' Circular Economy and Sustainability, 1st edition, Volume 2: Environmental Engineering, Elsevier, 297-317, https://doi.org/10.1016/B978-0-12-821664-4.00020-0

Liu, X. Yu, J. Okubo, K. Sato, M. and Aoki, T. (2021), 'Case Study on the Efficiency of Recycling Companies' Waste Paper Collection Stations in Japan,' Sustainability 13(20), 11536, https://doi.org/10.3390/su132011536

Maekawa, T. Kanaya, H. Suzuki, S. and Asada, M. (2016), 'Oscillation up to 1.92 THz in resonant tunneling diode by reduced conduction loss,' Appl. Phys. Express, 9, 024101.

Miles, R. E. Harrison, P. and Lippens, D. (2001), Terahertz Sources and Systems. Springer, Dordrecht, https://doi.org/10.1007/978-94-010-0824-2

Minister of the Environment of Japan (2021), Result of Separation and Collection of Waste Containers and Packaging in Each Municipals in 2019, www.env.go.jp/press/109333.html, accessed on August 8, 2021.

Naftaly, M. Vieweg, N. and Deninger, A. (2019), 'Industrial applications of terahertz sensing: state of play,' Sensors, 19, 4203.

National Supermarket Association of Japan (2020), Supermarket Annual Statistical Survey of Supermarkets in Japan, 2020, www.super.or.jp/wp-content/uploads/2020/10/2020nenji-tokei.pdf, accessed on August 8, 2021.

Newman, S. Watkins, E. Farmer, A. Brink, P. and Schweitzer, J. P. (2015), 'The Economics of Marine Litter.' In: Bergmann, M. Gutow, L. and Klages, M. (eds) Marine Anthropogenic Litter. 367-394. Springer, Cham.

Nishizawa, J. (2004), 'Development of THz wave oscillation and its application to molecular sciences,' Proceedings of the Japan Academy Ser B 80, 74–81, doi.org/10.2183/pjab.80.74.

Nishizawa, J. and Watanabe, Y. (1958), 'High frequency properties of the avalanching negative resistance diode,' Sci. Rep. Res. Inst. Tohoku Univ. B, 10, 91–108.

Nishizawa, J. Płotka, P. Kurabayashi, T. and Makabe, H. (2008), '706-GHz GaAs CW fundamental-mode TUNNETT diodes fabricated with molecular layer epitaxy,' Phys. Stat. Sol. C, 5, 2802–2804.

Okubo, Kazuaki. Yu, Jeongsoo. Osanai, Shiori and Serrona, Kevin, R. B. (2021), 'Present issues and efforts to integrate sustainable development goals in a local senior'

Oyama, Y. Zhen, L. Tanabe, T. and Kagaya. M (2009), 'Sub-terahertz imaging of defects in building blocks,' NDTandE International, 42, 28–33.

Suto, K. Sasaki, T. Tanabe, T. Saito, K. Nishizawa, J. and Ito, M. (2015), 'GaP terahertz wave generator and THz spectrometer using Cr:Forsterite lasers,' Review of Scientific Instruments, 76, 123109, https://doi.org/10.1063/1.2140223

Takahashi, S. Tamano, H. Nakajima, K. Tanabe, T. and Oyama, Y. (2014), 'Observation of damage in insulated copper cables by THz imaging,' NDTandE International, 61, 75–79.

The Japan Containers and Packaging Recycling Association (2021a), Annual Report 2021, www.jcpra.or.jp/Portals/0/resource/association/report/pdf/report2021.pdf, accessed on August 8, 2021.

The Japan Containers and Packaging Recycling Association (2021b), 'Estimated Emissions of containers and packaging,' https://www.jcpra.or.jp/recycle/related_data/tabid/678/index.php#kamipla, accessed on August 8, 2021.

The Japan Containers and Packaging Recycling Association, www.jcpra.or.jp/, accessed September 1, 2021.

Tonouchi, M. (2007), 'Cutting-edge terahertz technology,' Nature Photonics, 1, 97–105. https://doi.org/10.1038/nphoton

Ushijima, T. Tanaka, S. Suzuki, Y. Yukioka, S. Wang, M. Nabetani, Y. Fujii, S. and Takada, H. (2018), 'Occurrence of microplastics in Digestive Tracts of Fish with Different Modes of Ingestion in Japanese Bays and Lake Biwa,' Journal of Japan Society on Water Environment, 41, 4 (in Japanese). https://doi.org/10.2965/jswe.41.107

World Bank Group (2021), 'Market Study for the Philippines: Plastics Circularity Opportunities and Barriers,' Marine Plastics Series, East Asia and Pacific Region. Washington, DC.

World Population review (2021), 'Plastic Pollution by Country 2021,' https://worldpopulationreview.com/country-rankings/plastic-pollution-by-country, accessed September 1, 2021.

15 Marine Pollution and Ecosystem Health

Challenges for Developing Sustainable Blue Economy

Md. Nazrul Islam,[1*] Sahanaj Tamanna,[2] and
Al Rabby Siemens[1]
[1] Department of Geography and Environment, Jahangirnagar
University, Savar, Dhaka, Bangladesh.
[2] Bangladesh Environmental Modeling Alliance (BEMA), Non-Profit
Research and Training Organization, Mirpur, Dhaka, Bangladesh
*Corresponding author: Md. Nazrul Islam.
E-mail: nazrul_geo@juniv.edu

CONTENTS

DOI: 10.1201/9781003184287-15

15.1   INTRODUCTION

Developing the sustainable fisheries, aquaculture and the blue economy the pollution free marine ecosystems are currently the most important issues. But currently the marine pollution threats to marine biodiversity and to make the barriers for enhancing the blue economy at regional to globally as well (Riera et al. 2014; Bond, 2019; Voyer and van Leeuwen, 2019). In general, marine pollution is a combination of chemicals, inflow nutrients and trash, most of which comes from land and river-estuaries sources and is washed or blown into the ocean (Landrigan et al. 2020; National Geographic, 2022). This pollution results in damage to the marine environment and biodiversity, to the health of all organisms, and to economic structures, for example, blue economy worldwide (Patil et al. 2018; Alam and Xiangmin). There are many types of marine debris include various plastic items like shopping bags and beverage bottles, along with cigarette butts, bottle caps, food wrappers, and fishing gear etc. (Andrady, 2011Naik Mayur et al. 2021). Plastic waste is particularly problematic as a pollutant because it is so long-lasting. Most of the plastic items can take hundreds of years to decompose at the bottom of the oceans (Cuthbert et al. 2014). Many scientific research findings shows that an estimated 8 million tonnes of plastic waste and 1.5 million tonnes of microplastics enter the oceans every year (Abbas, 1973; National Geographic, 2022). This threatens marine ecosystems and the communities relying on the seas for their livelihoods (Kamal and Khan, 2009). Cousteau, 2022). From an international point of view, negligible data on socio-economic and environmental estimates of the ocean-based sector have been available in the scientific research domain (Sumaila et al. 2020). According to Forbes (1995) there are a few assessment procedures tools to monitor the marine pollution that have been started to evaluate on the basis of ocean based economic ideas since 1970 (Shepherd and Jackson, 2013; Wenhai et al. 2019). The lack of available marine pollution information and data evaluation limits the intentions of connected stakeholders, complicating the real assessment of those stakeholders in terms of economic success from a regional, national and global perspective (Delgado, 2003; Sidell and O'Brien, 2006). The blue economy encompasses a diverse mix of resources and assets that have aided in the growth of the regional and global economy. As an example, according to a basic calculation, the blue economy contributes between 3% and 5% of world GDP (Patil et al. 2018; FAO, 2021). Considering the largest population structure and food necessity in Asia it is quite important to enhance the blue economy in Asian Countries especially in South Asia (Ahmed, 1976; Sarker et al. 2016). The Bay of Bengal is surrounded by several south Asian states. India borders it on the west, northwest, and east, Myanmar borders it on the east, Sri Lanka borders it on the southwest, and Indonesia borders it on the southeast (Shahriar, 2020; Mukherji, 2021). This may open up possibilities and generate a lot of future ideas for the blue economy to grow, and they can use the riches and materials they have in their designated maritime zones to do so (Dietz and O'Neill, 2013).

In South Asia, residents in coastal regions rely mostly on fish and fisheries for their livelihoods, which account for 5% to 8% of total income (Aye et al. 2019). Because the Bay of Bengal is such a valuable resource in South Asia, India gets around half of the seafood it produces (1.2 million tonnes per year), and Myanmar (Burma) receives approximately 1.1 million tons (Funge-Smith et al. 2012; Anderson, 2014). If it looks at some of the countries involved by this phenomenon, it can see that Maldives, India, Bangladesh, Myanmar, Sri Lanka are acquiring 0.6 million tonnes, 0.12 million tons, and 0.16 million tonnes seafood's annually, respectively (Alharthi and Hanif, 2020). Considering these production scenarios, the management and processes of seafood cultivation activities must be carried out carefully to guarantee that quality requirements are satisfied while contributing to higher GDP in South Asian countries (Bari, 2017; Bond, 2019). Particularly, India has a coastline with six other countries, providing great prospects for blue economic growth through maritime expansion, commerce and the use of natural resources such as fisheries and aquaculture, minerals as well as energy to suit local needs (Agarwala, 2021). Through building a solid and long-term framework for maritime advancement, India may achieve financial and other economic success.

Five nations in south Asia (India, Bangladesh, Pakistan, Maldives and Sri Lanka) comprise less than 2% of the world's entire coastline (Bari, 2017). Coastal areas account for 40% of all trade in such nations' diverse regions and supply most of their financial foundation (Huang, 2016). South Asia's coastline neighborhoods are quite diverse. The beach tourism industry in the area has grown at an annual pace of 8% approximately. India had the most tourists in the area in 2014, while Bangladesh had the fewest (Naazer, 2018). If we look at some perceptions over the last few years, the concept becomes apparent in Table 15.1.

TABLE 15.1
Some Important Conferences and Declarations Made on Global Blue Economy Organized by Various Organizations from 2011–2021

Name	Date and year	Place	Specifications	Source
Funding the Sustainable Blue Economy (Hybrid Event)	11 Nov 2021 to 11 Dec 2021	Glasgow, UK	This event highlighted that while financial institutions are key enablers of this transition, public support is essential to close the ocean finance gap.	COP26 (2021)
Virtual Conference on Blue Economy and Blue Finance	10–11 November 2020	University of Wollongong, Australia	It focuses was on the blue economy and blue finance, including related governance planning, sectoral management, and risk management.	Tirumala and Tiwari (2020)
Declaration of the Indian Ocean Rim Association	2–3 September 2015	Pointe aux Piments, Republic of Mauritius	The development of the blue economy holds immense promise for the Indian Ocean region. The Indian Ocean is the world's preeminent seaway for trade and commerce.	Rogerson (2020)
The IORA Blue Economy Core Group Workshop	4–5 May 2015	Durban, South Africa	Promoting fisheries and aquaculture and maritime safety and security cooperation in the Indian Ocean region.	Bohler-Muller (2017)
The Indian Ocean Region Workshop	26–27 July 2015	Bali, Indonesia	Exploration and development of seabed minerals and hydrocarbons: current capability and emerging science needs.	Yamin et al. (2021)
The Changwon Declaration	2012	Changwon City, Republic of Korea	It underlined the need of expanding East Asian seas through a blue economy dependent on the ocean (EAS).	Upadhyay and Mishra (2020)
The East Asian Seas Congress	2012	Changwon City, Republic of Korea	It defined the notion of a blue economy and announced the dawn of a new era for the world.	Mallin et al. (2020)
The Xiamen Declaration	2014	Xiamen, China	It became a key cornerstone for activists and allowing them to identify new blue economy strategies to expand cooperation and coordination.	Silver et al. (2015)
The Abu Dhabi Declaration Conference	19 - 20 January 2014	Abu Dhabi, UAE	Its helps us in understanding the importance and insights of blue economy as well as Seychelles and the United Arab Emirates collaborated to host the Conference.	Sadally (2018)

To ensure the long term sustainability of coastal resources attention must be paid to integrate eco- friendly material with developing new infrastructures and paradigms (Eric Jordán-Dahlgren Rosa, 2007). Changes in institutions and regulations need to be also explored, with the main goal of assuring marine environmental sustainability and conserving the coastline landscapes, the ecosystem, marine elements, and aquatic resources all receive important considerations (Iftekhar, 2006). The blue economy and its intertwined relationship with productivity growth are commonly misunderstood, and additional knowledge is still required (Geisendorf and Pietrulla, 2018). Marine biotechnology as well as the manufacturing process of seafood and salt are two more related industries that employ ocean-based assets (Smith-Godfrey, 2016). Ship and boat building and repair, as well as marine tourism and the advancement of maritime legislation and management, are all part of the blue economy (EU, 2016). In South Asia and African countries, we have a huge manpower with low labor costs so that the ship and boat buildings are remarkably suitable to establish in this region. The blue economy also encompasses functions such as maritime exploration and invention (Eikeset et al. 2018). This also includes all major ocean-related industries, such as coast guards and security forces. The blue economy is an essential idea for nations with coastal areas, and this bigger image of the blue economy provides us with clarity into the notion as a whole and its value. No economy wants to waste potential growth prospects, thus the blue economy is an essential concept (Rudge, 2021). The value of the blue economy is now widely acknowledged across the world. Every policymaker, researcher, and scientist recognizes the importance of blue economy and marine pollution concern in the relevance of the Pacific Small Island Developing States (SIDS) and the Southern African area (Techera, 2018). Coastal states are fighting to keep their study rights. They also seek to make use of their marine riches (Alharthi and Hanif, 2020).

Because of this, the Seychelles and the United Arab Emirates teamed up in 2014 to convene the Abu Dhabi Declaration Conference in Abu Dhabi, which emphasized the importance of adapting to and coping with global warming (Houghton et al. 2001; Howel et al. 2012). This conference also focused on maritime environmental preservation, as well as the creation and gratification of the blue economy (Grilli et al. 2021). The notion of a blue economy has acquired momentum in more than the world's major countries. But it can be seen, emerging and third-world countries, such as African nations and small island developing states, have also been influenced (SIDS) at this sector (Robinson, 20220; Allam and Jones, 2021). Fishing has always been a source of income for South Asian countries. They have a lot of advantages and a lot of possibilities for building their blue economies because of their closeness to the Indian Ocean (NOAA, 2017). However, without the infrastructure in place to successfully manage and regulate the blue economy, it is highly unlikely that it will reach its full potential.

But currently marine pollution especially the micro-plastic has threatened these potential resources. Fisheries resources are declining day by day and on the other hand the fish consumption rate has increased dramatically in recent years. India's per capita fish intake grew by 4.3% between 1985 and 1997 (Ohlan, 2016). Many South Asian countries also showed a 3.3% increase in fish consumption over this time period (Delgado et al. 2003). Fisheries is the fastest growing industry that provides answers to issues including resource exploitation and degradation while also providing employment possibilities. In comparison to other regions, South Asian countries have seen tremendous expansion in the fishing sector (Funge-Smith et al. 2012). Figure 15.1 is a functional flowchart of the implementation of local, regional, global coastal planning for enhancing global blue economy.

Sri Lanka and Maldives have significant coastlines, which means that the biodiversity of the ocean offers more chances for expansion (Bari, 2017). Governments such as these are on the lookout for new methods to contribute and grow their economies. Nations with coastal territories have the option, according to UN Resolutions, to use whatever maritime resources they can, including mining and fishing (United Nations Convention for the Law of the Sea [UNCLOS] Article 56) (NOAA, 2017). Other UN articles allow coastal countries to dig, explore, and use the sea's riches (UNCLOS

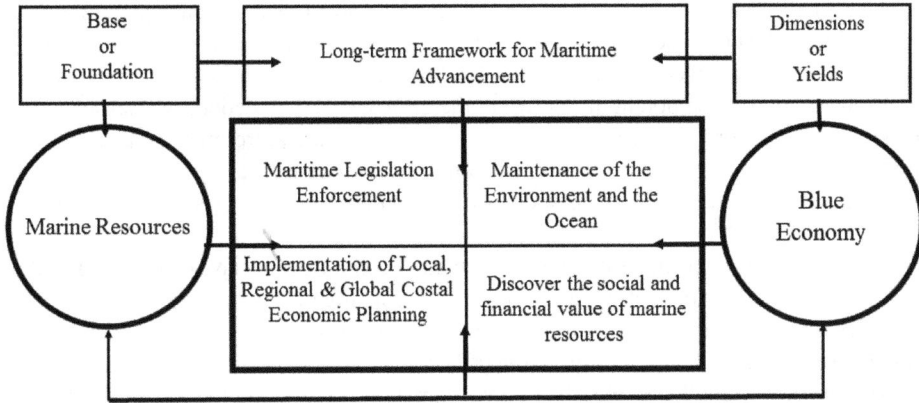

FIGURE 15.1 Functional flowchart that outlines of local, regional, global coastal planning for enhancing global blue economy.

Article 77) (Schoolmeesters et al. 2009; Smith, 2017). Because of the enormous impact that maritime sectors have on the Global market, China has recognized the need for developing the blue economy. According to experts Zhao et al. (2014) and Colgan and Judith (2013), China's blue economy related sectors contributed around US$ 240 billion to the GDP. The Chinese blue economy employs about nine million people, and according to Jiang et al. (2014), the sector's contribution to the national economy expanded from 6.46% to 13.83% from 2000 to 2011.

The World Bank defines sustainable use of marine assets as the use of ocean resources for financial growth, enhanced livelihoods, and jobs while maintaining the integrity of the ocean environment for future expansion and economic benefits (EU, 2016). According to information provided by the Commonwealth:

- The global marine economy is estimated to be worth roughly one and a half trillion US$ every year.
- 80% trade by volume is handled by the ocean, and fisheries support 350 million jobs worldwide.
- Offshore oil resources are expected to account for 34% of crude oil output by 2025.
- Aquaculture is the fastest growing food sector, accounting for nearly half of the fish consumed by humans (Commonwealth, 2022).

However, economic benefits and growth can only continue to be achieved if the resources which fuel the development can be protected and used sustainably (Bhanawat, 2021). Plastic waste in the world's seas actually costs up to US$ 2.5 trillion every 12 months, based on a research reported in the Marine Pollution Bulletin (Waste360, 2019). Not only that, projects such as offshore wind turbines, oil mining and waste management may not take the health of the ocean into account. Therefore, the challenges for blue economy are man-made, created in the name of development which is short-sighted, the negligence of the people, and the irresponsibility of the world leaders to take accountability (Fattah, 1979; UNECA, 2016). These challenges must be dealt with before marine pollution causes irreparable damage.

15.2 MARINE POLLUTION AND ITS IMPACT

15.2.1 Marine Pollution and the Loss of Marine Biodiversity

According to a report in 2015 by World Wildlife Fund, there has been about 50% decline in populations of marine life between the years 1970 and 2012 (Briggs, 2016). It is not a hyperbole to estimate that the statistics have been only increasing due to ever-increasing marine pollution (Goldberg, 2011). The ocean ecosystem's resilience to tolerate climate shocks is being weakened by the continued loss of coastal ecosystems. Because of the economic importance of marine variety, the climate change issue has the potential to prevent the growth of a blue economy (Hasan et al. 2018). Species-specific behavioral patterns emerge because of the temperature increase caused by human activity. Certain species respond to variations in temperature, while others relocate to the poles or to new locations, a phenomenon known as species invasion (WASA Group, 1998). This can also have negative consequences.

For instance, deadly algae endemic to the Indian and Pacific Oceans have diffused throughout the Mediterranean, displacing local plants and depleting marine life of food and shelter (Thelma, 2019). Other species that are unable to adjust to the changing environment may become endangered. In 2015, the International Union for Conservation of Nature (IUCN) reported the extinction of 15 identified marine species (Ceballos et al. 2015). Environmental degradation has the potential to bleach and kill some corals. This has an immediate effect on marine species that have calcareous skeletons or shells. Table 15.2 shows some sources and impacts of marine debris pollution and its mitigation process to build a sustainable blue economy.

Erosion and floods have a negative impact on ocean life in coastal regions, notably in some coastal ecosystems like estuaries and sea grass beds. Fish is a main provider of animal nutrition for at least 1 billion people on the planet, according to UNESCO (Lincoln et al. 2021). The seafood industry will be heavily impacted by the massive loss of biodiversity if the marine pollution continues to increase. This biodiversity reduction also means that the genes and compounds that could be useful in medical research and industrial applications will be hampered (National Geographic , 2019). All of these observations are relevant indicators of marine pollution pointing to a loss of potential commercial marine trades for the long-term blue economy (Sarker et al. 2018).

15.2.2 Marine Pollution and the Collapse of Fish Stocks from Overfishing

Fishing is one of the main drivers of decreases in sea natural life populations. Taking fish isn't innately awful for the sea, except when vessels take fish more quickly than stocks can renew, something many refer to as overfishing (Gershwin, 2013; Forrest et al. 2019). The quantity of overfished stocks internationally has significantly increased in 50 years and today 33% of the world's surveyed fisheries are at present pushed past their natural cutoff points (Willette et al. 2017). Table 15.3 shows marine pollution effects and global projected fish production scenarios through 2030 (live weight equivalent). Overfishing involves taking too many fish too quickly, so that the reproducing population is unable to recuperate (Jackson and Seeger, 2013). Overfishing regularly goes along with inefficient kinds of business fishing that take in large numbers of unwanted fish or other sea creatures, which are then disposed of (Wilhelmsson et al. 2013).

The table shows the amount of fishing and the increase day by day. Overfishing jeopardizes sea environments and the billions of individuals who depend on fish as an essential provider of protein (Davies and Baum, 2012; Dais et al. 2019). Without practical administration, our fisheries face breakdown and we face a food emergency (Srinivasan et al. 2010; Sumaila and Tai, 2019). The harm done by overfishing goes past the marine climate. Billions of individuals depend on fish for protein, and fishing is the chief business for many individuals all over the planet. (Sumaila and Tai, 2019). All over the planet, numerous fisheries are impacted by decisions that exacerbate the issue or that might be unprincipled.

15.2.3 Fertilizers and Waste from Humans and Animals

Human trash pollutes the oceans in a variety of ways (Awuchi and Awuchi, 2019). Waste and plastics detritus (packs, syringes, cutlery, plastic bottles, and so on) pose a significant danger to

TABLE 15.2

Sources and Impacts of Marine Debris Pollution and Its Mitigation Process in Building the Sustainable Blue Economy

Materials	Proportion	Impact	Mitigation Process	References
Plastics	37	Hundreds of marine species have been ingested, suffocated, killed, or entangled.	So, that plastics elements do not increase enormously, it is necessary to minimize output and recycle.	IUCN (2021)
Glass	09	Sharp edges endanger species, and it can take years for them to be ground down into 'beach glass.' and is referred to as garbage.	To avoid trash from being thrown away, a new tech-based manufacturing system is required.	Nature (2019)
Rubber	08	Tires release microscopic plastic polymers when the rubber wears down, which typically end up as contaminants in seas and streams.	We should minimize the overall production of rubber and also to reduce the recycling mechanisms of rubber if we want to see the clean ocean.	Binnemans et al. (2013)
Wood or Processed timber	18	Timber processes, such as the use of fertilizers and insecticides, and pulp paper mill trash management techniques also contribute to water pollution in terms of erosion, chemical pollution, and the damage of marine lines.	To move production away from the water's edge and decrease deforestation	Anh et al. (2010)
Metal	7	Sediments in ecosystems with chronic metal inputs, such as seaports or other industrialized coastal regions, are heavily polluted. This trait has raised concerns about the ecological consequences of poor sediment quality.	Many treatment approaches, such as mechanical, biochemical, and biological, were proposed to clean up heavy metal contamination in the ocean.	Ansari (2004)
Clothing and textiles	17	Clothes composed of synthetic materials and chemical compounds shed small plastic fibers that wind up in the environment when created, cleaned, and worn.	To reduce the consumption of harmful technologies and look at possibilities for creating recycled textiles so that ocean water remains safe.	NRDC (2015)
Others	4	Abandoned and derelict vessels or paper and cardboard.	To conserve the maritime ecology by reducing pollutant waste, mitigating vehicle pollution, and using less energy.	Sayed et al. (2021)

Source: Modified and adopted from Binnemans et al. 2013; Pavičić, 2015; Sayed et al. 2021.

the existence of maritime fauna. These forms of pollution are eaten, cause entanglement, and even death by asphyxia (Mason and Folkerts, 2013; Awuchi and Awuchi, 2019). Because many varieties of plastic do not float, they end up deep in the water (Gameiro, 2019). Other polymers tend to gather in subtropical gyres, forming massive waste patches. Biodegradable polymers, on the other hand, often only degrade at specific temperatures that aren't prevalent in oceans (Goel et al. 2021). Swimmers can contribute sunscreen, moisturizer, repellents, oils, beauty products, and cosmetics

TABLE 15.3

Marine Pollution Effects and Global Projected Fish Production Scenarios through 2030 (Live Weight Equivalent)

Projected Fish Production Scenarios 2030 (live weight equivalent)

Regions	2018	2030	2030 growth vs. 2018 growth
	(1,000 tones)		In percentage (%)
Asia	1,22,404	1,45,850	+ 19.2
Africa	12,268	13,820	+ 12.7
Europe	18,102	19,290	+ 06.6
North America	6,536	6,981	+ 06.8
South America and Caribbean	17,587	16,730	- 04.9
Oceania	1,617	1,750	+ 08.2
Developed countries	29,233	30,730	+ 05.1
Developing countries	1,35,096	1,73,691	+ 28.6
World	1,78,529	2,04,421	+ 14.5

Source: Kobayashi et al. 2015; FAO, 2020.

to water bodies (Awuchi and Awuchi, 2019). This is significant since much of this occurs because of tourism. As a result, even though it is profitable, actions must be taken to ensure tourism's long-term viability (Adkins, 2017). These chemicals harm phytoplankton, sea anemones, fish, and other animals in the water, as well as coral reefs, all of which are popular tourist destinations (Häder et al. 2007). Oil from cars on the road is washed away and ends up in the water. Boats have also dumped oil directly into the ocean.

Plastic is the most common man-made substance, and it has long been scrutinized by environmentalists (Vegter et al. 2014; Lee et al. 2020). However, there is a scarcity of reliable worldwide data, notably concerning its end-of-life destiny. A first global assessment of all large plastics ever conducted has taken place, detecting and synthesizing disparate data on thermoplastics, synthetic fabrics, and chemicals manufacturing, use, and end-of-life management (Geyer et al. 2017).

a) **Oiling and Lubrication**

  ✓ When oil causes physical and chemical damage to a plant or animal, it is known as fouling or oiling. A bird's wings, for example, may be oil-coated, rendering it impossible to fly or removing the insulating characteristics.

  ✓ The amount of greasing an animal receives might affect its prospects for survival.

b) **Sensitivity to Oil**

  ✓ Oil has a variety of harmful substances that can result in serious health issues such as heart disease, growth retardation, immune response impacts, and even mortality.

c) **Runoff from Farmland and Aquaculture**

  ✓ Farmers' ammonia fertilizers and insecticides wash down into waterways and into the sea.

  ✓ The aquaculture sector has also been accused of dumping unfinished foodstuff, chemicals, and parasites into bodies of water.

  ✓ Sewage and sanitary systems don't always operate well, and they don't always eliminate enough nitrogen and phosphorus before dumping the waste into the rivers.

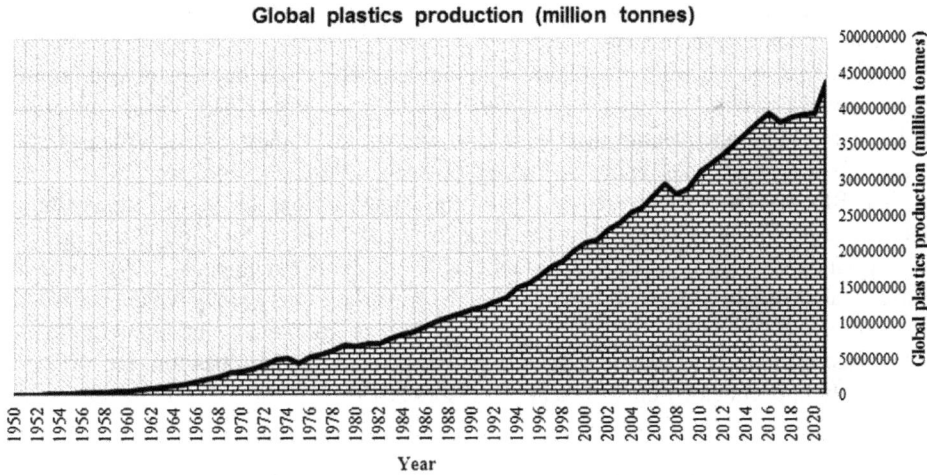

FIGURE 15.2 Increase of global plastic production by year (modified and regenerated by the author from Geyer et al. 2017; Goel et al. 2021).

Now, if we look at three occurrences that occurred in the years 2000, 2010, and 2020, we can see how oil toxicity, or an oil spill may have a severe influence on the ocean and its surroundings (Table 15.4).

d) Pollution from Industry
  ✓ Pollutants such as nuclear debris, arsenic, copper, chloride, and cyanide are found in commercial waste.
  ✓ As they are discharged into the water without being properly treated, they can pollute the water and damage the local ecology.

e) Nutrient Enrichment (Eutrophication)
  ✓ Eutrophication is caused by a shortage of oxygen incorporated in the water and an abundance of nutrients, mostly nitrates and phosphates, in coastal regions.
  ✓ This pollution is caused by runoff from household and industrial wastewater treatment plants, as well as manufacturing agricultural regions.

f) Increase of CO$_2$
  ✓ The acidity of the seas is increasing. This is because carbon dioxide is increasing because of air pollution, and carbon dioxide dissolves in water and forms carbonic acid.
  ✓ Corals as well as shellfish may be harmed by increased acidity.

g) Sound Pollution
  ✓ Several marine creatures rely substantially on their sensitivity to sound to survive.
  ✓ The noise produced by cargo vessels, submarines, oil drilling and mining, fishing industry, and leisure jet skis disturbs this.

h) Manure
  ✓ These are used to improve manufacturing quality and quantity. Fertilizers have a variety of negative consequences on the maritime ecosystem resulting from their widespread use.
  ✓ Runoffs can occur from both home gardens and industrial farms.

Manures must be delivered in the proper quantities, at the proper times and locations, and in the proper manner. Their effects include:

i) Oxygen depletion caused by commercial pesticides that include nutrients which promote the development and proliferation of microorganisms (Kako et al. 2014). This lowers the amount of available oxygen in marine habitats that can lead to suffocation in many aquatic animals (Islam and Tanaka, 2004). The water quality may be harmed because of these deceased species.

ii) Algal Blooms can be triggered by fertilizers. Many various forms of marine life can be poisoned by large concentrations of growing algae. Algae also takes a lot of oxygen to grow, which causes oxygen deprivation (Joyce, 2000; Islam et al. 2012; Wurtsbaugh et al. 2019). Algae as well as its toxins can cause an ecosystem to shut down. Dead zones emerge when typical marine life can no longer survive (Islam et al. 2013). Over 400 such dead zones may now be found throughout the world's coasts. It might take years for them to go back to their former, healthy state. These zones may potentially impact nearby environments (Jackson, 2019).

15.2.4　Marine Pollution Subsequent Impacts on Human Health and Well-being

The oceans are essential to human health and wellbeing. The oceans are particularly important to the health and well-being of people in small island nations (Landrigan et al. 2020). Most of the oceans and sea contamination are increasing because many nations inadequately controle marine pollution. It is a complicated combination of poisonous metals, plastics, man-made synthetic substances, petrol, metropolitan and modern wastes, pesticides, composts, synthetic drug substances, rural overflow, and sewage (Landrigan et al. 2020). Over 80% emerges from land-based sources (Landrigan et al. 2020). It arrives at the seas through waterways, spillover, the air and direct releases. It usually appears to be heaviest around the coastlines, and is most prevalent all along coastal areas of low and middle-income countries (Barbier and Cox, 2003). Plastic is a quickly expanding and exceptionally apparent part of sea contamination, and an expected 10 million metric tonnes of plastic waste enter the oceans every year (Landrigan et al. 2020). Additionally, all four regional seas in Europe have a large-scale contamination problem, ranging from 96 % of the assessed area in the Baltic Sea and 91 % in the Black Sea, to 87 % in the Mediterranean and 75 % in the North-East Atlantic Ocean. The magnitude of coverage of the assessed area is mostly good, but it varies considerably between the four seas and remains limited in the offshore waters of the Mediterranean Sea (Zhongming et al. 2019).

Mercury is the most noteworthy metal contamination in the seas. Mercury pollution of ocean environments comes from two primary sources of coal burning and limited scope gold mining (Rallo et al. 2012). Worldwide spread of industrialized horticulture with expanding utilization of synthetic compost prompts expansion of Harmful Algal Blooms (HABs) to previously unaffected districts (Roberts et al. 2021). Synthetic toxins are universal and pollute oceans and marine creatures from the high Arctic to the deep depths (Schnurr et al. 2018). The plastic waste in the ocean can upset endocrine flagging, decrease male fertility, harm the sensory system, and increase the risk of disease (Maes et al. 2021). HABs produce strong poisons that collect in fish and shellfish. When consumed, these toxins could induce severe brain damage as well as rapid death.

HAB poisons can likewise become airborne and cause respiratory sickness. Pathogenic marine microorganisms cause gastrointestinal illnesses and significant problems (Livesay et al. 2021). Regarding climate change and increased contamination, the risk of Vibrio infections, such as cholera, increasing, recurring and spreading to new areas is considerable (Maes et al. 2021). All the negative effects of sea pollution on human health fall disproportionately on vulnerable populations in the Global South, resulting in global ecological injustice.

TABLE 15.4
Three Selective Case Studies Showing the Events and Aftermath of Oil Spill in the Marine Ecosystem

Case study: 1	Case study: 2	Case study: 3
MV Treasure oil incident	*Gulf of Mexico oil incident*	*Oil incident in Mauritius*
Year: 2000	*Year:* 2010	*Year:* 2020
Location: Occurred at the coast of South Africa	*Location:* Occurred at Mexican Gulf	*Location:* Occurred at southeast of Mauritius
About the Event:	*About the Event:*	*About the Event:*
On June 23, 2000, the MV Treasure sunk six miles off the coast of South Africa while carrying iron ore from China to Brazil, causing an oil disaster.	As workers on an offshore drilling rig attempted to shut off an exploratory oil well deep in the Gulf of Mexico, a blast of gas surged up, crushing the drill pipe.	A Japanese bulk tanker ran straight into the ground on such a coral reef southeastern part of Mauritius on July 25, 2020, dumping almost 1,000 tons of gasoline oil into the island's lovely, clear waters (AFP, 2022)
The vessel was transporting fuel oil, which spilled into the sea (Crawford et al. 2002)	The 'blowout protector,' an alternative valve meant to seal the well in the event of a collision, collapsed, and gas rushed the drill site, causing an accident that killed 11 crew members (Borunda, 2020)	One of the world's largest bulk carriers capsized while travelling across the Indian Ocean on its route from China to Brazil (Laurette and Takada, 2021)
Aftermaths:	*Aftermaths:*	*Aftermaths:*
It put two of the most significant mating groups of the threatened African, or jackass, penguins, which live off the coastline of South Africa and eats largely fish, in jeopardy.	More than a third of federal gulf waterways were closed to fishing during the peak of the spill (Pallardy, 2010).	Poisonous elements of the oil spill were exposed to tens of thousands of individuals who participated for the clean-up, especially members of local fishing industry.
A thousand individuals labored 24 hours a day for more than 3 months to save the creatures. Over 19,000 penguins were engulfed in oil at Robben Island and the rescue's resources already stretched thin, the slick travelled towards Dassen Island, host to a breeding population of 50,000 more birds.	Numerous species died because of the dangerous spill, ranging from algae to whales, with a range of consequences including reduced breeding, limited growth, lesions, and illness.	It harmed the island's food system and diminished its tourism appeal.
Dassen's 19,500 vulnerable penguins were picked up by rescuers and moved 500 km away, where they were free to find their way back by ocean (Martel, 2010).	Scientists have devoted a decade to examining the influence on Gulf marine life after the accident (Boyle, 2020).	Apparently over 50 whales and dolphins were washed up, dead on the island's beaches; the administration has yet to reveal the complete findings of the animals' autopsy, which would clarify whether the oil leak was to blame for their deaths (Sandooyea and Steele, 2021)

Source: Modified and adopted from Crawford et al. 2002; Martel, 2010; Pallardy, 2010; Boyle, 2020; Borunda, 2020; Sandooyea and Steele, 2021; Laurette and Takada, 2021; AFP, 2022.

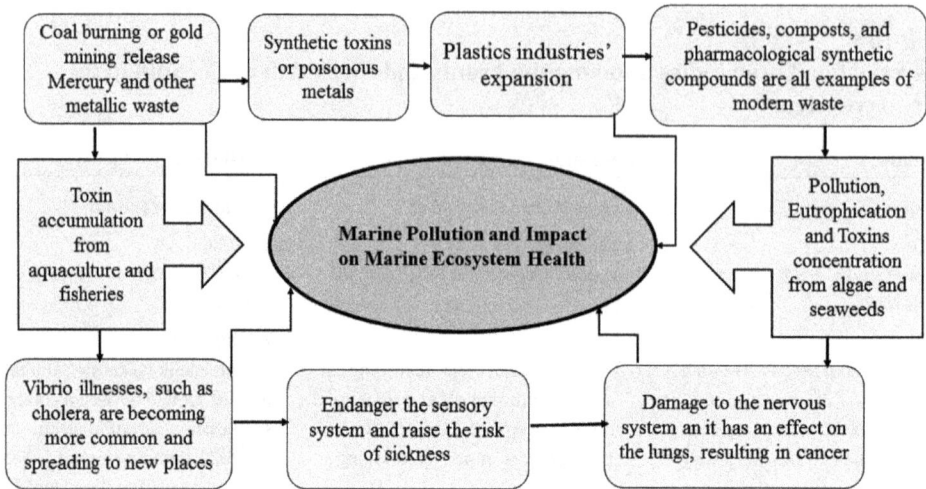

FIGURE 15.3 The approximation of contamination scenarios of marine pollution, its impact, and the state of marine ecosystem health.

15.3 TRACKING THE IMPACT ACHIEVED WITH THESE PROTECTED AREAS

To reduce pollution, some areas are specifically being protected. In the ocean, these are known as marine protected areas, or MPAs. Three types are:

1) *Fully protected areas which means that mining and processing operations are not permitted.*
2) *Areas which are severely guarded are those where no industrial and only a little leisure activity is permitted.*
3) *Partially protected areas which find a balance between human activities and protection.*

Protecting such areas always brings benefits. Properly protected regions can boost total marine life biomass by more than 400% on average (Grorud-Colvert and Lubchenco, 2017). Fish tend to grow larger and reproduce more. Even partially protected areas can provide some benefits, even though these are much less than those of strongly protected areas (Pittman et al. 2014). The MPAs also provide the species with a greater facility to combat environmental changes. When a low-oxygen event destroyed many abalones in the Gulf of California, threatening the local fisheries, the abalones in the maritime reserve were the first to revive and start to refill the region (Magazine, 2017).

15.4 OCEAN BASED POLLUTION SCENARIOS AND BLUE ECONOMY PERCEPTIONS

15.4.1 SOUTH ATLANTIC OCEAN SCENARIOS

The southwestern Atlantic Coastline is dominated by Brazil, Argentina, and Uruguay, whereas the eastern Atlantic Coastline is dominated by Cameroon, Congo, Democratic Republic of Congo, Angola, Namibia, Sao Tome and Principe, and South Africa (Miloslavich et al, 2011). Steep canyons run the length of the continental slope, connecting the shelf and deep waters. The heads of the submarine canyons have a significant benthic diversity, including 50% of the species associated with the shelf-break community (Bertolino et al. 2007; Schejter et al. 2016). With white sand, mangrove forests, rocky shores, marshes, and coral reefs in the north and sandy shorelines, mangrove swamps, rocky coastlines, lagoons, and coral reefs in the south, Brazil's coasts are host to a diverse spectrum of aquatic, estuarine, and coastal life. (Miloslavich et al. 2011). The population density of these countries is quite variable and mostly lives at the coast. Due to a lack of comprehensive land-use management and strong preparation mechanisms, these hazards disproportionately affect third world and emerging nations in the world (Hatje et al. 2021).

The equator divides the Atlantic Ocean into two halves, as we all know. The anticlockwise central subtropical gyre of surface and intermediate seas runs near to the coastlines of South America as well as South Africa and characterizes South Atlantic waters, with increasingly complicated currents forming on both continents' coasts (Campos et al. 1995; McDonagh and King, 2005). Between the equator and 40° S is a zone with a circumference of around 4,500 km (Piontkovski et al. 2000). A portion of the worldwide circulation is known as the Atlantic Meridional Overturning Circulation. Its eastern (African) branch transports hot water to the northern hemisphere, while the western (South American) branch transports cold water from the northern (Hunt et al. 2016). The sub Antarctic branch of the Antarctic Circumpolar Current closes the gyre to the south (West Wind Drift). When compared to surrounding locations, we can determine that the zone is a biogeographic zone with distinct biological constituents (Longhurst, 1998).

Oil and gas drilling, as well as mining, are key sources of revenue for nations on both sides of the South Atlantic (Ayuk et al. 2020). Anglo and Latin America, Greenland, Brasil, Europe, and Africa are all connected through these nations. The countries named are world-class producers of oil, fossil fuels, and minerals (Gonzalez-Silvera et al. 2004; Spalding et al. 2016). The overall amount of oil extracted, and the gas production was enormous. In January 2020, it was estimated as 3,120 million barrels/day and 139 million cubic meters, respectively (Hatje et al. 2021). Coastal sites, on the other hand, provided for 97% and 81% of national oil and gas supply. Mining has always been a significant business in Brazil's shoreline and sovereign sea. (Polovina et al. 2008; Henson et al. 2010).

The ocean is being viewed as a garbage sink from a global viewpoint, owing to humans' misguided belief that dilution is the cure to pollution (Boschi, 2010). Manufacturing activities and industrialized economies' usage of fossil fuels are the primary sources of pollutants in the North Atlantic (Huang et al. 2014). However, during the South America conference for the Decade of Ocean Science, the lack of adequate sewage technology and water systems was identified as the major source of pollution (Hu et al. 2004). Developed countries treat 70% of all industrial effluent in the globe. Only 8% of people in third world nations obtain effluent treatment. As a result, it is estimated that over 80% of all sewage is released without treatment worldwide and may accumulate in the environment (Anon, 2013). Contaminants reach coastal waters from a variety of sources, the majority of which are land-based. Surface runoff, waste dumping, agricultural and industrial operations, and inadequate waste treatment are among them, while others are related to the ocean, such as transportation, industrial and recreation fishing, oil drilling, extraction, and industrial emissions (Abreu, 2013). We also forget about noise and light pollution from time to time. Due to their negative effects on aquatic animals, both are becoming a significant global problem (Smith and Eckert, 1991; Anon, 2013). Rapid urbanization and development of coastal regions, commercial watercraft, geophysical and mining activities in the ocean all contribute to these types of pollution (Boschi, 2010).

15.4.2 NORTH ATLANTIC OCEAN SCENARIOS

Aside from natural variations, manmade emissions of greenhouse gases (GHG) or aerosols have an impact on global temperatures, and the global average temperature will continue to fluctuate (Chen and Dong, 2019). With increasing greenhouse gas emissions estimated in the later decades of the twenty-first century, all sections of the world are expected to warm, except for a few limited marine areas in some cases. During the last decade of the twentieth century and the first decade of the twenty-first century, the global mean temperature is expected to climb by 1.4° C to 5.8° C. (Wang and Swail, 2001) cover the complete range of forcing scenarios from the ATCSG's Specialized Assessment on Emissions Scenarios. The resulting rise in ocean level is expected to be between 0.09 and 0.88 meters by 2100 (Wang and Swail, 2001; Webb and Howard, 2011).

Almost every country is experiencing global warming as a result of this and scientists are afraid that the Earth's weather conditions will continue to become more powerful and unpredictable (Leiserowitz, 2007). One of the impacts, it has been proposed, may be a particularly severe summer in the United States. In late June of 2020 temperatures in the generally temperate Pacific Northwest reached record highs, with Portland, Oregon, hitting 116 degrees F (Chang and Bonnette, 2016; Vasquez, 2022). For the first time in history (30th June 2021), the temperature in Canada exceeds 120 degrees Fahrenheit (Nicholas, 2021; Samenow, 2021). The heat waves also took a devastating human toll, likely killing hundreds of people. Wildfires have also scorched millions of acres in the western U.S. with California on pace to surpass its record-shattering 2020 season in terms of the number of fires (Nicholas, 2021; Samenow, 2021). For the 2022 hurricane season, the National Oceanographic and Atmospheric Administration (NOAA) is forecasting a likely range of 14 to 21 named storms (winds of 39 mph or higher), of which 6 to 10 could become hurricanes (winds of 74 mph or higher), including 3 to 6 major hurricanes (category 3, 4 or 5; with winds of 111 mph or higher). NOAA provides these ranges with a 70% confidence (Blackwell, 2022). The research scholars of NOAA have predicted that the frequency of hurricane in Atlantic ocean will increase in future years (Reimann, 2021).

The Gulf of Mexico is a basin in the Atlantic Ocean, surrounded by the gulf coast of the United States, Mexico and Cuba (Turner and Rabalais, 2019). The dead zone here is one of the largest in the world. Its waters are full of nitrogen and phosphorous that come from major farming states in USA. The presence of these chemicals frequently turns Gulf of Mexico waters hypoxic, or low in oxygen (Altieri and Diaz, 2019). The North Atlantic Garbage Patch was first documented in 1972 and is entirely composed of man-made marine debris floating in the North Atlantic Gyre (Rabalais et al. 2002). Scientists estimate that the North Atlantic Garbage Patch is hundreds of kilometers in size and has a density of 200,000 pieces of trash per square kilometer in some places (Kaplan, 2021).

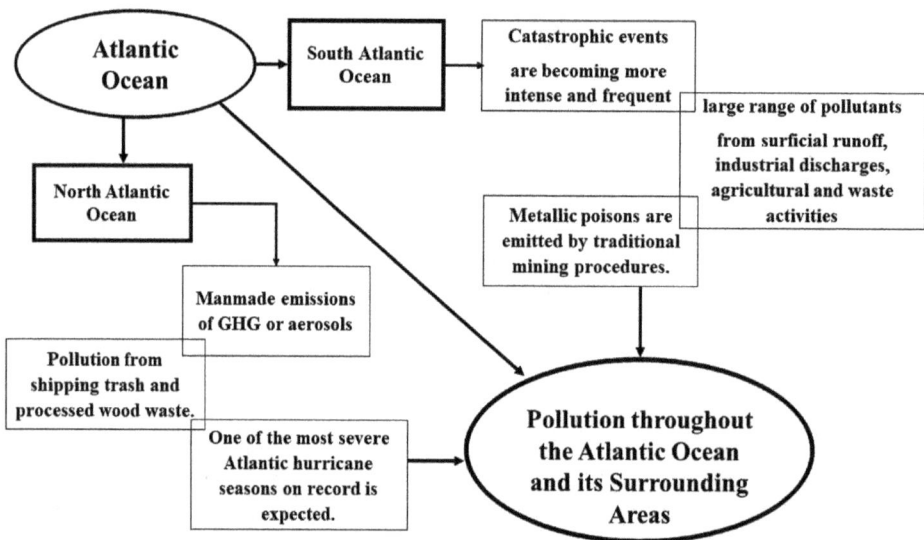

FIGURE 15.4 Scenarios of marine pollution through the North and South Atlantic Ocean and blue economy perceptions.

15.4.3 Pacific Ocean Pollution Status

The Great Pacific Garbage Patch (GPGP) is a collection of marine debris in the North Pacific Ocean (GPG, 2019) Also known as the Pacific trash vortex, the garbage patch is actually two distinct collections of debris bounded by the massive North Pacific Subtropical Gyre (Harse, 2011; Lebreton et al. 2018). The GPGP is a swath of ocean litter. It's taken from the North Pacific Ocean. Plastic use has now surpassed 320 million tons annually, with more plastic being created in the last century than ever before (NOAA, 2017). In the ocean plastic might remain on the surface waters and it can end up building up in remote parts of the world's oceans (FAO, 2014). Researchers identify and estimate the GPGP, is a significant agglomeration of plastic in the ocean zone established in subtropical areas and spanning from California to Hawaii. Approximately 60% of the plastic manufactured has a density lower than that of seawater (Leberton, 2017). Plastic can be carried by ocean trade winds once it is incorporated into the marine ecosystem (GPGP). It is either reclaimed by coastlines that have been damaged by the sun, temperature changes, waves, and marine life, or it loses stability and sinks (Leberton, 2017). On the other hand, most of these floating plastics make their way onshore and into marine gyres. A substantial concentration zone for floating plastic has been observed in the eastern part of the Northern Pacific Subtropical Gyre (Goldstein and Goodwin, 2013). The research team GPGP were able to predict whether the observed values from sea surface sweep literature are inside or outside the GPGP zone by creating a flexible GPGP window that accounts for seasonal and inter-annual fluctuations. They looked at the ten-year evolution of microplastic mass concentrations (kg km^2) in and around Los Angeles using our calibrated model (GPGP) (Lebreton et al. 2018).

15.4.4 Indian Ocean Pollution Rate

Scientists found that 414 million plastic garbage items totaling 238 tons are poisoning the Indian Ocean environment. The enormous plastic development included around 25% single-use or disposable plastics, packages, drink bottles, straws, plastic cutlery, sacks, toothbrushes and shoes, of the 414 million items of plastic toxins (Pattiaratchi et al. 2021). An expected 384 million buried items waste was viewed covering the surface at up to 10 cm depth (Lavers et al. 2019; Moore, 2020). Around 60% of this contained miniature garbage that was 2 - 5 mm in size and could present critical difficulties to untamed life and biodiversity (Moore, 2020).

15.4.5 Arctic Ocean Pollution Scenario

The biggest source of pollution in the Arctic waters isn't monetary activity. The growing presence of military weapons structures in the neighborhood raises concerns about contamination spreading farther (Schor, 2011). The gatherings resulted in an agreement to conduct a preliminary examination of public legislation and to discern differences in climate-change mitigation measures (Biesbroek et al. 2011). The review of explicit threats of marine pollution in the Arctic is mostly governed by public legislation in coastal nations, albeit they take into account current worldwide rules (Regan, 2021). 92% detected of microplastics were microscopic, manufactured fibers, the majority of which were polyester ('Pollutants in the Arctic', 2021).

Despite its remote location, the Arctic is intimately linked to our houses, our clothing, and our shopping proclivities across the rest of the globe, according to Ross, because roughly 66% of our clothing is made from manufactured materials such as polyester-nylon-acrylic ('Pollutants in the Arctic', 2021). The eastern strands were also half the length of the western strands and appeared to be more recent and fresher, suggesting that most strands seen in the Arctic Ocean originated in the Atlantic (Strand et al. 2020). Plastic particle concentrations were many times greater in the Eastern Arctic (above Western Europe and the North Atlantic Ocean) than in the Western Arctic (above the Western Canadian shoreline or more Alaska) in terms of microplastic concentrations (Halsband and Herzke, 2019).

15.5 IMPACT ON OCEAN HEALTH AND THE BLUE ECONOMY

The Global Organization for the Blue Economy has organized various discussions and seminars on the subject recently. This is crucial because a big part of economic activity happens in associated ocean economic activities rather than core industries (Fujita and Krugman, 2004). Fisheries, maritime trade and shipping, fuel, recreation, environmental services, and marine monitoring and surveillance are among the 26 marine economic functions that may be recognized (Potgieter, 2018). According to the European Commission's 2019 blue economy survey (excluding seabed extraction), marine living assets (namely, the fishing industry, commercial fishing, and fish production and marketing), the coastal tourist industry, maritime transport, port facilities, building and construction and repair, and the marine refining of crude oil, gas, and metals, directly employed approximately 4 million people in 2017 (up 7.2% from 2009) and produced €180 billion in gross value added (GVA) (EU, The EU Blue Economy Report, 2019).

All these roles take into consideration the supply chains that arise as a result of the possibilities for infrastructure growth in a range of sectors (Coe et al. 2017). These will provide jobs and aid in reducing poverty by allowing residents in coastal areas to participate in social and economic activities (MoFA, 2020). To achieve long-term success, however, solid political duties, in-depth analyses, cultural sensitivity, and a good attitude are essential.

15.5.1 WATER QUALITY AND OCEAN HEALTH

The Salish Sea's seawater condition is critical for all living things that rely on it. It has an impact on the number of phytoplankton in surface water, which is important since phytoplankton absorb and store vast amounts of free carbon while also releasing oxygen into the environment (Miner et al. 2018). As marine water quality impacts the capability of fish and marine mammals to flourish and breed, it has an impact on people as well. It further has an impact on the quality and amount of food that we may collect from the water. It impacts us both directly and passively when we visit a beach or enjoy the fauna and landscape of the place (EPA, Marine Water Quality, 2019).

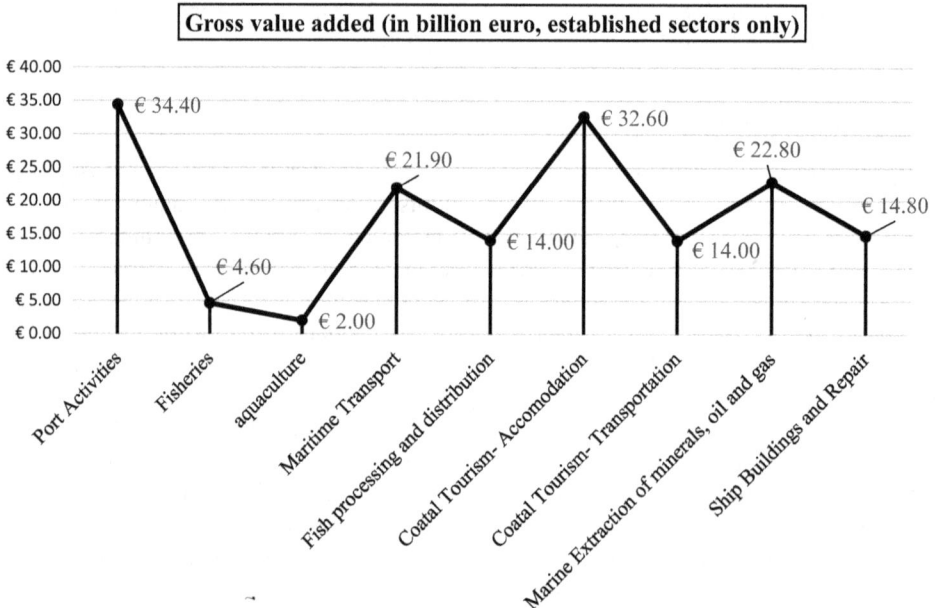

FIGURE 15.5 The marine portion expansion (gross value added by currency) (the EU Blue Economy, 2018).

TABLE 15.5
Comparing Calculated Data of pH Value of Marine Water with Lower and Upper Uncertainty pH Value of Marine Water

Year	Calculated data of pH value of marine water	Lower uncertainty limit of pH value of marine water	Upper uncertainty limit of pH value of marine water	Comment
1985	8.109	8.108	8.110	The ocean accumulates around 30% of the
1990	8.102	8.101	8.103	carbon dioxide that is released into the
1995	8.096	8.095	8.097	environment, raising ocean acidity. As the pH
2000	8.087	8.086	8.088	range is logarithmic, the pH of surface marine
2005	8.079	8.078	8.080	waters has decreased by 0.1 pH units during
2010	8.071	8.070	8.072	the industrial revolution, corresponding to a
2015	8.062	8.061	8.063	30% increase in acidity.
2018	8.057	8.056	8.057	

Source: Chou et al. 2016; Alvarez et al. 2020; FAO, 2020.

Table 15.5 lists soluble oxygen levels in coastal water over time. The coastal water score, established by the Washington State Department of Ecology, is just another approach to track marine water condition in Puget Sound, and is an indicator of the overall condition of marine waters in the Salish Sea (Wong and Rylko, 2014). It is offered here as an additional indicator of the quality of seawater. This sign also highlights a new concern to marine water health in the Salish Sea and throughout the world's oceans: micro-plastics (EPA, Marine Water Quality, 2019). There has been some research on the amount of heavy metal pollution, its origins, and its effects on marine ecosystems along various coastlines, such as the Indian Ocean and the Pacific Ocean, two of the world's largest oceans (Jayaraju et al. 2009).

15.5.2   Aquatic Biodiversity Loss and Reduction of the Blue Economy

Marine biodiversity has significant importance for enhancing the blue economy in the coastal and ocean systems. It is critical for the ocean environment to be able to withstand climatic disruptions, mitigate climatic changes, and fulfill its role as a global ecosystem (Cavicchioli et al. 2019). Ocean biodiversity, as a climate controller, must be conserved and restored (Whechel et al. 2018). Millions of species live in the marine water and ecosystems as well. Marine biodiversity is critical to the health of the seas. Climate control relies heavily on oceanic life. Marine animals are directly influenced by global warming resulting from human activities (Cavicchioli et al. 2019).

Temperatures rise causes animals of various kinds to behave in different ways. If their association with the microalgae that they shield and depend on is interrupted, some corals can quickly bleach and die (Dikens, 2018). Phytoplankton, crustaceans, and molluscs are amongst the marine animals whose calcareous skeletons or shells are directly affected by ocean acidification, which is a result of rising CO_2 absorption (Smith, 2016).

Climate changes wreak havoc on marine species in coastal locations, particularly in marshes and sea grasses, which are vital reproductive grounds and CO_2 capturing zones, and erosion and floods are two additional examples of how strong weather events devastate resources (Abisha et al. 2022). Because of the accumulated consequences of multiple changes in coastal environments, oceans are becoming increasingly sensitive to climate change (MoEF, 2001). Ocean habitats are being degraded, making them less able to respond to climate change. The significance of this crisis has received much too little appreciation.

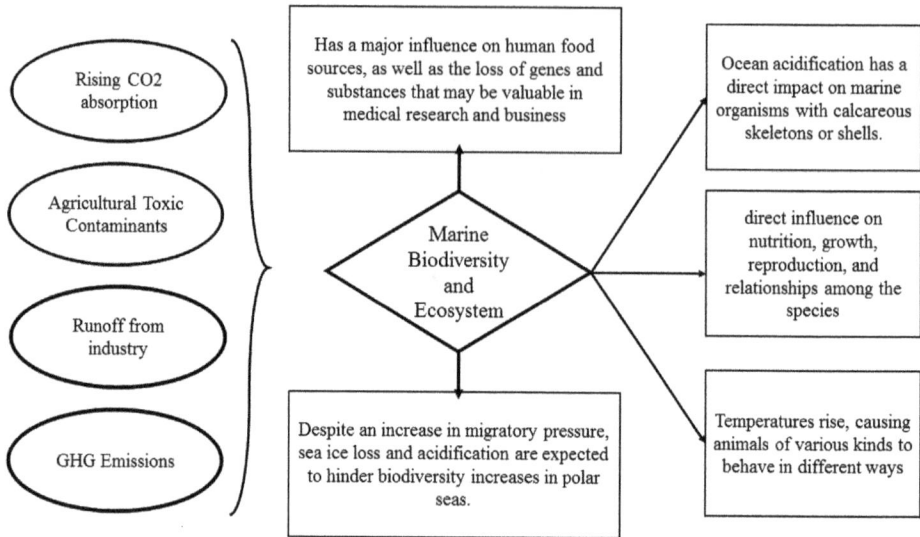

FIGURE 15.6   Marine biodiversity is critical to the health of the seas due to climatic and anthropogenic impacts.

On a worldwide basis, one third of fish populations are overfished, while another 60% of the total (Gullestad et al. 2020) are fished to their utmost sustainable levels. The impacts of growing acidification of the oceans may be felt across the world due to the global nature of ocean and circulation patterns. (Verberk, 2016). Climate change is a third factor contributing to the dwindling marine biodiversity, since it is predicted to reduce ocean net primary output by 3–10% and fish biomass by 3–25% by the end of the century (Bryndum-Buchholz et al. 2019). Coral reefs in the tropics are expected to experience more intense warming events with shorter recovery periods, resulting in massive bleaching events with significant death rates (Kubiak, 2019).

15.5.3   IMPACT OF THE NAVIGATION HAZARD ON THE BLUE ECONOMY

There are many natural disasters involving ocean processes include phenomena such as rain, tropical storms, tsunamis, storm surges, blooms of toxic algae, pathogen contamination of coastal waters, and recurring as well as long-term climate variability (National Research Council, 1999). The navigational hazards are things you could run into on the water. However, the person at the helm may constitute a 'danger to navigation' (Van Erve and Bonnor, 2006). When a boat collides with something while in motion, it is one of the most costly and dangerous claims that can arise in Marine Boat Insurance (Maloney, 2017). Investigations have shown that these tragedies are virtually always preventable if common sense and basic skills are used (Rutherford, 2017). Thousands of damaged and derelict ships litter harbors, rivers, and ports around the globe Bamford et al. (2008), posing a threat to our oceans, beaches, waterways, and Great Lakes by impeding nautical channels, inflicting environmental damage, and lowering financial and cultural value (Little, 2018). Our ocean, beaches, waterways, and Great Lakes are all under peril. This debris obstructs navigational channels, harming the ecosystem and lowers the financial and cultural value of the area. Lost and abandoned vessels can linger for years, wreaking havoc on protected ports and bays and scattering debris (Hopley, 2004). This work outlines how each coastal state deals with abandoned and derelict vessels in an effort to pull together facts and offer a thorough look at this issue (Bilkovic, 2019).

Natural disasters and maritime accidents can increase the marine trash issue. Natural catastrophes of many kinds, from hurricanes to tsunamis to floods and landslides, have devastating effects on human life and property (Shaw, 2006). Because of tremendous winds, torrential rainfall, floods, and tidal waves, extreme occurrences can carry anything as little as a cigarette butt or as massive as the top of a two-story home far out to ocean (Fradkin, 2005). Even medical waste, such as syringes and needles, may be washed into the ocean and coastal waterways during storms or other periods of strong winds or high waves. This can lead to floods and make it difficult or risky to get to people in an emergency (Pawar et al. 2016). Large volumes of marine debris can also be formed in a single incident at sea. In addition to blocking waterways, large items like complete vessels and cargo containers can create accidents that put human health and safety in jeopardy (Sheavly and Register, 2007). In order to reduce the impacts from marine debris created during a disaster, there is the NOAA Marine Debris Program (Blog, 2021). These instructions for coastal states and territories include information that can be used during a marine debris emergency. The information in these handbooks explains how different levels of government deal with maritime debris in the wake of a disaster and how to deal with dangerous material that can obstruct shipping lanes (Bhanawat, 2021).

This lost equipment can provide a threat to boats and navigational safety through tangling around engines and thrusters, or by snagging active fishing gear. Sunken and massive marine debris also serves as a hazard to boat movement (Brennan et al. 2018). When the seas are rough or the weather is unfavorable, removing abandoned fishing gear from a boat can be difficult and even dangerous. The greatest Great Lake, Lake Superior, is plagued by a ghost net issue (Sheavly and Register, 2007). Gill nets, a long-lasting form of net often employed in Lake Superior, can be destroyed by storms, wind, moving ice, waves, and other boats (Galili et al. 2010). In addition to knotting with active fishing gear and propellers, nets that have been cut loose will float for years below the water's surface, posing a danger to boaters who may not know how to free their craft from derelict gear (Tudela, 2004).

15.6 A POLLUTION MANAGEMENT APPROACH FOR A BLUE ECONOMY

15.6.1 Implementation Law and Policy

The Law of the International Maritime organization sets a fundamental legal framework within which all activity in the oceans must occur (Molenaar and Oude Elferink, 2009). It establishes the powers and obligations of states in the areas of navigation, living and non-living assets, marine ecological preservation and protection, and the development and transfer of marine technologies (Conference, 2006). Members of the World Trade Organization (WTO) have trading commitments and responsibilities that may affect the blue economy (Kohl et al. 2016). These were created to strengthen current institutional structures, but they also provide a challenge in terms of uniformity and integration (Table 15.6).

The United Nations General Assembly's recently decided (Resolution 69/292 of June 19, 2015) to establish a regional and international, valid signatory device under UNCLOS on sustainable development (Sung, 2016). It is also forms the mechanism of clean energy use in coastal ecosystems in areas beyond sovereign territory and it is a significant development in terms of investments in areas beyond national jurisdiction (Pauli, 2010; Tessnow-von Wysocki and Vadrot, 2020; Caroli et al. 2022). Africa's minimal participation in the Kyoto Protocol's Clean Development Mechanism (CDM) and carbon trading systems thus far reflects the continent's technical, geopolitical, and organizational competency limitations (Vogler, 2011; Hassan and Alam, 2019). Barriers to knowledge may be overcome with perseverance, and governments can use global legal measures to aid in the development of effective domestic Blue Economy administrative frameworks and policy recommendations.

Increasing the capacity of states to begin negotiating fair and substantial agreements requires that the essential legislative and judicial policy direction is framed by specific national circumstances

TABLE 15.6
Challenges and Opportunities for a Sustainable Blue Economy Sector

Challenges	Sectors	Opportunities
✓ Yet to be a fossil-fuel-based transportation system ✓ To ensure proper health, safety, security, and environmental protection ✓ To gain social and political support	*Maritime transportation* (KIM, 2019)	✓ To establish itself as a global marine trading centre ✓ Increased capacity and infrastructure, particularly for coastal shipping and inland waterways ✓ To improve the maritime environment's efficient approach
✓ Acquisition and management of maritime reserves, as well as conservation incentives ✓ To provide a worldwide framework or concern for ocean resource protection ✓ For mitigating the negative effects of global warming on the marine ecosystem.	*Biological variety in the sea* (Hayward, 2015; Broderick, 2015)	✓ To make money by using vast maritime resources as assets ✓ Using them as a source of nutrients for humanity ✓ To make the maritime environment more resistant to climate change's impacts
✓ To improve the foundation for sustainable fisheries management and aquaculture development ✓ Increased stakeholder involvement in the fisheries industry will help to increase openness in selection at all levels.	*Sustainable aquaculture and fishing* (FAO, 2021)	✓ To fulfill rising food and nutrient needs in terms of quantity and quality ✓ To successfully integrate aquaculture with other agricultural operations
✓ To make the complicated and varied nature of the situations encountered by maritime converter technology more understandable. ✓ To address and comprehend these loading situations, as well as the difficulties and possibilities that they provide	*Marine renewable energy* (Sánche, 2021)	✓ To create a massive supply of energy for commercial advantage while also supporting the whole global society ✓ Making a nation or zone more adaptable through new energy derived from marine resources and making them fluent or comfortable with it
✓ To drastically minimize the risk of marine pollution ✓ To maintain appropriate administration of national and international policy standards, as well as other concerns ✓ To secure the sector's long-term viability by providing fundamental financial assistance	*Coastal and marine tourism* (EU, Challenges and Opportunities for Maritime and Coastal Tourism in the EU, 2016)	✓ Coastal and maritime tourism has emerged as one of the most prominent touristic sub-sectors ✓ To give a nation's economy a substantial financial boost ✓ To assist in the long-term development of coastal regions that are more distant

(Keohane et al. 2009). It suggests that while improving institutional and legal capabilities for the Blue Economy, the following factors should be considered (Pauli, 2010; Okemwa, 2019):

- Implementing undone agreements by states based on marine policy.
- Relating ocean governance instruments to freshwater bodies and other natural resources.
- Making adaptations of legislation and policies regarding marine environment.
- Negotiating conflicts as well as marine boundaries.

- Training officials on economic sectors including international law and legislation.
- Developing integrated maritime strategies.
- Establishing 'Academic Centers of Excellence' on the Blue Economy in universities as think tanks (Pauli, 2010; Hassan and Alam, 2019).
- Establishing aquatic and marine related departments/institutions.
- Enhancing dialogues and consultations regionally to aggregate views and facilitate technologies on marine resources.
- Strengthening states' ability to execute fair and strong contractual agreements on a global scale.
- Developing environmental rules and regulations for marine and coastal ecosystem conservation, management, preservation, and sustainability (Hassan and Alam, 2019).
- Improving blue economy training, research, building capacity, and skills enhancement across the country.
- Ensuring involvement in agreements and conceptualization of a reasonable view relating the global advancement and ability to understand under UNCLOS.

The above cases highlight the significance of establishing a road map and vision for the maritime economy's long-term growth (Pauli, 2010). It focuses on the necessity to build a set of regulations for the maritime sector, as well as other institutional methods (Jakobsen et al. 2022). Establishing a blue economy ministry, coordination of the blue economy at a high office level, such as the president's or prime minister's office, or the formation of an operational coordination mechanism, are only a few examples of methods to develop (Voyer et al. 2018). The case and discourse let us see the value of long-term strategy, supervision, practical operations, and assessment in the blue economy.

15.6.2 FAIR TRADE POLICY

The fisheries sector employs 350 million people worldwide, with 90% of fisherman living in under-developed nations (Barnes-Mauthe et al. 2013). Fish is projected to be worth US$ 25 billion in emerging regions, making it the most expensive single trade product exchanged by these nations (Belton et al. 2018). Because aquaculture supplies 58% of all fish sold in global markets, revitalizing this sector can help to improve food security while also promoting social and economic inclusion for some of the world's poorest populations (Table 15.7) (FAO, 2016; Taylor et al. 2019).

TABLE 15.7
Fishermen and Fish Farmers' Employment across the World, Broken Down by Zone (Thousands)

Fishermen and fish farmers' employment across the world, broken down by zone (thousands)					
	Fisheries and aquaculture				
Region	1995	2000	2005	2010	2015
Africa	2,812	3,348	3,925	4,483	5,067
Americas	2,072	2,239	2,254	2,898	3,193
Asia	31,632	40,434	44,716	49,427	49,969
Europe	476	783	658	648	453
Oceania	466	459	466	473	479
Total	37,456	47,263	52,019	57,930	59,161

Source: Food and Agriculture Organization (FAO), 2020.

It also has the potential to promote food security, which is connected to economic growth, steady earnings, and a reduction in risk and vulnerability. If a producer earns more money, it indicates that he or she has more money to spend on food and investment (Fisher, 2018). Another key element of fair trade is that the producer is able to utilize environmentally friendly growing methods that safeguard the environment while also providing workers with safe, healthy, and humane working conditions (Sinclair, 2021). Significant signals are sent to the market by big merchants' commitments to purchase from sustainable sources. In addition, in certain situations, these commitments may yield direct money that can be used to fund sustainable practices (Potts et al. 2016).

The sustainability criteria of fresh fish's ability to support a blue economy, as well as its participation in it, is highly dependent on a few factors, including the characteristics of individual proposals, the ideological culture in which the proposals are carried out, and the overall economic environment in which they are incorporated (Potts et al. 2016).

15.6.3 USING SUSTAINABLE ENERGY AND THE BLUE ECONOMY

The ocean most likely contains enough energy in the form of heat, currents, waves, and tides to supply the world's need for electricity many times over (Owusu and Asumadu-Sarkodie, 2016). Further research and the implementation of renewable energy from the ocean is required for a broad, holistic, and sustainable energy strategy, since marine renewable energy will surely improve the climate by substituting fossil fuel plants and cutting carbon pollution (Pelc and Fujita, 2002; Al-Shetwi, 2022). Oceans are a significant source of renewable energy. For gathering that energy, different devices use different methodologies. The following are the primary sources of ocean energy (Ponta and Jacovkis, 2008):

- Tidal streams
- Thermal energy
- Currents in the ocean
- Range of tides
- Waves
- Gradients of salinity

As with any promising but new technology, it is recommended that research efforts continue but to move cautiously, prioritizing the health of the marine environment while creating clean energy as a top priority (Pelc and Fujita, 2002; Wilberforce et al. 2019).

So, as a result, it is reasonable to conclude that the future of such sustainable energy is bright, as it provides multiple advantages while having very little detrimental effect (Panwar et al. 2011). It also has the potential to become a major energy source in the long term, as well as a significant contributor to the blue economy (Ebarvia, 2016).

15.6.4 POLITICAL COMMITMENT

Political forces can have a significant impact on the pollution management approach for the blue economy. They handle pollutants both inside and externally (Voyer and van Leeuwen, 2019). Political leaders play a significant role in a nation and are, in most circumstances, policymakers (Cosgrove and Loucks, 2015). They must adopt a variety of criteria and circumstances in order to reduce pollution in the blue economy and make it long-term viable (Bond, 2019). Therefore, people are more likely to obey their instructions. Emphasizing the blue economy on a national level would have a great impact in the future economic development of a country (Bonstra et al. 2018).

At the Ocean Foundation a unique investment alternative is provided for those who have elected to withdraw from fossil fuels as part of the boycott movement, by default. As of 2019 they've raised roughly US$ 25 million in funding and they've been running for more than three and a half years

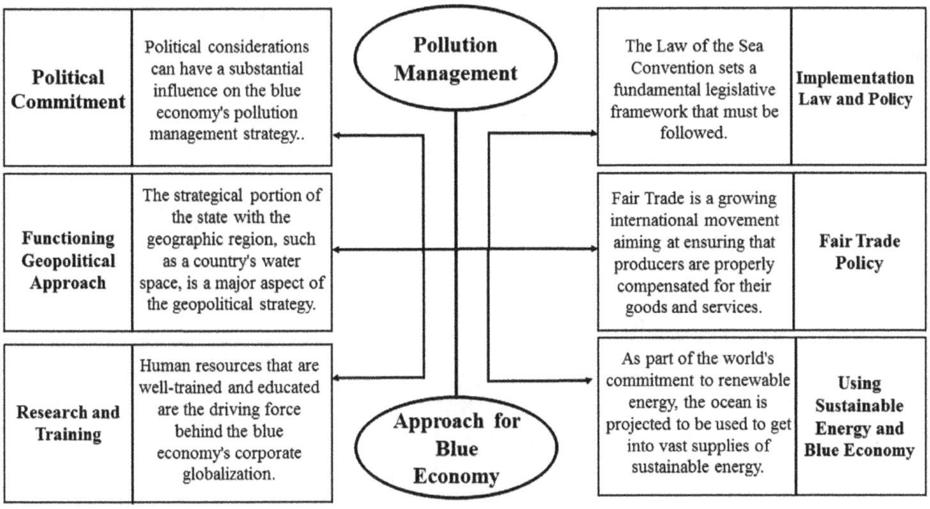

FIGURE 15.7 Political forces can have a significant impact on the pollution management approach of the blue economy.

(Cordier and Uehara, 2019). At the Ocean Foundation, every company in the portfolio is evaluated, as well as any new firms under discussion for admission, to ensure that the 'good for the ocean' statement is accurate (Cordier and Uehara, 2019). In October 2015, the very first workshop was held in Monterey, California. The conference's title is 'The Oceans in National Income Accounts (Attri and Bohler-Mulleris, Eds. 2018). Seeking Consensus on Definitions and Standards,' it gathered up 39 participants from 10 nations to discuss how to assess both the ocean economy and the (new) blue (sustainable) economy using national accounting categories (Silver et al. 2015; Colgan, 2016). The maritime economy of these ten countries was quite similar.

It will take a protracted effort to:

i) Develop a consistent set of classifications for measuring the ocean's market system, as well as a regular, adequate, and well-defined regions and.
2) Look for methods to quantify ecological integrity, which shows whether socioeconomic growth is long-term viable, and start working on a financial spreadsheet for ocean resources (Boer, 2000).

In the fall of 2016, China hosted a second summit with coastal states. Since then, The Ocean Foundation and the House of Sweden have co-sponsored and co-chaired a conversation on the Blue Economy (Spalding, 2016). The delegates at the conference discussed methods to improve transatlantic cooperation and collaboration, such as how the blue growth agenda might help reverse the present loss in ocean health while simultaneously producing employment and money. However, it protects ocean ecological services, which are difficult to quantify. The delegates at the conference can also assist with the assessment of natural resources. They teamed up with JetBlue to look at the true value of natural systems in the Caribbean (clean beaches, thriving reefs, and healthy mangroves) (Spalding, 2016).

15.6.5 Functioning Geopolitical Approach

The geopolitical approach is basically a method of understanding, explaining, and predicting international political behavior through researching foreign policy (Choudhury, 1964; Scholvin and

Wigell, 2018). Small Island authorities have meaningful ocean resources at their disposal, particularly in comparison to their mainland, which displays a huge potential for boosting economic growth while attempting to address unemployment, food security, and deprivation (Mustafa et al. 2019). The ocean based economy is more than just a perspective and looks at the ocean economy as a revenue source (Fusco et al. 2022). Therefore, managing pollutants is very essential for the blue economy. The policymaker of a country could come up with some steps or regulations to prevent the pollutants from entering a water body, for example, by reducing degradable wastes from the land water, using fewer plastic materials and not disposing of these in any water.

15.6.6 RESEARCH AND TRAINING

Well-trained, skilled, and knowledgeable human resources are the motivating power behind the development of a blue economy capable of participating in transnational corporations and the ensuing technology revolution (Farazmand, 2004). Progressive and long-term expansion is impossible without a skilled staff (Rostoka et al. 2019). When it comes to sustainability in the maritime sector, how can we explain its 'rebound' effect? Increased efficiency can result in financial savings that can be used to buy additional items and activities, therefore offsetting the ecological benefits of increased efficiency (Herring and Roy, 2007). Our Rockefeller Marine Initiative is still a financial offering that focuses on the ocean and includes investments. It is made up of companies who produce or provide goods or services that are 'good for the ocean' (Spalding, 2016).

15.7 CONCLUSION

A blue economy may considerably assist financial development provided the global blue assets in the ocean are efficiently mapped as well as incorporated within a sound institutional structure as well as based on particular laws and studies. As a result, there is potential for infrastructural growth. This will help to alleviate unemployment by introducing coastal inhabitants into the mainstream, in addition to producing jobs. According to the conclusions of this analysis, several countries have the potential to develop a blue economy that would contribute to regional economic success, but they will need strong political intentions, practical research, societal understanding, and favorable attitudes. In other words, if many countries properly utilize their advantages, they will advance more quickly. On the other hand, long-term success necessitates a strong governmental dedication, thorough study, cultural awareness, and a constructive attitude toward the blue economy. To secure the long-term viability of coastal resources, global collaboration is required, and this will assist us better in understanding the blue economy's role in encouraging socioeconomic progress.

ACKNOWLEDGEMENTS

I would like to express my gratitude to all of the anonymous authors and contributors whose articles I reviewed on many occasions to produce this scientific chapter. I have also used many websites, free domains, blogs and other sources for reviewing the literature, concepts and perceptions to build the scenarios of marine pollution and challenges for developing a sustainable blue economy. Also, I would like to show my gratitude to the Ministry of Science and Technology, Government of Bangladesh that have provided the funding support *(NST Research Allocation Project 2019-2020)* to the first author of this chapter for continuing his research on the global blue economy and seafood production and policies.

REFERENCES

Abbas, B. M. (September 1973). Water Resources Development. In: *Water for Human Development. Proceedings of the First World Conference on Water Resources*, Chicago, 24–29.

Abisha, R. Krishnani, K. K. Sukhdhane, K. Verma, A. K. Brahmane, M. and Chadha, N. K. (2022). Sustainable development of climate-resilient aquaculture and culture-based fisheries through adaptation of abiotic stresses: a review. *Journal of Water and Climate Change*, *13*(7), 2671-2689.

Abreu, F. V. (2013). O Pré-sal Brasileiro e a legislação do novo Marco regulatory: Uma avaliação geoeconômica dos recursos energéticos do pré-sal. *Revista de Geologia*, 26(1), 7–16.

Adkins, S. (2017). From Disposable Culture to Disposable People: Teaching about the Unintended Consequences of Plastics (Doctoral dissertation, Antioch University).

AFP. (2022). Oil spill in Mauritius in 2020: the captain of the Wakashio found guilty. Retrieved from africanews, Retreievd on August 7, 2022 from the https://www.africanews.com/2021/12/22/oil-spill-in-mauritius-in-2020-the-captain-of-the-wakashio-found-guilty/

Agarwala, N. (2021). Powering India's Blue Economy through ocean energy. *Australian Journal of Maritime and Ocean Affairs*, 1-27.

Ahmed, N. (1976). *An Economic Geography of Bangladesh*. Vikas Publishing Home, New Delhi.

Al-Shetwi, A. Q. (2022). Sustainable development of renewable energy integrated power sector: Trends, environmental impacts, and recent challenges. *Science of The Total Environment*, 153645.

Alam, M. and Xiangmin, X. (2019). Marine Pollution Prevention in Bangladesh: A Way Forward for Implement Comprehensive National Legal Framework. *Thalassas: An International Journal of Marine Sciences*, *35*(1), 17-27.

Alharthi, M. and Hanif, I. (2020), "Impact of blue economy factors on economic growth in the SAARC countries", *Maritime Business Review*, Vol. 5 No. 3, pp. 253–269. https://doi.org/10.1108/MABR-01-2020-0006

Allam, Z. and Jones, D. (2019). Climate change and economic resilience through urban and cultural heritage: The case of emerging small island developing states economies. *Economies*, *7*(2), 62.

Altieri, A. H. and Diaz, R. J. (2019). Dead zones: oxygen depletion in coastal ecosystems. In World seas: An environmental evaluation (pp. 453–473). Academic Press.

Álvarez, M. Fajar, N. M. Carter, B. R. Guallart, E. F. Pérez, F. F. Woosley, R. J. and Murata, A. (2020). Global ocean spectrophotometric pH assessment: consistent inconsistencies. *Environmental Science and Technology*, *54*(18), 10977–10988.

Anderson, R. C. (2014). Cetaceans and tuna fisheries in the Western and Central Indian Ocean. *International Pole and Line Federation Technical Report*, 2, 133.

Andrady, A. L. (2011) Microplastics in the marine environment. *Mar. Pollut. Bull.* 62, 1596–1605.

Anh, P. T. Kroeze, C. Bush, S. R. and Mol, A. P. (2010). Water pollution by intensive brackish shrimp farming in south-east Vietnam: causes and options for control. *Agricultural Water Management*, 97(6), 872–882.

Anon. (2013). *Report of the South-Eastern Atlantic Regional Workshop* UNEP/CBD/RW/EBSA/SEA/1/4.

Ansari, T. M. Marr, I. L. and Tariq, N. (2004). Heavy metals in marine pollution perspective-a mini review. *Journal of Applied Sciences*, *4*(1), 1-20.

Attri, N. V. and Bohler-Mulleris, N. eds (2018). *The Blue Economy Handbook of the Indian Ocean Region*, 15–16.

Awuchi, C. G. and Awuchi, C. G. (2019). Impacts of plastic pollution on the sustainability of seafood value chain and human health. *International Journal of Advanced Academic Research*, 5(11), 46–138.

Aye, W. N. Wen, Y. Marin, K. Thapa, S. and Tun, A. W. (2019). Contribution of mangrove forest to the livelihood of local communities in Ayeyarwaddy region, Myanmar. *Forests*, 10(5), 414.

Ayuk, E. Pedro, A. Ekins, P. Gatune, J. Milligan, B. Oberle, B. ... and Mancini, L. (2020). *Mineral Resource Governance in the 21st Century: Gearing extractive industries towards sustainable development*. International Resource Panel, United Nations Envio, Nairobi, Kenya.

Bamford, H. A. et al. (2008). *Interagency Report on Marine Debris Sources, Impacts, Strategies and Recommendations*.

Barbesgaard, M. (2018). Blue growth: savior or ocean grabbing? *The Journal of Peasant Studies*, 45(1), 130–149.

Barbier, E. B. and Cox, M. (2003). Does economic development lead to mangrove loss? A cross-country analysis. *Contemporary Economic Policy*, 21(4), 418–432.

Bari, A. (2017). Our oceans and the blue economy: opportunities and challenges. *Procedia Engineering*, 194, 5–11.

Barnes-Mauthe, M. Oleson, K. L. and Zafindrasilivonona, B. (2013). The total economic value of small-scale fisheries with a characterization of post-landing trends: An application in Madagascar with global relevance. *Fisheries Research*, 147, 175–185.

Belton, B. Bush, S. R. and Little, D. C. (2018). Not just for the wealthy: Rethinking farmed fish consumption in the Global South. *Global Food Security*, 16, 85–92.

Bertolino, M. Schejter, L. Calcinai, B. Cerrano, C. and Bremec, C. (2007). Sponges from a submarine canyon of the Argentine Sea. In: Custódio, M. R. Hajdu, E. LôboHajdu, G. and Muricy, G. (Eds), *Porifera Research: Biodiversity, Innovation, Sustainability*. Museu Nacional, Rio de Janeiro, pp. 189–201.

Bhanawat, C. O. (May 2021). *How To Use Parallel Indexing Techniques For Ship Navigation*. Retrieved on August 7, 2022, from marineinsight: https://marineinsight.com/marine-navigation/parallel-indexing-tec hniques-for-ship-navigation/

Biesbroek, R. Klostermann, J. Termeer, C. and Kabat, P. (2011). Barriers to climate change adaptation in the Netherlands. *Climate Law*, 2(2), 181–199.

Bilkovic, D. M. Mitchell, M. M. Davis, J. Herman, J. Andrews, E. King, A. ... and Dixon, R. L. (2019). Defining boat wake impacts on shoreline stability toward management and policy solutions. *Ocean and Coastal Management*, 182, 104945.

Binnemans, K. Jones, P. T. Blanpain, B. Van Gerven, T. Yang, Y. Walton, A. and Buchert, M. (2013). Recycling of rare earths: a critical review. *Journal of Cleaner Production*, 51, 1–22.

Blackwell, Jasmine (2022) NOAA predicts above-normal 2022 Atlantic Hurricane Season-Ongoing La Niña, above-average Atlantic temperatures set the stage for busy season ahead, National Oceanographic and Atmospheric Administration (NOAA), Retrieved on August 6, 2022, www.noaa.gov/news-release/noaa-predicts-above-normal-2022-atlantic-hurricane-season

Blog, M. D. (2021). U.S. Federal government. Retrieved from the NOAA Marine Debris Program: Retrieved on August 7, 2022, from https://marinedebris.noaa.gov/

Boer, G. J. G. Flato, M. C. Reader, and D. Ramsden (2000). A transient climate change simulation with greenhouse gas and aerosol forcing: Experimental design and comparison with the instrumental record for the twentieth century. *Climate Dyn*, 16, 405–425.

Bohler-Muller, N. (2017) The IORA Blue Economy Core Group initiative: keeping the future in mind. *Indian Ocean RIM Association Magazine*. April:24–25.

Bond, P. (2019). Blue Economy threats, contradictions and resistances seen from South Africa. *Journal of Political Ecology*, 26(1), 341–362.

Boonstra, W. J. Valman, M. and Björkvik, E. (2018). A sea of many colours–How relevant is Blue Growth for capture fisheries in the Global North, and vice versa? *Marine Policy*, 87, 340–349.

Borunda, A. (2020). We still don't know the full impacts of the BP oil spill, 10 years later. Retrievedon August 7, 2022, from nationalgeographic: https://nationalgeographic.com/science/article/bp-oil-spill-still-dont-know-effects-decade-later

Boschi, E. E. (2010). Crustáceos decápodos. In: M.B. Cousseau (Ed.). Peces, crustáceos, y moluscos registrados en el sector del Atlántico Sudoccidental comprendido entre 34ºS y 55ºS, con indicación de las especies de interés pesquero. INIDEP Serie Informe Técnico, 5, Mar del Plata, pp. 65–78.

Boyle, L. (April 15, 2020). Deepwater Horizon oil spill still affecting fish in Gulf a decade later. Retrieved on August 7, 2022 from independent: https://independent.co.uk/climate-change/news/deepwater-horizon-oil-spill-gulf-mexico-fish-pollution-a9466761.html

Brennan, M. L. Cantelas, F. Elliott, K. Delgado, J. P. Bell, K. L. Coleman, D. ... and Ballard, R. D. (2018). Telepresence-enabled maritime archaeological exploration in the deep. *Journal of Maritime Archaeology*, 13(2), 97–121.

Briggs, J. C. (2016). Global biodiversity loss: exaggerated versus realistic estimates. *Environmental Skeptics and Critics*, 5(2), 20.

Broderick, A. C. (2015). Grand challenges in marine conservation and sustainable use. *Frontiers*.

Bryndum-Buchholz, A. Tittensor, D. P. Blanchard, J. L. Cheung, W. W. Coll, M. Galbraith, E. D. ... and Lotze, H. K. (2019). Twenty-first-century climate change impacts on marine animal biomass and ecosystem structure across ocean basins. *Global Change Biology*, 25(2), 459-472.

Campos, E. J. D. Gonçalves, J. E. and Ikeda, Y. (1995). Water mass characteristics and geostrophic circulation in the South Brazil Bight: Summer of 1991. *J. Geophys. Res.* 100(C9), 18537–18550, doi: 10.1029/ 95JC01724.

Caroli, M. et al. (2022). Recommendations on Complementary Feeding as a Tool for Prevention of Non-Communicable Diseases (NCDs)—Paper Co-Drafted by the SIPPS, FIMP, SIDOHaD, and SINUPE Joint Working Group. *Nutrients, 14*(2), 257.

Cavicchioli, R. et al. (2019). Scientists' warning to humanity: microorganisms and climate change. *Nature Reviews Microbiology*, *17*(9), 569-586.

Ceballos, G. Ehrlich, P. R. Barnosky, A. D. García, A. Pringle, R. M. and Palmer, T. M. (2015). Accelerated modern human–induced species losses: Entering the sixth mass extinction. *Science advances*, *1*(5), e1400253.

Chang, H. and Bonnette, M. R. (2016). Climate change and water-related ecosystem services: impacts of drought in California, USA. *Ecosystem Health and Sustainability*, 2(12), e01254.

Chen, W. and Dong, B. (2019). Anthropogenic impacts on recent decadal change in temperature extremes over China: relative roles of greenhouse gases and anthropogenic aerosols. *Climate Dynamics*, 52(5), 3643–3660.

Chou, W. C. Gong, G. C. Yang, C. Y. and Chuang, K. Y. (2016). A comparison between field and laboratory pH measurements for seawater on the E ast C hina S ea shelf. *Limnology and Oceanography: Methods*, *14*(5), 315-322.

Choudhury, A. M. (1964). *Working Plan for the sunderban Forest Division for the period 1960–61 to 1979–80*. East Pakistan Government Press, Dhaka.

Coe, N. M. Hess, M. Yeungt, H. W. C. Dicken, P. and Henderson, J. (2017). 'Globalizing'regional development: a global production networks perspective. In *Economy* (pp. 199-215). Routledge.

Colgan, C. S. and Kildow Dr, J. T. (2013). Understanding the Ocean economy within regional and national contexts. *Presentations*. 1. Retrieved on August 7, 2022, from https://cbe.miis.edu/cbe_presentations/1

Colgan, C. S. (2016). Measurement of the ocean economy from national income accounts to the sustainable blue economy. *Journal of Ocean and Coastal Economics*, 2(2), 12.

Commonwealth, T. (2022). *Home > Blue Economy*. Retrieved from August 7, 2022, from https://thecommo nwealth.org/

Conference, I. L. (2006). *Maritime Labour Convention, 2006*.

COP26. (2021). *COP26 Event: Funding the Sustainable Blue Economy, 11 Nov 2021 to 11 Dec 2021*, Glasgow, UK. Retrieved on August 7, 2022, from https://unepfi.org/events/cop26-event-funding-the-sustainable-blue-economy/

Cordier, M. and Uehara, T. (2019). How much innovation is needed to protect the ocean from plastic contamination? *Science of the Total Environment*, 670, 789–799.

Cosgrove, W. J. and Loucks, D. P. (2015). Water management: Current and future challenges and research directions. *Water Resources Research*, 51(6), 4823-4839.

Cousteau, Jacques-Yves (2022) Clean oceans and the blue economy overview 2022, European Investment Bank 98-100 boulevard Konrad Adenauer L-2950 Luxembourg. Retrieved on 28 July 2022 from https://www.eib.org/en/publications/clean-oceans-and-blue-economy-overview-2020.

Crawford, R. J. M. et al. (2000). Initial impact of the Treasure oil spill on seabirds off western South Africa. *South African Journal of Marine Science*, *22*(1), 157-176.

Cuthbert, R. J. Cooper, J. and Ryan, P. G. (2014). Population trends and breeding success of albatrosses and giant petrels at Gough Island in the face of at-sea and on-land threats. *Antarctic Science* , 26(2), 163–171.

Davies, T. D. and Baum, J. K. (2012). Extinction risk and overfishing: reconciling conservation and fisheries perspectives on the status of marine fishes. *Scientific Reports*, 2(1), 1–9.

Dias, M. P. et al. (2019). Threats to seabirds: a global assessment. *Biological Conservation*, 237, 525–537.

Dietz, R. and O'Neill, D. (2013). *Enough Is Enough: Building a Sustainable Economy in a World of Finite Resources*. Routledge, New York, USA.

Dikens, J. (2018). The Decline of Marine Biodiversity. *Ocean and Climate Platform*. Paris.

Delgado, C. L. (2003). Fish to 2020: Supply and demand in changing global markets (Vol. 62). *WorldFish*.

Ebarvia, M. C. M. (2016). Economic assessment of oceans for sustainable blue economy development. *Journal of Ocean and Coastal Economics*, 2(2), 7.

Eikeset, A. M. Mazzarella, A. B. Davíðsdóttir, B. Klinger, D. H. Levin, S. A. Rovenskaya, E. and Stenseth, N. C. (2018). What is blue growth? The semantics of "Sustainable Development" of marine environments. *Marine Policy*, *87*, 177-179.

EPA. (December 2019). *Marine Water Quality*. Retrieved from United States Environmental Protection Agency: Retrieved on August 7, 2022, from https://epa.gov/salish-sea/marine-water-quality

Eric Jordán-Dahlgren Rosa, E. M. (2007). *The Atlantic Coral Reefs of Mexico*. ICML, Universidad Nacional Autónoma de Mexico, Mexico.

EU. (2016). *Challenges and Opportunities for Maritime and Coastal Tourism in the EU*. Brussel, EU.

EU. (2019). *The EU Blue Economy Report* . Luxembourg, EU.

FAO. (2014). *The State of World Fisheries and Aquaculture*. Retrieved on August 7, 2022, from http://fao.org/3/a-i3720e.pdf.

FAO. (2020). The State of Food Security and Nutrition in the World 2020. Transforming food systems for affordable healthy diets. Rome, FAO. Retrievd on July 16, 2022, from https://doi.org/10.4060/ca9692en.

FAO. (2021). *Future Challenges for the CWP*. FAO.

FAO (2016). FAOSTAT. Food and Agriculture Organization of the United Nations, Rome, Italy. Retrieved on August 7, 2022, from http://faostat.fao.org/default.aspx

Farazmand, A. (2004). Innovation in strategic human resource management: building capacity in the age of globalization. *Public Organization Review*, 4(1), 3–24.

Fattah, Q. A. (November 27–29, 1979). *Protection of Marine Environment and Related Ecosystems of St. Martin's Island*. Proceedings of the National Seminar on Protection of Marine Environment and Related Ecosystems, Dhaka, 104–108.

Fisher, E. (2018). Solidarities at a distance: Extending Fairtrade gold to east Africa. *The Extractive Industries and Society*, 5(1), 81-90.

Forbes, J. M. (1995). Tidal and planetary waves. *The Upper Mesosphere and Lower Thermosphere: A Review of Experiment and Theory, Geophys. Monogr. Ser*, 87, 67-87.

Forrest, A. Giacovazzi, L. Dunlop, S. Reisser, J. Tickler, D. Jamieson, A. and Meeuwig, J. J. (2019). Eliminating plastic pollution: how a voluntary contribution from industry will drive the circular plastics economy. *Frontiers in Marine Science*, 6;627. doi: 10.3389/fmars.2019.00627

Fradkin, P. L. (2005). *The great earthquake and firestorms of 1906: how San Francisco nearly destroyed itself*. Univ of California Press, USA

Fujita, M. Krugman, P. (2004). The new economic geography: Past, present and the future. In: Florax, R.J.G.M. Plane, D.A. (eds) Fifty Years of Regional Science. Advances in Spatial Science. (pp. 139–164). Springer, Berlin, Heidelberg. https://doi.org/10.1007/978-3-662-07223-3_6

Funge-Smith, S. Briggs, M. and Miao, W. (2012). Regional overview of fisheries and aquaculture in Asia and the Pacific 2012. *RAP Publication (FAO)*.

Fusco, L. M. Knott, C. Cisneros-Montemayor, A. M. Singh, G. G. and Spalding, A. K. (2022). Blueing business as usual in the ocean: Blue economies, oil, and climate justice. *Political Geography*, 98, 102670.

Galili, E. Rosen, B. and Sharvit, J. (2010). Artifact assemblages from two roman shipwrecks off the Carmel Coast. *Atiqot*, 63, 61–110.

Gameiro, E. S. C. (2019). Humanity is being driven ashore: a juridical and political essay on marine plastic pollution (Doctoral dissertation).

Geisendorf, S. and Pietrulla, F. (2018). The circular economy and circular economic concepts a literature analysis and redefinition. *Thunderbird International Business Review*, 60(5), 771-782.

Gershwin, L. A. (2013). *Stung! On Jellyfish Blooms and the Future of the Ocean*. University of Chicago Press. **Chicago, IL 60637 U.S.A.**

Geyer, R. Jambeck, J. R. and Law, K. L. (2017). Production, use, and fate of all plastics ever made. *Science Advances*, 3(7), e1700782.

Goel, V. Luthra, P. Kapur, G. S. and Ramakumar, S. S. V. (2021). Biodegradable/bioplastics: myths and realities. *Journal of Polymers and the Environment*, 29(10), 3079–3104.

Goldberg, O. (2011). Biodegradable plastics: a stopgap solution for the intractable marine debris problem. *Tex. Envtl. LJ*, 42, 307.

Goldstein, M. C. and Goodwin, D. S. (2013). Gooseneck barnacles (Lepas spp.) ingest microplastic debris in the North Pacific Subtropical Gyre. *PeerJ*, 1, e184.

Gonzalez-Silvera, A. Santamaria-del-Angel, E. Garcia, V. M. Garcia, C. A. Millán-Nuñez, R. and Muller-Karger, F. (2004). Biogeographical regions of the tropical and subtropical Atlantic Ocean off South America: classification based on pigment (CZCS) and chlorophyll-a (SeaWiFS) variability. *Continental Shelf Research*, 24(9), 983-1000.

GPGP. (2019). Caryl-Sue, National Geographic Society. *Great Pacific Garbage Patch. Retrieved on August 7, 2022, from* https://nationalgeographic.org/encyclopedia/great-pacific-garbage-patch/

Grilli, G. Tyllianakis, E. Luisetti, T. Ferrini, S. and Turner, R. K. (2021). Prospective tourist preferences for sustainable tourism development in Small Island Developing States. *Tourism Management*, 82, 104178.

Grorud-Colvert, K. and Lubchenco, J. (2017). *Do Ocean Preserves Actually Work?*

Gullestad, P. Sundby, S. and Kjesbu, O. S. (2020). Management of transboundary and straddling fish stocks in the Northeast Atlantic in view of climate-induced shifts in spatial distribution. *Fish and Fisheries*, *21*(5), 1008-1026.

Häder, D. P. Kumar, H. D. Smith, R. C. and Worrest, R. C. (2007). Effects of solar UV radiation on aquatic ecosystems and interactions with climate change. *Photochemical and Photobiological Sciences*, 6(3), 267–285.

Halsband, C. and Herzke, D. (2019). Plastic litter in the European Arctic: what do we know? *Emerging Contaminants*, 5, 308–318.

Harse, G. A. (2011). Plastic, the great pacific garbage patch, and international misfires at a cure. UCLA J. Envtl. L. and Pol'y, 29, 331.

Hassan, D. and Alam, Md. A. A. (2019) "Institutional Arrangements for the Blue Economy: Marine Spatial Planning a Way Forward," Journal of Ocean and Coastal Economics: Vol. 6: Iss. 2, Article 10. DOI: https://doi.org/10.15351/2373-8456.1107

Hasan, M. M. Hossain, B. S. Alam, M. J. Chowdhury, K. A. Al Karim, A. and Chowdhury, N. M. K. (2018). The prospects of blue economy to promote Bangladesh into a middle-income country. *Open Journal of Marine Science*, 8(03), 355.

Hatje, V. Andrade, R. L. Oliveira, C. C. Polejack, A. and Gxaba, T. (2021). Pollutants in the South Atlantic Ocean: sources, knowledge gaps and perspectives for the decade of ocean science. *Frontiers in Marine Science*. 8:644569. doi: 10.3389/fmars.2021.644569

Henson, S. A. et al. (2010). Detection of anthropogenic climate change in satellite records of ocean chlorophyll and productivity. *Biogeosciences*, *7*(2), 621-640.

Herring, H. and Roy, R. (2007). Technological innovation, energy efficient design and the rebound effect. *Technovation*, 27(4), 194–203.

Hopley, D. (2004). Geology and hydrogeology of carbonate islands. *Developments in Sedimentology*, 835–866.

Huang, Y. (2016). Understanding China's Belt and Road initiative: motivation, framework and assessment. *China Economic Review*, *40*, 314-321.

Houghton, J. T. Y. Ding, D. J. Griggs, M. Noguer, P. J. van der Linden, X. Dai, K. Maskell, and C. A. Johnson (Eds) (2001). *Climate Change 2001: The Scientific Basis*. Cambridge University Press, 881.

Howell, E. A. Bograd, S. J. Morishige, C. Seki, M. P. and Polovina, J. J. (2012). On North Pacific circulation and associated marine debris concentration. *Mar. Pollut. Bull*. 65, 16–22.

Hu, C. Montgomery, E. T. Schmitt, R. W. and Muller-Karger, F. E. (2004). The dispersal of the Amazon and Orinoco River water in the tropical Atlantic and Caribbean Sea: Observation from space and SPALACE floats. *Deep-Sea Research Part II: Topical Studies in Oceanography*, v. 51, issue 10-11, p. 1151-1171, Elsevier Publication, Netherlands.

Huang, Y. (2014). Quantification of global primary emissions of PM2. 5, PM10, and TSP from combustion and industrial process sources. *Environmental Science and Technology*, 48(23), 13834–13843.

Hunt Jr, G. L. et al. (2016). Advection in polar and sub-polar environments: Impacts on high latitude marine ecosystems. *Progress in Oceanography*, *149*, 40-81.

Iftekhar, M. S. (2006, August). Conservation and management of the Bangladesh coastal ecosystem: overview of an integrated approach. In *Natural resources forum* (Vol. 30, No. 3, pp. 230-237). Oxford, UK: Blackwell Publishing Ltd.

Islam, M. S. and Tanaka, M. (2004). Impacts of pollution on coastal and marine ecosystems including coastal and marine fisheries and approach for management: a review and synthesis. *Marine pollution bulletin*, 48(7-8), 624-649.

Islam, Md. Nazrul, Kitazawa, D. Kokuryo, N. Tabeta, S. Honma, T. Komatsu, N (2012) Numerical modeling on transition of dominant algae in Lake Kitaura, Japan, *Ecological Modelling*, 242: 146-163, Elsevier publication, DOI: 10.1016/j.ecolmodel.2012.05.013

Islam, M. Kitazawa, D. Hamill, T. and Park, H. D. (2013). Modeling mitigation strategies for toxic cyanobacteria blooms in shallow and eutrophic Lake Kasumigaura, Japan. *Mitigation and adaptation strategies for global change*, 18(4), 449-470.

IUCN. (2021). *Marine Plastic Pollution*. International Union for Conservation of Nature.

Jackson, A. (April 30, 2019). How Do We Pollute? 9 Types of Ocean Pollution. *Spring Power and Gas*. Retrieved December 15, 2021, from https://springpowerandgas.us/how-do-we-pollute-9-types-of-ocean-pollution/

Jackson, J. and Seeger, P. (2013). Overfishing: A Powerful Agent of Ecosystem Change. *Stung! On Jellyfish Blooms and the Future of the Ocean*, 111.

Jakobsen, S. E. Uyarra, E. Njøs, R. and Fløysand, A. (2022). Policy action for green restructuring in specialized industrial regions. *European Urban and Regional Studies*, 29(3), 312-331.

Jayaraju, N. Sundara Raja Reddy, B. C. and Reddy, K. R. (2009). Metal pollution in coarse sediments of Tuticorin coast, Southeast coast of India. *Environmental Geology*, 56(6), 1205-1209.

Jiang, G. Keller, J. and Bond, P. L. (2014). Determining the long-term effects of H2S concentration, relative humidity and air temperature on concrete sewer corrosion. *Water research*, 65, 157-169.

Joyce, S. (2000). The dead zones: oxygen-starved coastal waters. *Environmental health perspectives*, 108(3), A120-A125.

Kako, S. Isobe, A. Kataoka, T. and Hinata, H. (2014). A decadal prediction of the quantity of plastic marine debris littered on beaches of the East Asian marginal seas. *Mar. Pollut. Bull.* 81, 174–184.

Kamal, A. H. M. and Khan, M. A. A. (2009). Coastal and estuarine resources of Bangladesh: management and conservation issues. Maejo International Journal of Science and Technology, 3(2), 313-342.

Kaplan, S. (August 5, 2021). A critical ocean system may be heading for collapse due to climate change; study finds. Retrieved from washingtonpost: Retrieved on Janaury 14, 2022 from https://washingtonpost.com/climate-environment/2021/08/05/change-ocean-collapse-atlantic-meridional/

Keohane, R. O. Macedo, S. and Moravcsik, A. (2009). Democracy-enhancing multilateralism. *International Organization*, 63(1), 1–31.

KIM, S. (2019). *Challenges and Opportunities of Sustainable Shipping in the Asia-Pacific Region*. Bangkok.

Kobayashi, M. Msangi, S. Batka, M. Vannuccini, S. Dey, M. M. and Anderson, J. L. (2015). Fish to 2030: the role and opportunity for aquaculture. Aquaculture economics and management, 19(3), 282-300.

Kohl, T. Brakman, S. and Garretsen, H. (2016). Do trade agreements stimulate international trade differently? Evidence from 296 trade agreements. The World Economy, 39(1), 97-131.

Kubiak, L. (2019). Marine biodiversity in dangerous decline, finds new report. Retrieved from NRDC: Retrieved on August 8, 2022, from https://nrdc.org/experts/lauren-kubiak/marine-biodiversity-dangerous-decline-finds-new-report

Laurette, B. Valayden, R. Steele, M. and Takada, H. (2021). *Mauritius One Year after Oil Disaster*. Greenpeace.

Landrigan, P. J. et al. (2020). Human health and ocean pollution. *Annals of global health,* 86(1):151. doi: 10.5334/aogh.2831. PMID: 33354517; PMCID: PMC7731724.

Lavers, J. L. Dicks, L. Dicks, M. R. and Finger, A. (2019). Significant plastic accumulation on the Cocos (Keeling) Islands, Australia. *Scientific Reports*, 9(1), 1-9.

Lebreton, L. Van Der Zwet, J. Damsteeg, J. W. Slat, B. Andrady, A. and Reisser, J. (2017). River plastic emissions to the world's oceans. *Nature communications*, 8(1), 1-10.

Lebreton, L. et al. (2018). *Evidence that the Great Pacific Garbage Patch Is Rapidly Accumulating Plastic.* Nature.

Lebreton, L. et al. (2018). Evidence that the Great Pacific Garbage Patch is rapidly accumulating plastic. Scientific reports, 8(1), 1-15.

Lee, K. H. Noh, J. and Khim, J. S. (January 31, 2020). The Blue Economy and the United Nations' Sustainable Development Goals: Challenges and opportunities. *Environment International* 137, 105528. Retrieved December 10, 2021, from https://sciencedirect.com/science/article/pii/S0160412019338255

Leiserowitz, A. (2007). Communicating the risks of global warming: American risk perceptions, affective images, and interpretive communities. *Creating a Climate for Change: Communicating Climate Change and Facilitating Social Change*, 44–63.

Lincoln, S. et al. (2021). A regional review of marine and coastal impacts of climate change on the ROPME sea area. *Sustainability*, 13(24), 13810.

Little, D. I. (2018). Mangrove restoration and mitigation after oil spills and development projects in East Africa and the Middle East. C. Makowski, C. W. Finkl (eds.), In *Threats to Mangrove Forests* (pp. 637–698). Springer, Cham.

Livesay, H. N. Vance, P. H. Trevino, E. and Weissfeld, A. S. (2021). Algae-associated illnesses in humans and dogs and presence of algae on buildings and other structures. *Clinical Microbiology Newsletter*, 43(2), 9–13.

Longhurst, A. R. (1998). *Ecological Geography of the Sea*. Academic Press, San Diego.

Maes, T. et al. (2021). From Pollution to Solution: A Global Assessment of Marine Litter and Plastic Pollution.

Magazine, S. (2017). Do Ocean Preserves Actually Work? *Smithsonian Magazine*. Retrieved December 14, 2021, from https://smithsonianmag.com/science-nature/momentum-grows-ocean-preserves-how-well-do-they-work-180961690/

Mallin, F. and Mads, Barbesgaard. "Awash with contradiction: capital, ocean space and the logics of the Blue Economy Paradigm." *Geoforum* 113 (2020): 121-132.

Maloney, B. P. (2017). Interpleader of Maritime Claims: A Collision Between Double Vexation and Admiralty Jurisdiction. *Tul. L. Rev. 92*, 1063.

Martel, S. (2010). *Lessons from Another Oil Spill*. The National Wildlife Federation.

Mason, W. and Folkerts, G. (2013). Pollution Destabilizes Ecosystems. *Stung! On Jellyfish Blooms and the Future of the Ocean*, 187.

McDonagh, E. L. Bryden, H. L. King, B. A. Sanders, R. J. Cunningham, S. A. and Marsh, R. (2005). Decadal changes in the South Indian Ocean thermocline. *Journal of Climate, 18*(10), 1575-1590.

Miloslavich, P. et al. (2011). Marine biodiversity in the Atlantic and Pacific coasts of South America: knowledge and gaps. *PloS one, 6*(1), e14631.

Miner, C. M. et al. (2018). Large-scale impacts of sea star wasting disease (SSWD) on intertidal sea stars and implications for recovery. *PLoS One, 13*(3), e0192870.

Ministry of Foreign Affairs (MoFA, 2020). (n.d.). Retrieved December 10, 2021, from https://mofa.gov.bd/site/page/8c5b2a3f-9873–4f27–8761–2737db83c2ec/OCEAN/BLUE-ECONOMY--FOR-BANGLADESH

MoEF. (2001). 'Ministry of Environment and Forest,' Bangladesh: State of Environment. Retrieved on May 7, 2022, from http://moef.gov.bd/html/state_of_env/state_of_env.html

Molenaar, E. J. and Oude Elferink, A. G. (2009). Marine protected areas in areas beyond national jurisdiction-the pioneering efforts under the OSPAR convention. *Utrecht Law Review, 5*(1), 5–20.

Moore, T. (2020). Plastic pollution: Aldabra in Indian Ocean has most waste ever seen on any island, say scientists. *Sky News*. Retrieved December 15, 2021, from https://news.sky.com/story/plastic-pollution-aldabra-in-indian-ocean-has-most-waste-ever-seen-on-any-island-say-scientists-12068045

Mukherji, M. J. (2021). Embracing Curzon's Political Vision to Secure India's Cultural and Political Borders. ELECTRONIC JOURNAL OF SOCIAL AND STRATEGIC STUDIES, 2, 47-58.

Mustafa, S. Shapawi, R. and Estim, A. (2019). Higher Education for Blue Growth, Employment and Sustainable Development. *Blue Growth and Blue Economy in the Context of Development Policies and Priorities in Malaysia*, 92.

Naazer, M. A. (2018). Politics, tourism and regional cooperation in South Asia. *Pakistan Journal, 54*(1), 20-42.d

Naik Mayur, S. Supnekar Santosh, P. and Pawar Prabhakar, R. (2021). Assessment of marine debris and plastic polymer types along the Panvel Creek, Navi Mumbai, West Coast of India. *Intern. J. Zool. Invest, 7*(1), 278-293.

National Geographic. (September 20, 2019). Tires: The plastic polluter you never thought about. Retrieved from nationalgeographic: Retrieved on 22 July 22, 2022, from https://nationalgeographic.com/environment/article/tires-unseen-plastic-polluter

National Geographic (2022) Marine Pollution. National Geographic Headquarters, 1145 17th Street NW, Washington, DC 20036. Retrieved on 28, July 2022 from https://education.nationalgeographic.org/resource/marine-pollution

National Research Council. (1999). Climate and weather, coastal hazards, and public health. In *From monsoons to microbes: Understanding the ocean's role in human health*. National Academies Press, USA.

Nature Hub. (November 27, 2019). Is Glass Packaging Actually Sustainable? Retrieved on 21 May 2022 from medium: https://medium.com/naturehub/is-glass-packaging-actually-sustainable-3b06ac1b16b3

NOAA. (2017). Historical El Nino/La Nina episodes (1950–present). *Oceanic Niño Index (ONI)*. Retrieved on March 31, 2017, from http://origin.cpc.ncep.noaa.gov/products/analysis_monitoring/ensostuff/ONI_v4.shtml

NRDC. (2015). *Encourage Textile Manufacturers to Reduce Pollution*.

Ohlan, R. (2016). Dairy economy of India: Structural changes in consumption and production. *South Asia Research, 36*(2), 241-260.

Okemwa, E. M. A. (2019). *Harnessing the Potentials of the Blue Economy for Kenya's Sustainable Development.*

Owusu, P. A. and Asumadu-Sarkodie, S. (2016). A review of renewable energy sources, sustainability issues and climate change mitigation. *Cogent Engineering, 3*(1), 1167990.

Pallardy, R. (2010). Deepwater Horizon oil spill environmental disaster, Gulf of Mexico. Retrieved from britannica: Retrieved August 5, 2022, from https://britannica.com/event/Deepwater-Horizon-oil-spill

Panwar, N. L. Kaushik, S. C. and Kothari, S. (2011). Role of renewable energy sources in environmental protection: A review. *Renewable and sustainable energy reviews*, 15(3), 1513-1524.

Patil, P. G. Virdin, J. Colgan, C. S. Hussain, M. G. Failler, P. Vegh, T. (2018) Toward a Blue Economy: A Pathway for Bangladesh's Sustainable Growth. World Bank, Washington, DC. World Bank. https://openknowledge.worldbank.org/handle/10986/30014 License: CC BY 3.0 IGO."

Pattiaratchi, C. et al. (2021). Plastics in the Indian Ocean—sources, fate, distribution and impacts. Retrived on August 8, 2022, from https://os.copernicus.org/preprints/os-2020-127/os-2020-127.pdf

Pauli, G. A. (2010). *The Blue Economy: 10 Years, 100 Innovations, 100 Million Jobs*. Paradigm publications, *Boulder, Colorado, USA*

Pavičić, M. (2015). Fishing for litter activities in the fishing port Vira, island of Hvar, Croatia.

Pawar, P. R. Shirgaonkar, S. S. and Patil, R. B. (2016). Plastic marine debris: Sources, distribution and impacts on coastal and ocean biodiversity. *PENCIL Publication of Biological Sciences*, 3(1), 40-54.

Pelc, R. and Fujita, R. M. (2002). Renewable energy from the ocean. *Marine Policy*, 26(6), 471-479.

Piontkovski, S. A. and Castellani, C. (2009). Long-term declining trend of zooplankton biomass in the Tropical Atlantic. *Hydrobiologia*, 632(1), 365-370.

Pittman, S. J. et al. (2014). Fish with chips: tracking reef fish movements to evaluate size and connectivity of Caribbean marine protected areas. *PLoS One*, 9(5), e96028.

Pollutants in the Arctic. Npolar. no. (2021). Retrieved December 15, 2021, from https://npolar.no/en/themes/pollutants-in-the-arctic/.

Polovina, J. J. Howell, E. A. and Abecassis, M. (2008). Ocean's least productive waters are expanding. *Geophysical Research Letters*, 35(3).

Ponta, F. L. and Jacovkis, P. M. (2008). Marine-current power generation by diffuser-augmented floating hydro-turbines. *Renewable Energy*, 33(4), 665–673.

Potgieter, T. (2018). Oceans economy, blue economy, and security: notes on the South African potential and developments. *Journal of the Indian Ocean Region*, 14(1), 49-70.

Potts, S. G. et al. (2016). Safeguarding pollinators and their values to human well-being. *Nature*, 540(7632), 220-229.

Rabalais, N. N. Turner, R. E. and Wiseman Jr, W. J. (2002). Gulf of Mexico hypoxia, aka" The dead zone". Annual Review of ecology and Systematics, 235-263.

Rahman. (2006). A study on coastal water pollution of Bangladesh in the Bay of Bengal. Retrieved on August 7, 2022, from http://dspace.bracu.ac.bd/xmlui/bitstream/handle/10361/215/05268007.pdf?sequence=3andisAllowed=y

Rallo, M. Lopez-Anton, M. A. Contreras, M. L. and Maroto-Valer, M. M. (2012). Mercury policy and regulations for coal-fired power plants. *Environmental Science and Pollution Research*, 19(4), 1084–1096.

Regan, H. (2021). From Norway to Canada, the Arctic Ocean is being polluted by tiny plastic fibers from our clothes. *CNN*. Retrieved December 15, 2021, from https://edition.cnn.com/style/article/arctic-polyester-fibers-study-intl-hnk-trnd/index.html

Reimann, N. (August 5, 2021). Major Atlantic current may be on the verge of collapse, scientists warn. Retrieved on May 10, 2022 from Forbes: https://forbes.com/sites/nicholasreimann/2021/08/05/major-atlantic-current-may-be-on-the-verge-of-collapse-scientists-warn/?sh=61b5294847aa

Riera, R. Becerro, M. A. Stuart-Smith, R. D. Delgado, J. D. and Edgar, G. J. (2014). Out of sight, out of mind: Threats to the marine biodiversity of the Canary Islands (NE Atlantic Ocean). Marine pollution bulletin, 86(1-2), 9-18.

Roberts, C. Yost, M. Ransom, C. and Creech, E. (2021). *Effects of Irrigation, Herbicide, and Oat Companion Crop on Spring-Seed Alfalfa*. In Paper presented at the Western Society of Crop Science Annual Meeting (WSCS).

Robinson, S. A. (2020). A richness index for baselining climate change adaptations in small island developing states. *Environmental and Sustainability Indicators*, 8, 100065.

Rogerson, C. M. (2020). Coastal and marine tourism in the Indian Ocean Rim Association states: overview and policy challenges. *Geo Journal of Tourism and Geosites*, 29(2), 715–731.

Rostoka, Z. Locovs, J. and Gaile-Sarkane, E. (2019). Open innovation of new emerging small economies based on university-construction industry cooperation. *Journal of Open Innovation: Technology, Market, and Complexity*, 5(1), 10.

Rudge, P. (2021). Beyond the blue economy: creative industries and sustainable development in small island developing states. In written by Peter Rudge, ISBN 9780367820251, Routledge, CRC press, NY, USA

Rutherford, D. (June 2017). Navigational Hazards or a Hazard to Navigation? Retrieved from Boatus: https://boatus.com/expert-advice/expert-advice-archive/2017/june/navigational-hazards-or-a-hazard-to-navigation

Sadally, S. B. (2018). IORA's Policy Framework on the Blue Economy. *The Blue Economy Handbook of the Indian Ocean Region*, 190.

Samenow, J. (2021) Hard to comprehend': Experts react to record 121 degrees in Canada, Retrieved on August 06, 2022, from https://www.washingtonpost.com/weather/2021/06/30/canada-record-heat-experts-react/

Sánche, S. O. (April 30, 2021). *The Challenges and Opportunities of Marine Renewables*. University of Strathclyde. **Glasgow**, Scotland, UK

Sandooyea, S. Takada, H. and Steele, M. (February 6, 2021). The devastation of the Mauritius oil spill is still unaddressed. Retrieved on March 15, 2022 from aljazeera: https://aljazeera.com/opinions/2021/2/6/we-must-make-sure-the-mauritius-oil-spill-does-not-repeat

Sarker, J. Md, Patwary, S. A. Md, Uddin, A. M. M. B. Hasan, M. Md, Tanmay, M. H. et al. (2016). Macrobenthic Community Structure—An Approach to Assess Coastal Water Pollution in Bangladesh. Retrieved on August 8, 2022, from https://ej-geo.org/index.php/ejgeo/article/view/133

Sarker, S. Bhuyan, M. A. H. Rahman, M. M. Islam, M. A. Hossain, M. S. Basak, S. C. and Islam, M. M. (2018). From science to action: Exploring the potentials of Blue Economy for enhancing economic sustainability in Bangladesh. *Ocean and Coastal Management*, *157*, 180-192.

Sayed, E. T. Wilberforce, T. Elsaid, K. Rabaia, M. K. H. Abdelkareem, M. A. Chae, K. J. and Olabi, A. G. (2021). A critical review on environmental impacts of renewable energy systems and mitigation strategies: wind, hydro, biomass and geothermal. *Science of the Total Environment*, 766, 144505.

Schejter, L. et al. (2016). Namuncurá Marine Protected Area: an oceanic hot spot of benthic biodiversity at Burdwood Bank, Argentina. *Polar Biology*, *39*(12), 2373-2386.

Schnurr, R. E. et al. (2018). Reducing marine pollution from single-use plastics (SUPs): A review. *Marine pollution bulletin*, *137*, 157-171.

Scholvin, S. and Wigell, M. (2018). Power politics by economic means: Geoeconomics as an analytical approach and foreign policy practice. *Comparative Strategy*, *37*(1), 73-84.

Schoolmeesters, A. Eklund, T. Leake, D. Vermeulen, A. Smith, Q. Force Aldred, S. and Fedorov, Y. (2009). Functional profiling reveals critical role for miRNA in differentiation of human mesenchymal stem cells. *PloS one*, *4*(5), e5605.

Schor, J. B. (2011). *True Wealth: How and Why Millions of Americans Are Creating a Time-Rich, Ecologically Light, Small-Scale, High-Satisfaction Economy*. Penguin, City of Westminster, London, England.

Shahriar, S. (2020). India's Economic Relations with Myanmar: A Study of Border Trade. Journal of Borderlands Studies, 1-23.

Shaw, R. (2006). Indian Ocean tsunami and aftermath: need for environment-disaster synergy in the reconstruction process. *Disaster Prevention and Management: An International Journal*, *15*(1), 5-20.

Sheavly, S. B. and Register, K. M. (2007). Marine debris and plastics: environmental concerns, sources, impacts and solutions. *Journal of Polymers and the Environment*, 15(4), 301–305.

Shepherd, C. J. and Jackson, A. J. (2013). Global fishmeal and fish-oil supply: inputs, outputs and markets. *Journal of fish Biology*, 83(4), 1046–1066.

Sidell, B. D. and O'Brien, K. M. (2006). When bad things happen to good fish: the loss of hemoglobin and myoglobin expression in Antarctic icefishes. *Journal of Experimental Biology*, 209(10), 1791–1802.

Silver, J. J. Gray, N. J. Campbell, L. M. Fairbanks, L. W. and Gruby, R. L. (2015). Blue economy and competing discourses in international oceans governance. *The Journal of Environment and Development*, *24*(2), 135-160.

Sinclair, C. (2021). Review of Sarah Hayes (2021). Postdigital positionality: Developing powerful inclusive narratives for learning, teaching, research and policy in higher education. *Postdigital Science and Education*, *3*(3), 1051-1055.

Smith-Godfrey, S. (2016). Defining the blue economy. *Maritime affairs: Journal of the national maritime foundation of India*, *12*(1), 58-64.

Smith, G. W. and Eckert, K. L. (1991). WIDECAST Sea Turtle Recovery Action Plan for Belize-CEP Technical Report 18.

Smith, J. (2016). *The effects of ocean acidification on zooplankton: Using natural CO_2 seeps as windows into the future* (Doctoral dissertation, University of Plymouth).

Smith, L. H. (2017). To accede or not to accede: An analysis of the current US position related to the United Nations law of the sea. *Marine Policy*, *83*, 184-193.

Spalding, Mark J. (2016). The new blue economy: the future of sustainability. *Journal of Ocean and Coastal Economics* 2(2), 8.

Srinivasan, U. T. Cheung, W. W. Watson, R. and Sumaila, U. R. (2010). Food security implications of global marine catch losses due to overfishing. *Journal of Bioeconomics*, 12(3), 183–200.

Strand, E. Bagøien, E. Edwards, M. Broms, C. and Klevjer, T. (2020). Spatial distributions and seasonality of four Calanus species in the Northeast Atlantic. *Progress in Oceanography*, *185*, 102344.

Sumaila, U. R. and Tai, T. C. (2019). *Ending Overfishing Can Mitigate Impacts of Climate Change*. Institute for the Oceans and Fisheries Working Paper Series, 5.

Sumaila, U. R. et al. (2020). *Ocean Finance: Financing the Transition to a Sustainable Ocean Economy*. World Resources Institute. *Washington, D.C. , USA*

Sung, W. K. (2016). How did they become law: a jurisprudential inquiry about the outcome principles of historic united nations environmental conferences? *Ga. J. Int'l and Comp. L.* 45, 53.

Taylor, S. F. Roberts, M. J. Milligan, B. and Ncwadi, R. (2019). Measurement and implications of marine food security in the Western Indian Ocean: an impending crisis? *Food Security*, *11*(6), 1395-1415.

Techera, E. J. (2018). Supporting blue economy agenda: fisheries, food security and climate change in the Indian Ocean. *Journal of the Indian Ocean Region*, *14*(1), 7-27.

Tessnow-von Wysocki, I. and Vadrot, A. (2020). The voice of science on marine biodiversity negotiations: a systematic literature review. *Frontiers in Marine Science*, 7, 1044.

Thelma, J. (2019). Problems of Invasive Species. *Marine Pollution: Current Status, Impacts and Remedies*, *1*, 344-365.

Tirumala, R. D. and Tiwari, P. (2020). Innovative financing mechanism for blue economy projects. *Marine Policy*, 139:104194.

Tudela, S. (2004). *Ecosystem effects of fishing in the Mediterranean: an analysis of the major threats of fishing gear and practices to biodiversity and marine habitats* (No. 74). Food and Agriculture Org.

Turner, R. E. and Rabalais, N. N. (2019). The Gulf of Mexico. In World seas: An environmental evaluation (pp. 445-464). Academic Press.

UNECA. (2016). *Africa's Blue Economy: A Policy Handbook*. Economic Commission for Africa, Addis Ababa.

Upadhyay, D. K. and Mishra, M. (2020). Blue economy: Emerging global trends and India's multilateral cooperation. *Maritime Affairs: Journal of the National Maritime Foundation of India*, *16*(1), 30-45.

Van Erve, P. and Bonnor, N. (2006). Can the shipping-aviation analogy be used as an argument to decrease the need for Maritime Pilotage? *The Journal of Navigation*, 59(2), 359–363.

Vasquez, T. (2022). An Unprecedented Pacific Northwest Heat Wave Rings Alarm Bells. Weatherwise, 75(1), 22-27.

Vegter, A. C. et al. (2014). Global research priorities to mitigate plastic pollution impacts on marine wildlife. *Endangered Species Research*, 25(3), 225–247.

Verberk, W. (March 3, 2016). Water pollution makes river biodiversity more vulnerable to climate warming. Retrieved on 7 August, 2022 from freshwaterblog: https://freshwaterblog.net/2016/03/03/water-pollut ion-increases-makes-river-biodiversity-more-vulnerable-to-climate-warming/

Vogler, J. (2011). The challenge of the environment, energy, and climate change. *International relations and the European Union*, 349–79.

Voyer, M. Quirk, G. McIlgorm, A. and Azmi, K. (2018). Shades of blue: what do competing interpretations of the Blue Economy mean for oceans governance? *Journal of environmental policy and planning*, 20(5), 595-616.

Voyer, M. and van Leeuwen, J. (2019). Social license to operate' in the Blue Economy. *Resources Policy*, 62, 102–113.

Wang, X. L. and Swail, V. R. (2001). Changes of extreme wave heights in Northern Hemisphere oceans and related atmospheric circulation regimes. *J. Climate*, 14, 2204–2221.

WASA Group. (1998). Changing waves and storms in the Northeast Atlantic? *Bull. Amer. Meteor. Soc*, 79, 741–760.

Waste360. (2019). Study Puts Economic, Social Cost on Ocean Plastic Pollution. Retrieved on August 7, 2022 from waste360: https://waste360.com/plastics/study-puts-economic-social-cost-ocean-plastic-pollution

Webb, M. D. and Howard, K. W. (2011). Modeling the transient response of saline intrusion to rising sea-levels. Groundwater, 49(4), 560-569.

Wenhai, L. et al. (2019). Successful blue economy examples with an emphasis on international perspectives. *Frontiers in Marine Science*, 6, 261.

Whelchel, A. W. Reguero, B. G. van Wesenbeeck, B. and Renaud, F. G. (2018). Advancing disaster risk reduction through the integration of science, design, and policy into eco-engineering and several global resource management processes. *International journal of disaster risk reduction*, 32, 29-41.

Wilberforce, T. El Hassan, Z. Durrant, A. Thompson, J. Soudan, B. and Olabi, A. G. (2019). Overview of ocean power technology. *Energy*, *175*, 165-181.

Wilberforce, T. El Hassan, Z. Durrant, A. Thompson, J. Soudan, B. and Olabi, A. G. (2019). Overview of ocean power technology. *Energy*, *175*, 165-181.

Wilhelmsson, D. Thompson, R. C. Holmström, K. Lindén, O. and Eriksson-Hägg, H. (2013). Marine pollution. *Managing Ocean Environments in a Changing Climate: Sustainability and Economic Perspectives; edited by* Kevin J. Noone, Ussif Rashid Sumaila, Robert J. Diaz*Elsevier: Amsterdam, The Netherlands*, 127-169.

Willette, D. A. Simmonds, S. E. Cheng, S. H. Esteves, S. Kane, T. L. Nuetzel, H. ... and Barber, P. H. (2017). Using DNA barcoding to track seafood mislabeling in Los Angeles restaurants. *Conservation Biology*, *31*(5), 1076-1085.

Wong, C. and Rylko, M. (2014). Health of the Salish Sea as measured using transboundary ecosystem indicators. *Aquatic Ecosystem Health and Management*, 17(4), 463–471.

Wurtsbaugh, W. A. Paerl, H. W. and Dodds, W. K. (2019). Nutrients, eutrophication and harmful algal blooms along the freshwater to marine continuum. *Wiley Interdisciplinary Reviews: Water*, 6(5), e1373.

WWF. (2021). *Forests Forward*. WWF.

Yamin, S. Cedillo, D. Sikes, N. Sitaraman, S. and Wilkins, K. (2021). Strategic Competition, Cooperation, and Accommodation: Perspectives from the Indian Ocean Region. Retrieved on August 8, 2022, from https://apcss.org/wp-content/uploads/2021/03/N2548-Yamin-Strategic-Competition-Indian-Ocean-1.pdf

Zhao, R. Hynes, S. and He, G. S. (2014). Defining and quantifying China's ocean economy. *Marine Policy*, 43, 164–173.

Zhongming, Z. Linong, L. Xiaona, Y. Wangqiang, Z. and Wei, L. (2019). Contamination of European seas continues despite some positive progress. Retrieved on August 6, 2022, from https://www.eucc-d.de/beitrag/contamination-of-european-seas-continues-despite-some-positive-progress.html.

16 Seaweed Farming Potential in India

An Assessment and Review

Muthuswamy Jaikumar,[1]* Ramadoss Dineshram,[2] Temjensangba Imchen,[2] Sourav Mandal,[3] and Kannan Rangesh[4]

[1] Gujarat Institute of Desert Ecology, Bhuj, Kachchh, Gujarat. India
[2] Biological Oceanography Division, CSIR—National Institute of Oceanography, Dona Paula, Goa, India
[3] Ocean Engineering Division, CSIR—National Institute of Oceanography, Dona Paula, Goa, India
[4] Department of Marine and Coastal Studies, Madurai Kamaraj University, Tamil Nadu, India
*Corresponding author: Muthuswamy Jaikumar;
E-mail: jaikumarmarine@gmail.com

CONTENTS

DOI: 10.1201/9781003184287-16

16.1 INTRODUCTION

Seaweeds are a vital part of coastal ecosystems, which offer indispensable ecosystem services as well as socio-economic values for the lives of various marine forms. The seaweed industry has huge economic value and significantly contributes to the sustainable development of rural coastal provinces. Seaweeds are cultivated in many Asian countries, and the production of phycocolloids from seaweeds is extensive across the globe. Seaweed resources are abundant in India, particularly on the west coast and the southeast coast of India. India has an 8100 km stretch of coastline, and it includes an Exclusive Economic Zone (EEZ) around 2 million sq km, consisting of coastal and island ecosystems, and these are same as 66% of the total mainland area. About 30% of the people depend on these areas for utilizing a large variety of seaweed populations (Ganesan et al. 2020). The highest seaweed diversity was reported on the Southeast Coast of Tamil Nadu, Gujarat, Lakshadweep islands, and Andaman and Nicobar Islands.

In India, seaweed cultivation is being taken up on a large scale, particularly in the coastal districts of Tamil Nadu, because of suitable environmental conditions such as the shallow nature of the bottom and a lesser tidal influence (Figure 16.1). This favours the cultivators who can do more cycles of cultivation every season, especially in three coastal districts. In Gujarat, only a pilot scale of cultivation has been tried in a few districts because in this region, tidal amplitude is very high with high wave action. Such conditions are not favourable to do more cycles of cultivation. A pilot scale cultivation has been carried out in the state of Andhra Pradesh, for example, in Visakhapatnam district in cages and in the Krishna district, Nayalanka, in PVC rafts pipes.

On the South East coast of India, the coastline of Tamil Nadu extends from Tiruvalluvar district in the north to Kanyakumari district in the south, a total length of about 1076 km (Ramesh et al. 2008; Mantri et al. 2019a). The coastal districts of Tamil Nadu such as Pudhukottai, Ramanathapuram, Tuticorin and Kanyakumari have a rich diversity of seaweeds. Seaweed species like *Kappaphycus alvarezii*, *Gracilaria dura* and *G. edulis* are cultivated on a commercial scale in the coastal region of India, particularly in Tamil Nadu's Pudhukottai, Ramanathapuram and Tuticorin. Carragenophyte *Kappaphycus alvarezii* is cultivated extensively in this region. *K. alvarezii* is cultivated by vegetative propagation as it grows faster. The crop can be harvested within 40 days of cultivation. *Gracilaria edulis* and *G. dura* cultivation period extends to 60–75 days, and the revenue is less compared to *Kappaphycus* cultivation. The fishermen in this coastal region are actively engaged in *Kappaphycus* cultivation due to its shorter cultivation time and it is an additional source of income.

FIGURE 16.1   Seaweed locations in India (Government of Gujarat) (https://gaic.gujarat.gov.in/writeread ata/images/pdf/26-seaweed-culture.pdf).

16.2   DISTRIBUTION OF SEAWEEDS

Seaweeds are primitive macrophytic benthic marine algae found in the tidal regions of the seas/ oceans. Green, brown and red seaweeds are found in the intertidal, tidal and subtidal regions, respectively. Green seaweeds are more prevalent in the intertidal zone. Green seaweeds include the species of sea lettuce (*Ulva*), green string lettuce (*Enteromorpha*), *Codium, Chaetomorpha* and *Caulerpa*. Brown seaweeds live in the upper subtidal zone or tidal zone. They are *Sargassum, Turbinaria, Laminaria* and *Dictyota*. Red seaweeds mostly grow in subtidal waters, and they include *Gracilaria, Eucheuma, Gelidiella, Ceramium, Acanthophora*.

More than 100 species of seaweeds have been recorded between Dhanuskodi and Kanyakumari (Table 16.1) Kaliaperumal et al (1998). The surveys from the Central Salt and Marine and Chemical

TABLE 16.1
Seaweed Distribution: Dhanuskodi and Kanyakumari in Tamil Nadu

Name of Species	Number of Species
Chlorophyta	20
Phaeophyta	18
Rhodophyta	61
Cyanophyta	1
Estuarine Species	
Hypnea valentiae	3
Gracilaria verrucosa	
G. arcuate	

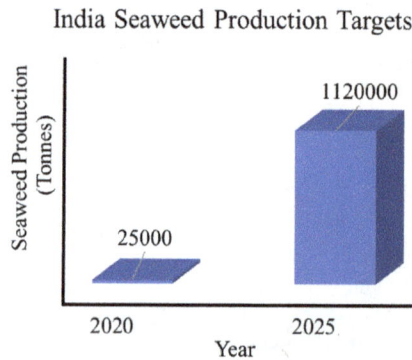

FIGURE 16.2 Seaweed production targets in India.

Source: Ministry of Fisheries, Animal Husbandry and Dairying, Government of India (2020)
Pradhan Mantri Matsya Sampada Yojana: Operational Guidelines, (http://nfdb.gov.in/PDF/
PMMSY-Guidelines24-June2020.pdf).

Research Institute (CSMCRI), the Central Marine Fisheries Research Institute (CMFRI) and fishery-based organizations showed that South India's coastal belt has a large number of seaweed resources. In Gujarat's West Coast, there are huge amounts of seaweed resources along the subtidal and inter-tidal regions. The coasts of Tamil Nadu, Gujarat, Andaman and Nicobar Islands, and Lakshadweep have a rich diversityk of seaweeds and standing biomass.

There are about 700 species of marine algae from the Indian coast, that is, from both deep-water and inter-tidal regions, and of these, about 60 species have commercial significance (Khan and Satam, 2004; Mantri et al., 2019b). Agar production from *Gelidiella acerosa* and *Gracilaria* sp. (red seaweeds) is performed all year round, whereas alginate production from *Sargassum* and *Turbinaria* (brown algae) is performed in the southern coast between the months of August and January. The bays and creeks in most of the Indian coastal regions are reported to have immense potential for the cultivation of seaweeds (Ayyappan, et al. 2006)

16.3 THE IMPORTANCE OF SEAWEEDS

16.3.1 Economic Significance

There are about 221 economically important seaweeds, of which 145 species are used for food and 110 species are used for the production of phycocolloids (Mohammed, 2015). In 2018, seaweed

TABLE 16.2
Global and Indigenous Production of Seaweed

Country/Area	Wet Weight in Tones	Share of World Total (%)
China	20351442	56.86
Indonesia	9962900	27.84
Korea, Republic	1821475	5.09
Philippines	1500326	4.19
Korea, Dem. People's Rep	603000	1.68
Japan	412300	1.15
Malaysia	188110	0.53
India	5300	0.01
Chile	427508	1.19
Peru	36348	0.10
Canada	12655	0.04
Mexico	7336	0.02
United States of America	3394	0.01
Norway	163197	0.46
France	51683	0.14
Ireland	29542	0.08
Russian Federation	19544	0.05
Iceland	17533	0.05
United Republic of Tanzania	106069	0.30
Morocco	17591	0.05
South Africa	11155	0.03
Madagascar	9665	0.03
Solomon Islands	5600	0.02
Papua New Guinea	4300	0.01
Kiribati	3650	0.01
Australia	1923	0.01

Source: FAO. 2021c. Fishery and Aquaculture Statistics. Global production by production source 1950–2019 (FishStatJ),
Mantri et al. 2022.

farming constituted 97.1 % by volume of the total of 32.4 million tonnes of the cultivated combination of aquatic algae and wild seaweed (Table 16.2) (Chopin and Tacon, 2021). In India, 5,300 tonnes of seaweed was produced through seaweed farming (FAO, 2020; Chen and Xu, 2005). On the other hand, about 25,000 tonnes of wild seaweeds was collected from the wild, accounting to a market value of about ~Rs or 300 - 500 crores (Raghuvanshi et al. 2021). Under the Pradhan Mantri Matsya Sampada Yojana (PMMSY) program, the Department of Fisheries, Government of India, has set a target to achieve 11.2 Lakh MT of fresh seaweed production in the next five years (2020 - 2025) (Shenoy and Rajpathak, 2021) (Figure 16.2). A study showed that India could meet only 25% of the seaweed-based bio-stimulant market demand, while the remaining 75% is imported from Europe and North America (Ferdouse et al. 2018).

16.3.2   The Significance of Seaweed Farming

To fulfil the burgeoning market of seaweeds and their products, seaweed farming is venturing towards open ocean, however, its development is still at a rudimentary stage. Seaweed culture can have noteworthy constructive environmental impact. The nutrient-absorbing properties of seaweeds can help improve coastal water quality. In addition, it can also help to treat the effluent wastewater and control eutrophication.

16.3.3 The Significance of Seaweed in Diet and Health

Seaweeds are popularly known as sea vegetables. They are a good source of vitamins and minerals. Seaweeds have become a great source of nutraceutical supplements with many health benefiting properties. They are a good source of vitamins A, B1, B12, C, D, E, niacin, folic acid, pantothenic acid, riboflavin, and minerals such as Ca, P, Na, and K. They have essential amino acids needed for metabolism and health. There are about 54 trace elements needed for the human body's physiological functions, and these vital elements are colloidal, chelated and balanced forms and hence they are bioavailable. The presence of biologically active compounds such as carotenoids, phlorotannins, fucoidan, alginic acid, etc. can preclude certain diseases like inflammation, cancer, diabetes, arthritis, hypertension, and cardiovascular ailments, by consuming seaweeds. Seaweeds are used as an ingredient in food formulations for their proteins, antioxidant, antimicrobial and anti-inflammatory compounds, and fibres. Seaweeds also serve as fodder because of high nutritional value. The seaweed phytochemical products like agar, alginate and carrageenan are used as stabilizers (alginate), in biotechnology (agar), in pharmaceuticals and as an emulsifier in the food industry (carrageenan). They also serve as manure for the cultivation of agricultural crops due to their micronutrients and growth promoting property.

16.4 THE POTENTIAL OF SEAWEED CULTIVATION IN INDIA

16.4.1 Seaweed as a Cultivated Crop

Seaweed cultivation commands a billion-dollar industry with an ecologically sustainable development. In a tropical country like India, seaweed cultivation is blessed with the gifts of nature, namely, sunlight and seawater. Seaweed cultivation has the potential to generate wealth with a quick return on investment due to its short cultivation cycle. Therefore, seaweed cultivation can bring a revolutionary change in the economy and overall prosperity to India. Harvested seaweeds from coastal regions are used for alginate and agar production in the Gulf of Mannar. Local people of this region are the victims of past tsunami who lost their livelihood and properties. The people, especially women, depend on wild seaweed collection for their daily earnings. The meagre income can meet the basic needs for survival. However, if the coastal community practice and adopt scientific farming of seaweed, they can generate a minimum amount of Rs.25,000 to 35,000 per month. India has a long coastline, and it has huge potential to develop seaweed aquaculture. It is estimated that the seaweed market has a potential to achieve 60 billion US$ globally by 2030. The standing stock of about 844 seaweeds is assessed to be 58,715 tonnes wet weight in the Indian seas (Banerjee et al. 2020.).

In India, the commercially harvested seaweeds like *G. acerosa* and *G. edulis* are used for agar production, whereas *Sargassum* and *Turbinaria* are produced for alginate production in India (Krishnamurthy, 1971). Pilot scale cultivation of *Kappaphycus* was initiated for fisherwomen's economy development in Mandapam. Later, it was extended to Kanyakumari and Tuticorin. The production from these areas showed that cultivation of seaweed is a profitable venture. The demand for *Kappaphycus alvarezii* is high in both local and global markets. In India, *K. alvarezii* is well recognized for its plant bio-stimulant/fertilizer property (Shanmugam and Seth, 2018.). However, the commercial potential of seaweed farming is not fully capitalized in many littoral countries. As of now, only *K. alvarezii* is commercially cultivated and this accounts for 30% of the requirement and the rest, *Gracilaria* sp., *Gelidiella* sp. and *Sargassum* sp., is harvested from naturally occurring areas. The State Bank of India (SBI) and NGO and seaweed companies has started financing seaweed cultivation to SHGs in 2004 (Krishnan and Narayanakumar, 2013). About 2000 families of coastal poor folk have been rehabilitated in the southern coastal districts of Tamil Nadu with a maximum between Mandapam and Rameswaram.

16.4.2 SUSTAINABLE LIVELIHOOD TO FISHERMEN THROUGH SEAWEED CULTIVATION

Seaweed is easy to cultivate. It is completely eco-friendly in cultivation, harvesting and use. The chosen varieties are non-invasive species, and they do not alter the water or affect other marine organisms. The vocation of seaweed cultivation will ameliorate the living conditions of marginal fishermen and earn additional income. The assorted benefits of seaweed farming are as follows: it offers job opportunities to coastal people, provides an uninterrupted supply of essential material for the seaweed industry, provides good quality raw material for industrial purposes, and conserves naturally occurring wild seaweed populations.

16.5 THE POTENTIAL OF COMMERCIAL SEAWEED CULTIVATION IN INDIA

Commercial seaweeds come under the categories of agarophytes, carrageenophytes, alginophytes, and edible seaweeds as shown in Figure 16.3 and Table 16.3.

Sargassum   Kappaphycus   Ulva

Gracilaria   Padina

FIGURE 16.3   Some commercial seaweeds of India.

TABLE 16.3
Commercial Seaweeds of India

S. No.	Seaweed Category	Representative Species
1.	Agarophytes	*Gelidiella acerosa, Gracilaria corticata, G. edulis, G. crassa, G. folifera, Gracilariopsis megaspore*
2.	Carrageenophytes	*Hypnea valentiae, H. musiformis, Kappaphycus* alone is cultivated at present on a commercial scale in Palk Bay since returns are attractive with an annual income exceeding Rs.100,000 / head
3.	Alginophytes	*Sargassum* sp, *Turbinaria conoides, Cystoseira trinodis, Hormophysa triquetra* for algin and liquid seaweed fertilizer
4.	Edible seaweeds	*Ulva flexuosa, Ulva lactuca, Caulerpa racemosa, C. lactervirens, Acanthophora spicefera*

16.6 POTENTIAL OF *KAPPAPHYCUS* CULTIVATION

16.6.1 THE CHARACTERISTICS OF *KAPPAPHYCUS* SEAWEED

The *Kappaphycus* is a red alga and belongs to Rhodophycea family. *K. alvarezii* comprises a cylindrical axis with enlarged maximally irregular branches that extend beyond the basal structure towards the light. The halli looks darkish brown in intense light and reddish in the deeper water or shade because of the large amount of phycoerythrin. Pale yellow thalli occur in bright light conditions. *K. alvarezii* perform phototrophy in intense light conditions. The growth condition depends on water motion. If there is high wave action, more damage is possible to plants. Clear water is desirable for growth. The site should be away from the lowest tidal level. Care is required to avoid very shallow water near the shore since the crop is destroyed due to high temperature in summer. *K. alvarezii* grows abundantly in the sandy bottom of the sand, and optimum salinity ranges from 29 to 34 ppt. About 150 g of seed plants grow in 45 days up to >600 g in Palk Bay as it has calmer conditions. It needs sunlight, no turbid seawater and lower wave action for refilling the bottommost nutrients. It has also been shown that *Kappaphycus* seaweed can grow up to 1 kg in the open sea water in the monoline net-bag culture method where the wave action is higher.

16.6.2 THE ADVANTAGES AND BENEFITS OF *KAPPAPHYCUS* SEAWEED

- *Kappaphycus* is a resourceful plant and can grow almost everywhere in marine environment.
- *Kappaphycus* is the chief source of Kappa carrageenan.
- It is propagated vegetatively by 'cloning' from bud cuttings and it is easy to multiply.
- The growth cycle is short, namely, ~45 days
- It has simplified cultivation technology and is eco-friendly.
- It does not require fertilizer or any other chemicals used in agricultural crops.
- Seeding, harvesting, the extraction of biofertilizer and drying can be done on the shore.
- *Kappaphycus* can assist in sustainable mariculture production.
- 2–3 members of a family can earn Rs. 25,000 to 35,000 per month. The techno-economic viability was established in Tamil Nadu.
- *Kappaphycus* cultivation can be valuable to the environment as it absorbs CO_2, controls pollution and enhances biodiversity.

16.7 METHODS OF SEAWEED CULTIVATION

16.7.1 POPULAR METHODS OF SEAWEED FARMING IN TAMIL NADU AND GUJARAT

The methods for seaweed farming in India were developed by Central Salt and Marine Chemicals Research Institute- Marine Algal Research Station (CSMCRI-MARS), a constituent laboratory of CSIR. The methods include the floating raft method (bamboo frame with nylon ropes and stone anchor) for calm and shallow water, the monoline method (wooden poles connected with nylon ropes) for shallow water with small to moderate waves, tube net method (3 to 7 m long tube nets with an eye size of 85 - 90 mm) for rough water, and net bag method (nylon net bags attached to a long rope) for rough and deep water. The application of a method depends on the geography and weather conditions of the site (Table 16.4).

16.7.2 THE SINGLE ROPE FLOATING RAFT METHOD

The single rope floating raft (SRFR) method developed by CSMCRI is suitable for culturing seaweeds in greater depths and over a wide area (Figure 16.4). A long rope made up of polypropylene (10 mm diameter) is attached to two wooden stakes with two synthetic fibre anchor cables and kept afloat with synthetic floats. The length of the cable is twice the depth of the sea (3 to 4 m).

TABLE 16.4
Cultivation Methods of Seaweed in India

S. No.	Site Type	Preferred Farming Method
1.	Very calm and very shallow	Pole monoline system, where seaweed lines are attached to poles hammered into the seabed or Bamboo raft system
2.	Very calm waters and slightly deep	Bamboo raft system or regular monoline system
3.	Moderately calm waters	Regular monoline system
4.	Rough and deep waters	HDPE pipe ladder system with tube nets

FIGURE 16.4 Single rope floating raft (SRFR) method (Image Courtesy: https://twocircles.net/2021dec23/444480.html?amp).

Each raft is kept afloat by means of 25-30 floats. A long polypropylene rope of 10 mm diameter is attached to 2 wooden stakes with 2 synthetic fibre anchor cables and kept afloat with synthetic floats. The length of the cable is twice the depth of the sea (3 to 4 m). The lower end of the cultivation rope is attached to the stone to keep this in a vertical position. Normally, ten fragments of *Gracilaria edulis* are bound on each rope. The distance between two rafts is kept at 2 m. Floating raft technology (FRT) is recommended for the Kerala coast for agarophyte cultivation. Some areas in the Gulf of Kutch are suggested as appropriate for seaweed cultivation in deep water. Besides, CMFRI also developed and improved the techniques for culturing *Gracilaria edulis*, *Gelidiella acerosa*, *Acanthophora spicifera* and *Hypnea musciformis*.

16.7.3 THE FLOATING BAMBOO RAFT METHOD

A raft of 1.8 to 2.4 m bamboo is used for the purpose. 15-18 lines of rope is used in a single raft, and the seedlings are tied to the ropes. The raft is anchored with stones or steel anchors (Shanmugam et al. 2017) (Figure 16.5).

16.7.4 THE POLE MONOLINE METHOD

The monoline method is suitable for shallow regions such as nearshore shallow waters (Figure 16.6). In the pole monoline method, the poles where rope lines are attached are hammered into the seabed. A 7 m monoline contains 20 twines at 10 cm intervals. An average of 5 kg of seed material is attached in each monoline, and it is tied with 12 mm nylon rope.

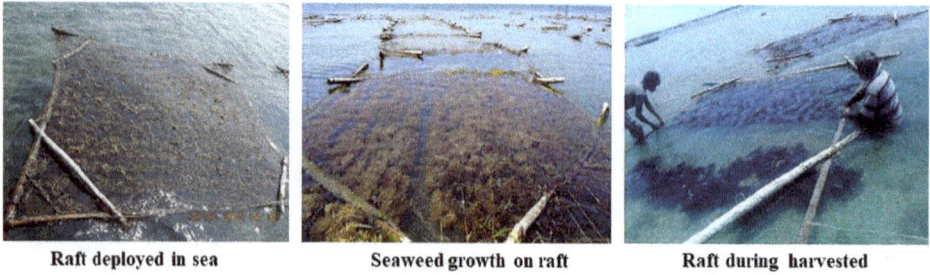

Raft deployed in sea   Seaweed growth on raft   Raft during harvested

FIGURE 16.5 Floating bamboo raft method.

FIGURE 16.6 Pole monoline method.

16.7.5 THE TYING HDPE PIPE LADDER/BAMBOO TUBE NET/TIE-TIE METHOD

This method is very similar to the monoline method, except in the seeding practice (Figure 16.7). It requires a fabrication of tube net with a mesh size of 45 mm, and the diameter of the tube will be 90 mm which is fixed in the permanent structure of a HDPE/bamboo raft, holding 20 seeding lines, each 3 m length with 2 kg seed/lines.

The seed material is not fastened with ties in this method but held compactly. The seed material of algae will be loaded into the tube-nets with the help of a small plastic pipe with smaller diameter than tube-net. A raft prepared from bamboo of 7 to 10 cm diameter and 30 × 30 cm in size will be used for fabrication of the main frame in the tube-net method of cultivation. Here bamboo supports the angular portions diagonally and is fixed using supporting bamboo of 4′ long to keep the raft structure together. 10–15 cm diameter plastic tube-net having 3–4 cm mesh size is cut into 3 m lengths.

16.7.6 THE PVC LADDER SEAWEED TYING METHOD

This method is similar to the tube net method. It uses 1 m × 2 m length PVC tubes (Figure 16.8).

16.7.7 THE POLE LONG LINE TUBENET METHOD (VERY SHALLOW WATER)

In this method, longline tubenet is set afloat by using plastic pet bottles used as floaters in which 10 mm polypropylene rope forms the monoline. The pole end of the monoline is anchored (Figure 16.9).

70X100M 150 monoline/sector/site

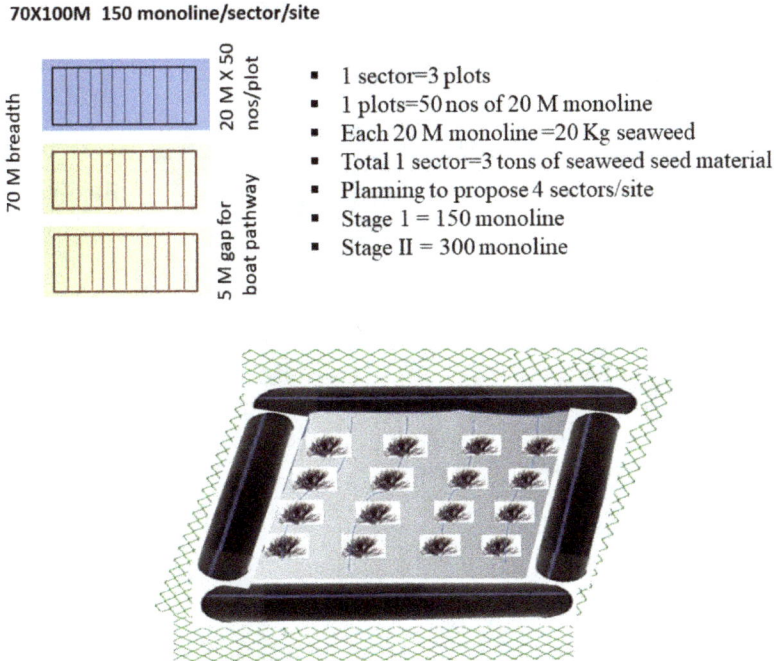

- 1 sector=3 plots
- 1 plots=50 nos of 20 M monoline
- Each 20 M monoline=20 Kg seaweed
- Total 1 sector=3 tons of seaweed seed material
- Planning to propose 4 sectors/site
- Stage 1 = 150 monoline
- Stage II = 300 monoline

FIGURE 16.7 Sector monoline method.

FIGURE 16.8 PVC ladder method.

FIGURE 16.9 Pole long line tube net method.

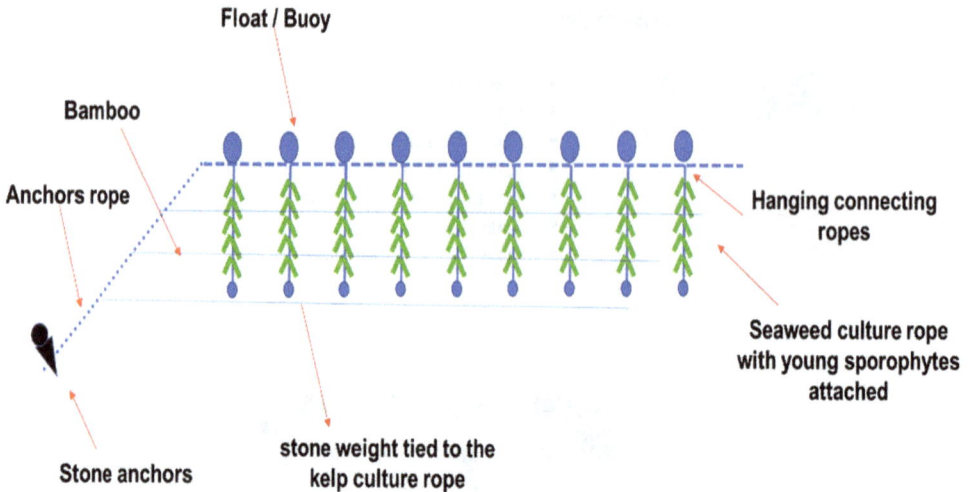

FIGURE 16.10   Long monoline method.

16.7.8 LONG-MONOLINE METHOD (DEEPER WATER)

In this method,10 mm polypropylene rope forms the monoline, which is anchored at both ends and kept afloat with the help of floaters tied to the monoline (Figure 16.10). 100 g of seed material is used for each planting. The seed material is tied to the monoline using soft plastic fabric (Doty and Alvarez, 1975), similar to the tie-tie method. A monoline of 7 m contains 40 seedlings at intervals of 15 cm. The harvest is collected after 45 days of seeding.

16.7.9 THE NET BAG METHOD

The mesh size of the net bag is about 0.64 cm. The fish net is prepared by cutting a net of 90 × 80 cm. The net is folded towards its 90 cm side and the bottom and the side are sewn (Figure 16.11). The net bag's size after sewing will be 45 cm and 80 cm long. The bottom side is bundled and tied with a 3 mm polyethylene rope (FAO, 2018). This net bag method is used for *Kappaphycus alvarezii* cultivation in rough waters. Generally, 0.5 kg to 1 kg of seed material is kept in the net bag for the cultivation method. After the seeding, it is monitored at 3-day intervals to wash off the mud and clean the attached animals on the surface of the net bag. After 30 - 45 days of culture, every net bag of 1 kg of seedlings will grow up to 5 kg/net bag.

16.7.10 THE CULTIVATION OF *GRACILLARIA EDULIS* USING HDPE PIPE

Gracilaria edulis is cultivated on a commercial level across the year in the Gulf of Mannar region and between June and September in Palk Bay. The new raft method is best suited to the bay region, where the waves are less or the depth is shallow (Figure 16.12). Production is about 120 t/ha (wet wt.) per year. A threefold increase is noted within 60 days (Johnson et al. 2020).

Raft design: One-meter HDPE pipe with 12 mm holes on both ends for tying the rope. The raft consists of 18 to 20 monolines each 4.5 m long. Bottom fencing is used to protect from fish bites. The bottom fencing also allows *Gracillaria* spores to settle in the net, ensuring a supply of seeding material for the next cycle of cultivation. Each line is stitched with a 4 mm tie rope. Approximately 0.065 g/line or 50 g/system of seed material is tied to the line and it takes about 50–60 days to harvest.

FIGURE 16.11 Net bag method.

FIGURE 16.12 HDPE pipe method.

16.8 THE COLLECTION OF WILD SEAWEED IN INDIA

The seaweed is collected from the wild during spring tides in five districts of Tamil Nadu (Tanjore, Pudukottai, Ramnathapuram, Tuticorin and Kanyakumari). A study by Ganesan et al. (2019) showed that in the Gulf of Mannar region, about 1270 women and 285 men of the fishing community, representing 14 coastal villages are engaged in harvesting seaweed. The report also showed that about 460 women and 210 men from the Palk Bay coast harvest seaweed. Their report indicates that around 2000 active men and women are engaged in the collection of wild seaweed. Most of the wild collection has been reported from the Gulf of Mannar region in the Ramanathapuram district (Table 16.5) (Johnson et al. 2017). *Gelidiella, Gracilaria, Sargassum* and *Turbinaria* are the major seaweeds collected from the wild in this region. However, in recent years, the collection from the wild has declined due to the restriction and subsequent ban by the department of forests in the Gulf

TABLE 16.5
List of Cultivable Species in India

S. No	Seaweed species	District	Season
1.	*Gelidiella acerosa*	Ramanathapuram	February - April
2.	*Gracilaria edulis*	Ramanathapuram	February - April
3.	*G.dura*	Ramanathapuram	February - April
4.	*Kappaphycus alvarezii*	Pudhukottai, Ramanathapuram, Tuticorin	Pudhukottai (October - April)
			Ramanathapuram (February - September)
			Tuticorin (May-October)

of Mannar and Palk Bay area. Despite this, some people are still engaged in illegal collection in the Gulf of Mannar region.

A survey during 1980 - 90 by CSMCRI (Central Salt and Marine Chemicals Research Institute) showed that the estimated total of seaweed standing stocks was 97,400 tonnes wet wt. (Kaliaperumal and Kalimuthu, 1997) from the coast of Tamil Nadu, 7,500 tonnes wet wt. from the Andhra Pradesh coast, and ~19,345 tonnes wet wt. from the Lakshadweep islands. The production of *Gelidiella acerosa* decreased from 500 tonnes (dry wt.) in 2005 to 280 tonnes in 2017 (Mantri et al. 2019a; Ganesan et al. 2019). During the same period, *Gracilaria edulis* decreased from 250 to 50 tonnes. Similarly, the SNAP Natural Products and Alginates (P) Ltd) (SNAP, 2020) showed that the wild collection of *Sargassum wightii* decreased from 2700 tonnes in 2005 to 1550 tonnes in 2017. The collection of different species of *Turbinaria conoides, T. ornata* and *T. decurrens* from the Gulf of Mannar islands and coastal mainland of Southeast India from Rameswaram to Kanyakumari also witnessed a drastic decline from 100 tonnes in 2005 to 10 tonnes in 2017. However, there is a steady increase in food grade agar production; every year 300 - 400 dry tonnes is produced for agar production (SNAP, 2020). Their report showed that ~15,000 MT of *Sargassum wightii* wet weight per year was collected from the natural beds at Ramanathapuram.

In three districts of Tamil Nadu, namely, Tuticorin, Ramnathapuram and Pudukottai, more than 2000 active men and women are engaged in *Kappaphycus alvarezii* cultivation. In the Mandapam region of Ramanathapuram district, *Gracilaria dura* and *G. edulis* cultivation are also being taken up, and around 50 members are actively involved in this cultivation (Table 16.6). Due to restrictions in wild collection, fishermen are looking for alternative income and they are now entering into seaweed cultivation in this region. Government funding bodies and private companies are supporting the local fisherfolk to take up *Kappaphycus alvarezii* cultivation.

16.9 THE PROBLEMS AND PROSPECTS OF SEAWEED CULTIVATION

Recent problems associated with the seaweed industry are overexploitation of raw materials, low quality of stocks and lack of labour. Besides, most seaweeds are prone to epiphytism and colonized by epibionts such as bacteria, protest, algae and invertebrates.

16.9.1 SEAWEED-ASSOCIATED BACTERIA AND THEIR INTERACTIONS

Kappaphycus alvarezii crop failure is mostly due to various environmental issues and using the same vegetative strains since its introduction in our Indian waters. Seaweed-associated bacteria play a major role in seaweed growth and morphogenesis (R.P Singh and CRK Reddy. 2014). Microbial communities from the phyla Proteobacteria and Firmicutes are the most abundant on seaweed tissue. Bacterial communities associated with seaweed produce plant growth-regulators, bioactive compounds, signalling molecules for quorum sensing, and other effective molecules which

TABLE 16.6
List of Seaweed Species Available in Palk Bay and Gulf of Mannar

S. No.	Palk Bay	Gulf of Mannar
1.	*Gelidiella acerosa*	*Gelidiella acerosa*
2.	*Gracilaria edulis*	*Gracilaria edulis*
3.	*Gracilaria lichenoides*	*G. verrucosa*
4.	*Sargassum spp.*	*G.salicornia*
5.	*Turbinaria ornata*	*G.crassa*
6.	*T. deccurens*	*G.foliifera*
7.	*T. conoides*	*Sargassum wightii*
8.	*Kappaphycus alvarezii*	*Turbinaria ornate*
9.	*Hydropuntia edulis*	*T. deccurens*
10.	*Hypnea* sp.	*T. conoides*
11.		*Hypnea musciformis*
12.		*Sarconema filiforme*

are responsible for the regular morphology, growth and development of seaweeds. Besides, bioactive compounds of associated bacteria control the presence of another group of bacterial strains on seaweeds and safeguard the host from harmful entities found in the pelagic realm.

16.9.2 SEAWEED-ASSOCIATED FAUNA

The most common problems encountered by seaweed farmers are fish bite in Mandapam region, and barnacles and ascidian attachments in Tuticorin cultivation sites (FAO, 1988; Miller and Hay, 1998). Research is being undertaken to address the problem of epiphytism and grazers in seaweed aquaculture (Table 16.7) (Ganesan et al. 2006). The outcome of these studies is expected to bring relief to the fisherfolk who are engaged in seaweed cultivation and increase their source of income.

16.9.3 EPIPHYTISM ON SEAWEED AND PREVENTIVE MEASURES

Most marine organisms are prone to epiphytism, and it is a common occurrence on seaweed (Chirapart et al. 2018; Imchen et al. 2021). They are colonized by epibionts such as bacteria, protest, algae (micro- and macro-forms) and invertebrates (Wang et al. 2022). Epiphytic organisms use the seaweed as a host for support, food and protection. Most of the epiphytes are facultative, providing a mutualistic association, while some share an obligate relationship (Wahl and Mark, 1999; Stachowicz and Whitlatch, 2005; El-Din et al. 2015).

16.9.4 THE ROLE OF EPIPHYTES

Epiphytes play an important role in energy and nutrient conversion in addition to their role as pollution indicators (Davies, 2009). Many factors affect the distribution and the load of epiphytes, including seasonal and geographical variation. Environmental factors such as nutrient availability, DO, pH, salinity, water quality, temperature, light, and other biochemical processes affect the epiphytic load (Wear et al. 1999; Frankovic et al. 2006; Amsler et al. 2005; El-Din et al. 2015). Study also shows that the complex architecture of seaweed provides an ideal substratum for the settlement of epiphytes and epifauna (Wuchter et al. 2003; Ferreiro et al. 2013). However, the infestation by epiphytes can cause great economic loss to the seaweed industry due to inferior quality. Studies

TABLE 16.7
Seaweed-Associated Fauna

S. No.	Epiphytes/Grazers	Images	District Affected by Epiphytes (month)
1.	Crabs, rabbit fishes, and Puffer fish	*Siganus canaliculatus* *Lagocephalus lunaris*	Tuticorin and Ramnad (July to September)
2.	Ascidians	*Botryllus leachii*	Tuticorin and Ramnad (November to January)
3.	Micro bivalve	*Lectroma alacorvi*	Tuticorin (July to August)
4.	Barnacles		Tuticorin, Kottaipattinam (November to December)

TABLE 16.7 (Continued)
Seaweed-Associated Fauna

S. No.	Epiphytes/Grazers	Images	District Affected by Epiphytes (month)
5.	Isopods		Mandapam (October to November)
6.	Macroalgae		Mandapam *Enteromorpha intestinalis, Chaetomorpha crassa* (May to July)

Source: mediastore house

showed that the decline has been due to disease and pest infestation, intensified by monocrop aquaculture activity in seaweed aquaculture farms (Gachon et al. 2010; Loureiro et al. 2015).

16.9.5   THE IMPACT OF EPIPHYTES ON SEAWEED FARMING

The first major outbreak of epiphytic infestation was recorded on *Kappaphycus* during the last quarter of the 20th century (Sahu et al. 2020). The incidence of disease drastically affects the seaweed productivity. The major effect of the dense growth of epiphytes on seaweed is competition for the same resources such as light and nutrients. The epiphytes cover the photosynthetic surface area of the host, affecting the photosynthetic capacity. The outbreak of epiphytes also weakens the host, which in turn becomes susceptible to infection by pathogens. This affects the productivity of biomass and causes economic loss due to decreased market value of the crop. Experience has shown that carrageenan quality of seaweed species like *Kappaphycus* and *Eucheuma* spp. is significantly reduced due to infestation of epiphytic filamentous algae (EFA) (Tsiresy et al. 2016; Ward et al. 2020). This lowers the quality of seaweed and causes economic loss to the farmers. The epiphytic *Polysiphonia* infestation on *Kappaphycus alvarezii* has been a major cause in the decline of productivity (Vairappan, 2006; Yang et al. 2017; Ingle et al. 2018). The loss due to disease and pests in *Saccharina japonica* was 25 - 30% (Wang et al. 2017) and up to 20% in *Pyropia* sp. (Kim et al. 2017). The thallus whitening and discoloration of eucheumatoid seaweed known as ice-ice syndrome is a major problem in the seaweed industry. This syndrome has been identified to be caused by a combination of both abiotic (variation in temperature and salinity) and biotic factors (microbial pathogens) (Ward et al. 2020; Wahl et al. 2012; Kambey et al. 2021). Vairappan et al. (2008) demonstrated that epiphytic infestation damages the cortex of *Kappaphycus* making the host susceptible to bacterial infection.

16.9.6   THE CHALLENGES AND IMPACT OF EPIPHYTIC FAUNA OR GRAZER IN *KAPPAPHYCUS ALVAREZII* CULTIVATION

Seaweed cultivation is diminished due to various issues faced by the cultivators. Particularly, fish bite was the most common issue in the Mandapam region and barnacles and ascidians in the Southern part of the cultivation sites from Tuticorin to Kanyakumari. Research and development activities commenced to control the epiphytic and grazing activities during the cultivation period, which will help the fisherfolk to improve the benefits of cultivation and improve their source of income through prospective cultivation techniques.

16.9.7 MITIGATION AND PREVENTIVE MEASURES

Studies on the effect of epiphytism showed that it significantly affects the growth and survival of seaweed by restricting the uptake of carbon, oxygen and nutrients, reduces the amount of light for photosynthesis, shading affects the sporulation, decreases the flexibility of thalli, and increases the palatability for grazers (Wahl et al. 1997; Wang et al. 2017). To prevent and mitigate the effect of disease and pest infestation in commercially cultivated seaweeds, various strategies have been proposed in recent years.

The acid washing of *Pyropia* blades had been effective in controlling many diseases (Kim et al. 2017). Wang et al. (2014) showed that modifying the culture conditions such as exposure to light and salinity can reduce the severity of infection in *Saccharina japonica*. Hand picking and removal of the epiphytic pest is a common practice for controlling EFA. Ask and Azanza (2002) demonstrated that immediate removal of infected stock and replacing with an uninfected planting material is an effective technique in containing the spread. On the other hand, the bioactive extract from brown algae has been shown to protect *Pyropia* spp. against *Olpidiopsis* infection (Prado et al. 2017; Qiu et al. 2019).

Some bacterial epiphytes are known to support the growth and defence of seaweeds (Albakosh et al. 2016). However, these procedures could be too expensive for small-scale farmers. This indicates that preventive methods like quarantine, and improved aquaculture practices is an ideal strategy as suggested by Campbell et al. (2019). A natural defence mechanism to control epiphytism is produced by seaweed itself through biochemical compounds. However, the most effective strategy for monitoring and controlling disease and pest infestation lies on having an improved understanding of the causal agents such as biotic and abiotic factors. The understanding of epidemiology will facilitate an early detection and removal of the pest. This will significantly reduce the impact of pest infestation. On the other hand, the availability of seaweed strains against epiphytism could provide an effective tool in developing a sustainable seaweed aquaculture. Therefore, it is imperative to have a strong biosecurity guideline with a low-cost monitoring system to assess the potential risk and provide management of the pests and diseases.

16.9.8 HYDRODYNAMIC MODELLING FOR OFFSHORE SEAWEED FARMING

Currently seaweed farms are placed in sheltered locations. However, there is a strong need to expand seaweed cultivation towards offshore for large-scale farming. The productivity of the culture area is closely related to the hydrodynamics of the culture field. The wave energy dissipation and changes in the flow pattern due to the farming structure have significant effects on seaweed health. Wave-current induced sediment transport adds more complexities. The suitable positioning of framing structures based on the simulated flow behaviour and sediment dynamics contributes to predicting the vulnerabilities of the floating frame structures and designing the mitigation measures. A feasibility study should be carried out to assess the farming structure's performance and the seaweed's productivity for the local prevailing wave-flow conditions and sediment behaviour. A coupled wave-flow hydrodynamic model is set up to simulate the flow patterns and wave conditions at the model domain. The nutrient dynamics and water quality modelling can help to understand the health of the site's ecosystem. Finally, the wave-current-farming structure interaction study through model testing at laboratory scale and numerical simulation under different (extreme) wave-current conditions and structural configurations gives insight into the structural response, integrity and fatigue failure.

FIGURE 16.13   Seaweed farming in a wind farm (Image Courtesy: © Denis Lacroix, Ifremer and Malo Lacroix).

The model testing also gives a vivid understanding of the load on the mooring system and dynamics of the floater. Based on the literature review, ecological parameters, model simulated flow behaviour, and wave-interaction with the farming structure, an offshore seaweed farming location is selected, keeping in mind low wave and flow conditions with sufficient nutrient supply for the growth of seaweed. Thereafter, a high-resolution hydrodynamic modelling study is carried out before and after placing farming structures in the area of interest to understand changes in detailed local hydrodynamic behaviour due to the farming structure. Figure 16.13 illustrates conceptual seaweed farming inside a wind farm. In the near future, the presence of nearshore/offshore wind farms will increase due to their significant contribution to the generation of cleaner energy technologies. Further, this approach can be extended to other marine infrastructures.

16.10   CONCLUSION

To improve the living conditions of marginal farmers and fisherfolk in the coastal regions, seaweed cultivation is a promising avenue to earn an additional income due to the short growth cycle of seaweeds. However, to derive optimum benefits, efforts are required to improve the aquaculture practices of seaweed cultivation, especially the enhanced growth and harvesting strategies and elimination of invasive species. Sophisticated technology is needed to improve large scale seaweed cultivation in offshore seawater. To encourage the expansion of seaweed aquaculture, surveys must be conducted on large spatial scales to identify suitable locations.

REFERENCES

Albakosh, M. A. Naidoo, R. K. Kirby, B. and Bauer, R. (2016) Identification of epiphytic bacterial communities associated with the brown alga *Splachnidium rugosum. Journal of Applied Phycology* 28:1891–1901.

Amsler, C. D. Okogbue, I. N. Landry, D. M. Amsler, M. O. McClintock, J. B. and Baker, B. J. (2005) *Potential Chemical Defenses against Diatom Fouling in Antarctic Macroalgae*. Walter de Gruyter.

Ask, E. I. and Azanza, R. V. (2002) Advances in cultivation technology of commercial eucheumatoid species: a review with suggestions for future research. *Aquaculture* 206:257–277.

Ayyappan, S. Jena, J. K. Gopalakrishnan, A. and Pandey, A. K. (2006) *Handbook of fisheries and aquaculture*. Indian Council of Agricultural Research.

Banerjee, K. Turuk, A. S. and Paul, R. (2020) Seaweed: The Blue Crop for Food Security as Mitigation Measure To Climate Change. *Plant Archives* 20(1):2045–2054.

Campbell, I. et al. (2019) The environmental risks associated with the development of seaweed farming in Europe-prioritizing key knowledge gaps. *Frontiers in Marine Science* 6:107.

Chen, J. and Xu, P. (2005). Culture aquatic species information programme–Porphyra spp. Departamento de pesca y acuicultura de la FAO (en línea). Roma. Disponible en www.fao.org/fishery/culturedspecies/Porphyra_spp/es.

Chirapart, A, Praiboon, J, Boonprab, K, and Puangsombat, P (2018) Epiphytism differences in the commercial species of *Gracilaria*, *G. fisheri*, *G. tenuistipitata*, and *G. atissimi*, from Thailand. *Journal of Applied Phycology* 30:3413–3423.

Chopin, T. and Tacon, A. G. J. (2021) Importance of seaweeds and extractive species in global aquaculture production. *Reviews in Fisheries Science and Aquaculture* 29(2):139–148, doi: 10.1080/23308249.2020.1810626

Davies, O. A. (2009) Epiphytic diatoms growing on *Nypa fructican* of Okpoka Creek, Niger Delta, Nigeria and their relationship to water quality. *Research Journal of Applied Sciences, Engineering and Technology* 1:1–9.

Doty, M. S. and Alvarez, V. B. 1973. Seaweed Farms: A New Approach for U. S. Industry. Mar. Technol. Soc. 9th Annu. Conf. Paper. pp. 701–708.

El-Din, S. N. Shaltout, N. Nassar, M. and Soliman, A. (2015) Ecological studies of epiphytic microalgae and epiphytic zooplankton on seaweeds of the Eastern Harbor, Alexandria, Egypt. *American Journal of Environmental Sciences* 11:450.

FAO (May 1988) *Report on the Training Course on Seaweed Farming*. ASEAN/UNDP/FAO Regional Small-Scale Coastal Fisheries Development Project (RAS/84/016) and Regional Seafarming Development and Demonstration Project (RAS/86/024). Manila, Philippines, 2–21.

FAO. 2020. The State of Food and Agriculture 2020. Overcoming water challenges in agriculture. Rome. https://doi.org/10.4060/cb1447en

Ferdouse, F. Holdt, S. L. Smith, R. Murúa, P. & Yang, Z. (2018). The global status of seaweed production, trade and utilization. Globefish Research Programme, 124, I.

Ferreiro, N. Giorgi, A. and Feijoó, C. (2013) Effects of macrophyte architecture and leaf shape complexity on structural parameters of the epiphytic algal community in a Pampean stream. *Aquatic Ecology* 47:389–401.

Food and Agriculture Organization (2018). *Fishery and Aquaculture Statistics. Global Aquaculture Production 1950–2016 (FishStatJ)*. Rome, Italy. Retrieved from: FAO Fisheries and Aquaculture Department. www.fao.org/fishery/statistics/software/fishstatj/en

Frankovich, T. A. and Fourqurean, J. W. (1997) Seagrass epiphyte loads along a nutrient availability gradient, Florida Bay, USA. *Marine Ecology Progress Series* 159:37–50.

Gachon, C. M. Sime-Ngando, T. Strittmatter, M. Chambouvet, A. Kim, G. H. (2010) Algal diseases: spotlight on a black box. *Trends in Plant Science* 15:633–640.

Ganesan, M. Thiruppathi, S, Nivedita, S. Rengarajan, N. Veeragurunathan, V. and Jha, B. (2006). In situ observations on preferential grazing of seaweeds by some herbivores. *Current Science* 91:1256–1260.

Ganesan, M. Trivedi, N. Gupta, V. Madhav, S. V. Reddy, C. R. K. and Levine, I. A. (2019) Seaweed resources in India—status of diversity and cultivation: prospects and challenges. *Botanica Marina* 62(5):463–482.

Imchen, T. (2021) Nutritional value of seaweeds and their potential to serve as nutraceutical supplements. *Phycologia*, 60(6):534–546.

Ingle, K. N. Polikovsky, M. Chemodanov, A. and Golberg, A. (2018) Marine integrated pest management (MIPM) approach for sustainable seagriculture. *Algal Research* 29:223–232.

Johnson, B. Jayakumar, R. Nazar, A. K. A. Tamilmani, G. Sakthivel, M. Rameshkumar, P. Anikuttan, K. K. and Sankar, M. (2020) Prospects of Seaweed farming in India. *Aquaculture Spectrum,* 3(12):10–23.

Johnson, B. Narayanakumar, R. Nazar, A. A. Kaladharan, P. & Gopakumar, G. (2017). Economic analysis of farming and wild collection of seaweeds in Ramanathapuram District, Tamil Nadu. *Indian Journal of Fisheries*, 64(4), 94-99.

Kaliaperumal, N. and Kalimuthu, S. (1997) Seaweed potential and its exploitation in India. *Seaweed Research and Utilisation,* 19(1and2):33–40.

Kaliaperumal, N. Chennubhotla, V. S. Kalimuthu, S. Ramalingam, J. R. Pillai, S. K. Muniyandi, K. and Subbaramaiah, K. (1998) Seaweed resources and distribution in deep waters from Dhanushkodi to Kanyakumari, Tamilnadu. Seaweed Research and Utilisation, 20(1 & 2):141–151.

Kambey, C. S. et al. (2021) Seaweed aquaculture: a preliminary assessment of biosecurity measures for controlling the ice-ice syndrome and pest outbreaks of a *Kappaphycus* farm. *Journal of Applied Phycology* 33:3179–3197.

Khan, S. I. Satam, S. B. (2004) Seaweed mariculture: scope and potential In India. *Aquaculture Asia* 8 (4): 26–29.

Kim, J. O. et al. (2017) A survey of epiphytic organisms in cultured kelp *Saccharina japonica* in Korea. *Fisheries and Aquatic Sciences* 20:1–7.

Krishnamurthy, V. (1971) Seaweed resources of India and their utilization. *Seaweed Res.* Utiln. 1:55–67.

Krishnan, M. and Narayanakumar, R. (2013) Social and economic dimensions of carrageenan seaweed farming. FAO Fisheries and Aquaculture Technical Paper, 580:163–184.

Loureiro, R. Gachon, C. M. and Rebours, C. (2015) Seaweed cultivation: potential and challenges of crop domestication at an unprecedented pace. *New Phytologist*, 206(2), 489–492.

Mantri, V. A. Kavale, M. G. and Kazi, M. A. (2019) Seaweed biodiversity of India: Reviewing current knowledge to identify gaps, challenges, and opportunities. *Diversity*, 12(1), 13. https://doi.org/10.3390/d12010013

Mantri, V. A. Ganesan, M. and Gupta, V. (2019) An overview on agarophyte trade in India and need for policy interventions. *Journal of Applied Phycology* 31:3011–3023. https://doi.org/10.1007/s10811-019-01791-z

Mohammed (2015) *Current Trends and Prospects of Seaweed Farming in India*. CMFRI, 78–84.

Miller, M. W. and Hay, M. E. (1998) Effects of fish predation and seaweed competition on the survival and growth of corals. *Oecologia*, 113(2), 231–238.

Prado, S. Vallet, M. Gachon, C. M. M. Strittmatter, M. and Kim, G. H. (2017) *Prevention or Treatment in Algae of Diseases Induced by Protistan Pathogens*. Pub. No.: WO/2017/125775 [patent].

Qiu, L. Mao, Y. Tang, L. Tang, X, and Mo, Z. (2019) Characterization of Pythium chondricola associated with red rot disease of *Pyropia yezoensis* (Ueda) (Bangiales, Rhodophyta) from Lianyungang, China. Journal of Oceanology and Limnology 37:1102–1112.

Raghuvanshi, M. S. Patil, N. G. Daripa, A. Naitam, R. K. Kharbikar, H. L. Malav, L. C. and Pandey, L. (2021) Seaweeds as future resource of livelihood options in India. *Food and Scientific Reports* 2(8):26–30.

Ramesh, R. Mammalvar, R. and Gowri, V. S. (2008) *Database on Coastal Information of Tamil Nadu*. Institute for Ocean Management, Anna University, Chennai.

Sahu, S. K. Ingle, K. N. and Mantri, V. A. (2020) Epiphytism in seaweed farming: causes, status, and implications. *Environmental Biotechnology* 1(44):227.

Shanmugam, M. Sivaram, K. Rajeev, E. Pahalawattaarachchi, V. Chandraratne, P. N. Asoka, J. M. & Seth, A. (2017). Successful establishment of commercial farming of carrageenophyte Kappaphycus alvarezii Doty (Doty) in Sri Lanka: Economics of farming and quality of dry seaweed. Journal of Applied Phycology, 29(6):3015–3027.

Shanmugam, M. and Seth, A. (2018). Recovery ratio and quality of an agricultural bio-stimulant and semi-refined carrageenan co-produced from the fresh biomass of Kappaphycus alvarezii with respect to seasonality. Algal research, 32:362–371.

Shenoy, L. and Rajpathak, S. (2021) *Sustainable Blue Revolution in India: Way Forward*. CRC Press.

SNAP (2020) https://snapalginate.com/home/seaweed-harvesting/

Stachowicz, J. J. and Whitlatch, R. B. (2005) Multiple mutualists provide complementary benefits to their seaweed host. *Ecology* 86:2418–2427

Tsiresy, G. Preux, J. Lavitra, T. Dubois, P. Lepoint, G. and Eeckhaut, I. (2016) Phenology of farmed seaweed *Kappaphycus alvarezii* infestation by the parasitic epiphyte Polysiphonia sp. in Madagascar. *Journal of Applied Phycology* 28:2903–2914

Vairappan, C. S. (2006) Seasonal occurrences of epiphytic algae on the commercially cultivated red alga *Kappaphycus alvarezii* (Solieriaceae, Gigartinales, Rhodophyta). Journal of Applied Phycology 18:611–617

Vairappan, C.S. Chung, C. S. Hurtado, A. Soya, F. E. Lhonneur, G. B. and Critchley, A. (2008) Distribution and symptoms of epiphyte infection in major carrageenophyte-producing farms. *Journal of Applied Phycology* 20:477–483

Wahl, M. and Mark, O. (1999) The predominantly facultative nature of epibiosis: experimental and observational evidence. *Marine Ecology Progress Series* 187:59–66

Wahl, M. Goecke, F. Labes, A. Dobretsov, S. and Weinberger, F. (2012) The second skin: ecological role of epibiotic biofilms on marine organisms. *Frontiers in Microbiology* 3:292

Wahl, M. Hay, M. E. and Enderlein, P. (1997) Effects of epibiosis on consumer—prey interactions. *Hydrobiologia* 355: 49–59. https://doi.org/10.1023/A:1003054802699

Wang, G. Ren, Y. Wang, S. *et al.* (2022) Shifting chemical defence or novel weapons? A review of defence traits in *Agarophyton vermiculophyllum* and other invasive seaweeds. *Mar Life Sci Technol* 4:138–149. https://doi.org/10.1007/s42995-021-00109-8

Wang, S. Wang, G. Weinberger, F. Bian, D. Nakaoka, M. and Lenz, M. (2017) Anti-epiphyte defences in the red seaweed *Gracilaria vermiculophylla*: non-native algae are better defended than their native conspecifics. *Journal of Ecology* 105:445–457.

Wang, X. et al. (2014) Assimilation of inorganic nutrients from salmon (Salmo salar) farming by the macroalgae (*Saccharina atissimi*) in an exposed coastal environment: implications for integrated multi-trophic aquaculture. *Journal of Applied Phycology* 26:1869–1878.

Ward, G. M. et al. (2020) A review of reported seaweed diseases and pests in aquaculture in Asia. *Journal of the World Aquaculture Society* 51:815–828

Wear, D. J. Sullivan, M. J. Moore, A. D. and Millie, D. F. (1999) Effects of water-column enrichment on the production dynamics of three seagrass species and their epiphytic algae. *Marine Ecology Progress Series* 179:201–213

Wuchter, C. Marquardt, J. and Krumbein, W. E. (2003). The epizoic diatom community on four bryozoan species from Helgoland (German Bight, North Sea). *Helgoland Marine Research* 57:13–19

Yang, L. E. Lu, Q. Q. and Brodie, J. (2017) A review of the bladed Bangiales (Rhodophyta) in China: history, culture and taxonomy. *European Journal of Phycology* 52:251–263

17 Modeling of Marine Policy Regime Creation for Enhancing Blue Economy in Global to Regional Aspects

Md. Nazrul Islam
Department of Geography and Environment, Jahangirnagar University, Savar, Dhaka, Bangladesh

CONTENTS

DOI: 10.1201/9781003184287-17

17.1 INTRODUCTION

The marine ecosystems are the storehouses of the natural resources to use for achieving food security for the people of the world and for the various species in the oceans. The oceans are unique, extraordinary, and vital elements of our earth, covering more than of seventy percent of its surface. The productivity of marine ecosystems sustain life by generating oxygen, absorbing carbon dioxide from the atmosphere, regulating climate and temperature (Hale et al. 2009). Moreover, increasing a sustainable blue economy requires a model to create employment in coastal communities while ensuring that the oceans remain healthy and keep clean (Bennett et al. 2019). Nowadays, the world is pushing for extra food production and many people rely on the ocean for food, for jobs, for transport, and for recreation (Spalding, 2016; Hasan et al. 2018; Bennett et al. 2021). Considering the importance of thes, all the stakeholders should stand by the ocean to realize its vastness, its power and it's potential. Currently, marine ecosystems and coastal environments are subject to cumulative impacts from human activities, and those impacts know no boundaries (Day et al. 2015). It is necessary to develop a sustainable management strategy to solve this problem. A system managing each sector independently and allowing sectors to ignore each other is therefore inadequate (Kappel et al. 2012; Boero et al. 2016). It is necessary to integrate across sectors and across borders to enrich the marine environment globally and regionally (Annan-Diab and Molinari, 2017). According to the 1996 London Protocol to the Convention on the Prevention of Marine Pollution by Dumping of Wastes and Other Matter (Verlaan, 2011; Ruiz, 2018) prohibits all wastes, except for those identified on the 'reverse list'. This protocol is intended to be more protective of the marine environment (A-Khavari, 1997). To prevent or reduce ocean dumping, it is essential that we all do our part in cleaning up the mess. And, in order to make that happen, a clean-up drive could be organized to clear the shoreline (Ruhl, 1997; Wood, 2018). If the shore can be properly cleaned of the waste, ocean dumping rates can be reduced significantly. Heavy rains and floods wash trash and debris into the water (Rode, 2013). Human waste and sewage water that has been partially treated or untreated goes into the ocean. This is called 'garbage dumping' and is one of the world's leading causes of ocean pollution (Muthaiyah, 2020).

The scientific evidence shows that the current problems in the marine ecosystem and coastal zones are unplanned coastal development, illegal and overfishing, resource and habitat degradation (Chong et al. 2010). All of these are reflecting the lack of effective resource management activities in these areas (Baine et al. 2007). There is no integration of marine policies and protocols are implemented to overcome this issue. Some key challenges that are facing fisheries managers and scientists include (Peterman, 2004; Tyre and Michaels, 2011): (1) dealing with lots climate change uncertainties and the resultant risks, (2) estimating probabilities for those uncertain quantities, (3) recognizing and dealing with changes in parameters over time, (4) comprehensively evaluating management options, and so forth. (Cox, 2012).

Due to global warming and climate change the ocean is warming and the sea surface temperature is increasing, marine fish and other invertebrates have shifted their distributions to access their preferred temperatures (Perry et al. 2005; Dulvy et al. 2008; Poloczanska et al. 2013; Pinsky et al. 2013). In general, this has resulted in shifts towards the poles and into deeper waters. At a mean rate of 72 km per decade, marine species have been moving an order of magnitude faster than terrestrial species (Poloczanska et al. 2013). These distribution shifts are already generating management challenges (Pinsky et al. 2018). For example, a 'mackerel war' erupted in 2007 when the northeast Atlantic mackerel stock shifted from waters managed by the European Union, Norway and Faroe Islands into Icelandic and Greenland waters. Disagreements over the drivers of the shift, the expected duration of the shift, and appropriate catch reallocations resulted in the stock becoming increasingly overfished (Spijkers and Boonstra, 2017).

Marine ecosystems are critical for human existence and a foundation of the global economy. According to the Paulo de Bolle (2022) Global Senior Director, Financial Institutions Group, IFC, *'the more than 3 billion people rely on the ocean for their food, jobs, and livelihoods. Yet the impacts of climate change, overfishing, and pollution are putting our oceans at great risk, imperiling prosperity for millions in developing countries. Protecting the blue economy from these mounting threats* *isn't just a moral imperative it's a growing financial opportunity. Managed appropriately, it can create a win-win for the environment and the health in emerging market economies. Prior to the COVID-19 pandemic, the ocean economy was expected to double from 2010 to 2030 to reach $3 trillion and employ 40 million people.'*

According to ARTICLE 193 of The United Nations Convention on the Law of the Sea (UNCLOS) confers upon member states the sovereign right to exploit their natural resources pursuant to their environmental policies and in accordance with their duty to protect and preserve the marine environment (Schrijver, 1995). Similarly, ARTICLE 193 of UNCLOS also states that member states should take all the measures consistent to prevent, reduce and control pollution of the marine environment from any sources mentioned in this convention. They have also complied with this duty to protect the marine environment and the obligation to take measures to prevent and control marine pollution from regional to global contexts.

The 1969 International Convention applied to casualties involving pollution by oil. In view of the increasing quantity of other substances, mainly chemical, carried by ships, some of which would, if released, because serious hazard to the marine environment, the 1969 Brussels Conference recognized the need to extend the Convention to cover substances other than oil (Bergesen et al. 2018). The International Convention for the Prevention of Pollution of the Sea by Oil 1973 (MARPOL) as modified by the protocol of 1978. The six annexes of MARPOL (Annex 1 to 6) that are in force global from 2002 cover pollution by oil, chemicals, harmful substances in packaged form, sewage, and garbage (Vaneeckhaute and Fazli, 2020). But there is no comprehensive national legislation for the enforcement of these conventions (Hongdao and Mukhtar, 2017). The UNEP's 'Sustainable

TABLE 17.1

The Major Scales and Policy Implementation of Marine Policies Considering the Global Blue Economy Guidance

Global Blue Economy Guidance				
Scale	Institution	Policy	Policy implementation	Reference
Global	United Nations Environmental Programme (UNEP) and Others	Green economy in the blue world; blue economy; sharing success stories to inspire change	Widely used in this policy instrument	Gachingiri (2015)
	World Wildlife Fund (WWF)	Principles for sustainable blue economy; Reviving the ocean economy; the case for action (2015)	Not enough funds for the developing and least developed countries	Hoegh-Guldberg (2015)
Continental	United Nations Economic Commission for Africa (UNECA)	Africa's Blue Economy: A policy handbook	Widely used in African countries	Hagy and Nene (2021)
	African Union (AU)	2050 Africa's integrated Maritime strategy (2050 AIM Strategy)	Widely used in African countries	Union (2012)
Regional	United Nations Conference on Trade and Development (UNCTAD)	The Ocean Economy: Opportunities and Challenges for Small Island Developing States	Widely used in Small Island Developing States (SIDS)	Salpin et al. (2018)
	Western Indian Ocean Science Association (WIOMSA)	Building the blue economy in the WID Region	Widely used in the Indian Ocean region	Conand and Muthiga (2016); Wenhai et al. (2019)
	WWF International	Reviving the Western Indian Ocean Economy: actions for a sustainable future	Widely used in the Indian Ocean region	Obura et al (2017)

Blue Economy Initiative' (Table 17.1) aims to facilitate sustainable ocean-based economic, social and environmental benefits within the planetary boundaries of oceans and coasts (Sumaila et al. 2020). It is mentioned that engaging with countries, regional seas and many partners, it seeks to enhance decision-making, enabling conditions and capacities to develop the blue economy (Waite et al. 2015). Also they have tried to implement sustainable, climate-resilient and inclusive blue economy polices, strategies and solutions that reduce human impacts and support the sound use of marine and coastal ecosystems and their many services (Sumaila et al. 2020).

Similarly, WWF International recently released, *'Principles for a Sustainable Blue Economy,'* a 2015 briefing that aims to clarify what a *'blue economy,'* a label now commonly used in the contexts of economics, agriculture, and conservation, truly means (Voyer et al. 2020). As there remains no specific definition of the term, in WWF's view, the ambiguity may pose a danger. Thus, the principles have been developed to 'fill this gap in shared understanding about what characterizes a sustainable blue economy, and to help ensure that the economic development of the ocean contributes to true prosperity, today and long into the future (Hadjimichael, 2018).

The global blue economy strategy is consolidated, based on the following five detailed thematic technical reports that are annexes to this strategy:

* Fisheries, aquaculture, conservation, and sustainable aquatic ecosystems
* Shipping/transportation, trade, ports, maritime security, safety, and enforcement
* Coastal and maritime tourism, climate change, resilience, environment, infrastructure
* Sustainable energy and mineral resources and innovative industries
* Policies, institutional and governance, employment, job creation and poverty eradication, innovative financing

17.2 THE HISTORY OF MARITIME LAW AND BLUE ECONOMY ISSUES

Maritime law, also known as admiralty law, is a body of laws, conventions, and treaties that govern private maritime business and other nautical matters, such as shipping or offenses occurring on open water (Rogers, 2019). In the mission of the sustainable development goals, considerable effort is needed to ensure that the proposed goals, targets and indicators are in alignment with the vision, principles, guiding framework and criteria set out at the global and regional level (Waage et al. 2010). While the SDGs are not legally binding, governments are expected to take ownership and to establish national frameworks for the achievement of the SDGs through recurrent national and sectoral development planning (Banik, 2019). It must be done in a sustainable manner designed to ensure the conservation of individual species for current and future generations.

International rules, governing the use of the oceans and seas, are known as the law of the sea. Maritime law is a mix of common practices and laws that have been adopted by seafaring nations for generations (Smith, 2017). Since 1974, the maritime attorneys at the Law Offices of Charles D. Naylor have been compassionately and aggressively representing injured workers from the maritime trades, including seamen, and longshore workers. There is documentation of individual judgments and the establishment of common rules that pre-date European admiralty doctrines (Naylor, 2013). Additionally, the customs of early Egyptians, Phoenicians, and Greeks played a major role in the development and establishment of maritime law (Hassanali, 2020). As in the continental perspective of the blue economy, for the regional blue economies UNCTAD supports developing countries in improving their trade policies to ensure the sustainable use of the oceans, seas and coasts for economic growth and to improve livelihoods and jobs, while preserving the health of the ocean ecosystem (Islam and Mostaque, 2016; Ayilu et al. 2022).

The Western Indian Ocean Marine Science Association (WIOMSA) is inviting proposals from senior consultants to prepare and develop a status report, regional action plan and roadmap for 'The Blue Economy and Coastal Cities of the Western Indian Ocean Region' (Table 17.2) (Wambiji, 2019).

TABLE 17.2
Global Zoning of Potential Marine Areas, Their Characteristics and Major Environmental Concerns of This Particular Zone

Regions	Specific Marine Areas and Characteristics	Major Environmental Concern
West Asia	The Mediterranean, the Persian Gulf, the Arabian and the Red Sea	In the GCC countries, the challenges are from oil related industries and desalination plants
North America	Almost 25% of Canada's and about 55% of the United States' populations live in coastal areas	Harmful Algal Blooms (HABs) and toxins produced in the marine ecosystems
Latin America and the Pacific	South America, Meso-America, and the Caribbean	Habitat destruction and overexploitation,
Europe	The Adriatic, Mediterranean, Black, Azov, North Sea, Caspian, Baltic and White seas and the northeast Atlantic coast	Many of the 200 nuclear power plants operating throughout Europe are in coastal regions
Asia and the Pacific	Australia and New Zeeland, Central Asia, South Asia, Northwest Pacific and East Asia, South Asia, South Pacific	Marine resource issues are very critical, especially in developing countries
Africa	The African coastal zone is the island states Madagascar, Mauritius, Reunion and Seychelles. Many big cities are situated along the coast: Alexandria, Tripoli, Benghazi, Tunis, Algiers and Cairo	Coastal and marine habitats are being physically eroded and biologically degraded through unsustainable rates of resource extraction
The Polar region a) The Arctic b) The Antarctic	The Arctic marine environment covers approximately 20 million km² and includes the Arctic Ocean and several adjacent water bodies	Radioactive contamination, former nuclear weapons testing, and the Chernobyl accident
	The Southern Ocean represents 10% of the world's oceans	Hydrocarbon contamination

Source: After modifying and adopted from Islam and Kitazawa, 2011.

The Western Indian Ocean Marine Science Association and UN Habitat, in partnership with the Nairobi Convention recently launched a portfolio of 6 reports on 15th December 2021 (Wright et al. 2017).

In the United States, Maritime Law or Admiralty Law came into effect when the Judiciary Act of 1789 gave federal district courts jurisdiction over admiralty law cases, which made the U.S. Supreme Court the final authority on admiralty issues (Brown 1993; Casto, 1993). Although all cases of admiralty and maritime jurisdiction were put under federal jurisdiction, there is still a 'saving' clause that allows state courts to hear some maritime cases (Table 17.3). Although not all of the original principles of Admiralty Law still apply, there are principles that are used today such as Maintenance and Cure, Marine Insurance, General Average, and Salvage (Stevens, 1950).

These concerns lead to the passage of maritime laws making ship owners 'strictly liable' for damage caused by oil spills, and to new work rules for seamen on tankers and other types of vessels.

Also, the Basel Convention focuses on the regulation of the transboundary movement of hazardous wastes to protect developing countries from importing such wastes that they are unable to manage in an environmentally sound manner (Choksi, 2001). However, Basel does not establish a system for ship recycling, rather this has been dealt with in the Hong Kong International Convention for the Safe and Environmentally Sound Recycling of Ships 2009 (Mikelis, 2010). Though Bangladesh is the 3rd largest ship recycling country, it has not ratified the Hong Kong Convention and the observance of this convention has not been mentioned in Ship Breaking and Hazardous Waste Management Rules 2010 (Mikelis, 2010).

TABLE 17.3
Marine Environmental Relevant Laws, International Regime, Protocol and Amendments

Major Marine Regimes	Marine Acts, Laws, Protocol and Amendment
Antarctic Regime (1959–1998)	✓ Antarctic Treaty (1959–1980)
	✓ Protocol on Environmental Protection (1989/91–1998)
Baltic Sea Regime (1974–1998)	✓ Regulations for all Sources of Marine Pollution (1974–1992) (1992–1998)
Barents Sea Fisheries (1975–1998)	✓ Norwegian-Russian Cooperation on Fisheries in the Barents Sea Region (1975–1998)
Biodiversity Regime (1992–1998)	✓ Convention on Biological Diversity (1992–1998)
Climate Change Regime (1992–1998)	✓ Kyoto-Protocol (1997–1998)
	✓ The UN Framework Convention (UNFCCC) (1997–1998)
Lakes Management Regime 1972–1998	✓ Great Lakes Water Quality and Ecosystem Management (1978–1998)
London Convention Regime 1972–1998	✓ Wastes and Substances the Dumping of which is Prohibited (1972–1991) (1991–1998)
	✓ Regulation of Incineration at Sea (1978–1991) (1991–1998)
North Sea Regime 1972/74–1998	✓ North Sea Conferences (1984–1998)
Oil Pollution Regime 1954–1998	✓ The OILPOL Convention in 1954
	✓ The MARPOL (1973/78–1998) Conevention
Ramsar regime on Wetlands 1971–1998	✓ Ramsar Convention (1971–1987) (1987–1998)
Regime of Black Sea 1992–1998	✓ Bucharest Convention and Protocols (1992–1998)
	✓ Black Sea Strategic Action Plan (1996–1998)
S. Pacific Fisheries Regime 1979–1998	✓ General Management of Fisheries (1979–1982) (1982–1995/97) (1995/97–1998)

Source: After modifying and adopted from Islam and Kitazawa, 2011.

17.2.1 Instruments of Marine Pollution at International, Regional and National Levels

A large number of instruments at international, regional and national levels have been adopted to tackle marine environmental pollution issues (Boyes and Elliott, 2014). These instruments comprise conventions, agreements, regulations, strategies, action plans, programs, and guidelines (Figure 17.1). They contain specific management measures that are either compulsory or voluntary (Chen, 2015). The corresponding regional or national instruments transposed from international ones include: i) the European Union (EU) PRF Directive, ii) Annex IV of the Helsinki Convention, iii) the United States (US) Marine Plastic Pollution Research and Control Act, iv) the United Kingdom (UK) Merchant Shipping (Prevention of Pollution by Sewage and Garbage from Ships) Regulations 2008, and v) various other national legislations (Boteler and Coastal, 2014). The second type comprises instruments, which are not explicitly transposed into regional or national schemes. These instruments mostly serve as global guiding instruments encouraging regional bodies or countries to follow the actions proposed therein, or as a platform for the states concerned to engage in coordination and cooperation in marine litter issues (Long, 2011). The most prominent examples are perhaps a series of initiatives developed by the United Nations Environment Programme (UNEP).

17.2.2 The United Nations Convention on the Law of the Sea (UNCLOS)

The United Nations Convention on the Law of the Sea (UNCLOS) incorporates in its part XII the first comprehensive statement of international law for the ocean environment. The United Nations

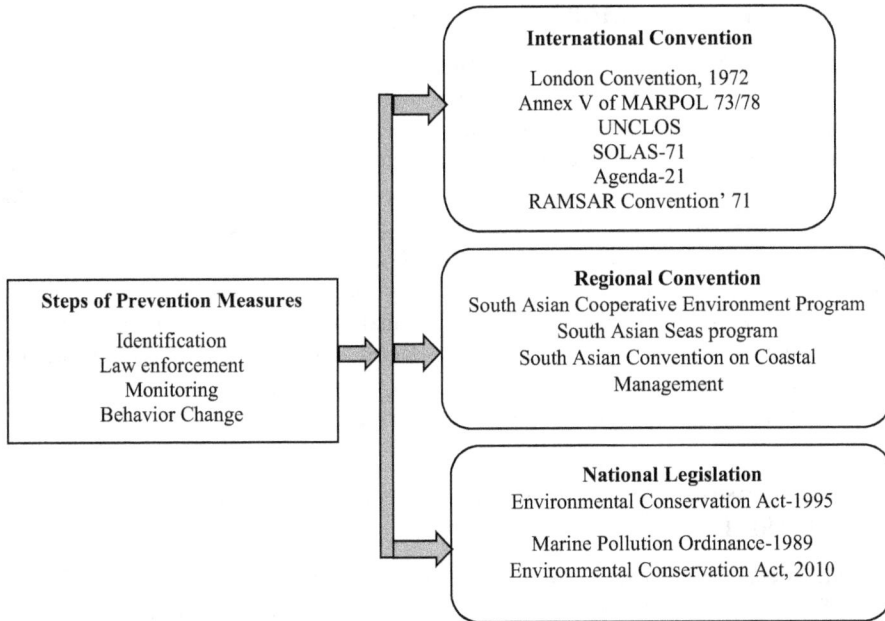

FIGURE 17.1 A thematic diagram of the number of instruments at international, regional, and national levels have been adopted to tackle marine environmental pollution issues.

Convention on the Law of the Sea (Jamilah and Disemadi, 2020) provides specifically for marine scientific research, protection of the marine environment, and the promotion of research centers ((Jamilah and Disemadi, 2020) .

- *Article 192 - States have the obligation to protect and preserve the marine environment*
- *Article 193 - States have the sovereign right to exploit their natural resources pursuant to their environmental policies and in accordance with their duty to protect and preserve the marine environment.*
- *Article 194 - States shall take, individually or jointly as appropriate, all measures consistent with this convention that are necessary to prevent, reduce and control pollution of the marine environment from any source, using for this purpose the best practicable means at their disposal and in accordance with their capabilities, and they shall endeavor to harmonize their policies in this connection* (Reeve, 2012; Karim, 2016).

17.2.3 INTERNATIONAL CONVENTION FOR THE PREVENTION OF POLLUTION (MARPOL), 1973/78

The International Convention for the Prevention of Pollution from Ships (MARPOL) is the main international convention covering the prevention of the pollution of the marine environment by ships from operational or accidental causes (Bergmeijer, 1992). The MARPOL Convention was adopted on 2nd November 1973 at IMO. The Protocol of 1978 was adopted in response to a spate of tanker accidents in 1976 - 1977. As the 1973 MARPOL Convention had not yet come into force, the 1978 MARPOL Protocol absorbed the parent Convention (Bergmeijer, 1992; Sheavly and Register, 2007). The combined instrument entered into force on 2nd October 1983. In 1997, a Protocol was adopted to amend the Convention and a new Annex VI was added which came into force on 19th May 2005. MARPOL has been updated by amendments through the years.

17.2.4 Salient Features of MARPOL 73/78 and Annex V

The MARPOL Annex V regulates the prevention of pollution by garbage from ships. Amendments to Annex V were adopted in 2011 (resolution MEPC.201 (62)) and came into force on 1st January 2013 (Campara et al. 2018). Under the amended MARPOL Annex V, discharge of all garbage is now prohibited, except as specifically permitted in the regulations of MARPOL Annex V. (Before these amendments, the discharge of garbage was generally allowed unless provided for otherwise in MARPOL Annex V, depending on the nature of the garbage and defined distances from shore.) Regulation 7 provides limited exceptions to the MARPOL Annex V restrictions in emergency and non-routine situations (Table 17.4) (Carpenter, 2019; Sovacool et al. 2021). Generally, discharge is restricted to food wastes, identified cargo residues, animal carcasses, and identified cleaning agents and additives and cargo residues entrained in wash water which are not harmful to the marine environment (Amon et al. 2020; Sovacool et al. 2021). It is recommended that ships use port reception facilities as the primary means of discharge for all garbage.

17.2.5 International Convention for the Safety of Life at Sea (SOLAS), 1971

Additional measures for tanker safety were incorporated into the 1978 Protocol to the International Convention for the Safety of Life at Sea (SOLAS), 1971 (Zhang et al. 2021). The SOLAS Convention in its successive forms is generally regarded as the most important of all international treaties concerning the safety of merchant ships (Zhang et al. 2021). The main objective of the SOLAS Convention is to specify minimum standards for the construction, equipment and operation of ships, compatible with their safety (Engtrø, 2022). Flag States are responsible for ensuring that ships under their flag comply with its requirements, and several certificates are prescribed in the Convention as proof that this has been done (Campara et al. 2018). Control provisions also allow Contracting Governments to inspect ships of other Contracting States if there are clear grounds for believing that the ship and its equipment do not substantially comply with the requirements of the Convention. This procedure is known as port State control (Gagatsia, 2007).

17.2.6 Marine Protected Areas: Territorial Water and Maritime Zones Act, 1974

The important legislation in the South and Southeast Asia region relates to marine parks and reserves and continues the tradition of dealing with them in the context of fisheries and environmental legislation (Agardy et al. 2011). The conservation, use, and exploitation of marine resources is provided for under the Territorial Water and Maritime Zones Act, 1974. According to the provisions in this Act, conservation zones may be established to protect marine resources from indiscriminate exploitation, depletion, or destruction (Islam et al. 2017). At present, there is no legal provision for the management of coastal zones.

17.2.7 The Need to Develop the Marine Environment Conservation Act 2004

As example considering the regional scale shows that in 2004, the government of Bangladesh drafted a Marine Environment Conservation Act, which were placed in the national parliament assembly for consideration after necessary scrutiny (Karim, 2019; Berkey and Williams, 2018). According to the Bangladesh Maritime Zones Act, 2018 (Draft) and under section 21, the owner or master or agent or the occupier of a ship is bound to report about discharges of oil or pollutants to the relevant authorities (Edu, 2011; Karim, 2016). Many different government Departments and devolved administrations have responsibility for regulating different activities and protecting the marine environment. In particular, the waters around Scotland, Wales and Northern Ireland out to 12 nautical

TABLE 17.4
The International Convention for the Prevention of Pollution from Ships (MARPOL) Is the Main International Convention Covering Prevention of Pollution of the Marine Environment by Ships from Operational or Accidental Causes

Type of Garbage	Ships Outside Special Areas	Ships Within Special Area	Offshore Platforms and All Ships Within 500 m of Such Platforms
Food wastes comminuted or ground	Discharge permitted ≥3 nm from the nearest land and en route	Discharge permitted ≥12 nm from the nearest land and en route	Discharge permitted ≥12 nm from the nearest land
Food wastes not comminuted or ground	Discharge permitted ≥12 nm from the nearest land and en route	Discharge prohibited	Discharged prohibited
Cargo residues not contained in wash water	Discharge permitted ≥12 nm from the nearest land and en route	Discharge prohibited	Discharge prohibited
Cargo residues contained in wash water	Discharge permitted ≥12 nm from the nearest land and en route	Discharge only permitted in specific circumstance and ≥12 nm from the nearest land and en route	Discharge prohibited
Cleaning agents and additives' contained in cargo hold wash water	Discharge permitted	Discharge only permitted in specific circumstanced and ≥12 nm from the nearest land and en route	Discharge prohibited
Cleaning agents and additives' contained in deck and external surfaces wash water	Discharge permitted	Discharge permitted	Discharge prohibited
Animal carcasses	Discharge permitted as far from the nearest land as possible and en route	Discharge prohibited	Discharge prohibited
All other garbage including plastics, domestic wastes, cooking oil, incinerator ashes, operational wastes, and fishing gear	Discharge prohibited	Discharge prohibited	Discharge prohibited

Source: Bergmeijer, 1992; Sheavly and Register, 2007; Campara et al. 2018; Carpenter, 2019.

miles are territorial seas and are the responsibility of the respective devolved administrations (Simas et al. 2015).

17.2.8 The Coastal Zone Policy 2005

The Coastal Zone Policy (CZP) 2005 with its eight development objectives forms a comprehensive framework for ensuring environmental friendly activities in coastal areas along with sustainable development (Rey-Valette et al. 2007). The Coastal Zone Policy (CZP) builds on the relevant segments on coastal issues and explicates them in a manner that provides direction for realizing the objectives of the policy (Rahman, 2006). The policy also prescribes provisions to control pollution on the coast to minimize environmental degradation.

17.3   CHALLENGES TO MARINE REGIME CREATION FOR THE MITIGATION OF CLIMATE CHANGE IMPACT

Climate change is reducing the productivity of marine fisheries globally. Regional impacts are especially pronounced, with some regions experiencing large gains in productivity whilst others are experiencing large losses (Hoegh-Guldberg et al. 2019). Climate change is significantly altering the ability for marine fisheries to provide food and income for people around the world (Shukla et al. 2019). These changes are commonly viewed as occurring through impacts on either the distribution of fish stocks (namely, where fish can be caught and by whom) or the productivity of fish stocks (namely, how much fish can be caught) (Bakun and Broad, 2003). In general, productivity is predicted to decrease in tropical and temperate regions and increase toward the poles (Lotze et al. 2019) as marine organisms shift their distributions to maintain their preferred temperatures (Pinsky et al. 2013; Poloczanska et al. 2013; Poloczanska et al. 2016). These regional shifts in productivity, range and fishing opportunity are likely to result in regional discrepancies in food and profits from fisheries (Lam et al. 2016), with tropical developing countries and small island developing states exhibiting the greatest vulnerability to the climate change (Allison et al. 2009; Blasiak et al. 2017; Guillotreau et al. 2012).

- **Ocean warming** is expected to raise mortality rates and lower productivity for higher-trophic-level species (bivalves, finfish, and crustaceans) (Lacoue-Labarthe et al. 2016).
- **Sea level rise** will increase the intrusion of saline water into deltas and estuaries compromising brackish-water aquaculture (Bricheno et al. 2021) and shifting shoreline morphology could reduce habitat availability (bivalves, finfish, crustaceans).
- **Increasing storm strength and frequency** pose risks to infrastructure (De Silva, 2012), and increased weather variability has been associated with lower profits (bivalves, finfish, crustaceans) (Li et al. 2014).
- **Ocean acidification** impedes the calcification of mollusc shells (Gazeau et al. 2013) resulting in reduced recruitment, higher mortality (Barton et al. 2012; Green et al. 2013) and increased vulnerability to disease and parasites (bivalves).
- **Increasing rainfall** will raise the turbidity and nutrient loading of rivers, potentially causing more harmful algal blooms (HABs) (Islam et al. 2012) that reduce production and threaten human health (bivalves, finfish, crustaceans) (Himes-Cornell et al. 2013; Rosa et al. 2014).

Every climatic event can have a range of harmful effects on the resources of coastal areas. Every year, many types of natural disaster occur in coastal regions (Mirza, 2003; Wenhai et al. 2019)). It brings both economic and environmental losses to countries. Every coastal country's peoples earn their livelihood using the ocean resources and this also enlarges the blue economy (Barbesgaard, 2018). In a particular season of annual cyclones in the ocean there is damage to the coastal infrastructure, human casualties and degradation of the environmental balance. Melting of the ice impacts sea level rise, thus increasing the possibility of regular flooding and the degradation of ecosystem services in mangrove areas (Bax et al. 2022). Similarly, regular erosion of land in coastal areas results in the loss of flora and fauna. Sometimes cyclones produce tidal surges in coastal areas (Sigren et al. 2014). As a result of these tidal surges, saline intrusion increases on coastal soil and water. Saline intrusion impacts on crops and aquaculture. Floods also create damage by way of tidal surges. Table 17.5 shows the key sources of marine regime creation challenges, environmental forcing factors and the magnitudes of impacts from natural and anthropogenic interference to marine ecosystems.

Of the natural resources in the marine environment, the second-largest ocean-related economic sector was tourism in 2010, next to offshore oil and gas (Hussain et al. 2014; OECD, 2016). The blue economy sectors and ocean tourism are projected to be the top contributor to ocean industry by 2030 in terms of production value, at which point it will account for 26% of the ocean-based

TABLE 17.5
Key Source of Marine Regime Creation Challenges, Environmental Forcing Factors and Magnitudes Impacts from Natural and Anthropogenic Interference to the Marine Ecosystems

Key Sources	Environmental Forcing Factors (EFF)	Magnitude of Impacts
Nuclear and power transmission	Tailings and water pollution	Marine ecosystem threats and species thrashing
Oil and gas exploration, dredging	Seismic effects, high level sound, oil spills	Contain toxic agents, namely, methanol
Microplastic increasingly leads	Disturbance of the oceanic environment	Adversely affects environmental and human health
Commercial marine shipping	Waste dumping, oil spills and ship groundings	Water pollution, toxicity, species missing
Overfishing and Aquaculture	Imbalance in productivity	Eutrophication and algal blooms
Highway and transportation	Habitat destruction and noise pollution	Marine fish and mammals shifted and killed
Hydroelectric power generation	Sound pollution and fish/mammals lost	Endangered fish and mammals
Military step and navigation research	Ranges and training areas conflict, hazards	Geopolitical crisis and cold war, pollution
Oceanographic survey and experiments	Habitat stress and noise pollution	Benthic fauna endangered, fish killed
Sonar and air guns for seismic surveys	Habitat dispersal and species movement	Habitat destruction and species loss
Laying submarine cables	Habitat disturbance and broad-spectrum dispersal	Harmful algal blooms and toxin produced
Underwater habitat and sea lab	Habitat disturbance and full dispersal of species	Long term marine ecosystem effects
Underwater Robotics (AUV, ROV)	Disturbance and harassment	Species threats and loss, fish killed
Subsurface and seabed aquaculture	Pollution and toxicity increased	Mammals and benthic species threats
Tourism and recreational activities	Human interference and habitat reduction	Ecosystems threats and species overwhelmed
Pipelines, tunnel borers and drilling	Chemical pollution, noise and sediment erosion	Continuing effects and biodiversity threats

Source: After modifying and adopted from Islam and Kitazawa, 2011.

economy, compared with 21% for oil and gas (OECD, 2016). Particularly in Asia and Africa, ocean tourism dwarfs the contribution of industrial capture fisheries, which constitute only 1% of ocean-based industries' production value (not accounting for artisanal fisheries, which are a critical component of the economies of Asia and Africa) (Raheem, 2022). Ocean resource exploitative activities include beach tourism, recreational fishing, swimming, diving, whale watching, and taking cruises, amongst. These, and possibly other blue economy sectors will be the emerging economic sectors in near future, in the world economy (Table 17.6) (Hall, 2001). Ocean tourism's global direct added value was estimated at US$390 billion in 2010, directly providing seven million full-time jobs. In addition, the ocean is a source of recreation for millions of people in the developed and developing worlds (Ghermandi and Nunes, 2013; Arlinghaus et al. 2019).

Ocean tourism directly supports the livelihoods of millions of people and the economies of the developing tropics and many small island developing states (Lee et al. 2015). For example,

TABLE 17.6

Climate Change Extreme Events with Their Impacts on Marine Resources and Options for Resilience Building to Climate Change and Enhancing Blue Economy

Climate Change Events	Impacts on Marine Resources	Options for Resilience Building for Enhancing Blue Economy
Warming	Coral bleaching, species migration, biodiversity loss, altered species life style, disruption in marine food chain	Mangrove plantation and restoration, sea grass, salt marsh and mussel bed conservation, coral reef protection and oyster reef development
Cyclone	Loss of coastal resources, degradation of coastal habitats, loss of infrastructure facilities	
Sea Level	Reduction in photosynthesis, disruption in the mangrove ecosystem	
Droughts	Crop loss	Plantation and crop insurance
Erosion	Clogging of air bladder of fish, mortality of the species, loss of coastal resources, degradation of coastal habitats, loss of infrastructure facilities	Mangrove plantation and restoration, sea grass, salt marsh and mussel bed conservation, coral reef protection and oyster reef development
Tidal surge	Loss of coastal resources, degrade coastal habitats, loss of infrastructure facilities	Mangrove plantation and restoration, sea grass, salt marsh and mussel bed conservation, coral reef protection and oyster reef development
Saline water intrusion	Crop damage, shift of species habitat	Mangrove plantation and restoration, crop insurance
Flood	Crop damage, loss of infrastructure, loss of habitats	
Change in precipitation	Crop loss	Plantation and crop insurance
Ocean acidification	Biodiversity loss, species migration, biodiversity loss, altered species life style, disruption in marine food chain	Mangrove plantation and restoration, marine spatial planning and marine protected area declaration

Source: Adopted from Techera, 2018; Sarker et al. 2019; Karani and Failler, 2020; Bax et al. 2022.

coral reef tourism alone contributes over 40% of the gross domestic products of Maldives, Palau, and St. Barthelemy (Spalding et al. 2017). Despite the importance of ocean tourism in the economy, data and research on the impacts of climate change in the tourism sector are limited (Scott et al. 2012).

17.4 POLICY OPTIONS AND THEIR APPLICATION FOR BLUE ECONOMY SECTORS

International maritime law stands on four strong pillars, namely, the Law of Sovereignty of Nations, the Law of Freedom of the High seas, the Law of Freedom of Contract, and the Legal Personality of a Ship (Khobragade et al. 2021). Each country is sovereign within its own political boundaries, in which its laws apply. Maritime law, also known as admiralty law, is a body of laws, conventions, and treaties that govern private maritime business and other nautical matters, such as shipping or offenses occurring on open water (Collins and Hassan, 2009; Rogers, 2019).

International rules, governing the use of the oceans and seas, are known as the Law of the Sea (De Lucia, 2019). With one of the industry's main concerns being the safety of crew and personnel onboard vessels, SOLAS 'Safety of Life at Sea' is generally regarded as the most important of all international Conventions (Table 17.7) (Joseph and Dalaklis, 2021).

TABLE 17.7
The SDGs Goal 14-Conserve and Sustainably Use the Oceans, Seas and Marine Resources for Sustainable Blue Economy and Others Development

SDGs Goal 14 and Sustainable Global Blue Economy: Policy Options	References
14.1 By 2025, prevent and significantly reduce marine pollution of all kinds, from land-based activities, including marine debris and nutrient pollution.	Cordova and Nurhati (2019) Smail et al. (2020)
14.2 By 2020, sustainably manage and protect marine and coastal ecosystems to avoid significant adverse impacts, including by strengthening their resilience, and act for their restoration in order to achieve healthy and productive oceans.	Segui et al. (2020) Halkos and Gkampoura (2021) Virto (2018)
14.3 Minimize and address the impacts of ocean acidification, including through enhanced scientific cooperation at all levels.	Tilbrook et al. (2019) Scott (2018)
14.4 By 2020, effectively regulate harvesting and end overfishing, illegal, unreported, and unregulated fishing and destructive fishing practices and implement science-based management plans, to restore fish stocks in the shortest time feasible, at least to levels that can produce maximum sustainable yield as determined by their biological characteristics.	Hurd et al. (2018) Friess et al. (2019) Singh et al. (2018) Nash et al. (2020)
14.5 By 2020, conserve at least 10% of coastal and marine areas, consistent with national and international law and based on the best available scientific information.	Dudley et al. (2017) Friess et al. (2019)
14.6 By 2020, prohibit certain forms of fisheries subsidies which contribute to overcapacity and overfishing, eliminate subsidies that contribute to illegal, unreported, and unregulated fishing and refrain from introducing new such subsidies, recognizing that appropriate and effective special and differential treatment for developing and least developed countries should be an integral part of the World Trade Organization fisheries subsidies negotiation.	Schmidt (2018) Oh (2018) Yingying (2017)
14.7 By 2030, increase the economic benefits to small island developing States and least developed countries from the sustainable use of marine resources, including through sustainable management of fisheries, aquaculture, and tourism	Nisa et al. (2022) Singh et al. (2018) Griffin et al. (2019)
14. a. Increase scientific knowledge, develop research capacity and transfer marine technology, adopting the Intergovernmental Oceanographic Commission Criteria and Guidelines on the Transfer of Marine Technology, to improve ocean health and to enhance the contribution of marine biodiversity to the development of developing countries, in particular Small Island Developing States and least developed countries.	Nisa et al. (2022) Singh et al. (2018) Griffin et al. (2019) Morgera and Ntona (2018)
14.b. Provide access for small-scale artisanal fishers to marine resources and markets.	Zelasney et al. (2020)
14.c Enhance the conservation and sustainable use of oceans and their resources by implementing international law as reflected in the United Nations Convention on the Law of the Sea, which provides the legal framework for the conservation and sustainable use of oceans and their resources, as recalled in paragraph 158 of 'The future we want'.	Cormier and Elliott (2017) Virto (2018) Mustafa et al. (2018) Rickels et al. (2019)

17.5 ACHIEVEMENTS FOR THE BLUE ECONOMY UNDER THE LONDON CONVENTION AND ITS PROTOCOL

The unregulated dumping and incineration activities that developed in the late 1960s and early 1970s have been halted (Louis, 2004). Parties to the Convention agreed to control dumping by implementing regulatory programs to assess the need for, and the potential impact of, dumping (Vare et al. 2018). They eliminated dumping of certain types of waste and, gradually, made this regime more restrictive by promoting sound waste management and pollution prevention (Ray,

2008). Prohibitions are in force for the dumping of industrial and radioactive wastes, as well as for incineration at sea of industrial waste and sewage sludge (Fytili and Zabaniotou, 2008).

- Guidance has been developed for the development of action lists and action levels for dredged material which assists regulators and policy makers on the selection of action lists and the development of action levels for dredged material intended for disposal at sea (Apitz and Agius, 2013). It is also remarked on the potential regulatory outcomes from changes to chemistry protocols in the Canadian disposal at sea program (Selin, 2013). An action list is a set of chemicals of concern, biological responses of concern, or other characteristics that can be used for screening dredged material for their potential effects on human health and the marine environment (Heise et al. 2020). On the other hand, the action levels establish thresholds that provide decision points to determine whether sediments can be disposed of at bottom of the ocean.
- Advice is available concerning the management of spoilt cargoes onboard vessels, best management practices for the removal of anti-fouling coatings from ships, and placement of artificial reefs (VanderZwaag, 2015).
- A technical co-operation and assistance programme has been established to assist with capacity building for waste assessment and management in marine system (Alam and Faruque, 2014), and in developing national regulations to comply with and implementation of the London Protocol and other maritime regimes (Stokke, 2018).
- As a remarkable mater is that the contracting parties to the London Protocol of marine regime have recently taken ground-breaking steps to mitigate the impacts of increasing concentrations of carbon dioxide in the atmosphere by amending the Protocol to regulate carbon capture and sequestration in sub-sea geological formations (Ringbom et al. 2018).
- London Convention and Protocol Contracting Parties have also adopted an 'Assessment Framework for Scientific Research Involving Ocean Fertilization' to guide Parties on how to assess proposals for ocean fertilization research which provides detailed steps for completion of an environmental assessment, including risk management and monitoring (Broder, 2017).
- Parties have developed a wealth of experience regarding marine pollution prevention issues, interpretation of the Convention and Protocol, licensing, compliance, and field monitoring activities (Gulas et al. 2017).

17.6 MAJOR PRIORITIES FOR A SUSTAINABLE GLOBAL BLUE ECONOMY

For much too long, the ocean has been out of sight, out of mind and out of luck. Attention has been scant from governments, funding agencies, financial institutions, food-security organizations, and the climate-mitigation community (Konar and Ding, 2020; Lubchenco et al. 2020). Nations usually manage their waters sector by sector, or issue by issue. The resulting hodgepodge of policies fails to consider collective impacts. It has been mentioned by US Secretary of State, John Kerry (2014) (Goldenberg, 2014), that *'Protecting our oceans is not a luxury. It is a necessity that contributes to our economy, our climate, and our way of life. Working together, we can change the current course and chart a sustainable future.'* It is essential to standardize our methods and definitions for valuing the coasts and ocean. The world's population depends upon the oceans for its very existence (Rockström and Klum, 2015). The ocean regulates our climate and our weather. It generates half of the oxygen we breathe (Andrews et al. 2014). It provides food and income for billions of people. Covering almost three quarters of the planet, the ocean is the life support system for Planet Earth. They express this life support as 'eco-system services' provisioning (for example, food, oxygen, and water), regulating (for example, climate/temperature regulation, and coastal stabilization), supporting (for example, pollution filtration, waste processing, and transportation of goods), and cultural services (for example, aesthetics, recreation, fun, and inspiration) (Raheem, 2022).

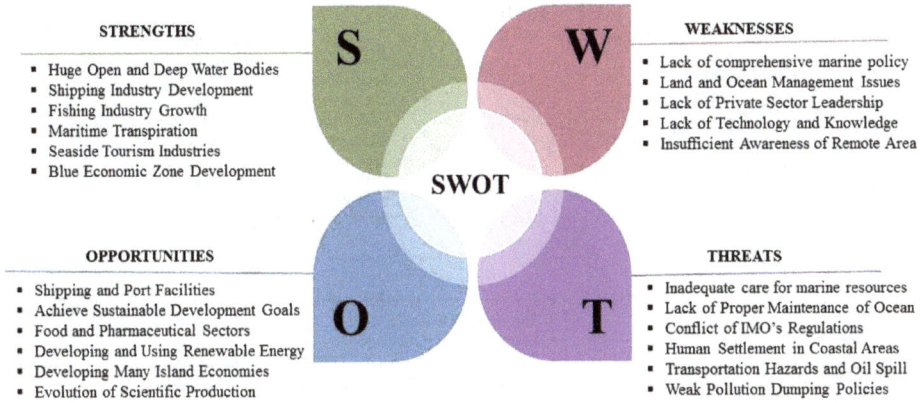

FIGURE 17.2 The SWOT Analysis shows the prospective Strengths, Weaknesses, Opportunities, and Threats for enhancing global blue economy in the marine environment.

Figure 17.2 is a SWOT Analysis that shows the prospective Strengths, Weaknesses, Opportunities, and Threats for enhancing Global Blue Economy in the marine environment.

Climate change impacts on the marine ecosystem will differ by the sea and oceans locations, territory, characteristics and dynamics of the oceans. By exploring climate change impact at the national level for fisheries, aquaculture and reef tourism, countries can assess what they stand to gain or lose due to climate change and understand how they might make use of these predictions to inform their investments and actions (Hoffman, 2005; Zougmoré, et al. 2016). Implementing certain key strategies will help build socioecological resilience to climate change and ensure the continued, or improved, provision of functions and services from the ocean, especially for the most vulnerable coastal nations (Hertel and Rosch, 2010). These strategies include the following:

- **A focus on equity:** Climate change is likely to cause and exacerbate global inequities, reducing resilience and thereby likely to worsen outcomes under all climate change scenarios (Lomborg, 2020). It will thus be profoundly important to examine the equity implications of all new and existing management decisions across all three sectors.
- **Looking forward:** The future of the ocean economy is expected to drastically change given climate change, and the nature and magnitude of these changes can be highly variable (Sellers and Ebi, 2018). Each of these three sectors will need to work to understand risks and anticipate changes, and to build precautionary and adaptive strategies into their management decisions (Figure 17.3).
- **Cooperating across boundaries:** As suitable habitats shift and change, marine species will move across jurisdictional boundaries and regional, national and international cooperative agreements will be necessary to ensure that these species are well-managed, and that the benefits are fairly distributed during and after the transitions (Roberts et al. 2017).

The steps for mitigation of and adaptation strategies for a sustainable blue economy are:

- It is needed to address the current impacts of climate change on the ocean including significantly cutting emissions, up-scaling proper protection for marine ecosystems to retain resistance and rebuild resilience (Wilson and Forsyth, 2018) , as well as implementing sustainable practices for all industries and uses across the oceans.
- The coastal states are well positioned to make use of MPAs for ecosystem-based adaptation and mitigation as a 'no-regret' climate change strategy (Lo, 2016). The Integrated Coastal

DPSIR Model
(Marine Regime and Blue Economy

Driving Forces

- Population Pressure
- Overfishing and Exploitation
- Marine Pollution
- Habitat destruction
- Climate Change Impact
- Invasive Alien Species

Responses

- Maintaining Marine Biodiversity
- Maintaining Species Distribution
- Awareness Building Program for Marine Issues
- Improving Awareness on Marine Issues
- Ensuring NGO and Community Involvement
- Develop the Marine Regime and Policies
- Innovate the London Protocol of Ocean Dumping

State

- Threatening Marine Population
- Disrupting Marine Habitats
- Eutrophication and Toxicity
- Oil Spills and Affected Species

Pressures

- Illegal and Unreported Fishing
- Agricultural Runoff to the Ocean
- Ocean Based Human Activities
- Climate Change Pressure
- Biologically Unsustainable Levels
- Spread of Invasive Organisms

Impacts

- Marine Biodiversity Loss
- Degradation of Salt Marshes
- Marine Recourse Depletion
- Environment Pollution
- Severe Environmental Impacts
- Decreased Water Quality

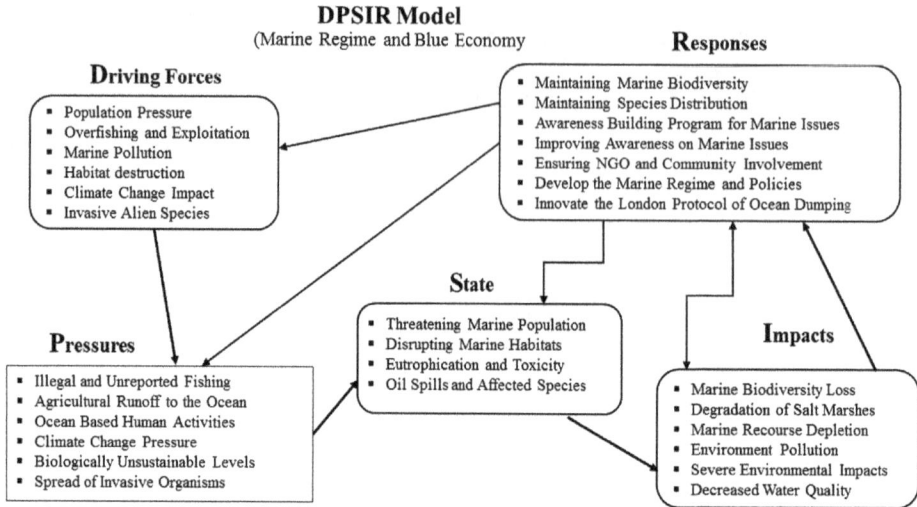

FIGURE 17.3   A DPSIR framework for the integrating marine and sustainable global blue economy.

Zone Management (ICZM) and Marine Spatial Planning (MSP) can be used by countries to improve the management of MPAs and help meet multiple objectives (Ramieri et al. 2019), including sustainable development, biodiversity conservation as well as climate change adaptation and mitigation.

- Adaptation strategies, including National Adaptation Action Plans (NAPA), as well as mitigation efforts such as REDD+ (Reducing Emissions from Deforestation and Forest Degradation) are needed for implementation in developing countries like Bangladesh. Also, the Nationally Determined Contributions (NDCs) under the Paris Agreement, provide opportunities to use MPAs as an implementation tool for ecosystem-based adaptation and mitigation (Bädeker, 2021) and are very useful to overcome climate impact challenges.
- Climate finance mechanisms enable increased support for the implementation of marine and coastal ecosystem-based adaptation and mitigation (Wamsler et al. 2016). For example, the Green Climate Fund (GCF) offers an opportunity for developing countries to receive support for mitigation and adaptation efforts, with a focus on biodiversity conservation and protected area management.

17.7   MODEL BUILDING AND POLICY PROCESS ANALYSIS

There are many prospects and challenges to the perspective of the blue economy such as frequent floods, marine pollution including ocean acidification and blue carbon, lack of trained personnel, plans and laws, poor ocean governance and political support, and so forth (Isoard and Winograd, 2013; Bir et al. 2020). The blue economy has diverse components, including established traditional ocean industries such as fisheries, tourism, and maritime transport (Spalding, 2016), but also new and emerging activities, such as offshore renewable energy, aquaculture, seabed extractive activities, and marine resources management (Voyer and van Leeuwen, 2019; Baltov, 2020). Considering these complex issues, the proposed Marine Environmental Management and Adaptation Strategies (MEMAS) Model could be the pathway to overcoming the hurdles of the marine regime creation challenges to enhance the global blue economy sectors (Table 17.8).

TABLE 17.8

Marine Regime Creations Factors and Challenges for Sustainable Development Blue Economy and Marine Management

Factors	Challenges
Leadership	Yes, US (NGO pressure and after accidents)
Science	Some debate, but not really an issue in negotiations
Problem framing	Shared sense of environmental problem and visibility of pollution, industry concerns could be overcome
Governance principles	No, only technical cooperation for implementation in developing countries
Incentives participation	International activity, competitiveness and threat of unilateralism

Source: Cohen et al. 2019; Farcy et al. 2019.

17.7.1 MARINE ENVIRONMENTAL MANAGEMENT AND ADAPTATION STRATEGIES (MEMAS) MODEL

Climate change and ocean acidification continue to have serious, adverse impacts on the marine environment with significant implications for people (Guinotte and Fabry, 2008). Several mitigation approaches are being developed and implemented in order to achieve a significant and rapid decrease in GHG emissions (Koch et al. 2013). Coastal ecosystem-based mitigation activities can be used alongside other land use changes and forestry activities to reduce ongoing emissions and to sequester new carbon (Kroeker et al. 2013). Effectively, apply carbon and coastal management, conservation and restoration strategies, including REDD+, NAMAs, MPAs and integrated coastal zone management, to increase the world's natural coastal carbon sinks (mangroves, salt marshes and seagrasses) and support other ecosystem functions and services (Dupont et al. 2008). It is to build a model, in collaboration with field studies, to develop tools for improving and enhancing management plans, including optimal scenarios for carbon allocation, CO_2 uptake and carbon management schemes.

17.7.2 CODING AND CONCEPTUALIZING THE PARAMETERS FOR THE MEMAS MODEL

To build the MEMAS model, Table 17.9: Identification and Processes of Major Parameters for developing the proposed model, is helpful.

In Table 17.9 are shown the identification, definitions and processes of the major parameters for the MEMAS Model. Application of the model based ecosystem approach will help to reach a balance in the conservation of marine biodiversity to enhance the global blue economy. It is based on the application of appropriate scientific methodologies focused on levels of selective major and minor influencing parameters which encompass the essential processes, functions and interactions among organisms and the marine environment. In Table 17.10 is shown the Selective Parameters Process and Functions: Marine Policy-Regime Creation Challenges for Enhancing the Blue Economy in Global to Regional Aspects to build a Marine Environmental Management and Adaptation Strategies (MEMAS) Model (Table 17.11). In this model, six major indicators are selected, that are highly challenging, to manage the marine ecosystem for developing a sustainable blue economy. Additionally, 30 minor influencing parameters are selected for the MEMAS model as the in-situ challenges to overcoming marine ecosystem management.

TABLE 17.9
Identification, Definitions and Processes of Major Parameters for MEMAS Model

Definition of Major Parameters	Coding	Process
Constraint of National Sovereignty	CNS	A fundamental principle of international law is that a state can generally control all activities within the territory over which it has sovereignty
Collective Action and Free Riding Problem	CAF	The free-rider problem because a person can enjoy the benefits of the good without having to pay for it
Conflict of Developed and Developing Countries	CDD	The main concept of conflict is, indeed, synonym of political, economic, and social instability
Rule Making in the Absence in the Government	RMA	An absence or lack of regulation
Affects Different Countries and Actors Differently	DCA	Working differently in fragile and conflict-affected situations
Lack of Integration of Marine Regime and Policies	IMR	Understanding regime shifts is thus critical for marine natural resource management

TABLE 17.10
Selective Parameters Process and Functions: Marine Policy-Regime Creation Challenges for Enhancing the Blue Economy in Global to Regional Aspects

Code	Parameters process, definition and functions	Parameters/Criteria/Factors
CNS	*Constraint of National Sovereignty*	*Major Influencing Factor*
CNSa	Regional Coordination Mechanisms	Minor Influencing Factor
CNSb	Pan African Fisheries Policy	Minor Influencing Factor
CNSc	Limitation of International Marine Law	Minor Influencing Factor
CNSd	Negotiations Imposed by Political Practice	Minor Influencing Factor
CNSe	States must Abide by International Law and Conventions	Minor Influencing Factor
CAF	*Collective Action and Free Riding Problems in Ocean*	*Major Influencing Factor*
CAFa	Climate Change Impact on Marine Ecosystem	Minor Influencing Factor
CAFb	Biodiversity Loss and Resource Depletion	Minor Influencing Factor
CAFc	Waste Accumulation and Environmental Degradation	Minor Influencing Factor
CAFd	Dumping of Plastic Wastes into the Ocean	Minor Influencing Factor
CAFe	Utilize Public Goods without Paying for Them.	Minor Influencing Factor
CDD	*Conflict of Developed and Developing Countries*	*Major Influencing Factor*
CDDa	Worsening State Services	Minor Influencing Factor
CDDb	Widening access to finance and business	Minor Influencing Factor
CDDc	Establishing investment support agencies or services	Minor Influencing Factor
CDDd	Growth of the economy or improved police force	Minor Influencing Factor
CDDe	Conflict with meaningful opportunities	Minor Influencing Factor
RMG	*Rule Making in the Absence in the Government*	*Major Influencing Factor*
RMGa	Environment data collection and testing.	Minor Influencing Factor
RMGb	MoU between the sub regional institutions	Minor Influencing Factor
RMGc	Clean up of oil and chemical spills	Minor Influencing Factor
RMGd	Scientific research on species, plant life and reefs	Minor Influencing Factor
RMGe	Conservation and restoration of coastal and ocean habitats	Minor Influencing Factor
ADC	*Affects Different Countries and Actors Differently*	*Major Influencing Factor*
ADCa	Significantly influence their foreign policy behavior	Minor Influencing Factor
ADCb	Actors are promoted international relations	Minor Influencing Factor
ADCc	Role of state and non-state marine actors	Minor Influencing Factor
ADCd	Increasing international interdependence and relations	Minor Influencing Factor

TABLE 17.10 (Continued)
Selective Parameters Process and Functions: Marine Policy-Regime Creation Challenges for Enhancing the Blue Economy in Global to Regional Aspects

Code	Parameters process, definition and functions	Parameters/Criteria/Factors
ADCe	Allocate funds for their participation	Minor Influencing Factor
IMR	*Lack of Integration of Marine Regime and Policies*	*Major Influencing Factor*
IMRa	Cross-cutting issues in policymaking	Minor Influencing Factor
IMRb	Encouraging Member States for Enhancing Blue Economy	Minor Influencing Factor
IMRc	Policy integration is a matter of coordination.	Minor Influencing Factor
IMRd	Implementation of administration and evaluation	Minor Influencing Factor
IMRe	Policy violations by the stakeholders	Minor Influencing Factor

TABLE 17.11
Matrix of the Marine Environmental Management and Adaptation Strategies (MEMAS) Model

Major/Minor Parameters	CNS	CAF	CDD	RMG	ADC	IMR
CNSa	CNS × CNSa	CAF × CNSa	CDD × CNSa	RMG × CNSa	ADC × CNSa	IMR × CNSa
CNSb	CNS × CNSb	CAF × CNSb	CDD × CNSb	RMG × CNSb	ADC × CNSb	IMR × CNSb
CNSc	CNS × CNSc	CAF × CNSc	CDD × CNSc	RMG × CNSc	ADC × CNSc	IMR × CNSc
CNSd	CNS × CNSd	CNS × CNSd	CDD × CNSd	RMG × CNSd	ADC × CNSd	IMR × CNSd
CNSe	CNS × CNSe	CAF × CNSe	CDD × CNSe	RMG × CNSe	ADC × CNSe	IMR × CNSe
CAFa	CNS × CAFa	CAF × CAFa	CDD × CAFa	RMG × CAFa	ADC × CAFa	IMR × CAFa
CAFb	CNS × CAFb	CAF × CAFb	CDD × CAFb	RMG × CAFb	ADC × CAFb	IMR × CAFb
CAFc	CNS × CAFc	CAF × CAFc	CDD × CAFc	RMG × CAFc	ADC × CAFc	IMR × CAFc
CAFd	CNS × CAFd	CAF× CAFd	CDD × CAFd	RMG × CAFd	ADC × CAFd	IMR × CAFd
CAFe	CNS × CAFe	CAF × CAFe	CDD × CAFe	RMG × CAFe	ADC × CAFe	IMR × CAFe
CDDa	CNS × CDDa	CAF × CDDa	CDD × CDDa	RMG × CDDa	ADC × CDDa	IMR × CDDa
CDDb	CNS × CDDb	CAF × CDDb	CDD × CDDb	RMG × CDDb	ADC × CDDb	IMR × CDDb
CDDc	CNS × CDDc	CAF × CDDc	CDD × CDDc	RMG × CDDc	ADC × CDDc	IMR × CDDc
CDDd	CNS × CDDd	CAF × CDDd	CDD × CDDd	RMG × CDDd	ADC × CDDd	IMR × CDDd
CDDe	CNS × CDDe	CAF × CDDe	CDD × CDDe	RMG × CDDe	ADC × CDDe	IMR × CDDe
RMGa	CNS × RMGa	CAF × RMGa	CDD × RMGa	RMG × RMGa	ADC × RMGa	IMR × RMGa
RMGb	CNS × RMGb	CAF × RMGb	CDD × RMGb	RMG × RMGb	ADC × RMGb	IMR × RMGb
RMGc	CNS × RMGc	CAF × RMGc	CDD × RMGc	RMG × RMGc	ADC × RMGc	IMR × RMGc
RMGd	CNS × RMGd	CAF × RMGd	CDD × RMGd	RMG × RMGd	ADC × RMGd	IMR × RMGd
RMGe	CNS × RMGe	CAF × RMGe	CDD × RMGe	RMG × RMGe	ADC × RMGe	IMR × RMGe
ADCa	CNS × ADCa	CAF × ADCa	CCD × ADCa	RMG × ADCa	ADC × ADCa	IMR × ADCa
ADCb	CNS × ADCb	CAF × ADCb	CCD × ADCb	RMG × ADCb	ADC × ADCb	IMR × ADCb
ADCc	CNS × ADCc	CAF × ADCc	CCD × ADCc	RMG × ADCc	ADC × ADCc	IMR × ADCc
ADCd	CNS × ADCd	CAF × ADCd	CCD × ADCd	RMG × ADCd	ADC × ADCd	IMR × ADCd
ADCe	CNS × ADCe	CAF × ADCe	CCD × ADCe	RMG × ADCe	CNS × ADCe	IMR × ADCe
IMRa	CNS × IMRa	CAF × IMRa	CCD × IMRa	RMG × IMRa	ADC × IMRa	IMR × IMRa
IMRb	CNS × IMRb	CAF × IMRb	CCD × IMRb	RMG × IMRb	ADC × IMRb	IMR × IMRb
IMRc	CNS × IMRc	CAF × IMRc	CCD × IMRc	RMG × IMRc	ADC × IMRc	IMR × IMRc
IMRd	CNS × IMRd	CAF × IMRd	CCD × IMRd	RMG × IMRd	ADC × IMRd	IMR × IMRd
IMRe	CNS × IMRe	CAF × IMRe	CCD × IMRe	RMG × IMRe	ADC × IMRe	IMR × IMRe

17.8 CONCLUSION

Many causes of pollution including sewage and fertilizers contain nutrients such as nitrates and phosphates. In excess levels, nutrients overstimulate the growth of aquatic plants and algae. Excessive growth of these types of organisms consequently uses up dissolved oxygen as they decompose and blocks light to deeper waters in the ocean. The rapidly increasing variety and number of offshore uses and the potential for conflicts between competing interests operating in the same area will increase the need for information concerning the nature and extent of offshore activities. In today's highly interdependent world, efforts to ensure national security, maintain environmental quality and manage the use of marine resources will require unprecedented awareness of activities, trends, and anomalies in the maritime domain, including those that may require some intervention. Politicians must think of sustainable development rather than economic expansion. Conservation strategies must become more widely accepted, and people must learn that energy use can be dramatically diminished without sacrificing comfort.

ACKNOWLEDGEMENTS

We have reviewed many books, articles, blogs and websites, I would like to express my gratitude to all the anonymous authors and contributors

In addition, I would like to express my thanks to the SUMITOMO Foundation, Tokyo, Japan for providing financing assistance for this study on the global blue economy and seafood production practices observed in Japanese tradition.

Also, I would like to thank the Faculty of Social of Science, Jahangirnagar University, Savar, Dhaka-1342, and the University Grants Commission (UGC) of Bangladesh for their financial support to continue my research on the seafood cultivation mechanisms and enhancing the blue economy in Bangladesh.

REFERENCES

A-Khavari, A. (1997). 1996 Protocol to the 1972 convention on the prevention of marine pollution by dumping of wastes and other matter. *Asia Pac. J. Envtl. L.* 2, 201.

Agardy, T. Di Sciara, G. N. and Christie, P. (2011). Mind the gap: addressing the shortcomings of marine protected areas through large scale marine spatial planning. *Marine Policy*, 35(2), 226–232.

Alam, M. N. et al. (2021). Blue technology for sustainability of small and medium fish firms: a study on small and medium fish firms of Bangladesh. *Environ Dev Sustain*, 23, 635–646. https://doi.org/10.1007/s10 668-020-00599-z

Alam, S. and Faruque, A. (2014). Legal regulation of the shipbreaking industry in Bangladesh: The international regulatory framework and domestic implementation challenges. *Marine Policy*, 47, 46–56.

Allison, R. J. Goodwin, S. P. Parker, R. J. De Grijs, R. Zwart, S. F. P. and Kouwenhoven, M. B. N. (2009). Dynamical mass segregation on a very short timescale. *The Astrophysical Journal*, 700(2), L99.

Amon, D. J. Kennedy, B. R. Cantwell, K. Suhre, K. Glickson, D. Shank, T. M. and Rotjan, R. D. (2020). Deep-sea debris in the central and western Pacific Ocean. *Frontiers in Marine Science*, 7:369. doi: 10.3389/fmars.2020.00369

Andrews, A. et al. (2014). CO_2, CO, and CH_4 measurements from tall towers in the NOAA Earth System Research Laboratory's Global Greenhouse Gas Reference Network: Instrumentation, uncertainty analysis, and recommendations for future high-accuracy greenhouse gas monitoring efforts. *Atmospheric Measurement Techniques*, 7(2), 647–687.

Annan-Diab, F. and Molinari, C. (2017). Interdisciplinarity: practical approach to advancing education for sustainability and for the Sustainable Development Goals. *The International Journal of Management Education*, 15(2), 73–83.

Apitz, S. E. and Agius, S. (2013). Anatomy of a decision: potential regulatory outcomes from changes to chemistry protocols in the Canadian Disposal at sea program. *Marine Pollution Bulletin*, 69(1–2), 76–90.

Araújo, R. Vásquez Calderón, F. Sanchez Lopez, J. Azevedo, I. Bruhn, A. Flunch, S. Garcia-Tasende, M. Ghaderiardakani, F. Ilmjärv, T. Laurans, M. MacMonagail, M. Mangini, S. Peteiro, C. Rebours, C. Stefánsson, T. and Ullmann, J. (2021). Emerging sectors of the blue bioeconomy in Europe: status of the algae production industry. *Frontiers in Marine Sciences*, 7, 626389 doi: 10.3389/fmars.2020.626389.

Arlinghaus, A. Bohle, P. Iskra-Golec, I. Jansen, N. Jay, S. and Rotenberg, L. (2019). Working Time Society consensus statements: evidence-based effects of shift work and non-standard working hours on workers, family and community. *Industrial Health*, 57(2), 184–200.

Ayilu, R. K. Fabinyi, M. and Barclay, K. (2022). Small-scale fisheries in the blue economy: review of scholarly papers and multilateral documents. *Ocean and Coastal Management,* 216, 105982.

Bädeker, J. C. (2021). Planning for Seaweed Optimism-Exploring Ecosystem-Based Marine Spatial Planning with the Case of the Norwegian Blue Kelp Forests (Doctoral dissertation).

Baine, M. Howard, M. Kerr, S. Edgar, G. and Toral, V. (2007). Coastal and marine resource management in the Galapagos Islands and the Archipelago of San Andres: issues, problems and opportunities. *Ocean and Coastal Management*, 50(3–4), 148–173.

Bakshi, A. (2019). Oceans and small island states: Prospects for the blue economy. *International Research Journal of Human Resources and Social Sciences*, 6, 43–60.

Bakun, A. and Broad, K. (2003). Environmental 'loopholes' and fish population dynamics: comparative pattern recognition with focus on El Niño effects in the Pacific. *Fisheries Oceanography*, 12(4–5), 458–473.

Baltov, M. (2020). Circular economy features to the blue growth domains. Economic *Science, Education and the Real Economy: Development and Interactions in the Digital Age*, 1, 352–361.

Banik, D. (2019). Achieving food security in a sustainable development era. *Food Ethics*, 4(2), 117–121.

Barbesgaard, M. (2018). Blue growth: savior or ocean grabbing? *The Journal of Peasant Studies*, 45(1), 130–149.

Barton, A. et al. The Pacific oyster, Crassostrea gigas, shows negative correlation to naturally elevated carbon dioxide levels: Implications for near-term ocean acidification effects. *Limnology and oceanography*, 57(3), 698–710.

Barton, A. et al. (2015). Impacts of coastal acidification on the Pacific Northwest shellfish industry and adaptation strategies implemented in response. *Oceanography*, 28(2), 146–159.

Bax, N. et al. (2022). Ocean resource use: building the coastal blue economy. *Reviews in Fish Biology and Fisheries*, 32(1), 189–207.

Bennett, N. J. Blythe, J. White, C. S. and Campero, C. (2021). Blue growth and blue justice: ten risks and solutions for the ocean economy. *Marine Policy*, 125, 104387.

Bennett, N. J. et al. (2019). Towards a sustainable and equitable blue economy. *Nature Sustainability*, 2(11), 991–993

Bergesen, H. O. Parmann, G. and Thommessen, Ø. B. (2018). International convention relating to intervention on the high seas in cases of oil pollution casualties (intervention convention). In *Yearbook of International Cooperation on Environment and Development 1998–99* (pp. 374). Taylor and Francis, Routledge. New York, USA.

Bergmeijer, P. (1992). The international convention for the prevention of pollution from ships. Editor(s): Antony J. Dolman and Jan Van Ettinger, In *Ports as Nodal Points in a Global Transport System* (pp. 65–76). Pergamon, Robert Maxwell, Paul Rosbaud, Oxford, UK

Berkey, C. G. and Williams, S. W. (2018). CALIFORNIA INDIAN TRIBES AND THE MARINE LIFE PROTECTION ACT. *American Indian Law Review*, 43(2), 307–351.

Bhuyan, M. S. et al. (2021). *Blue Economy Prospect, Opportunities, Challenges, Risks, and Sustainable Development Pathways in Bangladesh.*

Bir, J. Golder, M. R. Al Zobayer, M. F. Das, K. K. Zaman, S. Chowdhury, L. M. D. and Paul, P. C. (2020). A review on blue economy in Bangladesh: prospects and challenges. *Int. J. Nat. Soc. Sci*, 7, 21–29.

Blasiak, R. Spijkers, J. Tokunaga, K. Pittman, J. Yagi, N. and Österblom, H. (2017). Climate change and marine fisheries: least developed countries top global index of vulnerability. *PLoS One*, 12(6), e0179632.

Boero, F. et al. (2016). *CoCoNet: Towards Coast to Coast Networks of Marine Protected Areas (from the Shore to the High and Deep Sea), Coupled with Sea -Based Wind Energy Potential.*

Boteler, B. and Coastal, M. W. (2014). European maritime transport and port activities: identifying policy gaps towards reducing environmental impacts of socio-economic activities. *Ecologic* (https://www.ecologic.eu/sites/default/files/presentation/2014/european-maritime-transport-and-port-activities_0.pdf

Boyes, S. J. and Elliott, M. (2014). Marine legislation—The ultimate 'horrendogram': international law, European directives and national implementation. *Marine Pollution Bulletin*, 86(1–2), 39–47.

Bricheno, L. M. Wolf, J. and Sun, Y. (2021). Saline intrusion in the Ganges-Brahmaputra-Meghna megadelta. *Estuarine, Coastal and Shelf Science*, 252, 107246.

Broder, S. P. (2017). 12 International Governance of Ocean Fertilization and other Marine Geoengineering Activities. In *Ocean Law and Policy* (pp. 305–343). Brill Nijhoff.

Brown, J. R. (1993). Admiralty judges: flotsam on the sea of maritime law. *J. Mar. L. and Com.* 24, 249.

Čampara, L. Hasanspahić, N. and Vujičić, S. (2018). Overview of MARPOL ANNEX VI regulations for prevention of air pollution from marine diesel engines. In *SHS Web of Conferences*, 58, 01004. EDP Sciences.

Carpenter, A. (2019). Oil pollution in the North Sea: the impact of governance measures on oil pollution over several decades. *Hydrobiologia*, 845(1), 109–127.

Carver, R. (2019). Resource sovereignty and accumulation in the blue economy: the case of seabed mining in Namibia. *Journal of Political Ecology*, 26(1), 381–402.

Casto, W. R. (1993). The origins of federal admiralty jurisdiction in an age of privateers, smugglers, and pirates. *The American Journal of Legal History*, 37(2), 117–157.

Chen, CL. (2015). Regulation and Management of Marine Litter. In: Bergmann, M. Gutow, L. Klages, M. (eds) Marine Anthropogenic Litter. Springer, Cham. https://doi.org/10.1007/978-3-319-16510-3_15.

Choksi, S. (2001). The Basel Convention on the control of transboundary movements of hazardous wastes and their disposal: 1999 Protocol on Liability and Compensation. *Ecology LQ*, 28, 509.

Chong, V. C. Lee, P. K. Y. and Lau, C. M. (2010). Diversity, extinction risk and conservation of Malaysian fishes. *Journal of Fish Biology*, 76(9), 2009–2066.

Cohen, P. J. et al. (2019). Securing a just space for small-scale fisheries in the blue economy. *Frontiers in Marine Science*, 6, 171.

Collins, R. and Hassan, D. (2009). Applications and shortcomings of the Law of the Sea in combating piracy: a South East Asian perspective. *J. Mar. L. and Com.* 40, 89.

COM. (2018). 773 Final. *A Clean Planet for All—A European Strategic Long-Term Vision for a Prosperous, Modern, Competitive and Climate Neutral Economy.*

Conand, C. and Muthiga, N. (2016). *The Sea Cucumber Resources and Fisheries Management in the Western Indian Ocean: Current Status and Preliminary Results from a Wiomsa Regional Research Project (Grey Lit).*

Cordova, M. R. and Nurhati, I. S. (2019). Major sources and monthly variations in the release of land-derived marine debris from the Greater Jakarta area, Indonesia. *Scientific Reports*, 9(1), 1–8.

Cormier, R. and Elliott, M. (2017). SMART marine goals, targets and management—is SDG 14 operational or aspirational, is 'Life Below Water' sinking or swimming? *Marine Pollution Bulletin*, 123(1–2), 28–33.

Couto, G. Castanho, R. A. Pimentel, P. Carvalho, C. Sousa, Á. and Santos, C. (2020). The impacts of COVID-19 crisis over the tourism expectations of the Azores archipelago residents. *Sustainability*, 12(18), 7612.

Cox Jr, L. A. (2012). Confronting deep uncertainties in risk analysis. *Risk Analysis: An International Journal*, 32(10), 1607–1629.

Day, J. C. Laffoley, D. and Zischka, K. (2015). Marine protected area management. Editors: G. L. Worboys, M. Lockwood, A. Kothari, S. Feary, I PulsfordIn: Protected Area Governance and Management (pp. 609–650).

de Bolle, P. (2022) Blue Finance: How can this innovative wave finance the blue economy? World Bank Blogs, Published on Development and a Changing Climate, Retrieved on March 16, 2022 from https://blogs.worldbank.org/climatechange/blue-finance-how-can-innovative-wave-finance-blue-economy

De Lucia, V. (2019). Ocean commons, law of the sea and rights for the sea. Canadian Journal of Law and Jurisprudence, 32(1), 45-57.

De Silva, D. D. Rapior, S. Hyde, K. D. and Bahkali, A. H. (2012). Medicinal mushrooms in prevention and control of diabetes mellitus. *Fungal Diversity*, 56(1), 1–29.

De Silva, S. S. (2013). Climate change impacts: challenges for aquaculture. In *Proceedings of the Global Conference on Aquaculture 2010. Farming the waters for people and food* (pp. 75-110). FAO/NACA.

Diz, D. et al. (2018). Mainstreaming marine biodiversity into the SDGs: the role of other effective area-based conservation measures (SDG 14.5). *Marine Policy*, 93, 251–261.

Doussineau, M. Haarich, S. Gnamus, A. Gomez, J. and Holstein, F. (2020). *Smart Specialisation and Blue Biotechnology in Europe*, EUR 30521 EN, Publications Office of the European Union, Luxembourg, ISBN 978-92-76-27753-8, doi:10.2760/19274, JRC122818.

Dudley, N. Ali, N. Kettunen, M. and MacKinnon, K. (2017). Editorial essay: protected areas and the sustainable development goals. *Parks*, 23(2), 10–12.

Dulvy, Nicholas K. et al. "You can swim but you can't hide: the global status and conservation of oceanic pelagic sharks and rays." *Aquatic Conservation: Marine and Freshwater Ecosystems* 18.5 (2008): 459-482.

Dupont, S. Havenhand, J. Thorndyke, W. Peck, L. and Thorndyke, M. (2008). Near-future level of CO_2-driven ocean acidification radically affects larval survival and development in the brittlestar Ophiothrix fragilis. *Marine Ecology Progress Series,* 373, 285–294.

Edu, K. (2011). A review of the existing legal regime on exploitation of oil and the protection of the environment in Nigeria. *Commonwealth Law Bulletin*, 37(2), 307–327.

Ehlers, P. (2016). Blue growth and ocean governance—how to balance the use and the protection of the seas. *WMU Journal of Maritime Affairs*, 15(2), 187–203.

Elisha, O. D. (2019). The Nigerian blue economy: prospects for economic growth and challenges. *Int J Sci Res Educ*, 12(5), 680–699.

Engtrø, E. A discussion on the implementation of the Polar Code and the STCW Convention's training requirements for ice navigation in polar waters. *J Transp Secur* **15,** 41–67 (2022). https://doi.org/10.1007/s12198-021-00241-7

European Commission (2017), Directorate-General for Research and Innovation, *Study on lessons for ocean energy development: final report*, Publications Office, 2017, https://data.europa.eu/doi/10.2777/389418

FAO. (2020). *Regional Plan of Action to Prevent, Deter and Eliminate Illegal, Unreported and Unregulated (IUU) Fishing in WECAFC Member Countries (2019–2029)*. Rome. https://doi.org/10.4060/ca9457t

Farcy, P. et al. (2019). Toward a European coastal observing network to provide better answers to science and to societal challenges, the JERICO research infrastructure. *Frontiers in Marine Science,* 6, 529.

Friess, D. A. Aung, T. T. Huxham, M. Lovelock, C. Mukherjee, N. and Sasmito, S. (2019). SDG 14: life below water–impacts on mangroves. *Sustainable Development Goals*, 445, 445–481.

Fytili, D. and Zabaniotou, A. (2008). Utilization of sewage sludge in EU application of old and new methods—A review. *Renewable and Sustainable Energy Reviews*, 12(1), 116–140.

Gachingiri, A. (2015). Effect of leadership style on organisational performance: A case study of the United Nations Environment Programme (UNEP), Kenya. *International Academic Journal of Innovation, Leadership and Entrepreneurship*, 1(5), 19-36.

Gagatsia, E. (2007). *Review of Maritime Transport Safety and Security Practices and Compliance Levels: Case Studies in Europe and South East Asia*. In European Conference of Transport Research Institutes, Young Researchers Seminar (pp. 1–14).

Garai, K. Verghese, P. B. Baban, B. Holtzman, D. M. and Frieden, C. (2014). The binding of apolipoprotein E to oligomers and fibrils of amyloid-β alters the kinetics of amyloid aggregation. *Biochemistry*, 53(40), 6323–6331.

Gazeau, F. et al. (2013). Impacts of ocean acidification on marine shelled molluscs. *Marine biology*, 160(8), 2207-2245.

Getting to Zero Coalition. (2020). *The First Wave: A Blueprint for Commercial-Scale Zero-Emission Shipping Pilots and Capgemini Invent*, Fit for net-zero: 55 Tech Quests to accelerate Europe's recovery and pave the way to climate neutrality.

Ghermandi, A. and Nunes, P. A. (2013). A global map of coastal recreation values: Results from a spatially explicit meta-analysis. *Ecological Economics,* 86, 1–15.

Goldenberg, Suzanne (2014) Oceans: John Kerry launches global effort to save world's oceans 'under siege', The Guardian, Retrieved on 12 August 2022 from https://www.theguardian.com/environment/2014/jun/11/john-kerry-ocean-conservation-summit?RefID=Em2378_Email_EM1_Main

Green, T. C. Zaller, N. Palacios, W. R. Bowman, S. E. Ray, M. Heimer, R. and Case, P. (2013). Law enforcement attitudes toward overdose prevention and response. *Drug and Alcohol Dependence*, 133(2), 677–684.

Griffin, W. Wang, W. and de Souza, M. C. (2019). The sustainable development goals and the economic contribution of fisheries and aquaculture. *FAO Aquaculture Newsletter,* 60, 51–52.

Guillotreau, J. et al. (2012). Robotic partial nephrectomy versus laparoscopic cryoablation for the small renal mass. *European Urology*, 61(5), 899–904.

Guinotte, J. M. and Fabry, V. J. (2008). Ocean acidification and its potential effects on marine ecosystems. *Annals of the New York Academy of Sciences*, 1134(1), 320–342.

Gulas, S. Downton, M. D'Souza, K. Hayden, K. and Walker, T. R. (2017). Declining Arctic Ocean oil and gas developments: Opportunities to improve governance and environmental pollution control. *Marine Policy*, *75*, 53-61.

Hadjimichael, M. (2018). A call for a blue degrowth: unravelling the European Union's fisheries and maritime policies. *Marine Policy,* 94, 158–164.

Halkos, G. and Gkampoura, E. C. (2021). Where do we stand on the 17 Sustainable Development Goals? An overview on progress. *Economic Analysis and Policy,* 70, 94–122.

Hall, C. M. (2001). Trends in ocean and coastal tourism: the end of the last frontier? *Ocean and Coastal Management*, 44(9–10), 601–618.

Hassanali, K. (2020). CARICOM and the blue economy–Multiple understandings and their implications for global engagement. *Marine Policy*, 120, 104137.

Heidkamp, C. P. Garland, M. and Krak, L. (2021). Enacting a just and sustainable blue economy through transdisciplinary action research. *The Geographical Journal.*

Heise, S. Babut, M. Casado, C. Feiler, U. Ferrari, B. J. and Marziali, L. (2020). Ecotoxicological testing of sediments and dredged material: an overlooked opportunity? *Journal of Soils and Sediments*, *20*(12), 4218-4228.

Hertel, T. W. and Rosch, S. D. (2010). Climate change, agriculture, and poverty. *Applied Economic Perspectives and Policy*, 32(3), 355–385.

Himes-Cornell, A. K. Hoelting, C. Maguire, L. Munger-Little, J. Lee, J. Fisk, R. Felthoven, C. Geller, and P. Little. 2013. Community profiles for North Pacific fisheries - Alaska. U.S. Dep. Commer. NOAA Tech. Memo. NMFS-AFSC-259, Volume 4, 287 p.

Himes-Cornell, A. K. Hoelting, C. Maguire, L. Munger-Little, J. Lee, J. Fisk, R. Felthoven, C. Geller, and P. 2013. Community profiles for North Pacific fisheries - Alaska. U.S. Dep. Commer. NOAA Tech. Memo. NMFS-AFSC-259, Volume 5, 210 p.

Hoegh-Guldberg, Ove (2015). Reviving the Ocean Economy: the case for action. P.60; Geneva, Switzerland: World Wide Fund and The University of Queensland. Report available at http://assets. worldwildlife.org/publications/790/files/original/Reviving_Ocean_Economy_REPORT_low_res. pdf?1429717323and_ga=1.187

Hoegh-Guldberg, O. et al. (2019). The human imperative of stabilizing global climate change at 1.5, C. *Science*, 365(6459), eaaw6974.

Hoffman, A. J. (2005). Climate change strategy: the business logic behind voluntary greenhouse gas reductions. *California Management Review*, 47(3), 21–46.

Hongdao, Q. and Mukhtar, H. (2017). Joint development agreements: towards protecting the marine environment under international law. *JL Pol'y and Globalization,* 66, 164.

Hossain, M. S. Chowdhury, S. R. Navera, U. K. Hossain, M. A. R. Imam, B. and Sharifuzzaman, S. M. (2014). *Opportunities and Strategies for Ocean and River Resources Management.* Dhaka: Background paper for preparation of the 7th Five Year Plan. Planning Commission, Ministry of Planning, Bangladesh.

Hurd, C. L. Lenton, A. Tilbrook, B. and Boyd, P. W. (2018). Current understanding and challenges for oceans in a higher-CO2 world. *Nature Climate Change*, 8(8), 686–694.

Hussain, M. G. Failler, P. and Sarker, S. (2019). Future importance of maritime activities in Bangladesh. *Journal of Ocean and Coastal Economics*, 6(2), 3.

Hussain, M. G. Failler, P. Karim, A. A. and Alam, M. K. (2017). Review on opportunities, constraints and challenges of blue economy development in Bangladesh. *Journal of Fisheries and Life Sciences*, 2(1), 45–57.

Intergovernmental Panel on Climate Change (IPCC) (2019). Special Report on the Ocean and Cryosphere in a Changing Climate.

Intergovernmental Science-Policy Platform on Biodiversity and Ecosystem Services (IPBES) (2019). Summary for Policymakers of the Global Assessment Report on Biodiversity and Ecosystem Services.

Iqbal, M. K. (2020). Evolving concept of Blue Economy. *Blue Economy and Marine Spatial Planning (Blue Economy of Challenges)*, 1. Retrieved from Blue economy and marine spatial planning.

Islam, M. N. and Kitazawa, D. (April 2011). Underwater constructions: Challenges for ocean regime protocol in 21st century. In *2011 IEEE Symposium on Underwater Technology and Workshop on Scientific Use of Submarine Cables and Related Technologies* (pp. 1–10). IEEE.

Islam, M. N. Kitazawa, D. Kokuryo, N. Tabeta, S. Honma, T. and Komatsu, N. (2012). Numerical modeling on transition of dominant algae in Lake Kitaura, Japan. *Ecological Modelling*, *242*, 146-163.

Islam, M. and Mostaque, L. Y. (2016). Blue economy and bangladesh: lessons and policy implications. *Economics*, 2(2), 1–21.

Islam, M. K. Rahaman, M. and Ahmed, Z. (2018). Blue economy of Bangladesh: opportunities and challenges for sustainable development. *Advances in Social Sciences Research Journal*, 5(8).168-178

Islam, M. M. Shamsuzzaman, M. M. Mozumder, M. M. H. Xiangmin, X. Ming, Y. and Jewel, M. A. S. (2017). Exploitation and conservation of coastal and marine fisheries in Bangladesh: Do the fishery laws matter? *Marine Policy*, 76, 143–151.

Isoard, S. and Winograd, M. (2013). Adaptation in Europe. *Addressing Risks and Opportunities from Climate Change in the Context of Socio-Economic Developments.*

Jamilah, A. and Disemadi, H. S. (2020). Penegakan Hukum Illegal Fishing dalam Perspektif UNCLOS 1982. *Mulawarman Law Review*, 29-46.

Joseph, A. and Dalaklis, D. (2021). The international convention for the safety of life at sea: highlighting interrelations of measures towards effective risk mitigation. *Journal of International Maritime Safety, Environmental Affairs, and Shipping*, 5(1), 1–11.

Kappel, C. V. Halpern, B. S. and Napoli, N. (2012). *Mapping Cumulative Impacts of Human Activities on Marine Ecosystems.* SeaPlan: Boston, MA.

Karani, P. and Failler, P. (2020). Comparative coastal and marine tourism, climate change, and the blue economy in African Large Marine Ecosystems. *Environmental Development*, 36, 100572.

Karim, S. (2009). Implementation of the MARPOL Convention in Bangladesh. *Macquarie J. Int'l and Comp. Envtl. L.* 6, 51.

Karim, S. (March 2016). *Marine pollution in the Bay of Bengal: In Search of a Legal Response.* In Seminar.

Khobragade, J. W. Kumar, A. Binti Abd Aziz, S. N. and Maurya, D. (2021). The Anti-Maritime Piracy Law in India and Malaysia: An Analytical Study. *Journal of International Maritime Safety, Environmental Affairs, and Shipping*, 5(4), 208–219.

Koch, M. Bowes, G. Ross, C. and Zhang, X. H. (2013). Climate change and ocean acidification effects on seagrasses and marine macroalgae. *Global Change Biology*, 19(1), 103–132.

Konar, M. and Ding, H. (2020). A sustainable ocean economy for 2050. Approximating its benefits and costs. *High Level Panel for a Sustainable Ocean Economy. World Resources Institute.*Retrieved on 12 August 2022 from https://www.investableoceans.com/blogs/library/a-sustainable-ocean-economy-for-2050-approximating-its-benefits-and-costs

Kroeker, K. J. et al. (2013). Impacts of ocean acidification on marine organisms: quantifying sensitivities and interaction with warming. *Global Change Biology*, 19(6), 1884–1896.

Lacoue-Labarthe, T. et al. (2016). Impacts of ocean acidification in a warming Mediterranean Sea: An overview. *Regional Studies in Marine Science*, 5, 1-11.

Lam, W. V. Macrae, M. L. English, M. C. O'halloran, I. P. Plach, J. M. and Wang, Y. (2016). Seasonal and event-based drivers of runoff and phosphorus export through agricultural tile drains under sandy loam soil in a cool temperate region. *Hydrological Processes*, 30(15), 2644–2656.

Lee, D. Hampton, M. and Jeyacheya, J. (2015). The political economy of precarious work in the tourism industry in small island developing states. *Review of International Political Economy*, 22(1), 194-223.

Li, D. Bou-Zeid, E. and Oppenheimer, M. (2014). The effectiveness of cool and green roofs as urban heat island mitigation strategies. *Environmental Research Letters*, 9(5), 055002.

Lo, V. (2016). Synthesis report on experiences with ecosystem-based approaches to climate change adaptation and disaster risk reduction. *Technical Series*, 85.

Lomborg, B. (2020). Welfare in the 21st century: increasing development, reducing inequality, the impact of climate change, and the cost of climate policies. *Technological Forecasting and Social Change,* 156, 119981.

Long, R. (2011). The Marine Strategy Framework Directive: a new European approach to the regulation of the marine environment, marine natural resources and marine ecological services. *Journal of Energy and Natural Resources Law*, 29(1), 1–44.

Louis, G. E. (2004). A historical context of municipal solid waste management in the United States. *Waste Management and Research*, 22(4), 306–322.

Lubchenco, J. Haugan, P. M. and Mari E. P. (2020). Five priorities for a sustainable ocean economy. *Nature*, 588, 30–32.

Mahmood, S. and Rashid, H. (2018). The conception, planning and implementation of integrated coastal and ocean management for sustainable blue economy in Bangladesh.

Maritime Alliance for fostering the European (2019) Blue Economy through a Marine Technology Skilling Strategy. MATES is a 'Strategy baseline to bridge the skills gap between training offers and industry demands of the Maritime Technologies value chain, September 2019—MATES Project.'

Marques Santos, A. Madrid, C. Haegeman, K. and Rainoldi, A. (2020). *Behavioural Changes in Tourism in Times of Covid-19*, EUR 30286 EN, Publications Office of the European Union, Luxembourg, ISBN 978–92–76–20401–5 (online), doi: 10.2760/00411.

Midlen, A. (2021). What is the Blue Economy? A spatialised governmentality perspective. *Maritime Studies*, 20 (1): 1–26.

Mikelis, N. (2010). *The Hong Kong International Convention for the Safe and Environmentally Sound Recycling of Ships*. In UNCTAD Multi-Year Expert Meeting on Transport and trade Facilitation.

Miller, D. Roe, J. Brown, C. Morris, S. Morrice, J. and Ward Thompson, C. (2012). *Blue Health: Water, Health and Wellbeing*. Centre of Expertise for Waters, James Hutton Institute, pp. 1-53, Aberdeen. www.crew. ac.uk/publications. Available online at: www.crew.ac.uk/publications

Mirza, M. M. Q. (2003). Climate change and extreme weather events: can developing countries adapt? *Climate Policy*, 3(3), 233–248.

Mouton, S. F. (2019). *An Assessment of the Blue Economy in Namibia: A Case Study of Swakopmund and Walvis Bay*. Master's thesis, Faculty of Commerce.

Mussa, W. M. Jing, Z. Machochoki, A. S. and Bakari, S. J. (2019). Towards growth of the blue economy in Zanzibar: potentials and challenges, International Journal of Scientific Advances, 2(3): 308–315, Available Online: www.ijscia.com DOI: 10.51542/ijscia.v2i3.13.

Muthaiyah, N. P. (2020). Rejuvenating Yamuna River by wastewater treatment and management. *International Journal of Energy and Environmental Science*, 5(1), 14–29.

Morgera, E. and Ntona, M. (2018). Linking small-scale fisheries to international obligations on marine technology transfer. *Marine Policy*, 93, 295–306.

Muir, J. F. and Nugent, C. G. (December 1995). *Aquaculture Development Trends: Perspectives for Food Security*. FAO: Rome (Italy). Fisheries Department Government of Japan, Tokyo (Japan). International conference on the sustainable contribution of fisheries to food security. Kyoto (Japan), 4–9.

Mustafa, S. Estim, A. Shaleh, S. R. M. and Shapawi, R. (2018). Positioning of aquaculture in blue growth and sustainable development goals through new knowledge, ecological perspectives and analytical solutions. *Aquacultura Indonesiana*, 19(1), 1–9.

Nagy, H. and Nene, S. (2021). Blue Gold: Advancing Blue Economy Governance in Africa. *Sustainability*, 13(13), 7153.

Nash, K. L. et al. (2020). To achieve a sustainable blue future, progress assessments must include interdependencies between the sustainable development goals. *One Earth*, 2(2), 161–173.

Naylor, C. D. (2013) Cruise Injury Attorney Recognized as Certified Specialist in Admiralty and Maritime Law, Retrieved on August 2, 2022 from https://www.prweb.com/releases/cruise-injury-attorney/maritime-law yer/prweb10322812.htm; Contact Information Holly Naylor Law Offices of Charles D. Naylor. www. NaylorLaw.com, USA.

Nikčević, J. and Škurić, M. (2021). A Contribution to the sustainable development of maritime transport in the context of blue economy: the case of Montenegro. *Sustainability*, 13(6), 3079.

Nisa, Z. A. Schofield, C. and Neat, F. C. (2022). Work below water: The role of scuba industry in realising sustainable development goals in small island developing states. *Marine Policy*, 136, 104918.

NOAA. (2014). *National Centers for Environmental Information*. State of the Climate: Global Climate Report for Annual 2014, published online January 2015. Available from https://ncdc.noaa.gov/sotc/global/ 201413

Novaglio, C. et al. (2022). Deep aspirations: towards a sustainable offshore blue economy. *Reviews in Fish Biology and Fisheries*, 32(1), 209–230.

Obura, D. Burgener, V. Owen, S. and Gonzales, A. (2017). Reviving the Western Indian Ocean economy: Actions for a sustainable future (p. 64). *Gland, Switzerland: WWF International*.

OECD, I. (2016). *Energy and Air Pollution: World Energy Outlook Special Report, 2016*.

Oh, S. Y. (2018). World Trade Organization (WTO). *Yearbook of International Environmental Law*, 29, 492–494.

Østhagen, A. Rottem, S. V. Inderberg, T. H. J. Colgan, C. and Raspotnik, A. (2022). *Blue Governance Governing the Blue Economy in Alaska and North Norway AlaskaNor Work Package V*.

Patil, P. G. Virdin, J. Colgan, C. S. Hussain, M. G. Failler, P. and Vegh, T. (2018). *Toward a Blue Economy: A Pathway for Bangladesh's Sustainable Growth*. Environment and Natural Resources Global Practice, South Asia Region, World Bank.

Perry, A. L. Low, P. J. Ellis, J. R. and Reynolds, J. D. (2005). Climate change and distribution shifts in marine fishes. *Science*, 308(5730), 1912–1915.

Peterman, R. M. (2004). Possible solutions to some challenges facing fisheries scientists and managers. *ICES Journal of Marine Science*, 61(8), 1331–1343.

Pinsky, M. L. Worm, B. Fogarty, M. J. Sarmiento, J. L. and Levin, S. A. (2013). Marine taxa track local climate velocities. *Science*, 341(6151), 1239–1242.

Pinsky, M. L. Reygondeau, G. Caddell, R. Palacios-Abrantes, J. Spijkers, J. and Cheung, W. W. (2018). Preparing ocean governance for species on the move. *Science*, 360(6394), 1189–1191.

Poloczanska, E. S. et al. (2013). Global imprint of climate change on marine life. *Nature Climate Change*, 3(10), 919–925.

Poloczanska, E. S. et al. (2016). Responses of marine organisms to climate change across oceans. *Frontiers in Marine Science*, 3: 62.

Prideaux, B. Thompson, M. and Pabel, A. (2020). Lessons from COVID-19 can prepare global tourism for the economic transformation needed to combat climate change. *Tourism Geographies*, 22(3), 667–678.

Rey-Valette, H. Damart, S. and Roussel, S. (2007). A multicriteria participation-based methodology for selecting sustainable development indicators: an incentive tool for concerted decision making beyond the diagnosis framework. *International Journal of Sustainable Development*, 10(1–2), 122–138.

Raheem, S. (2022). The blue economy's entrepreneurial potential and its poverty mitigative powers in Nigeria. Edited by: Abayomi Al-Ameen, Vasilii Erokhin, Nima Norouzi, Saheed Nurein, Olawale Paul Olatidoye, Seda Yıldırım, Poshan Yu In *Implications for* Entrepreneurship and Enterprise Development in the Blue Economy, 185–217. Hershey, Pennsylvania.

Rahman, M. M. (2006). A study on coastal water pollution of Bangladesh in the Bay of Bengal (Doctoral dissertation, BRAC University).

Ramieri, E. Bocci, M. Markovic, M. (2019). Linking Integrated Coastal Zone Management to Maritime Spatial Planning: The Mediterranean Experience. In: Zaucha, J. Gee, K. (eds) Maritime Spatial Planning. Palgrave Macmillan, Cham. https://doi.org/10.1007/978-3-319-98696-8_12

Raudsepp-Hearne, C. et al. (2010). Untangling the environmentalist's paradox: why is human well-being increasing as ecosystem services degrade? *BioScience*, 60(8), 576–589.

Ray, A. (2008). Waste management in developing Asia: can trade and cooperation help? *The Journal of Environment and Development*, 17(1), 3–25.

Reeve, L. L. N. (2012). Of whales and ships: impacts on the great whales of underwater noise pollution from commercial shipping and proposals for regulation under international law. *Ocean and Coastal LJ*, 18, 127.

Rickels, W. Weigand, C. Grasse, P. Schmidt, J. and Voss, R. (2019). Does the European Union achieve comprehensive blue growth? progress of EU coastal states in the Baltic and North Sea, and the Atlantic Ocean against sustainable development goal 14. *Marine Policy*, 106, 103515.

Ringbom, H. Bohman, B. and Ilvessalo, S. (2018). Combatting eutrophication in the Baltic Sea: legal aspects of sea-based engineering measures. *Brill Research Perspectives in the Law of the Sea*, 2(4), 1–96.

Roberts, C. M. et al. (2017). Marine reserves can mitigate and promote adaptation to climate change. *Proceedings of the National Academy of Sciences*, 114(24), 6167–6175.

Rockström, J. and Klum, M. (2015). Big World, Small Planet. Yale University Press: New Haven, CT

Rode, S. (2013). Integrated sewage treatment and coastal management in Mumbai metropolitan region. *Management Research and Practice*, 5(2), 31.

Rogers, R. (2019). The sea of the universe: how maritime law's limitation on liability gets it right, and why space law should follow by example. *Ind. J. Global Legal Stud.* 26, 741.

Rosà, R. et al. (2014). Early warning of West Nile virus mosquito vector: climate and land use models successfully explain phenology and abundance of Culex pipiens mosquitoes in north-western Italy. *Parasites and Vectors*, 7(1), 1–12.

Rosa, R. et al. (2014). Differential impacts of ocean acidification and warming on winter and summer progeny of a coastal squid (Loligo vulgaris). *Journal of Experimental Biology*, 217(4), 518–525.

Ruhl, J. B. (1997). Thinking of environmental law as a complex adaptive system: how to clean up the environment by making a mess of environmental law. *Hous. L. Rev.* 34, 933.

Ruiz, J. J. (2018). Ocean options for climate change mitigation: Disposal of greenhouse gases at the sea under the 1996 London Protocol. In *Desafíos de la acción jurídica internacional y europea frente al cambio climático* (pp. 43–56). Atelier.

Salpin, C. Onwuasoanya, V. Bourrel, M. and Swaddling, A. (2018). Marine scientific research in pacific small island developing states. *Marine Policy,* 95, 363–371.

Sarker, S. Ara Hussain, F. Assaduzzaman, M. and Failler, P. (2019). Blue economy and climate change: Bangladesh perspective. *Journal of Ocean and Coastal Economics,* 6(2), 6.

Schmidt, C. C. (2018). *Issues and Options for Disciplines on Subsidies to Illegal, Unreported and Unregulated Fishing.* Fisheries Subsidies Rules at the WTO, 53.

Schrijver, N. (1995). *Sovereignty over Natural Resources: Balancing Rights and Duties in an Interdependent World.* University Library Groningen: Groningen, 240–241.

Schutter, M. S. and Hicks, C. C. (2019). Networking the Blue Economy in Seychelles: pioneers, resistance, and the power of influence. *Journal of Political Ecology,* 26(1), 425–447.

Scholaert F. (2020). *The Blue Economy: Overview and EU Policy Framework.* European Parliamentary Research Service (EPRS), 22, 29.

Scott, D. Simpson, M. C. and Sim, R. (2012). The vulnerability of Caribbean coastal tourism to scenarios of climate change related sea level rise. *Journal of Sustainable Tourism,* 20(6), 883–898.

Scott, K. N. (2018). Ocean acidification and sustainable development Goal 14: a goal but no target? In: Nordquist M. H. Moore J. N. (Eds.), *The Marine Environment and United Nations Sustainable Development Goal 14* (pp. 323–341). Brill Nijhoff. Koninklijke Brill NV, Leiden, The Netherlands.

Seabed Mineral Deposits in European Seas: Metallogeny and Geological Potential for Strategic and Critical Raw Materials (MINDeSEA). GeoERA European Union's Horizon 2020 research and innovation programme under grant agreement No 731166.

Segui, L. Smail, E. DiGiacomo, P. M. VanGraafeiland, K. Campbell, J. and von Shuckmann, K. (2020, December). Leveraging earth observation data at the science-society interface: tools to help countries monitor and mitigate marine pollution. In *AGU Fall Meeting Abstracts* (Vol. 2020, pp. SY035–0014).

Selin, H. (2013). Minervian Politics and international chemicals policy. In *Leadership in Global Institution Building* (pp. 193–212). Palgrave Macmillan, London.

Sellers, S. and Ebi, K. L. (2018). Climate change and health under the shared socioeconomic pathway framework. *International Journal of Environmental Research and Public Health,* 15(1), 3.

Sha, J. (2019). The emerging blue economy: its development and future propsects. *Liberal Stud.* 4, 61.

Sheavly, S. B. and Register, K. M. (2007). Marine debris and plastics: environmental concerns, sources, impacts and solutions. *Journal of Polymers and the Environment,* 15(4), 301–305.

Shiiba, N. Wu, H. H. and Huang, M. C. (2020). *Proposing Regulatory-Driven Blue Finance Mechanism for Blue Economy Development.*

Shukla, P. R. J. Skea, E. Calvo Buendia, V. Masson-Delmotte, H. O. Pörtner, D. C. Roberts, P. et al. (2019). IPCC, 2019. *Climate Change and Land: An IPCC Special Report on Climate Change, Desertification, Land Degradation, Sustainable Land Management, Food Security, and Greenhouse Gas Fluxes in Terrestrial Ecosystems.*

Shyam, S. S. and Elizabeth James, H. (2018). *Integrating Climate Change and Blue Economy in Fisheries Research and Education in India.*

Sigren, J. M. Figlus, J. and Armitage, A. R. (2014). Coastal sand dunes and dune vegetation: restoration, eroskion, and storm protection. *Shore Beach,* 82(4), 5–12.

Simas, T. et al. (2015). Review of consenting processes for ocean energy in selected European Union Member States. *International Journal of Marine Energy,* 9, 41–59.

Singh, G. G. et al. (2018). A rapid assessment of co-benefits and trade-offs among Sustainable Development Goals. *Marine Policy,* 93, 223–231.

Smail, E. VanGraafeiland, K. and Ghafari, D. (2020). *Methodology, Processing, and Application Development in Support of Sustainable Development Goal 14.1: Chlorophyll Global Analysis and Metrices.*

Smith, L. H. (2017). To accede or not to accede: An analysis of the current US position related to the United Nations law of the sea. *Marine Policy,* 83, 184–193.

Sovacool, B. K. Bazilian, M. Griffiths, S. Kim, J. Foley, A. and Rooney, D. (2021). Decarbonizing the food and beverages industry: a critical and systematic review of developments, sociotechnical systems and policy options. *Renewable and Sustainable Energy Reviews,* 143, 110856.

Spalding, M. J. (2016). The new blue economy: the future of sustainability. *Journal of Ocean and Coastal Economics*, 2(2), 8.

Spijkers, J. and Boonstra, W. J. (2017). Environmental change and social conflict: the northeast Atlantic mackerel dispute. *Regional Environmental Change*, 17(6), 1835–1851.

Stevens, T. F. (1950). Erie RR v. Tompkins and the Uniform General Maritime Law. *Harv. L. Rev.* 64, 246.

Stokke, O. S. (2018). Beyond dumping? The effectiveness of the London convention. *Yearbook of International Cooperation on Environment and Development 1998–99*, 39-49.

Sumaila, U. R. et al. (2020). *Ocean Finance: Financing the Transition to a Sustainable Ocean Economy*. World Resources Institute.

Stithou, M. (2017). Considerations of Socio-Economic Input, Related Challenges and Recommendations for Ecosystem-Based Maritime Spatial Planning: *A Review. Journal of Ocean and Coastal Economics*, 4(1), 1.

Stuchtey, M. R. et al. (2020). *Ocean Solutions that Benefit People, Nature and the Economy*. Report, World Resources Institute.

Spalding, A. K. and de Ycaza, R. (2020). Navigating shifting regimes of ocean governance: from UNCLOS to Sustainable Development Goal 14. *Environment and Society*, 11(1), 5–26.

Spalding, M. J. (2016). The new blue economy: the future of sustainability. *Journal of Ocean and Coastal Economics*, 2(2), 8.

Spalding, M. Burke, L. Wood, S. A. Ashpole, J. Hutchison, J. and Zu Ermgassen, P. (2017). Mapping the global value and distribution of coral reef tourism. *Marine Policy*, 82, 104–113.

Sunny, A. R. Mithun, M. H. Prodhan, M. H. Ashrafuzzaman, M. Rahman, S.M.A. Billah, M. M. Hussain, M. Ahmed, K. J. Sazzad, S. A. Alam, M. T. Rashid, A. and Hossain, M. M. (2021). Fisheries in the Context of Attaining Sustainable Development Goals (SDGs) in Bangladesh: COVID-19 impacts and future prospects. *Sustainability*, 13, 9912. https://doi.org/10.3390/su13179912

Techera, E. J. (2018). Supporting blue economy agenda: fisheries, food security and climate change in the Indian Ocean. *Journal of the Indian Ocean Region*, 14(1), 7–27.

Teh, L. C. L. and Sumaila, U. R. (2013). Contribution of marine fisheries to worldwide employment. *Fish and Fisheries*, 14(1), 77–88. doi:10.1111/j.1467–2979.2011.00450.x.

Tilbrook, B. et al. (2019). An enhanced ocean acidification observing network: from people to technology to data synthesis and information exchange. *Frontiers in Marine Science*, 6, 337.

Tyre, A. J. and Michaels, S. (2011). Confronting socially generated uncertainty in adaptive management. *Journal of Environmental Management*, 92(5), 1365–1370.

Union, A. (2012). 2050 Africa's integrated maritime strategy (2050 AIM Strategy). *2011-12-16)[2018-03-20]*. *https://au. int/en/documents-38.*

Upadhyay, D. K. and Mishra, M. (2020). Blue economy: Emerging global trends and India's multilateral cooperation. *Maritime Affairs: Journal of the National Maritime Foundation of India*, 16(1), 30–45.

van den Burg, S. W. K. Rockmann, C. Banach, J. L. and van Hoof, L. (2020). Governing risks of multi-use: seaweed aquaculture at offshore wind farms. *Front. Mar. Sci.* 7, 60. doi: 10.3389/fmars.2020.00060.

VanderZwaag, D. L. (2015). The international control of ocean dumping: navigating from permissive to precautionary shores. In *Research handbook on international marine environmental law* (pp. 132–148). Edward Elgar Publishing: Cheltenham and Camberley.

Vaneeckhaute, C. and Fazli, A. (2020). Management of ship-generated food waste and sewage on the Baltic Sea: A review. *Waste Management*, 102, 12–20.

Vare, L. L. et al. (2018). Scientific considerations for the assessment and management of mine tailings disposal in the deep sea. *Frontiers in Marine Science*, 5, 17.

Verlaan, P. (2011). Current Legal Developments London Convention and London Protocol. *International Journal of Marine and Coastal Law*, 26(1), 185–194.

Virto, L. R. (2018). A preliminary assessment of the indicators for Sustainable Development Goal (SDG) 14 'Conserve and sustainably use the oceans, seas and marine resources for sustainable development.' *Marine Policy*, 98, 47–57.

Voyer, M. Farmery, A. K. Kajlich, L. Vachette, A. and Quirk, G. (2020). Assessing policy coherence and coordination in the sustainable development of a Blue Economy. A case study from Timor Leste. *Ocean and Coastal Management*, 192, 105187.

Voyer, M. and van Leeuwen, J. (2019). 'Social license to operate' in the Blue Economy. *Resources Policy*, 62, 102–113.

Voyer, M. Quirk, G. McIlgorm, A. and Azmi, K. (2018). Shades of blue: what do competing interpretations of the Blue Economy mean for oceans governance? *Journal of Environmental Policy and Planning*, 20(5), 595–616.

Waage, J. et al. (2010). The Millennium Development Goals: a cross-sectoral analysis and principles for goal setting after 2015: Lancet and London International Development Centre Commission. *The Lancet*, 376(9745), 991–1023.

Waite, R. Kushner, B. Jungwiwattanaporn, M. Gray, E. and Burke, L. (2015). Use of coastal economic valuation in decision making in the Caribbean: Enabling conditions and lessons learned. *Ecosystem Services*, 11, 45–55.

Wambiji, N. (2019). *WIOMSA Annual Report, 2018*. Western Indian Ocean Marine Science Association.

Wamsler, C. et al. (2016). Operationalizing ecosystem-based adaptation: harnessing ecosystem services to buffer communities against climate change. *Ecology and Society*, 21(1).

Wenhai, L. et al. (2019). Successful blue economy examples with an emphasis on international perspectives. *Frontiers in Marine Science*, 6, 261.

Wilson, A. M. W. and Forsyth, C. (2018). Restoring near-shore marine ecosystems to enhance climate security for island ocean states: aligning international processes and local practices. *Marine Policy*, 93, 284–294.

Winder, G. M. and Le Heron, R. (2017). Assembling a Blue Economy moment? Geographic engagement with globalizing biological-economic relations in multi-use marine environments. *Dialogues in Human Geography*, 7(1), 3–26.

Wood, M. C. (2018). Atmospheric trust litigation: securing a constitutional right to a stable climate system. *Colo. Nat. Resources Energy and Envtl. L. Rev.* 29, 321.

World Resources Institute. (2020). The High Level Panel for a Sustainable Ocean Economy. *Transformations for a Sustainable Ocean Economy: A Vision for Protection, Production and Prosperity*. Report, World Resources Institute.

Wright, G. Schmidt, S. Rochette, J. Shackeroff, J. Unger, S. Waweru, Y. and Müller, A. (2017). *Partnering for a Sustainable Ocean: The Role of Regional Ocean Governance in Implementing SDG14*. PROG: IDDRI, IASS, TMG and UN Environment.

Yingying, W. U. (2017). Negotiation on fisheries subsidies within the framework of the WTO-special and differential treatment for developing members. *China Oceans L. Rev.* 22, 22–49.

Yoshioka, N. Wu, H. H. Huang, M. C. and Tanaka, H. (2020). *Proposing Regulatory—Driven Blue Finance Mechanism for Blue Economy Development*.

Zelasney, J. Ford, A. Westlund, L. Ward, A. and Riego Peñarubia, O. (2020). *Securing Sustainable Small-Scale Fisheries: Showcasing Applied Practices in Value Chains, Post-Harvest Operations and Trade* (Vol. 652). Food and Agriculture Organization of the United Nations.

Zhang, W. et al. (2021). Governance of global vessel-source marine oil spills: Characteristics and refreshed strategies. *Ocean and Coastal Management*, 213, 105874.

Zielinski, S. and Botero, C. M. (2020). Beach tourism in times of COVID-19 pandemic: critical issues, knowledge gaps and research opportunities. *International Journal of Environmental Research and Public Health*, 17(19), 7288.

Zougmoré, R. et al. (2016). Toward climate-smart agriculture in West Africa: a review of climate change impacts, adaptation strategies and policy developments for the livestock, fishery and crop production sectors. *Agriculture and Food Security*, 5(1), 1–16.

Index